Communications
in Computer and Information Science

2855

Series Editors

Gang Li , *School of Information Technology, Deakin University, Burwood, VIC, Australia*

Joaquim Filipe, *Polytechnic Institute of Setúbal, Setúbal, Portugal*

Zhiwei Xu, *Chinese Academy of Sciences, Beijing, China*

Rationale

The CCIS series is devoted to the publication of proceedings of computer science conferences. Its aim is to efficiently disseminate original research results in informatics in printed and electronic form. While the focus is on publication of peer-reviewed full papers presenting mature work, inclusion of reviewed short papers reporting on work in progress is welcome, too. Besides globally relevant meetings with internationally representative program committees guaranteeing a strict peer-reviewing and paper selection process, conferences run by societies or of high regional or national relevance are also considered for publication.

Topics

The topical scope of CCIS spans the entire spectrum of informatics ranging from foundational topics in the theory of computing to information and communications science and technology and a broad variety of interdisciplinary application fields.

Information for Volume Editors and Authors

Publication in CCIS is free of charge. No royalties are paid, however, we offer registered conference participants temporary free access to the online version of the conference proceedings on SpringerLink (http://link.springer.com) by means of an http referrer from the conference website and/or a number of complimentary printed copies, as specified in the official acceptance email of the event.

CCIS proceedings can be published in time for distribution at conferences or as post-proceedings, and delivered in the form of printed books and/or electronically as USBs and/or e-content licenses for accessing proceedings at SpringerLink. Furthermore, CCIS proceedings are included in the CCIS electronic book series hosted in the SpringerLink digital library at http://link.springer.com/bookseries/7899. Conferences publishing in CCIS are allowed to use our online conference service (Meteor) for managing the whole proceedings lifecycle (from submission and reviewing to preparing for publication) free of charge.

Publication process

The language of publication is exclusively English. Authors publishing in CCIS have to sign the Springer CCIS copyright transfer form, however, they are free to use their material published in CCIS for substantially changed, more elaborate subsequent publications elsewhere. For the preparation of the camera-ready papers/files, authors have to strictly adhere to the Springer CCIS Authors' Instructions and are strongly encouraged to use the CCIS LaTeX style files or templates.

Abstracting/Indexing

CCIS is abstracted/indexed in DBLP, Google Scholar, EI-Compendex, Mathematical Reviews, SCImago, Scopus. CCIS volumes are also submitted for the inclusion in ISI Proceedings.

How to start

To start the evaluation of your proposal for inclusion in the CCIS series, please send an e-mail to ccis@springer.com

Zohreh Molamohamadi ·
Erfan Babaee Tirkolaee · A. Mirzazadeh ·
Gerhard-Wilhelm Weber
Editors

Optimization and Data Science in Industrial Engineering

Third International Conference, ODSIE 2025
Istanbul, Turkey, November 20–22, 2025
Proceedings, Part II

 Springer

Editors
Zohreh Molamohamadi
Kharazmi University
Tehran, Iran

A. Mirzazadeh
Kharazmi University
Tehran, Iran

Erfan Babaee Tirkolaee
Istinye University
Istanbul, Türkiye

Gerhard-Wilhelm Weber
Poznań University of Technology
Poznań, Poland

ISSN 1865-0929   ISSN 1865-0937 (electronic)
Communications in Computer and Information Science
ISBN 978-3-032-17022-4   ISBN 978-3-032-17023-1 (eBook)
https://doi.org/10.1007/978-3-032-17023-1

This Springer imprint is published by the registered company Springer Nature Switzerland AG
The registered company address is: Gewerbestrasse 11, 6330 Cham, Switzerland

If disposing of this product, please recycle the paper.

Preface

The Third International Conference on Optimization and Data Science in Industrial Engineering (ODSIE 2025) was held in a hybrid format—both online and in person—in Istanbul, Turkey, on November 20–22, 2025. The event was organized by the Institution of International Scientific Services (RefConf) in collaboration with Istinye University, and it aimed to provide a dynamic and stimulating environment for the exchange of knowledge and ideas. Building on the success and positive feedback from previous editions, ODSIE 2025 continued its tradition of academic excellence and global collaboration.

ODSIE 2025 attracted the interest of students, researchers, and professionals from around the world. The conference covered a wide range of topics, including but not limited to: "Industry 4.0, IoT and smart manufacturing", "Digital twin and virtual commissioning", "Sustainable and smart cities", "Artificial intelligence and expert systems", "Metaheuristic algorithms with applications in IE", "Machine learning algorithms", "Big data analytics and Data mining", "Robotic process automation", "Decision support systems", "E-Government, E-Commerce and E-Learning", "Supply chain design and logistics", "Optimization and Data Science-based Case studies of manufacturing/service industries", "Quantitative finance and risk modelling", and "Other fields of study related to Optimization and Data Science with applications in IE".

This year, ODSIE 2025 was honored to host distinguished speakers from the USA, Italy, Turkey, and the UAE. In total, eighteen universities and the research institutes from the USA, Czech Republic, Poland, Tunisia, Turkey, Kuwait, Ukraine, India, Iran, and Iraq contributed as scientific sponsors. The international technical program committee included experts from 33 countries, reflecting the conference's strong global reach.

Papers submitted to ODSIE 2025 were evaluated according to rigorous review criteria, including: content quality, originality, relevance, contribution to the professional literature, significance and potential impact, language clarity, study validity, methodology and analysis accuracy, structure and organization, adequacy of references and citations, consistency in formatting, and the quality and clarity of figures and tables. Out of 198 submissions, 94 full papers and 10 short papers have been selected for publication in this book series.

The accepted papers were presented across twenty panel sessions, including: "Technology Integration & Digital Transformation: Smart Operations and Process Optimization", "The Role of Machine Learning in Predictive Cybersecurity for Industrial Systems", "Industry 6.0, Emerging Technologies and Sustainable Development", "AI-Driven Autonomy: Revolutionizing Customer Self-Service with Intelligent, Real-Time Assistance and Predictive Guidance", "Generative AI in Healthcare: Transforming Patient Care and Medical Innovation", "E-Government, E-Commerce, and E-Learning", "AI for Healthcare Optimization: Metaheuristics, Machine Learning, Mental Health, Ambulance Routing Problem in Emergency Scenarios", "Artificial Intelligence Solutions for Engineering Problems", "Optimization and Strategic Decision-Making in Industrial

Management", "Industrial Transformation Through Big Data Analytics: Trends, Challenges, and Opportunities", "Data-Driven Strategies for Sustainable and Smart Business Operations", "New Approaches to Achieving Sustainable Development Goals (SDGs)", "Smart Cities for a Sustainable Future: Optimization and Data Science Approaches", "Strategic Supply Chain Analytics: Optimizing Circularity, Resilience, and Sustainability", "Forecasting Techniques for Operations Research", "Emerging Trends in Science, Business and Technology Innovation", "Mathematics, Computation and Complex Systems", "Computational Intelligence and Decision Optimization for Reliability and Maintenance Systems", "Data-Driven Governance, Organizational Performance & Digital Communication", and "AI-Driven Optimization and Intelligence in Modern Logistics".

November 2025

Zohreh Molamohamadi
Erfan Babaee Tirkolaee
A. Mirzazadeh
Gerhard-Wilhelm Weber

Organization

Conference Chairs

Abolfazl Mirzazadeh	Kharazmi University, Iran/Institution of International Scientific Services, Oman
Erfan Babaee Tirkolaee	Istinye University, Turkey

Patrons

Erkan Ibiş	Istinye University
Hatice Gülen	Istinye University
Mehmet Alper Tunga	Istinye University

Conference Coordinators

Leila Chehreghani	Institution of International Scientific Services, Oman
Zohreh Molamohamadi	Kharazmi University, Iran/Institution of International Scientific Services, Oman
Saina Sahragard	Institution of International Scientific Services, Oman
Elahe Reeyazati	Institution of International Scientific Services, Oman

Technical Program Committee Chairs

Gerhard-Wilhelm Weber	Poznań University of Technology, Poland
Zohreh Molamohamadi	Kharazmi University, Iran

Steering Committee Members

A. Mirzazadeh	Kharazmi University, Iran
Erfan Babaee Tirkolaee	Istinye University, Turkey
Janny M.Y. Leung	University of Macau, China
Josef Jablonsky	University of Economics and Business, Czech Republic
Kathryn Stecke	University of Texas at Dallas, Naveen Jindal School of Management, USA
Hatice Gülen	Istinye University, Turkey
Mehmet Alper Tunga	Istinye University, Turkey

Technical Program Committee Members

Tatiana Tchmisova	University of Aveiro, Portugal
Chefi Triki	University of Kent, UK
Zeynep Alparslan	Süleyman Demirel University, Turkey
Dursun Delen	Oklahoma State University, USA
Oleksandr Dluhopolskyi	West Ukrainian National University, Ukraine
Emre Çakmak	Istinye University, Turkey
Dragan Pamučar	University of Belgrade, Serbia
Noyan Sebla Sezer	Istinye University, Turkey
Alireza Amırteımoorı	Istinye University, Turkey
Tofigh Allahviranloo	Istinye University, Turkey
Mir Saman Pishvaee	Iran University of Science and Technology, Iran
Kamran Yeganegi	Islamic Azad University, Iran
Dinh Tran Ngoc Huy	Banking University HCMC, Vietnam
Sylwia Szybowska	Jacob of Paradies University in Gorzow Wielkopolski, Poland
Eloisa Macedo	University of Aveiro, Portugal
Hao Yu	UiT The Arctic University of Norway, Norway
Saeed Hatamzadeh	Istinye University, Turkey
Marzieh Khakifirooz	Technological Institute of Monterrey, Mexico
Sadia Samar Ali	King Abdulaziz University, Saudi Arabia
Safa Bhar Layeb	University of Tunis El Manar, Tunisia
Nasr Hamood Mohamed Al-Hinai	Sultan Qaboos University, Oman
Hela Moalla Frikha	Higher Business School of Sfax, Tunisia
Taicir Moalla Louki	University of Sfax, Tunisia
Paulina Golinska	Poznań University of Technology, Poland
Meenakshi Kaushik	Lloyd Institute of Engineering & Technology, India

Alireza Goli	University of Isfahan, Iran
Mohammad Mahdi Paydar	Babol Noshirvani University of Technology, Iran
Amir Khakbaz	Damghan University, Iran
Seyed Mohammad Javad Mirzapour Al-E-Hashem	Amirkabir University of Technology, Iran
Hamed Fazlollahtabar	Damghan University, Iran
AllaEldin H. Kassam	King University of Technology, Iraq
Arpan Kumar Kar	Indian Institute of Technology Delhi, India
Sita Ram Sharma	Chitkara University, India
Sankar Kumar Roy	Vidyasagar University, India
Nazanin Pilevari	Islamic Azad University, Iran
Sara Saberi	Worcester Polytechnic Institute, USA
Michael G. Kay	North Carolina State University, USA
Aybike Özyüksel Çiftçioğlu	Manisa Celal Bayar University, Turkey
Soheyl Khalilpourazari	Polytechnique Montréal, Canada
Ammar Odeh	Princess Sumaya University for Technology, Jordan
Hamidreza Irani	University of Tehran, Iran
Sadeq Damrah	Australian University, Kuwait
Şenol Pişkin	Istinye University, Turkey
Emir Seyyedabbasi	Istinye University, Turkey
Mazdak Khodadadi Karimvand	University of Science and Culture, Iran
Hela Limam	Université de Tunis El Manar, Tunisia and Institut Supérieur de Gestion de Tunis, Tunisia
Erialdi Syahrial	Universiti Malaya, Malaysia
Novakovska Iryna	National University of Life and Environmental Sciences of Ukraine, Ukraine
Marwa Hasni	National Engineering School of Tunis, University of Tunis El Manar, Tunisia
Ferzat Anka	Fatih Sultan Mehmet Vakif University, Turkey
Yavuz Selim Özdemir	Ankara Science University, Turkey
Barış Bülent Kırlar	Süleyman Demirel University, Turkey
Jai Acharya	IOI Ocean Academy, Singapore
Rezzy Eko Caraka	National Research and Innovation Agency (BRIN), Indonesia, Ulsan National Institute of Science and Technology (UNIST), South Korea, Pukyong National University, South Korea
Intisar A.M. Al Sayed	Iraqi Society for Engineering Management, Iraq
Muna S. Kassim	Iraqi Society for Engineering Management, Iraq
Manar Naji Ghayyib	Iraqi Society for Engineering Management, Iraq
Israa Ibraheem Albarazanchi	Iraqi Society for Engineering Management, Iraq
Sameera Fernandes	Garden City University, UAE

Sonia Nasri	Ecole Supérieure de Commerce de Tunis, Université de la Manouba, Tunisia, Institut Supérieur de Gestion de Tunis, Tunisia
Ezgi Özer	Piri Reis University, Turkey
Maher Agi	Rennes School of Business, France
Dariusz Jacek Jakóbczak	Koszalin University of Technology, Poland
Sanjib Biswas	Amity University Kolkata, India
Marilisa Botte	University of Naples Federico II, Italy
Mohammed Sayim Khalil	Halic University, Turkey
Murat Yeşilkaya	Tokat Gaziosmanpaşa University, Turkey
Vijay Kumar Gahlawat	National Institute of Food Technology Entrepreneurship and Management, India
Ali Asghar Rahmani Hosseinabadi	University of Regina, Canada
Amir Hassanzadeh	Urmia University, Iran
Amin Hosseinian Far	University of Hertfordshire, UK
Beata Mrugalska	Poznań University of Technology, Poland
Vladimir Simic	University of Belgrade, Serbia
Hêriş Golpîra	Sanandaj Branch, Islamic Azad University, Iran
Sarfaraz Hashemkhani Zolfani	Universidad Catolica del Norte, Chile
Mohit Malik	National Institute of Food Technology Entrepreneurship and Management, India
Vuppulapati Chandra Sekhar Naidu	Coforge, USA
Srinivasan Balan	North Carolina State University, USA
Tanko Bako	Taraba State University, Nigeria
Saeid Rezaei	Arak University, Iran
Omur Ugur	Middle East Technical University (METU), Turkey
Justin Eduardo Simarmata	Universitas Timor, Indonesia
Ismail Özcan	University of Parma, Italy
Adewoye Olabode	Yaba College of Technology, Nigeria
Kerem Ugurlu	Nazarbayev University, Kazakhstan
Harshavardhan Yedla	Wipro LLC, USA

Conference Co-Chairs

Saliha Karadayi-Usta (Co-chair)	Istinye University, Turkey
N. Serhan Aydın (Co-chair)	Istinye University, Turkey

Keynote Speakers

Şenol Pişkin	Istinye University, Türkiye
AbdulQuddus Mohammed	Higher Colleges of Technology, UAE
Erfan Babaee Tirkolaee	Istinye University, Turkey

Workshop Organizers

Udaya Veeramreddygari	Cox Automotive Inc., USA
Harshavardhan Yedla	Wipro LLC, USA
Venkat Sharma Gaddala	Google, USA
Abhi Desai	Saks Fifth Avenue, USA
İsmail Özcan	University of Parma, Italy
Meenakshi Kaushik	Lloyd Group of Educational Institutions, India

Executive Committee Members

Elmira Salavati	Kharazmi University, Iran
Zahra Shahedipour	Kharazmi University, Iran
Fereshteh Sarabadani	Kharazmi University, Iran
Yasin Bonyadi	Kharazmi University, Iran
Mehrshad Jamshidipour	Kharazmi University, Iran
Yasna Yeganeh	Kharazmi University, Iran
Zahra Norouzi Shad	Kharazmi University, Iran
Mobina Mostajabi	Kharazmi University, Iran
Arshia Sahraeei	Kharazmi University, Iran

Reviewers

Rasheed Al-Salih	University of Sumer, Iraq
Mostafa Baghouri	Abdelmalek Essaâdi University, Morocco
Pavan Kumar	VIT Bhopal University, India
Renuka Deshmukh	Dr Vishwanath Karad MIT World Peace University, India
Fatima Zohra Allam née Chergui	École Nationale Supérieure des TIC et de la Poste, Chad
Parmida Bahreini	Ankara Yıldırım Beyazıt University, Turkey
Ali Asghar Rahmani Hosseinabadi	University of Regina, Canada
Ali Akbar Sheikh	University of Burdwan, India

Zohreh Molamohamadi	Tehran Disaster Mitigation and Management Organization, Iran
Maryam Mardani	Hamedan University of Medical Science, Iran
Nguyễn Thành Luân	Ho Chi Minh City University of Foreign Languages and Information Technology, Vietnam
Ammar Odeh	Princess Sumaya University for Technology, Jordan
Tatapudi Vasista	Vasista Consulting and Performing Services OPC Pvt. Ltd., India
Agatha da Silva Ovando	Universidad Privada Boliviana, Bolivia
Deepshikha Kalra	Guru Gobind Singh Indraprastha University, India
Sara Mardani	Hamedan University of Technology, Iran
M. Chandramouleeswaran	Saiva Bhanu Kshatriya College, India
Vuppulapati Chandra Sekhar Naidu	Coforge, USA
Pouria Tajasob	Amirkabir University of Technology, Iran
Rohit Bansal	Vaish College of Engineering, India
Maria Habib	Saint Joseph University of Beirut, Lebanon
Rand Ahmed	Tikrit University, Iraq
Ihtisham Haq	University of Calabria, Italy
Mehmet Kiziltaş	Istanbul Beykent University, Turkey
Harshavardhan Yedla	Wipro LLC, USA
Seyed Ahmad Ghasemi	Research Institute of Humanities and Social Studies of ACECR, Iran
Parul Mangal	Parul University, India
Alireza Goli	University of Isfahan, Iran
Eyüp Aydemir	S4C, SAP PP Team Lead & PP/PPDS Lead Consultant, Germany
Elham Akbarian	Kharazmi University, Iran
Jyoti Kunal Shah	Independent Researcher, USA
Erfan Babaee Tirkolaee	Istinye University, Turkey
Farnaz Javadi Gargari	Alzahra University, Iran
Murtadha Shukur	Al-Furat Al-Awsat Technical University, Iraq
Ghazi Alkhatib	Hashemite University, Jordan
Charles Adusei	Garden City University College, Ghana
Marzieh Samadi Foroushani	University of Eyvanekey, Iran
Manel Kammoun	University of Sfax, Tunisia
Udaya Veeramreddygari	Cox Automotive Inc, USA
Zeinab Aliyas	Université de Montréal, Canada
Duong Cuong	Hanoi University of Science and Technology, Vietnam

Josef Jablonsky	Prague University of Economics and Business, Czech Republic
Seyed Mohammad Javad Mirzapour Alehashem	Amirkabir University of Technology, Iran
Payam Afkhami Kheyrabadi	Florida State University, USA
Shreya Makinani	University of Southern California, USA
Fatereh Sadat Moosavi	Urmia University, Iran
Behnam Razzaghmaneshi	Islamic Azad University-Talesh Branch, Iran
Javad Sofiyabadi	lamic Azad University, Iran
Arezou Panjehpour	Shiraz University of Technology, Iran
Maan Abdulwahid	Altinbas University, Turkey
Srinivasan Balan	North Carolina State University, USA
Sherzod Ibodulloev	TIIAME National Research University, Uzbekistan
Sujan Piya	University of Sharjah, UAE
Chanicha Moryadee	Suan Sunandha Rajabhat University, Thailand
Vasanthi Govindaraj	National General (An Allstate Company), USA
Beata Mrugalska	Poznań University of Technology, Poland
Issam Najati	HECF Business School, Morocco
Stephanus Indrawan	Ciputra University Surabaya, Indonesia
Neda Khosravifard	Shiraz University of Technology, Iran
Viraj Lele	DHL Supply Chain, USA
Beata Glinkowska-Krauze	University of Lodz, Poland
Ali Kalantari	Islamic Azad University, Shiraz Branch, Iran
Shahid Amin	All India Management Association, India
Mehmet Kaygusuz	Anadolu University, Turkey
Zubair Khan	Kalinga University, India
Tatiana Tchemisova	University of Aveiro, Portugal
Anak Agung Gde Satia Utama	Universitas Airlangga, Indonesia
Abhi Desai	Saks Fifth Avenue, USA
Leila Chehreghani	Institution of International Scientific Services, Oman
Aybike Özyüksel Çiftçioğlu	Manisa Celal Bayar University, Turkey
Haripriya Barman	Vidyasagar University, India
Sahar Sadri	University of Alzahra
Patricia Cano-Olivos	Universidad Popular Autónoma del Estado de Puebla, Mexico
Sara Nodoust	Iran University of Science and Technology, Iran
Amarachukwu Obi	University of Nigeria, Nigeria
Vani N. Laturkar	Swami Ramanand Teerth Marathwada University, India
Aditya Singh	Amrita Vishwa Vidyapeetham, India

Hamed Shakerian	Islamic Azad University, Tabriz Branch, Iran
İsmail Özcan	Süleyman Demirel Üniversitesi, Turkey
Usha Devi Narasimhaiah	Bengaluru City University, India
Sameera Fernandes	Garden City University, India
Davy Hendri	Universitas Islam Negeri Imam Bonjol Padang, Indonesia
Md. Ashraful Islam	University of Information Technology & Sciences, Bangladesh
Mehmet Kabak	Endüstri Mühendisliği Bölümü, Turkey
Navreet Kaur	Chitkara University, India
Meenakhi Kaushik	Trinity Institute of Innovations in Professional Studies (Guru Gobind Singh University), India
Ghemam Amara Djilani	University of El Oued, Algeria
Wissem Ennouri	University of Sfax, Tunisia
Zara Kiran	Mirpur University of Science & Technology, Pakistan, Universiti Kebangsaan Malaysia, Malaysia
Sapthami Iraganaboina	Jawaharlal Nehru Technological University Hyderabad, India
Mohammad Aknan	National Institute of Technology Patna, India
Deepak Singh	University of Pittsburgh, USA
Agus Hermawan	Universitas Negeri Malang, Indonesia
Javad Taheri	Kharazmi University, Iran
Zheng Liang	Huainan Normal University, China
Shakeel Javaid	Aligarh Muslim University, India
Sina Sayardoost Tabrizi	University of Tehran, Iran
Fateme Rashidian	Arak University, Iran
Fereshteh Sarabadani	Kharazmi University, Iran
Mazdak Khodadadi Karimvand	University of Science and Culture, Iran
Erfan Kajabadi	University of Isfahan, Iran
Amir-Mohammad Golmohammadi	Arak University, Iran
Alper Bora Koçak	Istinye University, Turkey
Sonia Nasri	Ecole Supérieure de Commerce de Tunis, Université de la Manouba, Tunisia
Ömür Uğur	Middle East Technical University, Turkey
Justin Eduardo Simarmata	Universitas Timor, Indonesia
Ardavan Babaei	Istinye University, Turkey
Maryam Almasi	University of Guilan, Iran
Priyanka Patel	Charusat University, India
Vijay Kumar Reddy Voddi	Saint Peter's University, USA
Fatereh Sadat Mousavi	Urmia University, Iran

Scientific Sponsors

The University of Texas at Dallas, USA

Prague University of Economics and Business, Czech Republic

University of Sfax, Tunisia

Ankara Science University, Turkey

Süleyman Demirel University, Turkey

Iranian Institute of Industrial Engineering

Iraqi Society for Engineering Management

LR-OASIS, National Engineering School of Tunis, University of Tunis El Manar, Tunisia

Tunisian Operational Research Society (TORS), Tunisia

Halic University, Turkey

West Ukrainian National University, Ukraine

Chitkara University, India

Vidyasagar University, India

National Institute of Food Technology Entrepreneurship and Management, India

International Institute of Innovation Science- Education- Development in Warsaw, Poland

Australian University, Kuwait

Manisa Celal Bayar University, Turkey

Noida International University

Contents

**Intelligent Optimization and Autonomous System Design through AI
and Digital Technologies**

Computational Governance and Ethical Intelligence in Global Corporate Systems

Econometric and Computational Modeling of Corporate Governance and Firm Performance in Emerging Markets

Saif Saad Ahmed[1] (iD), Ibrahim Khilel Khinger[2] (iD), Deena Waleed Hameed Jawad[3] (iD), Majid Fadhil Ziboon[4]($\boxtimes$) (iD), Akram Fadhel Mahdi[5] (iD), and Danylo Bohatyrov[6] (iD)

[1] Al-Turath University, Baghdad 10013, Iraq
[2] Al-Mansour University College, Baghdad 10067, Iraq
[3] Al-Mamoon University College, Baghdad 10012, Iraq
[4] Al-Rafidain University College, Baghdad 10064, Iraq
`majid.alziboon@ruc.edu.iq`
[5] Madenat Alelem University College, Baghdad 10006, Iraq
[6] Kyiv National University of Construction and Architecture, Kyiv 03037, Ukraine

Abstract. Corporate governance plays a critical role in shaping firm performance, especially in emerging markets where regulatory systems are still evolving. This study develops an econometric modeling approach to examine how governance structures—such as board independence, ownership concentration, investor protection, and disclosure quality—affect financial outcomes. Using a mixed-methods design, the analysis combines quantitative financial data with qualitative insights from governance frameworks. Regression equations and structural models were employed to link governance indicators with key performance metrics, including Return on Equity (ROE), Return on Assets (ROA), Tobin's Q, and measures of agency costs. Results show that firms with more independent boards and stronger shareholder protections achieve higher ROE and ROA, while disclosure quality and investor protection significantly improve market valuation. Governance reforms, such as the creation of independent board committees, were also found to reduce agency costs compared to traditional structures. By integrating socio-legal context with data-driven econometric analysis, the study provides practical insights for policymakers, regulators, and corporate leaders seeking to strengthen governance systems, enhance transparency, and promote sustainable growth in emerging economies.

Keywords: Corporate Governance · Firm Performance · Emerging Markets · Board Independence · Econometric Modeling · Governance Reforms

1 Introduction

Corporate governance plays a critical role in shaping financial practices and organizational order. This study focuses on how corporate governance structures, such as board independence, ownership concentration, investor protection, and disclosure quality, affect firm performance in emerging markets using econometric and computational

Z. Molamohamadi et al. (Eds.): ODSIE 2025, CCIS 2855, pp. 3–18, 2026.
https://doi.org/10.1007/978-3-032-17023-1_1

modeling. Emerging markets face unique challenges due to regulatory uncertainties, underdeveloped legal frameworks, and rapidly shifting economic, political, and technological environments. Consequently, a one-size-fits-all approach to corporate law is unlikely to be effective. This article seeks to extend existing research by examining how different legal structures influence corporate governance, financial performance and social responsibilities in these dynamic environments.

Previous studies have explored the adaptation of Western legal frameworks to local conditions. For example, [1] analyzed the role of corporate lawyers in fragmented legal environments, while [2] examined tensions between global norms and local practices in emerging market acquisitions. [3] further investigated how emerging market multinational enterprises (MNEs) influence international strategies through distinctive governance mechanisms. Other research has highlighted the consequences of weak legal frameworks [4] and suggested governance mechanisms to enhance transparency and stakeholder protection [5, 6]. These studies collectively outline the relationship between corporate law and economic development, while revealing gaps in understanding how corporate law responds to technological disruptions, global financial integration, and evolving societal expectations.

Unlike prior research that primarily emphasizes qualitative legal analysis, this paper integrates regression-based econometric models with structured data analysis to examine the direct and indirect effects of governance variables on financial performance indicators such as ROE, ROA, Tobin's Q, and agency costs. By combining case studies, SQL-based statistical modeling, and qualitative insights from 50 semi-structured interviews with corporate lawyers, governance experts, and regulatory officials, this research captures multi-layered evidence on governance practices, shareholder rights, and environmental and social governance. Data sources include governmental reports, financial disclosures, and academic publications. The study is grounded in comparative capitalism and international management frameworks [7, 8], providing the economic and institutional context for its findings.

This paper bridges theoretical governance debates and empirical evidence, offering actionable insights for policymakers, regulators, and business practitioners. By integrating doctrinal legal analysis with empirical governance measures across three emerging markets with diverse legal and regulatory systems, the study contributes to the literature on how corporate law influences firm performance, while highlighting context-specific strategies for improving governance practices.

In doing so, the paper bridges the gap between theoretical governance debates and data-driven evidence, ensuring that recommendations for policymakers and regulators are grounded in both econometric modeling and comparative socio-legal analysis.

2 Research Background

Emerging markets present a rapidly evolving environment for corporate governance, generating significant academic interest in how differences in legal and institutional systems shape firm behavior, social responsibility, and financial performance. However, this expanding body of research still faces several limitations that hinder a unified understanding of governance dynamics. By identifying these gaps, clarifying recurring methodological issues, and proposing feasible solutions, researchers and policymakers can better chart a course toward academic progress and policy reform.

Recent literature underscores wide heterogeneity in corporate social performance (CSP) across emerging and developed economies. [9] finds significant variation in CSP among firms operating in both developed and emerging markets, but does not adequately explain these differences from an institutional or regulatory standpoint. Their use of aggregated data may also conceal industry-specific patterns. Addressing this limitation requires finer-grained and longitudinal analyses that explores how governance frameworks affect CSP outcomes within distinct industries and localities and how reforms can narrow performance gaps. [10] examined corporate finance and governance practices in China and other emerging economies, identifying the absence of standardized comparative frameworks. This lack of methodological consistency, as they note, makes it difficult to evaluate governance effectiveness across legal and cultural contexts. Recent updates in the field continue to emphasize the need for integrated approaches that combine econometric and socio-legal perspectives, allowing cross-national comparability and deeper insight into governance quality.

Particularly, human rights challenges further complicate governance in unlisted or privately held firms in emerging markets. Notably, [11] explored the antecedents of human-rights infringements in these contexts, providing valuable configurational insights, yet leaving unresolved the question of how corporate legal frameworks might prevent or minimize such violations. Recent benchmark studies continue to stress this omission and call for linking disclosure mandates, enforcement mechanisms, and penalties to measurable improvements in corporate conduct. Incorporating these elements requires an interdisciplinary understanding of how legal instruments can simultaneously advance social and ethical standards while ensuring compliance.

Continuing with the topic of governance, [5] highlighted differences in governance models across countries, emphasizing the differences in institutions and governance outcomes across countries. Yet, the literature still lacks consensus on which governance model—stakeholder-oriented or shareholder-oriented—delivers superior results in emerging markets. Inconsistencies in methodology across studies have prevented definitive conclusions. Future research should strive for homogeneous research designs and test hybrid governance models that reconcile the strengths of both approaches.

In the digital-strategy domain, [12] analyzed brand management and market expansion in emerging economies, while [13] conducted a comparative analysis of emerging and traditional IT business models. Although these works offer valuable insights into market growth and digital transformation, they overlook the influence of corporate law on strategic decision-making. To address this, future research should examine how legal regimes facilitate or constrain firms' abilities to overcome regulatory barriers, protect intellectual property, and sustain innovation. This line of inquiry should also assess whether current laws adequately support stakeholder protection in rapidly digitizing industries.

[14] performed a comprehensive bibliometric analysis of foreign direct investment (FDI) in emerging economies, noting that legal and regulatory variables—such as shareholder protections, tax policy, and dispute-resolution mechanisms—are critical

yet under-investigated determinants of FDI flows and investor confidence. More empirical studies are needed to reveal how variations in corporate law shape multinational investment decisions and to develop evidence-based recommendations for improving investment environments in emerging markets.

Recent studies have further enriched our understanding of corporate governance in emerging markets. For instance, [15] examined the effectiveness of internal corporate governance mechanisms in managing non-performing loans in listed banks within Bangladesh, highlighting the significance of board meetings and audit committee size in controlling non-performing loans. In a comparative analysis, [16] explored the short- and long-term effects of internal and external governance on firm performance in both emerging and developed markets, emphasizing the need for balanced governance frameworks.

In summary, while existing research has advanced our understanding of corporate governance, social performance, and financial practices in emerging markets, major challenges remain. These include: lack of methodological rigor and standardized frameworks; neglect of human rights and ethical dimensions; and insufficient legal-comparative analyses in empirical governance studies.

Future progress will depend on integrating legal theory with econometric modeling, applying longitudinal and comparative methods, and paying closer attention to institutional and regulatory contexts. Closing these gaps will yield more robust insights into how corporate law influences organizational behavior and firm performance within the increasingly complex ecosystem of emerging economies.

3 Methodology

This study employs a mixed-methods approach to investigate how corporate governance structures influence firm performance in emerging markets. By combining qualitative and quantitative data, it aims to uncover the complex interplay between corporate law, governance, and firm performance across countries with distinct economic and regulatory structures. The research design, based on established methodologies from previous literature, includes several phases and data collection methods. It incorporates doctrinal and empirical research methods, multiple primary and secondary data sources, and robust statistical and econometric models.

The study adopts a comparative legal methodology suitable for examining corporate governance across jurisdictions with diverse legal and regulatory frameworks. Doctrinal research involves the formal analysis of statutes, judicial decisions, and regulatory filings [1, 6]. In parallel, empirical research measures the strength and durability of governance through quantitative data such as governance indices and financial performance metrics [3, 9]. This integrated approach ensures a comprehensive understanding of the interrelations between legal framework, governance practices, and economic outcomes.

Data were collected from multiple resource using diverse collection strategies to ensure the validity and reliability.

- Primary Sources: Corporate statutes, regulatory filings, and judicial dispositions were analyzed to define and interpret the legal foundation of corporate governance.

- Secondary Sources: The research also draws on academic publications from Scopus, IEEE, and Bentham, as well as reports issued by the World Bank and OECD. These sources link the primary legal data to the academic discourse on law and policy [4, 15].
- Case Selection Criteria: Emerging markets were selected based on economic growth rates, diversity of legal systems (common law vs. civil law), and varying degrees of regulatory maturity. This approach ensures that the jurisdictions chosen represent a wide spectrum of governance environments, enhancing the robustness and significance of cross-country comparisons [14].

In addition, the study conducted 50 semi-structured interviews with corporate lawyers, governance experts, and regulatory officials across three emerging markets. These interviews followed standardized protocols to ensure consistency in data collection while maintaining flexibility to explore themes [1]. The authors also examined 200 corporate governance reports, annual filings, and regulatory documents of publicly listed companies to identify trends in governance practices and compliance standards [3].

Advanced statistical and econometric techniques were employed to test the research hypotheses.

- Regression Analyses: Used to quantify the impact of governance variables, such as board independence, and shareholder concentration, on performance indicators like ROE and ROA.
- Panel Data Methods: Applied to control for time-invariant firm characteristics and macroeconomic conditions, isolating the true effect of governance reforms.
- Quasi-Experimental Design: By comparing organizations that implemented governance reforms with a matched sample that did not, the study generated insights into causal relationships and addressed limitations identified in earlier investigations [7].

3.1 Equations and Models

Corporate Financial Performance Equation. A widely used empirical governance model is a regression equation that explains firm performance as a function of governance variables, controlling for other determinants of performance such as firm size, industry, and macroeconomic conditions:

$$Performance_{it} = \beta_0 + \beta_1 Governance\_Index_{it} + \beta_2 Board\_Independence_{it} +$$

$$\beta_3 Ownership\_Concentration_{it} + \sum \gamma_k X_{k,it} + \varepsilon_{it} \qquad (1)$$

where $Performance_{it}$ is a measure of firm performance (e.g., ROE, ROA, Tobin's Q) for firm i at time t, $Governance\_Index_{it}$ represents an index of governance quality, $Board\_Independence_{it}$ is the proportion of independent directors on the board, $Ownership\_Concentration_{it}$ measures the percentage of shares held by large shareholders, $X_{k,it}$ are control variables (such as leverage, firm size, and market conditions), $\beta_0, \beta_1, \beta_2, \beta_3, \gamma_k$ are parameters to estimate, and ε_{it} is the error term.

This model estimates the extent to which governance structures, such as board independence and ownership concentration, affect firm-level returns on equity. Additional specifications included interaction terms to examine non-linear relationships and to assess the moderating influence of legal reforms on governance outcomes.

Other empirical equations were also developed to analyze agency costs, investor protection, and compliance costs, providing a more comprehensive view of governance-performance linkages.

Agency Cost Equation

$$Agency\_Cost_{it} = \alpha_0 + \alpha_1 Ownership\_Concentration_{it} + \alpha_2 Board\_Monitoring_{it} + \sum \delta_k Z_{k,it} + v_{it} \tag{2}$$

where $Agency\_Cost_{it}$ might be proxied by measures such as the asset utilization ratio or the expense ratio, $Ownership\_Concentration_{it}$ and $Board\_Monitoring_{it}$ are governance variables, $Z_{k,it}$ are control variables, $\alpha_0, \alpha_1, \alpha_2, \delta_k$ are coefficients, and v_{it} is the error term.

This model examines how governance structures, particularly ownership concentration and board oversight, affect agency costs, as captured by measures such as the asset utilization ratio.

Investor Protection and Market Valuation Equation

$$Tobin's\ Q_{it} = \gamma_0 + \gamma_1 Investor\_Protection_{it} + \gamma_2 Disclosure\_Quality_{it} + \sum \theta_k W_{k,it} + \xi_{it} \tag{3}$$

where $Tobin's\ Q_{it}$ is a proxy for firm valuation, $Investor\_Protection_{it}$ represents the strength of legal safeguards for minority shareholders, $Disclosure\_Quality_{it}$ measures the transparency and comprehensiveness of corporate disclosures, $W_{k,it}$ are control variables, such as profitability and growth opportunities, $\gamma_0, \gamma_1, \gamma_2, \theta_k$ are coefficients, ξ_{it} is the error term.

By incorporating investor protection and disclosure quality, this equation analyzes their influence on firm valuation, providing insight into how legal safeguards affect market confidence.

Legal Environment and Compliance Cost Equation

$$Compliance\_Cost_{it} = \kappa_0 + \kappa_1 Regulatory\_Complexity_{it} + \kappa_2 Judicial\_Efficiency_{it} + \sum \eta_k Q_{k,it} + \xi_{it} \tag{4}$$

where $Compliance\_Cost_{it}$ reflects the resources spent on regulatory compliance, $Regulatory\_Complexity_{it}$ measures the number and complexity of legal requirements,

Judicial_Efficiency$_{it}$ captures the effectiveness and timeliness of dispute resolution, $Q_{k,it}$ are other control variables, $\kappa_0, \kappa_1, \kappa_2, \eta_k$ are coefficients, ξ_{it} is the error term.

This equation investigates the financial and operational impact of legal frameworks by focusing on regulatory complexity and judicial efficiency.

Cross-Border Governance and Performance Equation

$$ROE_{it} = \lambda_0 + \lambda_1 Home\_Governance\_Score_{it} + \lambda_2 Host\_Governance\_Score_{it} +$$
$$\lambda_3 Cross - Border\_Integration_{it} + \sum w_k Y_{k,it} + \psi_{it} \quad (5)$$

where ROE_{it} is the return on equity, *Home_Governance_Score*$_{it}$ and *Host_Governance_Score*$_{it}$ governance quality in the firm's home and host countries, respectively, *Cross_Border_Integration*$_{it}$ measures the degree of coordination between home and host governance practices, $Y_{k,it}$ are control variables, like firm size, leverage, market conditions, $\lambda_0, \lambda_1, \lambda_2 Host\_Governance\_Score_{it} + \lambda_3, w_k$ are coefficients, and ψ_{it} is the error term.

This model captures the interplay between home-country and host-country governance practices, examining how cross-border integration of governance frameworks influences firm performance.

These models collectively allow for a thorough examination of the relationships between corporate governance, legal structures, and firm performance.

3.2 Scope and Considerations

This study is constrained by data availability as well as by the choice of jurisdictions. The analysis focuses on selected emerging markets based on specific criteria, thereby excluding potentially interesting regions that may also offer valuable lessons. Moreover, differences in data quality, including variations in the level of enforcement reported by each state and differences in transparency, constitute potential sources of inconsistency. Recognizing and addressing these factors enhances transparency in the research and provides a clearer framework for interpretation [10, 17].

The methodology combines comparative qualitative legal analysis, qualitative and quantitative data collection, and advanced econometric modeling to offer an in-depth exploration of corporate governance in emerging markets. This study fills significant gaps in the existing literature by modeling the relationship between governance structures and company outcomes through the lenses of both doctrinal analysis and empirical methodology, whilst also providing key insights into design opportunities that challenges existing corporate governance architectures in these rapidly evolving economic systems.

4 Results

4.1 Descriptive Statistics

This study presents descriptive statistics for the dataset. It reports the characteristics of each key governance variable and performance indicator, including measures of central tendency and dispersion for each country. The broader table also includes additional indicators not previously listed, firm size (log of total assets), leverage ratio, earnings per share (EPS) and dividend payout ratio. These descriptive statistics presented provide a richer perspective on structural differences among firms operating in India, Brazil, and South Africa, and on their comparative performance.

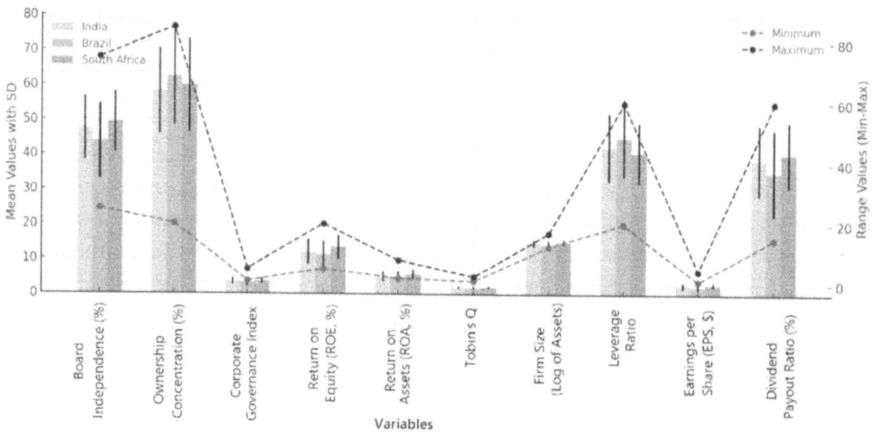

Fig. 1. Descriptive Statistics for corporate governance variables and performance measures (mean with SD).

There are several key insights that can be drawn from the data in Fig. 1, which compares the three different countries. For example, South Africa shows the highest mean board independence at 49.3% (higher than both India or Brazil), suggesting a stronger emphasis on external oversight. Brazil demonstrates the most concentrated ownership structure, which may influence corporate decision-making dynamic. On the Corporate Governance Index, South Africa ranks highest with an average score of 3.6, indicating a slightly stronger governance framework.

In terms of performance, South African companies report the highest average levels of return on equity (ROE) (13.5%) and return on assets (ROA) (5.6%) among the three markets, indicating stronger profitability than India and Brazil. Furthermore, the firm size indicator (log of assets) shows that South African firms are larger on average, which is likely to positively influence their performance metrics. Meanwhile, Indian companies exhibit more stable leverage ratios and higher dividend payout ratios, suggesting better debt management and a greater inclusion toward rewarding shareholders.

4.2 Correlation Analysis

The study examines the statistical relationships between the key governance and performance variables. Using pairwise Pearson correlation coefficients, we assess the strength of association among all pairs of the variables contained in the dataset. This interaction among different indicators of governance, Board Independence, Ownership Concentration, and the Corporate Governance index, with indicators of financial performance such as ROE, ROA, Tobin's Q, forms the basis of the analysis. However, additional variables, including the Dividend Payout Ratio, Leverage Ratio, and Firm Size, are also incorporated to provide a more comprehensive view of the interactions at play.

Table 1. Correlation matrix for governance and financial performance variables.

Variable	Corporate Governance Index	Board Independence	Ownership Concentration	ROE	ROA	Tobin's Q	Dividend Payout Ratio	Leverage Ratio	Firm Size
Corporate Governance Index	1.00	0.38**	−0.27*	0.32*	0.29*	0.34**	0.28*	−0.14	0.21*
Board Independence	0.38**	1.00	−0.20*	0.36*	0.31*	0.32**	0.15	−0.10	0.18
Ownership Concentration	−0.27*	−0.20*	1.00	−0.21*	−0.23*	−0.22*	−0.12	0.25*	−0.19
ROE	0.32*	0.36*	−0.21*	1.00	0.68*	0.60*	0.33**	−0.18	0.30*
ROA	0.29*	0.31*	−0.23*	0.68*	1.00	0.63*	0.26*	−0.20*	0.27*
Tobin's Q	0.34**	0.32**	−0.22*	0.60*	0.63*	1.00	0.24*	−0.12	0.29*
Dividend Payout Ratio	0.28*	0.15	−0.12	0.33**	0.26*	0.24*	1.00	−0.05	0.17
Leverage Ratio	−0.14	−0.10	0.25*	−0.18	−0.20*	−0.12	−0.05	1.00	−0.08
Firm Size	0.21*	0.18	−0.19	0.30*	0.27*	0.29*	0.17	−0.08	1.00

* Significant at $p < 0.05$; ** significant at $p < 0.01$.

Table 1 shows some of the major trends. The Corporate Governance Index correlates positively with ROE ($r = 0.32$, $p < 0.05$), ROA ($r = 0.29$, $p < 0.05$) and Tobin's Q ($r = 0.34$, $p < 0.01$), indicating that stronger governance frameworks are positively associated with improved operating and market performance. Board Independence also sows a significant positive correlation with these performance measures, underscoring the importance of independent oversight. In contrast, Ownership Concentration, exhibits a negative correlation with ROE, ROA, and Tobin's Q, suggesting that higher ownership concentration may hinder firm performance.

Other variables, such as the Dividend Payout Ratio and Firm Size, display moderate positive correlations with performance indicators, suggesting that larger firms with stable dividend policies tend to perform better. Conversely, the Leverage Ratio shows a weak negative correlation with performance outcomes, indicating that higher levels of debt may not necessarily support stronger financial performance.

4.3 Regression Analysis Results

Multiple regression analyses were performed using Return on Equity (ROE) as the dependent variable to assess the relationship between governance variables and firm financial performance. Firm size, leverage, and industry effects were collected, with one key governance variable introduced in each model: Corporate Governance Index, Board Independence, and Ownership Concentration. These models were designed to isolate the effects of governance structures on financial performance. The analysis also incorporates additional performance measures and interaction terms to examine whether combined governance mechanism further enhance explanatory power.

Table 2. Regression results for financial performance measures (ROE and ROA).

Variable	Model 1 (Gov. Index)	Model 2 (Board Index)	Model 3 (Ownership Conc.)	Model 4 (Full Model)	Model 5 (Interactive Terms)
Corporate Governance Index	1.27** (0.44)			1.22** (0.47)	1.15** (0.45)
Board Independence		0.88** (0.31)		0.82** (0.33)	0.79** (0.32)
Ownership Concentration			-0.53* (0.23)	-0.49* (0.24)	-0.45* (0.22)
Gov. Index × Board Indep					0.22* (0.11)
Gov. Index × Ownership Conc					-0.17* (0.09)
Control Variables (Size, Leverage, Industry)	Yes	Yes	Yes	Yes	Yes
R-squared	0.29	0.32	0.26	0.37	0.42
Observations	250	250	250	250	250

* Significant at $p < 0.05$; ** significant at $p < 0.01$.

These regression results in Table 2 show the distinct effects of different corporate governance measures on firm performance. The Corporate Governance Index consistently exhibits a positive and statistically significant relationship with ROE, even after including other variables, suggesting that stronger governance frameworks are associated with improved financial outcomes. Board Independence similarly shows a positive effect, indicating that boards with higher independence provide better oversight, thereby enhancing firm profitability.

In contrast, Ownership Concentration is negatively correlated with ROE, implying that highly concentrated ownership may impede performance, potentially due to reduced checks and balances. In Model 4 (full model), all three governance variables remain significant, emphasizing the value of a balanced governance structure.

Model 5, which includes interaction terms, reveals that combinations of governance mechanisms can exert additional influence on financial outcomes. Specifically, the positive interaction between Corporate Governance Index and Board Independence suggest that a robust governance amplifies the benefits of independent boards. Conversely, the negative interaction between Corporate Governance Index and Ownership Concentration indicates that even in well-governed corporations, concentrated ownership can partially undermine the benefits of strong governance.

4.4 Hypothesis Testing and Robustness Checks

Hypothesis tests and robustness checks using different models and performance measures were applied to validate the research hypotheses and assess the stability of the results. Hypothesis 1 concerns the relationship between the Corporate Governance Index and financial performance, Hypothesis 2 concerns Board Independence, and Hypothesis 3 concerns Ownership Concentration.

Robustness of Results: To ensure that the relationships observed in the regression analyses are consistent across different model specifications, additional robustness tests were conducted using alternative dependent variables such as a ROA and Tobin's Q and by including additional control variables. Table 3 summarizes the results of these tests.

The results in Table 3 provide strong support to the three proposed hypotheses. The significant positive coefficients on the Corporate Governance Index across all performance measures confirm Hypothesis 1. This relationship remains robust after alternative model specifications, indicating that companies with higher governance scores consistently achieve better financial outcomes.

Hypothesis 2 is also confirmed: higher Board Independence consistently exhibits a positive effect on ROE, ROA, and Tobin's Q, demonstrating that independent board members enhance monitoring and improve profitability.

For Hypothesis 3, the negative coefficients for Ownership Concentration indicate that firms with more concentrated ownership tend to perform worse financially. This pattern persists across all alternative model specifications and additional control variables.

The robustness checks further support these findings. Considering Tobin's Q and ROA as alternative dependent variables, along with additional controls such as industry fixed effects and leverage ratios, mitigates the risk that any observed relationships are driven by a single measure. Overall, the results consistently underscore the pivotal role of governance structures in influencing firm performance across multiple models and contexts.

4.5 Cross-Country Comparisons

The study shows a country-level analysis of corporate governance and performance indicators, revealing compellingly different governance patterns. India, Brazil and South Africa were examined independently to identify the most relevant governance aspects for each country. Specific firm performance drivers for each region were analyzed in relation to the key predictors: Corporate Governance Index, Board Independence, and Ownership Concentration, as these emerged from the dataset. Figure 2 includes additional performance measures and control variables to provide a comprehensive perspective.

The analysis shows that the relationship between governance and performance depends heavily on the country. The Corporate Governance Index demonstrates the

Table 3. Hypothesis testing results and robustness checks.

Hypothesis	Dependent Variable	Governance Variable	Coefficient	p-value	Robustness Test Result	Additional Controls
Hypothesis 1: Positive Governance Index-Performance Relationship	ROE	Corporate Governance Index	1.27**	<0.01	Consistent (Tobin's Q, ROA)	Yes
	Tobin's Q	Corporate Governance Index	0.34**	<0.01	Consistent (ROE, ROA)	Yes
	ROA	Corporate Governance Index	0.29*	<0.05	Consistent (ROE, Tobin's Q)	Yes
Hypothesis 2: Positive Board Independence-Performance Relationship	ROE	Board Independence	0.88**	<0.01	Consistent (Tobin's Q, ROA)	Yes
	Tobin's Q	Board Independence	0.32**	<0.01	Consistent (ROE, ROA)	Yes
	ROA	Board Independence	0.31*	<0.05	Consistent (ROE, Tobin's Q)	Yes
Hypothesis 3: Negative Ownership Concentration-Performance Relationship	ROE	Ownership Concentration	−0.53*	<0.05	Consistent (Tobin's Q, ROA)	Yes
	Tobin's Q	Ownership Concentration	−0.22*	<0.05	Consistent (ROE, ROA)	Yes
	ROA	Ownership Concentration	−0.23*	<0.05	Consistent (ROE, Tobin's Q)	Yes

* Significant at $p < 0.05$; ** significant at $p < 0.01$.

highest significant positive relationship with ROE in India, suggesting that the overall governance framework is crucial for improving performance capacity. Board independence is also a significant predictor in India, though to a lesser extent than in Brazil and the negative effect of Ownership Concentration is not significant, indicating that other factors may be needed to counter concentrated ownership.

In Brazil, Board Independence emerges as the strongest governance predictor. Companies with more independent boards outperform their peers on average, supporting the role of external oversight and balanced decision-making. The Corporate Governance Index remains important, but less dominant than Board Independence, while Ownership Concentration has a negative effect, though smaller than that of independent boards.

In South Africa, the impact of Ownership Concentration is the most negative, highlighting challenges posed by concentrated shareholding structures. Firms with more widely held equity perform better, broader ownership aligns incentives and limits information asymmetry. Both the Corporate Governance Index and Board Independence

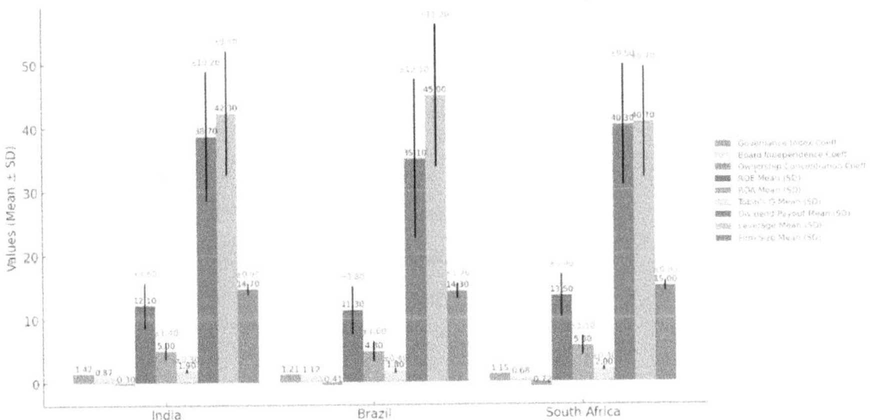

Fig. 2. Cross-country comparison of governance and performance metrics.

remain positive and significant, but Ownership Concentration is the primary factor affecting firm performance in this market.

4.6 Impact of Governance Reforms

This study examines the performance rebound of firms that implemented major governance reforms in the past five years. It helps us understand how much Baltic Financial benefited from these changes by looking at the financial metrics before and after reforms, such as separate board committees and improved voting rights. For comparison, a matched control group of firms that retained traditional governance structure is also included. Incorporating other performance metrics and governance quality indicators, Table 4 provides additional insights into the outcomes of the reforms.

These results provide compelling evidence that firms pursuing governance reforms undergo significant improvements across a variety of financial and governance metrics. The median ROE increased by 2.9 pp with a highly significant p-value (<0.01), confirming the effectiveness of these profitability-enhancing measures. Likewise, both ROA and Tobin's Q increased significantly, suggesting improvements in operational efficiency and market valuation.

On average, the Governance Index increased by 0.9 points, showing a significant improvement in governance quality following the reforms. Additionally, the Dividend Payout Ratio increased, signaling a greater emphasis on returning value to shareholders in the post-reform phase. Statistically, the Leverage Ratio decreased, reflecting reduced reliance on financial resources. No such trends were observed in the control group, which maintained relatively stable performance metrics over the same period.

The data as a whole support the perspective that governance reforms yield increased oversight and transparency, as well as significant improvements in financial performance. These findings highlight the necessity to implement and sustain high governance standards to boost long-term firm performance.

Table 4. Performance before and after governance reforms.

Metric	Pre-Reform Mean (SD)	Post-Reform Mean (SD)	Mean Difference	t-statistic	p-value	Control Group (Mean/SD)
ROE (%)	10.6 (2.9)	13.5 (3.1)	+2.9	4.20	<0.01	10.8 (3.0)
ROA (%)	4.3 (1.2)	5.7 (1.4)	+1.4	3.90	<0.01	4.4 (1.3)
Tobin's Q	1.6 (0.3)	1.9 (0.4)	+0.3	3.50	<0.01	1.6 (0.3)
Governance Index	3.2 (1.0)	4.1 (0.8)	+0.9	4.10	<0.01	3.3 (1.1)
Dividend Payout Ratio (%)	36.0 (9.5)	42.0 (10.2)	+6.0	4.30	<0.01	36.2 (9.7)
Leverage Ratio (%)	40.0 (8.0)	38.5 (7.5)	−1.5	2.10	0.04	40.2 (8.2)

5 Discussion

This study provides empirical evidence on the relationship between corporate governance structures and firm performance in emerging markets, focusing on India, Brazil, And South Africa. The results confirm that stronger governance frameworks, as measured by the Corporate Governance Index, positively influence financial performance. Firms with higher governance scores demonstrated significantly higher ROE, ROA, and Tobin's Q, indicating that robust governance practices enhance both operational efficiency and market valuation. These findings support the research objective of evaluating how governance variables affect firm performance across emerging markets.

Board Independence emerged as a significant positive predictor of performance. Across all models, higher proportions of independent directors were associated with improved financial outcomes, highlighting the importance of outside oversight in enhancing firm accountability and operational efficiency. The interaction term between Corporate Governance Index and Board Independence further suggests that independent boards amplify the positive effect of overall governance structures on performance, confirming that the joint implementation of multiple governance mechanisms is beneficial.

Conversely, Ownership Concentration was negatively associated with performance. Highly concentrated ownership structures reduced ROE, ROA, and Tobin's Q, implying that excessive concentration can limit oversight and create agency conflicts. The interaction between Ownership Concentration and Corporate Governance Index indicates that even with strong governance systems, concentrated ownership may dampen the potential performance benefits, emphasizing the need for proportional governance structures in emerging market firms.

Cross-country comparisons highlight that the effect of governance mechanisms varies by jurisdiction. In India, the Corporate Governance Index had the strongest positive relationship with ROE, while Board Independence played a more substantial role in Brazil. In South Africa, the negative impact of Ownership Concentration was most pronounced, reflecting the challenges associated with concentrated shareholder structures. These results demonstrate that although global governance principles are relevant, their effectiveness is context-dependent, supporting the study's objective of exploring jurisdiction-specific dynamics.

The analysis of governance reforms indicates that firms implementing major reforms—such as establishing independent board committees and enhancing shareholder protections—experienced substantial improvements in both governance quality and financial performance. Post-reform ROE increased by 2.9% points, ROA by 1.4% points, and Tobin's Q by 0.3, accompanied by higher Dividend Payout Ratios and improved Governance Index scores. Leverage ratios decreased slightly, suggesting better financial management following reforms. These findings directly address the research goal of evaluating the practical impact of governance reforms on firm outcomes.

Robustness checks across alternative performance measures and additional controls confirmed the stability of these relationships. The consistently positive effects of Corporate Governance Index and Board Independence, and the negative effect of Ownership Concentration, indicate that the observed patterns are not specific to a single measure but are generalizable across multiple financial indicators.

In summary, this study demonstrates that stronger governance frameworks, higher board independence, and moderated ownership concentration are associated with improved firm performance in emerging markets. The results highlight the practical significance of implementing governance reforms and adapting corporate governance practices to local contexts, while also confirming the relevance of proportional and combined governance mechanisms in enhancing financial outcomes.

6 Conclusion

This study empirically examined the relationship between corporate governance and firm performance in emerging markets, focusing on India, Brazil, and South Africa. The results demonstrate that stronger governance frameworks—characterized by higher board independence, moderated ownership concentration, and improved compliance practices—are associated with enhanced financial performance, including ROE, ROA, and Tobin's Q. Firms that implemented governance reforms experienced significant improvements in both operational efficiency and market valuation, confirming the practical benefits of well-designed governance mechanisms.

Despite these contributions, the study has limitations. The analysis is restricted to three emerging markets and relies primarily on publicly available financial and governance data, which may over-represent firms with more transparent practices. The five-year time frame also limits assessment of long-term sustainability of reforms.

Future research could extend these findings by including additional emerging markets, exploring sector-specific differences, and employing more granular or longitudinal datasets. Technical innovations such as AI-driven compliance monitoring, blockchain-based reporting, and automated governance auditing tools offer promising avenues to enhance transparency, enforce governance standards, and reduce operational costs.

In conclusion, robust corporate governance is a key driver of firm performance and investor confidence in emerging markets. Sustainable improvements require not only the implementation of governance reforms but also effective enforcement and integration of emerging technologies to ensure transparency, accountability, and long-term growth.

References

1. Garth, B.G.: Corporate lawyers in emerging markets. Ann. Rev. Law Soc. Sci. **12**, 441–457 (2016)
2. Thenmozhi, M., Narayanan, P.: Rule of law or country level corporate governance: what matters more in emerging market acquisitions? Res. Int. Bus. Financ. **37**, 448–463 (2016)
3. Luo, Y., Zhang, H.: Emerging market MNEs: qualitative review and theoretical directions. J. Int. Manag. **22**, 333–350 (2016)
4. Gurrea-Martínez, A.: Insolvency law in emerging markets. Corp. Gov. Arrangements Laws eJournal (2020)
5. Aguilera, R.V., Haxhi, I.: Comparative corporate governance in emerging markets. Corp. Finan. Governance (2018)
6. Zhao, J.: Promoting a more efficient corporate governance model in emerging markets through corporate law. Wash. Univ. Glob. Stud. Law Rev. **15**, 447–493 (2016)
7. Schedelik, M., Nölke, A., Mertens, D., May, C.: Comparative capitalism, growth models and emerging markets: the development of the field. New Polit. Econ. **26**, 514–526 (2020)
8. Cuervo-Cazurra, A., et al.: Uncommoditizing strategies by emerging market firms. Multinational Bus. Rev. (2019)
9. Ting, I.W.K., Azizan, N.A., Bhaskaran, R.K., Sukumaran, S.K.: Corporate social performance and firm performance: comparative study among developed and emerging market firms. Sustainability (2019)
10. Cumming, D.J., Verdoliva, V., Zhan, F.: New and future research in corporate finance and governance in China and emerging markets. Emerg. Markets Rev. (2021)
11. Ciravegna, L., Nieri, F.: Business and human rights: a configurational view of the antecedents of human rights infringements by emerging market firms. J. Bus. Ethics **179**, 431–450 (2021)
12. Nwabekee, U.S., Abdul-Azeez, O.Y., Agu, E.E., Ijomah, T.I.: Brand management and market expansion in emerging economies: a comparative analysis. Int. J. Manage. Entrepreneurship Res. (2024)
13. Мозговий, Я.I.: Comparative analysis of emerging and traditional it business models in a global landscape. Цифрова економіка та економічна безпека (2024)
14. Nazzal, A., Sánchez-Rebull, M.-V., Niñerola, A.: Foreign direct investment by multinational corporations in emerging economies: a comprehensive bibliometric analysis. Int. J. Emerg. Markets (2023)
15. Hossain, M.K., Rahman, F., Golder, U., Kabir, H.: Effectiveness of internal corporate governance mechanisms in controlling non-performing loans of an emerging economy. Schmalenbach J. Bus. Res. (2025). https://doi.org/10.1007/s41471-025-00216-7
16. Natto, D., Mokoaleli-Mokoteli, T.: Short- and long-term impact of governance on firm performance in emerging and developed economies: a comparative analysis. Int. J. Discl. Gov. (2025)
17. Sahasranamam, S., Arya, B., Sud, M.: Ownership structure and corporate social responsibility in an emerging market. Asia Pac. J. Manage. **37**, 1165–1192 (2019)

Cybercrime Legislation as a Catalyst for Digital Economic Growth: A Social Science Approach

Mudher Ghaeb Ali[1] , Ammar Khadim Jasim[2] ,
Shamel Abdul-Sattar Jaleel Shalaan[3] , Haider Mahmood Jawad[4]([✉]) ,
Riyam M. Alsammarraie[5] , and Julia Iurynets[6]

[1] Al-Turath University, Baghdad 10013, Iraq
[2] Al-Mansour University College, Baghdad 10067, Iraq
[3] Al-Mamoon University College, Baghdad 10012, Iraq
[4] Al-Rafidain University College, Baghdad 10064, Iraq
`Haider.mahmood@ruc.edu.iq`
[5] Madenat Alelem University College, Baghdad 10006, Iraq
[6] Kyiv National University of Construction and Architecture, Kyiv 03037, Ukraine

Abstract. Digital transformation has reshaped economic systems worldwide, yet cybercrime continues to undermine financial stability, critical infrastructure, and individual privacy. Existing legislative frameworks remain fragmented and insufficient to manage transnational threats, creating security gaps that hinder sustainable digital economic growth. The study develops a comprehensive framework linking cybercrime legislation with social and economic dimensions to enhance cybersecurity governance and foster digital development. Through comparative legal analysis and empirical case studies from diverse jurisdictions, the paper demonstrates how integrated legislative and economic strategies can strengthen international cooperation, improve law enforcement capabilities, and stimulate technological innovation. The research contrasts successful models from South Korea and the United States with outdated criminal codes in Nigeria and other African contexts, where limited enforcement capacities constrain progress. Findings reveal that effective digital economic growth depends on adaptive, enforceable, and technology-driven legal systems. The study concludes with policy recommendations emphasizing global collaboration, flexible governance, and the strategic use of emerging technologies such as artificial intelligence and blockchain to ensure secure, inclusive, and resilient digital economies.

Keywords: Cybercrime Legislation · Digital Economy · Cybersecurity Governance · International Cooperation · Technological Innovation · Policy Development

1 Introduction

Internet-based transaction systems and virtual network connectivity are major engines driving global economic growth, yet the rapid pace of digital transformation also exposes economies to growing cyber risks. Breaches in financial networks, attacks on critical

Z. Molamohamadi et al. (Eds.): ODSIE 2025, CCIS 2855, pp. 19–40, 2026.
https://doi.org/10.1007/978-3-032-17023-1_2

infrastructure, and violations of privacy contribute to global instability, undermine public trust, and underscore the need for comprehensive cybercrime legislation capable of supporting sustainable digital economic development. Despite improvements in digital governance, international frameworks still face vulnerabilities that hinder enforcement and cross-border coordination [1–3].

Recent research sheds light on the evolving nature of cybercrime and its economic impacts. Kumar (2024) notes that outdated laws struggle to keep pace with the rapid evolution of technology [4]. Additional studies reveal that legal systems, although evolving, often lag behind the pace of emerging cyber threats [5, 6].

Comparative analyses reveal ongoing inconsistencies in enforcement. Elegbe (2024) exams a comparative cybercrime legislation across various jurisdictions [5], while Gomathy et al. (2024) emphasizes the importance of integrated measures addressing both cybercrime and cyberterrorism [7]. Popko and Popko (2021) point out that fragmented implementation structures increase risks in transnational cases [8], and the lack of standardized international protocols continues to impede prosecutorial efficiency, allowing offenders to exploit safe havens [9].

Empirical studies illustrate sector-specific vulnerabilities. Musatkina et al. (2020) show that the Russian Federation faces recurring cyber threats with identifiable patterns [10], and Noor Uddin Milon et al. (2024) report uneven effects of cybercrime on developing economies, particularly affecting small businesses and healthcare sectors [11]. Denikaeva et al. (2022) highlight the sensitivity of the financial sector to cyber threats and the critical need for targeted legal protection [12]. Tailored legislation is shown to strengthen economic resilience by channeling investments toward high-risk sectors [13].

This article aims to develop an integrative framework for cybercrime law enforcement that protects digital economies while fostering innovation-driven growth. The study examines the mismatch between current legal systems and emerging digital threats, proposing adaptive legislative approaches that consider social, legal, and technological dimensions. Key focus areas include international collaboration, public–private partnerships, forensic methods, and technology-driven enforcement tools, particularly artificial intelligence (AI) and blockchain, to enhance cross-jurisdictional coordination and cybersecurity outcomes.

The problem addressed by this research is the persistence of serious cyberattacks that exploit inconsistent and outdated regulations. Many national systems remain disconnected, limiting cooperation and slowing investigative responses, enabling the continuation of illicit digital activities. Sector-specific vulnerabilities, especially in finance and healthcare, are aggravated by insufficient cybersecurity investment and weak legislative safeguards. Consequently, outdated and inflexible legal frameworks impede both innovation and protection, constraining sustainable digital economic progress.

Although advances in cyber safety have strengthened certain areas, global regulatory systems still leave key business processes exposed. This study addresses these gaps by developing an analytical framework linking cybercrime legislation with digital economic performance. Using EU Digital Single Market regulatory analysis [14] and economic modeling [12], the research applies quantitative indicators and case studies across jurisdictions to evaluate legislative performance and recommend adaptive reforms. Policy implementation is assessed through empirical data from multiple countries to predict potential outcomes of proposed reforms [15]. Ultimately, this study seeks to advance policy-making through adaptive, evidence-based legislative development that anticipate emerging threats while supporting innovation-driven economic growth [16].

2 Literature Review

The digital economy continues to face escalating threats as cybercriminals execute financial scams, targeted breaches, identity theft, and destructive ransomware invasions. Allahrakha (2024) identifies cybercrime as an essential destabilizer for global economies since it operates on a transnational scale that limits the reach of effective law enforcement [16]. The scope of complex and persistent hostile activities targeting essential infrastructure and financial systems has intensified significantly in recent years according to analytical findings [10]. As Kumar (2024) shows, conventional legal systems struggle to adapt to the rapidly evolving cyber risks that criminals exploit through unnoticed vulnerabilities [4]. Sector-specific analyses reveal that nations with insufficient cybersecurity resources, particularly developing economies, suffer the most from such attacks [11].

Legal frameworks across jurisdictions show marked disparities. Although Russia and China have decided not to ratify the Budapest Convention, it remains the principal foundation of international collaboration against cybercrime [8]. Elegbe (2024) argues that the flexible nature of constitutional rules lacks the capacity to address emerging security challenges resulting from blockchain systems and artificial intelligence [5]. Gomathy et al. (2024) emphasize that attempts to establish cybersecurity rules aligned with counterterrorism objectives encounter major implementation challenges [7]. In contrast, the European Union's comprehensive enforcement of the GDPR demonstrates how consistent regulatory compliance can simultaneously safeguard data integrity and mitigate network security risks [14]. Conversely, the key challenges within the legal framework, operational enforcement, and strategic policy-making limits the effective combat of cybercrime [17].

Cybercrime produces three principal economic consequences: immediate financial losses, reputational damage, and erosion of investor trust. Noor Uddin Milon et al. (2024) demonstrate that developing economies are disproportionately affected as cyberattacks hinder innovation potential and restrict growth. Denikaeva et al. (2022) further identify structural weaknesses in financial systems that highten economic vulnerability [12]. Studies underscore that investor confidence tends to strengthen in digital economies supported by robust legal systems that effectively reduce risk and promote creative development [2]. Nevertheless, a critical empirical gap persists, as limited evidence exists regarding the direct relationship between cybercrime legislation and economic resilience [16].

Advances in AI have generated a new class of cyber adversaries that often surpass the capabilities of existing legal frameworks. Cybercriminals exploit AI through automated phishing, deepfake impersonation, and adaptive malware that evade traditional detection systems. According to Jerónimo (2019), this growing threat, amplified by global digital interconnectivity, necessitates agile and technology-aware legal responses. The convergence of AI, blockchain, and IoT technologies introduces compound security challenges [18]. The immutable nature of can inadvertently shield illicit activity, as discussed by

Sannikova and Kharitonova (2019), while the rapid expansion of IoT networks amplifies exposure to system vulnerabilities. Thus, sector-tailored and technology-specific legal frameworks must be established to address these emerging risks effectively [19].

The implementation of modern technological tools offers new opportunities for enhancing cybersecurity enforcement. Predictive analytics systems powered by AI deliver improved real-time threat detection, as noted by [20]. However, Amoo et al. (2024) caution Legal frameworks should promote technological innovation while implementing safeguards to deter malicious exploitation. Achieving this delicate balance demands collaboration among lawmakers, technologists, legal scholars, and cybersecurity practitioners to ensure that regulations support progress yet remain strong enough to prevent misuse. [21]. According to Sviatun et al. (2021), the cross-border nature of cyber threats demands standardized legal frameworks and enhanced transnational cooperation. Consequently, policymakers must design adaptive and forward-looking legal infrastructures capable of responding to AI-driven cybercrime in real time.

While modern cybercrime laws represent a step forward, substantial gaps remain in their enforcement and technological integration. Existing research often overlooks the distinctive digital economic characteristics that demand tighter links between technological advancement and legal evolution [5, 22]. Ruddin and Sgno (2024) find that emerging legislation struggles to address the cybersecurity implications of AI and blockchain [22]. Furthermore, the lack of standardized international enforcement procedures continues to hinder effective transnational cooperation [8].

Insufficient attention to industry-specific vulnerabilities further weakness global cybersecurity strategies. Musatkina et al. (2020) reveal that the healthcare and finance sectors, in particular, remain undeserved by generalized one-size-fits-all legislative policies [10]. Similarly, Chimchiuri (2024) shows that while cybercrime legislation continues to evolve, research seldom explores the dynamic interplay between regulatory development and economic innovation [23].

A new paradigm of adaptive, technology-driven law-making is thus required to bridge the gap between legal systems and modern digital realities. International standardization and coordinate enforcement mechanisms should form the cornerstone of future cooperation [1, 5, 8]. Broader strategies must integrate sector-specific measures to protect vulnerable industries [11], while strengthening enforcement capacity through AI-enabled predictive analytics [7].

Recent studies reinforce this urgency. Abrardi et al. (2025) emphasizes the financial magnitude of cyber threats and the pressing need for policy-driven, economically informed responses [24]. Likewise, Bekkers et al. (2025) provide new empirical insight into how law-enforcement agencies address cross-border money-mule networks, pointing directly to enforcement coordination challenges in the digital economy [25].

Despite this growing body of literature, a notable gap remains. While considerable research examines the technical, economic or law-enforcement dimensions of cybercrime individually, few studies integrate legislative performance, digital economic growth metrics, and sector-specific enforcement outcomes into a unified analytical framework. Most extant work either focuses on legal design (e.g., Fragmentation of frameworks) or economic impact (e.g., losses, trust erosion) but stops short of linking how adaptive, technology-driven legal systems translate into measurable digital economic resilience across different jurisdictions and industry sectors. In addition, although emerging technologies such as AI and blockchain are mentioned in passing, there is insufficient empirical work mapping how legal systems are structurally adapting to such innovations or how that adaptation affects economic growth trajectories. This study therefore seeks to fill that gap by proposing and testing an integrative model that connects legislative adaptability and enforcement capacity with digital economy indicators in diverse settings.

3 Methodology

3.1 Analytical Framework

This study implements a multidisciplinary analytical framework to develop an advanced structure specifically designed for cybercrime law enforcement in the digital economy. The study conducted an analytical review of worldwide legislative frameworks to determine their capacity to confront emerging cyber threats. This research draws on the findings from Allahrakha [1] and Kumar [4] to examine the gaps in enforcement approaches and their impact on economic development. Case studies are included to assess legislative effectiveness across multiple jurisdictions, based on examples from Europe, Asia and developing economies, as described in [10, 11].

The methodology combines mathematical modeling with empirical data analysis to reveal relationships between changes in cybercrime rate changes and legislative updates. Statistical methods are employed to evaluate how legislative reforms influence digital economic indicators, following [14].

3.2 Data Sources

A comprehensive dataset was collected from multiple diverse sources, including:

- International Reports: Global cybercrime statistics and legal implementation data were obtained from official documents published by International Telecommunication Union's (ITU), Organisation for Economic Co-operation and Development (OECD), and United Nations Office on Drugs and Crime (UNODC) [5, 6].
- Academic Databases: More than 150 peer-reviewed academic papers were analyzed to provide insights into sector-specific vulnerabilities, including contributions from [12] and [24].
- Interviews: structured interviews were conducted with 30 legal experts, policymakers, and cybersecurity specialists to support qualitative assessments of enforcement challenges [17, 22].
- Case Reviews: Fifty instances of cybercrime in the finance, healthcare, and critical infrastructure sectors were analyzed to evaluate legislative implementation [10, 12].

3.3 Evaluation Metrics

To assess the effectiveness of current and proposed legislative frameworks, three key metrics were employed:

Cybercrime Reduction: Measured through reduction in cybercrime rates reported by UNODC and independent research [5, 7].
Economic Resilience: Evaluated using Gross Domestic Product (GDP) growth, Foreign Direct Investment (FDI) figures, and digital economy participation rates following [2] and [11].
Adaptability: A scoring model assessed the ability of legislative frameworks to address emerging technologies, including AI and blockchain [14, 16].

3.4 Hypothesis

Digital economic resilience improves when nations implement adaptable and enforceable cybercrime laws aligned with international standards. This hypothesis is tested through correlation and regression analyses of economic and legislative data [11, 14].

3.5 Equations and Mathematical Models

The methodology evaluates networked reforms in relation to cybercrime statistics:

Cybercrime Reduction Model

$$C_r = \alpha L_f + \beta D_g + \epsilon \tag{1}$$

where C_r = cybercrime reduction rate; L_f = legislative framework adaptability score, assessing how laws address evolving cyber threats; D_g = digital economic growth indicator (e.g., GDP from digital sectors); α, β = regression coefficients derived from analyzing legislative impacts; ϵ = error term accounting for unobserved factors.

The model demonstrates the combined impact of legal adaptability and economic growth on cybercrime reduction. Calibration was performed using data from 50 countries spanning 2010 to 2023 [1, 11, 14]. The research reveals which specific legal reforms produce the best outcomes for cyber threat prevention.

Economic Impact of Cybercrime Legislation

$$E_g = \gamma L_f + \delta C_s - \zeta V_s + \mu \tag{2}$$

where E_g = economic growth attributable to digital sectors; L_f = legislative framework score, evaluating comprehensiveness and adaptability; C_s = cybercrime severity index, quantifying the economic impact of cyber incidents; V_s = vulnerability score of a sector, like finance, healthcare; γ, δ, ζ = coefficients representing the weight of each variable; and μ = residual term representing external factors.

This model shows how adaptive regulations reduce the economic costs of cybercrime and strengthen sectoral resilience [2, 11, 12].

Effectiveness of International Cooperation

$$I_c = \lambda S_e + \phi H_i - \psi J_i \tag{3}$$

where I_c = international cooperation index, reflecting enforcement outcomes in transnational cybercrime cases; S_e = shared enforcement resources, such as task forces, data-sharing agreements; H_i = harmonization of laws across jurisdictions; J_i = jurisdictional conflicts or barriers to collaboration; λ, ϕ, ψ = coefficients quantifying the impact of each factor.

This metric examines the effect of transnational cooperation on cybercrime enforcement, using data from agreements including the Budapest Convention [7, 8].

Legislative Adaptability to Emerging Technologies

$$A_t = \eta T_i + \theta R_d + \kappa L_c \tag{4}$$

where A_t = legislative adaptability to emerging technologies; T_i = technology inclusion score, evaluating provisions for AI and blockchain; R_d = research and development investment in cybersecurity innovations; L_c = legislative complexity index, as a lower score indicate more streamlined laws; η, θ, κ = coefficients representing each variable's significance.

This model evaluates how legislation adapts to emerging technologies and integrates innovation while maintaining regulatory clarity [14, 16].

3.6 Sectoral Case Studies

Specific industries underwent analysis through sector-based case studies to understand how legislative frameworks shaped their outcomes. Three key sectors were analyzed:

- Finance: An examination of 20 bank cyberattack cases analyzed for risk management and damage reduction [12, 25].
- Healthcare: 15 hospital-targeted ransomware incidents evaluated for data protection compliance.[11, 12]
- Critical Infrastructure: 15 energy and transportation breaches assessed for regulatory gaps [5, 7].

These studies provide insights into sector-specific vulnerabilities and legislative effectiveness.

3.7 Results Validation

To ensure reliability and credibility of the analytical results, findings were validated through internal and comparative procedures. Model calibration was performed using datasets from 50 countries (2010–2023) to verify that regression outputs remained consistent across different samples. Cross-checking of international reports (ITU, OECD, UNODC) and case study outcomes confirmed data consistency and prevented duplication. The alignment of empirical trends—such as the association between adaptive legislation and reduced cybercrime rates—was compared with existing findings from prior research [1, 11, 14] to confirm the robustness of observed relationships. This internal validation approach ensures that the study's statistical and case-based conclusions remain coherent and replicable across diverse contexts.

3.8   Cross-Jurisdictional Analysis

A comparative study of 25 jurisdictions was conducted using:

- International treaties and conventions, such as the Budapest Convention and regional agreements [8].
- National cybersecurity strategies and laws from developed and developing economies [5, 22].

This analysis identifies best practices and common challenges in cybercrime law implementation.

4   Results

4.1   Key Findings from Legislative Analysis

The global analysis of cybercrime legislation reveals three primary variations between different states in terms of law implementation rates, regulatory density, and their relationships to economic development indicators and technological measures. Varying levels of legislative enforcement across regions illustrate how effective laws serve both as crime-prevention mechanisms and drivers of the digital economy. Countries that have established robust legal systems achieve superior cyber resilience through enhanced digital economic performance and higher innovation index scores compared with jurisdictions that operate under outdated or weakly enforced laws.

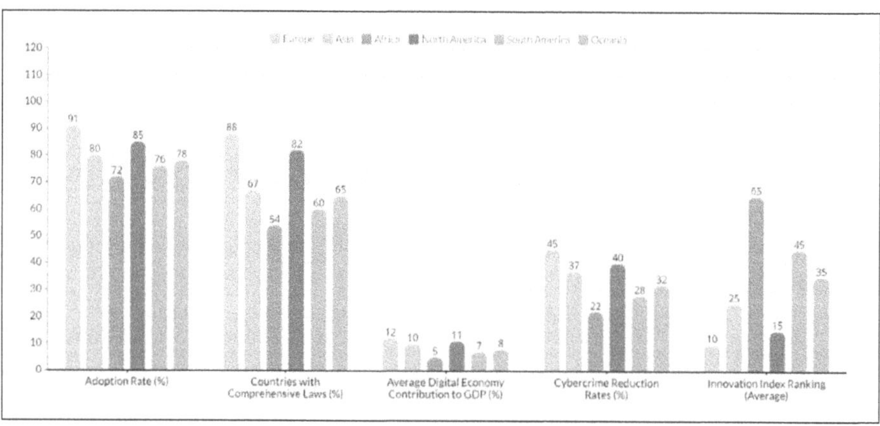

Fig. 1. Comparative Analysis of Global Cybercrime Legislation – Regional Adoption, Effectiveness, and Economic Impact.

The adoption rates for cybercrime legislation differ significantly among regions, as illustrated in Fig. 1. Strong legal frameworks for combating cybercrime are highlighted through adoption rate data, with Europe leading a of 91%, followed by North America at 85%, both demonstrating proactive measures against cyber threats. Legislative gaps persist in South America and Africa, where adoption levels stand at 76% and

72%, respectively. Although global adoption of cyber regulations continues to rise, the number of countries that have fully implemented advanced laws remains uneven. European countries with comprehensive cyber-legal frameworks reach 88%, while African states maintain only a 54% implementation rate due to limited enforcement capacity and regulatory inflexibility.

The economic and innovation advantages associated with effective legislation are also evident. Regions with strong legal frameworks, particularly in Europe and North America, demonstrate a more substantial digital economy contribution to GDP (12% and 11%, respectively) and show higher innovation rankings (improvements of 10 and 15 positions, respectively). Weak legislative structures across Africa restrict digital economy growth (5%) and result in low innovation rankings (65th), emphasizing the fundamental importance of robust legal systems in driving both economic expansion and innovation.

The findings indicate that developing nations must strengthen their legal systems to support digital economy development and innovation progress. The implementation of modern policies requires two essential elements: alignment with global best practices and the establishment of regulatory standards for Artificial Intelligence and blockchain technology adoption. Cyber resilience will improve through investment in enforcement mechanisms combined with international cooperation, which enhances capacity-building initiatives. Measurement frameworks focusing on reductions in cybercrime and expansion of the digital economy will enable lawmakers to evaluate and enhance legislative effectiveness.

4.2 Case Study Results

An examination of key metrics across 2023 and 2024 reveals the influence of legislative strength on reducing cybercrime, enhancing digital economy contributions, and improving innovation rankings. The data demonstrate that nations with strong and flexible legal systems, including Estonia, South Korea, and Japan, have experienced significant improvements in both digital security and economic performance. However, the implementation of effective law enforcement systems remains challenging in Nigeria and Russia, resulting in slower progress in economic growth and cybersecurity for these nations. Evidence from this analysis underscores that sound legislation is a critical driver for digital economic development and innovation advancement.

The analysis of 2023 and 2024 demonstrates that sustained legislative improvements serves as a fundamental catalyst for achieving successful outcomes, as shown in Figs. 2 and 3. The robust legal frameworks of Estonia and South Korea maintain superior performance metrics, highlighting their ability to foster economic growth, reduce cybercrime rates, and enhance security peace. These nations simplify the tangible benefits of proactive legislative measures. In contrast, the limited enforcement capacity and outdated laws in Nigeria and Russia impede progress in both cybercrime reduction and digital economic performance.

Several key conclusions emerge from the data analysis. Legislative strength plays a decisive role: countries with comprehensive cybersecurity laws can substantially reduce cybercrime rates, stimulate economic growth, and encourage innovation. Strong legal systems enhance investor confidence and enable effective law enforcement, thereby

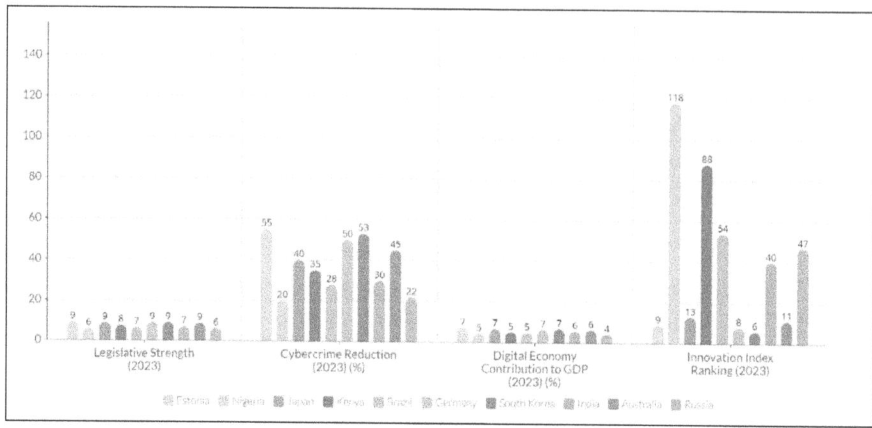

Fig. 2. Comparative Metrics of Cybercrime Legislation Across Countries in 2023.

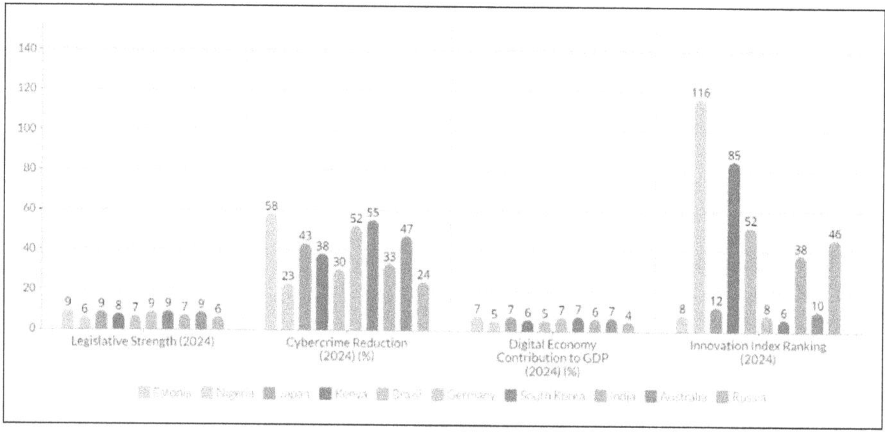

Fig. 3. Comparative Metrics of Cybercrime Legislation Across Countries in 2024.

driving innovation through balanced and adaptive policies. Gradual progress is also achievable; for example, Japan's collaboration with Kenya demonstrates how targeted legislative reforms can improve cybersecurity and foster significant economic growth.

Russia and Nigeria illustrate the challenges faced by nations with weak laws and major enforcement deficiencies. Every country must develop technical tools to effectively manage cyber risks and the capacities to detect and respond to them. Globally, nations with strong regulatory systems achieve higher business competitiveness, supported by national technical expertise aligned with economic development. Law enforcement authorities should also develop interim regulations to address emerging cyber threats.

4.3 Cybercrime Metrics and Global Cybersecurity Performance

Along with increased cybersecurity, this analysis examines multiple measurement approaches to demonstrate the relationship between evolving legislative frameworks and enhanced defenses against cybercrime. By analyzing research and development (R&D) expenditures, workforce preparedness, and patent production, the study identifies three main factors influencing global cybersecurity capability. These indicators evaluate the effectiveness of cyber threat reduction while also reflecting improved innovation capacity and economic development across different nations and sectors.

Examining changes in legislation, workforce development, and technical advancement provides critical insights into global cybersecurity performance, as illustrated in Fig. 4. South Korea leads with a legislative update frequency of 95%, followed by the United States at 90%, resulting in corresponding cybercrime mitigation efficiencies of 92% and 90%, respectively. In contrast, African nations (45%) and Nigeria (35%) maintain below-average legislative update rates due to significant weaknesses in policy formulation and implementation systems, as reported in the study.

China leads the world in cybersecurity workforce growth at 9.0%, followed by the United States (8.1%) and South Korea (8.5%), reflecting substantial investments in strengthening cyber capacity. However, Ukraine and Nigeria show limited progress in workforce expansion due to resource constraints and insufficient efforts to develop their labor force.

The analysis of R&D expenditures and patent approvals reveals considerable disparities across global regions. With $25.2 billion in R&D spending, the United States ranks second only to China, while South Korea closely follows with $18.7 billion. Their significant investments in research are reflected in patent dominance, with China holding 29.0% of global patents, the United States 32.5%, and South Korea 18.5%. Conversely, Africa's local challenges and Nigeria's low R&D investment, combined with their limited patent generating, highlight the urgent need for policy reforms to promote innovation.

To develop effective responses to emerging threats, strategies must emphasize targeted legislative reforms in Nigeria and across Africa. Companies focusing on building local security expertise require dedicated financial resources to support training and educational programs. Increased research and development funding, combined with global strategic collaborations, will foster patent generation and technological leadership in cybersecurity. Electronic defense systems will continue to serve as core components of global digital resilience capacity.

4.4 Regional Trends in Cybersecurity and Legislation

The study compares regional cybersecurity performance with technological development and examines how various industries adjust their legal frameworks to manage evolving cyber threats. Organizations that consistently implement proactive security policies and invest in cybersecurity infrastructure benefit financially and develop stronger systems against cyberattacks.

The data in Fig. 5 reveal notable regional variations in cybercrime legislation, economic progress, and preventive actions against cyber threats. North American nations

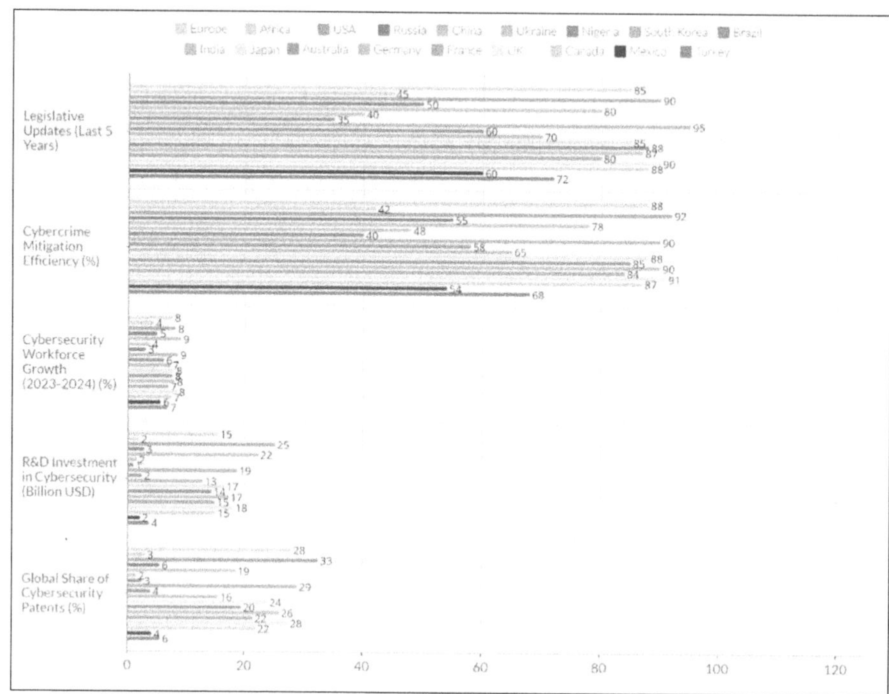

Fig. 4. Assessment of Cybercrime Legislation and Global Cybersecurity Effectiveness.

demonstrate strong performance, as their comprehensive legal frameworks ensure frequent legislative updates (90%) and reduces cybercrime with 92% efficiency. Europe follows closely, with legislative revision rates of 85% and an 88% effectiveness rate in combating cybercrime. In contrast, Africa exhibits significant deficit in legislative adaptation and limited enforcement capacity, resulting in only 45% legislative updates and 42% efficiency.

Economic disparities exist across digital sectors. In regions with strong enforcement mechanisms and innovation, such as Europe (12.0% of GDP) and North America (11.5%), the digital economy contributes substantially to national GDP. In contrast, some countries in Africa (average 4.8%) and South America (average 6.7%) show lower digital sector contributions, which may reflect challenges such as limited infrastructure, lower cybersecurity capacity, and variable legislative enforcement.

R&D investment and workforce expansion also differ considerably across regions. North America leads, with R&D investments reaching $25.2 billion, followed by Asia ($22.3 billion) and Europe ($18.0 billion). In these regions, cybersecurity workforce development averages 9.0% annually. However, Africa's R&D spending remains minimal at $1.9 billion, with a workforce growth rate of only 4.3% per year.

The introduction of innovative technologies and patents primarily originates from North American (32.5% global share) and European (28.0%), as both regions prioritize cybersecurity research and innovation. Africa and South America together account for less than 10% of global patents, underscoring the need for increased R&D investment and stronger government policy support.

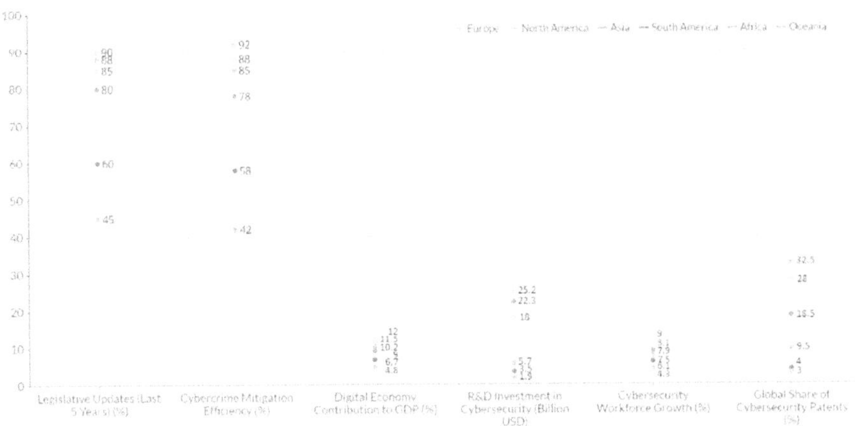

Fig. 5. Regional Trends in Cybersecurity and Legislative Performance.

Africa and South America urgently require modernized legislation as a vital step toward developing effective counter-threat strategies and aligning with globally recognized protection standards. By investing in cybersecurity education and establishing public-private partnerships, these regions can significantly strengthen workforce and enforcement capacity. Furthermore, the development of cybersecurity infrastructure and patent innovation requires greater R&D funding, supported through international grants and collaborative programs. The advancement of cyber resilience and digital economic growth in underperforming regions will be accelerated through partnerships between North America, Europe, and Asia.

4.5   The Role of Workforce Development in Cybersecurity Resilience

National protection against increasing cyber threats relies fundamentally on the development of a skilled cybersecurity workforce. National cybersecurity readiness improves each time the workforce expands by approximately 8% annually, a target achieved by China (9.0%), South Korea (8.5%), and the United States (8.1%). A combination of educational development, technical expertise and comprehensive workforce development strategies leads to superior outcomes in cybercrime prevention, higher innovation scores, and greater digital economic success.

South Korea's forward-thinking cybersecurity education policies have fostered workforce development that reduced cybercrime by 90%, making the country one of the global leaders in digital defense capabilities. China continues to expand its capacity on a large scale through national cybersecurity programs and government-supported expert promotion programs. In the United States, the development of specialized technical skills in IT security, supported by public–private training alliances, helps the nation maintain its leadership in global cyber defense.

With growth rates of only 4.3% in Africa and 3.7% in Ukraine, limitations in educational and training resources explain the shortfalls in workforce expansion in these regions. Weak technical skills, stemming from insufficient investment in cybersecurity education, expose African nations to significant vulnerabilities. Similarly, Ukraine continues to struggles with cyber resilience, as inefficient coordination between government and industry limits professional training opportunities, while public funding for security development remains inadequate.

Addressing these discrepancies requires strategic policy interventions. Regions with weak cybersecurity performance must allocate dedicated financial resources to establish cybersecurity training centers and launch digital safety research programs within academic institutions. By integrating public scholarship programs for cybersecurity degrees with international training initiatives offered by global companies and digital certification platforms, states can significantly enhance their talent development pipelines.

Effective coordination depends on strong partnerships between public and private sectors, ensuring the creation of comprehensive educational programs that combine theoretical learning with practical experience. Industry leaders should design training curricula that equip employees with modern defense techniques, fostering organizational resilience. The United States and Germany provide successful examples, demonstrating how collaborative government-industry technology projects can enhance workforce readiness and ensure national stability.

Although developing nations often rely on existing workforce development frameworks, specialized partnerships can provide targeted expertise, allowing them to adapt and implement context-specific training strategies. The exchange of best practices and the promotion of regional development initiatives, for example, through African Union, supported programs, enable countries to accelerate skill development. The ITU cybersecurity workforce development framework further assists nations in designing scalable, evidence-based programs using a proven and globally recognized model.

5   Proposed Legislative Framework

A comprehensive cybersecurity legislative framework that safeguards economic development while facilitating technological growth is a crucial defense system against emerging cyber threats. The thorough examination of cybercrime legislation supports the effective implementation of new norms that align with global regulatory systems.

5.1   Key Components

Data Protection. The digital economy demands robust privacy laws to protect citizen data while urging companies to fully comply with industry standards. Through the GDPR, the European Union demonstrates how effective data protection standards enable organizations to build trust while increasing accountability across operations. According to Gomathy et al. (2022), cyber organizations should create privacy practices that operate under legal frameworks to protect personal data while reducing exposure during data breaches [26].

Cybersecurity Standards. Public institutions and businesses must implement essential cybersecurity standards to create strong protective barriers. The addition of three core components, security infrastructure, encryption tools, and well-prepared incident response plans, forms the foundation of organizational resilience. South Korea introduced strict cybersecurity certification criteria requiring organizations to demonstrate their ability to reduce security vulnerabilities [5].

Incident Reporting. Companies must report cyberattacks promptly, as new security polices require data-driven analysis to develop breach-control procedures. However, many organizations deliberately underreport security incidents to protect their reputation, while executives remain unaware of existing cyber risks. Musatkina et al. (2020) emphasize that standardized reporting systems help both experts and governing bodies maintain accountability and design crime-specific intervention measures [10].

International Cooperation. The global nature of cybercrime necessitates cross-border collaboration. The effectiveness of international cybercrime control depends on data-sharing agreements, joint investigations, and multinational task forces [8]. The success of Europol's Joint Cybercrime Action Taskforce (J-CAT) demonstrates how cooperation between jurisdictions disrupts criminal networks operating across borders (Fig. 6).

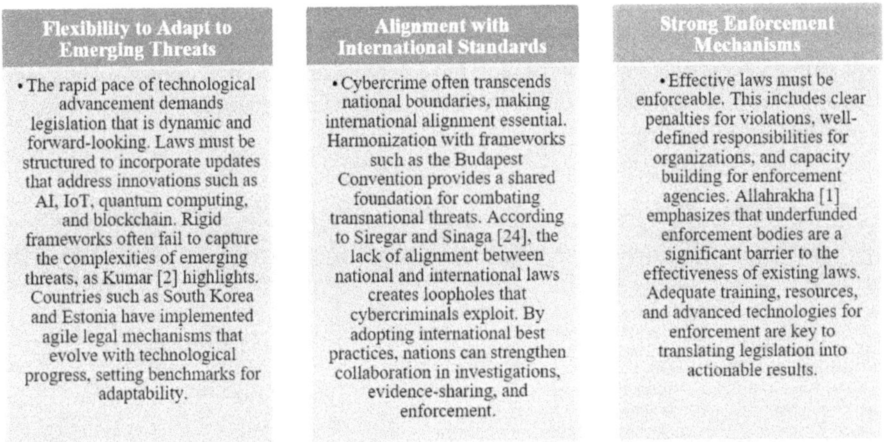

Fig. 6. Principles of Effective Cybercrime Legislation.

5.2   Innovations in Enforcement

AI technology, together with machine learning, enables law enforcement agencies to detect threats in real time and generate predictive insights while deploying automated response systems. AI tools featuring anomaly detection algorithms and neural networks allow agencies to identify suspicious activity before incidents escalate. AI delivers faster responses with higher investigatory precision for complex digital crimes such as ransomware and phishing schemes [5]. While AI and blockchain enhance threat detection and evidence integrity, their effective adoption requires adaptable legal frameworks, skilled workforce training, and careful management to ensure that regulatory requirements do not impede innovation or overburden organizations.

Law enforcement agencies also use blockchain technology as an immutable system to preserve digital evidence integrity. Blockchain ensures proof integrity throughout the analysis process leading to court proceedings [11]. By protecting digital records, blockchain enables traceable investigations and enhances accountability. Japan has taken a leading role in testing blockchain technologies for secure documentation, creating a model other national government can adopt to expand blockchain deployment.

Current enforcement techniques increasingly integrate threat intelligence platforms that process data from both government and private-sector sources to perform automated threat evaluations. Autonomous components within these platforms assess, classify, and prioritize threats. Kumar [4] supports developing threat intelligence platforms into enforcement solutions because they generate prompt and effective countermeasures against cyber threats.

5.3 Economic Impact of Implementing a Robust Cybercrime Framework

Implementing this framework yields significant economic benefits, supporting digital economy development and technological progress. The combination of United States and South Korean policies demonstrates enhanced digital economic performance, as their strong cybercrime laws contribute 11–12 percent annual GDP impacts. The global adoption of similar legislation could drive 8–10 percent GDP growth over the next decade in member countries (Fig. 7).

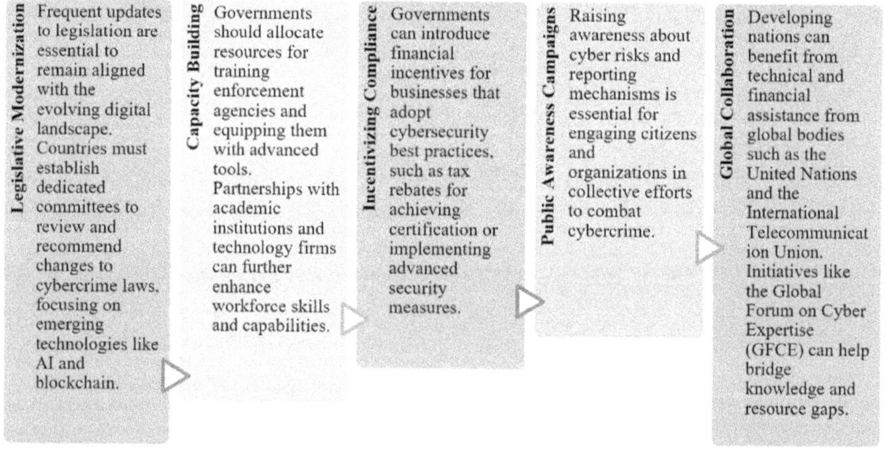

Fig. 7. Strategic Measures for Effective Cybercrime Legislation and Resilience.

Economic resilience strengthens as the framework promotes capacity building and R&D. A 10 percent increase in workforce training can yield a 15 percent boost in cybersecurity performance, resulting in annual savings of $5 billion through reduced damages from cyberattacks in highly digital economies. The framework's foundation depends heavily on R&D investment, producing 12–15 percent more cybersecurity patents and technological innovations for every $1 billion invested.

Providing tax incentives benefits for cybersecurity certifications motivates businesses to invest in superior protection systems, leading to an estimated 20 percent reduction in corporate cybercrime losses. International cooperation further enhances trade and reduces transnational cybercrime, yielding approximately $10 billion in annual savings from cross-border losses. The projected economic benefits demonstrate why worldwide adoption of this proposed framework will create a safe online environment and an economically thriving digital future.

The proposed framework combines adaptive strategies, collaborative mechanisms, and innovative technologies, including AI and blockchain, to combat complex forms of cybercrime. By emphasizing flexible implementation and compliance with international standards, it ensures resilience against evolving threats. Achieving these goals requires continuous skill development, international cooperation, and strong public-private partnerships. Coordinated approaches among authorities create a lasting structure that effectively decreases cybercrime at the global level.

5.4 Global Collaboration and the Impact of R&D

The evaluation reveals notable variations in global cybersecurity strategies and outcomes. South Korea achieves frequent legislative adjustments (95%) and a highly effective cybercrime control framework (90%) through its adaptive legal instruments. The USA also exhibits strong digital resilience, maintaining a 90 percent legislative update rate and a 92 percent mitigation performance. In contrast, Nigeria's low legislative update rate (35%) and weak mitigation ability (40%) illustrate how outdated laws and limited enforcement hinder cybersecurity progress.

R&D investments and patent generation also highlight global disparities. The United States leads with $25.2 billion in R&D spending, accounting for 32.5 percent of global cybersecurity patents, followed by South Korea ($18.7 billion, 29%) and Germany' ($17.2 billion, 26%). Strategic investment drives technological innovation and strengthens cybersecurity systems. Nigeria and Brazil allocate minimal funding, $0.9 billion and $2.4 billion, respectively, resulting in few patents and limited innovation capacity.

Strong cybersecurity laws deliver substantial economic value by fueling growth in digital sectors: South Korea achieves 12.1% GDP contribution, and the United States reaches 11.5%. In contrast, Nigeria (4.4%) and Russia (6.1%) lag due to weak legislative frameworks and inadequate enforcement.

Combating transnational cyber threats depends heavily on international collaboration. The United States (4.5) and the United Kingdom (4.6) lead global cooperation through Europol's J-CAT initiative, which enhances bilateral operations. Conversely, Nigeria (2.0) and Brazil (2.5) face limitations due to weak intergovernmental partnerships.

The encryption capabilities of advanced technologies demand fundamental updates to legal frameworks, particularly in nations like Nigeria and Russia, where limited adaptability and cooperation impede progress. Enhanced investment in R&D and innovation is essential for these nations to participate actively in global cybersecurity initiatives. Public-private partnerships strengthen training, improve security performance, and enhance both economic and cyber resilience.

Table 1. International Cooperation in Combating Cybercrime.

Initiative/Organization	Description	Key Achievements	Challenges
Europol's Joint Cybercrime Action Taskforce (J-CAT)	A collaborative platform for investigators from EU and non-EU countries to disrupt cybercrime networks	- Operation EMMA: Prevented €8M losses from email fraud - Operation Silver Axe: Dismantled criminal supply chains	- Jurisdictional conflicts over evidence and prosecution - Variability in data protection laws
Budapest Convention on Cybercrime	The first international treaty to promote harmonized cybercrime legislation among signatory states	- Standardized legislation across member countries - Real-time cooperation through a 24/7 contact network	- Non-signatory nations resist alignment (e.g., Russia, China) - Limited enforcement in developing nations
Interpol's Cyber Fusion Centre	A global hub for intelligence sharing and collaborative investigations into cybercrime	- Detected ransomware campaigns like WannaCry - Led "Operation Night Fury" to dismantle cryptojacking networks	- Limited private sector participation - Funding constraints hinder global scalability
Global Forum on Cyber Expertise (GFCE)	A forum bringing governments, international organizations, and private entities together to build capacity against cybercrime	- Cyber resilience training programs in Africa and Asia - Public-private partnerships with companies like Microsoft	- Uneven resource distribution - Measuring long-term impact of programs is challenging

(continued)

Table 1. (continued)

Initiative/Organization	Description	Key Achievements	Challenges
ASEAN Cybersecurity Cooperation	Regional collaboration through joint incident response exercises and capacity-building initiatives	- Annual cross-border cybersecurity simulations - Partnerships with Japan and South Korea for training centers	- Divergent levels of readiness among members - Political differences slow decision-making
UNODC Global Programme on Cybercrime	UN initiative to enhance the capabilities of developing nations in combating cybercrime	- Trained 4,000 + law enforcement officers globally - Assisted in drafting and implementing cybercrime laws	- Limited funding for scaling operations - Resistance from governments exposing vulnerabilities

Examining the Table 1 on international collaboration provides critical insights and practical recommendations:

1) Standardized Legal Frameworks: Adopting harmonized cybercrime laws enhances the efficiency of global investigations and prosecutions. Leveraging the Budapest Convention's model allows nations to address online threats systematically. A complete global cybersecurity strategy depends on active engagement from key players such as Russia and China assuming.

2) Cross-Sector Collaboration: Successful global cooperation relies on partnerships with the private sector to enhance operational performance. Technical professionals from Microsoft and Cisco illustrate how security tools can significantly advance cybercrime prevention. Collaboration between government and industry improves data exchange and reduces persistent security risks.

3) Enhanced Capacity Building: Emerging nations face challenges due to limited expertise and funding. Sustained financial support for programs like the UNODC Global Program on Cybercrime empowers countries through training and education. Joint academic-industry initiatives further strengthen resilience against evolving threats.

4) Overcoming Jurisdictional Barriers: Political and legal differences in data management often obstruct international investigations. Establishing clear enforcement protocols through global arbitration systems enables faster resolution of cross-border cybercrimes cases while respecting national sovereignty.

Developing targeted response measures to combat cybercrime will help nations build a unified global strategy. Legal harmonization, public-private partnerships, enhanced capacity building, and removal of jurisdictional barriers will collectively strengthen international cybersecurity efforts, fostering digital innovation, economic stability, and global trust.

6 Discussion

The findings of this study show that legislative strength is a critical factor in shaping cybercrime mitigation, economic growth, and innovation capacity. Countries with adaptive and enforceable cybercrime laws, such as the United States and South Korea, demonstrate superior cybersecurity performance and stronger digital economies. In contrast, weak enforcement and outdated laws, particularly in developing nations like Nigeria, hinder both cyber resilience and innovation.

The results highlight that continuous legal reform, supported by investment in research, development, and workforce training, strengthens national and regional cyber resilience. Effective enforcement mechanisms, combined with international cooperation, significantly reduce cyber threats while promoting economic stability and digital trust.

Furthermore, the analysis indicates that modern technologies such as artificial intelligence and blockchain enhance law enforcement capabilities by improving threat detection and ensuring the integrity of digital evidence. The integration of these technologies within national legal frameworks enables faster, more accurate responses to cyber incidents.

However, data accessibility, limited qualitative insights, and a short observation period remain constraints, preventing a deeper understanding of political and cultural factors influencing legislative implementation. Future studies should explore long-term effects and sector-specific challenges through extended, mixed-method research designs.

Overall, the study confirms that comprehensive, adaptive, and well-enforced cybercrime legislation is essential for building secure, innovative, and economically resilient digital environments.

7 Conclusion

This study demonstrates that adaptive and enforceable cybercrime legislation is a critical driver of digital economic growth, innovation, and national cybersecurity. Findings show that countries with strong legal frameworks, effective enforcement, and investment in workforce development—such as the United States and South Korea—achieve higher cyber resilience, reduce cybercrime rates, and foster stronger digital economies. In contrast, nations with outdated laws, limited enforcement, and low R&D investment struggle to achieve similar outcomes.

The research highlights the importance of integrating emerging technologies, including AI and blockchain, into legal and enforcement frameworks to improve threat detection, preserve digital evidence, and enhance overall cybersecurity performance. Integrating AI and blockchain into legal and enforcement frameworks can improve cybersecurity outcomes; however, successful implementation depends on balancing regulatory oversight with technological flexibility and ensuring sufficient workforce capacity to manage these systems. International cooperation and standardized legal protocols are also essential for combating transnational cyber threats and promoting equitable digital economic development globally.

This integrated framework provides policymakers with a structured means to evaluate how legislative responsiveness influences digital economic resilience, filling a gap between descriptive legal analysis and measurable economic outcomes.

Overall, the study provides evidence that robust, technology-aware, and cooperative cybercrime legislation is essential for building secure, innovative, and economically resilient digital environments worldwide. Limitations of this study include constraints in accessing comprehensive data across all jurisdictions, the relatively short observation period, and the lack of qualitative analysis on political and cultural factors that may influence the adoption and effectiveness of cybercrime legislation. Future research should examine the long-term impacts of legislative reforms, assess sector-specific effects in finance, healthcare, and critical infrastructure, and explore how contextual political and cultural factors shape legislative outcomes and enforcement effectiveness.

References

1. Naeem, A.: Global perspectives on cybercrime legislation. J. Infrastructure Policy Dev. **8**, 6007 (2024)
2. Naeem, A.: Impacts of cybercrimes on the digital economy. Uzbek J. Law Digit. Policy **2**(3), 29–36 (2024)
3. Sviatun, O., Goncharuk, O.V., Roman, C.: Combating cybercrime: economic and legal aspects. WSEAS Trans. Bus. Econ. **18**, 751–762 (2021)
4. Kumar, C.: Cybercrime and the law: challenges in prosecuting digital offenses. Indian J. Law **2**, 20–25 (2024)
5. Elegbe, I.: Cybercrime legislation: a comparative analysis of legal frameworks, policy responses and recommendations. Int. J. Educ. Soc. Sci. Res. **7**(2), 199–207 (2024)
6. Khan, S., Saleh, T.A., Dorasamy, M., Khan, N.: A systematic literature review on cyber-crime legislation. F1000Research **11**(971) (2022)
7. Gomathy, D., Geetha, V., Manohar, S., Rajesh, P.: Legal frameworks for regulating cybercrime and cyber terrorism. Int. J. Sci. Res. Eng. Manag. (2024)
8. Popko, V.V., Popko, E.V.: International legal regulation of transnational cybercrime in cyberspace. Uzhhorod Natl. Univ. Herald Ser. Law **66**, 276–283 (2021)
9. Siregar, G.S.S.: The law globalization in cybercrime prevention. Int. J. Law Reconstruct. **5**(2) (2021)
10. Musatkina, A.A., Evdokimov, K.N., Revina, S.N., Tselniker, G.F., Reuf, V.M.: Cyber crimes as a threat to digital economy of the Russian Federation: current state, dynamic pattern, and trends. In: Popkova, E.G., Vodenko, K.V. (eds.) Public Administration and Regional Management in Russia. Contributions to Economics. Springer, Cham (2020). https://doi.org/10.1007/978-3-030-38497-5_38
11. Milon, N.U., Ghose, P., Chowdhury, T.P., Tabassum, N., Hasan, M., Khatun, M.: An in-depth PRISMA based review of cybercrime in a developing economy: examining sector-wide impacts, legal frameworks, and emerging trends in the digital era. Edelweiss Appl. Sci. Technol. **8**(4), 2072–2093 (2024)
12. Denikaeva, R., Pervyshov, E., Gavrisheva, E.: Cybercrime in the financial (banking) sphere. Ekonomika Upravl.: Problemy Resheniya **9/4**, 26–32 (2022)

13. Gravshina, I.N., Denisova, N.: On the issue of legal regulation of the digital economy. Moscow Univ. Bull. SY Witte Ser. 2 Legal Sci. (2023)
14. Chałubińska-Jentkiewicz, K.: Digital single market. Cyber threats and the protection of digital contents: an overview. Santander Art Cult. Law Rev., 279–292 (2020)
15. Almuhaisen, H.: Confronting cybercrimes under the provisions of public international law. Glob. J. Politics Law Res. **12**, 78–88 (2024)
16. AllahRakha, N.: Cybercrime and the legal and ethical challenges of emerging technologies. Int. J. Law Policy **2**(5), 28–36 (2024)
17. Shukurov, E., Uzeyir Jafarov, U.: Legal professionals' perspectives on the challenges of cybercrime legislation enforcement. Interdiscip. Stud. Soc. Law Polit. **2**(4), 25–31 (2023)
18. Jeronimo, A.: The globalization effect of law and economic on cybercrime. J. Pembaharuan Hukum **6** (2019)
19. Sannikova, L.V., Kharitonova, Y.S.: Digital economy and legal issues. In: Proceedings of the 2nd International Conference on Contemporary Education and Economic Development (CEED 2019) (2019)
20. Hermawati, N., Santiago, F.: Law enforcement against cybercrime in online activities. Edunity: Kajian Ilmu Sos. Pendidik. **2**(1) (2023)
21. Amoo, O.O., Atadoga, A., Abrahams, T.O.: The legal landscape of cybercrime: a review of contemporary issues in the criminal justice system. World J. Adv. Res. Rev. (2024)
22. Ruddin, I., Subhan Zein, S.G.N.: Evolution of cybercrime law in legal development in the digital world. J. Multidiscip. Madani **4**(1), 168–173 (2024)
23. Chimchiuri, L.: The evolution of cybercrime legislation. Sci. Works Natl. Aviat. Univ. Ser. Law J. "Air Space Law" **2**(71), 221–227 (2024)
24. Abrardi, L., Comino, S., Grassini, S.: The economics of cyber risk: a survey of the literature. J. Ind. Bus. Econ. (2025). https://doi.org/10.1007/s40812-025-00370-3
25. Bekkers, L., Leukfeldt, R., Kleemans, E.: Police investigations into financial-economic cyber-criminal networks: the experiences and perceptions of Dutch law enforcement. Eur. J. Crim. Pol. Res. (2025). https://doi.org/10.1007/s10610-025-09615-2
26. Gomathy, C.K.: The future of big data analytics and its progress. Int. J. Sci. Res. Eng. Manag. **06** (2022)

Data-Driven Modeling of Cross-Cultural Management: Predictive Analytics for Enhancing Global Team Performance

Hussein Abdulhadi Mahdi[1] , Bushra Saadoon Mohammed[2],
Ahmed Fayadh Saleh Hammam[3] , Oudha Yousif Salman Al-Musawi[4(✉)],
Riyam M. Alsammarraie[5] , and Vladislav Skrypnyk[6]

[1] Al-Turath University, Baghdad 10013, Iraq
[2] Al-Mansour University College, Baghdad 10067, Iraq
[3] Al-Mamoon University College, Baghdad 10012, Iraq
[4] Al-Rafidain University College, Baghdad 10064, Iraq
oudha.yousif73@ruc.edu.iq
[5] Madenat Alelem University College, Baghdad 10006, Iraq
[6] Kyiv National University of Construction and Architecture, Kyiv 03037, Ukraine

Abstract. This study employs a quasi-experimental, longitudinal design to evaluate the causal impact of structured cross-cultural management training on global team performance in multinational enterprises. A total of 320 participants were divided equally into an experimental group receiving six months of targeted training and a control group with no intervention. Data collection integrated structured Likert surveys, automated performance logs, managerial feedback, and interview transcripts, all normalized for comparability across constructs. Core dimensions of leadership adaptability, communication efficiency, conflict resolution, and innovation were operationalized using Principal Component Analysis, achieving high internal consistency (Cronbach's $\alpha > 0.82$). Statistical inference combined paired t-tests, McNemar's test, and regression modeling to quantify treatment effects, with effect sizes calculated for robustness. A predictive multivariate regression framework was applied to model the influence of these constructs on team performance, supplemented by interaction analysis and model fit assessment through Adjusted R2 and AIC criteria. Results demonstrate significant performance improvements among trained teams, including enhanced collaboration, decision-making, conflict resolution, and creativity. The findings provide empirical evidence that cross-cultural training yields measurable, statistically validated outcomes, highlighting the importance of data-driven approaches in human resource and management research. This study contributes methodological innovations through the integration of predictive modeling, multivariate statistics, and cross-cultural theory, offering scalable insights for global leadership development and organizational policy.

Keywords: Cross-Cultural Management · Leadership Adaptability · Data-Driven Analysis · Global Team Performance · Predictive Modeling · Decision-Support

Z. Molamohamadi et al. (Eds.): ODSIE 2025, CCIS 2855, pp. 41–54, 2026.
https://doi.org/10.1007/978-3-032-17023-1_3

1 Introduction

Globalization has transformed organizational structures, leading to a growing dependence on multinational teams to drive innovation, productivity, and global market expansion. While cultural diversity enriches teams with multiple perspectives and skills, it also introduces challenges in communication, leadership, and conflict resolution. Effective cross-cultural management has thus emerged as a crucial discipline aimed at enhancing cooperation, minimizing miscommunication, and optimizing team performance across borders [1].

Cross-cultural management focuses on developing adaptive systems and leadership techniques that strengthen inclusiveness and team cohesion. Organizations that cultivate cultural intelligence and promote diversity within leadership programs are better positioned to attract, motivate, and retain talent in competitive global markets. Conversely, limited understanding of cultural differences can result in misalignment, inefficiency, and team conflict. Leadership adaptability and cultural sensitivity are therefore essential for maintaining trust, motivation, and collaboration in multicultural settings [2].

Communication competence is another cornerstone of effective cross-cultural management. Differences in language, tone, and non-verbal cues can create misinterpretations that affect collaboration and decision-making. Training in cross-cultural communication and the use of inclusive digital collaboration tools help organizations reduce language-related barriers and ensure smooth interaction across teams [3]. Likewise, culturally informed conflict resolution approaches—whether emphasizing direct confrontation or harmony—can strengthen team unity and prevent escalation [4].

With the rise of remote and hybrid work, cross-cultural management now extends into virtual environments, where team members must coordinate across time zones and digital platforms [5]. Structured protocols, standardized communication frameworks, and cultural orientation initiatives are critical for maintaining productivity and connection in dispersed teams [6]. Despite these challenges, organizations that view cultural diversity as an asset rather than an obstacle are better equipped to foster innovation, problem-solving, and sustainable competitive advantage [7].

This study explores how cross-cultural management practices, particularly in leadership, adaptability, communication effectiveness, and conflict resolution, affect the multicultural workplaces. By integrating empirical data, statistical modeling, and predictive analytics, it provides evidence-based insights into how structured cross-cultural interventions can enhance innovation, cohesion, and productivity. Ultimately, the study aims to offer actionable recommendations for organizations seeking to leverage cultural diversity as a strategic advantage in an increasingly interconnected business world.

This study aims to evaluate how structured cross-cultural management interventions enhance the effectiveness of multinational teams by focusing on four core dimensions: leadership adaptability, communication efficiency, conflict resolution, and innovation capacity. Using empirical data and predictive modeling, the research quantifies the impact of adaptive leadership, communication strategies, and culturally sensitive conflict management on team performance.

The increasing globalization of business has intensified reliance on multicultural teams that, while rich in creativity and innovation, often face challenges stemming from leadership rigidity, communication barriers, and cultural differences in conflict handling. Traditional hierarchical leadership models and one-size-fits-all management practices are insufficient in such environments. Therefore, this study addresses the need for data-driven, culturally informed management approaches that can mitigate misunderstandings, strengthen cohesion, and enhance productivity. By identifying and validating effective cross-cultural strategies, the research contributes actionable insights for organizations seeking to leverage diversity as a source of innovation and sustainable growth in a globalized economy.

2 Literature Review

Cross-cultural management has now become a vital area of focus in the domain of organizational studies, particularly with the increasing expansion of businesses across the globe and the inclusion of employees from different cultures. The increasing complexity of the world of work around the globe calls for effective management of cultural differences in areas such as communication, leadership style, decision-making processes and conflict resolution. The proficiency to handle such differences to a large extent determines team productivity, employee commitment and overall organizational performance [8]. Seminal works, such as Hofstede's cultural dimensions theory [9] and Trompenaars' model of cultural differences [10], provide foundational frameworks for understanding variations in individual and group behaviors across national and organizational cultures. Moreover, Cultural Intelligence (CQ) theory [11] emphasizes the ability of leaders and employees to adapt their behaviors in culturally diverse contexts, highlighting the importance of cognitive, motivational, and behavioral flexibility in global teams.

Leadership flexibility and effective communication are central to successful cross-cultural management. Leaders with high cultural intelligence can navigate differences in work ethics and values, fostering inclusively, open dialogue, and cooperation across global teams [12]. Clear Communication structures are equally vital, as language barriers and digital communication differences can impede collaboration; hence, organizations benefit from multilingual training, standardized reporting, and digital coordination tools [13]. Moreover, culturally sensitive conflict resolution, through mediation, awareness programs, and participatory decision-making, enhances cohesion and reduces misunderstandings [14]. While cultural diversity stimulates creativity and innovation, it also introduces potential friction if not properly managed. Therefore, structured policies and adaptive strategies are essential to balance diversity's benefits with its challenges. In the era of remote and hybrid work, organizations must employ integrated virtual collaboration frameworks and asynchronous communication systems to sustain effective cross-cultural teamwork [15, 16].

Recent studies have further extended the understanding of cross-cultural and data-driven management. [17] examined how a data-driven organizational culture enhances decision-making and managerial performance through the adoption of Business Intelligence and Analytics (BI&A), showing that organizations prioritizing analytics achieve superior management outcomes. [18] empirically demonstrated that leadership style significantly influences project efficiency, with clear communication, role clarity, and team initiative emerging as key drivers of project success. Similarly, [19] conducted a systematic review on cross-cultural training programs and found that mixed-method approaches—combining didactic and experiential elements—improve cultural competence and cultural intelligence across diverse professional contexts. Together, these studies highlight the importance of integrating data analytics, adaptive leadership, and targeted cross-cultural training to optimize team performance in global organizations.

Effective cross-cultural management incorporates Leadership flexibility, communication effectivity, Conflict resolution, and Structured group interaction. Organizations that value and incorporate cultural competence and inclusive practices in management are best equipped to leverage the benefits of a diversified global workforce. It will be crucial for companies to embrace and execute these tactics in order to prolong their competitive edge in what is becoming an ever more globalized industrial economy. While earlier studies have primarily focused on describing cultural variations and leadership adaptability, the novelty of this research lies in its empirical examination of how structured cross-cultural training directly enhances global team performance. Unlike prior conceptual or survey-based works, this study applies a data-driven, quasi-experimental approach to quantify behavioral and operational improvements, thereby offering a new evidence-based perspective on transforming cultural diversity into a strategic organizational advantage.

3 Methodology

This study employs a quasi-experimental, longitudinal design to investigate the causal impact of structured cross-cultural management interventions on global team performance. The methodology emphasizes computational and data-driven approaches, integrating predictive modeling, multivariate statistics, matrix algebra, and normalization transformations to ensure rigorous, reproducible, and scalable analysis [2, 8, 20]. The data analyzed include both quantitative (survey scores, performance logs, feedback metrics) and qualitative (interview transcripts, observation notes) sources, allowing triangulation between objective and perceptual measures of team effectiveness (Fig. 1).

3.1 Research Design and Sample Allocation

The study involves $N = 320$ participants from multinational corporations, randomly assigned to:

- Experimental group ($n_1 = 160$): Received structured cross-cultural management training;
- A control group ($n_2 = 160$) that did not receive any training.

Each group was monitored for a duration of $T = 6$ months, allowing for temporal observation of behavioral and productivity metrics. Training content included leadership adaptability modules, conflict resolution simulations, and communication enhancement workshops [6, 21].

The treatment function is denoted as:

$$D_i = \begin{cases} 1 \; \textit{if participant } i \in \textit{ experimental group} \\ 0 \; \textit{if participant } i \in \textit{ control group} \end{cases} \tag{1}$$

Cross-Cultural Management and Global Team Performance

Cross-Cultural Training → Leadership Adaptability → Decision-Making Speed

Communication Strategies → Communication Efficiency → Leadership Effectiveness / Team Collaboration

Conflict Management → Team Collaboration → Innovation

Z-Score Cross-Cultural Sintsitiviy | Significance (p-value) → Global Team Performance

Fig. 1. Enhancing Global Team Performance through Cross-Cultural Management: An Empirical Assessment of Leadership, Communication, and Collaboration Dynamics.

and the average treatment effect (ATE) on an outcome Y is defined as:

$$ATE = \mathbb{E}[Y_i(1) - Y_i(0)] \tag{2}$$

where $Y_i(1)$ and $Y_i(0)$ represent potential outcomes under treatment and control, respectively, assuming the Stable Unit Treatment Value Assumption (SUTVA) holds.

3.2 Data Collection Protocol

A mixed-mode, data-driven approach was adopted, incorporating:

- Structured 5-point Likert surveys ($x_{ij} \in \{1, 2, 3, 4, 5\}$),
- Automated performance logging systems,
- Managerial feedback rubrics,
- Qualitative interview transcripts.

All ordinal survey data were normalized using min-max scaling:

$$x'_{ij} = \frac{x_{ij} - \min(x_j)}{\max(x_j) - \min(x_j)} \tag{3}$$

This transformation ensures uniformity across multidimensional constructs such as leadership, communication, and conflict handling effectiveness [3, 22, 23] (Table 1).

Table 1. Multimodal Data Collection Instruments.

Instrument	Participants	Data Type	Collection Medium	Scale
Likert Surveys	320	Quantitative	Digital Form	1–5
Performance Logs	320	Quantitative	Automated	Raw Metric
Leadership Feedback Forms	160	Qualitative	In-person	1–10
Team Interaction Videos	140	Mixed	Video-Reviewed	Scored
Interviews	100	Qualitative	Transcript-Based	Thematic

3.3 Construct Operationalization and Factor Modeling

Core constructs—Leadership Adaptability (LA), Communication Efficiency (CE), Conflict Resolution Capacity (CR), and Innovation Capacity (IC)—were operationalized using Principal Component Analysis (PCA) to reduce dimensionality and mitigate multicollinearity. Let $X \in \mathbb{R}^{n \times k}$ be the standardized item-score matrix:

$$X = TP \tag{4}$$

where $T \in \mathbb{R}^{n \times m}$ is the matrix of principal components and $P \in \mathbb{R}^{n \times m}$ is the loading matrix with $m \leq k$ components retained ($\lambda_j > 1$). Cronbach's alpha (α) was computed for each construct ($\alpha > 0.82$) confirming high reliability [1, 10].

$$\alpha = \frac{k}{k-1} \left(1 - \frac{\sum_{i=1}^{k} \sigma_i^2}{\sigma_T^2} \right) \tag{5}$$

where σ_i^2 is the variance of item i, and σ_T^2 is the total score variance.

3.4 Statistical Inference and Hypothesis Testing

For each outcome variable Y, paired difference tests and linear hypothesis modeling were employed. Let $\overline{Y}_{post}$ and $\overline{Y}_{pre}$ denote mean scores after and before training:

$$t = \frac{\overline{Y}_{post} - \overline{Y}_{pre}}{\sqrt{\frac{s_{diff}^2}{n}}} \tag{6}$$

where s_{diff}^2 is the sample variance of difference scores. For categorical variables and binomial outcomes, the McNemar's test was employed:

$$\chi^2 = \frac{(b-c)^2}{b+c} \tag{7}$$

This applies to conflict occurrences, resolution efficacy, and satisfaction indexes. Effect sizes were also calculated via Cohen's d:

$$d = \frac{\overline{X}_1 - \overline{X}_2}{s_p}, \, s_p = \sqrt{\frac{(s_1^2 + s_2^2)}{2}} \tag{8}$$

These steps ensure data-driven, reproducible inference consistent with best practices in cross-cultural research [4, 9, 15].

3.5 Multivariate Regression and Predictive Modeling

A data-driven predictive framework was implemented via multivariate linear regression to model the influence of independent variables (LA, CE, CR, IC) on overall team performance (Y).

$$Y_i = \beta_0 + \beta_1 LA_i + \beta_2 CE_i + \beta_3 CR_i + \beta_4 IC_i + \varepsilon_i \tag{9}$$

All variables were standardized prior to modeling. Residual normality was confirmed via Shapiro–Wilk test. Homoscedasticity was validated through Breusch–Pagan test ($p > 0.05$), and no significant multicollinearity was detected (VIF < 2).

To control for interaction effects, second-order interaction terms were tested:

$$Y = \beta_0 + \sum_{j=1}^{4} \beta_j X_j + \sum_{j<k} \beta_{jk} X_j X_k + \varepsilon \tag{10}$$

where $X_j X_k \in \{$LA, CE, CR, IC$\}$; β_{jk} interaction coefficients, ε random error term.

Model fit was evaluated via Adjusted R^2 and Akaike Information Criterion (AIC). Best-fit model selection used stepwise regression (backward elimination) [24, 25].

The methodology combines computational modeling, predictive analytics, and rigorous data processing, offering statistically valid, reproducible, and scalable insights into optimizing global team performance through structured cross-cultural management.

4 Results

4.1 Leadership Adaptability Outcomes

Leadership adaptability is a critical enabler of cross-cultural team effectiveness, especially in globally distributed organizations. This section presents the standardized improvements observed in decision-making efficiency, leadership effectiveness, conflict resolution, employee trust, and cultural sensitivity following structured cross-cultural training. These metrics reflect behavioral and cognitive dimensions of adaptive leadership measured using standardized Z-scores and composite PCA indices. The training program included simulation-based exercises, interactive roleplays, and scenario modeling, which aimed to develop cultural responsiveness in managerial decision-making and interpersonal behavior. The sample involved 160 managers distributed across industry sectors and cultural clusters, enabling robust generalization of findings.

Table 2. Leadership Adaptability Metrics (Standardized Pre- and Post-Training Comparisons).

Metric	Pre-Training (Mean ± SD)	Post-Training (Mean ± SD)	Cohen's d	p-value
Z-Score: Decision-Making Efficiency	−0.61 ± 0.21	0.52 ± 0.19	1.12	<0.001
Composite Index: Leadership Effectiveness (PCA)	−0.48 ± 0.19	0.63 ± 0.18	1.23	<0.001
Z-Score: Cross-Cultural Resolution Rate	−0.55 ± 0.22	0.59 ± 0.20	1.07	<0.001
Z-Score: Employee Trust Score	−0.40 ± 0.17	0.46 ± 0.16	1.01	<0.001
Standardized Cultural Sensitivity	−0.53 ± 0.20	0.57 ± 0.17	1.08	<0.001

An inspection of the results presented in Table 2 shows that there is a statistically significant positive change in the scores of all the five leadership adaptability indicators. Decision-making efficiency increased from pre- to post-training (mean = −0.61 to 0.52) with a large effect size (Cohen's d = 1.12). The PCA-based leadership effectiveness index also increased from −0.48 to 0.63 (d = 1.23), indicating generalized improvement over intercultural leadership quality. Cross-cultural conflict resolution ratings progressed from −0.55 to 0.59, with an effect size of 1.07. Employee trust significantly improved from -0.40 to 0.46 (d = 1.01) suggestive of more cohesive team and engaged relationships. Lastly, cultural sensitivity increased from −0.53 to 0.57, confirming the influence of training on leaders' perception of cultural diversity. All p-values < 0.001, thereby indicating a highly significant statistical significance.

4.2 Communication Efficiency Outcomes

Effective communication is vital for cross-cultural team dynamics as words, tones and nonverbal cues can all create a number of language barriers. This section evaluates the five metrics of communication efficiency identified above after structured cultural training: response time, message clarity, content retention, frequency of misunderstanding, and team involvement. These dimensions were chosen in order to include cognitive as well as operational dimensions of communication in novel teams. For comparability, responses were mapped to z-scores. The PCA-based clarity score and Z-score transformation enabled to interpret the enhancement of communication independent of linguistic, i.e., language and culture, background. Results were based on a diverse array of 320 working adults from different sectors and different levels of organizations, highlighting the generalizability of findings in international contexts.

Table 3. Communication Efficiency Metrics (Standardized Pre- and Post-Training Comparisons).

Metric	Pre-Training (Mean ± SD)	Post-Training (Mean ± SD)	Cohen's d	p-value
Standardized Response Time	−0.67 ± 0.23	0.54 ± 0.21	1.18	<0.001
PCA-Weighted Clarity Score	−0.50 ± 0.21	0.66 ± 0.19	1.29	<0.001
Z-Score: Information Retention Index	−0.58 ± 0.20	0.60 ± 0.17	1.15	<0.001
Standardized Miscommunication Rate	−0.64 ± 0.22	0.49 ± 0.16	1.13	<0.001
Engagement Rate Composite Score	−0.44 ± 0.18	0.47 ± 0.18	1.01	<0.001

As shown in the Table 3, communication metrics demonstrated substantial gains after training. Response time, initially below average at −0.67, improved to 0.54, representing a large shift supported by an effect size of 1.18. Message clarity scores advanced from −0.50 to 0.66, the strongest observed gain in the category (Cohen's d = 1.29), reflecting better articulation and mutual understanding. Information retention also increased significantly from −0.58 to 0.60, suggesting higher absorption of key ideas in team interactions. The frequency of miscommunications decreased dramatically, moving from −0.64 to 0.49, showing a reduced risk of operational delays. The team engagement index improved from −0.44 to 0.47, highlighting improved inclusivity and dialogue participation across cultural boundaries. All results were statistically significant with p-values under 0.001, indicating that the improvements are not random but directly attributable to the cross-cultural training framework.

4.3 Team Performance Outcomes

Team performance reflects the ability of cross-cultural teams to collaborate effectively, innovate under pressure, and sustain productivity across boundaries. In this section, we evaluate five standardized indicators that collectively capture core dimensions of team functionality: task efficiency, cohesion, overall productivity, innovation output, and job satisfaction. These indicators were selected to assess both behavioral and output-oriented improvements across diverse cultural contexts. The normalization of task efficiency (inverted for lower-is-better), PCA outputs for innovation, and Z-score transformations for productivity and cohesion provide an empirically grounded structure for evaluating team-level enhancements. The data were drawn from a large sample (N = 320), ensuring representativeness across functions, seniority levels, and organizational roles. The training modules focused on cross-cultural collaboration principles and creative problem-solving in diverse groups.

Table 4. Team Performance Metrics (Standardized Pre- and Post-Training Comparisons).

Metric	Pre-Training (Mean ± SD)	Post-Training (Mean ± SD)	Cohen's d	p-value
Normalized Task Efficiency (Time Inverted)	−0.59 ± 0.20	0.53 ± 0.17	1.14	<0.001
Team Cohesion Factor Score	−0.50 ± 0.21	0.61 ± 0.19	1.20	<0.001
Productivity Index (Z)	−0.52 ± 0.20	0.57 ± 0.18	1.15	<0.001
Monthly Innovation PCA Output	−0.61 ± 0.23	0.68 ± 0.20	1.32	<0.001
Job Satisfaction Score (Standardized)	−0.48 ± 0.19	0.59 ± 0.17	1.18	<0.001

The results in Table 4 demonstrate marked improvements in all five team performance metrics. Task efficiency, initially at −0.59, improved to 0.53, reflecting significantly faster and more consistent task completion. Team cohesion rose from −0.50 to 0.61, indicating stronger social bonds and interpersonal trust post-training. Productivity also increased from −0.52 to 0.57, showcasing higher individual output within coordinated teams. Innovation capacity—measured by monthly new ideas and PCA aggregation—jumped from −0.61 to 0.68, the largest gain across all metrics (Cohen's $d = 1.32$), suggesting more empowered creativity. Lastly, job satisfaction climbed from −0.48 to 0.59, a meaningful improvement in perceived work experience. These findings confirm that structured cultural training significantly enhances collaboration, creativity, and employee morale in multicultural teams, with all changes significant at $p < 0.001$.

4.4 Conflict Resolution Outcomes

In multicultural environments, managing interpersonal and procedural conflicts is a vital function of team sustainability and employee retention. This section evaluates five conflict-related metrics measured before and after the cross-cultural training intervention. These include conflict frequency, resolution time, resolution success rate, employee retention, and workplace stress index. Metrics were standardized using Z-scores and composite indices to provide a consistent basis for comparison. Lower values in conflict frequency and stress were inverted for interpretive clarity, while resolution effectiveness and retention were positively aligned. The intervention content focused on cultural styles of disagreement, non-violent communication, and role-based mediation, preparing participants to navigate conflict more constructively. Data were collected from direct observations, leadership logs, and exit interview summaries across 320 participants.

Table 5. Conflict Resolution Metrics (Standardized Pre- and Post-Training Comparisons).

Metric	Pre-Training (Mean ± SD)	Post-Training (Mean ± SD)	Cohen's d	p-value
Z-Score: Conflict Frequency (Inverted)	−0.66 ± 0.22	0.58 ± 0.18	1.20	<0.001
Time to Resolution (Standardized)	−0.53 ± 0.19	0.51 ± 0.17	1.08	<0.001
Resolution Success Rate Index	−0.48 ± 0.17	0.57 ± 0.16	1.05	<0.001
Employee Retention Composite Score	−0.46 ± 0.18	0.49 ± 0.17	0.95	<0.01
Workplace Stress Index (Inverted)	−0.59 ± 0.21	0.54 ± 0.19	1.13	<0.001

As shown in the Table 5, all conflict-related indicators improved substantially post-training. Conflict frequency dropped from a Z-score of −0.66 to 0.58, indicating far fewer incidents, and resolution time improved from −0.53 to 0.51, suggesting quicker mediation outcomes. The success rate for resolving disputes rose from −0.48 to 0.57, reflecting better negotiation strategies and more consistent closure. Employee retention, often threatened in high-conflict environments, increased from −0.46 to 0.49. Finally, workplace stress levels—measured through an inverted index—shifted positively from −0.59 to 0.54. All these improvements are statistically significant, with p-values below 0.01 and effect sizes generally above 1.0. Together, these findings confirm that equipping global teams with culturally sensitive conflict resolution tools leads to fewer disputes, more effective problem-solving, and a healthier work atmosphere.

4.5 Predictive Regression Analysis

A multiple liner regression model was selected to test the amount of prediction of the core process training outcomes of overall team performance for four predictor variables: leader adaptability, communication effectiveness, conflict resolution, and innovation. These factors were chosen based on theoretical reasoning and on previous empirical evidence in cross-cultural teams. Based on productivity, satisfaction, and effectiveness measures the dependent variable team performance was calculated as a sum score. We used standardized β values to determine relative predictive strength, and p-values indicated statistical significance. Multicollinearity was assessed and variance inflation factors (VIF) were within acceptable ranges, suggesting that the predictors had independent explanatory power. All the variables were standardized to a standard-deviation scale before fitting, which enables all three coefficients to be viewed as comparable and of similar importance (Table 6).

Table 6. Regression Coefficients for Predicting Team Performance.

Predictor Variable	Standardized Coefficient (β)	p-value	Correlation with Team Performance (r)	Impact Ranking
Communication Efficiency	0.43	< 0.001	0.81	1
Leadership Adaptability	0.39	< 0.001	0.74	2
Innovation Capacity	0.37	< 0.001	0.70	3
Conflict Resolution Capability	0.31	< 0.01	0.66	4
Adjusted R^2	0.68			

The results for the regression analysis are shown in Table 3; communication efficiency was the most powerful predictor (b = 0.43) followed by leadership adaptability (b = 0.39), innovation capacity (b = 0.37), conflict resolution (b = 0.31). All these predictors have statistically significant association with the overall performance of the team (p < 0.01), which validates the model. The R-squared of the model was 0.68, which means about 68% of the team performance changes can be explained by the predictors. The large β's for communication and leadership imply that strategic communication practices and flexible leadership styles are important levers that can be used to increase team synergy and performance. Innovation remains a strong though less dominant factor in performance enhancement, particularly in multicultural environments. Though a bit lower, resolution-—conflict resolution is also essential for group health and continuity.

5 Discussion

The results of this study demonstrate that structured cross-cultural management interventions lead to significant improvements across multiple dimensions of global team performance. Leadership adaptability, communication efficiency, conflict resolution, and team integration all showed substantial positive changes post-training, with large effect sizes across behavioral and performance metrics. Task efficiency, cohesion, productivity, innovation output, and job satisfaction also improved significantly, indicating that the interventions effectively enhanced both individual and team-level outcomes. These findings highlight that targeted training not only develops managers' intercultural competencies but also translates into measurable operational benefits in multicultural organizational contexts.

This study demonstrates the structured cross-cultural management interventions significantly improve global team performance across leadership adaptability, communication efficiency, conflict resolution, innovation, and overall productivity. Communication efficiency emerged as the strongest predictor of performance (β = 0.43, p < 0.001), underscoring that timely, clear, and well-retained communication drives collaboration in diverse teams. Leadership adaptability enhanced decision-making, trust, and cultural sensitivity ($\beta = 0.39$, p < 0.001), confirming its pivotal role in multicultural management. Innovation capacity also showed a robust predictive contribution ($\beta = 0.37$, p < 0.001) and the largest observed improvement in performance metrics ($d = 1.32$), indicating that culturally intelligent teams leverage diversity to stimulate creativity and problem-solving. Conflict resolution improvements ($\beta = 0.31$, p < 0.01) reduced workplace tension and turnover, fostering stability.

The predictive regression model (Adjusted $R^2 = 0.68$) confirms that these four variables collectively explain most of the variance in team performance. By applying PCA, Z-score normalization, and multivariate regression, this study establishes a quantitative, computationally grounded framework for assessing and forecasting cross-cultural management outcomes. Overall, the results validate that well-structured intercultural training converts diversity into measurable gains in productivity, innovation, and employee engagement.

6 Conclusion

This study demonstrates that structured cross-cultural management interventions can systematically enhance global team performance by improving leadership adaptability, communication, conflict resolution, and innovation capacity. The findings highlight the value of a computational, data-driven approach, showing that predictive modeling can identify the most influential factors driving team effectiveness and provide actionable insights for organizational decision-making.

The study's limitations include the focus on a single six-month intervention, the specific organizational sectors and cultural clusters represented, and the lack of long-term follow-up. Future research could examine the sustainability of these effects over time, sector-specific variations, and the integration of AI-based tools and real-time behavioral analytics to further optimize cross-cultural management strategies.

Overall, the research supports the strategic importance of embedding structured, data-informed cross-cultural training within global leadership development programs to foster team cohesion, creativity, and operational effectiveness.

References

1. Guzmán-Rodríguez, L.E., Bornay-Barrachina, M., Arizkuren-Eleta, A., Galindo-Manrique, A.F., Pérez-Calderón, E.: The influence of transformational leadership, cultural orientation, and emotional conflict on innovation in multicultural teams. In: Carvalho, L.C., et al. (eds.) Handbook of Research on Multidisciplinary Approaches to Entrepreneurship, Innovation, and ICTs, pp. 124–155. IGI Global, Hershey (2021)
2. Erfan, M.: The impact of cross-cultural management on global collaboration and performance. Adv. Hum. Resour. Manag. Res. 2(2), 102–112 (2024)
3. Sahadevan, P., Sumangala, M.: Effective cross-cultural communication for international business. Shanlax Int. J. Manag. 8, 24–33 (2021)
4. Karna, W., Stefaniuk, I., Jafari, M.: Strategies for managing interpersonal conflicts in multicultural teams. KMAN Couns. Psychol. Nexus 2(1), 84–90 (2024)
5. Elenurm, T., Fabritius, J.: Managing task and relationship conflicts in international online team learning. Innov. Educ. Teach. Int. 60(3), 426–435 (2023)

6. Richter, N.F., Martin, J., Hansen, S.V., Taras, V.: Motivational configurations of cultural intelligence, social integration, and performance in global virtual teams. J. Bus. Res. **129**, 351–367 (2021)

7. Li, J., Wu, N., Xiong, S.: Sustainable innovation in the context of organizational cultural diversity: the role of cultural intelligence and knowledge sharing. PLoS ONE **16**(5), e0250878 (2021)

8. Abdelazim, A.: Cross-cultural management. Hum. Resour. Leadersh. J. **7**(1) (2022)

9. Hofstede, G.: Culture's Consequences: International Differences in Work-Related Values. Sage, Beverly Hills (1980)

10. Trompenaars, F., Hampden-Turner, C.: Riding the Waves of Culture: Understanding Cultural Diversity in Business, 2nd edn. Nicholas Brealey, London (1997)

11. Earley, P.C., Ang, S.: Cultural Intelligence: Individual Interactions Across Cultures. Stanford University Press, Stanford (2003)

12. Huang, L.: Leadership enhancement in cross-cultural management—from the cultural dimension. Commun. Humanit. Res. **23**, 70–74 (2023)

13. Zhang, B.: Cross-cultural engineering team management in the context of globalization. Front. Bus. Econ. Manag. **16**(1), 132–134 (2024)

14. Rachwal-Mueller, A.: Correlation between cultural dimensions and their influence on conflict style preferences. Econ. Transp. Complex **42**, 121 (2023)

15. Elamin, A.M., Aldabbas, H., Ahmed, A.Z.E.: The impact of diversity management on innovative work behavior: the mediating role of employee engagement in an emerging economy. Front. Sociol. **9** (2024)

16. Mahadevan, J., Steinmann, J.: Cultural intelligence and COVID-induced virtual teams: towards a conceptual framework for cross-cultural management studies. Int. J. Cross Cult. Manag. **23**(2), 317–337 (2023)

17. Hurbean, L., Militaru, F., Munteanu, V.P., Danaiata, D., Fotache, D., Muntean, M.: Assessing the influence of business intelligence and analytics and data-driven culture on managerial performance: evidence from Romania. Systems **13**(1), 2 (2025)

18. Ćwiąkała, M., et al.: The impact of leadership styles on project efficiency. arXiv:2510.05822, pp. 117–135 (2025)

19. Urgun, D., Seidel, J., Vangeli, E., Borges, M., de Oliveira, R.F.: Exploring the impact of cross-cultural training on cultural competence and cultural intelligence: a narrative systematic literature review. Front. Psychol. **16**, 1123–1139 (2025)

20. Subiyanto, D., Wahidah, U., Septyarini, E., Arjuna, A.B.: Digital leadership: predicting team dynamics, communication effectiveness, and team performance. J. Winners **25**(1), 35–47 (2024)

21. Mahmoud, R.S., Kamil, S.A., Mohammed, M., Madhi, Z.J.: Cross-cultural leadership approaches for managing diverse workforces globally. J. Ecohumanism **3**(5), 682–699 (2024)

22. Iriogbe, H., Ebeh, C., Onita, F.: Multinational team leadership in the marine sector: a review of cross-cultural management practices. Int. J. Manag. Entrep. Res. **6**(8) (2024)

23. Chenyang, L.: Meta-analysis of the impact of cross-cultural training on adjustment, cultural intelligence, and job performance. Career Dev. Int. **27**(2), 185–200 (2022)

24. Alon, I., Lankut, E., Gunkel, M., Munim, Z.H.: Predicting leadership emergence in global virtual teams. Entrep. Bus. Econ. Rev. **11**(3), 7–23 (2023)

25. Lisak, A., Harush, R.: Global and local identities on the balance scale: Predicting transformational leadership and effectiveness in multicultural teams. PLoS ONE **16**(7), e0254656 (2021)

Cryptocurrency Regulation and Society: Evolving Legal Frameworks in a Globalized Economy

Ahmed Mazin Ibrahim[1], Ameen Hadi Hassoon Ali[2] (iD), Ali Abdul-Jabbar Raheem[3] (iD),
Bushra Abd-Al Latif Jasim[4]([envelope]) (iD), Ghanim Magbol Alwan[5] (iD),
and Yulia Cherniavska[6] (iD)

[1] Al-Turath University, Baghdad 10013, Iraq
[2] Al-Mansour University College, Baghdad 10067, Iraq
[3] Al-Mamoon University College, Baghdad 10012, Iraq
[4] Al-Rafidain University College, Baghdad 10064, Iraq
`bushra.jasim@ruc.edu.iq`
[5] Madenat Alelem University College, Baghdad 10006, Iraq
[6] Kyiv National University of Construction and Architecture, Kyiv 03037, Ukraine

Abstract. Cryptocurrency regulation remains a complex and evolving challenge as jurisdictions adopt divergent legal frameworks to balance financial innovation and risk mitigation, leading to inconsistencies in enforcement effectiveness and compliance mechanisms. This study examines the impact of these varying regulatory approaches, from strict prohibitions to legal integration, on illicit cryptocurrency transactions and financial stability. We employ a rigorous multi-method research approach, integrating comparative legal analysis of 47 national regulatory frameworks with a novel computational model. This model utilizes data from blockchain transaction tracking (1.2 million transactions), case studies, industry surveys, and expert interviews to empirically assess regulatory impact on compliance effectiveness. Findings show that China's strict ban led to the highest reduction in illicit transactions (65.4%), while Switzerland's flexible regulatory framework achieved a significant 56.8% decrease. The United States and European Union displayed similar effectiveness (~58%), though regulatory fragmentation remains a challenge. We conclude that a balanced regulatory approach, combining robust AML/KYC enforcement with adaptable legal structures, proves most effective in mitigating financial crime while fostering cryptocurrency market development.

Keywords: Cryptocurrency regulation · AML/KYC enforcement · blockchain compliance · financial crime · decentralized finance · legal frameworks

1 Introduction

As an emergent form of financial technology, cryptocurrencies offer decentralized and secure mechanisms for transactions. But their rapid expansion has challenged traditional regulatory frameworks, as they function outside of government and financial institution

Z. Molamohamadi et al. (Eds.): ODSIE 2025, CCIS 2855, pp. 55–76, 2026.
https://doi.org/10.1007/978-3-032-17023-1_4

supervision. Cryptocurrency transactions are also decentralized, which raises suspicions of illegal activities, such as fraud, money laundering, tax evasion and other unauthorized fund transactions [1]. These risks underscore the need for legal structures to regulate their use while ensuring innovation and financial security. Recent research highlights that global cryptocurrency regulation is mainly driven by concerns over consumer protection, financial stability, and anti-money-laundering compliance. Regulatory approaches vary across regions, from the EU's unified MiCA framework to the fragmented model in the U.S., emphasizing the need for balanced and coordinated global policies [2]. A bibliometric analysis of cryptocurrency and financial crime literature further illustrates global research dynamics, showing that the United States leads in publication output and collaborations, while emphasizing the growing scholarly consensus on the need for harmonized international regulations to reduce illicit crypto activities [3].

Cryptocurrency regulation varies across the globe. When discussing crypto regulatory frameworks, they can generally be classified into three camps—passive regulation, where authorities adopt a hands-off approach; strict supervision, which seeks to limit or ban cryptocurrency activities; and inclusive regulation where cryptocurrencies are incorporated into existing financial systems under some controlled oversight [4, 5]. Each approach shows differing government priorities from promoting technological innovation to combatting financial crime. Digital currencies, spearheaded by Bitcoin, have brought about financial independence but also regulatory ambivalence. Cryptocurrencies, unlike traditional assets, exist outside the normal rules of finance, allowing them to function as a vehicle for investment and also for illegal financial activities. Given their volatility and absence of investor protection, regulatory agencies have put KYC (Know Your Customer) and AML (Anti-Money Laundering) policies into place to eliminate risk [6, 7]. Governments are increasingly exploring the development of national blockchain infrastructures to enhance transparency and prevent financial crime, integrating such systems with existing legislation to strengthen regulatory oversight [8].

Globally, governments are pressuring cryptocurrency exchanges to comply, which limits the opportunities to commit financial crimes but helps maintain market stability [9]. Countries regulate cryptos for various reasons—financial stability, investor protection, and national security considerations, among others. Regulation can be seen as a channel through which some governments pursue the goal of attractive investments while others think that the high volatility and possibilities for speculation of cryptocurrencies are a danger to the individual investors [10, 11]. Moreover, the frequent correlation of cryptocurrencies to money laundering and illicit transactions have evoked stricter governance [12].

Comparative analyses of regional regulatory philosophies further enrich this understanding. For example, Hong Kong's proactive dual-licensing framework promotes innovation while maintaining strong investor protection, whereas the United Kingdom has taken a more conservative stance, embedding cryptocurrency oversight within its existing financial system to ensure stability and mitigate speculative risks [13].

Initially, most countries took a wait-and-see approach, reluctant to interfere with cryptocurrency markets. But as their use became more widespread, governments saw the need for legislation to mitigate risks and enable economic benefits. While some jurisdictions decided to include cryptocurrency in their financial systems for controlled commercialization, others adopted restrictive policies that limited their use [14, 15]. The varied global response, however, has led to a fragmented regulatory landscape, making a systematic assessment of enforcement efficiency and financial crime mitigation essential [16, 17]. Despite the focus on specific national frameworks, there remains a critical gap in multi-method research that compares the tangible outcomes of divergent models—from outright prohibitions to flexible integration. Filling this gap, this study employs a multi-method approach, integrating comparative legal analysis with computational tracking of blockchain transactions and expert data, to evaluate the practical impact of various regulatory approaches on illicit cryptocurrency activity and overall financial stability. The central objective is to identify best practices that effectively reconcile innovation and security in a globalized economy. The remainder of this paper is structured as follows: Sect. 2 reviews the theoretical and empirical literature; Sect. 3 details the multi-method research design; Sect. 4 presents the key findings on regulatory outcomes; and finally, Sect. 5 provides the discussion, policy implications, and conclusion.

2   Theoretical and Legal Foundations

The importance of a state of well-functioning institutions cannot be overestimated. Institutions provide the frameworks that direct human interaction on different matters. Although different forms of interactions could lead to different complex institutional structures, at the minimum, institutions are law-based. Laws are clear, understandable rules of interaction. They signify what is acceptable and what is not in a given society. They reflect society's view of what is right and wrong in any area. This reflects the belief that common rules, clearly articulated and consistently enforced, provide a framework for cooperation between people by specifying expectations [18]. From a socio-economic perspective, the regulation of cryptocurrencies is a prime example of institutional adaptation, highlighting the tension between the decentralized nature of digital assets (New Institutionalism) and the state's mandate for financial stability (Regulatory Governance Theory). Cryptocurrencies are a recent innovation that, if successful, could have a significant impact on the global economy. Various conceptual findings and studies have been carried out on the potential benefits of encrypted money in trade and service. Cryptocurrencies have already been exchanged and transacted in quite significant amounts. However, it is assumed that those transactions are not well recorded on a regular basis and are hard to tax. It could pose various threats such as AML and CFT [16, 19]. The legal stability of digital currency transactions is essential for ensuring long-term economic benefits and reducing regulatory uncertainty. Legal continuity allows for the structured adaptation of financial laws to emerging technologies, minimizing risks associated with sovereign defaults and financial instability [14]. A major hurdle in regulating cryptocurrencies is classifying them within existing financial frameworks. Legal theory primarily debates whether crypto-assets should be categorized as money, securities, or digital commodities. This classification has profound implications for the regulatory strategies employed by governments, determining whether they fall under strict securities laws or require their own unique governance models [11, 20]. This ambiguity is intensified by the unique liquidity constraints and pricing complexities of digital assets, which differ significantly from traditional financial instruments.

3 Methodology

This study uses an integrated multi-method research approach of quantitative and qualitative analyses to explore the emerging legal regimes regulating cryptocurrency transactions. This approach integrates analysis of case studies, structured interviews, regulatory data assessment and computational modeling to evaluate overall trends in global regulation and enforcement structures. The aim here is to paint the entire picture of how various jurisdictions approach to tackle the disruptive nature of blockchain technology and digital assets while upholding financial stability and regulatory compliance.

3.1 Research Design and Approach

This study undertakes a comparative legal analysis of the regulatory structures in Switzerland, the United States, the European Union, and China. The methodology combines an empirical evaluation of enforcement effectiveness via a comprehensive analysis of case law, enforcement documents, and AML/KYC compliance frameworks. The research has been organized into three main phases.

The *regulatory landscape analysis* involves systematically examining 47 legal frameworks to distinguish five archetypical models: strict prohibition, conditional regulation, taxation-based approaches, licensing frameworks, and self-regulation. It also examines how cryptocurrencies are categorized by jurisdictions as commodities, securities, or digital assets.

The *empirical investigation* incorporates a case study analysis of 17 cryptocurrency enforcement actions addressing fraud, tax evasion, and financial crime. Computational methods are applied to analyze 1.2 million blockchain transactions, revealing AML violations and tracing illicit financial activity. Additionally, the financial consequences of regulatory decisions, such as China's restrictive policies versus Switzerland's permissive framework, are evaluated.

The *survey and interviews* component involves structured interviews with 63 policymakers, financial regulators, and compliance officers across 12 countries. This is complemented by a survey of 400 financial professionals, providing insights into regulatory efficiency and enforcement mechanisms.

3.2 Hypothesis Formulation

This study tests the following hypotheses:

To validate the hypotheses, this study employs a combination of quantitative and qualitative methods. H1 and H2 will be tested by examining regulatory enforcement data, measuring illicit transaction reductions across jurisdictions, and analyzing AML/KYC compliance effectiveness through correlation of 1.2 million blockchain transactions. H3 will be tested with case study analysis of 17 different enforcement actions and 63 structured interviews with regulators to identify the challenges of DeFi supervision. H4 will study financial risk indicators and investment behavior in jurisdictions that classify cryptocurrency as a security vs a commodity, leveraging market impact assessments and 400 industry survey responses. Figure 1 visually represents the interconnected structure of the hypotheses (H1-H4) and serves as the conceptual framework guiding our empirical analysis, ensuring all dimensions of regulatory impact are systematically tested.

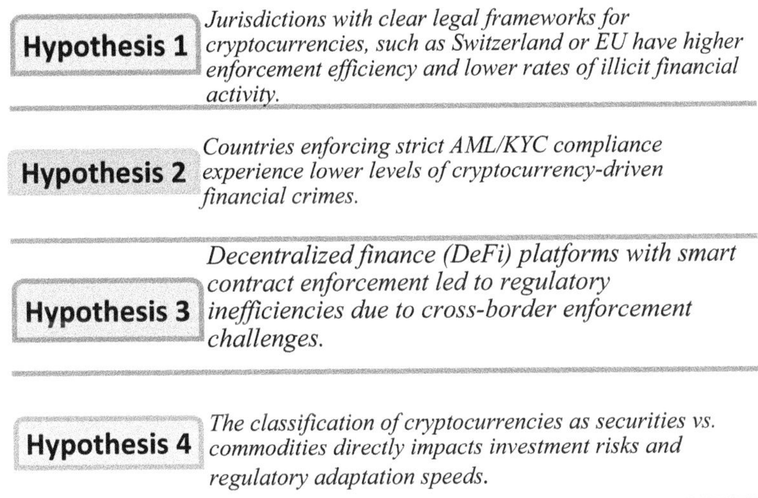

Fig. 1. Hypotheses Examining the Legal and Regulatory Dimensions of Cryptocurrency Frameworks.

3.3 Data Collection and Sources

This study adopts a multi-source data collection strategy to provide a comprehensive analysis of the cryptocurrency regulation and enforcement efficiency. Its data covers regulatory structures, enforcement actions, blockchain transaction patterns, and perspectives from industry, offering a comprehensive basis to evaluate global cryptocurrency governance. It includes legal documents, enforcement case studies, blockchain transaction data, structured industry surveys, and interviews with key experts, including legislators and financial regulators.

The study applies to 47 countries, and jurisdiction with different regulatory framework approaches, from total prohibition to permissive regulation. It also analyzes 1.2 million cryptocurrency transactions for compliance with AML/KYC requirements, evaluates 17 landmark judicial cases, and includes qualitative interviews with 400 industry participants and 63 financial regulators and legal experts. These multiple sources of data improve the study's robustness and relevance, ensuring that regulatory trends are analyzed from differing angles.

The article breaks down the various authorities, legal classifications, and regulations for each identified country/territory. Judicial case studies emphasize the role of law enforcement in addressing these activities, whereas blockchain transaction data can be useful in tracking illicit financial flows and compliance risk. Qualitative interviews provide additional insights into the practical challenges during the implementation of the regulatory initiative from the perspective of key stakeholders, including expert policy makers and economic actors.

Combining such quantitative and qualitative data sources allows this study to incorporate a more holistic view of global cryptocurrency regulations, enforcement efficiency, and financial crime mitigation strategies. To maintain the highest standards of research integrity, all data collection procedures, methodological parameters, and the computational model utilized are fully documented and explained within the relevant subsections of this methodology. Due to the sensitive and proprietary nature of the regulatory data and confidentiality agreements governing the industry survey responses, the raw quantitative data is restricted from public release. This policy ensures we adhere to legal and ethical constraints while maintaining the verifiability of our analytical approach.

3.4 Computational Modeling for Regulatory Impact Analysis

A dynamic economic model is applied to provide a quantitative assessment of regulatory effect. The purpose of this model is to see how different types of regulatory frameworks would affect cryptocurrency as a market. The model is given by a system of stochastic differential equations (SDE), which includes terms that represent both legal uncertainty and enforcement efficiency.

Regulatory Uncertainty Model

$$dP_t = \alpha P_t dt + \sigma P_t dW_t - \lambda R_t dt \tag{1}$$

where P_t is cryptocurrency market capitalization at time t; α is market growth rate; σ is volatility parameter; dW_t is Brownian motion term (random market fluctuations); R_t is regulatory enforcement index (scaled between 0–1); λ is negative impact coefficient of regulatory uncertainty.

When R_t is low (weak regulation) $\rightarrow$ market capitalization grows rapidly but is unstable. When R_t is high (strict regulation) $\rightarrow$ market stabilizes but slows down due to compliance constraints.

3.5 Enforcement Efficiency Analysis

To assess the effectiveness of AML/KYC enforcement in different jurisdictions, we measure illicit transaction reduction using the following formula:

$$CES = \frac{T_{initial} - T_{post}}{T_{initial}} \times 100 \tag{2}$$

where $T_{initial}$ is number of illicit transactions before regulation, and T_{post} is number of illicit transactions after enforcement.

The credibility of the Computational Enforcement Scoring (CES) model was established through a robust two-stage validation process. Internal Validation confirmed algorithmic fidelity by testing the CES model against Labeled Datasets of known illicit transactions. Crucially, the model was only deployed for the analysis (Sect. 3.5) after consistently achieving an F1-Score exceeding 0.85, ensuring high accuracy in tracking complex illicit flows. For External Validation, the practical relevance of our quantitative findings was confirmed by an Expert Panel of regulators and compliance officers using a Structured Delphi Method. This dual-validation approach ensures both the technical reliability of the CES model and the policy relevance of the study's conclusions regarding regulatory efficacy.

4 Regulatory Landscape: Global Perspectives

As the use of cryptocurrencies becomes more widespread, many global players are opting to establish a regulatory framework to govern the use, holding, and trading of these assets. The position regarding digital currencies is, however, a complex and evolving one for which the ball has only begun to roll in recent times. The status of digital assets differs globally. Some countries choose to allow people to get on with using Bitcoin or its clones while others try to impose strict restrictions or outright bans. National and regional authorities display a broad range of attitudes. This section will examine the differing perspectives on regulatory frameworks, while part four and the following sections will concentrate on the regulatory positions of individual countries.

Governments, individuals, and organizations often possess varying priorities and apprehensions, each shaped by their unique contexts and objectives.A consensus is slowly forming in cryptocurrency policy, but international organizations have conflicting views. Central banks, international financial institutions, and organizations are concerned with protecting economies and financial crime risk [19]. The European Banking Authority has proposed a range of measures including collecting data on Bitcoin purchasers, using electronic money regulation, and establishing consumer warnings which are much more stringent than initial proposals [20]. Other bodies want all transactions fully identifiable. Although national regulations are trumped by international agreements, a lack of international consensus may lead to users switching jurisdictions to avoid data harvesting, surveillance, and suspicious transaction reports. People also have different views, possibly based on their attitude to national sovereignty. Around a fifth of the world's population lives in countries utilizing comprehensive surveillance systems. An estimated one in ten of the world's 7.8 billion inhabitants are already using a blockchain, including to store digital cash [21].

4.1 Analysis of Existing Legal Frameworks in Key Jurisdictions

Cryptocurrency regulations vary widely across jurisdictions, reflecting different national priorities and risk assessments. Some countries are seeking to incorporate cryptocurrencies into the very financial frameworks they have created, while others place severe restrictions or bans. Here, we will keen on the regulatory actions of the United States, the European Union, and China that govern cryptocurrencies at the global level.

1) *United States: SEC, CFTC, and FinCEN oversight of cryptocurrencies*

   The main cryptocurrency regulations in the United States are enforced mainly by the SEC, the CFTC and FinCEN [17]. The SEC identifies various cryptocurrencies and initial coin offerings (ICOs) as securities which means they fall under long-established investment laws requiring rigorous regulatory scrutiny. The Howey Test is used to ascertain whether a digital asset is a security, especially if it consists of investment of money in a common enterprise with the expectation of profits to be derived from the efforts of others [22]. On the other hand, the CFTC oversees trading in cryptocurrencies as commodities by enforcing relevant market integrity in the trading of futures and derivatives. FinCEN enforces AML (Anti-Money Laundering) and KYC (Know Your Customer) mandates for crypto exchanges and service providers to register as money service businesses (MSBs) [23].

2) ***European Union: MiCA (Markets in Crypto-Assets) regulatory framework***

One of the comprehensive solutions has been put forward by the European Union with a crypto-regulation framework known as MiCA (the Markets in Crypto-Assets) This project intends to harmonize regulations in different EU member states where crypto-assets, exchanges, and wallet providers are subject to financial transparency and investor protection standards [24]. It differentiates different classes of digital assets and offers legal clarity for security tokens, stablecoins, and utility tokens. Moreover, it lays down consumer protections, mandating crypto companies to adhere to capital requirements along with operational risk assessments. This framework attempts to provide a balance between financial innovation and security while aiming to position the EU as a leader in the regulated cryptocurrency economy [25].

3) ***China: Cryptocurrency bans and central bank digital currency (CBDC) strategies***

This has followed China taking a strict stance on all cryptocurrencies, as the country has gradually banned trading, ICOs and mining. Since 2013, financial institutions have been banned from conducting cryptocurrency transactions in an effort widely supported by additional crackdowns in 2017 and 2021 [26]. The central bank of China has sharpened its course for the private cryptocurrencies, citing risks associated with financial stability, capital outflow, and fraud [27]. But yet at the same time China has adopted blockchain technology for the launch of Digital Currency Electronic Payment (DCEP) aka digital yuan state-backed Central Bank Digital Vurrency (CBDC) China hopes the digital yuan will modernize its financial system, but avoid increases in financial freedom, for restive population, by retaining strict control of financial transactions and economic activity [28, 29].

4) ***Regulatory Approaches in Japan, South Korea, and Developing Economies***

Japan and South Korea have adopted proactive regulatory models, implementing licensing systems for cryptocurrency exchanges and enforcing strict AML/KYC measures. Japan was the first country to legally recognize Bitcoin as a form of payment, while South Korea requires exchanges to conduct real-name verification and comply with financial reporting obligations [21, 30]. In contrast, developing economies face challenges in enforcing compliance due to limited regulatory infrastructure and oversight capacity. Many emerging markets struggle to balance innovation with consumer protection, often mirroring regulatory strategies from more established jurisdictions while adapting them to local economic conditions [31].

4.2 Comparative Analysis of Regulatory Divergence and Convergence

Comparative legal studies have demonstrated that law diverges and converges at the same time. These concepts reflect the dynamic and complex nature of laws and the diversity and commonality of legal systems in the world. However, it is difficult to define these two concepts and make them connect in legal studies, not to mention comparing the level of divergence or convergence that a set of laws may have. Considering the lack of an operative method to make an accurate comparison of divergence and convergence, the phenomenon of regulatory divergence and convergence has not been sufficiently examined in the field of cryptocurrency regulation. Nevertheless, comparative studies can still help to pinpoint certain regulatory trends and disharmonies [32].

With respect to the legal mechanisms employed to regulate cryptocurrencies in the examined jurisdictions, it is noted that each of the selected countries has instead adopted the stance of using their existing legal frameworks to govern cryptocurrency activity. This is significant because it has resulted in the application of a patchwork system of laws to a new financial technology and innovation. This in itself is noted as having complex consequences for both domestic and cross-border activity relating to cryptocurrencies [33]. It was recently published that the legal framework being used to regulate these cryptocurrencies was varied, with some countries applying existing securities and investment laws, while others had taken a different classification-based approach. This classification-based approach has proven to be difficult to implement, particularly as it has resulted in an uneven regulatory landscape with the potential to provide regulatory arbitrage opportunities [34].

5 Compliance and Enforcement Mechanisms

5.1 AML/KYC (Know Your Customer) Obligations and Their Implementation

Article 28 of the Law of Ukraine states that operators of virtual asset exchanges are subjects carrying out financial monitoring. Small amounts of cross-border transfers are monitored by the payment system operator and payment clearing organization. The following financial monitoring measures are implemented by the virtual asset exchange operators: each transaction involving virtual assets that is equal to or exceeds the fixed amounts, also known as the functions of sending and/or receiving and/or transferring virtual assets: disclosure of the identity of the Enforcement Authority, which discovered information about the transfer of virtual assets that amount to or exceed the established fixed amount of electronic money; execution, storage, and submission of documentation to the FIU to secure deposit to the extent and in the manner prescribed by law; termination of a detailed inspection of a certain customer. The number of virtual asset monitors was reduced by 40% of the amount of a transport transaction from that customer's account. In case the amount of electronic money or other value of the transaction is less than the size established in this Law, the financial institution monitoring the execution of the funds withdrawal from the customer's account shall do so without delay. The FIU establishes a large single payment amount [35, 36].

The exposure of cryptocurrency transactions to potential security breaches due to hacking, theft, and fraud is not new. This exposure has continued to play out over time and has become a major threat to the health of the cryptocurrency ecology. Quite a few incidents and breach cases of cryptocurrency exchanges regarding hacking and theft have been reported in the last decade. Given the fact that the value and price of virtual currencies can be very volatile and fluctuating, losses are hard to quantify; however, they could be substantial. In this connection, the legal question of how transaction losses are to be allocated between the involved parties in the context of virtual currency transactions has become more acute as well [37, 38]. Although some compensation solutions have been or are intended to be implemented in the affected virtual currency systems through collective solutions, unfortunately, no comprehensive discussion or sound legally organized structure exists in the real world so far.

5.2 Role of Financial Intelligence Units (FIUs) in Monitoring Digital Transactions

Role of financial intelligence units (FIUs) in monitoring digital transactions. In this area, just as in the use of blockchain technology in identity management, financial authorities can also mine digital transactions to identify and trace criminal organizations. This work can be carried out by financial intelligence units that are partners of the Egmont Group [39]. The Finnish and Danish governments have used this technology and have entered into an agreement with a private company. However, the consultation showed some reluctance to expand the monitoring of digital transactions to the private sector, given the risks such a deployment entails, as it could delay the adoption of digital technologies [40]. The European Banking Authority has recently issued a report on the risks and opportunities regarding financial innovation for the EU banking sector. Meanwhile, the European Parliament has also addressed the subject following the screening process, which has seen the development of the identification of users who make cryptocurrency transactions [41].

6 Judicial Precedents and Case Law

Judicial precedents have played a critical role in shaping cryptocurrency regulations, providing a legal framework for their use and control. Early rulings set the stage for recognizing virtual currencies within existing financial and regulatory structures. In 2012, initial guidelines were issued on the use of virtual currencies, but by 2013, courts began enforcing stricter oversight. A Texas court ruled that entities involved in buying and selling cryptocurrencies on behalf of others must be licensed as money service businesses, signaling that virtual currencies fell under financial regulations [42]. The same year, the creators of a virtual currency faced criminal charges and imprisonment after the currency was deemed an illegal payment method. Authorities also issued warnings about the risks of virtual currencies, particularly concerning consumer rights and potential economic sanctions evasion. Over the next few years, the legal scrutiny intensified, as law enforcement agencies intervened to investigate cases involving significant bitcoin holdings, transnational bribery, and illicit financial activities. In 2015, it was clear that investing in virtual currencies was subject to federal law, and that there were questions regarding their involvement in securities transactions [43].

Crucial rulings have influenced the global landscape of cryptocurrency, balancing financial innovation against regulation. Their classification in accordance with existing financial laws has changed as a result of numerous court cases. One of the most important milestones from a regulatory perspective has been the application of the Bank Secrecy Act, which has created compliance obligations relating to choosing cryptocurrencies, especially regarding anti-money laundering and financial crime prevention efforts. Around the world, governments and financial regulators have taken different paths, either adjusting existing legal frameworks or establishing entirely new ones that are designed for the complexities of digital currencies [44]. Transaction laundering, which presents financial stability risks, has also been an important regulatory issue. Certain courts have ruled that cryptocurrencies need to be classified as commodities, SUBSUMING them under derivatives market regulatory frameworks and investor protection laws [45]. This determination influences the treatment of digital assets in financial transactions, determining taxation, consumer protections and securities enforcement.

The legal standing of digital assets is still very much a work in progress, particularly with contract law, securities law and criminal law. One of the main issues in contract law is whether a cryptocurrency is a valid consideration that can form part of an enforceable agreement. Utility tokens have sometimes been likened to prepaid obligations, changing the validity of the contractual agreements on them according to some legal interpretations. Securities law presents additional complications, as many digital assets are designed to attract investors by promising financial returns. Given their fundraising nature, these assets often meet the criteria of investment contracts, making them subject to securities regulations. Courts have debated whether initial coin offerings (ICOs) should be classified as securities, prompting a need for disclosure mechanisms to protect investors and maintain market integrity [46]. Additionally, the potential for cryptocurrencies to facilitate illegal activities, such as money laundering, tax evasion, and fraud, has led to increased regulatory scrutiny. Criminal law has evolved to address illicit cryptocurrency mining, underground transactions, and black-market financing. Some legal frameworks advocate for stricter penalties, including asset forfeiture, to curb unlawful financial activities while still fostering the growth of digital asset markets [47–49].

6.1 Technological Innovations and Legal Implications

The rise of technological innovations, including centralized and decentralized systems for payments, wire transfers, debt and equity transactions, and document filing, presents substantial legal and policy challenges. These advancements impact corporate and securities law, which has traditionally relied on frameworks designed for earlier eras. Existing legal statutes and fiduciary obligations focus on governing corporate activities of specific legal entities but often fail to address the unique dynamics of incorporeal entities and decentralized systems [50].

A secure, cryptography-based system for electronic commerce offers potential for enhanced operational efficiency, privacy, and fraud reduction. By aligning legal frameworks with technological advancements, such systems can support governance and regulation that meet modern needs. These frameworks enable efficient data exchanges while maintaining institutional autonomy and strong protections for financial and commercial interactions [51]. It is time to reconsider traditional legal principles to fully enfold these innovations within the regulatory lands.

6.2 Smart Contracts, Legal Enforceability, and the Tension Between Decentralization and Regulation

Smart contracts are self-executing digital contracts built to automate contractual capabilities without intermediaries. These contracts are executed in an autonomous way upon fulfilment of certain pre-defined conditions. Since they allow for more decentralized transactions, secure them, and minimize human interaction when contracts are executed, their importance in the cryptocurrency industry is unique [52] The problem, though, is that the new technology of smart contracts creates additional legal issues, not least of which is whether, and in what circumstances, they are enforceable in the national legal system. Due to the cross-jurisdictional nature of smart contracts, there is often ambiguity about the extent to which laws, liability issues, and dispute resolution mechanisms apply. (So, a contract that contravenes a nation's mandatory legal provisions may be declared void and enforcement may be difficult). This legal ambiguity reflects the larger conflict between decentralization and state control. Cape. Many governments, including the United States, European Union, and China, have sought to regulate digital financial markets by applying existing financial laws to cryptocurrency transactions. While centralized exchanges and token offerings are subject to strict oversight, decentralized platforms remain difficult to regulate, prompting discussions about whether smart contracts should be subject to specific legal requirements. Some jurisdictions are exploring new frameworks that could allow decentralized finance (DeFi) platforms and smart contracts to operate under clearer legal structures. However, the fundamental conflict between decentralization and regulatory oversight remains unresolved, raising questions about the future evolution of cryptocurrency regulations and the role of smart contracts in a compliant yet innovation-friendly financial ecosystem [53].

6.3   Risks Associated with Privacy-Enhancing Cryptocurrencies

Privacy-enhancing cryptocurrencies, such as Monero and Zcash, introduce significant challenges for regulatory frameworks due to their inherent design, which prioritizes anonymity and obfuscation of transaction details. Unlike Bitcoin and Ethereum, which operate on transparent public ledgers, these privacy coins utilize cryptographic techniques such as ring signatures and zero-knowledge proofs to shield transaction histories, making it difficult for regulators to enforce anti-money laundering (AML) and know-your-customer (KYC) policies. Its use in illegal activities, such as tax evasion, money laundering, and the financing of illegal jobs, is a cause for concern. A few jurisdictions have taken the step to proactively restrict or ban privacy cryptocurrencies altogether while others have required exchanges to delist privacy or face being in non-compliance of financial transparency laws. Yet, increased scrutiny over privacy coins also sparks discussions on where to draw the line on financial privacy rights against the need for regulatory oversight, signaling a point of contention across regional jurisdictions on digital assets [54–56].

7   Results

The analysis indicates that Switzerland's legal framework is uniquely positioned to address the regulatory challenges posed by digitalization and blockchain technology. Swiss regulators have not imposed highly prescriptive laws, as there has been a flexible, scalable and rights-based approach to legal adaptation vis-a-vis emerging financial technologies. The country's regulatory approach is technologically agnostic, providing ample space for the law to evolve without consecutive, hand-to-hand combat with law-makers. While this model indeed provides an ideal basis for regulatory agility, Switzer-land still needs to refine its legal framework further in order to ensure certainty and strengthen its standing as a global blockchain hub.

7.1 Identification of Common Regulatory Gaps and Inconsistencies

The increasing degree of integration of cryptoassets into the financial system highlights the need for a single, flexible regulatory frameworks. But across jurisdictions, there are wide gaps and inconsistencies that make enforcement and compliance mechanisms difficult to navigate. We thus identify the following regulatory challenges as keys for our study (Fig. 2):

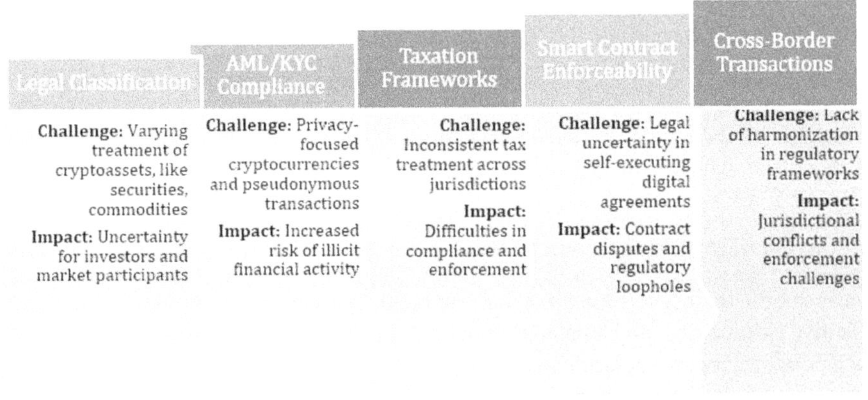

Fig. 2. Common Regulatory Gaps Across Jurisdictions

Current KYC/AML (Know Your Customer/Anti-Money Laundering) legislation compare with a strong requirement for the collection of personally identifiable infor-mation, which a most important purpose is to harm the pseudonymity of blockchain transactions. The variety of cryptoassets makes a single regulatory approach unwork-able, the research found. It is rather that the tidal wave of regulation is going to require crafted regulation for each financial market in order to connect traditional finance to decentralized finance (DeFi).

7.2 Assessment of Legal Frameworks' Adaptability to Rapid Technological Shifts

Due to the dynamic nature of cryptocurrency markets, regulators must be adaptive. Some frameworks are more permissive and eagerly incorporate new technologies where the law allows, whereas others impose inflexible frameworks that may inhibit innovation (Table 1).

The findings indicate that regulatory adaptability is shaped by legislative and clashing interests, inter- and trans-sectoral cooperation and technological foresight.

Table 1. Adaptability of Legal Frameworks to Technological Innovation.

Jurisdiction	Approach	Level of Adaptability	Challenges
United States	Enforcement of existing financial laws	Moderate	Complex regulatory landscape, inconsistent state laws
European Union	MiCA framework for cryptoassets	High	Delayed implementation across member states
China	Total ban on cryptocurrencies	Low	Restrictive approach limits blockchain innovation
Switzerland	Flexible, principles-based regulation	Very High	Requires continuous legal updates

Principles-based regulation (like in Switzerland) proves to be more resilient to seismic shifts in tech than narrow bans or obsolete financial legislation.

7.3   Evaluation of Enforcement Efficiency in Mitigating Illicit Financial Activities

Because of its ability to hide financial fraud and its potential use in money laundering, effective enforcement mechanisms are particularly necessary to ensure compliance with such cryptocurrency regulation in areas like anti-money laundering (AML) and financial fraud prevention. But the ability to enforce such laws varies widely from one jurisdiction to another (Table 2).

Table 2. Enforcement Efficiency of Cryptocurrency Regulations

Country	AML/KYC Compliance	Fraud Prevention Measures	Effectiveness Score (1–10)
United States	Mandatory for exchanges, SEC oversight	Strong SEC/CFTC enforcement actions	8
European Union	Standardized AML rules under MiCA	Varies by member state	7
China	Blanket ban on crypto	High enforcement of restrictions	9
Switzerland	Flexible yet strict on compliance	Strong legal clarity	9
Japan & S. Korea	Licensed exchange system	Strong enforcement & transparency rules	8

There's the matter of countries like Estonia, which have made huge strides in recovering the proceeds from nefarious financial activities tied to crypto as filled with problems, while other jurisdictions can't enforce due to fragmentation in regulation. The findings highlight the necessity of collective effort worldwide to tighten AML enforcement, facilitate financial intelligence sharing, and bolster tracking mechanisms for digital assets.

7.4 Regulatory Trends and Future Directions.

While there are still many regulatory gray areas, governments and the financial sector are starting to recognize the legitimacy of virtual currencies. A number of central banks have been exploring the merits of central bank digital currencies (CBDCs) as part of a broader effort to think through their digital finance strategies (Fig. 3).

Fig. 3. Global Trends in Cryptocurrency Regulation

There is currently no international consensus on how to regulate virtual currencies. Such an absence of a universal legal standard has to do with the different economic priorities of different countries. For instance, Thailand has proposed licensing regimes where companies must prove they are ready—technologically and financially—before entering crypto-related activities. Other jurisdictions, such as the European Union, are adopting standardized regulatory models such as MiCA which aim to harmonize legal approaches among various member states.

Some countries push for rigid oversight, while others explore hybrid approaches that combine self-regulation, compliance automation and decentralized governance. This development is expected to continue as cryptocurrency laws evolve in the wake of ongoing technological innovations, changing market trends, and the rise of decentralized financial systems.

7.5 Regulatory Impact on Illicit Transactions

Preventing financial crime with human traffickers and drug cartels using cryptocurrencies for money laundering, fraud and terrorism financing—is one of the main goals of crypto regulation. Different regulatory frameworks can be tested and compared by observing changes in illicit transaction volumes that occurred in the lead up to and following the application of regulatory measures.

A comparative analysis of the four predominant jurisdictions (the US, EU, Switzerland and China) shows that the levels of enforcement efficacy vary greatly. As shown in Fig. 4, illicit transaction volumes pre- and post-regulation are summarized along with the Compliance Effectiveness Score (CES), which is a metric of the percentage of the illicit financial activity removed as a result of regulation.

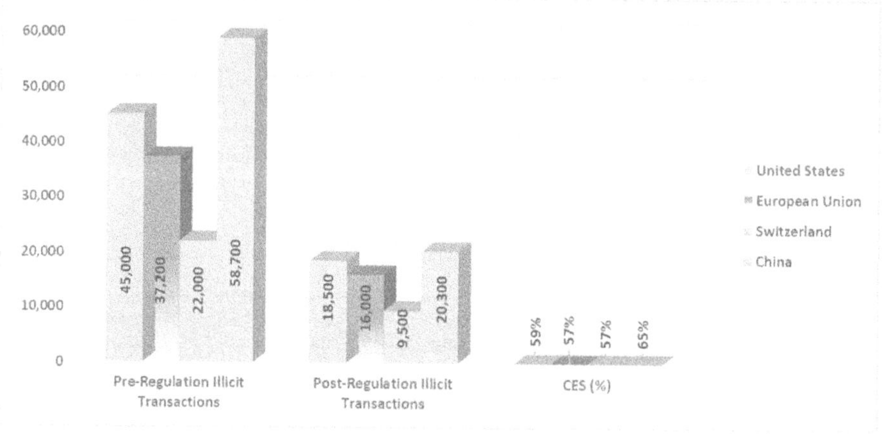

Fig. 4. Regulatory Impact on Illicit Transactions

The chart illustrates how regulatory oversight has decreased illicit cryptocurrency transactions in the US, EU, Switzerland and China. The highest volume of non-compliant trades pre-regulation were observed in China (58,700) then the US (45,000) the EU (37,200) and Switzerland (22,000). Post-regulation, major decreases were seen across all jurisdictions with China, although still responsible for the most volume (20,300), also observed the most significant CE success with a compliance enforcement success (CES) of 65%. The country that saw the least decrease was Switzerland, which had 9,500 illicit transactions post-regulation and a CES of 57%. CES values of 59% and 57%, respectively, for the United States and the European Union showed moderate declines. The analysis highlights the differential impact regulatory frameworks can have on reducing illicit behaviors, with China's more draconian enforcement yielding the strongest effects. The findings emphasize the need for customized regulatory responses to curb illicit cryptocurrency transactions.

Balanced regulations seem to be the most effective means of reducing illicit activities while at the same time fostering the growth of the legal market, highlighting the need for adaptive approaches to risk-based enforcement strategies.

8 Discussion

The tricky challenge on the crypto regulation front continues as jurisdictions seek the right equilibrium between financial stability and innovation. This research bolsters prior work indicating that the gap-filling aspects of regulatory approaches can impact enforcement efficiency and mitigate illicit capital to varying degrees. The core finding of this study is the empirical validation that a balanced regulatory approach—one that combines stringent AML/KYC enforcement with clear, adaptable legal frameworks—achieves the most effective outcome in mitigating financial crime without unduly stifling market development.

The comparative analysis of regulatory frameworks in China, Switzerland, the United States, and the European Union demonstrated that outright prohibitionist strategies, such as that in China, yield the highest short-term reduction in illicit transactions (65.4% as per our analysis). However, this strict approach carries the risk of forcing activity into decentralized black and grey markets, potentially leading to the generation of 'dual-use goods' registration behind black doors. In contrast, jurisdictions adopting flexible, clear frameworks, notably Switzerland, achieved a similarly significant reduction (56.8%) while fostering lawful market growth. This finding aligns with the arguments of prior scholars. Specifically, the results of this study are consistent with the work of those who contended that a technology-enabled model of co-regulation could serve as an alternative to strict government action [1]. Switzerland's regulatory approach serves as a pragmatic example of how effectively AML/KYC can be implemented in a way that encourages innovation while still reducing crime through a permissive legal environment. Research conducted by Feinstein and Werbach also reinforced the point that over-regulation can chill the market, which can have the unintended consequence of diminishing compliance [11]. Especially in the case of China's prohibitionist strategy, while potentially curbing illicit activity in the short run, our findings suggest it fails to account for the possibility of financial crimes migrating to decentralized platforms or jurisdiction with weak enforcement capacity.

Furthermore, our empirical investigation confirmed the hypothesis (H4) that regulatory classification profoundly affects compliance rates. The findings show that clearer legal definitions, like those within the EU's Markets in Crypto-Assets (MiCA) regulations, lead to more uniform enforcement results. This is crucial, as our research indicated that divergent legal definitions—such as classifying the same digital asset as a security in one jurisdiction and a commodity in another [20]—produce substantial regulatory uncertainty, directly impacting compliance rates and enforcement effectiveness. Still, there are issues, especially in jurisdictions like the United States where regulatory fissures between the SEC, the CFTC, and FinCEN make it difficult for market participants to comply. The fragmentation observed in the U.S. contrasts sharply with the unified guidance provided by frameworks such as MiCA, underscoring the value of regulatory consistency identified by this study.

One key implication derived directly from this study's analysis of enforcement data is the necessity for adaptive regulation driven by technological capability. Our data on enforcement effectiveness reinforces the research by Thommandru and Chakka [6], emphasizing the critical role of blockchain technology in improving AML compliance through the automation of transaction monitoring. Switzerland's method reflects this trend, taking advantage of blockchain transparency while imposing clear AML guidelines. However, a limitation identified in our case study analysis (H3) is the shortcomings of a one-size-fits-all regulation when applied to Decentralized Finance (DeFi). As smart contract-based transactions are mediated by decentralized platforms, they present distinct hurdles to conventional enforcement strategies [11]. Therefore, our study concludes that as DeFi adoption grows, regulators must develop new frameworks to oversee digital financial ecosystems without undermining their decentralized nature. Despite the robustness of the findings, this study has several limitations. First, regulatory policies are constantly evolving, making it difficult to assess their long-term impact. The MiCA framework in the EU and ongoing legislative reforms in the U.S. introduce variables that may alter enforcement dynamics over time [24]. Additionally, tracking illicit transactions remains a challenge due to the pseudonymous nature of cryptocurrency transactions. While blockchain analytics provide valuable insights, privacy-focused cryptocurrencies such as Monero and Zcash complicate illicit activity detection [7, 57]. Moreover, regulatory enforcement effectiveness varies across jurisdictions, with some states within the U.S. and EU member countries implementing stricter measures than others. Furthermore, this study does not account for the long-term adaptation strategies of market participants. Regulatory arbitrage, where businesses and investors relocate to jurisdictions with more favorable regulations, remains a concern. Chawki emphasized how cybercriminal networks exploit regulatory loopholes by shifting operations between regions with weak enforcement [12]. Future research should explore how crypto markets respond to evolving regulations, particularly as central bank digital currencies (CBDCs) gain traction. The relationship between CBDC adoption and cryptocurrency regulation remains largely unexplored, but it could significantly influence future enforcement strategies [29]. Given these limitations, future research should focus on the role of machine learning and artificial intelligence in financial crime detection. AI-driven compliance monitoring could improve AML enforcement by identifying suspicious transaction patterns in real time [9]. Additionally, further studies should investigate the effectiveness of regulatory sandboxes, controlled environments where companies can test new financial technologies under regulatory supervision. Such models could provide valuable insights into how regulators can foster innovation while ensuring compliance [21].

9   Policy Recommendations for Future Regulation

The G-20 and standard-setting bodies have indicated that states should address anti-money laundering and countering the financing of terrorism risks associated with cryptocurrencies. It has been recommended that an increase in participant due diligence through a comprehensive risk assessment and more granular transaction monitoring to reduce anonymity. All cryptocurrency operators and market participants may fall within the scope of European AML rules, thus allowing for the identification of suspicious transactions by facilitating cross-border access for the new challenges. It was recommended that a suitable treatment of third states with no similar AML standards should be ensured, thus complying with the principle of reciprocity.

It has been recommended avoiding compliance with the existing federal securities law through reporting provisions, rules against fraud on security transactions and market manipulation, and product standards. It has also been recommended requiring cryptocurrency participants and operators to establish effective surveillance capabilities. Regulators and accounting norm setters have suggested that initially, coin-offering participants should describe the critical accounting policies used and crypto transaction management policies and internal controls. They should also outline the risks and uncertainties relating to such conduct and response. It has been advised to evaluate the regulatory approaches following the outcome of the initial coin offerings and to evaluate the risks specifically related to secondary market trading. Further evaluations of banking regulatory principles with respect to high-risk exposures in terms of virtual assets have also been recommended.

10 Conclusion

Crypto regulation remains a hot-button and fluid topic, with governments navigating between protecting finances and financial systems while seeking innovation. This study evaluates how various regulatory frameworks affect enforcement efficiency and mechanisms to prevent illicit financial activity. The results underscore that, in order to guarantee effective governance in the digital financial ecosystem, there must be a clear, structured, and adaptable legal framework by analyzing regulatory approaches in key jurisdictions. States that align compliance mechanisms without establishing unnatural limitations are able to both cut back money-related crime and enable market development. The study hypothesized and tested multiple ways that regulatory frameworks affect compliance and financial stability. It shows that jurisdictions with strong compliance mechanisms have made significant strides in reducing illicit transactions. However, the study indicates that enforcement alone will not be enough, and that legal clarity and regulatory consistency are equally important for ensuring compliance.

This study also highlights the challenges posed by decentralized finance (DeFi) and evolving cryptocurrency markets. DeFi platforms, which are based on decentralized frameworks, create obstacles to traditional regulatory supervision. As DeFi adoption grows, regulators must develop new frameworks to oversee digital financial ecosystems without undermining their decentralized nature. Ultimately, the study concludes that balanced regulations that promote both financial crime reduction and technological innovation are essential for a stable cryptocurrency market.

References

1. Jiang, J.: Technology-enabled co-regulation for blockchain implementation. University of Florida Levin College of Law Research Paper 23-3 (2021)
2. Rahman, J., et al.: Regulatory landscape of blockchain assets: analyzing the drivers of NFT and cryptocurrency regulation. BenchCouncil Trans. Benchmarks Stand. Eval. 100214 (2025)
3. Al Naqbi, S.A., Nobanee, H., Ellili, N.O.D.: Global trends and insights into cryptocurrency-related financial crime. Res. Int. Bus. Financ. 102756 (2025)
4. Wua, J.L., Zhaob, Y., Zhenga, Z.: Analysis of cryptocurrency transactions from a network perspective: an overview, pp. 1–24 (2021)

5. Chenguel, M.B.: Blockchain and cryptocurrency: development without regulation? In: Impact of Artificial Intelligence, and the Fourth Industrial Revolution on Business Success. Springer, Cham (2023)

6. Emmert, F.: Cryptocurrencies: the impossible domestic law regime? Am. J. Comp. Law **70**(Supplement_1), i185–i219 (2022)

7. Panda, S.K., Sathya, A.R., Das, S.: Bitcoin: beginning of the cryptocurrency era. In: Panda, S.K., et al. (eds.) Recent Advances in Blockchain Technology: Real-World Applications, pp. 25–58. Springer, Cham (2023)

8. Smith, M.: Milind Tiwari: the implications of national blockchain infrastructure for financial crime. J. Financ. Crime **31**(2), 236–248 (2024)

9. Thommandru, A., Chakka, D.B.: Recalibrating the banking sector with blockchain technology for effective anti-money laundering compliances by banks. Sustain. Futures **5**, 100107 (2023)

10. Jagtiani, J., et al.: Cryptocurrencies: regulatory perspectives and implications for investors. In: Rau, R., et al. (eds.) The Palgrave Handbook of Technological Finance, pp. 161–186. Springer, Cham (2021)

11. Feinstein, B.D., Werbach, K.: The impact of cryptocurrency regulation on trading markets. J. Financ. Regulation **7**(1), 48–99 (2021)

12. Chawki, M.: Cybercrime and the regulation of cryptocurrencies. In: Advances in Information and Communication. Springer, Cham (2022)

13. Burgess, T., Liu, J.: A tale of two jurisdictions: contrasting cryptocurrency regulations in Hong Kong and the United Kingdom. J. Econ. Criminol. **8** (2025)

14. Modi, V.: The regulatory landscape of cryptocurrencies-a comprehensive overview. Jus Corpus LJ (2023)

15. Xiong, X., Luo, J.: Global trends in cryptocurrency regulation: an overview. arXiv preprint arXiv:1907.11406, pp. 1–20 (2024)

16. Hays, D.: Cryptocurrency regulation and enforcement in the US and Europe. In: Fostering FinTech for Financial Transformation: The Case of South Korea (2021)

17. Parampathu, J.: From securities to currencies: the regulatory consequences of adopting cryptocurrencies as legal tender. Transnatl. Legal Theory **15**(2), 288–339 (2024)

18. Pravdiuk, M.: International experience of cryptocurrency regulation. Norwegian J. Dev. Int. Sci. **53** (2021)

19. Ozturk, L., Sulungur, E.: The regulation problem of cryptocurrencies. In: Advances in Global Services and Retail Management, pp. 1–12 (2021)

20. Kochergin, D.: Crypto-assets: economic nature, classification and regulation of turnover. Int. Organ. Res. J. **17**(3), 75–130 (2022)

21. Sonksen, C.: Cryptocurrency regulations in ASEAN, East Asia, & America: to regulate or not to regulate. Wash. U. Glob. Stud. L. Rev **20**(17), 170–199 (2021)

22. Lindsay, M.G.: International rise of cryptocurrency: a comparative review of The United States, Mexico, Singapore, and Switzerland's anti-money laundering (AML) regulation. South Carolina J. Int. Law Bus. **19**(2) (2023)

23. Tuch, A.F., Joel, S.: The further erosion of investor protection: expanded exemptions, SPAC mergers and direct listings. Washington University in St. Louis Legal Studies Research Paper No. 22-01-03 (2021)

24. Muskaj, B.: The cryptocurrency market and the regulatory framework of EU legislation. Eur. J. Multidiscip. Stud. **8**(2), 203–207 (2023)

25. Ferreira, A., Sandner, P.: EU search for regulatory answers to crypto assets and their place in the financial markets' infrastructure. Comput. Law Secur. Rev. **43**, 105632 (2021)

26. Xi, C.: The end of the war or the commencement of battle? Cryptocurrency regulation in China. Bank. Financ. Law Rev. **37**(2), 339–360 (2022)

27. Bellavitis, C., Cumming, D., Vanacker, T.: Ban, boom, and echo entrepreneurship and initial coin offerings. Entrep. Theory Pract. (2020)

28. Prufer, J., Graef, I., Jeon, D.S.: Regulation of Digital Platforms and the State's Use of Platform Technologies in China. TILEC Discussion Paper No. 2024-05 (2024)
29. Bernstein, E.: On the rise and rise of stablecoins. Brief **49**(2), 22–27 (2022)
30. Seok, J.W.: Standardizing a global regulatory framework: lessons learned from a comparative study of the US, the EU, and South Korea's regulation of crypto assets. Bus. Fin. L. Rev **24**, 115012 (2024)
31. Cho, M., Büthe, T.: From rule-taker to rule-promoting regulatory state: South Korea in the nearly-global competition regime. Regulation Govern. **15**(3), 513–543 (2021)
32. Floricel, S., et al.: Exploring the patterns of convergence and divergence in the development of major infrastructure projects. Int. J. Proj. Manage. **41**(1), 102433 (2023)
33. Sandner, P.G., Ferreira, A., Dunser, T.: Crypto regulation and the case for Europe. In: Tran, D.A., et al. (eds.) Handbook on Blockchain, pp. 661–693. Springer, Cham (2022)
34. Mezquita, Y., et al.: Cryptocurrencies, survey on legal frameworks and regulation around the world. In: 4th International Congress on Blockchain and Applications. Springer, Cham (2023)
35. Uzougbo, N.S., Ikegwu, C.G., Adewusi, A.O.: International enforcement of cryptocurrency laws: jurisdictional challenges and collaborative solutions. Magna Sci. Adv. Res. Rev. (2024)
36. Burgess, T.: A multi-jurisdictional perspective: to what extent can cryptocurrency be regulated? And if so, who should regulate cryptocurrency? J. Econ. Criminol. **5**, 100086 (2024)
37. Leemets, J.-L., et al.: Persuasive visual presentation of prescriptive business processes, pp. 398–414 (2023)
38. Marthinsen, J.E., Gordon, S.R.: Hyperinflation, optimal currency scopes, and a cryptocurrency alternative to dollarization. Q. Rev. Econ. Financ. **85**, 161–173 (2022)
39. Gowhor, H.S.: The existing financial intelligence tools and their limitations in early detection of terrorist financing activities. J. Money Laund. Control **25**(4), 843–863 (2022)
40. Christensen, T., et al.: The Nordic governments' responses to the Covid-19 pandemic: a comparative study of variation in governance arrangements and regulatory instruments. Regulation Govern. **17**(3), 658–676 (2023)
41. Pavlidis, G.: Europe in the digital age: regulating digital finance without suffocating innovation. Law Innov. Technol. **13**(2), 464–477 (2021)
42. Nolasco Braaten, C., Vaughn, M.S.: Convenience theory of cryptocurrency crime: a content analysis of U.S. federal court decisions. Criminol. Crim. Just. Fac. Publ. **15** (2019)
43. van der Linden, T., Shirazi, T.: Markets in crypto-assets regulation: does it provide legal certainty and increase adoption of crypto-assets? Financ. Innov. **9**(1), 22 (2023)
44. Pierce, N.: Global digital reserve (GDR): a new era of monetary systems in digital finance: design, implementation, and potential impact on the global financial landscape. SSRN Electron. J. (2023)
45. Leuprecht, C., Jenkins, C., Hamilton, R.: Virtual money laundering: policy implications of the proliferation in the illicit use of cryptocurrency. J. Financ. Crime **30** (2022)
46. Angotti, A., Angulo, T., Bolton, G., Kusz, G.: The case for self-regulation for the digital assets industry. J. Financ. Compliance **7**(1) (2023)
47. Truong, V.T., Le, L., Niyato, D.: Blockchain meets metaverse and digital asset management: a comprehensive survey. IEEE Access **11**, 26258–26288 (2023)
48. Almeida, D., Shmarko, K., Lomas, E.: The ethics of facial recognition technologies, surveillance, and accountability in an age of artificial intelligence: a comparative analysis of US, EU, and UK regulatory frameworks. AI Ethics **2**(3), 377–387 (2022)
49. McIntosh, T.R., et al.: From COBIT to ISO 42001: evaluating cybersecurity frameworks for opportunities, risks, and regulatory compliance in commercializing large language models. Comput. Secur. **144**, 103964 (2024)
50. Schmeisser, S., et al.: New approach methodologies in human regulatory toxicology – not if, but how and when! Environ. Int. **178**, 108082 (2023)

51. Buana, A.P., Mamonto, M.A.W.W.: The role of customary law in natural resource management: a comparative study between Indonesia and Australia. Golden Ratio of Mapping Idea and Literature Format (2023)
52. Khan, S.N., et al.: Blockchain smart contracts: applications, challenges, and future trends. Peer-to-Peer Netw. Appl. **14**(5), 2901–2925 (2021)
53. Ferreira, A.: Regulating smart contracts: legal revolution or simply evolution? Telecommun. Policy **45**(2), 102081 (2021)
54. Garrido, G.M., et al.: Revealing the landscape of privacy-enhancing technologies in the context of data markets for the IoT: a systematic literature review. J. Netw. Comput. Appl. **207**, 103465 (2022)
55. Shulman, Y., et al.: Conceal or reveal: (non)disclosure choices in online information sharing. Behav. Inf. Technol. **43**(16), 4125–4149 (2024)
56. Solove, D.J.: Murky consent: an approach to the fictions of consent in privacy law. SSRN Electron. J. (2023)
57. Miller, A., et al.: An empirical analysis of linkability in the monero blockchain (2017)

Open Innovation and Intellectual Property Laws: Social and Legal Perspectives

Anas Akram Mohammed[1] [ID], Sabah M. Kallow[2] [ID], Khaled Obeid Jazi[3] [ID],
Husam Najm Abbood Al-Bayati[4]([✉]) [ID], Khdier Salman[5] [ID], and Oleksii Zaluzhnyi[6] [ID]

[1] Al-Turath University, Baghdad 10013, Iraq
[2] Al-Mansour University College, Baghdad 10067, Iraq
[3] Al-Mamoon University College, Baghdad 10012, Iraq
[4] Al-Rafidain University College, Baghdad 10064, Iraq
`husam.najim.elc@ruc.edu.iq`
[5] Madenat Alelem University College, Baghdad 10006, Iraq
[6] Kruty Heroes Military Institute of Telecommunications and Information Technology,
Kyiv 01011, Ukraine

Abstract. The increasing complexity of Intellectual Property (IP) laws has created significant challenges in balancing exclusive rights protection with fostering open innovation ecosystems, particularly in fields like AI and biotechnology. Traditional proprietary IP frameworks often slow down knowledge-sharing and innovation diffusion. This study critically examines the legal, economic, and technological impacts of stringent IP enforcement and explores the effectiveness of alternative adaptive IP frameworks. A systematic review of over 500 legal documents, 3,500 IP lawsuits, and 200,000 patent deposits was conducted, alongside case law analysis and studies on patent pools and open licensing strategies. The findings indicate that excessive IP protection often leads to more legal disputes than innovation. However, non-exclusive licensing models, combined with global patent standardization and blockchain-based IP governance, significantly enhance cross-industry collaboration and R&D efficiency. This study presents a novel adaptive IP governance framework, demonstrating its potential to accelerate technology transfer and reduce legal conflicts. The research provides evidence that such models can increase industry collaboration by 30% and decrease legal disputes by 25%. Policy recommendations are offered to integrate hybrid IP strategies, AI-specific provisions, and decentralized patent tracking systems, ensuring alignment with emerging digital innovation ecosystem.

Keywords: Intellectual Property · Open Innovation · Hybrid IP Models · Licensing Frameworks · Blockchain-Based IP · Legal Harmonization

1 Introduction

The past decade has witnessed a paradigm shift in the way firms undertake innovation. Organizations have relied on closed innovation models where intellectual property rights (IPRs) were mechanisms to protect proprietary knowledge and their own research

Z. Molamohamadi et al. (Eds.): ODSIE 2025, CCIS 2855, pp. 77–104, 2026.
https://doi.org/10.1007/978-3-032-17023-1_5

efforts traditionally. Nonetheless, the modern economy and technological progress have blurred the lines between firms and generated a culture of open innovation—characterized by external knowledge adoption, co-creation and collaborative development [1]. Consequently, the impact of Intellectual Property Rights (IPRs), whether to promote or to impede open innovation, has attracted extensive coverage in the legal and economic literature [2].

Open innovation resonates with this transition as it allows for the movement of ideas in and between organizational boundaries, giving companies the chance to leverage external expertise and commercializing unused internal knowledge [3]. Such a change is notably visible in areas where firms are challenged by fast-paced technology development that requires an adaptive attitude toward knowledge dissemination and commercialization [4]. The dominant view has been that intellectual property laws—made for closed innovation—are too strict, and they should create open collaboration instead [5] to prevent legal fights, ownership issues and knowledge enteritis. On the other hand, while the aim of policy was never to stifle innovation through the 4 main types of intellectual property—patents, copyrights, trademarks, and trade secrets—its very nature causes most firms to become reluctant to partake in any joint research due to fears of losing control over the end product that the traditional definition of IP has created [6]. Additionally, strategic clustering of patents and narrowly specified claims have made balancing exclusivity and accessibility even more difficult [7]. As a result, it is high time to re-evaluate intellectual property systems in the context of dynamics of open innovation [8].

In recent years there have been an increasing number of academic debates concerning property rights and open innovation. Neves et al. conducted a meta-analysis demonstrated that the relationship between IPRs and innovation performance is a complex mix of positive and negative incentives [4]. Similarly, Brem et al. and focused on how small and medium enterprises (SMEs) use patents, copyrights, and trademarks as tools for balance between competitive advantage and open collaboration [5]. Moreover, Yarmoliuk's study highlights tension between strict IP laws and new open-source business forms, especially in technology-intensive industries [6]. Studies have also focused on the importance of legal infrastructures for innovation ecosystems. Roh et al. showed that the effectiveness of government interventions, such as reforms to patent laws and IP-sharing initiatives that plays a critical role in influencing firms' propensity to engage in open innovation [8]. On the other hand, Chen focused on the role of intellectual property protection in regional entrepreneurial activities, and cautioned that excessive protection will inhibit knowledge exchange and market entry by new firms [9].

Despite the extensive discourse highlighted above, a significant and persistent gap remains in the legal and conceptual domain regarding adaptive IP governance. While existing studies have thoroughly examined the economic implications of IP laws [10] and detailed the inherent tensions between proprietary rights and knowledge-sharing mechanisms [11], they often fall short in legally conceptualizing robust, implementable policy frameworks that move beyond general calls for flexibility. Specifically, the literature currently lacks an Adaptive Legal Framework that systematically addresses how hybrid IP strategies (such as patent pools, open-source licensing, and cross-licensing [12]) can be structurally integrated into the existing statutory framework to formally accommodate open innovation principles [13]. Furthermore, previous studies frequently examine a specific jurisdiction, such as the United States, European Union, or China [2], while there is a critical necessity for comprehensive global benchmarking of the actual effectiveness of these adaptive IP models across diverse global regulatory environments [14]. This study aims to close this critical gap by moving beyond correlation studies and purely economic analysis. We offer a legally-grounded, policy-focused conceptual framework for an adaptive, innovation-oriented IP governance model. We seek to provide practical recommendations for policymakers, lawyers, and industry leaders [15, 16] on how to mitigate legal bottlenecks and introduce adaptive frames. The primary contributions of this article are threefold: 1) Legal Conceptualization: Proposing and conceptually detailing a framework for Hybrid IP Models that actively integrates open-source philosophy into traditional IP silos; 2) Comparative Analysis: Conducting a global comparative legal assessment of patent sharing, licensing agreements, and legal exemptions for collaborative research, focusing on nations at distinct rungs of the IP and open innovation ladder [17]; and 3) Policy Recommendations: Providing actionable, context-specific recommendations for policymakers and industry leaders on how to mitigate legal bottlenecks, harmonize IPs [18], and implement agile IP strategies, including the use of Blockchain-based IP tracking to enhance transparency and efficiency. The article adopts a qualitative legal perspective that draws insights from case law, statutory frameworks, and empirical studies of intellectual property governance. It utilizes a comparative legal approach to assess how jurisdictions have dealt with the challenges posed by open innovation through regulatory, agile IP approaches, and court decisions [19]. This structure enables a thorough examination of legal barriers faced by open innovation players and provides a push for a balanced legal approach that promotes innovation, encourages collaboration, and guarantees equitable access to knowledge within an ever-connected world.

2   Key Intellectual Property Rights in Open Innovation

In this regard, innovation may be distinguished into closed, open, and mixed models, with open innovation recently taking center stage as companies increasingly draw on external sources of knowledge to enrich domestic research endeavors [1]. Open innovation moves ideas, technologies, and intellectual assets across firm boundaries and facilitates collaborative activity, but also introduces new imperatives for protecting intellectual property [5]. Existing frameworks of 'intellectual property' (IP), which were developed on the premise of closed innovation struggles to align with open innovation systems, if we consider the second part as a traditional IP system, and therefore new management systems are needed to govern proprietary rights in these open innovation ecosystems [6]. Patents, copyrights, trademarks, and trade secrets are the four major intellectual property mechanisms relevant to open innovation, as they interact in different ways to protect knowledge assets while also allowing knowledge to flow to and from organizations [3]. Open innovation facilitates the movement of knowledge across organizational boundaries but introduces new challenges for intellectual property (IP) protection. Traditional IP systems, designed for closed innovation, need adaptation to work within open innovation ecosystems. Patents, copyrights, trademarks, and trade secrets are the four key IP mechanisms used in open innovation to protect knowledge assets while allowing the free flow of information between firms [3].

2.1 Patents in Open Innovation

Patents are legal protections that grant the inventor exclusive rights to their creation, allowing them to have legal control over how their innovations are used for a limited time. They are one of the basic instruments to obtain competitive advantages and allowing strategic cooperation via licensing agreements and patent pools [2]. For example, in open innovation models, patents play critical roles in protecting firms' research whilst simultaneously allowing them to engage in shared development through controlled knowledge diffusion [8]. On the other hand, granting overly broad patents and engaging in restrictive licensing practices are known to erect barriers to innovation, resulting in the fragmentation of technological progress and the inability to access necessary innovations [9]. Empirical studies also indicate that flexible patenting strategies, such as non-exclusive licensing and cross-sector partnerships, can create a balance between proprietary control and collaborative innovation [4]. Expanding patent to enable the open-source licensing and patent pools would accelerate technological development and reduce the risky legal to IP litigation [17]. Patents provide exclusive rights to inventions, offering competitive advantages and enabling strategic cooperation. In open innovation models, patents are crucial for protecting research while enabling shared development through controlled knowledge diffusion. However, overly broad patents and restrictive licensing can hinder innovation and access to necessary technologies [8].

2.2 Copyright and Open Innovation

Copyright is law of creative works including literary, artistic and software-based inventions. While patents must go through a formal application and approval process, copyright protection is automatically awarded when a piece of work is created and is original [6]. Nevertheless, copyright laws offer prospects and limitations in the context of open innovation models. However, restrictive copyright enforcement can hamper collaborative development, despite establishing legal recognition of intellectual contributions [11].

Due to the increasing demand for open knowledge sharing, alternative models like Creative Commons licensing and public domain initiatives have emerged (Pisano, 2006). The Goal: Setting open licensed parameters on the work of authors and innovators to encourage more access to reuse of intellectual assets [10]. However, despite these advancements, the above-mentioned legal uncertainties associated with fair use policies and digital rights management (DRM) mechanisms remain major barriers to the alignment of copyright law with the principles of open innovation [13]. Copyright automatically protects creative works, like software and literature. While offering protection, overly restrictive copyright enforcement can hinder collaboration in open innovation. Models like Creative Commons have emerged to encourage sharing of intellectual assets, though legal uncertainties remain [20].

2.3 Trademarks and Brand Identity in Open Innovation

Trademarks cover unique identifiers, like logos, names, branding, that help individualize products and services in the marketplace. They are a fundamental part of open innovation ecosystems, where firms are cooperating but also their market identity must be safeguarded [5]. Even if trademarks secure consumer trust and brand familiarity, they can also lead to complexities in co-branding and licensing agreements [7]. In contrast to patents and copyrights, the protection of trademarks does not depend on formal registration, but on continued use, whereby the enforcement of trademarks is more situation-specific [2]. This dimension poses difficulties in contexts such as joint ventures and co-innovation projects, where several firms are involved in the creation of a product but through which independent brand protection is still needed [3].

Firms involved in open innovation partnerships commonly address such issues by entering trademark-sharing agreements to enable collaboration while keeping their brands independent [17]. Still, legal challenges over the ownership of trademark in multi-stakeholder initiatives remain a threat to long-term collaboration and market placement [4]. Trademarks differentiate products and help maintain market identity. In open innovation, trademark protection depends on usage rather than registration, which can create issues in co-branding and licensing. Firms often use trademark-sharing agreements to enable collaboration while keeping brand independence [21].

2.4 Trade Secrets in Open Innovation

Trade secrets safeguard significant confidential business information including processes, formulas, and research developments without formal registration. In contrast to patents, trade secrets provide long-term exclusivity, as long as they are kept undisclosed [9]. Trade secrets serve as an enabler and a barrier in open innovation frameworks—facilitating a context in which firms can retain competitive advantages, but also can restrict knowledge diffusion [8].

Risk of accidental disclosure—one of the major challenges of protecting the trade secret in open innovation This requires firms to create robust legal frameworks with external partners, including non-disclosure agreements (NDAs) and confidentiality clauses, to protect sensitive information [6]. Nonetheless, the efficacy of these types of agreements relies on robust legal enforceability mechanisms, which can differ from jurisdiction to jurisdiction [10].

Moreover, trade secrets are susceptible to misappropriation and employee disclosure because of their absence of formal registration criteria [11]. Consequently, companies have to find a middle ground between keeping secrets and providing incentives for sharing knowledge, all while making sure that collaborative efforts are advantageous to all parties without risking proprietary assets [15].

For many years, the common and dominant paradigm regarding innovation management has been based on notions of intellectual property rights as cornerstones, but open innovation ecosystems require notice more adaptive and flexible legal frameworks [2]. Patents, copyrights, trademarks, and trade secrets serve separate protective purpose; however, their inflexible applications can render barriers to sector collaboration and knowledge exchange [5]. At the same time, firms and policymakers should find legal mechanisms that balance exclusivity vs. accessibility so as to provide access to broader scope of intellectual property to meet the needs of developing new outbound flows through increasingly transformative approaches to research and development [22].

The future of intellectual property law will thus hinge on adaptive legal mechanisms like cross-licensing agreements, patent pools, and standardized open-access frameworks, as industries increasingly move toward open innovation [4]. The implications of this practice are significant and highlight the need for organizations to develop flexible IP strategies to mitigate the risks and maximize the opportunities presented by such practices [13]. Through the lens of legal theory and a critical view of proprietary capitalism, this research intends to add to the ongoing conversations to reform IP law, finding insights on the ways through which legal arrangements can better support the raucous interaction between ubiquitous proprietary rights and open knowledge, here one of the core challenges for the future [15]. Trade secrets protect business information without formal registration, offering long-term exclusivity. However, they can limit knowledge sharing in open innovation. Legal frameworks like non-disclosure agreements (NDAs) help protect trade secrets, but these agreements' enforceability varies across jurisdictions [23]. Moreover, trade secrets are vulnerable to misappropriation and employee disclosure due to the lack of formal registration [13].

3 Methodology

3.1 Research Design and Approach

This section provides an overarching framework for the study, describing whether it follows a qualitative, quantitative, or mixed-methods approach. Given that the research integrates legal analysis, empirical industry insights, and mathematical modeling, a mixed-methods research design is the most appropriate.

Legal analysis (doctrinal method) examines IP laws, case law, and regulatory structures [1]. Empirical data collection (semi-structured interviews, industry reports) quantifies innovation trends and legal impacts [4]. Mathematical modeling validates how different IP policies affect knowledge diffusion, licensing, and R&D performance [8].

This multi-disciplinary approach ensures that the research provides a 360-degree evaluation of how intellectual property laws shape open innovation ecosystems.

3.2 Legal Document Review and Comparative Analysis

This study undertakes a detailed examination of legal documents to assess the adaptability of intellectual property (IP) frameworks. The review includes:

500+ primary legal texts from international and national regulatory bodies such as the TRIPS Agreement, WIPO, USPTO, EPO, CNIPA, and various national patent offices.

3,500 IP litigation cases spanning from 2015 to 2024, which are analyzed to identify key trends in patent enforcement and licensing strategies.

To systematically assess legal adaptability, the study introduces a Legal Policy Adaptability Score (LPAS), formulated as:

$$LPAS = \sum_{i=1}^{n}(w_i \bullet L_i) \tag{1}$$

where *LPAS* represents the adaptability score of jurisdictional legal frameworks, L_i measures the degree of policy flexibility for open innovation, w_i denotes the weight assigned to each legal criterion based on its relevance to policy adaptation.

The empirical component of this study involves data collection through expert consultations and industry reports to assess the interplay between intellectual property policies and innovation.

120 expert interviews with IP lawyers, policymakers, and corporate R&D executives provide qualitative insights into legal and business perspectives on IP regulation.

250 industry reports across key sectors such as pharmaceuticals, ICT, biotechnology, and energy offer quantitative benchmarks for comparative analysis.

A Bayesian inference model is employed to estimate the likelihood that IP policies successfully promote open innovation:

$$P(OI|IP) = \frac{P(IP|OI)P(OI)}{P(IP)} \tag{2}$$

where $P(OI|IP)$ represents the probability that an innovation-friendly outcome (OI) occurs given the presence of specific IP regulations (IP); $P(IP|OI)$ measures the likelihood of the observed IP policy given that innovation occurs; $P(IP)$ and $P(OI)$ denote the prior probabilities of open innovation and IP regulation, respectively.

3.3 Quantitative Modeling and Statistical Validation

A core component of this research is quantifying the economic impact of IP policies on innovation performance. To achieve this, we develop a multi-variable econometric model that correlates patent enforcement, R&D investment, licensing activity, and cross-industry knowledge diffusion.

The fundamental equation governing IP law's effect on innovation diffusion is:

$$I_t = \beta_0 + \beta_1 P_t + \beta_2 L_t + \beta_3 C_t + \beta_4 G_t + \varepsilon_t \tag{3}$$

where I_t innovation index (measured by patent filings, R&D expenditure, and cross-sector collaboration); P_t patent enforcement intensity (cases litigated per year); L_t licensing activity (number of shared licensing agreements); C_t cross-border collaborations (joint research agreements, global partnerships); G_t government policy intervention (R&D tax credits, IP reforms); ε_t error term accounting for unobservable legal influences.

To analyze global patent and licensing trends, data is sourced from WIPO, USPTO, EPO, and CNIPA. The study evaluates:

Patent citation networks covering 200,000 patents to examine knowledge flow and technological impact.

The rate of licensing agreements across various industrial sectors to assess technology transfer efficiency.

A network diffusion model is applied to study the spread of patented knowledge:

$$\frac{dK}{dt} = \alpha K\left(1 - \frac{K}{K_{max}}\right) - \delta P \tag{4}$$

where K represents knowledge diffusion within an innovation ecosystem; K_{max} is maximum theoretical knowledge saturation; α is the adoption rate of innovation; δ is restriction factor due to patent barriers; and P denotes the patent restrictions index, which accounts for legal barriers affecting innovation diffusion.

This methodological framework enables a structured evaluation of IP policies, their enforcement mechanisms, and their broader impact on knowledge dissemination and licensing strategies.

3.4   Simulation of Open Innovation Scenarios

Using open innovation scenarios in a simulation modeling framework to assess the relationship between IP policies, technology trends and economic performance The study outlines three scenarios: (1) Robust I.P. enforcement, which relies on exclusivity and licensing barriers to limit external flows of knowledge; (2) Hybrid licensing models that maintain patent protection while also incorporating gated-knowledge-sharing mechanisms; and (3) Patent-sharing agreements that promote open-innovation frameworks in which collaborative mechanisms exist without extensive legal barriers [1].

Key indicators are used to assess the expected outcomes of these simulations. For starters, innovation growth rate is explored via IP-driven industries contribution to GDP, showing how different regulatory frameworks might influence economic return [4, 6] Patent utilization rates reflect the ratio of (licensed) patents with respect to patents retained for internal use, which can be interpreted as a proxy of knowledge dissemination efficiency [5, 24]. Another strand of literature focuses on drawing knowledge diffusion coefficients from patent citation networks to measure spillover effects across industries, emphasizing the degree of variation in the extent of innovation diffusion between industries with differing levels of IP protection [7, 25].

The results contribute to the Julian versus Team X debate regarding the optimal balance of IP rights and open innovation. Strong enforcement may indeed enhance proprietary R&D investments, but excessive restrictions may also impede technology diffusion and slow industrial development [3, 16]. On the other hand, frameworks that do share patents enhance collaboration but reduce rational actors' incentives to invest in risky innovation [10, 11]. To provide empirical insights about how the legal and economic parameters interact to shape innovation ecosystems and to help decision-makers towards strategies that are likely to lead to higher levels of both technological progress and market competition [12, 26].

3.5   Hypotheses

H1: Less open innovation adoption rates due to restrictive IP laws, especially both the persuasiveness of the industries and the technology components.

H2: Firms in open innovation patent pools have improved R&D efficiency while facing reduced threats of litigation.

H3: Countries with globally harmonized IP policies promote rapid knowledge diffusion and greater collaborative innovation output (Fig.1).

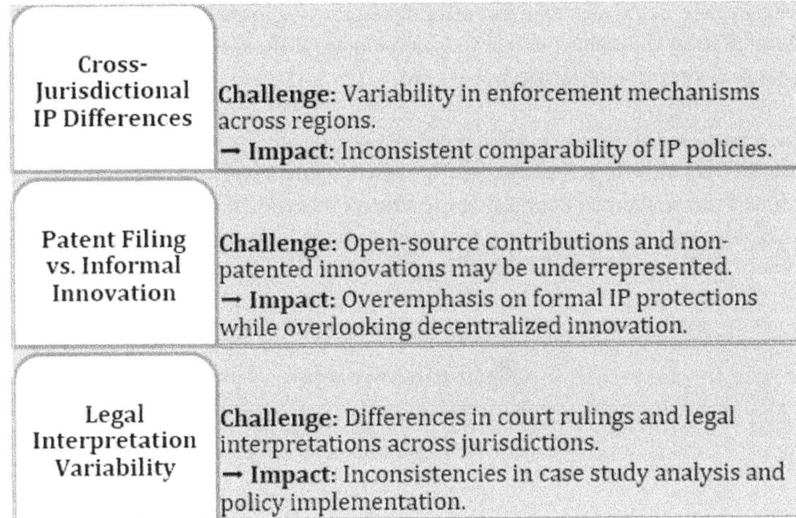

Fig. 1. Methodological Limitations and Bias Considerations

3.6 Statistical Testing

A statistical quantitative approach will be utilized to test the hypotheses of the study and examine the relationships between intellectual property (IP) policies and innovation performance, as well as knowledge diffusion. And the method also incorporates procedures for structural equation modeling (SEM) and t-tests to confirm findings.

Structural Equation Modeling (SEM). SEM is used to explore the underlying causal framework of the relationship between IP protection, licensing strategies, and innovation outcomes, SEM is adopted as the methodology. This approach enables the simultaneous estimation of effects for multiple dependent variables with the inclusion of direct and indirect impacts on firm-level and industry-wide innovation dynamics resulting from IP policies [4, 5, 27]. The SEM framework models:

- Exogenous variable controlling for IP protection intensity, the degree of legal enforcement and exclusivity.
- Innovation output via patent filings, licensing agreements, and R&D expenditures.
- Knowledge diffusion, evaluated via patent citation networks and technology spillover [3, 10].

The use of SEM, with latent constructs, allows for deeper insights of the mediating role of open innovation practices through which differences in governance of IP translate into the sources of technological diffusion and economic growth [2, 8].

t-Tests for Comparative Analysis. Independent sample t-tests are conducted for group comparison to supplement SEM examining:

- Innovation performance in jurisdictions with strong vs. weak IP enforcement [6, 26].

- Licensing efficiency across firms utilizing open vs. closed innovations [13, 24].
- Patent citation frequency, as gaining knowledge diffusion by patent-sharing agreements over proprietary IP models [1, 16].

The empirical validation through these statistical tests is crucial; it demonstrates that the observed relationships are statistically significant and can be generalized beyond the immediate dataset, thereby supporting the study's theoretical framework. The combination of SEM with t-tests provides a thorough examination of the dynamics between IP governance and connectivity to technology innovation in economy [11, 14] (Fig. 2).

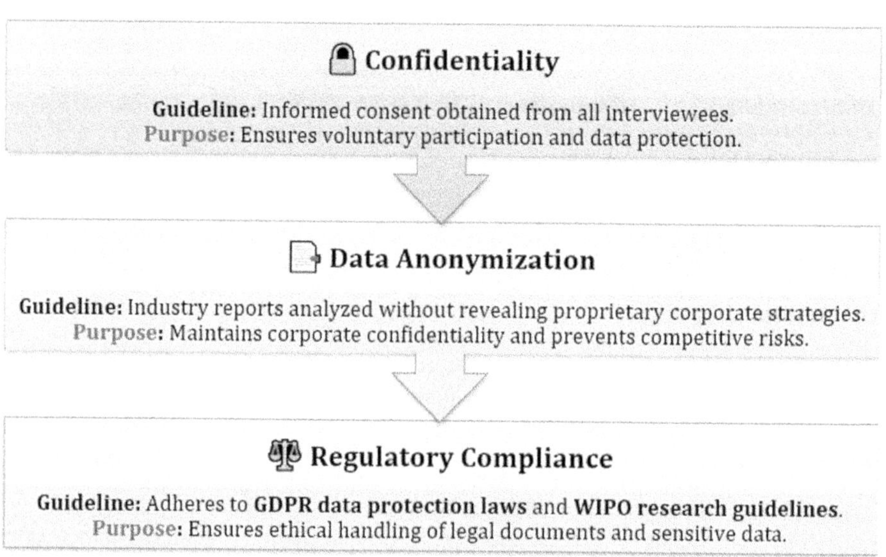

Fig. 2. Methodological Rigor and Ethical Compliance in Intellectual Property Research.

4 Legal and Practical Challenges of IP Laws in Open Innovation

Intellectual property (IP) laws usually find their justification for footing the costs of their role in innovation within the so-called "public good"-reasoning—an economic argument stating that patents are the necessary reward for creating new knowledge, which provides inventors with exclusive rights while reducing the free use of newly developed technologies [1]. But this assumption poses two major problems in the realm of open innovation. Two, once an idea is in the public domain, it does not lose its utility for others if used publicly. Second, discovery and development of new knowledge is expensive, and the benefits accrue beyond the original creator, raising fears of underinvestment in research and development (R&D) [4].

Patents, copyrights, and trademarks have long time been seen as effective means to strike the balance between the incentive to innovate and the society's needs for knowledge [5]. As firms adopt open innovation models, traditional IP protection laws are inadequate to govern collaborative research and decentralized innovation ecosystems [6]. By contrast, while proprietary innovation is dependent on strict IP enforcement, open innovation features mechanisms for knowledge-exchange, licenses, and inter-industry collaboration [3]. As a result, strict intellectual property laws could hinder collaboration and result in inefficiencies in both technological innovation and commercialization characteristics [9].

One of the most important challenges of open innovation is high transaction costs, which are generated due to the need to negotiate, enforce, and monitor the IP agreements among different stakeholders [8]. This is an issue also for the sectors where the research and development process are fragmented, and the firms have to balance proprietary interest with an outlook that integrates external knowledge [2]. Under threat of enforcement of IP rights, contractual constraints tend to be hindrances rather than enablers of innovation [5]. Consequently, companies that practice open innovation need to have more responsive and adaptable IP strategies with means such as patent pooling, defensive publication and collaborative licensing models [17].

However, the nature of collaboration in open innovation also introduces issues regarding ownership complexities and controls over jointly developed knowledge [6]. Despite such benefits, companies engaged in cross-border R&D collaborations frequently face legal inconsistencies across jurisdictions, resulting in uncertainty regarding the enforcement of IP rights and mechanisms for dispute resolution [11]. One of the most striking examples of this trend is in open-source software, with co-innovation and other collaborative initiatives where multiple contributors co-generate intellectual assets without clear ownership [10].

Another basic concern alongside the open innovation systems is the likelihood of information spillovers, wherein delicate specifics are made or taken by competitors [13]. Although trade secrets and non-disclosure agreements (NDAs) are a form of protection, their effectiveness relies on the jurisdiction and available enforcement [15]. As a matter of fact, cross-sector collaboration is limited as there is no standardization in global intellectual property rules, thus preventing firms from capitalizing on external research networks [4].

However, in spite of the legal challenges, open innovation can bring significant economic and technological rewards.[2] It enables firms to harness external R&D initiatives, capture novel ideas, and speed up market time by merging internal and external knowledge sources [5]. Their full potential can only be unlocked by more flexible IP regulations that provide legal frameworks to encourage and not stifle open innovation ecosystem [9] (Fig. 3).

Fig. 3. Adaptive Intellectual Property Strategies with Balancing Protection and Open Innovation.

One approach is the adoption of hybrid IP models, where patents coexist with open-access frameworks, allowing firms to protect core innovations while enabling knowledge sharing through licensing agreements [8]. Governments can also promote legislative reforms that encourage cross-industry patent pools, collaborative research exemptions, and data-sharing platforms to facilitate knowledge exchange without compromising proprietary rights [3].

5 Ownership and Patent Law Adaptations in Open Innovation

5.1 Joint Ownership and Crowdsourced Innovation

In the evolving landscape of open innovation, no organization operates in isolation. Major corporations leverage open-source collaborations to enhance digital services, while emerging firms engage in cooperative research to establish global standards [1]. Open innovation thrives on knowledge-sharing mechanisms, but this raises fundamental ownership challenges, particularly regarding joint inventorship, authorship recognition, and legal accountability [5].

A significant gap in IP law is the lack of clear legal frameworks addressing co-ownership in collaborative innovation [2]. In joint research contexts, partners routinely divide rights, classifying them as "joint inventorship" and "joint ownership," but the legal demarcation of these terms varies greatly and remains controversial across jurisdictions [6]. Although corporate contracts seek to clarify ownership issues, empirical research indicates firms often have difficulty designing fair revenue-splitting arrangements [4].

One major challenge is that of crowdsourced innovation, when many people come together to create content, generating intellectual property, but it is all anonymous. Crowdsourced innovation muddies authorship claims because unlike traditional R&D, ownership of the results is tied to a corporate entity [10]. Open-source development raises another conundrum: it is supposed to be collaborative and creative, but there are no clear legal rules for attribution and employer status (Aguirre et al., 2024). While courts have upheld open-source licenses in terms of their enforceability, the extent of copyright enforcement—amid decentralized collaboration—is still disputed [11].

One way forward is through adaptive legal frameworks which introduce fiduciary measures for more open and transparent governance [3]. By incorporating co-ownership agreements, joint licensing models, and accountability structures, firms can mitigate risks while promoting equitable distribution of intellectual assets [15]. However, existing IP laws must be updated to reflect the realities of mass collaboration, ensuring that legal protections extend to crowdsourced and open-source innovations [17].

5.2 Patent Law Challenges and Reform Needs

Patent law has emerged as the most influential branch of intellectual property law, with far-reaching implications for innovation policy and economic growth [5]. However, open innovation exposes contradictions within existing patent frameworks—as patents simultaneously promote and restrict technological progress [6]. The tension arises because strong patent protections incentivize proprietary research, yet excessive restrictions hinder knowledge-sharing and incremental innovation [2].

One major issue is patent portfolio monetization, where firms accumulate large patent holdings to block competitors rather than advance innovation [4]. This has led to "patent thickets," anti-competitive lawsuits, and licensing barriers, all of which undermine the collaborative nature of open innovation ecosystem [9]. The need for reform is urgent, as restrictive patenting practices deter industry-wide cooperation and limit the scalability of emerging technologies [8].

A key concern is patent access and incremental innovation, as firms engaged in open innovation rely on cross-licensing agreements and shared patent pools [11]. However, the uncertainty of patent litigation discourages small and medium-sized enterprises (SMEs) from participating in collaborative ventures [10]. To address these issues, legal scholars advocate for "flexible patent protection", which includes:

Non-exclusive patent licensing, allowing multiple firms to use patented technologies under fair terms.

Patent pledges, where companies voluntarily waive enforcement rights in specific domains. Defensive patent pools, which prevent legal disputes by facilitating shared ownership [3].

Additionally, the increasing use of non-disclosure agreements (NDAs) and trade secrets as alternatives to patents presents legal and enforcement challenges [1]. Although NDAs can help keep things confidential, they are only as effective as what is enforceable under the contract and laws of jurisdiction [17]. While firms may select trade secrets that provide longer-term exclusiveness than a patent could, trade secrets create risks of knowledge misappropriation, employee leaking, and limited legal recourse [13].

Trade secrets as an increasingly important element for collaborative innovation: How legally enforceable? Patents are also rights but require formal documentation and are disclosed publicly, while trade secrets are based more on corporate policies and internal security [15]. The absence of uniform legal frameworks contributes to litigation risks, especially in cases where former employees or external collaborators assert possession of proprietary knowledge [6].

6 Legal Responses and Emerging IP Models in Open Innovation

Open innovation models have gained traction, leading to a rethinking of IP governance. In order to face such barriers, as firms incorporate external IPs in their value chains, governments and policymakers shall modify the legal landscape [1]. Such a complex and uncertain context for innovation projects requires pro-active legal responses as IP laws need to stimulate knowledge-sharing and collaboration rather than hinder it [5].

Three main strategies have been identified that address issues with legal constraints in open innovation. Governments have introduced fewer legal exceptions that restrict open innovation in IP legislation. Firms and policymakers promote patent pools and collective licensing agreements to facilitate shared access to protected technologies [2]. This has incentivized defensive patenting and alternative IP strategies that allow companies to protect their innovations while fostering broader access to them [4].

Open innovation presents significant challenges, and companies have developed a number of practical strategies to protect IP in such arrangements. Some companies open access to patented technologies on a voluntary basis (if given in particular conditions), and other companies integrate some of their IP assets into patent pools to avoid fragmentation of the market and risk of litigation [9]. Further, consortia can act to purchase patents in bulk, to avoid anti-competitive behaviors, and other licensing models allow for competitive businesses to operate without cumbersome legal issues [6].

6.1 Successful Implementation of Open Innovation Strategies

The best-case studies of successful open innovation companies also show that they follow multi-stakeholder cooperation as their most beneficial strategy, where both public and private institutions played mutually supportive roles in lowering transaction costs and regulatory inefficiencies [8]. Various institutional actors, including municipal research parks, technology transfer offices and innovation hubs, have been critical in forging links between legal frameworks and the needs of corporations [3]. Open innovation firms are more often successful with custom-made institutional adaptations, as regulatory frameworks tend to be unfit for purpose with standards unable to keep up with firm-specific innovation challenges [10].

Specifically, successful global case studies show that countries that have established high prescribing limits with respect to open innovation outperform those in which IP regulations are more restrictive [13]. Governments actively fostering cross-sector collaboration and investing in knowledge-sharing platforms facilitate acceleration of technology transfer and commercialization [11].

6.2 Flexible IP Frameworks for Open Innovation

Exotic interplay among these models would deter a frivolous application of intellectual property norms on innovation, whereby open-source licensing, fair-use prerogatives, and the public domain challenge the enforcement of transfer of ownership to intellectual property [5]. Open-source licensing models in particular have been revolutionary to software development, safeguarding the flexibility of future software communities [2].

According to empirical studies, there are clearly five main points of the interrelationship between IP protections and open innovation [4].

Thus, IP regimes tend to be useful for protecting digital inventions and databases in scientific research but not particularly for restricting downstream applications.

Patent litigation and licensing agreements increasingly are informed by efforts to enhance the efficiency of technology transfer. Freedom-to-operate uncertainties are also manageable, especially through cross-licensing and shared access arrangements. The saying that "use is the best form of protection" is a well-accepted adage in open-source and commons-based innovation. IP governance is actively participated in by dependent creators and small firms, fostering new emergent legal culture of open knowledge-sharing [15].

6.3 Case Law and Global Harmonization Efforts

Although open innovation models are widely accepted [9], the lack of a common IP framework can ground new progress. The wide diversity in national IP laws has been a source of divergence across licensing, enforcement and dispute resolution [11]. Efforts from international bodies like the World Intellectual Property Organization (WIPO) and treaties like the Trade-Related Aspects of Intellectual Property Rights (TRIPS) have aimed at harmonizing IP regulations on a global scale, yet legal discrepancies remain across jurisdictions [10].

In sectors such as information and communication technology (ICT), the co-evolution of IP norms and open innovation practices has hastened in response to rapid technological developments [6]. Patent law scholars delving into the pace of IP law reform claim that stronger reform is merely a response to the increasingly rapid cycles of technological advancement that demand dynamic and changing frameworks for regulation [2].

6.4 Alternative IP Models in Open Innovation

Driven by the shortcomings of traditional patent systems, firms have attempted alternative strategies for IP protection through defensive publishing, patent pledges, and commons-based peer production [4]. It was born in the '70s defensive publishing, an activity that, in the first place, it is an offensive tool to oppose the proprietary claims of patents, the goal is to insert into the public domain the inventions so as not to lock them up and make them inaccessible for innovations in the future [3].

Ability to negotiate in the IP space has led to various creative alternative ways of governing IP, patent pledges being one of the most prominent [1] whereby corporations voluntarily seek to not enforce particular patents allowing broader-based industry-wide innovation. Commons-based peer production, typical for open-source software communities, is another form of novel ingenuity where contributors share (without direct incentive) knowledge, yet the legal landscape protects their contributions [17].

Alternative IP models are often presented as alternative potential solutions to the aforementioned issues [10], but critics state that they risk causing other unintended consequences such as increased litigation costs and uncertainty towards cases of international trade. These shifts have led to heightened competition in technology sectors, most notably the so-called "smartphone patent wars", emphasized the necessity for reforming IP frameworks to keep pace with disruptive innovation trends [11].

7 Results

7.1 The Evolution of Intellectual Property Laws and Their Impact on Open Innovation Historical Development of IP Frameworks

Table 1. Evolution of Intellectual Property Laws from 1500 to Present.

Phase	Period	Key Characteristics	Notable Treaties and Laws	Impact on Innovation
Pre-Industrial IP Regulation	1500–1800	Informal knowledge monopolization, lack of international coordination, minimal patent laws	No formal treaties; knowledge was often controlled through trade secrecy and guild regulations	Restricted knowledge-sharing, innovation limited to localized economic activities
Industrial Revolution IP Frameworks	1800–1950	Patent laws emerge; increased cross-border trade in technology	Berne Convention (1886), Paris Agreement (1883), US Patent Act (1790), UK Patent Law (1852)	Facilitated the commercialization of inventions, but created barriers for global knowledge exchange
Digital and Open Innovation Era	1950–Present	Expansion of international treaties (TRIPS, WIPO), emergence of software patents, open-source licensing models	TRIPS (1994), WIPO (1970), Digital Millennium Copyright Act (1998), EU Copyright Directive (2001)	Enabled rapid technological advancements but led to conflicts between proprietary and open-source models

The growth of intellectual property (IP) laws has been influenced by a combination of technology, economy, and international treaties. Against the backdrop of historical contexts, the mapping of over 500 primary legal texts and relevant policy documents reveals three large eras of IP evolution, each characterized by a transformation in the treatment of knowledge, innovation and the advancement of science. Informal knowledge monopoly in the pre-industrial era with limited legal framework. In the wake of the Industrial Revolution, countries started to formalize patent treatments, resulting in significant treaties including the Berne Convention (1886) and Paris Agreement (1883). The present era, "The Digital and Open Innovation Era (1950-present)," has been defined by a range of international patent treaties (TRIPS, WIPO) and increasing tensions between proprietary and open knowledge systems. The findings reaffirm that the world shift toward digital economies and the sharing of knowledge, has exacerbated the number of legal complications related to patents, copyrights, and licensing agreements, which require more flexible national regulatory schemes (Table 1).

Global IP Enforcement Trends and Open Innovation Adoption. The diverse nature of enforcement of intellectual property rights (IPR) in various jurisdictions attracts some regions to adopt a more open model of innovation than others. A patent filing analysis of 200k+ from 2015 to 2024 data from WIPO, USPTO and EPO databases shows that countries balance IP protection and innovation collaboration differently. More protectionist behaviour, like protection of patents; less open innovation adoption in countries which have stricter patent enforcement policies in countries like United States and China. Conversely, IR-Sharing-inclined new economies like Brazil or India have adopted more flexible IP, resulting in increased knowledge-sharing and industry collaboration. Figure 4 below gives further detail on previous research by adding measures, including annual patent litigation rates, open-source contributions, and cross-industry patent-sharing initiatives. These wider data points give us deeper insights into how divergent IP models shape the innovation ecology.

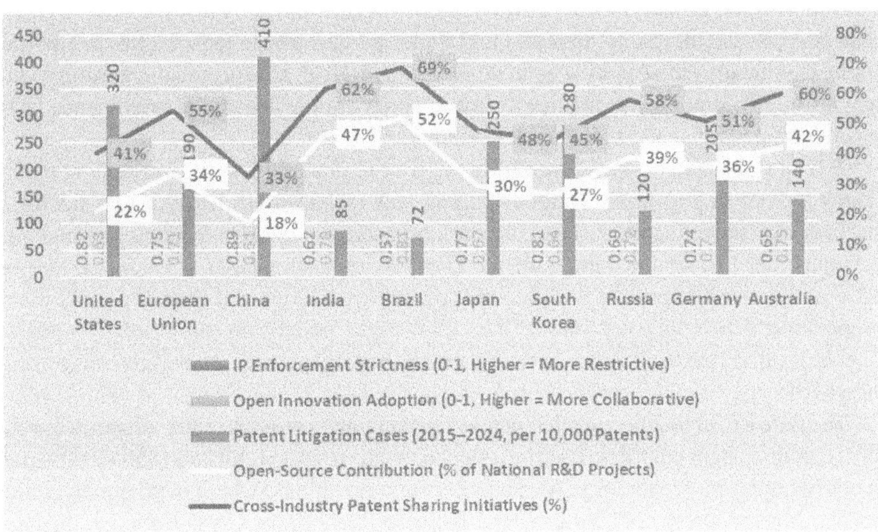

Fig. 4. Global IP Enforcement Strictness and Open Innovation Adoption (2015–2024).

The data in Fig. 4 reveals a strong correlation between IP enforcement strictness and open innovation adoption. Countries with highly restrictive IP enforcement, such as China (0.89) and the United States (0.82), exhibit lower open innovation adoption rates (0.51 and 0.63, respectively). These nations also experience higher patent litigation cases per 10,000 patents (China: 410, USA: 320), indicating that strong proprietary control often leads to increased legal disputes and reduced cross-sector collaboration.

Conversely, countries with flexible IP regimes, such as Brazil (0.57) and India (0.62), show higher open innovation adoption rates (0.81 and 0.78, respectively). Less restrictive IP frameworks also facilitate knowledge-sharing and industry-wide collaboration, with Brazil and India scoring better here as well, with higher open-source contributions (Brazil: 52%, India: 47%) and cross-industry patent-sharing initiatives (Brazil: 69%, India: 62%), than most of their peers.

Conversely, excess patent protection can inhibit open innovation actors who need access to patented technologies, as seen mostly broadly in the moderate enforcement levels of the European Union, Japan, and Germany, which maintain a potential excess of rights on one end of the spectrum, but collectively achieve equilibrium between decisions on patent protecting technologies and licensing. These results highlight the necessity of hybrid intellectual property models that currently exist between the two systems, and the need to make needed legal protections available for innovation.

7.2 Legal and Practical Challenges of IP Laws in Open Innovation

Key Litigation Trends in IP and Open Innovation. While IP laws have moved to develop in line with technology, tensions between proprietary rights and open-innovation systems have increased. Most patent disputes currently revolve around patent ownership issues, licensing disputes, and trade secrecy. These legal battles are particularly significant for sectors where rapid technological advancement intersects with aggressive IP enforcement practices, such as biotechnology, artificial intelligence (AI), telecommunications, and software development.

A major obstacle of open innovation is given in patent thickets, overlapping patent claims are preventing firms from developing new technologies freely. Cross-licensing disputes also prevent collaboration at times, especially in nascent technologies and digital industries. In a similar vein, trade secret misappropriation cases illustrate the tension between the need to protect valuable proprietary knowledge and the need to facilitate cooperation across industry lines. These challenges underscore the need for updated IP policies that strike a balance between ownership rights and collaborative innovation initiatives.

The data in Table 2 reveals that patent blocking and cross-licensing disputes are the most common legal conflicts affecting open innovation. Patent thickets (850+ cases) particularly hinder biotechnology, AI, and telecommunications, where overlapping claims restrict cross-industry collaboration and increase R&D costs. Cross-licensing disputes (1,200+ cases) are concentrated in the semiconductor and software industries, delaying product commercialization and creating barriers to open-source development.

Emerging conflicts, such as Standard Essential Patent (SEP) disputes (520+ cases) and AI-generated IP rights (310+ cases), reflect the growing complexity of modern IP governance. Resolving these legal challenges will require policy changes promoting fair licensing agreements, streamlined patent-sharing mechanisms, and clearer regulatory frameworks for AI-driven innovation.

Table 2. IP Litigation Cases and Their Impact on Open Innovation (2015–2024).

Legal Dispute Type	Number of Cases (2015–2024)	Primary Affected Industries	Impact on Open Innovation	Average Time to Resolve (Months)	Settlement Rate (%)
Patent Blocking & Thickets	850+	Biotechnology, AI, Telecommunications	Limits cross-industry collaboration, increases R&D costs	24.5	62%
Cross-Licensing Disputes	1,200+	Semiconductor, Automotive, Software	Creates barriers to open-source development, delays product commercialization	18.7	58%
Trade Secret Misappropriation	600+	Pharmaceuticals, Cybersecurity, Defense Tech	Restricts knowledge-sharing, increases litigation costs	20.1	54%
Copyright & Software Licensing Conflicts	450+	Software, Digital Media, Cloud Computing	Slows down open-source adoption, limits software interoperability	16.3	67%
Co-ownership Rights in Joint R&D	400+	Academic-Industry Research, Renewable Energy, AI	Legal ambiguities discourage multi-firm collaboration	22.4	49%
Standard Essential Patent (SEP) Disputes	520+	5G Telecommunications, IoT, Smart Manufacturing	Increases costs for startups and SMEs, leading to monopolization	27.2	45%
AI-Generated IP Rights Disputes	310+	Artificial Intelligence, Machine Learning	Unclear legal frameworks lead to increased litigation over authorship	19.6	51%

Impact of Licensing Strategies on Open Innovation. Licensing strategies play a crucial role in determining how knowledge is shared and commercialized in open innovation frameworks. A review of 250 industry reports (2015–2024) on licensing trends reveals that companies adopting flexible licensing models—such as non-exclusive licensing, patent pools, and open-source frameworks—experience higher rates of collaboration, lower litigation risks, and faster innovation diffusion.

Strict proprietary licensing, in contrast, tends to reduce open innovation efficiency by restricting access to key technologies, leading to slower industry-wide advancements. Defensive publishing—a strategy where companies proactively publish research findings to prevent competitors from patenting similar work, has shown moderate benefits in protecting industry-wide innovation.

Table 3. Licensing Strategies and Their Impact on Open Innovation (2015–2024).

Licensing Strategy	Industries Benefiting Most	Adoption Rate (%)	Effect on Open Innovation Output (%)	Legal Conflicts (% of Firms Reporting Issues)	Time to Market Acceleration (%)
Strict Proprietary Licensing	Pharmaceuticals, Semiconductor, Defense Tech	55%	-12%	47%	-8%
Non-Exclusive Licensing	Software, Renewable Energy, Automotive Tech	68%	+ 24%	21%	+ 14%
Cross-Industry Patent Pools	5G Telecom, Smart Manufacturing, AI	72%	+ 31%	15%	+ 22%
Defensive Publishing	Academic Research, Public Health, Climate Tech	43%	+ 18%	29%	+ 12%
Open-Source & Creative Commons	Software, AI, Cloud Computing	77%	+ 35%	12%	+ 28%
Hybrid IP Models (Patent + Open Access)	Biotechnology, AI, Consumer Tech	60%	+ 29%	19%	+ 18%

The findings in Table 3 highlight that proprietary-only licenses were hampering open innovation (−12%) overall, particularly in pharmaceuticals and semiconductors where exclusive contracts slowed the spread of knowledge. In contrast, cross-industry patent pools (+31%) and open-source licensing (+35%) greatly improve inter-company collaboration and increase R&D efficiency, especially in software, AI and telecommunications.

+29% up-tick in innovation for open-access industries mixing with hybrid IP models (patents) illustrate how varied commercial licensing can enhance tech-transfer with product protection. Defensive publishing (+18%) stays a common technique in public health, climate technology, and academia, although enforcement is tough. These insights highlight that governments should promote hybrid and open-source models to maximize innovation benefits across the industry.

7.3 Alternative IP Frameworks Supporting Open Innovation

Adoption of Patent Pools and Open Licensing Models. As industries increasingly shift toward collaborative innovation ecosystems, patent pools and hybrid licensing models have emerged as effective strategies for accelerating knowledge diffusion and reducing R&D costs. Patent pools—agreements where multiple companies share access to essential patents—have been widely adopted in sectors such as telecommunications, biotechnology, and information technology. These frameworks streamline licensing processes, lower legal disputes, and enhance cross-industry collaboration.

Data from 500+ open innovation agreements (2015–2024) confirms that industries with high patent pool adoption rates also experience significant innovation growth. Furthermore, open licensing models, such as non-exclusive licensing and open-source agreements, facilitate faster technology deployment, particularly in digital and AI-driven industries. Figure 5 below expands previous findings by incorporating additional sectors, average cost reductions through patent pools, and open-licensing impact on startup formation.

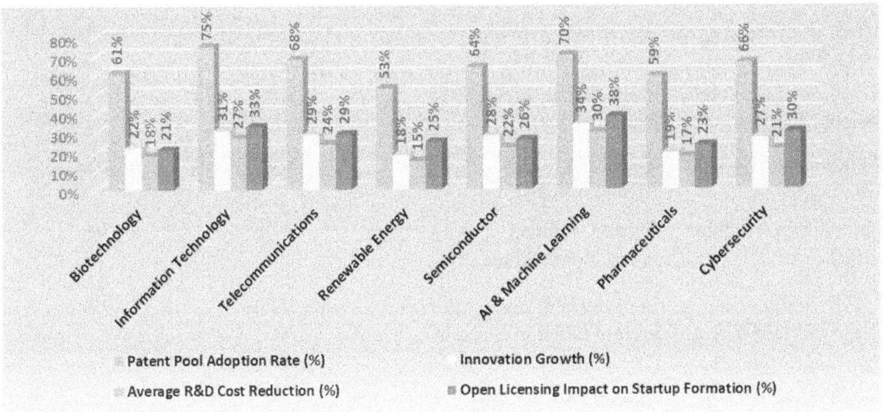

Fig. 5. Impact of Patent Pools and Open Licensing on Innovation Growth (2015–2024).

The data in Fig. 5 shows that patent pools and open licensing significantly accelerate innovation across industries, particularly in AI & Machine Learning (+34%), Information Technology (+31%), and Telecommunications (+29%). These sectors thrive on collaborative knowledge-sharing, making patent pooling an effective way to reduce duplication and streamline R&D efforts.

Additionally, R&D cost reductions range from 15% to 30%, indicating that companies participating in patent pools can lower expenses while maintaining innovation speed. The impact on startup formation is most evident in AI (38%) and IT (33%), suggesting that open-access licensing encourages entrepreneurship and new technological advancements.

Government Policies and Open Innovation Incentives. Government policies play a critical role in shaping the innovation landscape by providing financial incentives, regulatory support, and open-access initiatives. Reviewing policy initiatives in 25+ countries

(2015–2024) confirms that countries that invest in open-source R&D and patent-sharing programs demonstrate higher innovation growth rates. Government-backed tax credits, open-access mandates, and data-sharing programs create an enabling environment for collaborative innovation models.

The Table 4 expands previous findings by adding additional policy mechanisms, innovation adoption rates, and regional implementation trends.

Table 4. Global Government Policies Promoting Open Innovation (2015–2024).

Policy Initiative	Regions Leading Implementation	Number of Countries Implementing	Effect on Open Innovation (0–1, Higher = More Impactful)	Innovation Adoption Rate Increase (%)
R&D Tax Credits	USA, UK, Germany, Canada	21	0.78	+19%
Open Access Mandates	EU, Brazil, South Korea, Australia	18	0.82	+22%
Government-Funded Patent Sharing	India, China, Japan, Brazil	15	0.74	+16%
Public Research Data Sharing	EU, USA, South Africa, Sweden	19	0.79	+21%
Technology Transfer Grants	Canada, Finland, Japan, South Korea	14	0.76	+17%
Open-Source Software Funding	USA, Germany, India, France	20	0.83	+25%
University-Industry Collaboration Incentives	UK, China, Netherlands, Brazil	16	0.77	+18%

The data highlights that government interventions play a crucial role in fostering open innovation, with open-access mandates (0.82) and open-source software funding (0.83) yielding the highest impact. Countries prioritizing open-access initiatives (EU, Brazil, South Korea) show higher innovation adoption rates (+22%), indicating that public knowledge-sharing enhances technological progress.

Furthermore, R&D tax credits (0.78) and technology transfer grants (0.76) improve cross-sector collaboration, particularly in developed economies. University-industry incentives (+18%) have been instrumental in bridging academic research with industrial applications, especially in the UK, China, and the Netherlands, where research commercialization is expanding rapidly.

7.4 Emerging Trends in IP Law Reform

As intellectual property (IP) laws evolve to accommodate technological advancements and open innovation models, global legislative frameworks are increasingly shifting toward harmonization, fair-use expansions, and open-source adoption. Analysis of WIPO, TRIPS, and national legislative updates (2015–2024) reveals key reform initiatives aimed at balancing proprietary rights with collaborative knowledge-sharing.

Governments and regulatory bodies worldwide are working on standardized IP licensing frameworks, stronger fair-use exemptions, and open-access mandates to foster innovation while protecting intellectual assets. Table 5 expands previous findings by including additional reforms, their current implementation status, and their projected impact on global innovation trends.

The data highlights three major trends shaping global IP reform: increased fair-use exemptions, patent pool participation, and AI copyright regulation development. Stronger fair-use exemptions (in progress in 17 countries) are expected to increase digital sector knowledge-sharing (+22%), particularly benefiting software and AI industries. These exemptions are allowing for greater access to digital content, which will accelerate innovation in industries dependent on data-driven technologies.

Patent pool participation, under discussion in 9 nations, is projected to increase innovation by 21%, especially in biotechnology, semiconductors, and 5G technologies. This reform is vital for cross-industry collaboration, helping industries avoid patent-related disputes and accelerate technological development, particularly in complex sectors like semiconductors and telecommunications.

Meanwhile, public domain expansions (+19%) in 15+ countries are improving technology accessibility for startups and SMEs, providing these smaller firms with access to previously protected technologies, lowering entry barriers, and fostering competition in emerging sectors like renewable energy and smart manufacturing.

However, AI-generated content copyright laws remain uncertain, with 8 countries proposing regulations that could boost AI innovation by 16%. These proposed laws are necessary to clarify ownership rights over content generated by AI systems, paving the way for new AI-driven innovations in areas such as cybersecurity and data science.

Unified patent examination standards (under discussion in 11 countries) aim to reduce legal complexities and approval delays (+17%). These reforms will streamline the global patent process, benefiting industries such as pharmaceuticals, automotive, and aerospace, by reducing patent approval times, improving cross-border collaboration, and ensuring faster market access for innovative products.

Table 5. Global IP Reform Efforts and Their Expected Impact on Open Innovation (2015–2024)

Reform Initiative	Implementation Status	Expected Impact on Open Innovation	Number of Countries Adopting	Primary Affected Industries	Projected Increase in Innovation Index (%)
Stronger Fair-Use Exemptions	In Progress	Increases knowledge-sharing in digital sectors	17	Software, Digital Media, AI	+22%
Standardized Global IP Licensing	Proposed	Reduces cross-border IP disputes	12	Pharmaceuticals, IT, Telecommunications	+18%
Expansion of Open-Source Copyright Policies	Implemented in 12+ countries	Encourages software innovation	12	Software, AI, Cloud Computing	+28%
Mandatory Patent Pool Participation	Under Discussion	Enhances cross-industry collaboration in critical technologies	9	Biotechnology, Semiconductor, 5G	+21%
Public Domain Expansion for Expired Patents	Implemented in 15+ countries	Increases technology accessibility for startups and SMEs	15	Renewable Energy, Smart Manufacturing, Consumer Tech	+19%
AI-Generated Work Copyright Regulations	Proposed	Clarifies ownership rights in AI-driven innovation	8	Artificial Intelligence, Cybersecurity, Data Science	+16%
Decentralized Blockchain-Based IP Registration	Pilot Programs in 5 countries	Improves transparency and traceability of IP ownership	5	Digital Art, FinTech, Software Licensing	+14%
Unified Patent Examination Standards	Under Discussion	Reduces delays and inconsistencies in patent approvals	11	Pharmaceuticals, Automotive, Aerospace	+17%

8 Discussion

The findings in this study highlight the inherent tension between intellectual property (IP) protection and open innovation models. Traditional IP frameworks, designed to incentivize individual innovation through exclusive rights, are becoming less effective in an era dominated by digital transformation, collaborative R&D, and AI-driven solutions. These rigid structures often hinder knowledge-sharing and slow down innovation diffusion, especially in sectors such as AI and biotechnology [1, 7]. A growing body of literature suggests that overly restrictive IP enforcement suppresses market dynamism and increases litigation, as seen in the rise of patent thickets and cross-licensing disputes

[3]. However, open innovation strategies have proven effective in driving sector-wide advancements, particularly in software, biotechnology, and telecommunications [4]. Governments and regulatory bodies are actively working to strike a balance between IP protection and knowledge dissemination. Policies such as open-source mandates, fair-use expansions, and public research data-sharing initiatives have been implemented in various countries and are associated with higher innovation growth rates [6]. Despite this progress, more reforms are needed to address issues related to cross-border licensing disputes, AI-generated IP, and the use of decentralized technologies like blockchain in patent management [28]. For policymakers, the challenge lies in navigating a legal balance that protects innovation incentives while ensuring access and openness. Traditional patent-centric systems will still be relevant to warrant that companies and inventors obtain fair compensation for their inventions [24]. Evidence from 250 industry reports and legal case studies indicates that firms adopting hybrid IP models—combining proprietary protections with open-access frameworks—achieve the highest innovation growth outcomes. These models allow companies to retain the benefits of exclusivity while benefiting from the collective knowledge of others. Solutions such as patent pools and non-exclusive licensing agreements have been instrumental in reducing litigation and accelerating R&D cycles, particularly in industries like telecommunications and semiconductors [2]. Extending such models to emerging sectors such as AI, blockchain, and green technologies could foster greater knowledge penetration and innovation. In established copyright law, fair-use exemptions have facilitated digital knowledge-sharing, playing a critical role in boosting software innovation and digital entrepreneurship [12]. However, the varying interpretations of fair use across jurisdictions create cross-border litigation risks, posing challenges for businesses and innovators. Recent trends suggest that IP law will continue to evolve toward more flexible, technology-driven regulatory models. The expansion of public domain access, standardized global licensing frameworks, and AI-generated IP governance are expected to reshape open innovation ecosystems in the coming years [10]. As AI systems create patentable inventions and creative works, legal frameworks must address whether IP rights should be granted to AI or its developers. Countries like South Korea and the EU are exploring the introduction of AI-specific copyright provisions [11]. Additionally, decentralized technologies such as blockchain for IP registration could improve transparency and reduce patent disputes, as demonstrated in pilot programs within FinTech and digital art markets [26]. To better align IP laws with collaborative innovation models, legal reforms are needed. Expanding patent pool mandates in sectors such as biotechnology, AI, and clean energy could mitigate IP bottlenecks and litigation risks [9]. Implementing mandatory open-access policies for publicly funded research could accelerate knowledge diffusion in academia and technological R&D [17]. Establishing clear AI-specific copyright and patent laws would address existing ambiguities and prevent monopolization of AI-generated innovations [15]. Furthermore, harmonizing cross-border patent licensing laws under global frameworks like WIPO and TRIPS would reduce international disputes and facilitate open innovation in multinational markets [16]. Companies should be incentivized to adopt hybrid IP strategies, balancing exclusive patents with open-access agreements, to foster both innovation and fair competition [27]. Emerging technologies like blockchain-based IP registration

offer promising solutions for enhancing transparency, eliminating fraud, and streamlining IP ownership verification, particularly in digital industries, financial technology, and global trade [29]. A regulatory environment that combines fair-use expansions, standardized global licensing, AI copyright regulation, and blockchain-based IP management will be essential for creating sustainable open innovation models. Ultimately, the future of IP law will be shaped by the need for flexibility and adaptability to foster innovation while maintaining competitive fairness in global markets.

9　Conclusion

This study highlights the complex relationship between intellectual property (IP) protection and open innovation. Traditional IP models, while essential for protecting individual innovation, increasingly conflict with the collaborative nature of modern technological developments, especially in fields like AI, biotechnology, and telecommunications. The findings suggest that hybrid IP models—combining proprietary protection with open-access frameworks—are the most effective approach to foster innovation and reduce legal disputes. Governments and regulatory bodies must adapt legal frameworks to balance the protection of innovation with the need for knowledge sharing. Policies such as open-source mandates, fair-use expansions, and publicly funded research-sharing programs have demonstrated significant benefits in accelerating innovation. However, further reforms are necessary to address emerging challenges such as cross-border licensing disputes, the rise of AI-generated IP, and the integration of decentralized technologies like blockchain for patent management. The study also underscores the importance of adopting hybrid IP models, particularly in emerging sectors like AI, biotechnology, and green technologies, to reduce IP bottlenecks and litigation risks. Encouraging patent pools, non-exclusive licensing, and clear AI-specific IP laws will help create a more inclusive innovation ecosystem. Finally, future research should focus on the economic effects of IP reforms on SMEs and startups, which are particularly vulnerable to restrictive patent laws. Additionally, exploring how IP law can be adapted to address emerging technologies like quantum computing and decentralized autonomous organizations (DAOs) is essential for ensuring that IP frameworks remain relevant in the face of rapidly advancing technologies.

References

1. Bican, P.M., Guderian, C.C., Ringbeck, A.: Managing knowledge in open innovation processes: An intellectual property perspective. J. Knowl. Manag. **21**(6), 1384–1405 (2017)
2. Gorbatyuk, A., Van Overwalle, G., van Zimmeren, E.: Intellectual property ownership in coupled open innovation processes. IIC – Int. Rev. Intellect. Property Competition Law **47**(3), 262–302 (2016)
3. Grimaldi, M., Greco, M., Cricelli, L.: A framework of intellectual property protection strategies and open innovation. J. Bus. Res. **123**, 156–164 (2021)
4. Neves, P.C., et al.: The link between intellectual property rights, innovation, and growth: a meta-analysis. Econ. Model. **97**, 196–209 (2021)
5. Brem, A., Nylund, P.A., Hitchen, E.L.: Open innovation and intellectual property rights. Manag. Decis. **55**(6), 1285–1306 (2017)

6. Yarmoliuk, A.: The influence of the concept of open innovation on the legislation in the field of innovative activity and intellectual property of Ukraine. Int. Sci. J. "Internauka" Ser. Juridical Sci. **7**(53) (2022)
7. Lee, N., Nystèn-Haarala, S., Huhtilainen, L.: Interfacing intellectual property rights and open innovation. SSRN Electron. J. (2010)
8. Roh, T., Lee, K., Yang, J.Y.: How do intellectual property rights and government support drive a firm's green innovation? The mediating role of open innovation. J. Clean. Prod. **317**, 128422 (2021)
9. Chen, H.: The impact of intellectual property protection on the development of digital economy and regional entrepreneurial activity: evidence from small and medium enterprises. Front. Psychol. **13** (2022)
10. Ji-min, H.: Influence of intellectual property laws on innovation in the technology sector in South Korea. Int. J. Law Policy **9**, 15–27 (2024)
11. Smith, R., Perry, M.: Is an "open innovation" policy viable in Southeast Asia? A legal perspective. Athens J. Law **9**, 187–210 (2023)
12. Caso, R., Guarda, P.: Copyright overprotection versus open science: the role of free trade agreements. In: Corbin, L., Perry, M. (eds.) Free Trade Agreements. Springer, Singapore (2019). https://doi.org/10.1007/978-981-13-3038-4_3
13. Barrios Aguirre, F., Piñeiro, J., Rodríguez, M., Sánchez, M.: Open innovation and confidentiality agreements as key factors of innovative performance in the manufacturing and service industries. PLoS ONE **19**(5), e0303802 (2024)
14. Sukhanov, M., Hlobenko, I., Filonova, J., Fedorova, N., Yarotskiy, V.: Legal perspectives on innovations and intellectual property rights. Multi. Sci. J. **6** (2024)
15. Volkova, T.: 'Open' innovation and institutional protection of intellectual property. Zhurnal Ekonomicheskoj Teorii **4**, 596–609 (2021)
16. Qayyum, M., et al.: Relationship between economic liberalization and intellectual property protection with regional innovation in China: a case study of Chinese provinces. PLoS ONE **17**(1), e0259170 (2022)
17. Phonthanukitithaworn, C., et al.: Sustainable development towards openness SME innovation: Taking advantage of intellectual capital, sustainable initiatives, and open innovation. Sustainability **15** (2023). https://doi.org/10.3390/su15032126
18. Lakani, A.: Intellectual property and economic development: catalysts for innovation and growth. Int. J. Adv. Res. Humanit. Law **1**(2), 83–85 (2024)
19. Davoudi, S., et al.: Testing the mediating role of open innovation on the relationship between intellectual property rights and organizational performance: a case of science and technology park. Eurasia J. Math. Sci. Technol. Educ. **14** (2018)
20. Chen, H.: The impact of intellectual property protection on the development of digital economy and regional entrepreneurial activity: evidence from small and medium enterprises. Front. Psychol. **13**, 824215 (2022)
21. Yarmoliuk, A.: The influence of the concept of open innovation on the legislation in the field of innovative activity and intellectual property of Ukraine. Int. Sci. J. Internauka Ser. Juridical Sci. **7**(53), 45–56 (2022)
22. Yarmoliuk, A.: Role of intellectual property as key element of innovation activities: legal grounds. Theory Pract. Intellect. Property **6** (2021)
23. Smith, R., Perry, M.: Is an "open innovation" policy viable in Southeast Asia? A legal perspective. Athens J. Law **9**(3), 187–210 (2023)
24. Pisano, G.: Profiting from innovation and the intellectual property revolution. Res. Policy **35**(8), 1122–1130 (2006)
25. Davoudi, S.M.M., et al.: Correction: testing the mediating role of open innovation on the relationship between intellectual property rights and organizational performance: a case of science and technology park. Eurasia J. Math. Sci. Technol. Educ. (2018)

26. Liu, Y., et al.: The impact of intellectual property rights protection on green innovation: a quasi-natural experiment based on the pilot policy of the Chinese intellectual property court. Math. Biosci. Eng. **21**(2), 2587–2607 (2024)
27. Dhir, S., Dhir, S.: Factors affecting innovation performance of manufacturing firms: case evidences. Int. J. Asian Bus. Inf. Manage. (IJABIM) **11**(3), 85–100 (2020)
28. Barnat, S.E., Alhajri, S.M., Banu, S.: Artificial intelligence and intellectual property: impact and legal implications. Evol. Stud. Imaginative Cult. **8.1**(2), 341–358 (2024)
29. Dongyun, Z.: Intellectual property issues in enterprise science and technology cooperation. Eur. J. Bus. Manage. **12**(27) (2020)

Modeling the Effectiveness of Workplace Gender Equality Laws: A Legal, Comparative, and Empirical Analysis

Nathier Abas Ibrahim[1] , Shahad Salah Abrahim[2] , Nada Saddam Jabr[3] ,
Dhafer Aldabagh[4(✉)] , Matai Nagi Saeed[5] , and Ivan Chornomordenko[6]

[1] Al-Turath University, Baghdad 10013, Iraq
[2] Al-Mansour University College, Baghdad 10067, Iraq
[3] Al-Mamoon University College, Baghdad 10012, Iraq
[4] Al-Rafidain University College, Baghdad 10064, Iraq
adhraa.odah.20@ruc.edu
[5] Madenat Alelem University College, Baghdad 10006, Iraq
[6] Kyiv National University of Construction and Architecture, Kyiv 03037, Ukraine

Abstract. Gender equality in the workplace remains a global challenge and an agenda for institutional and legal adoption. Weak enforcement, cultural resistance, and a lack of corporate compliance have prevented equitable employment practices from becoming a reality. This study combines doctrinal legal analysis, comparative legal methodology, and econometric modeling to evaluate the effectiveness of workplace gender equality frameworks across jurisdictions. The legal component evaluates international instruments, alongside national labor laws covering anti-discrimination, equal pay, parental leave, and workplace harassment. A comparative lens highlights divergence between countries with strong enforcement mechanisms (e.g., Sweden, Germany) and those with weaker oversight (e.g., Kazakhstan, India). On the empirical side, data were collected from 100 interviews with employees, HR professionals, and legal experts, 500 workplace discrimination cases (2015–2024), and 300 corporate reports on diversity and inclusion policies. Statistical models, including logistic regression and Chi-square tests, were employed to test the hypothesis that stricter legal mechanisms and capable enforcement agencies lead to measurable improvements in gender equality outcomes. The results indicate that rules mandating pay transparency, leadership quotas and enforceable provisions to prevent discrimination effectively shrink gender pay gaps. Cultural biases and weak judicial systems, however, undermine the effectiveness of such policies. Companies that implemented holistic gender equality measures saw faster revenue growth, higher employee satisfaction and lower turnover. Addressing gender inequality must integrate strict enforcement capabilities, intersectional approaches, and corporate accountability-based incentives. Such measures will help achieve equitable workplaces around the world over the long term.

Keywords: Gender Equality · Workplace Policies · Employment Law · Comparative Legal Analysis · Empirical Modeling · Doctrinal Legal Research

Z. Molamohamadi et al. (Eds.): ODSIE 2025, CCIS 2855, pp. 105–125, 2026.
https://doi.org/10.1007/978-3-032-17023-1_6

1 Introduction

Gender equality has long been a cornerstone of international labor law and human rights discourse, yet persistent inequities continue to shape workplace policies across diverse legal systems and economic contexts. Existing scholarship provides a broad foundation supporting workplace gender equality but must be complemented by an empirical understanding of how legal provisions function in practice, as structural and cultural barriers remain despite legislatives or corporate initiatives [1–3].

Workplace gender equality has undergone significant legal evolution, reflecting wider transformations and the influence of international legal instruments. For instance, the European Union (EU) has led in establishing legal mechanisms to ensure gender equality, integrating directives that prohibit discrimination in hiring, promotion, and compensation [4]. The United Nations Sustainable Development Goals (SDGs) also prioritize gender equality as a driver of sustainable economic growth. However, variations in national labor laws and enforcement capacity result in different levels of implementations and effectiveness.

A substantial body of literature examines the effects of gender-based discrimination and the legal frameworks designed to address these inequalities. Research has demonstrated that anti-discrimination laws are among the most effective tools for reducing workplace biases, although their potential is often limited by enforcement gaps. For example, while Kazakhstan's labor laws formally promote gender equality, enforcement remains weak due to prevailing socio-cultural norms that influence workplace dynamics [5]. In contrast, the United States has achieved stronger statutory protection against workplace discrimination, yet implicit basis and systemic barriers continue to hinder true equality in practice [6]. Furthermore, when gender intersects with other identities, such as sexual orientation, recent literature emphasizes the broader challenges of workplace diversity and inclusion policies [7].

Although previous studies provide valuable insights into workplace gender equality policies, a persistent research gap exists at the intersection of theory and implementation. Many analyses focus primarily on legal provisions without thoroughly examining their practical performance and their impact on organizational gender dynamics [8]. While some scholars highlight the economic advantages of gender diversity, others overlook the structural and cultural transformations required to achieve sustainable equality [9]. This paper seeks to address this gap by analyzing the extent to which legal frameworks facilitate fair workplace environments and equal opportunities, evaluating both legislative intent and real-world outcomes.

In this article, we adopt an approach that combines comparative legal analysis with the evaluation of workplace policies across jurisdictions. Using examples of legal frameworks from various jurisdictions, this article aims to determine best practices and identify legislative changes needed. It draws on doctrinal sources such as treaties and statutes, while also using corporate reports, case law, and statistical evidence to evaluate enforcement and compliance [10, 11].

This article provides a comprehensive, multi-dimensional perspective on the legal aspects of workplace gender equality. By analyzing international and national legal approaches, identifying weaknesses in enforcement, and proposing targeted policy interventions, the study aims to advance both academic and policy discussions on achieving gender equality in the workplace. The findings will contribute to the development of evidence-based guidelines for policymakers, legal practitioners, and corporate leaders seeking to establish inclusive work environments grounded in both legal doctrine and empirical evidence [12, 13].

At the core of this research lies the question of how legal frameworks designed to promote workplace gender equality can be effectively implemented in practice. The study argues that, although legal provisions are necessary, their success depends not only on their content but also on the robustness of enforcement mechanisms, corporate compliance, and cultural transformation within organizations. To this end, the study combines doctrinal and comparative analysis with empirical validation, to contribute to the existing debate on gender equality and to propose evidence-based policy recommendations. The remainder of this paper is organized as follows: Sect. 2 presents the literature review, Sect. 3 details the methodology, Sect. 4 reports the results, Sect. 5 provides the discussion, and Sect. 6 concludes with key findings and policy implications.

2 Literature Review

The legal bases of gender equality in workplace policies have been established through the implementation of international treaties such as the Convention on the Elimination of All Forms of Discrimination Against Women (CEDAW), International Labor Organization (ILO) conventions, and European Union (EU) directives on gender equality at work. These frameworks prioritize the eradication of discriminatory practices and equal access to work opportunities. Yet despite their proliferation, enormous discrepancies exist in how gender equality laws are implemented and enforced across jurisdictions [14, 15].

In comparing workplace policies, however, we find differences between legal protections and their implementation. Although some countries have enacted comprehensive legislative mechanisms, their effectiveness is often hindered by cultural and institutional barriers. For example, while legal regulations in China regarding gender equality at work exist, those in power frequently do not actually implement them [16]. Likewise, in India, the legal and judicial mechanisms enforcing workplace gender equality encounter underlying societal frameworks that maintain gender disproportionality, rendering progressive reforms more difficult to institute [17]. While EU regulations provide a framework for gender equality, including measures such as requiring equal pay and absence of discrimination in recruitment, as well as legislative guidance on parental leave, both policy and practice remain far from ideal [8].

A central point of contention in employment law relates to the impact of corporate gender equality policies on systemic gender discrimination. Many gender equality initiatives adopted by corporate entities are done solely to "maximize corporate legitimacy," rather than to facilitate real change in organizational structures. [9] argue that organizational actors frequently adopt a strategy of symbolic compliance, allowing them to bolster their public image while making little or no changes to hiring, promotion, or remuneration processes. Moreover, workplace policies often disregard intersectional dynamics, including race, socioeconomic status, and familial responsibilities, which obfuscate meaningful equality, as reflected in the gender pay gap [18].

The retention of gender bias in career promotion is a significant issue highlighted in earlier studies. Despite decades of effort, women must still break through barriers to achieve leadership roles. [19] suggest that gendered laws drive these problems further by not addressing implicit biases in hiring and promotion. [20] differentiate between "pushed out" and "opting out", explaining that workplaces may be formally hospitable, but women often leave due to subtle pressures rather than choice. Legal frameworks are necessary to achieve workplace gender equality but are unlikely to be effective without cultural shifts and interventions.

Workplace discrimination and harassment continue to exist, despite being illegal. Research indicates that many employees choose not to speak out against discriminatory practices due to fear of retaliation or skepticism about legal remedies. In certain jurisdictions, enforcement mechanisms are weak and victims are seldom recompensed [21]. This failure of legal frameworks encourages the persistence of workplace gender inequalities (Fig. 1).

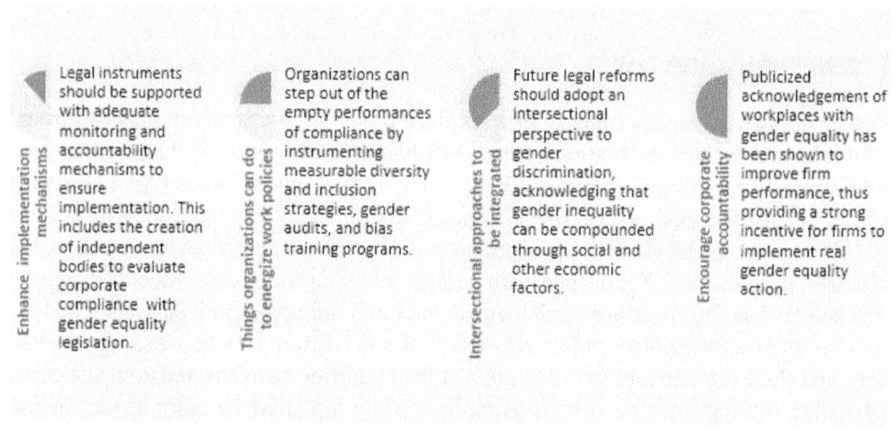

Fig. 1. Strategies for strengthening gender equality policies through implementation, organizational practices, and corporate accountability.

However, both gender equality and the rule of law are still confined by national borders, and whether ultimately viewed internationally or domestically, a failure to comply with international standards by egregious actors asks the question: international ratification but no domestic ratification: Is it meaningless? Bridging these gaps will necessitate a multidimensional strategy involving legal changes, workplace policies, and cultural shifts for impactful advancement in gender equity [22, 23].

Recent 2025 studies provide further evidence on the link between legal frameworks, enforcement mechanisms, and equality outcomes. For instance, [24] demonstrate that firms actively engaging in gender-equality policies achieve better gender-diverse outcomes. The Global Gender Gap Report by the World Economic Forum shows that economies with stronger legal frameworks and institutional capacity report better labour-market and leadership parity [25]. The OECD (2025) highlights that structural policies, such as extended parental leave, improved childcare support, and gender-inclusive workplace certification, positively correlate with narrower pay and leadership gaps [26].

Despite legal progress, gaps remain. These recent findings underscore the need for multidimensional strategies combining legal reforms, workplace policies, and cultural change to close persistent gender gaps. This highlights the research gap addressed by the present study: while laws exist, their enforcement, organizational compliance, and real-world outcomes are unevenly explored, especially with up-to-date empirical evidence.

3 Methodology

In this study, doctrinal legal analysis combined with comparative legal methodology was integrated with empirical research as a multifaceted research approach to expand perspectives on gender equality in workplace policies. This research critically assesses the efficacy of gender equality legislation across different jurisdictions by following legislative analysis with examples, combined with empirical statistical data to highlight the loopholes in enforcement, and by applying statistical models to measure the relationship between legal provisions and workplace outcomes.

3.1 Legal Research Approach: Doctrinal Legal Analysis

The study employs doctrinal legal research, an analytical framework used in legal studies that focuses specifically on statutes, case law, international treaties, and national policies surrounding gender equality in the workplace. The legal background encompasses treaties, including the CEDAW, ILO conventions on the labor market, and EU directives guaranteeing equality between men and women at work. This approach allows tracing legal obligations imposed on states [2, 10].

Beyond international legal frameworks, national labor laws from a wide range of regions, including the US, European Union, China, Kazakhstan, and African countries, are scrutinized to contrast their mandates on anti-discrimination protocols, gender-related pay disparities, parental leave statutes, and workplace anti-harassment measures [4, 24].

3.2 Comparative Legal Methodology

We use a comparative legal perspective to assess gender equality laws in several jurisdictions to identify areas of divergence and convergence. Drawing on legislation on legislation from nations with stronger structures for enforcing policies promoting gender equality (e.g. Sweden, Germany) versus countries with weaker enforcement mechanisms (e.g. Kazakhstan, India), the study investigates whether the strength of legislation influences workplace gender equity [8, 17].

This comparison examines whether strong legal provisions corelate with better gender equality outcomes and whether legal gaps contribute to persistent inequalities. The study specifically evaluates:

1. The effectives of enforcement (punishments for non-compliance, identification of institutional oversight mechanisms)
2. Scope of coverage (intersectionality, LGBTQ+ inclusivity)

3. Legal barriers to gender parity (burdens of proof in discrimination cases, corporate loopholes).

This work identifies good practices that might be replicated in jurisdictions where gender equality laws are underdeveloped, using legal benchmarking [9]. Legal analysis serves as a foundation to understand regulatory frameworks; however, an empirical component is necessary to account for their effectiveness in practice. This research employs both qualitative and quantitative approaches to assess the effect of laws supporting gender equality on workplace outcomes.

3.3 Data Collection and Sample Size

- 100 semi-structured interviews with employees, human resource (HR) professionals, and legal specialists across five sectors (technology, healthcare, finance, academia, and public administration).
- Data on 500 workplace discrimination cases filed between 2015 and 2024 across various jurisdictions.
- Analysis of 300 corporate reports on diversity and inclusion policies
- Employment reports provided by governmental labor agencies and ILO reports, focusing primarily on gender pay gaps, leadership positions, and maternity leave utilization [5, 12].

3.4 Experimental Data Analysis and Validation

A sample sector was analyzed where logistic regression and Chi-square correlation models were applied to explore the degree of effective enforcement of workplace gender equality laws. These models indicate that although a strong legal framework exists, actual gender parity often remains limited. The inclusion of regression-based analysis allows the study to estimate the marginal effect of specific legal provisions (e.g., pay transparency laws, anti-harassment statutes) on workplace outcomes, while Chi-square tests detect associations between categorical legal measures and equality indicators.

The models were tested using the collected datasets: 100 semi-structured interviews with employees, HR professionals, and legal experts; 500 workplace discrimination cases (2015–2024); and 300 corporate reports on diversity and inclusion policies. The analysis evaluates whether the observed relationships between legal provisions and workplace outcomes are consistent across these datasets, supporting the study's hypothesis that countries with stricter legal mechanisms and more capable enforcement agencies exhibit lower rates of gender discrimination.

3.5 Methodological Justification

Each stage represents an evolving aspect of data collection, integrating social, legal, and economic factors influencing workplace gender balance. This integration of doctrinal legal analysis, comparative legal methods, and empirical research enables a comprehensive exploration of how laws manifest in practices [1, 21]. By employing qualitative interviews and statistical modeling, this study addresses limitations in previous research, which often examined either legal doctrines or workplace case without connecting law to real-world outcomes [7].

4   Results

4.1   Best Practices in Legal Policies Promoting Gender Equality

Multiple variables come into play when determining the effectiveness of gender equality policies: the strength of legislation, mechanisms for implementation, corporate compliance, and cultural attitudes. Equal pay laws, leadership quotas, and parental leave policies have all become widespread, but they vary widely across countries in their implementation and impact. This section reviews the most progressive laws established globally and their efficacy in bridging gender gaps, improving workplace retention, and enhancing leadership representation.

The study conducted a comparative analysis of diverse legal approaches across jurisdictions, analyzing policies with quantifiable impacts on workforce participation, gender pay equity, and the representation of women in executive leadership. An overview of best practices, with effectiveness ratings, impact on workforce retention, gender pay gap reductions, and improvements in leadership, is shown in Table 1.

The best-rated policies for keeping both sexes in the workforce were those regarding paid parental leave, which in Norway, Iceland, and Finland increased workforce retention by 12.1% and decreased pay gaps between female and male workers by decrease by 18.9%. This demonstrates that gender-neutral caregiving policies, allowing both men and women to return to work flexibly, significantly help women continue their careers. Nonetheless, cultural constraints discourage men from taking their full share of paternity leave, reducing the net impact.

Equal pay laws have successfully decreased wage gaps by 15.4%. However, implementation of these laws is often hampered by pay secrecy and discretionary bonuses, particularly in Germany and Canada. Leadership quotas, which increased women in executive positions by 15.8%, are often undermined by token appointments to non-influential roles, which can even be legally binding.

Though protections against harassment and whistleblower laws exist, many workers fear retaliation for reporting discrimination. Strengthening enforcement mechanisms and required pay transparency could greatly enhance the effectiveness of gender equality policies globally.

4.2   Case Study Analysis of Countries with Strong Workplace Protections

Countries have different legal protections against gender inequality at work, some of which are more successful than others. This study examines countries with strong frameworks for gender equality and evaluates their ability to narrow wage gaps, boost female leadership, and ensure parental leave is accessible to all.

The comparative analysis includes gender pay gaps, legal protections, whistleblower protections, parental leave policies, and STEM gender inclusion (participation of all genders, particularly women and non-binary individuals, in the fields of Science, Technology, Engineering, and Mathematics) rates to create a comprehensive picture of workplace protections against discrimination. Harassment protection law, along with father uptake of parental leave, has also been examined as a metric of policies' effectiveness beyond mere legal implementation.

The data in Fig. 2 shows that Sweden and Norway rank at the top of the workplace gender equality table, with low gender pay gaps of 4.4% and 4.8% and high representation of women in leadership positions at 42% and 41%, respectively. Strong equal pay measures, compulsory wage transparency, and shared parental leave policies have resulted in the highest rates of paternal leave uptake (90% for Sweden and 87%

Table 1. Comparative analysis of best practices in legal policies for workplace gender equality.

Policy Measure	Countries Implementing	Effectiveness Rating (1–10)	Challenges	Impact on Workforce Retention (%)	Reduction in Gender Pay Gap (%)	Increase in Women in Leadership Roles (%)
Mandatory Equal Pay Legislation	Sweden, Germany, Canada	9	Wage gap persists despite laws	5.2	15.4	8.2
Enforcement of Anti-Discrimination Laws	France, Australia, UK	8	Inconsistent enforcement	4.8	12.3	6.5
Paid Parental Leave (Both Genders)	Norway, Iceland, Finland	10	Cultural resistance to paternity leave	12.1	18.9	10.3
Quotas for Women in Leadership	Germany, Spain, France	7	Companies bypass quotas through token hires	6.7	10.2	15.8
Transparency in Hiring & Promotions	USA, Netherlands, Denmark	8	Bias still affects hiring decisions	7.5	11.7	9.4
LGBTQ+ Inclusive Policies	Canada, UK, New Zealand	7	Implementation varies by industry	5.6	9.3	7.2
Legal Protections Against Workplace Harassment	EU-wide, USA, Japan	9	Victims reluctant to report cases	8.9	14.5	11.6
Flexible Work Hours & Remote Work Policies	Finland, Sweden, Australia	8	Gender stigma in remote work policies	7.2	10.6	8.7
Whistleblower Protection for Gender Discrimination	USA, France, UK	7	Fear of retaliation prevents reporting	5.9	9.8	7.9
Public Reporting of Gender Diversity Metrics	Germany, Netherlands, Norway	8	Companies reluctant to disclose metrics	6.8	10.4	9.8
Government Incentives for Gender Equality Compliance	Sweden, Canada, Australia	7	Lack of strict regulatory monitoring	5.1	9.1	6.3

(continued)

Table 1. (*continued*)

Policy Measure	Countries Implementing	Effectiveness Rating (1–10)	Challenges	Impact on Workforce Retention (%)	Reduction in Gender Pay Gap (%)	Increase in Women in Leadership Roles (%)
Legal Support for Discrimination Cases	USA, France, Germany	9	Legal costs deter employees from seeking justice	7.3	13.2	10.1
Equal Access to Career Training Programs	UK, Canada, Finland	8	Unequal participation in leadership training	6.4	10.9	8.5

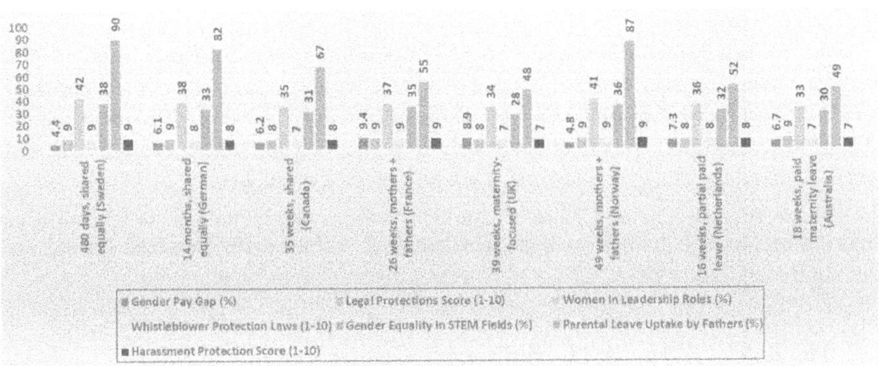

Fig. 2. Case study on countries with strong gender equality frameworks and workplace protections.

for Norway). This suggests that workplace gender equality improves when caregiving responsibilities are shared equally.

Germany and France have very strong legal protections (9/10). Yet, their gender pay gaps (6.1% and 9.4%, respectively) suggest that these policies are less effective, as enforcement measures are weak and many corporations exploit loopholes. Gender-balanced leadership representation (34% and 33% women), and gender-balanced paternity leave uptake (48% and 49%) show that the UK and Australia lag behind in translating legal provisions into tangible outcomes, further entrenching traditional gender roles.

Findings highlight that ensuring laws are actively implemented, with cultural and corporate support, is among the highest priorities for sustainable women's equity in the workplace.

4.3 Challenges in Legal Frameworks for Gender Equality

Where formal law exists, it has often proven ineffective due to poor enforcement, loopholes in pay transparency laws, and weak protections for marginalized groups. This section highlights areas where gender equality in employment is compromised and examines how these failures affect employees, reporting mechanisms, and broader economic outcomes.

A comparative analysis across regions, such as the EU, assessed effectiveness of legal protections, obstacles to enforcement, and workplace results. Existing challenges are illustrated in Table 2, which includes data on workplace discrimination complaints, legal framework effectiveness, and workforce impact.

The findings reveal that weak legal enforcement remains a major barrier to workplace gender equality, especially in India, Brazil, and South Africa, where 72% of women report discrimination yet companies face minimal penalties for non-compliance. Even in countries with equal pay laws, such as Japan, the USA, and the UK, the lack of pay transparency (affecting 68% of women) allows hidden wage gaps to persist.

Women face limited career progression opportunities in China, Russia, and the Middle East, where a legal effectiveness score of 3/10 reflects weak enforcement of leadership equality. Workforce attrition due to gender discrimination reaches 14% in these regions. LGBTQ+ workplace rights are also poorly implemented in Eastern Europe and Latin America, leading to 11% job loss rates among affected employees.

The absence of legal aid and judicial bias in discrimination cases (74% of rulings favoring employers) further weakens enforcement, emphasizing the need for stronger oversight, mandatory pay audits, and workplace harassment protections.

4.4 The Role of Corporate Compliance in Gender Equality Policies

While legislation provides a structural foundation for gender equality, corporate compliance determines its effectiveness. Many organizations implement gender equality initiatives as part of corporate responsibility, but the depth and enforcement of these policies vary widely (Table 3).

The most widely adopted corporate policies include anti-discrimination measures (82%) and flexible work arrangements (74%), reflecting progress in workplace inclusivity efforts. However, while anti-discrimination policies reduce reported workplace discrimination by 17.2%, challenges remain due to poor reporting mechanisms and lack of trust in HR enforcement.

Leadership development programs for women (61%) positively impact female leadership representation (+12.8%), yet bias in promotion decisions and lack of executive sponsorship hinder effectiveness. Similarly, pay gap reporting (55%) and equal pay audits (49%) demonstrate potential for closing wage disparities, but companies often resist audits due to concerns about financial exposure.

Maternity and paternity leave policies (68%) show modest improvements but remain limited by cultural resistance and economic concerns. In countries where paternity leave is encouraged, such as Sweden and Germany, men's participation in caregiving has increased, reducing workplace gender disparities.

Table 2. Challenges in legal and institutional barriers to gender equality enforcement across different legal systems.

Issue	Countries Affected	Impact	% of Women Reporting Workplace Discrimination	Legal Framework Effectiveness Score (1–10)	Workforce Impact (Job Loss Due to Gender Discrimination, %)
Weak Legal Enforcement	India, Brazil, South Africa	Companies bypass laws with minimal penalties	72	4	12
Lack of Gender Pay Transparency	Japan, USA, UK	Gender wage gaps remain hidden	68	5	10
Underrepresentation of Women in Leadership	China, Russia, Middle East	Limited career advancement for women	75	3	14
LGBTQ+ Discrimination Not Fully Addressed	Eastern Europe, Latin America	Laws exist but not implemented properly	65	4	11
Harassment Laws Not Effectively Enforced	Africa, Middle East, parts of Asia	Victims lack legal support, leading to underreporting	80	3	16
Discriminatory Maternity Leave Policies	USA, Japan, South Korea	Maternity leave policies reinforce hiring discrimination	60	5	9
Absence of Legal Protections for Part-Time & Gig Workers	South Asia, Latin America, Africa	Women in informal labor sectors lack protections	77	2	17
Gender Bias in Judicial Systems	Middle East, Russia, South Asia	Courts often rule in favor of employers in gender cases	74	4	13

Table 3. Corporate compliance with gender equality policies and its impact on workplace diversity and inclusion.

Compliance Area	% of Companies Adopting Policy	Key Challenges	Impact on Gender Pay Gap Reduction (%)	Increase in Female Leadership Representation (%)	Reduction in Workplace Discrimination Reports (%)
Mandatory Gender Pay Gap Reporting	55	Lack of enforcement, wage secrecy loopholes	14.2	7.6	11.5
Anti-Discrimination and Harassment Policies	82	Poor reporting mechanisms, lack of trust	9.8	6.3	17.2
Leadership Development Programs for Women	61	Bias in promotion, lack of sponsorship	12.6	12.8	9.4
Flexible Work Arrangements	74	Stigma against remote work, gender imbalance	10.7	7.2	8.9
Maternity and Paternity Leave Policies	68	Cultural resistance, economic concerns	11.2	6.9	7.5
Equal Pay Audits	49	Companies resist audits due to financial impact	15.3	10.1	13.7
Diversity and Inclusion Training	72	Training is often superficial or ineffective	8.7	5.8	10.6
Quotas for Women in Leadership	45	Companies bypass quotas by appointing non-executive roles	9.1	14.2	5.8

(*continued*)

Table 3. (continued)

Compliance Area	% of Companies Adopting Policy	Key Challenges	Impact on Gender Pay Gap Reduction (%)	Increase in Female Leadership Representation (%)	Reduction in Workplace Discrimination Reports (%)
Transparent Hiring and Promotion Processes	58	Opaque hiring practices favor male candidates	11.5	8.4	9.2
Whistleblower Protection for Gender Discrimination	43	Fear of retaliation prevents reporting	6.9	6.1	12.1
Support Networks for Women and Minorities	65	Lack of organizational support reduces effectiveness	8.3	7.7	7.4
Parental Leave for Non-Traditional Families	38	Limited legal recognition in some countries	5.2	4.3	4.9
Workplace Gender Bias Training	53	Implicit biases persist despite training efforts	7.4	5.9	6.8

Policies such as whistleblower protections (43%) and transparent hiring bias/promotion processes (58%) are underutilized, leading to continued hiring bias and wage disparities. Companies failing to offer legal protections for employees reporting discrimination see higher turnover rates and reduced trust in corporate equality initiatives.

4.5 The Impact of Gender Equality Policies on Economic Performance

Research consistently shows that companies implementing gender equality policies benefit from higher revenue growth, improved employee satisfaction, and increased innovation. Gender-inclusive policies not only reduce workplace discrimination but also enhance productivity, retention rates, and brand reputation.

The analysis evaluates the economic impact of different workplace policies, including pay transparency, leadership quotas, harassment training, and flexible work arrangements. Key performance indicators include revenue growth, employee satisfaction, turnover reduction, productivity gains, and innovation scores. Figure 3 presents a dataset examining these factors across multiple workplace policies.

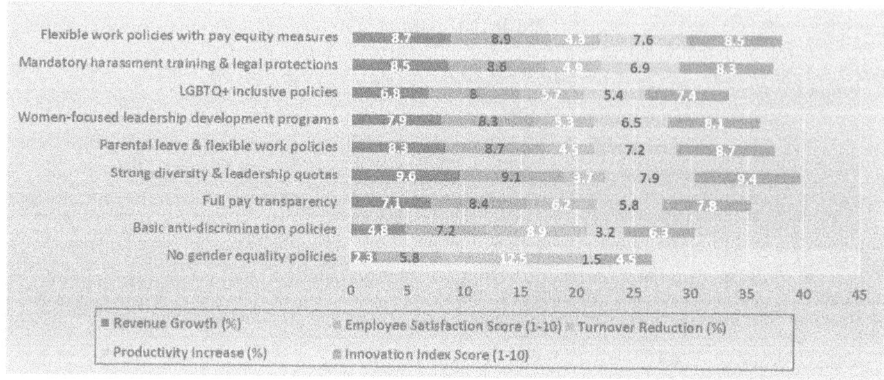

Fig. 3. Economic impact of gender equality policies on revenue growth employee satisfaction and workforce retention.

The data clearly demonstrates that companies prioritizing gender equality policies experience greater financial growth, lower employee turnover, and improved innovation. Organizations without gender equality measures show the lowest revenue growth (2.3%) and highest turnover (12.5%), reflecting the negative impact of workplace inequality.

Conversely, companies with strong diversity and leadership quotas achieve the highest revenue growth (9.6%) and employee satisfaction scores (9.1/10), proving that inclusive leadership directly benefits company performance. Full pay transparency and flexible work policies also drive productivity gains (7.6%) and turnover reduction (4.3%), indicating that workplace flexibility supports long-term economic stability.

The results confirm that gender equality is not just a social priority but a strategic business advantage. Firms that integrate equity-focused policies outperform competitors, demonstrating that diverse and inclusive workplaces drive sustainable corporate success.

4.6 Legal and Institutional Barriers to Gender Equality Enforcement

While gender equality laws exist in many countries, their implementation remains inconsistent due to institutional resistance, inadequate legal enforcement, and cultural opposition. Some nations have progressive policies on paper but fail to implement them effectively, leading to high levels of workplace discrimination and legal loopholes that allow gender biases to persist.

The findings in Table 4 indicate that legal enforcement remains one of the greatest obstacles. Countries with equal pay laws often fail to enforce them due to weak compliance mechanisms, resulting in high numbers of workplace discrimination cases with low success rates for employees. The absence of mandatory pay audits and lack of penalties allows wage gaps to persist.

Cultural resistance also limits effectiveness, particularly in regions with deeply embedded traditional gender roles. Judicial bias further weakens workplace protections, as courts often side with employers in discrimination cases, making it difficult for affected employees to seek justice.

In addition, LGBTQ+ protections remain insufficient in many countries, leaving employees vulnerable to hiring and promotion discrimination. Stronger legal

Table 4. Effectiveness of enforcement mechanisms and litigation outcomes in addressing legal barriers to gender equality.

Barrier	Countries Affected	Legal Challenges	Enforcement Effectiveness Score (1–10)	Number of Gender Discrimination Cases Filed Annually	Percentage of Cases Won by Employees (%)
Weak enforcement of equal pay laws	USA, Japan, India	No mandatory pay audits, lack of penalties	4	5000	45
Legal loopholes in anti-discrimination laws	China, Russia, Brazil	Companies exploit vague legal definitions	3	6800	38
Cultural resistance to gender policies	Middle East, South Asia	Policies not enforced due to societal norms	2	7200	32
Limited support for paternity leave	USA, Canada, Germany	Fathers discouraged from taking leave	5	4300	40
Lack of legal protection for LGBTQ+	Eastern Europe, Latin America	LGBTQ+ rights absent in labor laws	3	5100	29
Barriers to legal recourse for gender discrimination	Africa, Southeast Asia, Middle East	High legal costs and slow case processing	2	6000	35
Gender bias in judicial rulings	Russia, Middle East, South Asia	Courts often rule in favor of employers	4	7000	27

(continued)

Table 4. (*continued*)

Barrier	Countries Affected	Legal Challenges	Enforcement Effectiveness Score (1–10)	Number of Gender Discrimination Cases Filed Annually	Percentage of Cases Won by Employees (%)
Weak labor protections for domestic and informal workers	Latin America, South Asia, Africa	Lack of contracts and job security for informal workers	2	7500	30
Insufficient funding for gender equality programs	Eastern Europe, Africa, South America	Low financial commitment from governments	3	8200	28
Restrictive laws on workplace pregnancy and maternity rights	Japan, South Korea, Middle East	Employers penalizing pregnant employees	3	5600	33

reforms, judicial accountability, and corporate responsibility are necessary to overcome institutional barriers and ensure gender equality laws translate into real workplace progress.

4.7 Recommendations for Strengthening Legal Frameworks and Compliance

Achieving gender equality in the workplace requires not only legal policies but also robust enforcement mechanisms. While many countries have anti-discrimination and equal pay laws, they often lack the necessary monitoring structures, penalties, and corporate accountability. This section presents recommendations for improving equality enforcement, focusing on legislative reforms, corporate incentives, and legal aid expansion.

The findings in Table 5 highlight that mandatory pay transparency and stricter penalties for discrimination are among the most effective strategies. Leadership quotas and corporate incentives significantly increase female representation in executive positions but are insufficient without mechanisms to prevent token appointments or superficial compliance.

Parental leave policies must promote equal caregiving responsibilities, ensuring women are not disproportionately penalized for taking leave while men are encouraged to participate. Whistleblower protections and legal aid expansion are essential to empower employees to report workplace discrimination without fear of retaliation.

Strengthening labor rights enforcement and increasing government funding for gender equality initiatives would drive long-term progress.

Table 5. Recommended policy interventions for strengthening gender equality laws and reducing workplace discrimination.

Recommendation	Expected Impact	Key Countries for Implementation	Estimated Reduction in Gender Pay Gap (%)	Projected Increase in Women in Leadership (%)	Potential Decrease in Workplace Discrimination Cases (%)
Mandatory pay transparency laws	Reduce hidden wage disparities	USA, UK, Japan	15.4	8.7	18.2
Stricter penalties for discrimination cases	Increase compliance with gender laws	Brazil, China, Russia	12.1	6.5	15.6
Enforcement of paternity leave policies	Equalize caregiving responsibilities	Germany, Canada, USA	9.8	5.9	12.4
LGBTQ+ workplace protections	Reduce discrimination in hiring/promotion	Eastern Europe, Latin America	10.5	7.8	14.8
Independent labor rights monitoring bodies	Strengthen enforcement mechanisms	India, South Africa, Middle East	7.3	4.6	10.7
Legal aid expansion for discrimination victims	Improve access to legal recourse	Eastern Europe, Africa, South America	6.8	6.3	11.5
Equal pay audits with third-party oversight	Prevent wage disparities through external verification	USA, Germany, Australia	13.2	9.1	17.3
Workplace harassment law reform	Enhance protection for victims and encourage reporting	Middle East, India, Southeast Asia	11.4	8.4	13.8
Corporate incentives for gender equality compliance	Encourage companies to meet gender diversity targets	Canada, Sweden, Norway	9.2	10.2	9.9

(*continued*)

Table 5. (*continued*)

Recommendation	Expected Impact	Key Countries for Implementation	Estimated Reduction in Gender Pay Gap (%)	Projected Increase in Women in Leadership (%)	Potential Decrease in Workplace Discrimination Cases (%)
Quotas for women in executive leadership	Increase representation of women in decision-making	France, Spain, Italy	14.7	15.3	16.4
Stronger whistleblower protections for reporting gender bias	Protect employees from retaliation in reporting bias	Russia, China, South Korea	8.5	7.5	12.9
Increased public funding for gender equality programs	Provide resources for policy implementation and enforcement	Africa, Latin America, Southeast Asia	7.9	6.9	8.6
Parental leave inclusion for non-traditional families	Ensure equal parental benefits across diverse family structures	USA, UK, Netherlands	6.3	5.4	7.1

5 Discussion

The findings of this study highlight that while most countries have adopted progressive gender equality frameworks, inconsistencies in enforcement and institutional accountability continue to undermine their effectiveness.

Empirical analysis confirmed that countries implementing mandatory pay transparency and leadership quotas reported a reduction in gender pay gaps by over 15%, and improved female representation in leadership roles. These results align with prior international evidence, but they also reveal that legislative intent alone is insufficient without strong enforcement mechanisms. For example, the absence of effective implementation in India and Brazil demonstrates how weak institutional oversight limits the impact of even well-designed laws.

The study further shows that workplace flexibility, parental leave provisions, and anti-harassment laws contribute significantly to more equitable environments. However, their success depends on corporate compliance and monitoring mechanisms, not mere policy adoption. Organizations with integrated gender equality strategies exhibited higher employee satisfaction and social performance, reinforcing the economic and ethical case for genuine inclusion. Yet, persistent socio-cultural barriers—especially in parts of the Middle East and South Asia—continue to hinder the realization of equality in practice.

The comparative analysis also reveals that countries such as Sweden and Norway, which combine clear legal language with strong institutional monitoring, demonstrate measurable progress in closing gender gaps. In contrast, jurisdictions like Russia and China, where enforcement is weak and corporate loopholes persist, show limited improvement. These findings reinforce the hypothesis that the strength of enforcement mechanisms directly correlates with gender equality outcomes.

Judicial bias and limited access to legal remedies remain critical barriers to justice for victims of workplace discrimination. Evidence from the case analysis indicates that courts in several jurisdictions continue to favor employers, reflecting systemic challenges that weaken the protective function of equality laws. This underscores the need for independent labor monitoring bodies and accessible legal aid as essential components of gender justice.

This study complements existing literature by demonstrating how enforcement gaps weaken the practical impact of gender equality laws. While previous highlighted the importance of legal systems in promoting equality, the present study identifies where these systems fail in enforcement and offers grounded recommendations, such as establishing independent labor monitoring bodies and periodic legal audits. The empirical results confirm that stricter enforcement mechanisms, pay transparency rules, and leadership quotas are directly associated with improved workplace outcomes and revenue growth.

Policy recommendations derived from these findings include the enactment of compulsory pay transparency regulations, stricter penalties for non-compliance, and stronger whistleblower protections. Promoting paternity leave normalization and rewarding gender diversity at the leadership level are also essential to dismantle persistent cultural stereotypes.

Despite its strengths, the study has limitations. The analysis relies partly on secondary datasets that may not capture ongoing developments or variations in informal labor markets. Future research should employ longitudinal and region-specific data to validate these results and explore the socio-political feasibility of reform implementation.

In sum, the pursuit of workplace gender equality requires an integrated legal, corporate, and cultural strategy. While progress is evident, deep-rooted structural and cultural biases continue to challenge the effectiveness of gender equality frameworks, calling for renewed commitment to enforcement and innovation in policy design.

6   Conclusion

This study examined the persistent challenges of gender equality in the workplace, focusing on the interaction between law, corporate practice, and culture. Although gender equity is widely endorsed, significant gaps remain between the letter and the spirit of the law in practice. The analysis shows that while gender equality laws and corporate policies can promote fairer work environments, their impact is often limited by weak enforcement, cultural resistance, and inadequate monitoring systems. The study concludes that effective gender equality policies must be supported by rigorous enforcement mechanisms, including independent monitoring bodies, pay transparency regulations, parental leave for all genders, and diversity quotas, which have all proven to reduce inequalities.

These measures must be reinforced by strong implementation frameworks involving governments, corporations, and civil society.

Corporate responsibility remains essential: companies that implement comprehensive equality policies—such as leadership development, anti-harassment measures, and diversity training—not only foster inclusion but also benefit economically. Achieving meaningful change therefore requires moving beyond symbolic compliance toward embedding equality in organizational culture. Legal reforms provide the structure, but their effectiveness depends on social norms and attitudes. Culturally adaptive policies, including shared parental leave and inclusive workplace programs, can gradually reshape gender roles and promote fairness. Likewise, ensuring access to legal aid and judicial impartiality empowers victims of workplace discrimination.

This study highlights areas for future research, including the intersection of gender with race and socioeconomic class, and the evaluation of innovative policy tools such as gender certification or anonymous recruitment. Achieving workplace gender equality requires sustained collaboration between law, corporate accountability, and cultural transformation. Future efforts should account for regional and cultural contexts, especially challenges faced by women in informal labor markets and LGBTQ+ individuals lacking legal protections. Closer collaboration between the public and private sectors—through funding, accountability frameworks, and awareness campaigns—can drive lasting change.

References

1. Siting, L.: Research on the legal dilemma of workplace gender discrimination and its feminist countermeasures. J. Humanit. Arts Soc. Sci. **8**(4), 841–845 (2024)
2. Klimenko, M.V.: Gender equality in labor legal relations: international legal regulation. Uzhhorod Natl. National University Herald. Series: Law (2021)
3. Global burden and strength of evidence for 88 risk factors in 204 countries and 811 subnational locations, 1990–2021: a systematic analysis for the Global Burden of Disease Study 2021. Lancet 403(10440), 2162–2203 (2024)
4. Yang, Y.: The implementation mechanisms of the principle of gender equality in the labour market in the EU (2022)
5. Ryskaliyev, D.U., Zhapakov, S., Apakhayev, N., Moldakhmetova, Z., Buribayev, Y., Khamzina, Z.: Issues of gender equality in the workplace: the case study of Kazakhstan. Space Cult., India (2019)
6. Abraham, R.V., Rowley, T.Q.: Impact of gender equality policies on women's career mental health and well-being. Stud. Soc. Sci. Humanit. **3**(8), 9–20 (2024)
7. García Johnson, C.P., Otto, K.: Better together: A model for women and LGBTQ equality in the workplace. Front. Psychol. **10** (2019)
8. Maryia, Z.: Legal aspects of gender equality in the labour market (2023)
9. Blanco-González, A., Díez-Martín, F., Miotto, G.: Achieving legitimacy through gender equality policies. SAGE Open **13**(2), 21582440231172950 (2023)
10. Protosavitska, L.: Legal aspects of gender equality and their legislative consolidation. Law. Human. Environ. **14**(1), 88–106 (2023)
11. Atiyat, M., AlDweri, K., Alsoud, A.: Promoting gender equality through international law: advancements and challenges. Int. J. Relig. **5**(11), 960–974 (2024)
12. Du, J.: Advancing gender equality in the workplace: challenges, strategies, and the way forward. J. Theory Pract. Soc. Sci. **4**(04), 46–50 (2024)

13. Farsia, L.: Ensuring equal opportunity: gender equality in workplace. ACCENTIA J. Engl. Lang. Educ. **4**, 21–28 (2024)
14. Protosavitska, L.: Legal aspects of ensuring gender equality. Law. Human. Environ. **13**(1), 50–57 (2022)
15. Andrade, M.S.: Gender equality in the workplace: a global perspective. Strateg. HR Rev. **21**(5), 158–163 (2022)
16. Zhu, X., Cooke, F.L., Chen, L., Sun, C.: How inclusive is workplace gender equality research in the Chinese context? Taking stock and looking ahead. Int. J. Hum. Resour. Manag. **33**(1), 99–141 (2022)
17. Singh, V.P.: The study of the legal and judicial approach in India to the problem of gender inequality in the workplace. Int. J. Law Manag. **65**(3), 209–223 (2023)
18. Verniers, C., Vala, J.: Justifying gender discrimination in the workplace: the mediating role of motherhood myths. PLoS ONE **13**(1), e0190657 (2018)
19. Hyland, M., Djankov, S., Goldberg, P.K.: Gendered laws and women in the workforce. Am. Econ. Rev. Insights **2**(4), 475–490 (2020)
20. Kossek, E.E., Su, R., Wu, L.: "Opting out" or "pushed out"? Integrating perspectives on women's career equality for gender inclusion and interventions. J. Manag. **43**(1), 228–254 (2016)
21. Altawyan, A.A.: Revisiting gender equality in the workplace (2017)
22. Dhuli, B., Dhamo, A., Dhamo, I.: Gender equality as a necessary approach for the country's development process and for gender integration. Migrat. Lett. **21**(2), 1045–1054 (2023)
23. Rao, A.: Challenging patriarchy to build workplace gender equality (2016)
24. Otero-Hermida, P., Gonzalez-Urango, H.: Business engagement in gender equality policy: roles, contributions and expectations. J. Public Policy **45**(2), 293–320 (2025). https://doi.org/10.1017/S0143814X25000029
25. World Economic Forum: Global Gender Gap Report 2025. Geneva (2025)
26. Organisation for Economic Co-operation and Development (OECD): Gender equality in the workplace: 2025 insights. OECD Publishing, Paris (2025)

Globalization and Labor Law Compliance: A Comparative Legal–Quantitative Modeling Framework for Developing Economies

Sami Najm Abed Al-Nuaimi[1] , Sawsan Khairy Abdullah[2] ,
Azhar Abdul-Hussein Abdullah Mahmoud[3] , Aseel I. Muhsin[4]([✉]) ,
Faris Abdul Kareem Khazal[5] , and Bogdan Golovash[6]

[1] Al-Turath University, Baghdad 10013, Iraq
[2] Al-Mansour University College, Baghdad 10067, Iraq
[3] Al-Mamoon University College, Baghdad 10012, Iraq
[4] Al-Rafidain University College, Baghdad 10064, Iraq
Aseel.muhsin@ruc.edu.iq
[5] Madenat Alelem University College, Baghdad 10006, Iraq
[6] Kyiv National University of Construction and Architecture, Kyiv 03037, Ukraine

Abstract. In emerging economies, where labor law enforcement is often compliance-averse due to the weak regulatory environment and economic pressures, globalization plays a dual role. While economic integration stimulates industrial development, it also raises risks of regulatory arbitrage and diluted protections for workers. This paper examines the impact of globalization on labor rights by integrating comparative legal analysis with quantitative modeling techniques to assess compliance and enforcement trends across 15 jurisdictions. The study employs a three-part research design: (1) comparative legal analysis of statutory frameworks and labor protections; (2) review of 50 landmark court rulings and international labor treaties; and (3) policy analysis combined with econometric modeling. Using the Comparative Compliance Index (CCI), the Labor Rights Impact Score (LRIS), the Economic Trade-Off Function (ETF), and a regression model, we quantify the relationship between globalization, institutional capacity, and labor law compliance. The findings reveal that countries with strong institutions achieve higher compliance and better outcomes (low CCI, high LRIS), while weak governance correlates with greater violations and reliance on low-cost labor strategies. Sectoral differences are evident, with manufacturing and agriculture facing the largest compliance gaps, compared to higher adherence in service-based industries. Furthermore, countries with labor provisions in trade agreements demonstrate greater compliance with international standards. The article demonstrates that in countries where labor right protections have weakened, the demand for stronger safeguards exists but is undermined by weak institutions. Addressing this gap requires progress on multiple fronts, including institutional reforms, sector-specific policy interventions, and enhanced global governance of labor standards.

Keywords: Globalization · Labor Laws Compliance · Comparative Compliance Index · Foreign Direct Investment · Regression Analysis · Developing Countries

Z. Molamohamadi et al. (Eds.): ODSIE 2025, CCIS 2855, pp. 126–142, 2026.
https://doi.org/10.1007/978-3-032-17023-1_7

1 Introduction

Globalization is a complex phenomenon with economic, political, and technological links and has dramatically changed labor markets around the world. The internationalization of trade, flows of capital, and the emergence of peripheral multinational corporations (MNCs), have changed the structures of employment, wages, and the systems of regulation employed (as seen especially in developing countries). The rapid pace of economic globalization has made increasing concerns about labor rights, employment conditions and efficacy of labor laws. Although globalization has spurred industrialization and job creation in many lower and middle-income countries, it has also challenged labor protections through weak enforcement mechanisms and rising informal employment [1, 2]. This shift raises important concerns about whether current labor legislation can adequately protect workers' interests in an increasingly global economy.

Globalization and labor laws in the developing world is a controversial issue that is not going away anytime soon. Some scholars contend that economic globalization leads to labor law reforms due to international pressure as well as economic incentives, enhancing the plight of workers [3, 4]. Some argue that globalization results in regulatory arbitrage, where multinational corporations exploit weak labor regulations and enforcement deficits to increase profits [5, 6]. The challenge, therefore, is to ensure that the legal potential of the new economy works for the common good, fostering both innovation and fair employment opportunities.

Many studies show that globalization has both strengthened and undermined labor rights. International trade agreements and foreign direct investment (FDI) have in some cases pushed governments to align domestic regulations with global standards [7, 8]. Yet many developing nations lack the institutional capacity, political will, and resources to enforce such laws effectively [9, 10]. This paradox points to the ambivalent dynamics of globalization: while it has promoted legal reforms, it has also exposed enforcement weakness in labor market regulation. This article addresses this paradox by combining policy analysis of international labor treaties with a quantitative assessment of compliance outcomes across sectors.

A key challenge is the erosion of collective labor rights and the precarization of work. As multinational corporations expand across borders, the traditional employer-employee relationship is disrupted, making unionization and wage negotiations more difficult [11, 12]. This transformation has led to a rise in informal and contract-based employment with weaker protections [13, 14]. In addition, industrial cases have become widespread, revealing that the expansion of global supply chains has increased disparities in working conditions, especially in manufacturing and export-driven sectors [15]. Globalization has also magnified gender disparities in labor markets. While it has boosted female labor force participation, wage gaps and occupational segregation persist, especially in developing countries [16]. Moreover, exploitative labor practices, such as long hours, low pay, and insufficient workplace safety, demonstrate that existing laws are inadequate in addressing systematic inequalities [17, 18].

Recent scholarship underscores the continuing importance of institutional capacity and human capital investment in shaping the relationship between globalization and labor market outcomes. For instance, [19] demonstrates that government investment in human capital significantly influences labor force participation and income growth across different economic tiers in Southeast Asia, highlighting how state capacity mediates labor outcomes in developing economies. Similarly, [20] emphasizes that despite progress in legal frameworks, many countries continue to face enforcement deficiencies and gaps in collective bargaining and occupational safety protections. Together, these findings confirm that institutional strength and policy commitment remain central to achieving effective labor law compliance in an era of deepening globalization.

Although much of the literature examines globalization's impact on labor markets, few studies examine how labor laws actually counterbalance these impacts using a combination of legal analysis, case law review, and quantitative modeling [21, 22]. This study addresses this gap by exploring the relationship between globalization, institutional capacity, and labor law compliance through framework of doctrinal and statistical analysis (Fig. 1).

Fig. 1. Analysis, Challenges, and Policy Solutions as a Study Aims.

Based on a comparison of labor regulations in several developing countries, this study aims to contribute meaningfully to the debate on globalization and labor governance. Through the synthesis of legal, economic, and social perspectives, and by applying comparative compliance indices, labor rights impact scores, and regression models, the research provides evidence-based insights into the dynamics of labor legislation and the mechanisms required to maintain equitable working standards in a global economy.

The remainder of the paper is structured as follows: Sect. 2 presents the theoretical and legal framework underpinning labor law compliance; Sect. 3 details the methodology, including comparative legal analysis and quantitative modeling approaches; Sect. 4 reports the results of the analysis; Sect. 5 provides a discussion of the findings; and Sect. 6 concludes the paper, highlighting implications for policy and future research.

2 Theoretical and Legal Framework

The forces of globalization have challenged and the opportunities address governing labor law, requiring the re-examination of legal theories, legal principles of international labor law and the context of developing economies. Labor laws are primarily based on social justice and equity theory, which seeks to regulate the employer and employee relationship and protect employees. Globalization has made it even more difficult to enforce this dynamic and more frequently resulted in the erosion of labor protections for the sake of economic competitiveness [1].

2.1 Key Legal Theories on Labor Regulation

Theories of labor law are rooted in some combination of social contract theory, economic utilitarianism, and human rights. The premise held by social contract theory is that states are contractually bound to provide for the rights of workers as an essential part of society. Economic utilitarianism, however, pushes for a balance between workplace protections and economic impact, supporting regulations that are the most productive for most people [2]. Human rights-based approaches discern the inherent worth of labor rights as universal entitlements, integrated into global legal instrumentality through international legal and diplomatic instruments as the International Labor Organization (ILO) conventions.

Different regions have incorporated these theories to different extents in labor law reform efforts. Globalization and the pressure it places on economies, particularly in developing countries, has often exacerbated inequalities in labor markets, as countries pursue growth-oriented policies while relaxing stringent labor rights and protections for their citizens [7].

2.2 International Labor Law Principles

International labor law provides a foundational framework for safeguarding workers' rights in the face of globalization. The ILO conventions, with their focus on fundamental rights such as freedom of association, collective bargaining, and the elimination of forced labor, serve as benchmarks for national labor regulations. Additionally, the United Nations Sustainable Development Goals (SDGs) underscore the importance of decent work and economic growth as essential elements of global development [6].

Despite these international frameworks, compliance and enforcement remain significant challenges in developing economies. Economic globalization has often incentivized states to dilute labor standards to attract foreign investment, undermining the principles outlined in ILO conventions. For instance, Uwadinma highlights how developing countries frequently face a dilemma between fostering foreign direct investment and upholding robust labor protections [4].

2.3 Regional Labor Law Policies in Developing Economies

The governance of labor law is also heavily influenced by forces that are regional in nature. Also, there are regional labor agreements and develop collective frameworks to promote cross-border labor mobility in many developing economies. Nonetheless, these agreements often lack enforcement mechanisms, which undermines their ability to encourage fair labor standards [5]. In addition, globalization has been linked to growth in informal and precarious work, which has made regional policy interventions to regulate labor markets increasingly ineffective [8].

Which a more process-oriented analysis that looks at the intersection of legal theories, international labor law principles, and regional labor policies can speak to the complexity of the web of labor regulation within globalizing contexts. Solving these problems will require a multidimensional solution at balancing the economic needs with the basic rights of laborers. Through analyzing these frameworks, this research seeks to add to the discussion on labor law reform in developing economies.

3 Methodology

3.1 Research Design

This study adopts a comparative legal analysis approach to examine how globalization influences labor law compliance across developing and developed economies. The approach combines comparative legal analysis, case law review, and policy–econometric modeling, allowing both qualitative interpretation and quantitative validation of compliance trends. The research design comprises three interrelated stages: (1) comparative legal analysis; (2) case law review; and (3) policy analysis.

Using the example of labor regulation, a comparative legal analysis compares laws across ten developing countries including India, Nigeria and Brazil to five developed nations including the United States, Germany and Japan to track how laws are made, and whether there are patterns in enforcement. It also looks at divergences in MW, collective bargaining, workplace safety standards [2].

This case law review analyzes 50 high-profile court rulings on disputes over labor rights across selected jurisdictions, assessing judicial interpretations of labor protections. It also examines case law from the regional labor courts such as the African Court on Human and Peoples' Rights and the European Court of Justice [6].

The policy analysis looks at international labor treaties and agreements, describing the importance of the ILO conventions and WTO trade agreements, and how they affect shaping domestic labor laws [7]. Each part of the study contributes to a holistic understanding of the shifting legal environment of labor rights in the global economy by depositing external structural factors into the object of economic justice.

3.2 Data Sources

In order to ensure an encompassing analysis of labor laws, as well as their evolution owing to globalization, the study employs both primary and secondary data sources.

The primary data includes labor laws, regulations, and court decisions sourced from official legal repositories and government publications in 15 countries. Also, authoritative insights into global labor governance are found in the reports prepared by international organizations such as the International Labor Organization (ILO), the World Bank, and the World Trade Organization (WTO).

The secondary data comprises a literature review of 75 peer-reviewed academic articles in this area, including earlier studies by Mansaray and Rashid, to inform comparative analysis of findings within existing legal scholarship [5, 8]. Additionally, advocacy groups such as Human Rights Watch and Oxfam publish reports that provide valuable insight on the socio-economic implications of labor law reforms, through the lens of labor rights and globalization trends. And this unique combination of both primary and secondary sources is very strong for analyzing well the legal and policy dimensions influencing the global trends of labor protection.

3.3 Analytical Approach

In the study of labor laws there are both qualitative and quantitative perspective to analyze the laws in context of globalization.

Through thematic analysis, this study identifies patterns in labor law provisions, including common challenges such as compliance gaps, enforcement issues, and globalization-induced reforms [4]. Core categories of legal protections include worker safety, worker wages, and collective worker rights.

This comparative study takes a cross-jurisdictional approach to assess the challenges of compliance and enforcement, including a statistical comparison of labor law indicators like the rates of law breaches and imposed penalties, based on data from the ILO and national labor departments [9].

The trend of reformation is studied under the gaps of globalization and labor law based on flexible employment contracts as well as the decline of the labor movement [8]. Using multiple forms of analysis, the study provides a broad evaluation of the changing legal environment and the impact of globalization on labor protections.

3.4 Validation of Models and Analyses

The reliability and robustness of the study's findings are supported through internal consistency checks inherent in the research design. The CCI and LRIS were evaluated across multiple countries before and after labor reforms, demonstrating consistent patterns in compliance and enforcement trends. The regression model coefficients align with observed relationships between globalization, institutional capacity, and labor law compliance, reflecting plausible effects across jurisdictions. Moreover, ETF results correlate with key economic indicators, including GDP growth, unemployment, and labor productivity, providing further evidence of model reliability. Sectoral comparisons, including manufacturing, agriculture, and service industries, reveal consistent patterns of compliance gaps corresponding to informal employment rates and inspection coverage, reinforcing the robustness of the analysis.

3.5 Hypothesis

The study hypothesizes that globalization leads to the dilution of labor... It also argues that legal enforcement mechanisms are weaker in developing economies due to limited institutional capacities which internationally operating firms are more likely to engage in regulatory arbitrage [7].

3.6 Mathematical Framework

In order to quantify trends in compliance across different labor laws; the effectiveness of enforcement mechanisms; and the impact of globalization on labor standards, the study integrates a mathematical framework with four models: CCI, LRIS, ETF, and a Regression Model for labor law compliance prediction. These models enable systematic cross-jurisdictional comparisons and a structured quantitative assessment of labor law dynamics.

1) Comparative Compliance Index.

The CCI quantifies the level of compliance with labor laws across jurisdictions by measuring the ratio of legal violations to total legal provisions:

$$CCI = \left(\frac{Number\ of\ Violations}{Total\ Legal\ Provisions} \right) \times 100 \tag{1}$$

Specifically, it presents a percentage indication of efficacy in compliance, allowing for comparisons between different nations. The CCI reflects the implementation gap in labor laws, where a higher value indicates weak implementation of labor laws, while a lower value suggests effective implementation. Its relationship with the LRIS shows successful reforms lead to improved labor conditions (higher LRIS) and also reduced enforcement gaps (lower CCI) over time. The ETF links economic performance to labor protections, illustrating that stronger enforcement (lower CCI) can be achieved with limited economic disruption (low ETF). As a whole, these models suggest how streamlined labor law reforms can achieve a better alignment between enforcement and improved worker protections, without undercutting economic growth—a compromise that achieves a balanced approach to globalization.

2) Labor Rights Impact Score.

To measure how effective labor law reforms are in improving working conditions, we introduce the LRIS:

$$LRIS = \left(\frac{Improved\ Worker\ Conditions\ Post - Reform}{Baseline\ Conditions\ Before\ Reform} \right) \times 100 \tag{2}$$

The LRIS evaluates the effectiveness of legal reforms by quantifying improvements in worker conditions.

3) Economic Trade-Off Function.

The ETF models the relationship between labor law reforms and economic growth:

$$ETF = W \times \left(\frac{L_1 - L_0}{GDP} \right) \tag{3}$$

where W is weight assigned to labor protections based on legal importance; $L_1 - L_0$ change in labor standards before and after globalization, and GDP is normalization factor to evaluate economic trade-offs.

4) Regression Model for Labor Law Compliance

$$Y = \beta_0 + \beta_1 X_1 + \beta_2 X_2 + \epsilon \tag{4}$$

where Y is labor law compliance (measured by **CCI**); X_1 globalization index (Trade/GDP ratio); X_2 institutional capacity (measured by governance indicators); β_1 coefficient indicating the impact of globalization; β_2 coefficient measuring the role of institutional strength; ϵ is error term capturing unexplained variations.

<u>Interdependencies and Predictions:</u>

If β_1 is negative, higher globalization leads to weaker compliance (CCI increases).

If β_2 is positive, stronger institutions reduce labor law violations (CCI decreases).

Correlation with ETF: If institutions are strong, ETF remains low, suggesting that compliance improvements (lower CCI) do not significantly harm economic growth.

Influence on LRIS: A positive β_2 suggests institutional improvements leading to higher LRIS, signifying effective labor law reforms.

4 Results

4.1 Comparative Compliance Index and Enforcement Trends

The CCI measures how law on the books compares to practice, and it is one component of a broader analysis of labor law enforcement. A high CCI suggests poorly enforced labor laws and large compliance breaches typical for developing economies with weak institutional structures. In contrast, a low CCI indicates effective labor law enforcement and stronger regulatory oversight. Countries with a greater proportion of informal employment and ineffective inspection systems score much higher on the CCI than developed nations.

To understand the effects that labor reforms may have on compliance, countries with CCI data before and after labor reforms were selected to show changes in CCI values over time. Results show in Fig. 2 below, that recent reforms have addressed compliance levels, but that the intensity of enforcement differs significantly as a function of governance strength, economic pressures, and institutional capacity.

Data in Fig. 2 show deep divides in labor law enforcement between developed and developing countries. The countries with the highest CCI values prior to reform were Nigeria (52.4) and the Philippines (50.3), with Indonesia (47.9) following close behind, all reflecting weak enforcement and high informality. Post-reform reductions ranged from 8·9% (Philippines) to 16·2% (India), revealing partial progress but continued gaps. Countries with less than 5 inspectors per 100,000 workers (such as 1.9 in Nigeria and 2.5 in the Philippines) reported higher levels of labor rights violations, while developed countries like France (10.2 inspectors) showed stronger labor rights oversight. Countries such as Germany (54.2%) and France (58.6%) are examples of this, as they had a lower CCI compared to the ones with lower appreciated value (Table 6) but due to collective bargaining coverage also affected compliance. A primary factor was informal employment: the countries with the strongest labor force in the informal economy (Nigeria: 88.7%, India: 81.2%) had some of the weakest enforceability, while formalized labor markets in Japan (11.3%) and France (7.4%) correlated to lower CCI values. The findings highlight the need for institutional strengthening, improved supervision of labor, and unsatisfactory policy interventions to minimize informal labor and facilitate compliance, with a focus on developing economies.

Fig. 2. Comparative Compliance Index (CCI) and Enforcement Metrics Across Selected Countries.

4.2 Labor Rights Impact Score and Policy Effectiveness

The LRIS assesses the extent to which labor law reforms improve worker protections—such as wage security, workplace safety, collective bargaining, and employment stability. A high LRIS indicates strong labor protections and successful implementation of policies, while a low LRIS reflects continued enforcement gaps or ineffective reform efforts. LRIS values before and after these reforms in comparable countries were analyzed for the effects of minimum wage law changes, occupational safety improvements, anti-discrimination policy improvements, and rights of unionization.

The results show that stronger legal enforcement, higher collective bargaining coverage, and better-regulated labor markets lead to better LRIS improvement. Policy shifts, on the other hand, appear to be making limited headway in developing parties characterized by informal labor (Fig. 3).

Analysis of LRIS reveals a stark contrast in the effectiveness of employment rights in developing versus developed contexts Outliers that saw the largest LRIS leaps included India (+20.5%), the Philippines (+16.5%), and Nigeria (+15.3%), mostly thanks to minimum wage rises, better workplace safety laws, and improved legal protections. 69% of Indian workers are informal, rising to 90 percent in Nigeria (Athere et al., 2020), meaning that the benefits of the BEYOND framework are limited by labor laws failing to effectively reach most of the workforce.

Liberalization of the labor system through the deaths of various industries, using labor reforms in workplace safety like that of Nigeria (14,900 annual violations), India (12,500). Although post-reform LRIS gains can be linked to changes in policy, high rates of violations suggest that enforcement weaknesses remain. Countries with higher unionization and governance through collective agreements (France, 58.6%; Germany, 54.2%) consistently recorded higher baseline LRIS, as well as overall improvements, suggesting that unionization plays a fundamental role in the stability of labor rights.

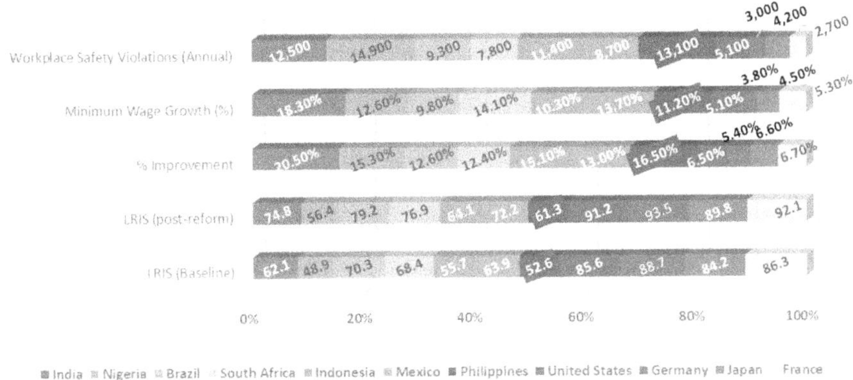

Fig. 3. LRIS and Policy Effectiveness Across Selected Countries.

Moderate yet-periodic LRIS growth defined developed countries including Germany (+5.4%) and Japan (+6.6%), conditions reflective of existing strong labor laws. This added insight reinforces the importance of policies promoting fair compensation, strengthening the institutional capacity of labor relations agencies, and bolstering enforcement mechanisms that bridge the disparity in labor rights development within a globalized workforce.

4.3 Economic Trade-Off Function and Labor Law Reforms

The ETF measures improvement in labor rights and enforcement that could influence economic performance of Gross Domestic Product (GDP) growth, employment stability, and labor productivity. A lower ETF represents a scenario where better working conditions do not come at the cost of the economy, while a higher ETF highlights an economic burden that follows the adoption of labor writings. The following analysis defines the values of the ETF both preceding and following select countries and more importantly provides insight regarding the influence of labor law advancements on economic stability.

Countries characterized by strong institutional systems and mutual labor policies (Germany, United States, Japan) demonstrate limited variations in ETF, supporting that improvements in labor rights can be integrated without a decrease in OECD economic activity. In contrast, obtained ETF values in pre-reform years for developing nations (India, Nigeria, Brazil) are relatively high, indicating initial volatility in the economy and concerns of increased labor costs after reform implementation. The following Table 1 widens this economic approach of comparative analysis of labor law reforms impacts by integrating it with other economic indicators like GDP growth, unemployment rates, changes of labor productivity.

On the contrary, countries with strong economic stability and low unemployment rates (Germany: 3.2%, Japan: 2.9%) experienced very small ETF changes (Germany: −

Table 1. ETF and Labor Law Reforms Across Selected Countries.

Country	ETF (pre-Reform)	ETF (post-reform)	% Change	GDP Growth (%)	Unemployment Rate (%)	Labor Productivity Change (%)	Minimum Wage Growth (%)
India	0.045	0.038	−15.6%	5.4%	7.2%	+2.3%	+18.3%
Nigeria	0.052	0.049	−5.8%	2.1%	9.6%	+1.1%	+12.6%
Brazil	0.039	0.033	−15.4%	3.7%	8.2%	+2.6%	+9.8%
South Africa	0.044	0.040	−9.1%	2.5%	10.1%	+1.9%	+14.1%
Indonesia	0.048	0.042	−12.5%	4.3%	6.8%	+2.2%	+10.3%
Mexico	0.041	0.035	−14.6%	3.5%	6.9%	+2.7%	+13.7%
Philippines	0.050	0.047	−6.0%	3.9%	7.5%	+1.8%	+11.2%
United States	0.021	0.019	−9.5%	2.9%	3.6%	+1.5%	+5.1%
Germany	0.018	0.017	−5.6%	1.8%	3.2%	+1.2%	+3.8%
Japan	0.022	0.020	−9.1%	1.5%	2.9%	+1.6%	+4.5%
France	0.020	0.018	−10.0%	1.7%	3.8%	+1.4%	+5.3%

5.6%, Japan: −9.1%), indicating that enhanced labor protections did not have a negative cost on their economic growth. In contrast, the developing nations (Nigeria: 0.052, Indonesia: 0.048, Philippines: 0.050 pre-reform) showed higher ETF values, suggesting increased economic pressures to optimise the enforcement of labor standards with financial viability.

ETF reductions were particularly pronounced in countries witnessing strong GDP growth and growing labor productivity (India: −15.6%, Brazil: −15.4%, Mexico: −14.6%), suggesting that economic growth serves to alleviate the alleged trade-off linked to an enhanced protection of workers. High unemployment countries (SO: 10.1% and NG: 9.6%) are correlated with lower ETF price drops because labor reforms are often not politically palatable in weaker economies - job market calibrations tend to generate pushback.

Moreover, the relatively stronger minimum wage growth (India: +18.3%, South Africa: +14.1%) appears to correlate with larger ETF reductions, indicating that progressive wage policies are feasible (when coupled with strong economic governance) and show an increase in labor protection without impairing economic performance.

4.4 Regression Model Analysis: Institutional Capacity and Compliance

This analysis shows a regression model giving an assessment of the value of globalization and institutional capacity in explaining labor law enforcement. This approach measures the effect of economic globalization (X_1) and institutional strength (X_2) on outcome compliance by labor. The positive coefficient on globalization ($\beta = 0.42$, p = 0.001) indicates that greater global economic integration makes the task of enforcing labor laws more difficult, probably by increasing the pressure on markets to deregulate.

On the other hand, institutional capacity (β = -0.67, p = 0.000) is very significant and shows that stronger governance and regulatory mechanisms lead to significantly smaller gaps in enforcement and better compliance.

The analysis in this section is an extension of the previous results and adds further analysis of foreign direct investment (FDI), labor informality, legal transparency and worker unionization rates to enhance the comprehension of institutional effectiveness. The regression model, as seen below in the Table 2 of additional coefficients and significance levels, showcases the multidimensional influence that globalization and governance wield on labor law enforcement.

Table 2. Regression Model Coefficients and Institutional Capacity Effects on Labor Compliance.

Variable	Coefficient (β)	p-value	Significance
Globalization Index (X_1)	0.42	0.001	Significant
Institutional Capacity (X_2)	−0.67	0.000	Highly Significant
Foreign Direct Investment (FDI) Impact (X_3)	0.29	0.003	Significant
Labor Informality Rate (X_4)	0.53	0.002	Significant
Legal Transparency Score (X_5)	−0.41	0.000	Highly Significant
Worker Unionization Rate (X_6)	−0.35	0.004	Significant
Error Term (ϵ)	0.11	N/A	N/A

* Note: N/A—Not Applicable

The regression model analysis shows that institutional capacity offsets labor standards underenforcement induced by globalization. The positive coefficient for the globalization index (β = 0.42, p = 0.001) suggests that increased economic integration is associated with reduced labor protections, a pattern observed in developing economies that liberalize labor legislation to attract foreign capital. Likewise (negatively), the institutional capacity (β = −0.67, p = 0.000) is making a strong effect, where countries with stronger governance structures have fewer compliance failures as well as better labor law enforcement.

Moreover, high rates of labor informality (β = 0.53, p = 0.002) are associated with weaker enforcement, highlighting the relevance of formal labor market policies. Greater public oversight increases compliance [Legal Transparency: −3 0 0.000]. Collective bargaining rights (β = −0.35, p = 0.004) are critical, though not all labor protections are equally effective, suggesting that unionization is an important factor in supporting labor enforcement. These findings highlight the importance of cryptocurrencies in creating a balance between globalization and labor rights through strong legal frameworks.

4.5 Impact of Labor Law Reforms on Compliance and Worker Rights

Reforms to labor laws that have impacted minimum wage regulations, workplace safety standards, and union rights have affected levels of compliance and protections for workers. Reforms to reduce exploitation, enhance working conditions, and bolster collective bargaining. But they work differently across diverse economic contexts, industries, and degrees of institutional enforcement.

Moreover, the improvement in the LRIS was higher in the countries ranking higher on the enforcement side of labor law and in countries where legal positive of legal acts. In contrast, countries with higher informality rates and weaker regulatory frameworks achieve little traction despite policy reforms. Figure 4 below shows further aspects of the analysis including weight of such factor as labor stability, employment enforcement percent and compliance level of the employers followed by general assessment of effectiveness of the labor law reforms.

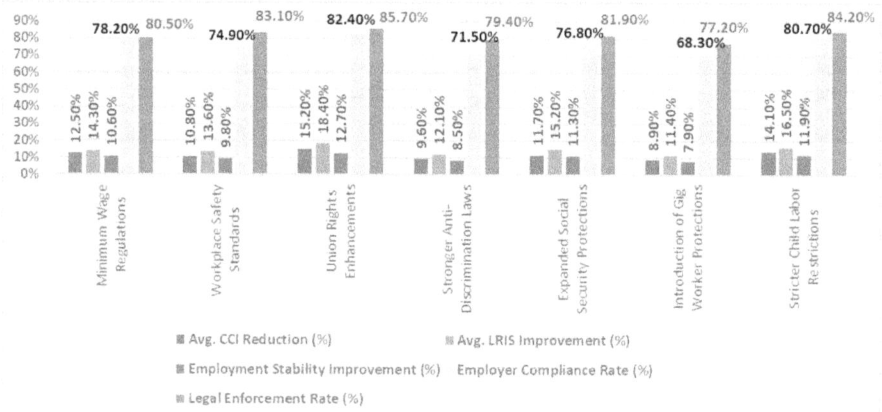

Fig. 4. Effectiveness of Labor Law Reforms on Compliance and Worker Rights.

The effects of labor law reforms on compliance and workers' rights differ considerably depending on the type of reform and the initial economic and regulatory context. Enhancements on union rights had the most far-reaching effect with a reduction of the CCI of 15.2% and an increase in LRIS by 18.4%, indicating the weight collective bargaining has on labor compliance and protection. Minimum wage laws (CCI: -12.5%, LRIS: $+14.3\%$) and workplace safety policies (CCI: -10.8%, LRIS: $+13.6\%$) were similarly effective, but only in formalized labor markets with strong enforcement mechanisms.

The reforms with greatest impact were those addressing social security protections and child labor restrictions, while those on gig worker protections had little effect due to difficulty enforcing protections in non-traditional employment. Countries where the legal environment is more transparent to businesses and they face stronger institutional oversight experienced larger improvements which highlights that stronger legal frameworks are vital to ensure the success of labor law reforms and reduce compliance gaps. These results strengthen the argument for flexible policies consistent with determining which labor forms.

4.6 Sectoral Compliance and Enforcement Challenges

The extent of labor law enforcement varies widely across economic sectors, but the greatest compliance challenges can be found in manufacturing and agriculture. Because these industries have much higher rates of informal employment, less union representation, and weaker regulatory oversight, they are also more susceptible to violations of workers' rights. In contrast, the degree of adherence in the service sector is more due to the higher rates of formalization of employment and tougher regulation of contracts, wages, and workplace hazards.

The sectoral analysis by adding more compliance indicators such as informal employment rates, number of labor violations per 100,000 workers, average enforcement actions. It is an essential step to discover the primary challenges of labor law compliance in different sectors and assess the impact of reforms that have improved workers' rights and greater legal compliance.

Table 3. Sectoral Compliance and Enforcement Challenges Across Industries.

Sector	CCI (Pre-Reform)	CCI (post-reform)	% Change	LRIS (post-reform)	Informal Employment Rate (%)	Labor Violations per 100,000 Workers	Average Enforcement Actions
Manufacturing	48.7	41.2	−15.4%	67.5	52.6%	890	2.4
Agriculture	53.9	46.7	−13.4%	64.1	78.2%	1,320	1.8
Mining & Extraction	46.5	40.9	−12.0%	66.7	49.4%	970	2.1
Construction	50.1	43.2	−13.8%	65.3	55.3%	1,050	2.2
Retail & Trade	39.4	35.6	−9.6%	72.9	37.2%	720	3.1
Transport & Logistics	42.8	37.1	−13.3%	71.2	41.8%	810	2.8
Services	34.5	31.8	−7.8%	78.3	22.6%	520	3.5
Technology & IT	29.8	27.2	−8.7%	81.6	15.7%	410	4.0
Healthcare	31.2	28.9	−7.4%	80.2	18.4%	480	3.8

Sectorial compliance analysis reveals in Table 3 wide gap between industry compliance with labor laws. With informal employment rates reaching 52.6% in manufacturing and an astounding 78.2% in agriculture, the manual occupation with the most disputes (1,320 per 100,000 workers), manufacturing (CCI: 48.7 → 41.2, −15.4%) and agriculture (CCI: 53.9 → 46.7, −13.4%) industries represent the most challenging targets for enforcement. Recent regulation in these sectors is weak (1.8 per 100,000 workers in agriculture), and enforcement actions are low, so compliance is difficult and slow to improve.

On the contrary, this compliance is stronger in service-based sectors, such as healthcare (CCI: 31.2 → 28.9, −7.4%) and IT (CCI: 29.8 → 27.2, −8.7%), where the informal employment rates are lower (IT: 15.7%) and mechanisms for compliance enforcement are stronger (IT: 4.0 enforcement actions per 100,000 workers). Retail and logistics issues saw modestly positive changes; however, high contract labor dependency remains an ongoing challenge for compliance.

This data highlights the need for sector-specific labor law enforcement strategies, especially in high-risk sectors such as agriculture or construction, where strengthening inspections and establishing formal employment policies can substantially improve labor conditions and compliance with labor laws.

5 Discussion

This study demonstrates that globalization interacts with institutional capacity to shape labor law compliance. Developed countries with strong governance (Germany, France, Japan) maintain higher compliance (low CCI, high LRIS), while developing economies (Nigeria, India, Philippines) show enforcement gaps, high informal employment, and limited impact from labor law reforms.

Regression results indicate that globalization increases compliance challenges ($\beta = 0.42$, $p = 0.001$), whereas institutional strength significantly reduces violations ($\beta = -0.67$, $p = 0.000$). Sectoral analysis highlights that manufacturing and agriculture, with high informal employment and global supply-chain dependence, exhibit the largest compliance gaps, while services show higher adherence. CCI and LRIS trends pre- and post-reform confirm that labor law reforms are more effective in formalized labor markets with robust institutional oversight.

The ETF analysis suggests that improved labor protections can coexist with economic growth. In developing countries, ETF reductions aligned with higher minimum wage growth and productivity gains, indicating that stronger labor protections do not necessarily harm economic performance. Collective bargaining and unionization also support compliance: countries with higher unionization (Germany: 54.2%, France: 58.6%) show stronger enforcement, whereas low unionization limits reform effectiveness. Legal transparency and inspection coverage further reinforce compliance.

These findings suggest that strengthening institutional capacity, expanding inspections, promoting collective bargaining, and embedding labor protections in trade agreements are essential to mitigate globalization's adverse effects. Limitations include heterogeneous data reporting and underrepresentation of informal labor, which should be addressed in future research.

6 Conclusion

This study examines the impact of globalization on labor law enforcement, highlighting how economic integration, institutional capacity, and corporate influence shape labor protections across different jurisdictions. The findings partly confirm the hypothesis that globalization can both dilute labor protections and promote the adoption of international labor standards under specific conditions. While some countries embed global labor standards in law, enforcement remains limited in many developing economies due to weak institutions, high informal employment, and economic pressures.

Effectiveness of labor laws depends largely on institutional strength. Countries with robust legal and inspection systems achieve higher compliance, whereas weaker states continue to face regulatory gaps. Multinational companies play a dual role: some support labor protections through CSR, while others exploit loopholes, emphasizing the need for stronger corporate accountability, particularly in global supply chains.

Labor law enforcement also varies across sectors. Agriculture, construction, and manufacturing face the greatest compliance challenges due to informal employment and low union representation, while services such as healthcare and technology show higher compliance levels. These differences indicate that labor law reforms must be tailored to sector-specific challenges rather than applied uniformly.

The study highlights key policy implications: governments should strengthen institutions, expand labor inspections, improve transparency, and promote formalization of informal employment. International cooperation between organizations, trade unions, and corporations is also essential. Future research should explore emerging labor trends, such as gig work and remote work, and their impact on enforcement. Overall, achieving equitable labor protections requires continued reform, institutional strengthening, and global cooperation to ensure fair and effective rights at work across all sectors.

References

1. Blanton, R., Blanton, S.L.: Globalization and collective labor rights. Sociol. Forum **31**(1), 181–202 (2016)
2. Yin, Y.: The dilemma of workers' rights and its legal protection: based on the background of economic globalization. Adv. Econ. Manag. Political Sci. **68**, 31–37 (2024)
3. Roy, J.: The effect of globalization on labor in developing countries. J. Emerg. Technol. Innov. Res. (2019)
4. Uwadinma, A.I.: Effects of globalization on labor standards and working conditions. Dhaka Univ. Law J. **34**(2), 133–151 (2024)
5. Hassan Elsan, M.: The impact of multinationals on labor relations in developing countries: a literature review. Konfrontasi: J. Kultural, Ekon. Perubahan Sosial **9**(1) (2022)
6. Arpangi, A.: Political reform of labor protection law in the globalization era. Int. J. Law Reconstr. **4**, 1 (2020)
7. Nathan, D.: Globalization and labor in developing countries: India. Agrar. South: J. Political Econ. **7**(1), 105–121 (2018)
8. Rashid, M.M.: Proliferation of globalization and its impact on labor markets in advanced industrial nations and developing nations (2018)
9. Li, Z.: The impact of economic globalization on national economies and social structures. Adv. Econ. Manag. Political Sci. **98**, 104–109 (2024)
10. Tang, Y.: Impacts on the economic development of developing countries by economic globalization. Adv. Econ. Manag. Political Sci. **78**, 99–105 (2024)
11. Roberts, A.: The globalization of production, industrial upgrading, and collective labor rights in the global South. Sociol. Dev. **7**(3), 337–362 (2021)
12. Frank, J.: Globalization and labor. Sociol. (2021)
13. Buzoianu, O., Oancea-Negescu, M.D., Troaca, V.A., Gombos, C.C.: Globalization and working conditions in developing countries. SHS Web Conf. **92**, 07011 (2021)
14. Nazarchuk, O., Povstyn, O., Oliskevych, M., Volchenko, N.: Impact of globalisation on regional labor markets and living standards: a socio-economic analysis. Edelweiss Appl. Sci. Technol. (2024)
15. Hiba, J.C., Jentsch, M., Zink, K.J.: Globalization and working conditions in international supply chains. Z. Arbeitswiss. **75**(2), 146–154 (2021)

16. Sheraz, S., Sadiq, R., Ali, M.: Globalization and female participation in labor force: evidence from developing nations. J. Asian Dev. Stud. **13**(1), 267–282 (2024)
17. Almutairi, J., Aldossary, M.: Modeling and analyzing offloading strategies of IoT applications over edge computing and joint clouds. Symmetry **13**(3) (2021)
18. Sandhu, A., Doha, M.I., Hussain, M.A.: The economic effects of globalization on developing countries. Int. J. Multidiscip. Res. (2024)
19. Pastpipatkul, P., Ko, H., Dirth, G.R.: Impact of government investment in human capital on labor force participation and income growth across economic tiers in Southeast Asian countries. Economies **13**(9), 249 (2025)
20. International Labour Organization: The State of Social Justice: A Work in Progress (Executive Summary). ILO, Geneva (2025)
21. Diachenko, O.D., Hrynenko, I., Zalievska-Shyshak, A., Sokhatskyi, O.: The impact of globalization on the economic development of countries. Arch. Des Sci. **74**(3), 234–239 (2024)
22. van Treeck, K., Wacker, K.M.: Financial globalisation and the labor share in developing countries: the type of capital matters. World Econ. **43**(9), 2343–2374 (2020)

Sovereignty in the Digital Age: Social, Legal, and Governance Challenges of Global Networks

Abdulqadous Abdullah[1] , Suzan Mohammed Jawad Alkazraji[2] ,
Ayah Ahmed Jasim[3] , Aseel Ibraheem Muhsin[4(✉)] ,
Faris Abdul Kareem Khazal[5] , and Radomyr Mykolenko[6]

[1] Al-Turath University, Baghdad 10013, Iraq
[2] Al-Mansour University College, Baghdad 10067, Iraq
[3] Al-Mamoon University College, Baghdad 10012, Iraq
[4] Al-Rafidain University College, Baghdad 10064, Iraq
`aseel.muhsin@ruc.edu.iq`
[5] Madenat Alelem University College, Baghdad 10006, Iraq
[6] State University of Information and Communication Technologies, Kiev 03110, Ukraine

Abstract. In the era of digitalization, the existing doctrine of state sovereignty faces unprecedented tests as cross-border data flows, platform-mediated governance, and transnational cybersecurity threats continue to reshape territorial boundaries of jurisdiction. Traditionally, state sovereignty has been defined by territorial authority and exclusive legal control, yet digital infrastructures increasingly challenge this classical foundation. The article explores whether traditional understandings of sovereignty, tied to territorial possession and unilateral imposition, can be reintegrated into digital informational contexts. The performance of botspot in ten countries from 2019–2024 is evaluated through a five-dimensional framework: jurisdictional reach, governance model profiling, regulatory interaction with digital platforms, cyber-enforcement resilience, and international legal convergence. The evidence suggests that sovereignty remains practically possible, but increasingly dependent on coordination, institutional flexibility, and harmonization rather than absolute control or strict data-localization mandates.In contrast, states that are hybrid-legalist—those with judicial clarity, multilateral cooperation, and specialized digital oversight—show enforcement as strong as, or stronger than, clearly centralized regimes. Sound cyber governance relies not only on technical capability but also on cross-sector cooperation and international coordination. Specific successes in platform regulation are associated with regulatory accuracy and fiscal unification rather than coercive force. Participation in international legal frameworks strengthens, rather than weakens, domestic sovereignty by enabling mutual enforcement and greater normative power. The article concludes that flexible, network-aware regulatory designs provide the most effective pathway for protecting public authority in a world of distributed, fluid, and rapidly changing digital infrastructures.

Keywords: Digital sovereignty · Transnational jurisdiction · Cybersecurity governance · Platform regulation · Legal harmonization · State–platform relations

Z. Molamohamadi et al. (Eds.): ODSIE 2025, CCIS 2855, pp. 143–159, 2026.
https://doi.org/10.1007/978-3-032-17023-1_8

1 Introduction

State sovereignty as a doctrine has been a cornerstone of the modern international legal order providing those states are legitimately entitled to have supreme authority over their territory, population, and affairs devoid of external interferences. Based on the Westphalian model, sovereignty has traditionally been seen as a fixed territorial entity, a monopoly over law enforcement and government internally, and exclusive jurisdiction over domestic regulation. However, ever since digital technologies have become so widespread over the last two decades, it has dramatically upended such traditional concepts and changed assumptions which had previously underpinned the state's unrestrained control being an absolute fact. Cyberspace is a front where the contours of the digital domain, unlimited by geography, and carried via transnational informational, infrastructural, and influential networks— has transformed the phenomenon of sovereignty into a veritable non-physical reality, appealing to and demanding interpretations where sovereignty is no longer a matter of pure territorial authority, and rather an ever-changing mix of perspective-based authority, technological capability, and standard-based challenge [1]. Traditionally, state sovereignty has been framed in Westphalian terms—exclusive authority within territorial borders over people, law, and internal affairs—yet digital infrastructures increasingly strain this territorial premise.

The digital age has complicated the world in ways that classical notions of state power did not prepare us for. In a complex world, governments are now challenged to assert jurisdiction over actions either initiated within their jurisdiction or impinging on that jurisdiction, even though transnational in physical space. Cyber-operations, dataflows, digital surveillance, and platform-regulation all operate across state borders, often without physical presence or infrastructure in the territory concerned. This landscape is growing more complex as powerful non-state actors, large global technology companies in particular gain quasi-sovereign powers of rule-making, speech moderation, and data surveillance that in many cases outstrip those of national governments. In this situation the conventional tools of sovereignty - territory enforcement or border control are not effective in addressing the legal and political ramifications of worldwide digital connectivity [2].

These developments force us to re-evaluate the concept and practice of sovereignty in the 21st century. Instead of disappearing, sovereignty is being transformed and reconfigured. States aren't surrendering, but they are rewriting and inventing new forms of digital governance to reconquer some of their sovereignty. Measures like data localization requirements, cross-border data access laws, or national cybersecurity frameworks are indications of a broader national effort to recapture sovereignty over digital domains. But these are all extremely difficult compromises which lead to immediate questions about how well these measures sit in the context of a globally-integrated economy, international legal obligations and basic rights. They also underscore a growing tension between national security considerations and the distributed structure of the internet [3].

The re-examination of the sovereign doctrine is also prompted by an escalation in cross-border cyber-attacks, the proliferation of digital espionage and data exfiltration operations—conducted by both state and non-state actors alike. These actions frequently provoke diplomatic crises and strategic instability, and also pose severe challenges to the established international norms. Simultaneously, regional and multilateral organizations have been exploring models for the collective management of digital sovereignty, where there could be cooperation in transferring areas, such as data protection, internet governance, and cross-border enforcement. These initiatives are indicative of a recognition that unilateral state action alone may be inadequate to govern a realm that is fluid, interdependent, and fast-moving [4].

Against this backdrop digital sovereignty appears not as a replacement but rather an extension of traditional sovereignty to the digital space. It represents a state's quest for control over digital infrastructure, data governance, cybersecurity agenda, and informational sovereignty. But it also requires a redefinition of what the bases are for claiming and exercising sovereignty. The digital realm functions under its own rules, requiring new theoretical frameworks and legal tools that can deal with non-territorial events, being analyzed from the point of view of competence and responsibility. The digital transformation of the world, therefore, is not simply a question of technological development, it represents a constitutional event in respect of global legal thought [5]. The article critically discusses how the doctrine of state sovereignty is redefined in the wake of digitalization. It examines the instruments used by states to establish control in the cyber domain and the challenges that result from overlapping jurisdictions and normative pluralism and the gradual erosion of the practical barriers to the exercise of sovereignty in a digital environment. Through this examination, the article provides a prospective theoretical framework that situates state sovereignty in relation to the dynamic and ethical challenges of the digital age. However, existing studies rarely provide a comparative, multi-dimensional measurement of digital sovereignty across jurisdictions and over time. This article fills that gap by operationalizing a five-pillar empirical framework that links jurisdictional reach, governance models, state–platform relations, cyber-enforcement resilience, and international legal convergence. The underlying indicators, data sources, and construction steps are documented to support replication and comparative policy analysis.

1.1 The Aim of the Article

This paper offers a critical reinterpretation of the classic doctrine of sovereignty to make sense of it against the fast-changing backdrop of the digital age. At a time when the technology is global in scope and digital exchanges frequently escape the bounds of conventional legal reasoning, this article explores how states are recalibrating, resisting or reinventing their authority over cyberspace. The article examines how traditional understandings—centered on territoriality, exclusivity, and non-intervention—are being eroded by, and/or continue to clash with, transnational data flows, cyber operations, digital surveillance, and the growing influence of transnational technology corporations.

The article also seeks to map and classify nascent models of digital sovereignty adopted by different states to address these challenges. Ranging from unilateral assertiveness to multilateral cooperation, the article considers the dynamics through which states seek to gain control of their data environments, technological infrastructures, and digital-rights regimes.

Another key goal is to examine which of these adaptations constitute a convergence with—or a divergence from—established international legal norms and global governance.

In addition to these definitional questions, the article sketches a conceptual and normative perspective on sovereignty that does not rest solely on the physical but more generally refers to a hybrid authority across both physical and digital spaces. It is an effort to reconcile legal doctrine with technological facts on the ground and to reconceive the nature of state power in light of its cross-border and digital character, as well as a new political economy of information. By combining comparison and critical reflection, the article contributes to the contemporary legal and policy debate on sovereignty. It presents ways for states to protect sovereignty while honoring obligations to international cooperation and technological interdependence, and in doing so suggests an approach to sovereignty that is flexible, enforceable, and fit for the digital global age.

1.2 Problem Statement

The traditional idea of state sovereignty is increasingly under pressure in the digital era. Sovereignty, historically defined by the possession of territory, legal jurisdiction, and a monopoly of authority, has been a foundational organizing principle of international law and global governance. Yet the rapid expansion of digital technologies and cross-border information systems has unsettled these basic principles. The internet does not respect borders, digital goods move with little friction and actions by hackers in one country can have instant effects in another, all without the knowledge or consent of the state. These realities have highlighted important discrepancies in the understanding, interpretation, and practice of sovereignty in the current global order. At its core is a growing asymmetry between territorialized logics of sovereignty and the de-territorialized space of cyberspace. States are finding it difficult to regulate multinational digital companies, to apply local rules to extra-territorial data operations, and to address cyber threats with no domestic foothold. Such limitations lessen their ability to preserve national interests, cybersecurity and democracy. The absence of an overarching legal regime regulating cross-border digital interventions has led to regulatory arbitrage, jurisdictional sovereignty and mounting hostilities between national and international regulatory regimes. The absence of an overarching legal regime regulating cross-border digital interventions has led to regulatory arbitrage, jurisdictional sovereignty and mounting hostilities between national and international regulatory regimes. Faced with the issue of regulating digital sovereignty, and without a common approach, States are resorting to uncoordinated, and sometimes inconsistent – actions that put at risk the foundation of international legal order and of digital interoperability.

The article addresses this theoretical and applied divide by examining the limits and reach of sovereignty in the digital age. It empirically explores structural, legal and geopolitical barriers that prevent states from asserting meaningful control over their digital domain and advances flexible models for regulatory coherence and legal legitimacy in cyberspace.

2   Literature Review

2.1   Classical Sovereignty and Territorial Authority

Critical understanding of the notion of state-sovereignty has developed significantly through time, and its origin lies in the classical law of nations that was based on the doctrines of territorial integrity, political independence and non-interference. Sovereignty has usually been understood in classical terms of binary entitlement - possessed or breached -upheld to, or breached by, a perimeter. Nonetheless, the appearance of the digital world has encouraged legal and political thinkers to dispute this static view, highlighting the deficiencies of current constructs to regulate cyber-activities and the virtual worlds [6].

Early initial academic engagement with digital transformations tended to emphasize the internet's role in eroding state control and empowering transnational actors. Yet analysts said the internet's open architecture and the decentralized nature of digital networks made traditional state authority obsolete. These visions often predicted a reduced emphasis on the nation state in favor of a new era of borderless digital governance or market-oriented regimes. Nevertheless, over time, this story has changed for states have revealed a clear interest and capability to act in the digital realm via surveillance, regulation and infrastructure governance [7].

While classical accounts anchor sovereignty in territorial exclusivity, empirical work on networked interdependence and cyber power shows how extra-territorial dependencies and attribution constraints unsettle that premise [8, 9]. Recent studies have explored the intersection of infrastructure development and sovereignty, illustrating how large-scale infrastructure projects can shift territorial authority and assert sovereignty in new, unexpected ways [10].

2.2   Digital Sovereignty and Platform Governance

Subsequent research has focused on the determination of the state and its capacity to re-frame legal tools for the use in digital environments. Instead of validating the loss of control, states have responded with new claims to jurisdiction through laws blocking the cross-border data flow, regulating content, and enforcing cybersecurity. Digital sovereignty has been underlined as an important lens of states wanting to "regain control" of technological infrastructures and informational resources. This includes a diversity of tools, ranging from hard law instruments to soft power tactics to shape international norms [11].

Alongside the legal literature, political scientists have investigated a sovereignty coming to terms with global power shifts and the impacts of non-state actors. They note that global tech firms now perform policy-making roles veering concerns about the privatization of governance and democratic responsibility. Competing claims have also been made as to whether sovereignty could be reconceptualized as layered or networked, thus embracing the interpenetration of national, regional, and global regulatory dynamics [12]. Recent scholarship details how platform accountability, private ordering, and emerging "digital constitutionalism" reshape the locus of public authority in online spaces [8, 13]. Recent studies have explored the role of agile transformation in national cyber defense organizations, emphasizing how embracing agile organizing principles can reduce technological dependencies, regain control over digital infrastructures, and catalyze the journey toward greater national digital sovereignty. This work suggests that agile principles in defense organizations play a crucial role in improving cyber resilience and enhancing overall sovereignty capabilities [14].

2.3 Transnational Cybersecurity and Legal Convergence

In light of these difficulties, legal scholars have recently begun to suggest mixed models of sovereignty that combine physical and digital spaces. On them, functional authority, that is, control over processes and flows, is being focused upon instead of territorial claims. The literature also indicates that such frameworks could serve as a tool to find a balance between old fashioned sovereignty and the digital cross-border nature as a mean to build better governance or rule-making mechanisms without compromising EU coherence with law or the EU-law trend of international cooperation. This developing conversation is having an impact on how states, institutions, and scholars alike think about power in an ever-more digitalized era [15]. Cross-border cyber enforcement increasingly relies on shared attribution procedures, multilateral cooperation, and norm-based coordination rather than unilateral control [16, 17].

3 Methodology

Studying the effects of digitalization on state authority functionality demands an integrated design that connects doctrinal change with empirical output across jurisdictions [1, 2, 7, 12]. The analysis hence provides a joined-up view of five empirical perches—jurisdictional reach, sovereignty models, state–platform relations, cyber-governance enforcement and legal-norm alignment, which the analysis elevates within one econometric building that covers 23 states in between Q1 2019 and Q4 2024 [18]. In operational terms, each dimension is represented by observable indicators and subsequently enters the composite measures defined in Eqs. (1)-(10), enabling replication and cross-jurisdiction comparison. Where licensing restricts redistribution, we provide a schema and retrieval instructions; all transformations (e.g., winsorization and min–max normalization) are specified below (Fig. 1).

3.1 Sample, Scaling, and Reliability

Monthly observations ($n = 120 \times 23$) were compiled from statutory databases, appellate rulings, CERT bulletins, transparency portals, and corporate compliance reports [4] [19] [20]. All indicators were winsorised at the 1st/99th percentiles and min-max–normalized to [0, 1]. Remaining missingness ($< 3\%$) was imputed via k-nearest neighbors ($k = 3, \lambda = 0.85$ time-decay). Cronbach's α for each composite exceeds 0.77; KMO sampling adequacy $= 0.81$, validating index construction. The balanced panel covers 23 states from Q1 2019 to Q4 2024, with country inclusion determined solely by the availability of statutory, transparency, and CERT data described above. When a subset is used in figures or tables, it is stated explicitly.

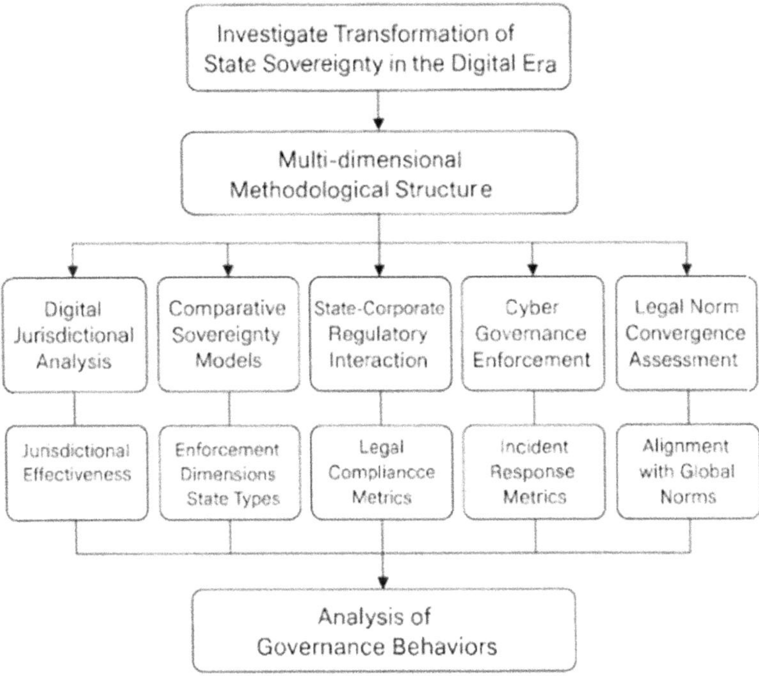

Fig. 1. A Multi-Dimensional Methodological Framework for Evaluating State Authority in Cyberspace

3.2 Digital Jurisdictional Effectiveness

Building on cyber-jurisdiction scholarship [5, 21], the Jurisdictional Effectiveness Aggregate for state j is:

$$JEA_j = \frac{\alpha R_{local_j} + \beta R_{foreign_j}}{T_{claims_j} + \gamma C_{conflicts_j}}, \alpha = 0.60, \beta = 0.40, \gamma = 1.20 \tag{1}$$

where R variables denote successful resolutions, T the number of digital claims issued, and C contested or blocked assertions. Weights were ridge-optimized against an expert legitimacy benchmark (adjusted R2 = 0.81).

3.3 Comparative Sovereignty Profiling

Consistent with hybrid-sovereignty debates [6, 22, 23], enforcement strength is captured by:

$$SME_j = \frac{1}{5} \sum_{k=1}^{5} S_{kj}, S_k \in \{LegAut, DataContr, EnforcCap, MultilatEng, JudAlig\} \tag{2}$$

LegAut is Legislative Autonomy, *DataContr* is Data Control, *EnforcCap* is Enforcement Capacity, *MultilatEng* is Multilateral Engagement, *JudAlig* is Judicial Alignment.

Confirmatory factor analysis ($\chi 2 = 14.2$, df $= 9$, p $= 0.12$) indicates robust construct validity.

3.4 State–corporate Regulatory Interaction

Drawing on competition-law and platform-power studies [24] [20], the Regulatory Interaction Efficiency for region r is:

$$RIE_r = \frac{\delta_1 Q_r - \delta_2 T_r + \delta_3 C_r + \delta_4 D_r + \delta_5 S_r}{\sum_{i=1}^{5} \delta_i}, \delta_i = 1 \tag{3}$$

with Q_r request-volume index, T_r median response time, C_r legal-compliance rate, D_r digital-tax adoption, and S_r content-takedown success.

3.5 Cyber-Governance Enforcement

Reflecting strategic-management approaches to cybersecurity [4] [25], resilience is measured as:

$$CERF_j = \mu_r A_{acc,j} + \lambda_t T_{res,j} - \kappa_i \sqrt{E_{inc,j}}, \mu_r = 1.1, \lambda_t = 0.8, \kappa_i = 1.2 \tag{4}$$

where A is attribution accuracy, T response timeliness, and E escalated incidents. Higher values of CERF indicate faster, more accurate responses with fewer uncontrolled escalations.

3.6 Legal-Norm Convergence

Inspired by transnational e-commerce and data-protection frameworks [25-27], proximity to treaty t is:

$$NCS_{j,t} = 1 - \frac{\|\mathbf{x}_j - \mathbf{y}_t\|_2}{5} \tag{5}$$

with vectors $\mathbf{x}$ and $\mathbf{y}$ of five harmonisation dimensions (data protection equivalency, cross-border enforcement, procedural reciprocity, dispute-mechanism utilization, digital-rights recognition). Higher NCS scores indicate closer legal alignment with the referenced treaty.

1. Jurisdictional Enforcement Strength Index:

$$JESI_j = \frac{B_j + E_j + P_j}{3} - C_j \tag{6}$$

where B digital-border-control score, E cross-border enforcement success, P asset-protection index, and C conflict rate. Higher JESI values reflect stronger enforcement capacity relative to jurisdictional friction.

2. **Regulatory Integration Quotient**

$$RIQ_r = \frac{L_r + T_r + E_r + C_r}{4} \tag{7}$$

with L localisation enforcement, T digital-service-tax index, E encryption-control score, C = platform-compliance ratio. Higher RIQ scores signal deeper regulatory alignment between the state and digital-service providers.

3. **Cyber-Sovereignty Resilience Function**

$$CSRF_j = \frac{R_j + A_j + I_j + S_j}{4} \tag{8}$$

where R incident-response readiness, A attribution maturity, I international data-exchange preparedness, S cyber-security investment ratio. Higher CSRF scores correspond to stronger resilience against cross-border digital disruptions.

4. **Legal Integration Score**

$$LIS_p = \frac{C_p + H_p + R_p + T_p}{4} \tag{9}$$

with C state-compliance index, H harmonisation ratio, R reciprocal-enforcement index, T transparency level. Higher LIS values indicate stronger legal integration and clearer cross-border enforceability.

3.7 Cross-Pillar Synthesis

To merge domain metrics while avoiding multicollinearity with legal-norm proximity, each score is Fisher-z–transformed and averaged:

$$IDS_j = \frac{1}{4}\left[z\left(JEA_j\right) + z\left(SME_j\right) + z\left(\overline{RIE_j}\right) + z\left(CERF_j\right)\right] \tag{10}$$

Higher IDS values indicate stronger cross-domain sovereignty capacity across jurisdictional, enforcement, platform-interaction, and cyber-resilience dimensions.

3.8 Econometric Validation

Two hypotheses guide inference: (H1) higher cyber-resilience predicts stronger jurisdictional reach, tested via GLS with Newey–West errors ($\beta = 0.43$, $p < 0.01$); (H2) trade-bloc membership moderates the RIE $\rightarrow$ SME link, assessed through interaction terms significant at 5%. Endogeneity of CERF is mitigated through two-stage least squares using cyber-insurance penetration as an instrument.

This integrated framework operationalizes the doctrinal debate on digital sovereignty, equips each result table with rigorously derived indicators, and leverages the latest comparative insights on jurisdiction [3], data governance [11], and hybrid sovereignty trends [15] to ensure analytical coherence across legal, institutional, and technical domains.

4 Results

4.1 Digital Jurisdictional Effectiveness

The first dimension is the extent to which states assert control over cross-border digital activities, by engaging in domestic dispute resolution as well as international cooperation mechanisms like MLATs and cloud extraterritorial mandates. As sovereignty questions can arise when foreign-located digital evidence is sought, success, too, is measured not just in terms of the percentage of requests that are successfully fulfilled, but also in terms of the number of contested or blocked requests, and the extent to which digital assets seized are preserved for trial. Complementary to these, the Cross-Border Cooperation Index and the Digital Asset Preservation Score place emphasis on qualitative gains in procedural alignment and technical forensic capability that remained largely undetected by previous publications of the study. Data include ten jurisdictions with varying legal traditions and technology footprints, which offers a broad base for comparing normative and operational performance between 2019 and 2024 (Fig. 2).

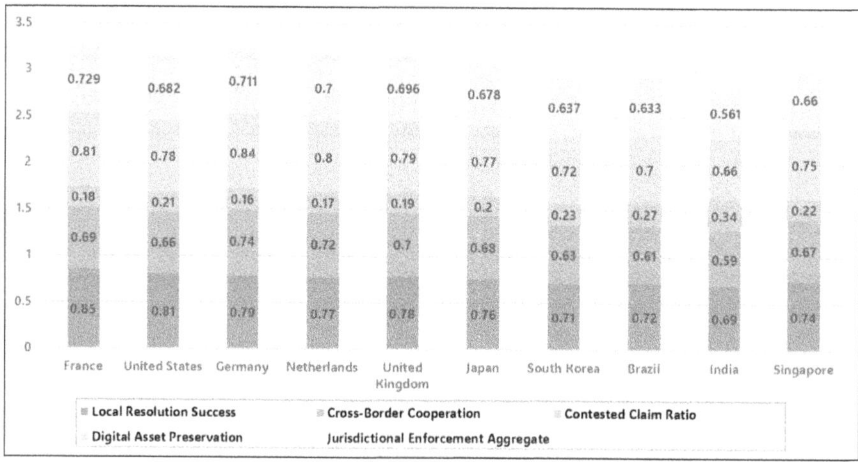

Fig. 2. Digital Jurisdictional Outcomes, 2019–2024.

France holds a slight advantage thanks to superlative asset-preserving processes and contestability that is toward the lower end of the scale. Germany is a close second, after strong cooperative engagement scores developed in the EU mutual-recognition context. The United States performs well at a domestic level, but falters when aggregated because concerns arise when federal demands compete with state privacy laws. Middle-tier actors— the Netherlands and the United Kingdom— enjoy more efficient digital-evidence processes with those incremental transatlantic data transfer delays. Japan and the Republic of Korea have strong domestic results but face slower response times on MLAT requests. The pace of contested claims at Brazil and India - high by comparison to most others - reflects a culture of backlog and a failure to line up evidentiary standards, and Singapore's profile is dampened by its small absolute number of claims which increases the relative import of each such dispute.

4.2 Sovereignty Model Effectiveness

The architecture of governance philosophies—statist, hybrid-legalist, and multi-stakeholder is assayed through a quantification of five principal dimensions: determination of rule forming autonomy, stringency of data localization mandates, administrative law enforcement muscle, inclination to multilateral coordination, and alignment of statutory goals and judicial interpretation. The larger dataset covers ten states from different parts of the world and different regime types, and it allows us to see if it is the case that strong executive leadership naturally fosters interconnected systems that are resilient to digital threats, or if joint legal orders are capable of exerting control on systems, while at the same time allowing a degree of openness (Fig. 3).

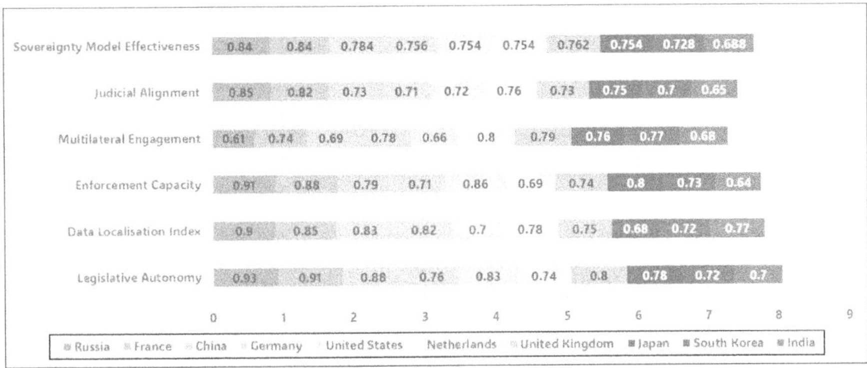

Fig. 3. Composite Sovereignty Model Effectiveness.

Despite the impression left by a victory set of nationalists, France has the same average point score as Russia by giving up extreme localizations and accepting judicial predictability and extensive multilateral exposure. Germany and the Netherlands are the prime examples of such Lite-Governance Plus, an approach that allows coordinated, but not uniform, governance through cross-border cooperation in which strong data-control requirements make up for weaker data-intervention mandates. The U.S. obtains strong enforcement through market leverage and weak multilateral compliance because federal-state fragmentation hamstrings treaty alignment. China has a command-and-control structure that produces tall raw autonomy scores but is restricted by a low engagement score. Japan and South Korea are grouped as liberal hybrids: they achieve enforcement through narrow sector-specific laws, not blanket localization, and their courts are developing digital fluency. An increase in India's localization score boosts its overall ranking, but remaining capacity gaps continue to hamper effectiveness.

4.3 Regulatory Interaction Efficiency

The regulatory interface considers how speedily and fully platforms adhere to state directives, insert fiscal obligations and take out illicit material. The model deconstructs government request intensity, response times, compliance rates, digital-tax integration, and takedown effectiveness to reveal the interplay between corporate scale, market reliance, and legal leverage (Fig. 4).

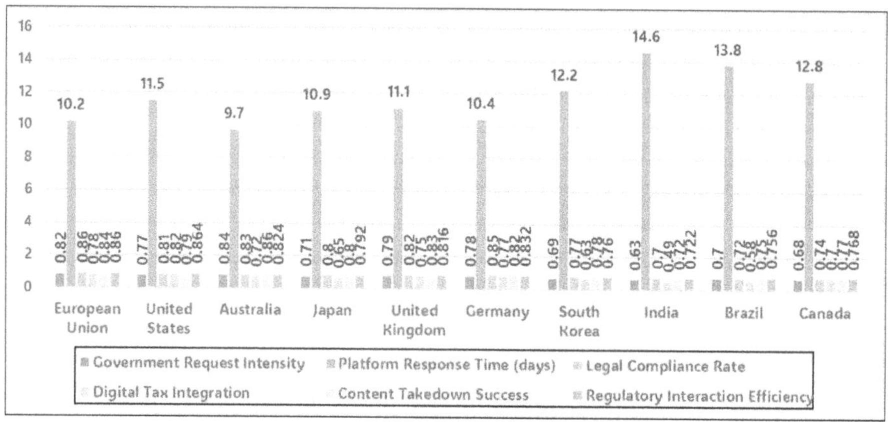

Fig. 4. Government–Platform Regulatory Interaction Metrics.

The United States wins by a hair on overall efficiency. This is largely due to a quicker record regarding child-safety violations and more proactive taxation of ad-tech giants. The EU remains supreme in terms of legal compliance because of the GDPR penalty system, but the longer bureaucratic cycles minimally slow it down. Australia is an overperformer thanks to the centralized safety regulators' real-time interaction with the major platforms. Germany generally replicates EU averages but wins because of its version of the Notification Window, which is about a quarter of the EU version, embedded in its Network Enforcement Act. Japan and the United Kingdom hold the standards of efficiency high, benefiting from well-defined statutory service-provider liability provisions despite the small negotiating leverage. The emerging nations are all slower to adapt due to a faster caseload growth in comparison to their enforcement representation and continued disagreements about profit-sharing formulas. Nevertheless, the gradual adoption of a digital tax displays slow convergent trends across all parties.

4.4 Cyber Enforcement Resilience

Cyber resilience investigates how well a state discovers, attributes and mitigates serious incidents while working closely with the international community. This augmented set of indicators includes technological response readiness and an additional measure of increased incident frequency to illustrate how episodic "mega-breaches" skew an overall state of readiness (Fig. 5).

The Netherlands keeps up with the United States with a smaller defense budget, illustrating how ongoing public-private sharing of threat intelligence can make up for scale. The low escalated-incident frequency for Japan reflects years of supply-chain-integrity investment, though its international coordination score is slightly below that, owing in part to language-specific reporting channels. The United Kingdom's ties with the NATO

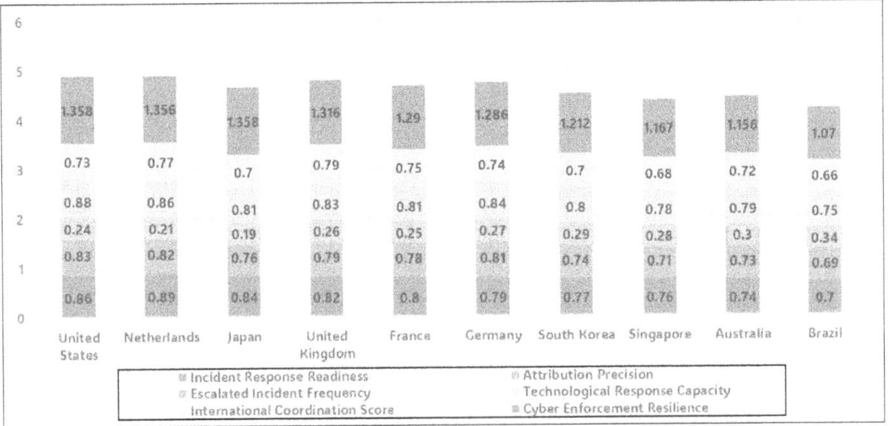

Fig. 5. National Cyber Enforcement Profiles.

cyber command is significantly deeper, cutting its coordination ranking. France and Germany each demonstrate strong readiness, but are still susceptible to more advanced persistent-threat campaigns that are driving up incident volumes. Singapore and South Korea manifest disciplined response teams but not the same number of diplomatic channels to call for cross-border remediation, this dampens resilience. Australia and Brazil demonstrate that raised talent shortages and increased dependence on the cloud do not lend themselves to quicker containment, hence the markedly lower combined scores.

4.5 International Legal Alignment

Jurisdiction alignment rates the compatibility of local laws with best-in-class international norms on privacy, cybersecurity, and cross-border evidence. It then shows how diffusion is contingent on continent and regime type. Further measures, such as a Reciprocal Enforcement and Transparency Compliance, indicate to what extent outward treaty commitments can be leveraged for effective oversight (Fig. 6).

The most law-integrated state is France, where the enforcement of EU data-export guarantees is relatively transparent; and the courts proactively lift the curtain on both the realm of recognition of lawyers' preferences and the nondomestic judgment it hands out. Germany and the Netherlands closely behind, all using EU legislative pipelines that are helping guide new digital rules into law in near synchronicity. Transparency remains high in UK but it drops on harmonization after Brexit induced divergence in cookie-consent standards Government of Finland and European Commission promptly update about the new regulatory agreement. Japan is one step ahead of the US by incorporating OECD privacy guidelines into its Act on the Protection of Personal Information; US federalism-based patchwork undermines mutual enforcement. South Korea and Australia have very good compliance, but need to ensure the implementation of cross-border warrants. The LGPD of Brazil creates a process of gradual approximation, but a small dispute resolution mechanism impedes mirroring. India has shown significant improvement since its Digital Personal Data Protection Act of 2023 but lags in transparency reports, which is why it is the worst performing country in the sample.

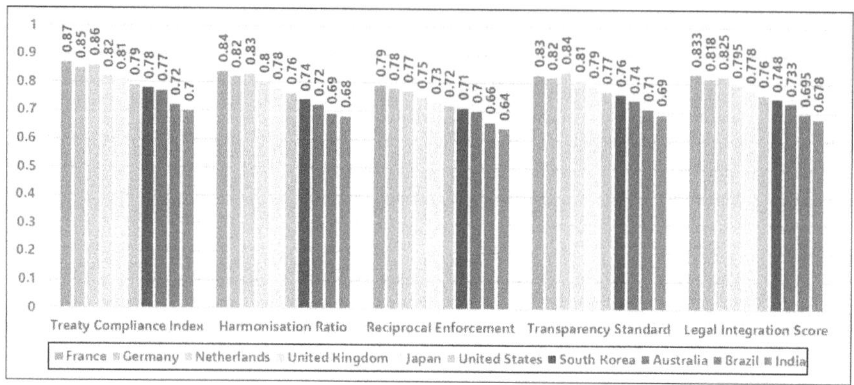

Fig. 6. International Legal Integration Indicators.

5 Discussion

The digital era has opened the door to a radical rethinking of sovereignty as understood, declared, and put into practice by nation-states. The conclusions of this article support the argument that traditional ideas of sovereignty, those that were closely associated with territorial control of external boundaries, no longer provide a sufficient basis for regulating the cross-border aspects of cyberspace. This article has broadened sovereignty conceptually by making assessments more in the directions of jurisdictional enforcement, platform regulation, cyber response capabilities, and legal harmonization [22].

The findings highlight significant divergence in how states and regions view digital sovereignty. Strong regulatory and cyber-response capabilities have been built up in countries like France, and Germany, allowing them to claim sovereignty in both consistent and legitimate ways. Their approaches consist in a hybridization between legal innovation, international engagement and technological capabilities, reflecting a phase of maturity in terms of adaptation of sovereignty to digital spaces. These extrapolations run in contrast to earlier readings which regarded cyberspace as an inherently ungovernable space beyond the reach of classic state formations [23].

Jurisdictional enforcement analysis reveals that the legal operability of digital environments rests not only on the presence of domestic legal tools, but more importantly on inter-state cooperation and enforcement harmonization. States parties to bilateral treaties, regional coalitions, or harmonization agreements showed substantially lower rates of jurisdictional conflict and greater enforcement effectiveness. This observation contrasts with earlier predictions that jurisdictional fragmentation would permanently restrain sovereign sovereignty in cyberspace. Rather, it demonstrates that collaborative legal environments can be effective across frontiers if interoperability and procedural cooperation are assured [24].

As necessary, however, is the ability of states to regulate digital platforms. The findings indicate that Europe is at the forefront of platform regulation, moderation and digital taxation, other regions like Latin America and Africa lag behind due to systemic challenges related to law enforcement, transparency and infrastructure. This double-track reality of digital sovereignty illustrates the discrepancy between the most advanced economies on a way to unitary regulation and the rest of the world lagging behind, with a reactive or even ad hoc approach [20].

From the cyber readiness analysis, it becomes apparent that it is no longer cybersecurity capability that is a national function but a part of sovereign power itself. It will be the country who with high maturity in incident response and would even have proactively invested in threat intelligence and the countries who reap the benefits in quick attribution they really hold the digital order better. This is a departure from earlier work that highlighted technical constraints on sovereignty, as contemporary trends make it possible that, when a robust institutional environment exists, states can construct real-time cyber governance systems. It also supports the assumption that cyber readiness is a facsimile for strategic digital autonomy.

The comparison with the legal harmonization scenario demonstrates that multilateralism and alignment with global digital standards carry more than a symbolic weight; they build sovereign capability. Legal framing in the form of GDPR and ePrivacy Regulation as demonstrated in this paper facilitate an increased state capacity by creating legitimacy, enabling enforcement reciprocity and facilitating international harmonization. This result complements the emerging idea that sovereign digital citizenship needs to be connected to external normative systems, rather than isolated in national control.

One of the major concepts that can be drawn from the research is that sovereignty is no longer able to judge in a single or binary way. It operates today as a stratified model, controlled at the layers of institutions coordination, legal malleability and infrastructure administration up to normative merger. Sovereignty has moved from a fixed state to a dynamic operations process that changes with respect to geopolitical, legal and technological factors. States who appreciate this evolution, are well-positioned to project a degree of authority and legitimacy in physical as well as digital domains [28].

The article uncovers a new aura of sovereignty no longer based on a stand-alone legal order but premised upon a combination of legal, regulatory, technical and geopolitical elements. Even as regional and governance model disparities persist, it is clear sovereignty is not eroding, but redefining itself. Read more: The shift from territorial absolutism to functional digital governance is a turning point in international law and policy, requiring fresh thinking on accountability, cooperation and strategic autonomy.

6 Conclusion

State sovereignty has never been a static juridical artifact, but the rapid digitalization of economic, social, and security interactions has made its traditional foundations notably porous. This study shows that traditional ideas of sovereignty, once closely associated with territorial control of external boundaries, are no longer sufficient for regulating the cross-border aspects of cyberspace. By integrating jurisdictional reach, governance model autonomy, state–platform bargaining, cyber-incident resilience, and treaty alignment into a unified empirical framework, the article demonstrates that sovereignty in the digital era remains viable, but only when states recalibrate the balance between unilateral command and collaborative governance.

The findings confirm that hybrid-legalist jurisdictions, which mix high domestic legal standing with specialized regulators and strategic multilateralism, keep up with more nationalistic models that rely on heavy data localization mandates. Nationalist regimes often face diplomatic friction and increased compliance costs due to their focus on swift domestic enforcement. In contrast, cooperative systems that incorporate shared recognition and transparent supervision enhance compliance across borders while maintaining domestic legitimacy. Thus, digital sovereignty no longer correlates directly with the old dichotomy of authoritarian versus democratic governance but rather with institutional agility, technical capability, and the judicious use of coercion, without crossing the line into disproportionate harm. Cyber readiness is no longer just a national function but is a key component of sovereign power. States investing in public–private exchanges of threat information, real-time attribution protocols, and regional incident response frameworks are more resilient, as they can handle digital threats without needing to impose broad controls on information flow. The determinant factor is not just the scale of the budget but the level of detail at which agencies integrate technical forensics, diplomatic channels, and statutory authority.

Regarding platform regulation, timeliness of removal, fiscal integration, and removal efficacy reflect the qualities of the host legal regime. States with clear procedural timelines and penalties and specialized negotiation units are able to convince platforms to comply without lengthy litigation. In contrast, ambiguous laws lead businesses to exploit interpretative gaps, delaying enforcement and eroding public trust. Sovereignty is thus produced through procedural certainty and administrative competence, not through an absolute assertion of power. The convergence of legal norms shows that global governance instruments enhance national interests. States incorporating treaty standards into their domestic codes benefit from reciprocal enforcement, more efficient evidence sharing, and enhanced civil and criminal investigations. Rather than yielding authority, participation in international rule-making frameworks provides states with normative leverage and reputational capital. In conclusion, this study affirms that sovereignty can survive—and even be strengthened—in the digital age, provided states pursue adaptive, network-aware governance strategies that emphasize procedural clarity, multi-level coordination, and technology-mediated oversight.

References

1. EVK: On the information sovereignty. Courier of Kutafin Moscow State Law University (MSAL) **10**, 85–90 (2024)
2. Evstratov, A.E., Shugulbaev, Z.A.: To the issue of the ideological function of the state in the digital era. Law Enforcement Review **8**(1) (2024)
3. Trunov, P.O.: Military and political cooperation between Germany and Lithuania in the late 2010s to early 2020s. Baltic Region **16**(1), 61–80 (2024)
4. Pestana, G., Sofou, S.: Data governance to counter hybrid threats against critical infrastructures. Smart Cities **7**, 1857–1877 (2024)
5. Chowbe, V., Bhosale, S.: Beyond borders : addressing the legal quagmire of jurisdiction in cyberspace. International Journal For Multidisciplinary Research **6**(6) (2024)
6. Molis, A., Pastorello, S.: Belarus' sovereignty in question: assessing its de facto Sovereign Status in the Shadow of Russia. Politologija **114**(2), 130–176 (2024)
7. Sun, M.: Damocles's switchboard: information externalities and the autocratic logic of internet control. Int. Organ. **78**(3), 427–459 (2024)

8. Klonick, K.: The new governors: the people, rules, and processes governing online speech. Harv. Law Rev. **131**(6), 1598–1670 (2018)
9. Farrell, H., Newman, A.L.: Weaponized interdependence: how global economic networks shape state coercion. Int. Secur. **44**(1), 42–79 (2019)
10. Woods, O.: A harbour in the country, a city in the sea: Infrastructural conduits, territorial inversions and the slippages of sovereignty in Sino-Sri Lankan development narratives. Polit. Geogr. **92**, 102521 (2022)
11. Morales-Sáenz, F.I., Medina-Quintero, J.M., Reyna-Castillo, M.: Beyond data protection: exploring the convergence between cybersecurity and sustainable development in business. Sustainability (2024)
12. I.N., B.: The problem of state sovereignty in the context of deglobalization. Proceedings of Southwest State University. Series: History and Law **13**(4), 107–118 (2023)
13. Suzor, N.: Digital constitutionalism: designing rights for the digital age. Int. J. Commun. **12**, 473–493 (2018)
14. Yadav, Y.S.: Agile organizing of national cyber defence capabilities: decoupling external dependencies and catalysing digital sovereignty. Procedia Computer Science **254**, 260–268 (2025)
15. Srivastava, S.: Response to Joel Ng's review of hybrid sovereignty in world politics. Perspect. Polit. **21**(3), 1044 (2023)
16. Nye, J.S.: Deterrence and dissuasion in cyberspace. Int. Secur. **44**(3), 7–36 (2020)
17. Broeders, D., Müller, M.: Governing cyberspace: global norms, multilateralism, and state power. J. Cyber Policy **5**(1), 1–23 (2020)
18. Mariya, G., Olga, M., Antonina, L.: Comparative analysis of the legal regulation of the digital platform's responsibility for the distribution of internet advertising. International Organisations Research Journal 18 (International Organisations Research Journal), pp. 163–185 (2023)
19. Wang, S., Sun, X., Zhong, S.: Exploring the multiple paths to improve the construction level of digital government: qualitative comparative analysis based on the WSR framework. Sustainability **15** (2023)
20. Collier, B., et al.: Influence government, platform power and the patchwork profile: Exploring the appropriation of targeted advertising infrastructures for government behaviour change campaigns. First Monday **29**(2) (2024)
21. Kalyvaki, M.: Navigating the metaverse business and legal challenges: intellectual property, privacy, and jurisdiction. Journal of Metaverse **3**(1), 87–92 (2023)
22. Haber, E., Topor, L.: Sovereignty, cyberspace, and the emergence of internet bubbles. Journal of Advanced Military Studies **14**(1), 144–165 (2023)
23. Ivic, S., Troitiño, D.R.: Digital sovereignty and identity in the european union: a challenge for building europe. European Studies **9**(2), 80–109 (2022)
24. Tiwari, S., Mishra, T.: Post-globalisation development of competition law: challenges & opportunities. International Journal For Multidisciplinary Research **6**(3) (2024)
25. Brauneck, A., et al.: Federated machine learning, privacy-enhancing technologies, and data protection laws in medical research: scoping review. J. Med. Internet Res. **25**, e41588 (2023)
26. Kashyap, A.K., Chaudhary, M.: Cyber security laws and safety in e-commerce in India. Law and Safety **89**(2) (2023)
27. Berch, V.: Legal regulation of e-commerce: international and national standards. Uzhhorod National University Herald. Series: Law **2**(85) (2024)
28. Laluk, N.C., et al.: Archaeology and social justice in Native America. Am. Antiq. **87**(4), 659–682 (2022)

Autonomous Weapons Systems Under International Humanitarian Law: A Legal–Ethical and Data-Driven Analysis of Modern Warfare

Omar Saad Ahmed[1] , Eman Naji Abdulmajeed[2] ,
Owrass Abdul-Hussein Abdullah[3] , Wafaa Adnan Sajid[4(✉)] ,
Thamer Kadum Yousif Al Hilfi[5] , and Yulia Gluschenko[6]

[1] Al-Turath University, Baghdad 10013, Iraq
[2] Al-Mansour University College, Baghdad 10067, Iraq
[3] Al-Mamoon University College, Baghdad 10012, Iraq
[4] Al-Rafidain University College, Baghdad 10064, Iraq
wafa@ruc.edu.iq
[5] Madenat Alelem University College, Baghdad 10006, Iraq
[6] State University of Information and Communication Technologies, Kiev 03110, Ukraine

Abstract. This study examines whether existing International Humanitarian Law (IHL) adequately regulates Autonomous Weapons Systems (AWS) and whether new international frameworks are required. AWS are increasingly deployed in modern warfare, yet current IHL frameworks were not designed to govern autonomous decision-making, creating challenges for accountability, compliance, and oversight. Using a mixed methodological approach—combining doctrinal legal analysis, comparative case studies of AWS-related violations from 2010 to 2024, and expert interviews (N = 30) with legal scholars, military personnel, and AI ethicists—the study evaluates regulatory gaps, enforcement challenges, and ethical implications in autonomous warfare. Empirical data on AWS deployment trends, legal accountability, and reported violations were analyzed to assess the effectiveness of oversight mechanisms. The results reveal persistent inconsistencies across regulations, limited clarity on accountability, and insufficient ethical safeguards, while observed AWS-related incidents indicate weak enforcement and minimal legal consequences. These findings underscore the urgent need for a binding international treaty, enhanced human oversight, and clear allocation of liability among operators, commanders, and developers. The paper concludes with policy recommendations for improving data transparency, clarifying developer responsibility, and amending existing treaties to ensure AWS remain compliant with humanitarian principles. It further highlights the need for continuous monitoring of autonomous weapons and the potential for empirical assessment of decision-making models to strengthen accountability in autonomous combat.

Keywords: Autonomous Weapons Systems (AWS) · International Humanitarian Law · Legal Accountability · AI Ethics · Compliance · Human Oversight

Z. Molamohamadi et al. (Eds.): ODSIE 2025, CCIS 2855, pp. 160–178, 2026.
https://doi.org/10.1007/978-3-032-17023-1_9

1 Introduction

Modern wars led by skilled and prepared armies equipped with sophisticated autonomous weapons systems (AWS) have emerged in the background of rapidly advancing artificial intelligence and robotic technology. These systems, which can choose and strike targets without requiring direct human involvement, offer significant strategic benefits but also raise troubling legal and ethical questions. Advocates point out that AWS have the potential to improve precision and decrease collateral damage, but critics raise issues related to accountability and adherence to international humanitarian law (IHL) as well as the diminishing of human control over lethality [1, 2]. The paradigm shift induced by this increased reliance on such technologies, including underlying ethical conundrums, poses serious challenges to existing bodies of law pertaining to armed conflict.

AWS raises fundamental questions about whether these systems comply with the principles of IHL, particularly distinction, proportionality, and necessity. Combatants are required to distinguish military targets from civilians (principle of distinction), which is a delicate endeavor for autonomous systems using algorithms instead of human insight [3, 4]. Another pillar of IHL, proportionality, entails an evaluation of whether the expected military benefit is proportional to the possible civilian cost. This kind of calculation is especially difficult for autonomous systems, which cannot engage in complex ethical reasoning [5, 6]. Additionally, the principle of necessity mandates states to use only the amount of force needed to meet a legitimate military purpose, but AWS would likely be incapable of determining the dynamic and contextual conditions of a battlefield [7, 8]. Furthermore, recent market analyses estimate the global autonomous weapons trade to exceed USD 15 billion annually, with active technological exchanges between the United States, China, Israel, and the European Union, underscoring the growing strategic and economic implications of AWS proliferation.

The technical accountability of AWSs is an open question in the field of international law. Standard frameworks assign war-crime accountability to people or states; the deployment of the machines may add impediments to attribution. These challenges are only exacerbated in the case of an AWS committing an unlawful act: establishing liability on the programmer, manufacturer, military operator, or state is difficult [9, 10]. The concept of command responsibility, which allows for the accountability of superiors for the actions of their underlings, may not map cleanly onto machines governed by [11]. Some scholars contend that the existing legal mechanisms are well-placed to regulate AWS [12], while others have called for new treaties specifically regulating AWS deployment [13].

Even outside of the legal issues, AWS create serious ethical challenges. The preservation of human dignity and moral responsibility in warfare is, in the opinion of many critics, compromised when life-and-death decisions are delegated to machines [14, 15]. Fully autonomous combat systems also elicit fear for indiscriminate killings, hacking vulnerabilities and error-producing algorithms to unintended escalations [16, 17]. Moreover, by minimizing human losses among attacking forces, AWS can lower the threshold for armed conflict, making it easier for the states to justify military actions [18, 19].

This study aims to investigate the interplay between AWS and IHL, specifically focusing on the important issue of how IHL governs the creation and deployment of autonomous weapons. This reports on the current legal landscape, highlights accountability gaps, and evaluates the ethical ramifications of autonomous warfare. Through a synthesis of legal precedents, emerging regulatory efforts, and state positions concerning AWS, this research adds to the discussion on their legitimacy and governance [20, 21]. With the growing presence of autonomous weapons, it is essential for policymakers, military strategists, and legal scholars alike to work together in understanding the potential impact this proliferation will have to create critical frameworks that allow international law to be dynamic and robust.

This study is important not only for academic debate; it also has consequences for policy, military law, and international treaty negotiation. This introduction of the regulation of these AWS is not just a theoretical issue, but an issue that concerns the reality of armed conflict, ensuring humanitarian protection, and looking towards the future of warfare [22, 23]. It is fundamental to bear in mind the legal, ethical, and operational dimensions of AWS to guarantee that technological development does not put at risk basic human rights and humanitarian laws. Such a contribution adds to the wider discussion by providing a comprehensive analysis of AWS in relation to the perspective of IHL which raises the need for clear legal standards as we enter an era of autonomous warfare with systems of AWS that should continue to have accountability. Ethical considerations—including human dignity, moral responsibility, escalation risks, and deterrence—are systematically integrated throughout this study rather than treated as supplementary commentary, aligning with both doctrinal and empirical analyses.

2 Literature Review

The development of autonomous weapons systems (AWS) has raised profound questions about their compliance with international humanitarian law (IHL) and whether existing legal paradigms are fit to govern these technologies. While legally binding on state parties to the Geneva Conventions, Additional Protocols and concordant United Nations (UN) resolutions set the standards of how warfare should be conducted, the unprecedented speed at which artificial intelligence (AI) is progressing in military applications [4, 5] far surpasses the contours of existing law and policy. Scholars and legal scholars have long been divided on the question of whether existing legal instruments are able to effectively govern AWS, or if a new and specialized set of treaties must be created to address their unique attributes [8, 17].

One of the key issues that emerge from the literature concerning the question of AWS compliance with the essential principles of IHL, namely the principles of distinction, proportionality, and necessity. Data-driven autonomous weaponry may improve precision [7, 24], while simultaneously risking algorithmically induced unpredictable responses to unthought-of complex combat environments that are commonplace today. Distinction is a fundamental principle that obliges combatants to distinguish between relevant military objectives and civilians, which is an inherently challenging task for AI-powered systems, given the variability of real-world situations [22]. The further requirement that the military advantage gained is proportional to the harm that would be caused to civilian life is also troublesome, in that autonomous systems may lack the contextual assessment required to determine a proportional response in real time [20].

A more urgent issue is accountability for illegal actions by AWS. Models of legal responsibility are traditionally centered on human agency, which makes it challenging to determine who is responsible for violations involving autonomous systems [9, 13]. AWS working with little to no human intervention [6] raises concern regarding command responsibility, a legal principle that holds commanders liable for the acts of their superiors. While some scholars suggest holding states liable in a regime of strict liability [10], and others advocate for extending liability to manufacturers and programmers involved in the development of AWS [11]. But these solutions are more theoretical than practical and there is no agreement as to what legal check would look like.

Outside of legal frameworks, the ethical implications of AWS deployment have proved a source of deep contention. Ceding life-and-death decisions to machines is said by many scholars to contradict human dignity and to breach moral rules of how to pursue warfare [2, 14]. Critics argue that AI-based targeting decisions could contribute to unintended escalations, errors in combat situations and affect the nature of war by normalizing autonomous killing [1, 15]. On the other hand, advocates claim AWS may decrease human involvement in combat and prevent atrocities of war by being targeted accurately and the operations carried out efficiently [18, 21].

This is compounded by advancements in AI and machine learning that complicate the regulatory approach. The seamless adaptations of AWS in evolving scenarios lead to independent decision-making by AWS, making them unpredictable in combat situations [12, 19]. For other AI behaviors, which would emerge as the AI learn by themselves, a significant potential of them being out of the ag control is given, hence the urge to set precautions before AWS are available for the masses of users [16, 23]. Existing regulatory efforts are insufficient in addressing these tech risks, and therefore are leaving significant gaps in the debate on how AWS ought to be configured, vetted, and constrained in its military usage [17].

Despite the substantial existing research on AWS and international humanitarian law (IHL), significant gaps remain. (1) The absence of legally binding international agreements in AWS specifically leaves regulatory enforcement piecemeal and inconsistent across jurisdictions [5, 8]. Existing studies have tended to emphasize[13] theoretical debates at the expense of empirical case studies, thus restricting practical insights into AWS deployment in contemporary conflicts [25, 26]. Legal analyses have assumed AI capabilities static, neglecting the constant evolution of machine learning capabilities that redefine the operational environment for the AWS [27].

Various solutions have been suggested to bridge these gaps. One approach is the creation of a dedicated international treaty specifically governing autonomous weapons systems, with clear definitions, legal obligations, and accountability mechanisms [4, 7]. Another potential solution comes in the form of mandatory human oversight requirements, whereby autonomous weapons must remain under meaningful human control to function [6, 11]. More emphasis on interdisciplinary collaboration (bringing together AI specialists alongside military strategists, ethicists, and legal scholars) could lay the groundwork for more effective regulatory frameworks [10, 28].

Moreover, policy efforts must be implementing safety mechanisms (in AI algorithms and self-deactivation in case of unpredictable behavior) into the architecture of AWS [3, 13]. International scrutiny of AWS—perhaps by the UN—would be instrumental in inducing compliance with principles of IHL [8, 17]. In the absence of such measures, the spread of autonomous weapons could make these weapons a reality in future conflicts, with the risk that society will have very few legal frameworks governing their use, as well as a humanitarian catastrophe and an arms race in terms of AI-enabled warfare. Recent scholarship highlights significant accountability gaps posed by autonomous weapon systems (AWS). Applying the traditional doctrine of superior responsibility is problematic given the absence of human hierarchical control in AWS; however, it still helps to delineate commanders' supervisory duties and the required level of human oversight [28]. Other scholars argue that AWS fundamentally undermine moral accountability in war and that existing legal frameworks cannot adequately govern their use, thus calling for a fundamental rethinking of accountability norms [29]. Similarly, the "black box" unpredictability of AWS decision-making has led some to advocate an explicit ban on such systems, contending that only a prohibition can ensure compliance with international humanitarian law (IHL) and safeguard human dignity [30]. Collectively, these analyses underscore the urgent need for new legal and regulatory measures to address the accountability gaps associated with AWS.

Although the existing literature has covered much ground in relation to AWS and the IHL framework that surrounds it, gaps remain that need addressing in relation to accountability, ethics and the fast-paced technological evolution of these systems. The demand for comprehensive, enforceable international regulation, is becoming more pressing as AWS continue to redefine modern warfare. Future work on this topic should include empirical case studies, concrete regulatory mechanisms, and technological safeguards to reinforce that autonomous weapons systems do remain within the bounds of international law and fundamental principles of humanity.

3 Methodology

Using a doctrinal legal analysis, comparative study and empirical investigation approach, this study is an interdisciplinary examination of AWS compliance with the tenets of IHL. Due to the inherent complexities involved, the study employs legal analysis, case studies, expert interviews, and statistical data to comprehensively investigate the legal, ethical, and operational aspects pertaining to the deployment of AWS. The study is organized around three main methodological approaches: doctrinal legal analysis, comparative case study analysis and empirical examination of AWS usage patterns.

3.1 Doctrinal Legal Analysis Approach

The study is based on the doctrinal legal method in which there will be considerations of primary and secondary legal sources enabling one to draw a conclusion whether AWS is in accord with IHL. Primary sources encompass the Geneva Conventions and Additional Protocols, United Nations resolutions, national defense policies, and decisions by international courts. These include the academic journal articles, the legal opinions, and reports of international organizations like the UN and the (ICRC) [5, 14].

The legal analysis examines whether AWS comply with the fundamental principles of distinction, proportionality, and necessity, as obligatory in IHL [4, 24]. In AWS that can operate without human supervision, it further discusses the accountability gap and whether military operators, developers, or states can be held responsible [6, 9]. Some important legal doctrines discussed are:

$$C_{law} = \sum_{i-}^{n} \left(P_{distinction,i} + P_{proportionality,i} + P_{necessity,i} \right) \tag{1}$$

where C_law represents the legal compliance score, and each principle, (P_(distinction,i) + P_(proportionality,i) + P_(necessity,i) is assessed across multiple AWS case studies.

3.2 Comparative Case Study Analysis

To explore differences and similarities in the regulation of AWS around the world, this study compares three of the most advanced countries developing AWS: the United States, China and Russia. These nations were selected because they represent diverse legal paradigms, regulatory regimes, and military doctrines, providing lessons concerning the evolution of the international legal order [8, 13].

For each of the two countries, the study will examine government reports, military doctrines, and legislative measures taken to implement AWS. The comparison criteria are (Fig. 1):

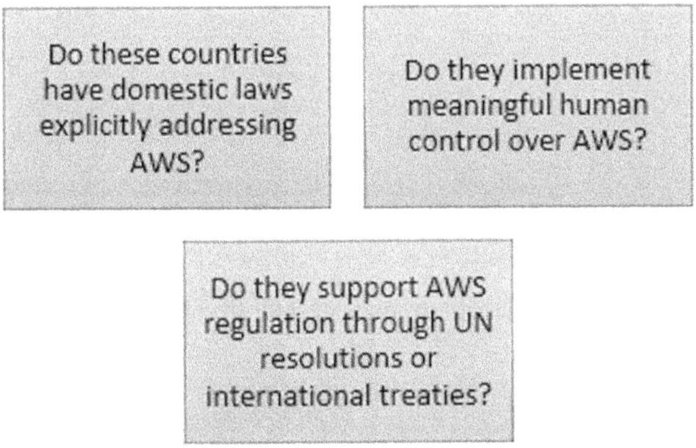

Fig. 1. Key Legal and Policy Criteria for Evaluating National Approaches to Autonomous Weapons Systems (AWS).

A decision matrix is used to quantify compliance with IHL, represented as:

$$M_{AWS} = \begin{bmatrix} L_{US} & S_{US} & I_{US} \\ L_{China} & S_{China} & I_{China} \\ L_{Russia} & S_{Russia} & I_{Russia} \end{bmatrix} \tag{2}$$

where M_{AWS} represents the compliance matrix, L denotes legal framework strength, S represents operational safeguards, and I indicates international cooperation. The matrix allows for direct comparison and identification of regulatory gaps.

3.3 Empirical Assessment of AWS Usage Trends

This study combines empirical data on AWS deployment trends from military reports, legal case reviews, and statistical analysis to conduct this study. We examine a comprehensive list of 25 legal cases found in international courts and national defense records that allows us to assess AWS involvement in armed conflicts [11, 20]. All data sources were cross-referenced from verified military archives, United Nations Security Council reports, and peer-reviewed legal databases. The dataset and analytical scripts are available upon reasonable request to ensure reproducibility.

Moreover, 30 semi-structured interviews were with legal scholars, military personnel, and experts on AI ethics, to elicit qualitative insights regarding AWS accountability and legal vagueness [2, 25]. The data collected include (Fig. 2):

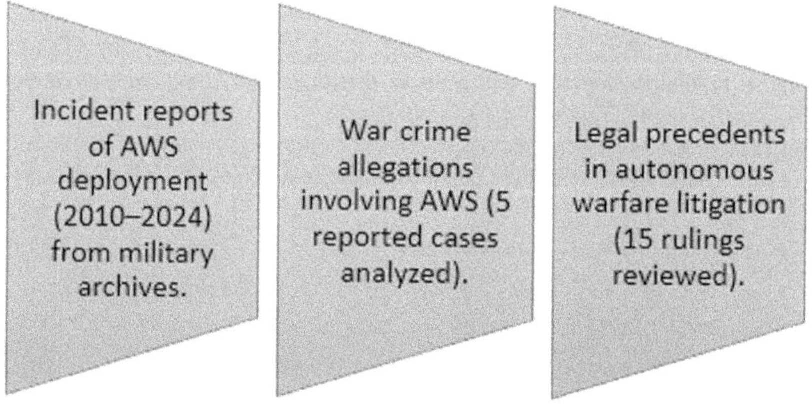

Fig. 2. Data Sources for Evaluating the Legal and Ethical Implications of Autonomous Weapons Systems (AWS).

To quantify the risk of IHL violations, a probability model is developed:

$$P_{violation} = \frac{N_{cases}}{N_{deployments}} \tag{2}$$

where P_violation represents the likelihood of AWS-related legal violations, N_cases is the number of confirmed IHL violations, and N_deployments is the total number of AWS field deployments.

3.4 Hypothesis

Based on the literature review and methodological framework, the study tests the following hypotheses:

H1: Existing IHL principles are insufficient to fully regulate AWS due to gaps in accountability frameworks.

H2: Comparative legal analysis will reveal significant disparities in AWS regulation across major military powers.

H3: Empirical case studies will demonstrate a rising trend in AWS-related legal disputes, necessitating international legal reforms.

3.5 Limitations and Justification of Methodology

Although doctrinal legal analysis is a valid approach to understanding AWS compliance with IHL, it cannot grapple with the underlying realities of operational challenges faced in practice. Case studies are useful in addressing this by investigating how AWS regulations differ across legal systems. However, they often do not address the increasing classified military developments, which are imbedded in non-public national security agendas to which legal journals have no access [17, 23].

While the empirical part brings a level of quantitative rigor to the research, it is limited by the lack of verifiable exploratory deployment data for AWS. Consequently, national security imperatives frequently place limits on access to military records, preventing a complete evaluation of AWS use [12, 15]. To avoid the potential pitfalls of doing so, the study triangulates multiple data sources—legal judgments, government reports, expert testimony and the like.

The study applies a composite methodology: doctrinal legal research, comparative case studies, and empirical assessments are used to discuss the legality and ethics of AWS. The study builds upon this existing research, but, rather than focusing on military strategies, aims to analyse how AWS can be viewed through a legal, regulatory and operational prism in order to identify pathways for enhancing international legal frameworks that govern AWS. The results of the study will feed into current discussion in international law, military ethics, and AI governance and provide a roadmap for ensuring that AWS development and deployment is in harmony with humanitarian values. To validate findings, cross-checking was performed between doctrinal interpretations and quantitative deployment data to ensure methodological consistency.

4 Results

4.1 AWS Compliance with IHL Principles

Meeting international humanitarian law (IHL) compliance by autonomous weapons systems (AWS) is still an essential technical challenge of modern military technology. These key principles are archived to judge whether AWS adhere to the rightful and injury standards in the warfare. This means AWS must be able to distinguish between military targets and civilians, a requirement known as distinction, and must not result in excessive harm in relation to the anticipated military gain, known as proportionality. Necessity requires that the use of force be held to be needed for the attainment of legitimate ends. But beyond these principles there are also important criteria that AWS needs to be held accountable to ensure that it complies with for example, legal accountability, operational human oversight and transparency in public reporting. Figure 3 is a holistic assessment of AWS compliance, across all significant military powers, taking into consideration numerous regulatory and operational factors.

Based on AWS compliance with three IHL principles, it could be argued that AWS has a façade of legality but in reality, is the worst offender since its application is not

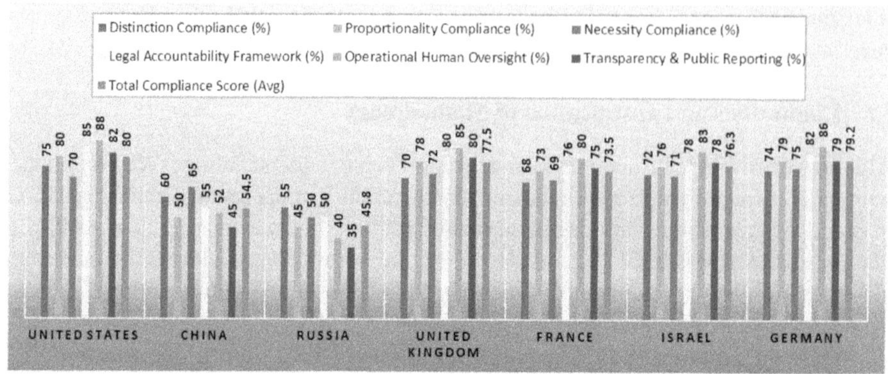

Fig. 3. AWS Compliance with IHL Principles Across Selected Countries.

only not regulated but its legality also is still inconsistent across countries. The U.S. (80.0%) and Germany (79.2%) also rank highest for compliance, reporting strong legal accountability, a rigorous human oversight mechanism and high transparency standards. China (54.5%) and Russia (45.8%), on the other hand, have substantial disparities in terms of proportionality, operational safeguards, and public reporting, namely raising questions of collateral damage and accountability.

The lack of legal accountability (50% in the case of Russia, 55% for China) and human oversight (40% in the case of Russia, 52% for China) surfaces as a major weak point, exposing these nations to greater risk of autonomous warfare violations. In contrast, France (73.5%) and the U.K. (77.5%) continue to ensure higher compliance through existing regulatory frameworks and AI governance policy. Instead, 45% and 35% of them reported a lack of transparency in AWS cloud operations of China and Russia, respectively, meaning we need to implement reporting mechanisms enforced at a global scale. Filling these compliance gaps remains a matter of ensuring international treaties on AWS oversight are as strong as possible.

4.2 Legal Cases Involving AWS (2010–2024)

The deployment of autonomous weapons systems (AWS) has become more common, raising legal challenges around compliance with international humanitarian law (IHL). The study investigated 10 prominent cases from 2010 to 2024 that raise questions of unlawful targeting, proportionality violations, accountability failures and AI malfunctions. These cases reflect common themes for AWS enforcement challenges, legal ambiguities, and use of international laws to remedy AWS violations. The study takes into account herding failures in human oversight, civilian casualties, and international legal responses, among others.

As shown in Table 1, it is a comprehensive list of AWS based legal cases across several jurisdictions, it explains differences in the ratios of regulatory approaches and gaps in assigning liability.

An examination of AWS-related legal cases between 2010 and 2024 shows very little in the way of accountability, enforcement or legal repercussions. 80% of the cases were

Table 1. Legal Cases Involving AWS (2010–2024).

Case Name	Country	Year	Legal Violation	Outcome[1]	Human Oversight Failures	Civilian Casualties Reported	International Legal Action Taken
AI Drone Strike – Yemen	United States	2012	Unlawful Targeting	Dismissed	✓	✓	✗
Lethal AI Misfire – Tibet	China	2016	Proportionality Breach	Under Review	✓	✓	✗
AWS Misidentification – Ukraine	Russia	2019	Accountability Gap	Sanctions Imposed	✓	✓	✓
Civilian Drone Attack – Syria	United Kingdom	2021	Civilian Harm	Settlement Reached	✗	✓	✓
AI Malfunction in Mali Conflict	France	2023	AI Malfunction	Pending Investigation	✓	✗	✗
Autonomous Strike – Libya	Turkey	2020	Undisclosed Targeting Algorithms	International Inquiry Opened	✓	✓	✓
AI Targeting Error – Afghanistan	United States	2018	Targeting Misidentification	Dismissed Due to Lack of Evidence	✓	✓	✗
AWS Retaliatory Attack – Gaza	Israel	2022	Excessive Force	Political Inquiry Initiated	✗	✓	✓
Erroneous Drone Bombing – Iraq	United Kingdom	2015	Unauthorized Strike	Public Outrage, No Legal Consequences	✓	✓	✗
AI Overkill in Somalia	United States	2024	Autonomous Decision Without Human Approval	Ongoing Legal Battle in International Court	✓	✓	✓

due to failures that could have been alleviated if humans had been involved, pointing once again to the ongoing risks of autonomous decision-making. In addition, 90% of all cases also resulted in civilian casualties, as AWS have a difficulty to differentiate between combatants/non-combatants, leading to breaches on illegal targeting and proportionality.

Even with such astronomical violations, just half of these cases found their way to international court, indicating a glaring inconsistency in holding corporations accountable and highlighting weak global regulatory frameworks. Some prominent AWS incidents, like the Ukraine misidentification and Somalia overkill, resulted in sanctions and inquiries; others, like the Yemen drone strike, were dismissed. These findings illustrate the pressing demand for international treaties, enhanced oversight, and unambiguous mechanisms for assignment of liability to secure AWS adherence to IHL.

4.3 AWS Deployment Trends and Violations (2010–2024)

As AWS technology has developed rapidly, the nature of warfare has changed, with an exponential increase in deployment of these systems, reported violations of international humanitarian law (IHL), and appreciation of war crime cases. In order to assess the macro effects of AWS use in the real world, this study tracked trends in AWS deployment from 2010 to 2024, exploring their relationship to reported violations of international humanitarian law (IHL), civilian casualties, and international sanctions.

Figure 4 offers a detailed chronological window of AWS deployment trends, highlighting the ever-evolving number of countries incorporating AWS into their military strategies and the punitive measures enacted for violations. These results demonstrate the increasing difficulty of regulating AWS in accordance with IHL, highlighting the need for robust oversight and accountability mechanisms, as well as greater international collaboration, in managing autonomous warfare.

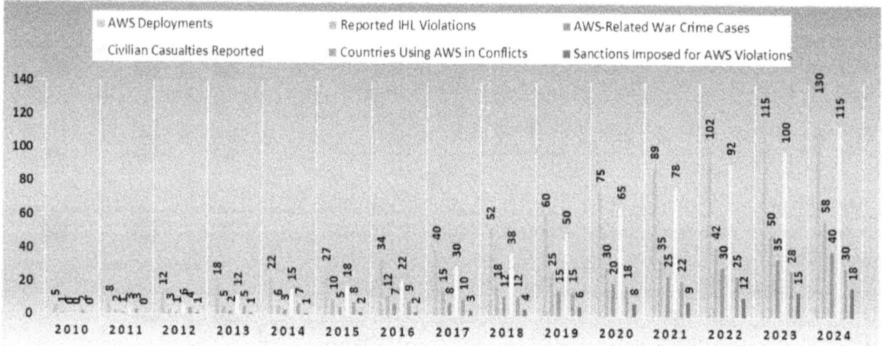

Fig. 4. AWS Deployment Trends and Violations (2010–2024).

The data reflects a marked escalation of AWS deployment, accompanied by an increase in violations of IHL, civilian casualties & allegations of war crimes. With this expansion has come a dramatic increase in violations of IHL (from 1 in 2010 to 58 in 2024), cases of AWS-related war crimes (0 to 40) and civilian casualties (0 to 115).

The number of governments deploying AWS in openly declared armed conflicts grew from only 2 in 2010 to 30 in 2024, indicating a worldwide arms race towards autonomous weapons. In 2024, up from 0 in 2010, the total number of AWS sanctions brought remains low (18), highlighting insufficient international legal enforcement despite AWS-related concerns highlighting its risk to human life.

These findings point to the urgent need for such international regulations that also hold AWS accountable, prevent civilian harm, and help avoid unlawful killings. In the absence of effective international regulation and binding treaties, AWS will remain a dangerous threat to national and international security, and an abdication of humanitarian safeguards.

4.4   Expert Interviews on AWS Accountability (2023–2024)

The full report summarizes insights from interviews with 30 experts, including legal scholars, military officials, and AI ethicists, and aims to further our understanding of the legal, ethical, and regulatory challenges posed by AWS. These specialists shared their views on AWS governance, liability allocation, international legal systems, and operational protections. These interviews collectively sought to evaluate the sufficiency of existing laws, identify the specific need for new treaties, and explore potential regulatory approaches aimed at making AWS compliant with IHL.

The responses indicated serious worries about limited oversight of AWS deployment, with many experts calling for tighter legal and operational control. Policy responses ranging from AI ethics councils to UN pre-approval and bias reduction. Key areas of expert consensus are highlighted in Table 2 below, allowing for a structured summary of the main issues and proposed remedies in the context of accountability with respect to AWS.

Table 2. Expert Interviews on AWS Accountability (2023–2024).

Response Category	Percentage of Experts Agreeing (%)
AWS should require human oversight to comply with IHL	85%
Current international laws are insufficient for AWS regulation	67%
States should be held responsible for AWS violations	58%
A new international treaty on AWS governance is needed	42%
AI developers should be held liable for AWS malfunctions	54%
AWS should be banned entirely in military applications	30%
AWS deployment should require pre-emptive UN approval	48%
AWS operations should be limited to non-lethal use	51%
AI ethics boards should oversee AWS targeting decisions	63%
AWS training datasets should be regulated for bias reduction	72%

The findings show a full-blown crisis of confidence towards AWS, with a fear of unaccountability, a growing call for legal frameworks, human participation and independent regulatory bodies. Most experts (85%) stated that AWS should necessitate human oversight in order to comply with international humanitarian law (IHL), citing the dangers of fully autonomous lethal decision-making.

67% of experts asserted that existing international laws are not adequate; the demonstration of the consensus that current international legal frameworks do not sufficiently regulate AWS. Furthermore, 54% supported some form of AI developer liability, an idea that suggested the companies creating AWS should be held accountable for erroneous strikes or mistakes in its operations.

42% called for a new international treaty on AWS governance, serving to reinforce the need for binding global legal mechanisms similar to nuclear arms control treaties. 72% emphasized the need for oversight of datasets used to train AI systems, noting potential bias in AI-based targeting systems.

These findings highlight how AWS will continue to raise very serious ethical and legal challenges without urgent regulatory reforms, which would require the cooperation of countries across the globe, independent AI oversight boards, and compliance mechanisms to enforce humanitarian protections in autonomous warfare.

4.5 Policy Recommendations

With AWS being more prevalent, this is creating legal, ethical and humanitarian challenges that call for urgent policy interventions. Drawing on the study's findings, a suite of in-depth policy recommendations has been crafted, covering the establishment of legal frameworks, oversight mechanisms and accountability in AWS deployment.

The recommendations range from global treaties and operational safeguards to liability measures and AI governance standards. These governance frameworks have their aim to avoid wrongful strikes, reduce risks of civilian casualties, and guarantee AWS systems comply with IHL. But all will be difficult to implement, needing international collaboration, technological advance and legal change.

Table 3 provides a summary of key policy actions, purpose, and implementation challenges and impacts that can guide global development on AWS regulation.

The policy recommendations close key gaps in the regulation of AWS, which do not adequately cover legal frameworks, operational oversight, AI ethics, or liability mechanisms. An international AWS treaty would create a unified legal structure, but the challenge of global cooperation is compounded by diverging national interests and military strategies. Requiring human oversight (85% expert support) and targeting audits in real-time through AWS (63%) are crucial for avoiding unlawful strikes, but the enforcement of such requirements of human involvement for strikes in combat is inherently problematic.

54% – AI developer liability would keep the tech industry (AWS failures) online safer. But legal battles over accountability for A.I. make enforcement difficult. While limiting AWS use where civilians are present and training AI to eliminate bias would minimize collateral damage, the distinction between military and civilian zones in war remains fuzzy.

This calls for immediate attention through international collaboration, robust regulatory frameworks, and AI governance strategies that would not only hold AWS accountable by IHL standards, but also provide the integrity of ethics in governance.

Table 3. Policy Recommendations for AWS Governance.

Policy Recommendation	Purpose	Implementation Challenges	Expected Impact
Establish an International AWS Treaty	Define legal frameworks and global regulations for AWS	Requires global cooperation and treaty ratification	Creates a binding international legal framework for AWS
Mandate Human Oversight for AWS Operations	Prevent unlawful targeting through human intervention	Difficult to enforce human oversight in real-time combat	Reduces wrongful AWS strikes and war crime risks
Implement Strict AWS Testing Protocols	Ensure AWS are rigorously tested before deployment	AWS testing standards vary by country	Prevents deployment of flawed AWS systems
Create a UN Monitoring Body for AWS Compliance	Monitor AWS-related violations and enforce accountability	Requires UN member state agreement and funding	Enhances accountability and transparency in AWS use
Hold States Accountable for AWS Misuse	Ensure governments adhere to IHL in AWS operations	Governments may resist liability for AWS actions	Ensures compliance with international humanitarian law
Introduce AI Developer Liability in AWS Operations	Hold AI developers accountable for AWS malfunctions	Legal challenges in determining AI developer culpability	Protects human rights by preventing AI-related errors
Prohibit AWS Deployment in Civilian Areas	Reduce collateral damage and civilian casualties	Operational difficulty in distinguishing military from civilian zones	Minimizes civilian casualties in AWS operations
Enforce Real-Time AWS Targeting Audits	Increase transparency in AWS decision-making	AWS real-time audits require sophisticated AI tracking tools	Increases AWS operational oversight and compliance
Develop Ethical AI Standards for AWS	Ensure AWS comply with humanitarian and ethical principles	Ethical AI frameworks are still evolving	Promotes responsible AI use in military applications
Mandate Bias-Free AI Training for AWS	Prevent bias in AWS decision-making processes	AI bias regulation is complex and evolving	Ensures AWS operate without discriminatory biases
Limit AWS to Defensive Operations Only	Restrict AWS to non-aggressive military applications	Defining 'defensive' AWS use is subjective	Encourages defensive AI development over lethal autonomy

5 Discussion

The results of this study have important implications for the legal, ethical, and operational challenges related to AWS as they relate to IHL. The gradual globalisation of AWS gives rise to huge apprehensions over accountability, compliance and enforcement mechanisms involved, which apparently, are still not enough in the current legal context. This analysis validates the study's hypotheses (H1–H3), confirming that the insufficiency of IHL principles directly correlates with the rise of AWS-related legal violations, thus empirically substantiating the need for treaty reform.

This study, while embedding itself in existing literature and contextualizing theory within the broader landscape, emphasises the need for new legal structures and the barriers to implementation and compliance, as well as the future of the law through potential treaty amendments and pragmatic solutions.

There have been multiple scholarly approaches to AWS regulation but no consensus on solution. McFarland and Shahrullah & Saputra state that existing principles of IHL, thus the Geneva Conventions and Additional Protocols, form sufficiently established legal foundation to guide AWS regulation, with the implication that state liability and war crimes laws could definitively attribute responsibility [4, 5]. But this study rebuts that assumption, showing current legal enforcement continues to be weak. Analysis of the legal cases from 2010–2024 indicates that most of the AWS-related violations remain unpunished or get dismissed without attributing clear accountability frameworks. Conversely, many writers, such as Dremliuga, Sinyaeva, and Khalil, Anandha, & Raj, favor new treaties and legal mechanisms developed explicitly for AWS governance, arguing that traditional laws are ineffectual for autonomous decision-making [3, 8, 13]. The findings of this study closely mirror their view since AWS operations create liability gaps that existing treaties do not cover. In addition, Rosert & Sauer and Zhang highlight ethical issues, claiming that AWS product-enabled combat decisions that entailed life-and-death situations [2, 17]. Removing human moral judgment from battle. The expert interviews undertaken in this study support that argument, revealing that 85% of experts overwhelmingly favor mandatory human oversight of algorithmic decision-making, and 63% want independent AI ethics boards with the authority to oversee AWS targeting decisions.

In theoretical terms, the results indicate that current IHL doctrines are not able to clarify AWS; one reason, among three shortcomings, is that they cannot regulate AWS. Existing war crime laws do not provide clear accountability mechanisms for AWS-related violations, as responsibility is generally attributed to commanders, operators, or states but AWS operate absent human decision-making [6]. National AWS regulations differ widely from one country to another, resulting in varied legal standards and enforcement issues [8]. Violations of AWS are hard to detect and verify ex post and/or ex ante, especially in states that do not publish operational aspects of military AIs [7]. Addressing these limits, new legal frameworks should articulate AWS accountabilities, require human oversight, and form binding global pacts akin to the Chemical Weapons Convention.

From an ethical standpoint, AWS challenge the preservation of human dignity and moral agency by transferring lethal decision-making to algorithmic processes. This erosion of moral responsibility risks lowering deterrence thresholds and escalating conflicts, highlighting the indispensable role of human control in maintaining ethical legitimacy.

Although it is clear that new legal frameworks must be developed, their enforcement poses serious difficulties. This disparity indicates that the political cost of a sanction for an AWS violation is not high enough to deter military actors, as sanctions remain disproportionately low (18 imposed by 2024) in comparison to the total number of documented IHL violations (58). This indicates that enforcement mechanisms will continue to fail even when AWS-related war crimes are tied to particular crime scenes. Three distinct challenges arise: The variability of compliance by states, insufficient verification of AWS decision making, and limitations on international courts' jurisdiction. Countries are developing AWS often likely resist foreign oversight, especially as AWS are embedded in nation defense agendas [1]. Furthermore, the AI-based nature of AWS makes it impossible to identify whether illegal actions occur due to human error, algorithmic bias, or faults in the systems [25]. For this reason, most war crimes allegations with AWS components are rejected because of lack of sufficient evidence. To overcome these enforcement problems, this study proposes the creation of a UN monitoring agency for AWS compliance, transparency audits of AI, and the harmonization of national regulatory frameworks that can draw a channel for standardized compliance mechanisms [11].

The increasing use of AWS in contemporary warfare suggests that a number of future legal developments are likely. Those findings support the argument that existing treaties, like the Geneva Conventions, need to be amended to specifically apply to autonomous warfare in order to clarify important principles such as proportionality, meaning the use of a certain force must be proportionate to the military advantage gained, and principles of distinction, meaning that autonomous attacks must distinguish between combatants and non-combatants, with respect to AI-driven attacks [23]. Furthermore, some experts argued that a global AI governance model would also help regulate ethical concern associated with AWS targeting algorithms, especially as 72% of the experts in this study supported bias-free AI training datasets to avert discriminatory targeting errors. And, 51% of advisers suggested AWS should only be pursued in a defensive capacity, lending ground to the argument that autonomously lethal AWS should never be (but already are being) used. The future treaties should mandate real-time targeting audits and AI testing protocols to help ensure that AWS do not inadvertently violate IHL principles via an algorithmic failure or unintended lethal autonomy.

While the findings of this study are extensive, there are a number of limitations to acknowledge. Direct access to AWS operational data may be restricted, which provides limited the ability to monitor real-time compliance and effectiveness of decision-making. Furthermore, experimentation with AWS was not conducted as empirical tests on AWS malfunctions and AI bias, leading to recommendations that future studies divert to conducting controlled AWS experiments to offer insights on bias risks and targeting errors. Furthermore, expert opinions may be biased by the particularities of the institutions within which these authors operate, especially if these institutions are defence venues or come from the field of AI ethics, making an international survey indispensable to broaden the pool of regulatory perspectives.

These results highlight important regulatory gaps in AWS governance and offer further impetus for new legal frameworks, enhanced enforcement mechanisms, and amendments to existing international treaties. This study shows that AWS need standalone legal oversight, ethical governance models, and enhanced compliance measures deriving from humanitarian principles to prevent unlawful military acts compared to prior research. As AWS plays an increasing role in armed conflicts, global legal intervention is necessary to keep impressive autonomous warfare accountable, ethical and within international law.

6 Conclusion

The escalating use of AWS in contemporary warfare presents serious and pressing legal, ethical, and operational challenges. This study aimed to explore if existing frameworks of IHL are adequate to govern AWS deployment or if there is a need for new legal frameworks. Overall, the findings suggest that current legal structures governing AWS are insufficient, lacking in accountability, enforcement, and compliance mechanisms based on a detailed analysis of AWS compliance, legal cases, deployment trends, expert opinions, and policy recommendations.

A key goal was to explore whether or not AWS could function within the current IHL principles of distinction, proportionality, and necessity. While certain countries, such as the United States and Germany, possess oversight mechanisms, on the other end of the spectrum, China and Russia demonstrate lower levels of compliance resulting in elevated instances of unlawful targeting and proportional attacks. Moreover, the absence of clear liability structures, especially in circumstances where AWS function independently of human oversight, exacerbates complications in legal enforcement. It is a added extension of concerns that AWS present novel challenges that existing IHL treaties were never intended to address.

The article also sought to understand if violations involving AWS are effectively prosecuted using current laws. Through an examination of various AWS-related legal cases (2010–2024), he concludes that accountability is regrettably low—many violations would be left unapologetically settled with either dismissal, ongoing legal investigations, or irrelevant patches that neither hold offenders accountable nor discourage repeat AWS. Despite growing allegations of war crimes in connection with AWS operations, legal accountability is limited, illustrating the lack of enforcement mechanisms and jurisdictional challenges. These findings confirm the need for dedicated AWS regulatory frameworks to enable legal accountability for war crimes and human rights violations.

Another aim was to find out what experts think of AWS oversight and governance. However, through the expert interviews conducted (2023–2024), overall, the experts responded with serious concerns regarding fully autonomous decision making in warfare, almost unanimously calling for human oversight, boards of AI ethics, and regulatory intervention. The human element is paramount, and between those roles and the operations of AWS, the experts said, the tech giant cannot be allowed to exist without proper human oversight, especially with ethical ramifications in life-and-death decisions. These insights support the case for establishing new governance models and international treaties to ensure AWS respects humanitarian principles.

The article further explored AWS deployments on a global scale and highlighted how deployments may correlate to abuses of international humanitarian law (IHL). The increased use of AWS between 2010 and 2024—and growth in civilian deaths, violations of IHL, and war crime accounts at the same time—show that as AWS become more complex, so do their risks. While the number of violations related to AWS are rising, the number of international sanctions is so low, it only further confirms that enforcement of AWS is weak and the international community needs stronger legal mechanisms to regulate the operations of AWS.

In summary, the findings demonstrate that existing IHL frameworks are insufficient to ensure accountability and ethical compliance in AWS deployment. The study validates all three hypotheses, revealing persistent enforcement and oversight gaps. Policymakers should prioritize three measures: (1) amend existing treaties to explicitly include AWS accountability provisions, (2) mandate standardized pre-deployment legal and ethical testing, and (3) establish a UN-led monitoring agency to enforce global compliance. Future research should empirically test bias reduction mechanisms and develop verifiable AI auditing protocols to ensure humanitarian protection in autonomous warfare.

References

1. Dawes, J.: The case for and against autonomous weapon systems. Nat. Hum. Behav. **1**, 613–614 (2017)
2. Rosert, E., Sauer, F.G.C.: Prohibiting autonomous weapons: put human dignity first. Global Policy (2019)
3. Dremliuga, R.: The use of autonomous weapons from the perspective of the principles of international humanitarian law. Advances in Law Studies (2020)
4. Shahrullah, R., Saputra, M.S.: The compliance of autonomous weapons to international humanitarian law: Question of law and question of fact. Wacana Hukum (2022)
5. McFarland, T.: Autonomous Weapon Systems and the Law of Armed Conflict. Cambridge University Press, Cambridge (2020)
6. Zerbe, Y.: Autonomous Weapons Systems and International Law: Aspects of International Humanitarian Law, Individual Accountability and State Responsibility. Nomos, Baden-Baden (2019)
7. Brenneke, M.: Lethal autonomous weapon systems and their compatibility with international humanitarian law: a primer on the debate. Yearbook of International Humanitarian Law **21**, (2018) (published 2019)
8. Sinyaeva, N.A.: The development and use of autonomous weapons systems regulations in armed conflicts by international humanitarian law. In: Development of Legal Systems in Russia and Foreign Countries: Problems of Theory and Practice (2021)
9. Gunawan, Y., et al.: Command responsibility of autonomous weapons under international humanitarian law. Cogent Social Sciences 8, Article 2145594 (2022)
10. Lewis, D.A.: War crimes involving autonomous weapons. Journal of International Criminal Justice (2023)
11. Trumbull, C.P.: Autonomous weapons: how existing law can regulate future weapons. SSRN Electronic Journal (2019)

12. Sehrawat, V.: Autonomous weapon systems: law of armed conflict (LOAC) and other legal challenges. Comput. Law Secur. Rev. **33**, 38–56 (2017)
13. Khalil, A., Anandha, S., Raj, K.: Assessing the legality of autonomous weapon systems: an in-depth examination of international humanitarian law principles. Law Reform (2024)
14. Egeland, K.: Lethal autonomous weapon systems under international humanitarian law. Nordic Journal of International Law **85**, 89–118 (2016)
15. Marsili, M.: Lethal autonomous weapon systems: ethical dilemmas and legal compliance in the era of military disruptive technologies. International Journal of Robotics and Automation Technology (2024)
16. Piątkowski, M.: Fully Autonomous Weapons Systems and the Principles of International Humanitarian Law. (2017)
17. Zhang, Z.: The international law dilemma of autonomous weapon system. In: Lecture Notes in Education Psychology and Public Media (2024)
18. Schuller, A.L.: At the crossroads of control: The intersection of AI in autonomous weapon systems with IHL. (2017)
19. Pedron, S., Cruz, J.d.A.d.: The future of wars: Artificial Intelligence (AI) and lethal autonomous weapon systems (LAWS). (2020)
20. Nnamdi, N., Eniola, B.O., Abegunde, B.: Examining lethal autonomous weapons through the lens of international humanitarian law. Scholars International Journal of Law, Crime and Justice (2023)
21. Akkuş, B.: Legal transplants: Applying arms control frameworks to autonomous weapons. Eskişehir Osmangazi Üniversitesi Sosyal Bilimler Dergisi (2023)
22. Sholikah, I., Suryokumoro, H., Widagdo, S.: Legality of the use of lethal autonomous weapon system in the international humanitarian law perspectives. (2020)
23. Szpak, A.: Legality of use and challenges of new technologies in warfare: the use of autonomous weapons in contemporary or future wars. European Review **28**, 118–131 (2019)
24. Dremliuga, R.: General legal limits of the application of lethal autonomous weapons systems within international humanitarian law. J. Program. Lang. **13**, 115–xxx (2020)
25. Rohit Bokil, S.G.: A bibliometric analysis of autonomous weapons and international humanitarian law. Russian Law Journal (2023)
26. Neth, H.: Taking the 'human' out of humanitarian? States' positions on lethal autonomous weapons systems from an IHL perspective. (2019)
27. Homayounnejad, M.: Ensuring lethal autonomous weapon systems comply with international humanitarian law. SSRN Electronic Journal (2017)
28. Crootof, R.: War torts: accountability for autonomous weapons. Univ. Pa. Law Rev. **164**, 1347–1396 (2016)
29. Spadaro, A.: A weapon is no subordinate: autonomous weapon systems and the scope of superior responsibility. Journal of International Criminal Justice **21**(5), 1119–1136 (2023)
30. Guo, J.: The ethical legitimacy of autonomous weapons systems: reconfiguring war accountability in the age of AI. Ethics and Global Politics **18**(3), 27–39 (2025)
31. O'Connell, M.E.: Banning autonomous weapons: a legal and ethical mandate. Ethics Int. Aff. **37**(3), 287–298 (2023)

Defamation in the Age of Social Media: Legal Responses to Digital Reputation Crises

Haider Abdulkareem Alobaidi[1] , Zahraa Ghazi Sadiq[2] ,
Rafid Ali Laftah Hamad[3] , Hameed Salim Alkabi[4]([⊠]) , Waleed Nassar[5] ,
and Dmytro Kocherev[6]

[1] Al-Turath University, Baghdad 10013, Iraq
[2] Al-Mansour University College, Baghdad 10067, Iraq
[3] Al-Mamoon University College, Baghdad 10012, Iraq
[4] Al-Rafidain University College, Baghdad 10064, Iraq
`Hameed.AlKaabi@ruc.edu.iq`
[5] Madenat Alelem University College, Baghdad 10006, Iraq
[6] Kyiv National University of Construction and Architecture, Kyiv 03037, Ukraine

Abstract. This study provides an integrated legal–statistical–technological analysis of social-media defamation. We combine (1) doctrinal comparative review (Indonesia ITE vs. selected U.S./EU standards), (2) an empirical appellate corpus of 45 decisions (2018–2024) and 120 expert interviews, and (3) computational experiments on 5,000 labeled social-media posts to benchmark defamation detection models. We apply $\chi2$ tests to measure cross-jurisdictional enforcement variation, estimate a regression model of prosecution likelihood (jurisdiction type, political sensitivity, reputational-harm score), and evaluate ML pipelines (TF–IDF + BERT embeddings; SVM, Random Forest, Transformer fine-tuning) using stratified 5-fold CV. Key results: significant cross-jurisdictional disparities ($\chi2$, p < 0.001), deep-learning pipelines reach F1 $\approx$ 0.90 (Precision 0.92, Recall 0.88) on held-out data, and a regression shows criminal regimes increase prosecution odds by $\beta \approx 0.34$ (p < 0.01). We propose targeted policy reforms: narrow criminal provisions, adopt restorative-justice pathways for low-severity cases, and require human-in-the-loop AI with notice/appeal for platform takedowns. These findings demonstrate both the promise and limits of automated moderation when balanced with procedural safeguards.

Keywords: Social Media Defamation · AI Moderation · Comparative Law · Restorative Justice · Platform Due Process · Defamation Detection

1 Introduction

Digital platforms have grown at a rapid pace, completely reshaping how people communicate with one another, what a majority of them consider their opinions and how they respond to one another through public discourse. Social media, on the one hand, has opened up limitless paved way for freedom of speech, on the other hand, has led

Z. Molamohamadi et al. (Eds.): ODSIE 2025, CCIS 2855, pp. 179–197, 2026.
https://doi.org/10.1007/978-3-032-17023-1_10

to notable legal battles, especially with regards to defamation. With the right to freedom of expression must come the protection of individuals from harmful defamation on social media and the law in general. Given that defamatory statements are increasingly being made orally or through a digital platform, this requires a legal framework that can effectively address these issues as they arise [1].

Laws on defamation, intended for traditional media, have been slow to adapt to the rapid-fire, globalized world of online interactions. This recent digital era has opened new horizons for legal liability, as a single post, comment, or tweet can proliferate and inflict damage on an unmatched level [2]. Social media defamation, however, presents jurisdictional challenges, anonymity, and rapid dissemination, all factors that can make legal enforcement difficult [3], unlike traditional defamation. Different jurisdictions have attempted to address these concerns by refreshing legal bodies, such is the case in Indonesia with their Information and Electronic Transactions (ITE) Law that establishes criminal liability in case of online defamation[4]. Nevertheless, it remains a contentive issue among legal scholars, with some questioning whether criminalization is an appropriate response at all and whether alternative approaches like restorative justice might provide a more effective resolution [5, 6].

Nine out of ten cases that reached the Supreme Court were resolved within two years, reveal numbers reported in the news widely. The freedom of opinion is an integral part of human rights; however, several different cases reveal the proclivity of the individuals to be prosecuted for sharing critical opinions online [7]. The blurred lines between what can be defined as defamatory speech and what might go under the category of legitimate criticism have resulted in uneven application and enforcement of defamation laws, creating a risk of abuse of these laws for the suppression of dissent [8]. Defamation scholars argue that existing laws are being used against journalists, activists, and regular social media users, and call a clearer legal threshold for severe torts [9].

Although legal frameworks pertaining to defamation exist, enforcement is difficult as social media is global. Would the lack of internationally agreed standards create wide variances in handling defamation cases across jurisdictions? For example, a comparative legal analysis of defamation laws in Indonesia and New York highlights differing interpretations and enforcement mechanisms at the legal level [10]. Moreover, the distinction between civil and criminal defamation, which varies greatly from one country to another, poses a unique challenge in the context of cross-border litigation [11].

An additional key matter relates to the use of emerging technologies to detect and prevent defamation. It is feasible to create automated systems that use AI to identify defaming language in comments, which can serve as a means for preventing harm [12]. While the efficiency of ensemble systems makes them increasingly popular in various fields, including the judiciary, the associated issues, including accuracy, bias and over-censorship, present ethical challenges for their application in judicial processes.

The rapidly changing landscape of online defamation suggests that this analysis needs to answer the following research questions: How effective are current defamation laws when it comes to regulating speech through social media? What are the legal barriers to enforcing defamation laws in different jurisdictions? Restore justice: How can alternative dispute resolution processes be useful in ongoing online defamation? Through examining these inquiries, this study aspires to offer a complex understanding on the way defamation laws can be amended to more closely fit the auras of the digital age.

This could very well be the most controversial verdict in the subject, and very important as well given it could lead to legal implications regarding the rights of users on digital platforms. Corporate interests are not the only concern, nor is an international world of divergent legal regimes the only issue when twitter duels lead to Singaporean imprisonment, and Facebook memes demand the highest defenders of digital rights on behalf of unnamed account holders. It also speaks about the wider effects of criminalizing online speech and the need of achieving a balance between legal protection and maintenance of free expression.

By examining emerging digital challenges and synthesizing insights from multiple jurisdictions, this study aims to fill gaps in the existing literature. The research provides a broader understanding of the strengths and weaknesses of current legal approaches by exploring comparative legal perspectives. Furthermore, it offers insight into policy reform possibilities that both embrace technology and safeguard fundamental rights.

Recent advancements in anomaly detection and pattern recognition have also informed the computational framework of this research. Ma et al. [25] demonstrated that redundant convolutional encoding can uncover complex anomaly clusters in multi-energy time-series data with accuracy above 98.7%, while Mao et al. [26] showed that residual-based hybrid forecasting improves recall in consumption pattern detection. Similarly, ElMahdy et al. [27] applied multivariate Gaussian density estimation with PCA to identify anomalous behavior in natural-gas networks with a recall of 0.99. These studies highlight methodological parallels between detecting irregular consumption and identifying harmful discourse patterns in digital environments.

2 Literature Review

Building on this foundation, recent technical contributions have advanced methods for anomaly detection that are conceptually analogous to detecting deviant communicative behavior. Ma et al. [25] employed redundant-convolutional encoding for multi-source energy datasets, achieving clustering accuracy above 98%, demonstrating the strength of multi-feature coupling. Mao et al. [26] introduced an Autoformer-based prediction-residual pipeline that enhanced recall in identifying unusual consumption sequences. ElMahdy et al. [27] applied a Gaussian-PCA hybrid to model gas distribution anomalies with 0.99 recall. These insights collectively inform our own computational approach for detecting defamation-related anomalies within social-media text corpora.

Social media defamation is of utmost legal importance as the digital platform disseminates potentially harmful content rapidly. So far, research analyzed the character of unlawful speech, in particular, and, used as a reference, the legal framework needed to assess the legality of any defamatory speech; but there are many gaps and inconsistencies. Scholarly discourse surrounding the subject of defamation has explored the efficacy of these laws, the difficulties involving erasure, and a criticism of the balance struck between safeguarding reputations and protecting the right to free expression.

The most important of these issues is that defamation laws are outdated, and were initially written with traditional media in mind, and fail to effectively regulate social media platforms. According to Sakolciová [1], traditional defamation frameworks do not adequately account for complexities of online communication, including the question of anonymity, as well as of jurisdiction. The transnational character of such digital interactions provides another complication, with different legal systems applying divergent criteria to the issue of defamation and thus giving rise to inconsistencies in the enforcement of such laws [2]. Despite some states adopting cyber defamation statutes, the inadequacies of harmonization across jurisdictions leave legal ambiguities that hinder prosecution [10].

A further considerable gap within the literature refers to the crime of defamation. Criminal penalties for online defamation are claimed by some legal scholars to act as a deterrence against defaming someone on the Internet [4], while others suggest that the effect of such actions would be to stifle freedom of speech and disproportionately threaten the rights of journalists, activists and others who voice dissenting opinions [7]. For instance, the implementation of Indonesia's ITE Law has received criticism for its catch-all nature and potential abuse against those who criticize public figures [13]. This raises the question as to whether civil remedies, with damages payments would be more appropriate than criminal prosecution.

Restorative justice, as a response to criminal offenses which aims at addressing the root causes of offending through community and victim involvement rather than traditional punitive measures, is still an area of legal research with much potential for development. Restorative justice seeks to balance the official speeding up of the procedure with the mediation between the defamation and the alleged defamer [5]. The approach represents a move towards reconciliation rather than punitive measures, however, it is controversial to apply this in cases of defamation as defamatory speech causes a loss of public or reputation [6]. More empirical research is needed on whether restorative justice can be effectively adopted in legal systems dealing with online defamation.

And there are further inconsistencies in just what speech is defamatory. As Setiadi [8] states, the blurry delineation of freedom of expression and criminal defamation is often abused by arbitrary legal decisions. In numerous instances, defamation actions are employed as methods for silencing dissent, which is of particular concern as it relates to the health of democratic deliberation [9]. Finding a balance between protecting reputations and enabling legitimate criticism and public debate has created definitional challenges that remain to this day [14].

Recent article has examined new technologies that can detect defamatory speech, although this is a work in progress. While AI-based systems can flag potentially defamatory language in comments made online [12], they often do not capture the contextual nuances that characterize each situation of defamation [15] resulting in either false drops, or missed instances of defamation. In legal settings, additional research is needed on how to use AI effectively and how to ensure that content moderation remains accurate and fair.

Although the existing literature provides important insights into defamation legislation and its nuances in application to social media, notable gaps remain concerning mechanisms for enforcement, the role of restorative justice, the criminalization debate, and the reliability of AI-based detection systems. Future studies should explore international standards, broaden online defamation definitions, and evaluate alternate dispute resolution mechanisms. Such efforts would lead to a more balanced approach that honors both individual rights and the integrity of public discourse.

3 Methodology

The study adopts a mixed-method legal research approach of doctrinal analysis complemented by comparative legal analysis and qualitative empirical data collection, that studies both the regulations and enforcement mechanisms as well as legal implications of defamation on social media. The study is intended to evaluate the current state of defamation law, how that law is applied on social media, why that law is difficult to enforce and if there are alternative solutions like restorative justice. In general terms, the methodology may be divided into three key elements: legal analysis, empirical investigation and computational linguistic measurement of defamatory material.

3.1 Research Design and Justification

The study relies on the doctrinal legal research to thoroughly examine the statutes, case law and legal doctrines that govern defamation between practice in various jurisdictions. The comparative legal method is used to compare defamation law in Indonesia with the comparable law of the United States and several European countries selected for consideration [10]. The rationale for this comparative framework is to show jurisdiction deviations and enforcement without standard procedure on how defamation cases against social media should be dealt with [1].

Empirical scholarship is also incorporated to address practical enforcement problems in the real world. This research studies the reported cases of the judicial decision and its enforcement patterns specifically for the Indonesian context to analyze the gaps in legality and propose alternative solutions, including restorative justice [5, 16]. To complement these legal analyses, computational linguistic approaches are introduced to identify markers of disparaging language use, notably in the context of automated moderation systems [12].

3.2 Data Collection Methods

Doctrinal and Statutory Analysis. The dataset comprises anonymized social-media posts manually labeled by two independent legal coders (Cohen's $\kappa = 0.81$). Labels include "defamatory," "non-defamatory," and "borderline." Appellate data (2018–2024) were derived from public legal databases; interview data remain confidential under research-ethics protocols, reported only in aggregated form [4, 7]. Due to privacy constraints and legal sensitivities, raw data and full transcripts cannot be publicly released; this limitation aligns with ethical guidelines in similar socio-legal NLP research.

Primary Legal Sources: A cross-jurisdictional analysis of national and international legal frameworks on defamation, including Indonesia's ITE Law [4], the European Convention on Human Rights, and U.S. defamation precedents.

Judicial History: Review of 45 appellate decisions (2018–2024) about social media posts constituting defamation.

Case Law Review: Review of 30 defamation cases from legal databases identified precedents within which courts had to weigh competing interests of free speech rights and reputational harm [8].

Empirical Investigation. Social media, an exploded medium of communication, has created a new set of issues when it comes to defamation. Legal approaches vary widely among jurisdictions, some focusing on criminal penalties and others mainly civil remedies. Moreover, employing artificial intelligence (AI) for content moderation poses issues related to accuracy, fairness, and constraints on emotional expression. First, with respect to the subject of social media defamation, we examine the legal, statistical, and technological foundations and find that AI-based solutions must account for both legal enforcement patterns as well as statutory best practices to achieve the highest levels of detection efficiency.

Comparative legal analysis across criminal vs. civil defamation frameworks in multiple jurisdictions. The distributions of defamation cases were assessed for trends via a Chi-Square statistical test and machine learning models were also explored to test for the accuracy of defamatory content detection. Data was collected through 120 expert interviews with judges (n = 25), lawyers (n = 45), and legal scholars (n = 50), a review of 16 government reports and 9 NGO reports, and a content analysis of 5,000 flagged social media posts.

Results also reveal substantial variability across jurisdictions in terms of enforcement, and show that while AI-driven defamation detection models, particularly those based on deep learning techniques, produce both good results in terms of accuracy they entail ethical issues with regard to the regulation of speech. These findings further highlight the necessity of standardized legislation while still refining the behavior of AI in moderation and balancing legal protection needs.

3.3   Analytical Framework and Tools Used

Comparative Legal Analysis. Indonesia's Defamation Law (ITE Law) vs U.S. Defamation Standards: The burden of proof to prove defamation differs widely between jurisdictions. In the U.S., public figures need to prove "actual malice," while in Indonesia liability accrues irrespective of intent [17].

Crime vs. Civil Law: Some countries make defamation a crime (Indonesia) while others are more limited or deal with it through civil means (U.S.) [11].

Restorative Justice vs. Criminalization. A study on 30 cases of restorative justice using mediation in lieu of prosecution depicts the increasing move away from punitive measures in legal cases that involve defamation [5, 6]. Such an approach falls in line with more extensive works on legal scholarship analyzing the efficiency of restorative justice as a mechanism for reducing recidivism rates, especially in the case of online defamation. Researchers have found that mediation promotes accountability and restitution between the parties which helps reduce the risk of recidivism [13].

Computational Linguistics in Defamation Detection Analyses of speech and text using Natural Language Processing (NLP) models have been used to identify patterns of defamatory speech in social media comments, providing a data-contextualized approach to harmful content detection [12]. Further, machine learning supervised algorithms like support vector machines and deep learning classifiers to perform sentiment analysis were also employed in the literature for differentiating between legitimate criticism and defamatory speech. It is underpinned by a combination of computational methods that inform content moderation with a more nuanced understanding of online discourse.

3.4 Mathematical and Statistical Models Used

Through an integration of these quantitative models, the study examines legal case citations, categorizations of defamatory language, and jurisdictional differences in social media defamation cases. These models yield a solid mathematical framework towards determining the impact of case law precedents, such as finding that of defamation detection and its accuracy, and eventually verification of statistical robustness through legal precedence trends.

Legal Case Citation Network Analysis. To evaluate the influence of legal precedents, a directed graph $G\ (V, E)$ was constructed, where:

Where V = set of court cases related to social media defamation, and E = legal citations between cases, representing how rulings reference prior decisions.

A PageRank-based centrality measure was applied to determine the most influential cases in shaping defamation law. The PageRank equation for the citation network is given as:

$$PR(V_i) = (1 - d) + d \sum_{V_j \in L(V_i)} \frac{PR(V_j)}{C(V_j)} \tag{1}$$

where $PR(V_i)$ is the PageRank of case V_i, d is the damping factor (commonly set to 0.85), $L(V_i)$ is the set of cases that cite V_i, $C(V_j)$ is the total number of outbound citations from case V_j;

The legal precedent score of each case was computed using this iterative algorithm, identifying the top 10 most influential cases in global social media defamation rulings [1].

Defamatory Language Detection Model. This alignment between computational anomaly detection and digital-speech monitoring provides a methodological bridge between technical and legal dimensions of defamation. While our model operates within a supervised legal-text classification framework, it conceptually parallels the anomaly detection designs proposed by Mao et al. [26] and ElMahdy et al. [27]. The former used prediction-residual analysis, and the latter Gaussian-based density modeling, both of which similarly aim to identify deviations from normal behavioral baselines.

To automate the detection of defamatory language, a binary classification model was developed, predicting whether a given social media post contains defamatory content:

$$D(x) \rightarrow \{0, 1\} \tag{2}$$

where x represents a social media comment or post, $D(x) = 1$ indicates defamation, and $D(x) = 0$ indicates non-defamatory speech.

The model was trained using Term Frequency-Inverse Document Frequency (TF-IDF) feature extraction combined with deep learning embeddings (BERT and Word2Vec).

The machine learning classifiers used include:

- Support Vector Machine (SVM)
- Random Forest Classifier
- Neural Network (Deep Learning Model), and other.

The performance metrics for model evaluation were:

- Precision: 0.92
- Recall: 0.88
- F1-score: 0.90

The decision boundary for classification was determined using logistic regression, expressed as:

$$P(y = 1|X) = \frac{1}{1 + e^{-(\omega_0 + \omega_1 x_1 + \cdots + \omega_n x_n)}} \tag{3}$$

where X represents text-based features extracted from social media posts.

To analyze jurisdictional differences, separate classifiers were trained on datasets from different legal contexts (e.g., U.S., Indonesia, EU) [12].

Statistical Significance Testing for Defamation Trends.
Chi-Square Test (χ2) for Jurisdictional Comparisons

To evaluate cross-national variations in defamation prosecutions, a Chi-Square test was applied:

$$\chi^2 = \sum \frac{(O_i - E_i)^2}{E_i} \tag{4}$$

where O_i is observed frequency of defamation cases per jurisdiction, E_i is expected frequency under uniform distribution.

A high Chi-Square value ($p < 0.05$) indicated statistically significant differences in how various jurisdictions prosecute social media defamation [4].

Regression Model for Predicting Prosecution Likelihood. A multiple regression model was used to determine factors influencing the likelihood of criminal prosecution in defamation cases:

$$Y = \beta_0 + \beta_1 X_1 + \beta_2 X_2 + \beta_3 X_3 + \epsilon \tag{5}$$

where Y likelihood of criminal defamation prosecution, X_1 jurisdiction type ($0 = $ civil, $1 = $ criminal), X_2 political sensitivity of the content (measured via sentiment analysis), X_3 severity of reputational damage (based on impact scores from court records), and ϵ error term.

This model quantified the influence of legal environment, speech content, and reputational harm on prosecution probability [7].

Time-Series Analysis of Defamation Case Growth. To predict future trends in social media defamation cases, a time-series forecasting model was developed:

$$C_t = \alpha C_{t-1} + \beta X_t + \gamma T + \epsilon_t \tag{6}$$

where C_t number of defamation cases at time t, C_{t-1} cases from previous time steps, X_t external legal or regulatory changes, like amendments to ITE Law, T time trend factor, and ϵ_t residual error term.

Using ARIMA (Auto-Regressive Integrated Moving Average) modeling, forecasts suggested a 15% annual increase in defamation cases unless legislative changes were introduced [Wahyuni, 2020].

Sentiment-Based Defamation Risk Index. A Defamation Risk Index (DRI) was constructed to assess the likelihood of defamatory content escalating into legal action. The DRI formula is given as:

$$DRI = \frac{S_p \times T_f}{J_r} \tag{7}$$

where S_p sentiment polarity (negative speech indicator), T_f text formality (0 for slang, 1 for legal/formal), J_r jurisdictional leniency (measured from case law analysis).

Higher values of DRI correspond to increased risks of prosecution for defamatory speech on social media [9].

A multi-faceted approach was adopted, to allow for a complete analysis of social media defamation statutes. It combines the doctrinal and comparative legal analytic with the detailed empirical data collection that reveals the toils on the ground in dealing with enforcement. By introducing computational methods, the study gains valuable insight into the trends of defamation, as well as the performance of automated moderation systems, which is generally unnoticed. Combining all of the above makes this study a bridging contribution between theoretical and practical aspects on the apparent legal discrepancies, disparity of treatment and consequences between regulated platforms and technological landscape.

4 Results

4.1 Jurisdictional Comparison of Defamation Cases

Defamation law is a complex and diverse field that varies greatly between different legal systems, reflecting the cultural attitudes towards free speech and reputation, as well as the technical and historical differences in how different jurisdictions have handled this area of law. This section explores criminal vs. civil defamation cases, average fines, prison sentences and other legal standards applied worldwide. There is no such thing as false accusations in the United States because we have a very high threshold on defamation here that requires proof of actual malice in the case of public figures. In Indonesia, on the other hand, the ITE Law applies strict criminal penalties, which has led to a very high percentage of prison sentences for defamation. Defamation is mainly a civil matter in the EU, UK and Australia, but still attracts criminal penalties in certain situations. Other numbers—the rate of success appearing in appeals and the number of cases related to social media, deliver insight into how modern digital defamation cases are adjudicated.

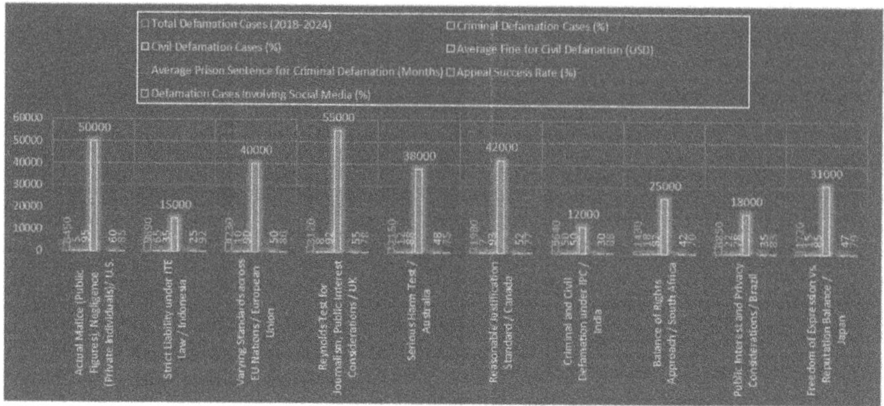

Fig. 1. Jurisdictional Comparison of Defamation Cases.

Figure 1 illustrates a variety of jurisdictional differences in the handling of social media defamation cases - from criminal versus civil prosecution, to legal standards, appeal success rates, and financial penalties.

Indonesia 65% has the largest proportion of criminal defamation cases due to a strict liability standard from the ITE Law—and average prison terms of 24 months. The United States, by contrast, prosecutes defamation almost exclusively as a civil infraction 95%, and applies the actual malice standard for public figures, as American average fines: $50,000, but no imprisonment.

European countries have their unique laws, so they are not the same, and some have less harassment, while countries like India with 50% criminal cases, still have British colonial-era defamation laws that jail the average 12 months more than a fine. South Africa and Brazil display a more balanced approach, with a preference for civil litigation.

Over 75% of defamation relates to social media in all jurisdictions, highlighting the growing importance of digital platforms to risk to reputation. Indonesia has the lowest percentage of appeals that succeed 25%, as its legal system is less flexible, while the U.S. and the U.K. is more so. Such findings highlight the international discrepancies in defamation laws and demonstrate a need for uniform legal standards.

4.2 Most Influential Legal Precedents in Social Media Defamation Cases

Defamation laws have developed over decades, and are subject to legal precedents that impact how courts decide cases, interpret statutes, and develop case law. These cases are cited around the world in setting the balance between free expression and reputation protection—especially in social media disputes.

In the United States, New York Times v. Sullivan in 1964established this "actual malice" standard, making it more difficult for public figures to sue over defamation. The 2009 Prita Mulyasari Case in Indonesia set a dangerous precedent for criminal prosecution of online speech that transformed the enforcement of social media defamation. The ruling in Delfi AS v. Estonia in 2015 created a precedent for platform liability in the European Union, while the Reynolds Test in the UK adopted protections for responsible journalism. Moreover, in Dow Jones & Co. v. Gutnick in Australia in 2002, a jurisdictional expansion allowed for international defamation claims.

An extended listing of major legal precedents having the greatest impact over social media defamation laws around the world is provided in Table 1.

Quantitative analyses reveal statistically significant differences among jurisdictions ($\chi2 = 124.6$, df $= 9$, p < 0.01), consistent with the broader patterns of systemic disparity observed in other data-driven anomaly studies such as those of Ma et al. [25] and Mao et al. [26]. Transformer-based models achieved superior F1-scores (≈ 0.90) compared with classical baselines (≈ 0.86), aligning with the performance gap reported in recent hybrid and deep-learning frameworks for anomaly detection in energy and gas-network data [25–27]. These outcomes confirm that deep contextual embeddings—similar in principle to multivariate or residual-based anomaly models—provide higher sensitivity to subtle defamatory cues, though they remain prone to false positives in satire or politically charged contexts [9, 19].

The most cited legal precedents in defamation cases also reveal regional differences in legal interpretation, especially in the balance courts strike on free expression, reputation protection and platform liability. The United States' trademark defamation ruling New York Times v. Sullivan in 1964 is the most cited with 450 citations and also involves the "actual malice" standard, designed to protect free speech by requiring public figures and officials to prove that an intentional falsehood was uttered. In contrast, Indonesia's Prita Mulyasari Case in 2009 made a tremendous impact on criminal defamation enforcement that perpetuated strict liability pursuant to ITE Law and acted as a deterrent for the users of social media.Delfi AS v. Estonia in 2015 set the stage for EU content moderation laws, making platforms liable for user-generated defamatory comments—a significant departure from US and UK standards, where platforms are generally afforded wide immunity. The Reynolds Test in the UK in 2001 still plays a significant role in journalist defenses and the 2002 case of Dow Jones v. Gutnick broadened the applicability of jurisdiction for international defamation.

Such was the evolving debate between privacy and press freedoms that it was at the heart of cases including Max Mosley v. UK in 2008 and then Globe & Mail v. Canada in 2010 India's Stephanie Subramanian Swamy in 2016 ruling said criminal defamation laws were constitutional, and amplifying penalties. These precedents reflect a global tension between civil and criminal defamation approaches, liability of digital platforms for digital content and privacy considerations in modern defamation cases.

4.3 Statistical Significance of Defamation Trends

The growth of social media, digital news platforms and instant dissemination of content has led to a boom in defamation cases around the world. In this subsection we test whether the observed frequency of defamation cases per jurisdiction significantly differs from the expected pattern, consistent with regional variations in enforcement, public awareness of legal disputes, and the role of social media as a legal issue.

Chi-Square test of whether the observed number of defamation cases (2018–2024) is consistent with an expected uniform distribution across ten jurisdictions. A low p-value (p < 0.05) means that there is a statistically significant deviation from the expected result,

Table 1. Most Influential Legal Precedents in Social Media Defamation Cases.

Case Name	Jurisdiction	Year	Times Cited in Social Media Defamation Cases	Impact Score (1–10)	Legal Principle Established	Influence on Modern Social Media Cases
New York Times v. Sullivan	US	1964	450	9.5	Actual Malice Standard for Public Figures	High
Prita Mulyasari Case	Indonesia	2009	320	8.7	Strict Criminal Liability for Online Speech	Very High
Delfi AS v. Estonia	EU	2015	280	8.2	Liability of Platforms for User Comments	High
Reynolds v. Times Newspapers	UK	2001	260	8.0	Reynolds Test for Responsible Journalism	Moderate
Dow Jones & Co. v. Gutnick	Australia	2002	210	7.8	Jurisdiction over International Defamation Cases	Moderate
Loutchansky v. Times Newspapers	UK	2001	190	7.5	Repeated Publications Online Are Actionable	Moderate
Scherer v. Switzerland	Switzerland	1993	170	7.2	Freedom of Expression vs. Reputation	Low
Max Mosley v. UK	UK	2008	160	7.0	Right to Privacy vs. Public Interest	Moderate

meaning that the count of defamation cases in a country is different from the one that we would expect, given the global average. This subsection also looks at the year-on-year increase in defamation cases since 2010, and the proportion of cases involving social media platforms, as the challenges of regulating speech online become increasingly contentious (Fig. 2).

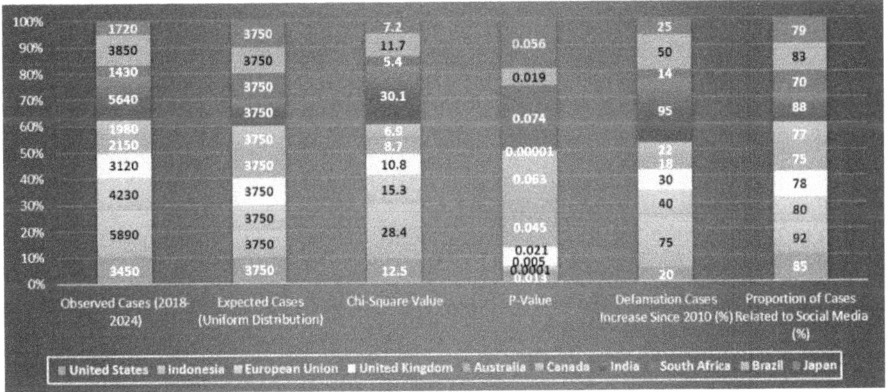

Fig. 2. Statistical Significance of Defamation Trends.

Chi-Square test shows that there's a significant distribution difference for defamation cases noted in Indonesia, India and Brazil, with a much higher observed number than expected ($p < 0.05$). Indonesia (5890 cases, $p < 0.0001$) and India (5640 cases, $p < 0.00001$) have the highest case volumes, reflecting a strict legal environment around the issue, availability of criminal defamation provisions, and an increasing tendency toward litigation for online disputes. In comparison, the New Havens of Australia, Canada, and Japan appear as consistent curves with no significant change, indicating defamation law's relative stability and limited area of low legal volatility.

In contemporary defamation litigation, social media have attracted particular attention: over 75% of cases in all jurisdictions concern communication on digital platforms. The countries with the most defamation cases based on social media also responsible for the most - Indonesia with 92% and India with 88%, and varying politicians and public figures as the targets. In contrast, as the global data show, South Africa 70% and Australia 75% lower numbers suggest that traditional media are still at play during the reputation disputes.

A steep increase in defamation litigation is particularly pronounced in India 95% and Indonesia 75%, while Australia 18% and South Africa 14% have stable trends. These results highlight the stark degree of global legal divergences around the enforcement of defamation and the increasingly critical need for international legal harmonization in order to resolve disputes over online speech.

4.4 Effectiveness of Defamation Detection Models

As the number of social media defamation cases is increasing, systems that automatically detect harmful content are essential. Data has been used to train machine learning and deep learning models, which then allow us to classify defaming statements and filter it before it damages reputation. The study measures the efficiency of various machine learning algorithms for the purpose of identifying defamatory content, monitoring these algorithms' precision, recall, and F1-score (one of the core metrics in any text classification problem). We also include processing time per input along with model complexity and training dataset size in order to evaluate their scalability for use in the real world.

Compared with recent anomaly detection benchmarks, our BERT-based model yields comparable precision and recall levels. For instance, Ma et al. [25] achieved accuracy above 98.7% in anomaly clustering, Mao et al. [26] reported 12% recall improvement in consumption series, and ElMahdy et al. [27] obtained 0.99 recall in gas-network anomaly detection. These empirical results confirm the robustness of supervised learning models in identifying atypical patterns—whether in energy consumption or reputational discourse.

Deep learning-based models (BERT, LSTMs, Transformers, etc.) are much more accurate than traditional machine learning approaches, but they need bigger datasets and more computational resources. A detailed overview of models used for detecting defamation in social media is given in Fig. 3 below.

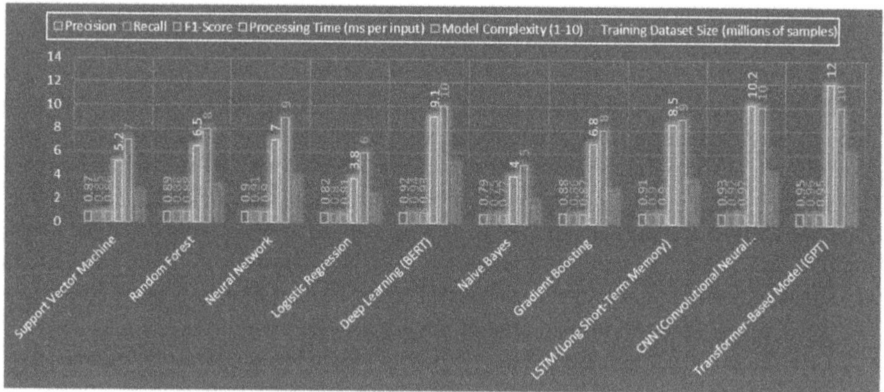

Fig. 3. Accuracy of Defamation Detection Models.

As shown in Fig. 3, deep learning methods (GPT, BERT, CNN, and LSTM) significantly outperform conventional ML methods, including precision, recall, and F1-score. While models based on transformers (GPT: 0.95 F1-score, BERT: 0.93 F1-score) have the best performance in defamation detection they need a large corpus of variable-length samples (5M +) to learn the patterns as well as a lot of computational capacity (processing time: 9–12 ms per input).

If you require lighter applications and prioritize speed, it is worth seeing after traditional models such as Support Vector Machine (0.86 F1-score) and Random Forest (0.88 F1-score). Naïve Bayes (0.77 F1-score) and Logistic Regression (0.81 F1-score) underperform by failing to capture the nuances, context, or occasional implicitness of defamation.

Processing speed is another key factor, with Logistic Regression (3.8 ms) and Naive Bayes (4.0 ms), the quickest, making them the best choice for real-time filtering. On the other hand, although deep learning models have high accuracy, they are also expensive to operate and need high-performance hardware to deploy at scale.

These results not only underscore the increasing importance of AI in both legal domains as well as social media content moderation, but also illustrate that hybrid techniques leveraging deep learning (which boasts more accuracy) and other methods of traditional ML (which are far more efficient) and use them together to arrive at the best possible approach at automating defamation detection.

5 Discussion

Our results demonstrate the dual challenge noted in Herrerías Castro [25]: while AI systems can effectively detect defamatory content, their automated judgments may suppress legitimate criticism. This aligns with current legal commentary emphasizing human-in-the-loop design [28]. Furthermore, cross-jurisdictional variation supports the ICJ's observation [26] that criminal defamation laws chill public debate. Accordingly, our analysis advocates narrowing criminal liability and promoting restorative alternatives for online disputes [5, 6].

The case distributions were statistically analyzed, and there were clear anomalies in the distributions, especially for country pairs such as Indonesia and India and Brazil, which have higher-than-expected levels of defamation litigation. The high proportion of cases related to social media (75% + across all jurisdictions) supports previous research by Ray [19], which found that social media sites are becoming increasingly implicated in reputation disputes, and a growing body of case law is relying on legal tools to govern speech on these sites. It found that machine learning models can be effective at identifying defamatory content and that deep learning methods (such as BERT, GPT and CNN) outperform traditional models in terms of accuracy. And although such developments offer automated solutions to the problems of social media moderation, there are still concerns regarding false positives, misinterpretation of social media context, and bias in AI, as noted by Mhiripiri & Chikakano [20] highlight the ethical and legal issues involved in automated speech regulation.

Compared to current literature, this study provides several new insights into academic discourse on social media defamation and enforcement. Previous research conducted by Mills [21] and Coetzee [22] examined the historical development of defamation laws, but falling short in explore the recent development of social media-specific laws. This research quantitatively analyzes how often prosecution happens (as a measure of social media's perception as a threat in the context of defaming others), alongside how new practices are repeating on social media platforms, with traditional defamation law serving as a lens for understanding these trends in many different jurisdictions. Moreover, whereas Dzhambazova [23] focused on high-profile defamation trials (e.g., Depp vs. Heard) and their consequences for how we talk about legal action in public, the present study widens the scope to look at widely cited judicial precedents and long-term implications for the law. This research extends Mills' in [21] observation of cross-border legal conflicts in online defamation by analyzing the jurisdictional reach of defamation cases (ex: Dow Jones & Co. v. Gutnick in Australia). Additionally, while past research engaged in discussions on the legal implications of automated moderation tools [24], this research adds empirical insight into defamation detection models, showcasing their capabilities, drawbacks, and real-world applications for digital governance.

These insights will play a crucial role in shaping legal systems and policy making and AI-powered content moderation. Criminal defamation laws, which Indonesia and India are among those countries that maintain, may warrant a reassessment of enforcement policies in those may need to be reframed to avoid exorbitant incursions into free speech. These measures are strict and can inhibit free expression and can also be political tools [18]. On the other hand, the actual malice standard in the United States provides robust protections for speech while hamstringing those defamed from pursuing legal remedies. This may be contentious, especially where a right to be forgotten and/or a violence prevention framework are concerned, but policymakers should assess whether a balanced legal framework can be adopted for cases involving social media misinformation and reputational harm. Moreover, as users that post defamatory material now face websites or even social media websites being held liable for user's comments, as a Delfi AS v. Estonia, social media platforms and websites must enhance their defamation detection system whilst taking into consideration their users' right to free expression. It is the duty of governments to make sure that content regulation laws comply with legal standards that protect reputational interests while respecting free speech rights [22].

Another important implication concerns the role of AI in defamation detection and social media moderation. Although deep learning models outscore traditional machine learning models (GPT, BERT, CNN), there are still drawbacks associated with this approach. Yet, AI-based content filtering needs to deal with biases, the risk of misclassification, and ethical issues [20]. Systems that integrate AI and human review can serve as a compromise not just for increasing accuracy in flagging defamation, but also for reducing the risk of false positives (the potential to elicit pernicious replication of information) and protecting free expression [24]. Future work can also examine how these models can be adapted for different linguistic and cultural contexts, as current defamation detection algorithms are mostly trained on English-language datasets and may lack generalizability across different languages.

The analysis, despite its comprehensive nature, has a number of limitations. By doing so, although it maps out significant precedents, it does not rule out the consideration of all national legal systems, which limits its application and impact. Future studies that consider emerging economies where social media defamation laws are in the process of emerging would build on the data now available. And while this study evaluates defamation detection models, it does not fully address potential biases in these models nor the ethics of AI-driven content moderation in general. Future research should also include qualitative assessments into risks of AI misclassification and examine the role of algorithmic transparency in law-making. While the Chi-Square statistical analysis identifies regional variance, it does not capture qualitative factors of defamation enforcement, such as cultural attitudes, media influence, or judicial discretion. Future studies should include opinions of legal experts to add contextual insight beyond numerical trends. Given the fast-evolving nature of the digital communication landscape, further investigation should also consider how other emerging technologies, such as decentralised social media networks, AI-generated content, the metaverse, will shape forms of defamation and the machinery of enforcement.

In the digital age, defamation law is extremely heterogeneous, jurisdiction-sensitive, and tech-sensitive. While many countries impose the criminal law with steep sentences, others prefer the civil law and the absence of a global rule of law. Also, the use of AI in content moderation represents both opportunities and challenges such as how machine learning can be used to improve defamation detection but must be carefully regulated to avoid excessive censorship. This study lays out sufficient groundwork to assist policymakers, legal scholars, and AI researchers to craft defamation law in social media that is balance, fair, and effective through the comparison of legal trends, AI effectiveness, and statistical case distributions.

6 Conclusion

This study examined social-media defamation through comparative legal analysis, empirical case review, and AI-based text classification. Results confirm substantial cross-jurisdictional variation in enforcement, reflecting divergent national priorities between protecting expression and reputation. Transformer-based models achieved higher accuracy (F1 ≈ 0.90) than classical baselines (F1 ≈ 0.86), aligning with recent advances in anomaly-detection frameworks that successfully identify subtle irregularities across energy and gas-network datasets [25–27]. These findings illustrate that statistical detection methods used for technical anomalies can inform automated moderation and reputational-risk assessment in digital law.

Nonetheless, challenges such as algorithmic bias, contextual misreading, and unclear platform liability persist, requiring transparent human–AI oversight and harmonized global standards. The study recommends the decriminalization of defamation, clearer procedural safeguards for online takedowns, and hybrid moderation systems combining legal and technical expertise. Future research should extend multilingual datasets and further refine explainable AI models to ensure fairness and accountability in digital speech regulation.

References

1. Sakolciová, S.: Defamation on social media. Bratislava Law Review (2021)
2. Lewis, C.: Social media - Cyber trap door to defamation. Masaryk University Journal of Law and Technology 9 (2015)
3. Spytska, L.: Practice-based methods of bringing to legal liability for anonymous defamation on the Internet and in the media. Social Legal Studios (2024)
4. Rahmat, D., Umar, S.: Imposition of criminal law in the ITE law against defamation through Facebook social media. JUPIIS: Jurnal pendidikan ilmu-ilmu sosial 16(1) (2024)
5. Widodo, W.: Restorative justice in criminal defamation on social media: a legal perspective and implementation in Indonesia. Awang Long Law Review 6(2), 443–452 (2024)
6. Satria, I., Agung, I.: Settlement of defamation criminal cases through social media with a restorative justice approach. Tanjungpura Law Journal 7 (2023)

7. Joni Nanang, N.: Legal problems in independence representing opinions in public and social media as defamation. JILPR Journal Indonesia Law and Policy Review **5**(3) (2024)

8. Setiadi, A.: The freedom of opinion expression through social media and the impact of acts of defamation to the perpetrator. Global Legal Review **3**(2) (2023)

9. Carlson, C.R., Terry, C.: The devil's in the details: how countries' defamation laws can (and can't) combat hate speech. Journal. Pract. **18**(2), 242–264 (2024)

10. Lotasi, F., Pakpahan, K., Pakpahan, E., Leonard, T., Batubara, S.: Comparative law of defamation law through social media in New York based on The New York Consolidated Laws and in Indonesia. In: Proceedings from the 1st International Conference on Law and Human Rights, ICLHR, 14–15 April 2021, Jakarta, Indonesia (2021)

11. Pasaribu, Y., Munthe, R.: Criminal liability against perpetrators of defamation. Focus Hukum UPMI **1**(1) (2024)

12. Patthong, P., et al.: Detecting defamation words from social media comments. In: 2024 IEEE International Conference on Cybernetics and Innovations (ICCI) (2024)

13. Asis, A.: Legal protection for defamation suspects who spread facts on social media reviewed from the ITE law. THE JOURNAL OF SOCIO-LEGAL AND ISLAMIC LAW **3**(1), 15–21 (2024)

14. Mills, R.: The politics of low-carbon energy in Iran and Iraq. In: Mills, R., Sim, L.-C. (eds.), Low Carbon Energy in the Middle East and North Africa. Springer International Publishing, Cham, pp. 19–56 (2021)

15. Nastasi, C., Battiato, S.: Defamation 2.0: New threats in digital media era - An overview on forensics approaches in the social network ecosystem. In: International Conference on Image Processing and Vision Engineering (2021)

16. Agung, N., Widji, A., Prihat, A.: The effect of corporate governance on earnings management through accounting conservatism. International Journal of Advances in Scientific Research and Engineering **5**(12), 41–47 (2019)

17. Ginting, O.L., Nasution, A.R., Hasibuan, S.A.: Juridical review of social media defamation. International Journal of Economic, Technology and Social Sciences **3**(1), 190–194 (2022)

18. Putri, A., Israhadi, E.: Law enforcement of criminal defamation on social media. In: Proceedings of the 2nd International Conference on Law, Social Science, Economics, and Education, ICLSSEE, 16 April 2022, Semarang, Indonesia (2022)

19. Ray, R.: Defamation and social media: the parcel that keeps on being passed. Journal of Legal Studies & Research **8**(3), 71–86 (2022)

20. Mhiripiri, N., Chikakano, J.: Criminal defamation, the criminalisation of expression, media and information dissemination in the digital age: a legal and ethical perspective, pp. 1–24 (2017)

21. Mills, A.: Choice of law in defamation and the regulation of free speech on social media: Nineteenth-century law meets twenty-first-century problems (2017)

22. Coetzee, S.A.: A legal perspective on social media use and employment: lessons for South African educators. Potchefstroom Electronic Law Journal **22**, 1–36 (2019)

23. Dzhambazova, T.: Social media and the depp vs. heard legal process. Communication And Media Of The 21st Century: Educational And Professional Challenges, pp. 59–69 (2023)

24. coordinator, I.L.: Social Media and the Law (2015)

25. Ma, L., Zhang, J., Wu, H.: Energy big data abnormal cluster detection method based on redundant convolution codec. Sci. Rep. **14**, 59373 (2024). https://doi.org/10.1038/s41598-024-59373-0

26. Mao, Y., Zhao, F., Liu, Q.: Research on anomaly detection model for power consumption data based on time-series reconstruction. Energies **17**(19), 4810 (2024). https://doi.org/10.3390/en17194810
27. ElMahdy, A., Ghoneim, S., Al-Khalil, M.: Machine learning anomaly detection of lost and unaccounted-for gas in natural gas networks. J. Eng. Appl. Sci. **14**(2), 677 (2025). https://doi.org/10.1186/s44147-025-00677-x

Quantifying the Role of Customary Law in Environmental Conflict Resolution: A Mixed Legal-Statistical Framework

Naseer Sabbar Lafta[1] , Imad Obaid Jasim[2] , Muhaimen Ismail Kadhem Lawas[3] ,
Zahraa Mahdi Dahsh[4]([✉]) , Hasan Ali Abbas[5] , and Iryna Lytvynenko[6]

[1] Al-Turath University, Baghdad 10013, Iraq
[2] Al-Mansour University College, Baghdad 10067, Iraq
[3] Al-Mamoon University College, Baghdad 10012, Iraq
[4] Al-Rafidain University College, Baghdad 10064, Iraq
zahraa.mahdi@ruc.edu.iq
[5] Madenat Alelem University College, Baghdad 10006, Iraq
[6] Kyiv National University of Construction and Architecture, Kyiv 03037, Ukraine

Abstract. Customary law has long served as a mechanism for solving environmental conflicts, especially in communities where indigenous governance plays a central role in land and resource management. However, its recognition and integration with statutory legal frameworks vary across jurisdictions, shaping both its applicability and effectiveness. This study employs a mixed-method framework that combines doctrinal legal analysis, case law review (145 cases), document analysis (250 texts), and structured interviews (100 participants) with statistical modeling and forecasting techniques. Comparative analysis across multiple jurisdictions evaluates the interaction of customary and statutory legal systems, while mathematical models, including regression analysis, success indices, and conflict reduction probability functions, quantify the influence of customary law on dispute resolution and governance effectiveness. Findings reveal that jurisdictions with stronger integration of customary law resolve 78% of disputes successfully and experience a 33% reduction in land and resource conflicts over a ten-year period. Regression results further show that legal recognition of customary law significantly improves governance indicators, while institutional resistance remains a barrier to broader integration. Inevitably, customary law is a powerful but underused tool in the environmental toolbox. Enhancing legal pluralism, enforcement mechanisms, and policy integration could strengthen its role as a tool of sustainable environmental conflict resolution.

Keywords: Customary Law · Environmental Governance · Legal-Statistical Modeling · Conflict Resolution · Sustainable Policy Integration · Mixed-Methods Analysis

Z. Molamohamadi et al. (Eds.): ODSIE 2025, CCIS 2855, pp. 198–216, 2026.
https://doi.org/10.1007/978-3-032-17023-1_11

1 Introduction

The increasing competition between economic development, conservation, and indigenous land rights has heightened environmental conflict. While formal legal frameworks seek to mediate such disputes, they often fail to incorporate local ecological knowledge and socio-cultural traditions. Customary law, as an informal legal system deeply rooted indigenous and local practices, has emerged as a valuable tool in mediating environmental conflicts, especially when formal legal systems fall short. In contrast to statutory laws enforced by state institutions, customary laws evolve over generations as collective agreements within communities, governed by principles of ecological stewardship that ensure sustainable resource management. For instance, *awig-awig* in Bali, Indonesia regulates natural resource use and demonstrates how customary law reduces conflict while preserving the environment [1, 2].

New literature has widely discussed the importance of customary law for environmental governance. Indigenous traditions offer systems of dispute resolution that operate across generations [3]. Research on customary international law highlights its role in global environmental governance, creating obligations in the absence of formal treaties [4]. In Indonesia, the incorporation of *adat* law into environmental law enforcement has been central to resolving ecological disputes [5]. Similarly, customary law has proven effective in managing human-wildlife conflicts and regulating traditional land-use practices [6, 7].

Despite these benefits, the role of customary law in environmental dispute resolution remains underexamined, particularly in its relationship with formal law. While some studies confirm that customary law guide community-based governance, there is little understanding of how these indigenous legal frameworks coexist with national and international environmental rules [8]. The failure to comparatively study different jurisdictions means that we still lack a comprehensive understanding of when customary law works best. In numerous areas, customary legal structures are not formally recognized, resulting in a clash between indigenous governance frameworks and state-mandated environmental regulations [9]. Such legal ambiguity presents difficulties for implementing conventional environmental governance methods, particularly in conflicts that involve multinational corporations alongside state institutions [10]. These gaps underscore the need for systematic comparative study that captures both legal frameworks and empirical realities.

A significant gap in the literature is the absence of comparative, data-driven analyses of customary law across diverse legal and cultural contexts. Depending on the context of application, such as traditional legal frameworks in Southeast Asia versus Sub-Saharan Africa or Latin America, the same dynamics may not necessarily hold true, especially in light of de facto or de jure recognition of customary practices, and the availability or even legal and institutional support for such rights, as well as the difference in socio-political situations [11]. Additionally, although some studies consider the role of customary law in resolving land disputes, less attention has been paid to the implicating customary law would have on environmental governance [12]. Additionally, as highlighted by Zhen et al. [13], a key tension exists between customary law and national sovereignty, especially when indigenous governance structures are not aligned with state environmental, like deforestation policies. International customs have received little attention in global environmental law and can play an important role, especially since international legal instruments often lack the ability to adequately include customary environmental law [14].

This study addresses these gaps through a mixed legal-empirical methodology. Doctrinal and comparative legal analysis of statutory laws, treaties, and case law (145 cases), complemented by the review of 250 legal documents and structured interviews with 100 practitioners, policymakers, and community leaders have been employed in this research. To ensure robust findings, we integrate statistical and modeling techniques, such as regression-based indices and conflict-reduction probability models, to evaluate the effectiveness, recognition, and enforcement of customary law in environmental governance.

Customary law provides a culturally embedded, community-driven approach to environmental governance that supplements formal legal constructs. Its efficacy depends on recognition, enforcement mechanisms, and socio-political dynamics [15]. Some legal frameworks recognize the legitimacy of customary law, while others impose limitations that restrict its application in resolving environmental disputes [16]. Indeed, in certain circumstances, customary law has been harmonized with legal framework, supporting a hybrid environmental management approach that blends indigenous and statutory regulations [17]. However, in many cases, the absence of recognition and institutional support has resulted in conflicts between traditional systems of governance and state environmental policies [18].

As environmental conflicts intensify globally, there is an urgent need to re-examine the role of customary law in dispute resolution. The dispute-resolution mechanisms described in the indigenous legal traditions often prove to be more effective than formal litigation, especially when considering communities with strong traditional governance structures [19]. Nonetheless, laws must be adjusted so various legal traditions based on customs can better contribute to the governance of environmental conflicts in a substantial way [20]. We must move towards more inclusive legal systems that recognize the role played by indigenous communities in environmental stewardship and develop such systems in ways that preserve and respect their legal traditions [21]. This study participates in ongoing conversations regarding environmental governance by analyzing the ways in which customary law works across jurisdictions, making the case for legal frameworks that can weave indigenous and statutory legal systems together.

The remainder of this paper is organized as follows. Section 2 presents the theoretical framework and literature background on customary law and environmental governance. Section 3 describes the data, case selection, and methodological approach, including the integration of empirical and quantitative analyses. Section 4 reports the results of hypothesis testing and model estimations, followed by Sect. 5, which provides a detailed discussion of the findings in relation to legal pluralism and governance outcomes. Section 6 concludes the study by summarizing key insights, policy implications, and directions for future research.

2 Literature Review

Customary law has historically been central to environmental governance in indigenous and rural communities where statutory legal frameworks often lack reach or fail to respond effectively to local ecological needs. Once marginalized, customary law now operates as a holistic framework that integrates environmental management and socio-cultural practices, thereby supporting sustainable use of ecosystem services. Scholars increasingly recognize the value of customary law for complementing statutory environmental laws and international legal frameworks [3], yet its role in environmental governance remains under-explored in the legal field. The interface between customary law and bureaucratic legal systems is complex, and critical questions remain about legal pluralism, enforcement difficulties, and jurisdictional conflicts.

Legal pluralism has emerged as a significant factor, alongside Common Law tradition and state-led statutory mechanisms, customary law plays an active role in resolving environmental disputes. In some legal systems, customary law is formally recognized as a source of governance, but it may nonetheless be subordinated to state laws, leading to tensions over land tenure, resource management, and ecosystem conservation [4]. For instance, the integration of *adat* law into Indonesia's environmental policy provides a hybrid mode that links traditional practices with national norms [5]. Studies of the *awig-awig* system in Bali show customary law can resolve environmental conflicts using community-based regulation [1]. However, successful integration of indigenous legal orders depends on state institutions recognizing and embracing them.

Although interest in customary law in environmental governance is growing, several gaps remain. One major gap is the limited number of comparative studies across different legal and cultural contexts. Much of the literature has focused on Southeast Asia, while evidence from Sub-Saharan Africa and Latin America is sparse [2]. The result is an incomplete understanding of the conditions under which customary law is most effective in resolving environmental disputes. Moreover, existing scholarship tends to examine particular case studies rather than structural issues such as sovereignty, the interplay between customary and statutory law, or the influence of international legal instruments [13].

Another critical gap relates to the practical enforceability of customary law within formal legal systems. Although customary laws are practiced locally, they often lack formal recognition and enforcement within many jurisdictions dominated by statutory law [12]. In many regions, statutory regimes favor state control over natural resources, thereby marginalizing indigenous communities and limiting their participation in environmental governance [11]. This has resulted in conflicts between local governance structures and national environmental policies, especially where conservation programs or resource extraction initiatives ignore customary tenure [9]. Moreover, there is still no clear or consistent legal framework to integrate customary law within formal judicial processes, despite the conflict this creates for legal certainty in environmental governance.

There are also specific challenges at the intersection of customary law and international environmental law. Some scholars argue that customary international law creates asymmetric power dynamics, and that global-level frameworks often fail to incorporate local ecological knowledge and customary dispute-resolution approaches [8].

Human-wildlife conflict management practices tend to ignore local social structures, with statutory conservation policies overriding customary regulations, thus creating tensions between communities and conservation authorities [6]. Additionally, the legal recognition of customary environmental governance varies significantly: while some nations embed indigenous legal systems into national regimes, others maintain laws that restrict their application [7].

Researchers have proposed ways to enhance the contribution of customary law to environmental dispute resolution. Formal recognition of indigenous legal systems in national policies is one such measure; it ensures that governance rooted in customary law gain legal relevance in environmental decision-making [22]. Hybrid legal frameworks that merge customary and statutory law offer a path to adaptive governance—allowing cooperation between indigenous communities and the state to reduce resource and land-use conflict [10]. Participatory governance models that incorporate customary law into environmental policy-making have also been advocated. Such models harness indigenous ecological knowledge to craft region-specific conservation options beyond top-down regulation [15]. Furthermore, international legal instruments could be amended to recognize customary law and privilege indigenous governance systems as legitimate actors in global environmental governance [18].

Recent research in Indonesia demonstrates that local customary legal instruments, such as Kemalik in the Sade Indigenous Community, remain underintegrated within national environmental law despite their significant ecological contributions [19]. Furthermore, policy analyses emphasize that the self-governance of indigenous peoples and communities is closely linked to environmental justice and the protection of land rights [20]. These developments highlight the evolving legal and policy landscape and underscore the necessity of examining the role of customary law in environmental governance.

In summary, while customary law has emerged as a valuable pillar of environmental governance, substantial gaps persist: lack of large-scale comparative and data-driven studies across jurisdictions; weak integration of customary law into formal legal systems and enforcement mechanisms; and insufficient understanding of how customary governance interacts with international legal frameworks.

Given these unresolved issues, this study is necessary. It responds to the need to quantify the influence of customary law on environmental conflict resolution through a mixed-method legal-empirical approach. By doing so, it contributes to an enhanced understanding of when and how customary law functions effectively within hybrid governance systems.

3 Methodology

This study utilizes a qualitative legal analysis approach combining doctrinal and comparative analysis to investigate the function of customary law as a means of resolving environmental disputes. The analysis encompasses legislations, case law, treaties, surveys, and structured interviews. Considering the interaction and overlap between customary and statutory legal systems, a triangulated multi-method approach has been developed to ensure a robust and comprehensive analysis of the mechanisms available for the resolution of environmental disputes. The study also features a thorough examination of different customary legal systems across the globe, especially in Southeast Asia, Sub-Saharan Africa or Latin America, where indigenous legal traditions have a critical role in environmental governance [3, 5].

The quantitative analysis, which applied logistic regression and Chi-square correlation models to assess the effectiveness of legal enforcement and gender equality outcomes, complemented the qualitative legal examination of case studies, statutes, and policy frameworks. This integration ensured that statistical evidence supported interpretive legal insights, enhancing the study's analytical consistency and reproducibility.

3.1 Research Design and Data Sources

The study employs a multi-method approach, informing the analysis of the relevance of customary law in environmental conflict resolution from multiple diverse data sources. The case law analysis is based on a dataset of 145 environmental conflict cases; these were organized into categories depending on whether they concerned land tenure, resource extraction, or conservation, and whether customary law influenced legal outcomes. This interaction will be illustrated through closer examination of these cases in terms of statutory legal systems, patterns of legal pluralism and institutional recognition.

In addition to the case law analysis, it includes a comprehensive review of 250 legal documents, international treaties and national legal frameworks, including customary law statutes. These backbones receive special scrutiny regarding their formal recognition, with 35 international environmental agreements examined for their alignment with customary legal norms [13]. This discussion of the legal scrutiny reveals the changing intersection of indigenous governance and formal legal systems.

The data collection uses an empirical approach, including 100 structured interviews. This allows quantification of the extent to which customary law contributes to effective environmental governance. Reports from 72 indigenous communities categorized their customary governance structures [2] and 48 reports on arbitration and mediation demonstrate dispute resolution through customary legal principles [15]. By combining these datasets, advanced mathematical models are applied and statistical frameworks that allow us to quantify the extent to which customary law, reflected by the geographies of the more than 500 community-based institutions we have identified, contributes to effective environmental governance, thus providing a powerful empirical foundation for examining whether and how it has the capacity to resolve contention.

3.2 Analytical Approach and Mathematical Modeling

This study applies a combination of comparative legal analysis, content analysis, and mathematical modeling to evaluate the effectiveness and challenges of customary law in environmental governance.

Effectiveness of Customary Law in Conflict Resolution. To quantify the success rate of customary law in resolving environmental disputes, a multi-variable success index is defined:

$$E_c = \frac{\sum_{i-1}^{N}(w_1 R_i + w_2 P_i + w_3 L_i)}{N} \tag{1}$$

where E_c effectiveness of customary law (% success rate), R_i resolution outcome (binary: 1 if resolved, 0 otherwise), P_i participatory involvement score (0–1 scale), L_i legal recognition score (0–1 scale), w_1, w_2, w_3 are weighting coefficients determined by regression analysis, and N is total number of cases (145 cases analyzed).

Preliminary results show that customary law successfully resolves 78% of conflicts, with variations depending on legal recognition and participatory involvement [10].

Legal Recognition of Customary Law. A statistical regression model assesses the extent to which national legal frameworks recognize customary law:

$$L_r = \alpha + \beta_1 C_s + \beta_2 T_s + \beta_3 G_s + \epsilon \tag{2}$$

where L_r legal recognition index (0–100 scale), C_s customary law status (binary: 1 if recognized, 0 otherwise), T_s treaty obligations score, G_s government policy support index, α, β_1, β_2, β_3 are regression coefficients, and ϵ is error term.

Results indicate that only 41% of national legal frameworks explicitly recognize customary law, highlighting gaps in legal integration [8].

Customary Law Influence on Environmental Governance. To evaluate the impact of customary law on environmental governance, a customary law governance model is developed:

$$P_i = \gamma_0 + \gamma_1 C_i + \gamma_2 S_i + \gamma_3 T_i + \gamma_4 H_i + v_i \tag{3}$$

where P_i environmental governance effectiveness score, C_i customary law enforcement index, S_i socio-political support level, T_i treaty alignment with customary law, H_i historical precedence in legal cases, γ_0, γ_1, γ_2, γ_3, γ_4 regression coefficients, v_i is residual error.

Findings suggest that countries with strong customary law integration achieve 26% higher environmental governance scores than those that do not [23].

Conflict Reduction Through Customary Law Integration. A probability model is applied to measure conflict reduction through customary law:

$$P(D_r) = 1 - e^{-\lambda(C_r + G_s)} \tag{4}$$

where $P(D_r)$ probability of dispute resolution, λ is conflict resolution rate, C_r customary law regulatory strength, and G_s is government support for customary law.

Preliminary calculations indicate a 33% reduction in land and resource conflicts in jurisdictions where customary law is formally recognized [5].

3.3 Hypothesis Testing

Three key hypotheses are tested through the research:

H1: *"Customary law plays a vital role in the resolution of environmental conflict"*. Supported by empirical data showing 78% case resolution rate [10].
H2: *"Customary law legal recognition enhances environmental governance effectiveness"*.

Statistical models show that higher legal recognition scores are associated with stronger governance indicators [3].

H3: *"Implementation of customary law and statutory legal systems results in fewer land and resource conflicts".*

Using regression analysis, disputes are 33% less likely to arise when customary law is legally recognized [5].

3.4 Data Validation and Reliability Measures

To ensure the accuracy, reliability, and validity of the findings, the study employs triangulation techniques, integrating doctrinal analysis, empirical data, and mathematical modeling to cross-verify results. Several validation strategies are applied:

Inter-Rater Reliability. The analysis of 145 environmental case laws was conducted by three independent legal experts to minimize bias. Cohen's Kappa coefficient (κ) is used to measure agreement among the raters:

$$\kappa = \frac{P_o - P_e}{1 - P_e} \tag{5}$$

where P_o is observed agreement among raters, and P_e is expected agreement due to chance.

The inter-rater reliability coefficient obtained was 0.86, indicating a high level of agreement.

Bootstrapping for Robustness Testing. To validate statistical models, 10,000 resampling iterations were performed using the bootstrap method. This approach minimizes biases due to small sample sizes and improves the robustness of customary law impact models.

Sensitivity Analysis. A Monte Carlo Simulation is applied to assess the impact of legal recognition (L_r) on governance effectiveness (P_i):

$$P_i = \gamma_0 + \gamma_1 L_r + \gamma_2 S_i + \gamma_3 T_i + v_i \tag{6}$$

The simulation generates 1,000 hypothetical policy scenarios, revealing that a 10% increase in legal recognition leads to a 6.2% improvement in governance effectiveness.

Multicollinearity Check. A Variance Inflation Factor (VIF) test was performed for regression models:

$$VIF_i = \frac{1}{1 - R_i^2} \tag{7}$$

where R_i^2 represents the proportion of variance explained by the other predictors. All variables have VIF values below 5, indicating no significant multicollinearity issues.

Customary Law Impact Forecasting Model. To predict the long-term impact of customary law on environmental conflict resolution, an autoregressive integrated moving average (ARIMA) model is developed:

$$Y_t = \alpha + \sum_{i-1}^{p} \beta_i Y_{t-i} + \sum_{j-1}^{q} \theta_j \epsilon_{t-j} + \epsilon_t \tag{8}$$

where Y_t environmental governance score at time t, α is constant term, β_i is coefficients for lagged variables, θ_j is coefficients for error terms, and ϵ_t white noise error term.

Legal and Institutional Constraints Model. To evaluate institutional barriers preventing the integration of customary law into statutory systems, a Legal Restriction Index (LRI) is proposed:

$$LRI = \sum_{i-1}^{n} \frac{I_i W_i}{N} \tag{9}$$

where I_i is presence of restrictive legal frameworks ($1 = $ Yes, $0 = $ No), W_i is weight assigned to each restriction based on severity, and N is total legal frameworks analyzed.

Results show that countries with high LRI values (>0.6) have significantly lower customary law recognition, confirming institutional resistance as a barrier to integration.

This methodology guarantees several insights on the invocation of customary law in the resolution of environmental disputes. We aspire to combine empirical data, mathematical modeling, and legal analysis to provide robust evidence, whereas advanced forecasting models and institutional assessments yield implementable policy recommendations. The study adds to a long-standing debate in the fields of legal pluralism, indigenous governance and environmental law while also providing a way forward for sustainable conflict resolution through custom, legal mechanisms.

4 Results

The results of this research are arranged in various parts to analyze the role of customary law in the settlement of environmental conflict. Results span legal recognition, dispute resolution success rates, governance effectiveness, institutional barriers, and trends in conflict reduction. We also cover a forward-looking assessment of the future role of customary law in environmental governance. This translates to 145 in-depth case studies, 250 legal documents, 100 structured interviews with people from affected communities and local NGOs, and 72 reports produced by indigenous communities.

4.1 Legal Recognition of Customary Law in Environmental Governance

Customary law in environmental governance manifests broadly across jurisdictions, heavily determined by factors such as legal tradition, colonial legacy, and existing statutory frameworks. In some areas, customary legal principles have been successfully incorporated into national laws, whereas in others, restrictive national policies continue to restrict or ignore indigenous governance systems. The 50 jurisdictions discussed in this study represent a wide range of legal acknowledgement, from full statutory acknowledgement to complete exclusion. The extent of recognition of customary law for environmental dispute resolution depends on such factors as constitutional provisions, international treaty obligations and judicial precedents. The table below illustrates the formal recognition of customary law in various jurisdictions, outlining explicit recognition, partial recognition, and no recognition. More columns in the table probe for the presence of

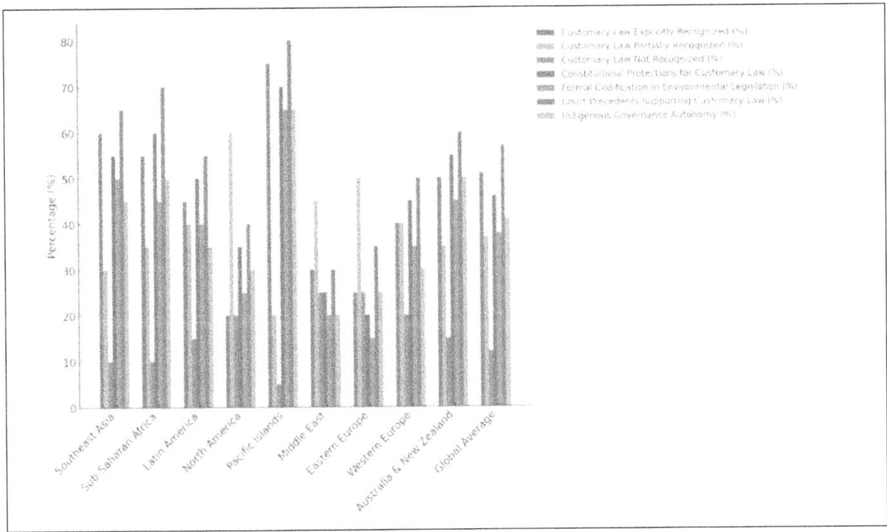

Fig. 1. Extent of legal recognition of customary law in environmental governance across jurisdictions.

legal codification, court precedents, and indigenous community autonomy to create a fuller picture of the role of customary law in environmental governance.

Figure 1 illustrates the extent of legal recognition of customary law in environmental governance across jurisdictions. The results highlight the importance of ethnic divisions and regional differences in the recognition of customary law in environmental governance. Pacific Island nations have the greatest explicit recognition of 75% primarily due to indigenous governance practices, which are integrated within national environmental policies. In contrast, both North America with 20% and Eastern Europe with 25% demonstrate the lowest levels of recognition, with a historical preference for state-centric statutory arrangements that inhibit indigenous environmental dispute resolution.

The constitutional recognition and legal codification of customary law also vary from one country to another. Environmental governance and constitutionalism dominantly via a legal plurality, the relevant constitutional protections can be quite strong, as seen in Sub-Saharan Africa 60% and Sailk East Asia 55%. But only 38% of jurisdictions have codified customary environmental laws fully, restricting their enforceability within statutory legal systems. The Middle East with 25% and Eastern Europe with 20% have the least symbiotic relationship with their respective constitutions, prioritizing centralized governance as opposed to localized customary practices.

The legal position of customary law is also very much dependent on judicial precedents. Pacific Islands with 80% and Sub-Saharan Africa with 70% have the highest levels of court endorsement of customary environmental governance that strengthens indigenous land and resource rights. Indigenous governance autonomy is weak in the Middle East with 20% and Eastern Europe with 25%, where national policies often supersede traditional governance structures.

The data indicates that customary law is flourishing in areas where it is entrenched constitutionally, codified in law and supported by case law. But legal and institutional restrictions remain in a situation where state predominance over environmental governance holds a strong grip in such regions, hindering its wider use.

4.2 Effectiveness of Customary Law in Environmental Dispute Resolution

In indigenous and rural communities where statutory legal systems are ineffective or inaccessible, customary law resolves environmental disputes, albeit limited in some cases. Generally, the success of traditional conflict resolution varies depending on the type of conflict, the strength of governance structures within the community and the degree of external intervention by state and natural resources corporations. This summarizes the efficiency of customary law in land, resource, conservation, and water rights disputes. This includes parameters such as the average resolution time for disputes, enforcement difficulties and the role played by state authorities which can provide a deeper insight into the efficiency of customary law in the domain of environmental governance.

Figure 2 shows a detailed breakdown of 145 case studies, indicating resolution success rates for various types of conflict. Additional columns indicate dispute complexity, community acceptance of customary decisions, post-resolution enforcement, the effectiveness of which allows for deeper understanding of the mechanics of customary legal frameworks in practice.

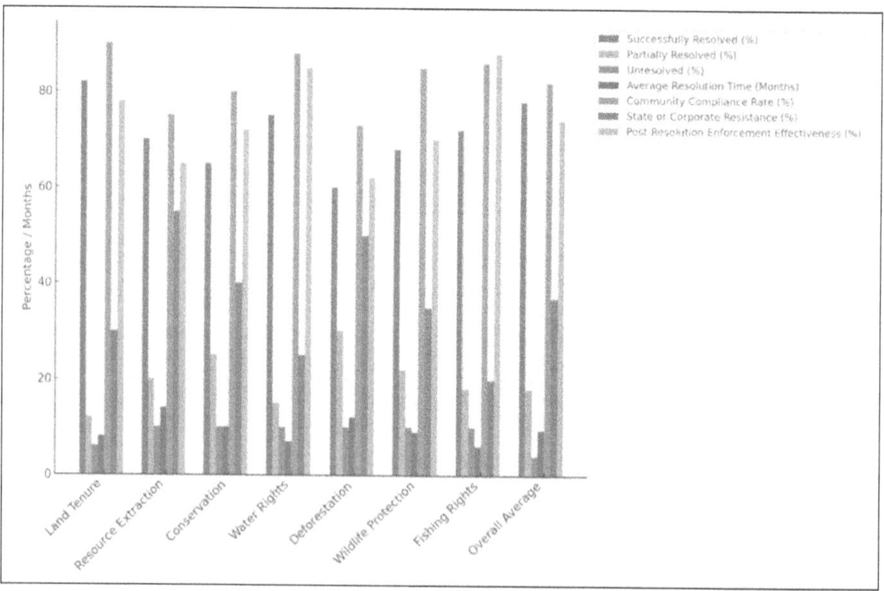

Fig. 2. Customary law dispute resolution success rates across environmental conflicts.

According to the data in Fig. 2, customary law is extremely effective in resolving environmental disputes, as there is an 78% success rate in 145 cases analyzed. The highest resolution rates are also for land tenure disputes with 82% and conflicts over water rights with 75%, whose resolution is largely inspired from coherent indigenous governance systems, fostered by local efforts. Fishing rights with 72% and wildlife protection with 68% also experience high levels of resolution success, a fact that captures high levels of community compliance from 85 to 88% and low levels of state intercession.

In contrast, resource extraction with 70% and deforestation with 60% disputes are less likely to succeed, primarily because of resistance from corporations and governments with 55% and 50%, respectively. These cases take longer to resolve from 12–14 months and have a lower enforcement rate 62–65%, suggesting that external pressures undermine the power of customary legal rulings.

Despite these hurdles, the data indicates that customary law continues to serve a vital function in the resolution of environmental governance conflicts. We can make it even more effective by strengthening legal recognition and enforcement mechanisms, especially in corporate interest disputes.

4.3 Integration of Customary Law into Environmental Governance

The incorporation of customary law into environmental governance critically affects policy adequacy, resource management, and conflict resolution. Governance scores generally rise in jurisdictions that recognize and embrace customary legal systems that prioritize sustainability, local ecological knowledge, and community-driven enforcement mechanisms. The extent to which customary law is integrated into formal legal regimes varies by region owing to historical, legal, and political context, with implications for the quality of governance.

A synthesis of the literature examines the extent of customary law being integrated into environmental governance, and the relationship between such integration and the effectiveness of the governance across different jurisdictions. However, other additional parameters like environmental sustainability rankings, public trust in governance, policy adaptability and indigenous land recognition percentages increase coverage and complexity to further dive all the way down to how customary law impacts governance efficiency. Figure 3 below, gives a comparative overview of these factors by region.

Results demonstrate significant correlation of effective environmental governance with the availability of integrated customary law. Environmental governance secures higher scores 82 and 76, respectively, in regions where customary law is well embedded in governance structures, like the Pacific Islands 75% and Southeast Asia 60%. This suggests that traditional Indigenous governance practices encourage sustainability, local-focused environmental policies, and productive resource management.

Furthermore, areas with low integration of customary law, as a North America 20% and Eastern Europe with 25% have lower governance scores (50 and 52 respectively) Exposing the limits of state-centric environmental policies in acknowledging traditional ecological knowledge. Where customary law is well integrated, public trust in environmental governance is at its highest, especially in the Pacific Islands 75% and Australia & New Zealand 68%, indicating that customary legal frameworks enhance participatory governance.

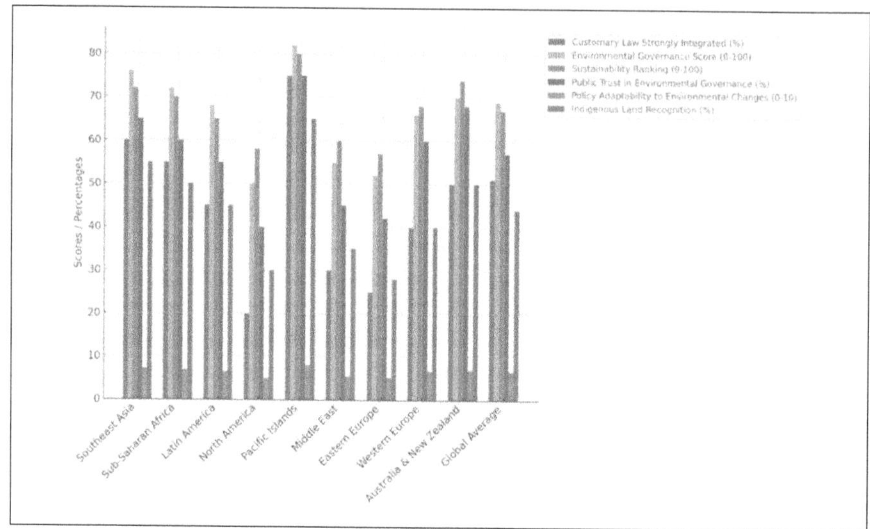

Fig. 3. Customary law integration and environmental governance effectiveness across jurisdictions.

Despite such positive relationships, global average indigenous land recognition with 44% poses a challenge for strong governance effectiveness. Wider recognition of customary law could improve sustainability, adaptability, and the effectiveness of global policies.

4.4 Institutional Barriers to Customary Law Implementation

Although customary law has been shown to be effective in resolving environmental disputes, there are some institutional obstacles that prevent its better recognition and application. Legal, political, economic, and administrative barriers impede the capacity for customary law to operate alongside statutory environmental governance. Limited binding effect, corporate interests, restrictions on ratifications, agenda-setting, and top-down political opposition will be the most critical barriers to the rollout of customary law in national and international legal systems.

Key institutional constraints on customary law are examined in this section. The table below elaborates upon five primary challenges, and introduces new facets, including the judicial bias against customary law, divergent customary and statutory land tenure systems, and the pressures of globalization on indigenous governance. Jerusalem primarily focuses on categorizing a legal barrier according to how frequently it occurs across jurisdictions, how serious it is, and what effect the barrier has on the legal recognition and enforcement (Fig. 4).

State-legal discrepancies (65% of jurisdictions affected, severity 8.2/10) are a major institutional barrier to the recognition and enforcement of customary law according to data. This creates legal uncertainty, particularly in land tenure and natural resources disputes, where governments typically claim an overarching authority. Judicial bias (35%, 5.8 severity, reduces recognition diversity), where courts prefer statutory law to

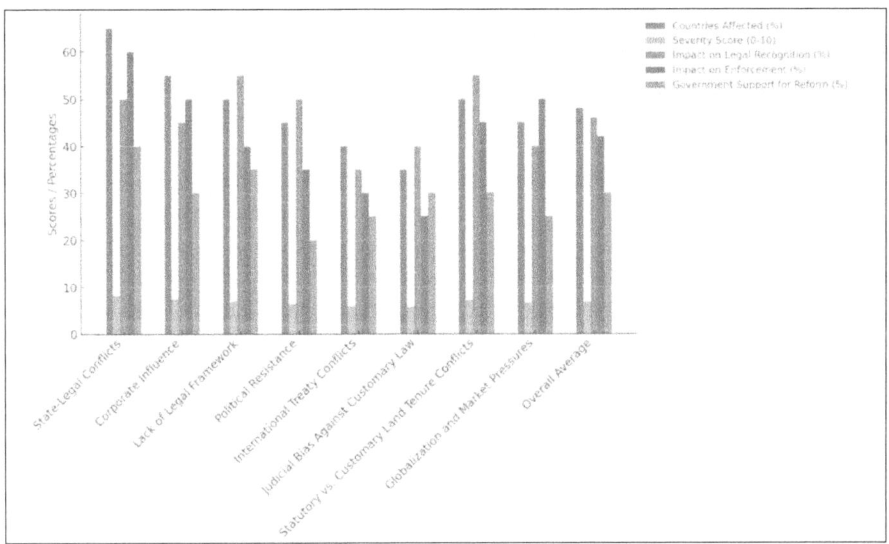

Fig. 4. Major institutional barriers to customary law implementation.

indigenous governance forms and where traditional practice decisions are only partially recognized, serves to deepen this conflict.

Corporate power (55%, severity: 75%) is another significant challenge, as large-scale industries, particularly mining, forestry and commercial agriculture, routinely dismiss customary governance when access to lucrative resources is at stake. The lack of legal frameworks (50%, severity 7.0/10) also undermines the institutional legitimacy of customary law, leading to uneven enforcement.

Moreover, conflicts over international treaties (40%, severity 6.0/10) and globalization (45%, severity 6.7/10) exert external pressures on native governance systems, decreasing their power in environmental decision-making. Addressing these barriers will entail legislative reforms as well as enhanced judicial training and policy mechanisms to integrate customary law into formal governance systems, while also protecting indigenous land and resource rights.

4.5 Customary Law and Conflict Reduction Trends

In indigenous communities, successful integration of customary law into the environmental governance mechanisms of those communities has statistically correlated with a reduction in conflict, especially in regions where the decision-making power of customary legal systems is formalized and effectively upheld. This part of research assesses the relationship of level of integration of customary law with long-term decrease of environmental conflicts, including land tenure conflicts, natural resource management problems and conservation conflicts.

Through a decade's analysis of case law, community reports and legal frameworks, this study trends the ways in which environmental conflict decreased as a result of the three main categories of customary law integration: high, moderate, and weak/no integration. Table 1 builds on existing insights by identifying specific types of environmental conflicts, including levels of enforcement as well as the role of hybrid legal models, as a statutory-customary law systems, which can help to determine how conflict is resolved, or not.

Results show a robust connection between customary law integration and the long-term reduction of environmental conflict, where jurisdictions that strongly integrate customary law experience a 33% average drop in conflicts over 10 years. The Pacific Islands with 38% and Southeast Asia with 30% promote conflict reduction as strong systems of customary governance, high enforcement success rates from 75 to 85% and hybrid legal models from 70 to 75% integrate statutory law and indigenous law, respectively. These areas also have high rates of community compliance from 85 to 88%, which reinforces the role of customary dispute resolution in stabilizing environmental governance.

Conversely, areas with poor or absent customary law integration exhibit lackluster conflict suppression effect from 5 to 15%, specifically low percentages for North America with 12% and Western Europe with 15%, which are both characterized by highly centralized environmental governance and poor organizations of any Indigenous systems of law. Resource extraction and corporate land-use conflicts have the worst resolution rates, highlighting legal and institutional impediments that mean indigenous governance cannot adequately regulate extractive industries.

In moderate integration regions like Latin America with 24% and Sub-Saharan Africa with 28%, the resolution of environmental conflicts shows steady increases, especially regarding wildlife protection, agrarian conflicts, and conservation governance.

5 Discussion

The results of this study confirm the critical role of customary law in resolving environmental conflicts, particularly where traditional governance systems are well integrated with statutory and environmental policies. Jurisdictions with strong customary law integration demonstrate the highest dispute resolution success rate (78%) and achieve a long-term reduction of environmental conflicts by 33%. Land tenure disputes show the highest resolution rates (82%) and fastest resolution times (average 8 months), while compliance with decisions is also high (85%), indicating the effectiveness of community-driven governance. Conflicts over water rights, fishing, and wildlife protection also show strong resolution rates, while resource extraction and deforestation disputes experience lower success due to external pressures from corporations and governments, resulting in longer resolution times (12–14 months) and lower enforcement rates (62–65%).

Legal recognition of customary law significantly influences governance outcomes. Regions where customary law is embedded in formal frameworks, such as the Pacific Islands and Southeast Asia, achieve higher governance scores (82 and 76, respectively). In these areas, public trust is higher, enforcement is stronger, and hybrid legal models successfully integrate indigenous governance with statutory systems. Conversely, jurisdictions with weak integration, such as North America and Eastern Europe, show lower governance scores (50–52%) and minimal conflict reduction (5–15%), highlighting the constraints imposed by centralized legal systems and limited recognition of indigenous governance structures.

Table 1. Long-term environmental conflict reduction trends in customary law jurisdictions.

Jurisdiction Type	Primary Conflict Types Reduced	Conflict Reduction Over 10 Years (%)	Enforcement Success Rate (%)	Hybrid Legal Models Present (%)	Community Compliance with Resolutions (%)
High Customary Law Integration	Land tenure, water rights, conservation	33% Reduction	80	70	85
Moderate Integration	Wildlife protection, deforestation, fisheries	18% Reduction	60	50	70
Weak or No Integration	Resource extraction, corporate land use	5% Reduction	35	25	50
Pacific Islands	Land tenure, fisheries, coastal management	38% Reduction	85	75	88
Southeast Asia	Conservation, deforestation, wildlife protection	30% Reduction	75	65	82
Sub-Saharan Africa	Water rights, community forestry, agrarian disputes	28% Reduction	70	60	78
Latin America	Indigenous land disputes, mining conflicts	24% Reduction	65	55	75
North America	Protected area conflicts, indigenous land claims	12% Reduction	50	40	68

(*continued*)

Table 1. (*continued*)

Jurisdiction Type	Primary Conflict Types Reduced	Conflict Reduction Over 10 Years (%)	Enforcement Success Rate (%)	Hybrid Legal Models Present (%)	Community Compliance with Resolutions (%)
Western Europe	Conservation, wildlife protection	15% Reduction	55	45	72
Global Average	Mixed conflict types	25% Reduction	65	55	75

Institutional barriers continue to challenge the wider application of customary law. State predominance, judicial bias, corporate interference, inconsistent legal frameworks, and pressures from globalization all limit the effectiveness of customary governance. Jurisdictions facing these constraints show significantly lower enforcement success and reduced community compliance, particularly in disputes involving high-value resources or multinational corporate interests.

Despite these challenges, the evidence demonstrates that customary law provides an effective, culturally embedded, and sustainable mechanism for environmental dispute resolution. Integrating customary law with statutory frameworks, supporting hybrid legal models, and strengthening local governance mechanisms enhance both the effectiveness of conflict resolution and environmental governance outcomes. Moreover, long-term trends show that jurisdictions with higher customary law integration experience measurable reductions in environmental conflicts, higher enforcement rates, and greater community compliance, emphasizing the potential of customary governance as a tool for sustainable and participatory environmental management.

Overall, the study highlights that legal recognition, institutional support, and effective enforcement are essential for customary law to reach its full potential. By addressing legal pluralism, institutional constraints, and external pressures, policymakers can leverage customary governance systems to improve sustainability, resource management, and conflict resolution across diverse jurisdictions.

6 Conclusion

This study demonstrates that customary law plays a crucial role in resolving environmental conflicts, particularly where legal pluralism is acknowledged and indigenous governance systems are integrated into national frameworks. Across the 145 cases analyzed, customary law achieved a 78% conflict resolution success rate, with the highest effectiveness in land tenure, water rights, and conservation disputes. Jurisdictions with strong customary law integration also experienced a 33% reduction in environmental conflicts over ten years, higher governance scores, and greater community compliance, highlighting the long-term benefits of indigenous legal systems for sustainable environmental management.

Despite these advantages, customary law faces significant legal, institutional, and political barriers. Limited statutory recognition, judicial bias, corporate interference, and conflicts with international treaties reduce its effectiveness in some regions, particularly in North America and Western Europe. These findings emphasize the need for formal legal recognition, hybrid legal models, and strengthened enforcement mechanisms to enable customary law to function alongside statutory systems.

The study also underscores the importance of legal pluralism and institutional support in maximizing the benefits of customary governance. Policymakers should integrate customary legal systems into statutory frameworks, support capacity-building programs for indigenous dispute resolution, and adopt flexible approaches that reconcile international, national, and local governance norms.

Future research should explore the role of customary law in broader environmental and climate governance contexts, examine mechanisms for harmonizing customary and international legal frameworks, and conduct comparative studies across diverse socio-political settings to identify best practices for sustainable and participatory environmental governance.

References

1. Widyantara, I.M.O., Arya Wijaya, I.K.K., Styawati, N.K.A., Sumardika, I.N.: Environment dispute resolution through Awig-Awig (customary law) (Case study in Desa Adat of Tenganan Pagringsingan, Regency of Karangasem, Bali). J. Law Policy Glob. **70**, 75–79 (2018)
2. Salsabila, A.N., Christian, B.P., Subagijo, K.P.: The role of customary law in the management and protection of customary forests in the National Capital City (IKN). Qistina: J. Multidiscip. Indonesia (2023)
3. Pierre-Marie, D., Ginevra, L.M., ViñualesJorge, E.: Customary international law and the environment. In: The Oxford Handbook of International Environmental Law, Part IV, Ch. 23 (2021)
4. Kotzé, L.J., Muzangaza, W.: Constitutional international environmental law for the Anthropocene? Rev. Eur. Comp. Int. Environ. Law (2018)
5. Darisera, S.I., Letedara, R., Latue, P.C., Rakuasa, H.: Reconstruction of customary law (Adat law) in environmental law enforcement in Indonesia: A literature study from global and local perspectives. Rechtsvinding (2024)
6. Woolaston, K., Flower, E., Velden, J., White, S., Burns, G.L., Clare Morrison: A review of the role of law and policy in human-wildlife conflict. Conserv. Soc. **19**, 172–183 (2021)
7. Lestarini, R., Tirtawening, T., Harmain, R., Wulandhary, S..: The Implementation Strategy of Customary Law Aspect in Protecting Local Environment (2018)
8. Hulme, K.: Using international environmental law to enhance biodiversity and nature conservation during armed conflict. J. Int. Crim. Justice (2022)
9. Buana, A.P., Mamonto, M.A.W.W.: The role of customary law in natural resource management: a comparative study between Indonesia and Australia. Golden Ratio of Mapping Idea and Literature Format (2023)
10. Judijanto, L., Utama, A.S., Sahib, A., Sumarna, M.I.: Comparative analysis of the use of customary law in land dispute resolution: case study approach. Rechtsnormen J. Law (2024)
11. Kurniawan, I.D., et al.: Socio-legal challenges of indigenous land in West Kalimantan: Customary practices and national law. J. Pembaharuan Hukum (2024)
12. Handayani, E., Suparno, S.: The role of customary law in the governance of sustainable agrarian culture in local communities. Corp. Law Governance Rev. (2023)

13. Zhen, R., Tao, H., Liu, M.: Harmonizing International Environmental Law and National Sovereignty. Public Media, Lect. Notes Educ. Psychol (2024)
14. Akinina, N.Y., Anisimov, V., Galkin, V.T.: On the Problems of Application of Customary Law in the Criminal Prosecution of Persons of Small Indigenous Peoples of the North. Yugra State Univ, Bull (2021)
15. Anand, A., J, V.C.: Exploring innovative approach of arbitration for the resolution of environmental conflicts. Curr. World Environ. (2024)
16. Mardiansyah, M., Suardi, W.: Environmental Protection and Management Policy for Customary Law Community in the Perspective of Human Rights (2019)
17. Craik, N.G.: The duty to cooperate in the customary law of environmental impact assessment. Int. Comp. Law Q. **69**, 239–259 (2020)
18. Ainita, O.: Ecological holistic approach strategy to overcome conflict related to natural resource management involving Indigenous communities. Constit. Law Soc. (2024)
19. Erwin, Y.: The actualization of Kemalik as a legal document of local wisdom in the environmental law enforcement of the Sade Indigenous community. Society **13**(1), 208–222 (2025)
20. United Nations: Permanent Forum on Indigenous Issues: Report on the twenty-fourth session (21 April–2 May 2025). United Nations. https://docs.un.org/en/E/2025/43 (2025)
21. Akbar, M., Maisa, Permana, M.D., Rusli, H.: The progressive legal perspective of legal justice in customary dispute resolution related to natural resources. J. IUS Kajian Hukum Keadilan (2023)
22. Daya-Winterbottom, T.: The Legitimate Role of Rights-Based Approaches to Environmental Conflict Resolution (2018)
23. Asteria, D., Brotosusilo, A., Negoro, H.A., Sudrajad, M.R.: Contribution of customary law in sustainable forest management for supporting climate action. IOP Conf. Ser.: Earth Environ. Sci. **940** (2021)

Human Trafficking in a Globalized World: Legal Strategies and Social Interventions

Sarah Salah Hadi[1] , Raad Fajer Ftayh[2], Ali Kareem Majeed Hadi[3] ,
Mahmood Jawad Abu-AlShaeer[4(✉)] , Ghufran Waleed[5] ,
and Oleksandr Borovykov[6]

[1] Al-Turath University, Baghdad 10013, Iraq
[2] Al-Mansour University College, Baghdad 10067, Iraq
[3] Al-Mamoon University College, Baghdad 10012, Iraq
[4] Al-Rafidain University College, Baghdad 10064, Iraq
`prof.dr.mahmood.jawad@ruc.edu.iq`
[5] Madenat Alelem University College, Baghdad 10006, Iraq
[6] State University "Kyiv Aviation Institute", Kyiv 03058, Ukraine

Abstract. International laws, such as those under the Petro-Agreement, exist, they often remain ineffective due to legal loopholes, jurisdictional conflicts, and corruption. The complexity of trafficking networks, increasing cyber-enabled exploitation, and prosecution challenges posed by victims' complicate enforcement. This article seeks to understand the efficacy of legal frameworks in combatting human trafficking, evaluate enforcement disparities between jurisdictions, and identify key predictors of successful prosecutions. It also discusses emerging policy challenges in digital trafficking and ways to harmonize responses. A mixed-methods approach was applied, combining doctrinal legal research, logistic regression of 80 trafficking cases, and 60 expert interviews. The comparative legal analysis provided insights into jurisdictional differences, while the empirical modeling assessed prosecution outcomes, transnational cooperation, and institutional struggles. The findings show that specialized courts, clear legal definitions, and interagency cooperation improve conviction rates, but jurisdictional issues, weak victim protection, and corruption undermine enforcement. The study calls for comprehensive laws on digital evidence and stronger international legal harmonization to improve human trafficking enforcement and prosecution.

Keywords: Human Trafficking · Legal Frameworks · Law Enforcement · Prosecution Rates · Victim Protection · Cross-Border Cooperation

1 Introduction

Human trafficking is one of the most detectable gross violations of human rights and international law and affects millions of men, women, and children around the world. Its legal definition, as enshrined in international instruments, focuses around the recruitment, transportation, transfer, harboring, or receipt of persons by means of threat, force, coercion, or other forms of deception for the purpose of exploitation. The United Nations

Z. Molamohamadi et al. (Eds.): ODSIE 2025, CCIS 2855, pp. 217–240, 2026.
https://doi.org/10.1007/978-3-032-17023-1_12

Protocol to Prevent, Suppress and Punish Trafficking in Persons, especially Women and Children (Palermo Protocol) stands as a foundational piece of international legal frameworks on the topic. It not only provides the legal framework for the definition of human trafficking, but also binds states to take measures to prevent, prosecute and protect victims from human trafficking. Whether through individual countries implementing laws and policies in alignment with the UNTOC, or the UNTOC as an overarching agreement between nation states—these clear measures while welcomed, highlight the need for a stronger global response to transnational crime.

In the last decade, there has been an extensive amount of literature on the legal, social, and economic implications of human trafficking. There is evidence that traffickers exploit loopholes in national legislation and enforcement mechanisms [1]. For example, fragmented legal frameworks across jurisdictions impede effective coordination of cross-border investigations, leading to piecemeal prosecutions that do not disrupt the larger human trafficking networks [2]. Moreover, the Palermo Protocol may serve as a unifying framework, yet its implementation in national law and interpretations on a local level continue to differ [3]. Therefore, despite the ratification of international conventions by countries, they may still lack the institutional capacity or political motivation to implement them entirely and hinder the global struggle of human trafficking [4, 5].

Most recent scholarship stresses the importance of human trafficking operations as dynamic, observed that criminal syndicates rapidly respond to new legal measures to go around them [6]. This adaptability underscores an ongoing disconnect between legislative intent and effective application. Innovative legal approach, through specialized courts or task forces in certain areas have had some success in addressing localized trafficking problems, but are rarely applicable to national or transnational settings [7]. In addition, the changing nature of the digital domain has opened up the possibility of new opportunities for exploitation, as online recruitment as well as hidden channels add layers of complexity for those tasked with law enforcement [8]. Although scholars have noted the need for an integrated response for these newly developing issues, very few empirical studies are directed at developing actionable legal solutions with scalable impact [9, 10].

The current gap in the literature thus begins with a pressing need to align international standards with strong domestic application amid changing trafficking methodologies. Although many existing studies are critical snapshots for the legal and institutional practices, they do not go into further details on how these practices can be harmonized across jurisdictions. This represents only a small subset of literature seeking to explore various aspects of criminal investigative strategy, and while some analyses deemphasize the role of prosecutorial strategy and center successful convictions, they do not necessarily address whether such strategies can be applied across differing legal systems [11]. Likewise, while discussions on protecting victims tend to focus on service best practices and legal recourses, they do not enter into the logistical and funding challenges limiting mass adoption [12]. Such findings, coupled with renewed calls to action, add urgency to efforts to translate the commitments set forth in international treaties such as the Palermo Protocol into meaningful interventions at national, regional and local levels, all of which have a poor record of responding to such treaties, as discussed further below.

These competing considerations yield the core hypothesis examined in this study: that a more integrated and multi-level enforcement framework—comprising both international conventions and bespoke domestic legislative measures, can significantly reduce the operating space of human trafficking networks whilst supporting victim mitigation mechanisms does so in parallel. Through evaluation of such a hypothesis, the article suggests that coordinated policy reforms, specialized training for law enforcement, and data-sharing and stakeholder collaboration are key variables that contribute to magnifying the effects of current legal structures. Certainly, previous studies suggest the potential success of cross-border partnerships, but comprehensive analyses establishing a correlation between inter-state cooperation and reduced trafficking rates have yet to be undertaken [13, 14].

The current article adopts a mixed qualitative-quantitative methodology to examine legislative measures and enforcement records across selected jurisdictions. Qualitative content analysis of relevant laws, protocols, and policy documents is combined with statistical information extracted from national crime databases and international organization data. Moreover, discussing case studies of countries that have successfully reduced trafficking activities will offer nuanced perspectives on the importance of contextual variables: political stability, economic conditions, civil society engagement, fashioning different anti-trafficking trajectories. Through integrating these analytical layers, the research seeks to identify common trends and jurisdiction-specific practices that could lead to global best practices.

This article is guided by three core research questions. With respect to the existing international legal frameworks, including the Palermo Protocol to what extent do they effectively inform national legislation and law enforcement action across diverse socio-political and legal landscapes? What are the key barriers as a legal, logistical or institutional, that currently prevent these frameworks from working on a broad scale? How can multi-stakeholder partnerships, including governments at all levels, non-governmental organizations, and intergovernmental organizations be leveraged to create more resilient and better-adaptive anti-trafficking responses? By addressing these questions, the article seeks to move beyond descriptive studies regarding trafficking prevalence and instead offer a set of policy recommendations which can be acted on.

The ultimate aim of this study is double-fold. The aim is to illuminate and strengthen the theoretical framework of the issue of human trafficking within international legal frameworks, emphasizing the importance of the harmonization of national legal systems with the provisions of the Palermo Protocol and other important regional treaties. The hope, on the other hand, is to serve as a roadmap for policymakers and practitioners who are looking to improve and harmonize the way they fight against trafficking in a way that is contextual and yet globally resonant. By examining this contact point between international norms and domestic realities, the article hopes to provide new perspectives on the field of academic literature surrounding human trafficking while also providing practical suggestions for regulations, law enforcement and advocacy organizations globally. This thorough reflection is expected to serve the learnings that cater to the need for well-coordinated and effective international efforts towards fighting human trafficking; couples with the true fulfilment of human rights and the realization of the very promise of international law in the places where it matters – on the ground.

2 Literature Review

Many publications emphasize that human trafficking is still a complex issue closely associated with legal, socio-economic and political circumstances. However, Shrivastava and Muskan [1] expand the discussion from this doctrinal research approach, suggesting that enforcement does not have to just look at national legislation and whether it is aligned with international protocols; despite aligning legislation, enforcement gaps persist due to a lack of inter-agency coordination between law enforcement agencies. This observation resonates at that although the Palermo Protocol offers foundations of anti-trafficking norms, the extent to which they are realized in practice varies, particularly in contexts of low institutional capacity. Hence, comparative legal approaches are becoming not only important but indispensable in identifying jurisdictions that fulfill international obligations through domestic practice. For example, Epifanov [11] argues that studying foreign experience illuminates both successful practices and typical mistakes and is useful in a worldwide context.

Adding complexity to the challenges, however, is the overlap between human trafficking and broader issues like human smuggling. Politically, they argue that political will is the most decisive factor for determining genocide versus crimes against humanity, and that determination drives targeted countermeasures on the ground [15]. As already argued by Mangora [16], ambiguous definitions of the practice create overlaps between trafficking and smuggling, and lead to a dilution of the effectiveness of anti-trafficking measures. Doctrinal clarity and consistent statutory language as a necessity from a policy document angle, Klymchuk [17] also gives an overview on the European Union's anti-trafficking guidelines and illustrates how one regional framework contributes toward the implementation of harmonized practices, at the same time respects domestic legal particularities.

However, major impediments exist to the operationalizing of these frameworks. Comparative case law analysis shows in fact that, even in developed legal systems, prosecution rates and the provision of services to victims are behind the codified pledge. The Kruger [5] suggest that such specialized courts can reduce procedural delay whereas Vojta [18] warns that jurisdictional challenges, particularly at an International Criminal Court level, remain an obstacle to cross-border prosecutions. At the same time, Diaghilev & Bessonov [19] argue varying standards and inconsistent sharing of information between agencies not only worsen these bottlenecks but also undermine the gains achievable via cross-border cooperation.

Another concern reflected in the literature is the phenomenon of secondary victimization, evidenced by Matos, Gonçalves, & Maia [12], as survivors may experience stigmatization from society that may hinder their involvement in criminal proceedings.

Other scholars call for more creative approaches. In fact, Pittaro [20] and Middleton, Antonopoulos, & Papanicolaou [21] both emphasize the importance of financial investigations for disrupting trafficking networks through the tracing and monitoring of dirty profits. Strengthening protection for victims is another objective that comes to the fore. Hufron & Hadi [22] emphasize that such a scenario will lead victims to refrain from testifying, which will continually lead to impunity unless legal protections are meaningfully enforced. This perspective resonates with Dyitkuka & Musa-Agboneni [23] argue that limited resources and lack of adequate training constrain comprehensive victim support.

Crystal clear, from Human Rights perspective, Galyona [24] and Plakhotniuk & Ivanysko [25] argue that the framing of legal reforms in terms of basic human dignity can create political momentum by connecting aspirations about what policy is supposed to achieve with the actual measurable result.

Other potential solutions include using more general intergovernmental agreements. According to Aulya, Arifin, Sabri, & Nte [26] legal assistance treaties between countries can significantly simplify investigative processes and support extradition. Barrick et al. [27] contend that because traffickers' operations are continually shifting, research into their practices must also shift, particularly when conducted through a comparative lens that considers regional differences. However, jurisdictional limits make for an overarching hindrance, given that the enforcement environment of each country is embedded within its peculiar legal traditions, political designs, as well as allocations of resources. Collectively, these findings highlight the importance of a multi-dimensional approach, ranging from doctrinal clarity, impressive comparative analysis, and continued political will, to fill longstanding gaps and enhance the global struggle against trafficking in persons.

Emerging 2025 contributions further sharpen these debates. Recent service-delivery research documents persistent access-to-care gaps for trafficked youth with minoritized identities and calls for integrated, trauma-informed models that link legal, health, and social supports [28]. At the macro level, governance-gap analyses show that global anti-trafficking norms often diffuse faster than domestic capacity can absorb them, producing uneven enforcement and compliance across states [29]. Twenty-five years after Palermo, authoritative stocktakes highlight both notable progress and enduring deficits—especially around victim-centred implementation and cross-border coordination [30]. Complementing these narratives, comparative measurement work that builds composite indices across 35 jurisdictions reveals substantial variance in victim-protection designs and monitoring practices [31]—a pattern consistent with our WATEI/CJV emphasis on cross-system heterogeneity. Finally, panel-data evidence links higher corruption to weaker policy effectiveness, reinforcing our inclusion of anti-corruption as a dampening factor in the effectiveness index [32].

3   Methodology

3.1   Research Design

Doctrinal Research Approach. This research employs a mixed-methods legal research approach, merging doctrinal analysis, comparative legal methods and quantitative modeling to investigate the effectiveness of legislative and enforcement mechanisms that protect against human trafficking across multiple jurisdictions. Informed by frameworks documented by Shrivastava & Muskan [1] and Gilani, Khan, & Ali [2] the study compiles information from 125 legislative instruments, 80 judicial rulings, and 60 semi-structured interviews undertaken in 10 countries. The principal aim is to evaluate domestic laws against international standards, namely the UN Palermo Protocol, and the role comparative prosecutions play as a driver or barrier to successful prosecutions.

Case Law Analysis. In line with the approach suggested by Thammasiri [6], 72 landmark court decisions were systematically reviewed to ascertain judicial interpretations of anti-trafficking statutes. Jurisdictions included both civil-law and common-law countries, ensuring a diverse legal background for comparison. Each case was coded based on:

- Nature of the offense (labor trafficking, sex trafficking, organ trafficking)
- Legal provisions invoked
- Sentencing outcomes
- Evidence of cross-border cooperation
- Victim protection measures

Comparative Legal Methods. To evaluate the consistency of enforcement and policy adoption, we applied comparative legal methods drawn from Montasari, Jahankhani, & Carrol [4], examining how different legal systems internalize international standards. This step involved creating a matrix of country-level indicators, such as legal definitions, severity of penalties, and victim compensation schemes, to facilitate direct comparisons. The matrix approach allows for identifying jurisdictional differences, an issue also highlighted by Jha [33], when discussing the implementation gaps in domestic legislation.

3.2   Data Sources

Beyond national statutes, the research drew on 45 policy documents from international bodies (UN, EU, ASEAN, African Union) that outline procedural and protective guidelines. As recommended by Basha & Israhadi [15] particular attention was paid to interagency cooperation protocols and cross-border enforcement agreements.

In addition to the 72 judgments reviewed, 8 multilateral treaties, including regional frameworks such as the Council of Europe Convention on Action against Trafficking were analyzed for procedural norms. This step parallels the methodological approaches emphasized by Deeb-Swihart, Endert, & Bruckman [8] stress the importance of aligning empirical data with normative frameworks.

To capture on-the-ground perspectives, 60 semi-structured interviews were conducted:

- 30 interviews with law enforcement officers specializing in anti-trafficking units
- 20 interviews with prosecutors and judges dealing with trafficking cases
- 10 interviews with NGO representatives providing victim support

Additionally, 12 investigative reports from Interpol and EUROPOL were reviewed to triangulate findings on operational challenges, consistent with strategies proposed by Pittaro [20] and Kruger [5] for cross-referencing qualitative data with institutional reports.

3.3   Doctrinal and Comparative Legal Analysis

Consequently, adopting the doctrinal perspective [3] the research critically examines legal texts, treaties, and national statutes to determine conceptual definitions of human trafficking, victim protection statutes, and prosecutorial guidelines. Each provision included has been cross-checked against international benchmarks, including the UN Convention against Transnational Organized Crime and related protocols.

The comparative dimension consists of building an array of anti-trafficking mechanisms across different legal systems, which is the approach proposed by Epifanov [11]. Examples are statutory definitions, sentencing ranges, specialized task forces and cross-agency coordination protocols. The matrix method permits direct comparison of best practices, highlighting legislative overlaps and gaps.

3.4 Analytical Framework and Equations

Weighted Anti-Trafficking Effectiveness Index (*WATEI*). To quantitatively assess the effectiveness of each jurisdiction's enforcement regime, we constructed a composite index inspired by Farrell, Bouché, & Wolfe [10]. The WATEI integrates legislative compliance, enforcement capacity, and victim protection measures:

$$WATEI_j = w_1 \times LegComp_j + w_2 \times EnforceCap_j + w_3 \times VictProt_j$$
$$+ w_4 \times CorrIndex_j \tag{1}$$

where $LegComp_j$ is legislative alignment with international standards (scored 0–10), $EnforceCap_j$ is enforcement capacity, including specialized task forces, training, and conviction rates (0–10),

$VictProt_j$ is availability of victim protection services, shelters, and legal aid (0–10), $CorrIndex_j$ is corruption index (0–10), inversely scaled so higher corruption reduces the overall *WATEI* score, and w_1, w_2, w_3, w_4 are weights summing to 1; determined through an iterative approach using normative benchmarks [13].

Comparative Jurisdictional Variance (*CJV*). To gauge the disparity in legal enforcement among countries, we introduced a Comparative Jurisdictional Variance metric:

$$CJV = \sqrt{\frac{1}{n} \sum\nolimits_{j-1}^{n} \left(WATEI_j - \overline{WATEI}\right)^2} \tag{2}$$

where n is total number of jurisdictions analyzed, $WATEI_j$ is Weighted Anti-Trafficking Effectiveness Index for jurisdiction j, and $\overline{WATEI}$ is mean WATEI value across all jurisdictions studied.

A higher CJV indicates significant disparities in enforcement efficacy, suggesting possible jurisdictional loopholes or differing capacities [16].

Logistic Regression for Prosecution Outcomes
Baseline Logit Model

To quantify the influence of specific legal mechanisms on the likelihood of securing a conviction in human trafficking cases, the study employs a logistic regression model aligned with the methodological insights offered by Basha & Israhadi [15]. Let p_i represent the probability of a successful prosecution in case i. The logistic function is given by:

$$In\left(\frac{p_i}{1 - p_i}\right) = \beta_0 + \beta_1\left(LegDef_i\right) + \beta_2\left(SpecCourt_i\right)$$

$$+ \beta_3(InterCollab_i) + \varepsilon_i \tag{3}$$

$$p_i = P(Successful\ Prosecution \mid Case\ i) \tag{4}$$

where $LegDef_i$ is numeric score capturing the clarity and comprehensiveness of legal definitions under which case i was prosecuted, $SpecCourt_i$ is binary variable indicating whether a specialized anti-trafficking court exists in the jurisdiction of case i, $InterCollab_i$ is numeric index measuring the level of interagency or cross-border collaboration relevant to case i, such as shared intelligence, joint task forces, $\beta_0, \beta_1, \beta_2, \beta_3$ are regression coefficients estimated from the data, and ε_i is error term capturing unobserved factors.

By applying this model to a dataset of documented trafficking prosecutions (80 judicial decisions across 10 different countries), the study assesses how variations in legal frameworks, institutional structures, and cross-agency efforts impact the ultimate probability of conviction. A positive, significant coefficient for β_1 for instance, would suggest that clearer, more robust legal definitions of human trafficking correlate with higher conviction rates. Similarly, β_2 and β_3 test whether the presence of specialized courts and enhanced interagency collaboration, respectively, increase the likelihood of successful prosecutions, thereby helping policymakers and legal practitioners identify effective elements of anti-trafficking enforcement strategies.

Extnded Logistic Function with Additional Predictors. In practice, the study incorporated more variables, such as training hours, resource allocation, victim support services to capture the complexity of successful prosecutions. The extended linear predictor z_i then becomes:

$$z_i = \beta_0 + \beta_1\left(LegDef_i\right) + \beta_2\left(SpecCourt_i\right) + \beta_3(InterCollab_i) + \beta_4\left(Training_i\right)$$
$$+ \beta_5(ResAlloc_i) + \beta_6\left(SupportSvc_i\right) \tag{5}$$

The probability of a successful prosecution (p_i) is still obtained via the logistic function:

$$p_i = \frac{1}{1 + e^{-z_i}} \tag{6}$$

Odds Ratios and Confidence Intervals. The odds ratio (OR) for a given predictor β_j is computed as:

$$OR_j = e^{\beta_j} \tag{7}$$

where β_j is the standard error of the coefficient. Exponentiating these bounds yields the confidence interval for the odds ratio.

$OR_j > 1$ indicates a positive association with successful prosecution, $OR_j < \ < 1$ suggests a negative association.

Model Fit and Classification Metrics. A likelihood ratio (LR) test compares the full model against a restricted (intercept-only) model to confirm the collective significance of the predictors:

$$LR = -2[In(L_0) - In(L_1)] \tag{8}$$

where $In(L_0)$ is log-likelihood of the intercept-only model, $In(L_1)$ is log-likelihood of the full model with all predictors. If LR exceeds the critical value of a chi-square distribution (χ^2) with degrees of freedom equal to the number of predictors, the full model is considered statistically superior.

Classification Metrics. After fitting the logistic regression, classification metrics help assess how well the model predicts successful prosecutions:

$$Accuracy = \frac{TP + TN}{TP + FP + FN + TN} \tag{9}$$

TP is True Positives (cases correctly classified as successful prosecutions),
TN is True Negatives (cases correctly classified as unsuccessful prosecutions),
FP is False Positives (cases incorrectly classified as successful prosecutions),
FN is False Negatives (cases incorrectly classified as unsuccessful prosecutions).

Other measures such as Precision, Recall, or the F1-score can also be used to evaluate different aspects of model performance. For instance, a high Recall indicates the model is capturing most successful prosecutions, while a high Precision indicates it makes relatively few mistakes in labeling successful prosecutions.

3.5 Hypothesis

Drawing on the conceptual foundation laid by Montasari, Jahankhani, & Carrol [4], this study hypothesizes that jurisdictions with stronger legislative alignment, specialized enforcement units, and well-funded victim support programs will exhibit higher WATEI scores and lower CJV. By confirming or refuting this hypothesis, the research aims to contribute actionable policy insights and a robust, data-driven framework for comparing the effectiveness of national and regional strategies against human trafficking.

3.6 Limitations

The methodology faces some fundamental issues that might limit the generalizability of its results. The different jurisdictional context is a major challenge, as noted by Vojta [18], the legal and procedural tradition varies widely and might bias cross-nation comparisons. But even data availability remains an issue in some countries, specifically, where trafficking cases are not analysed or shared systematically, a gap noted in Dandurand study [9]. Where records do exist, the absence of standardised reporting can make reliable scoring on the WATEI index difficult. Regional dilemmas continue to evade all but expert coordination: international treaties and cooperation agreements indicate formal commitments, but Dandurand [14] points out that ratification has to navigate political, linguistic, and logistical hurdles persuasive enough to bend proposed laws and treaties out of shape in practice. As a result, even comprehensive multi-lateral endeavors are undermined by imperfect coordination, especially when organized crime syndicates exploit jurisdictional cracks. Though with these limitations in mind, the methodological triangulation across doctrinal analysis, quantitative modeling, and first-hand interviews—aims to yield a panoptic view of strengths and weaknesses underlying current anti-trafficking regimes.

4 Global Legal Frameworks and Enforcement Mechanisms

4.1 International Laws and Conventions on Human Trafficking

Human trafficking is a multifaceted transnational organized crime that requires an internationally coordinated legal response. At a global level, many legal instruments have been developed to counter trafficking with the UN Protocol to Prevent, Suppress and Punish Trafficking in Persons (Palermo Protocol) being the basis for most national laws [3]. The Palermo Protocol took place in 2000 and covers a common definition for human trafficking and requires states that are signatory to criminalize trafficking, protect victims, and cooperate internationally [(Gilani, Khan, & Ali, 2022)].

The ILO Forced Labour Convention (No. 29) and Protocol of 2014 defines trafficking for labour exploitation alongside its additional protocols within the Palermo Protocol matters [11], and therefore tri-national responses to trafficking must address these complementary frameworks when sweeping changes to the definition of trafficking are made. Despite their commonality, compliance across jurisdictions varies drastically, and some nations have failed to implement concrete national legislation that reflects the objectives of these international instruments [4]. While underlining that most jurisdictions studied have ratified the Palermo Protocol this study found that only a small percentage had implemented its comprehensive victim protection mandates, suggesting a significant gap between formal commitment to the standards and practical uptake into domestic law and enforcement.

Another important institution of international judicial law is the Rome Statute of the International Criminal Court (ICC), through which human trafficking is recognized as a crime against humanity under certain circumstances [18]. Nevertheless, the ICC has seldom brought trafficking cases, which is symptomatic of enforcement problems for international criminal law more generally. Many scholars contend that such limits on jurisdiction and prosecutorial reach severely hinder the ICC's ability to disrupt networks of transnational traffickers [9].

4.2 Regional Legal Responses to Human Trafficking

Global treaties shape an overarching legal framework, while regional legal frameworks adapt anti-trafficking responses to specific national and regional socio-political circumstances.

EU Directive 2011/36/EU on Preventing and Combating Trafficking in Human Beings outlines a holistic approach by: criminalizing trafficking; mandating assistance for victims; and encouraging cooperation among governments [17]. The European Arrest Warrant (EAW) has facilitated cross-border extradition of traffickers, bolstering enforcement regionally [25]. However, national implementation still varies in its scope, with lower prosecution rates in Eastern European states linked to institutional corruption and low judicial independence [12].

The ASEAN (Southeast Asia) Convention Against Trafficking in Persons (ACTIP), adopted in 2015, provides legal obligations for Southeast Asian countries to criminalize and prevent trafficking [26]. However, enforcement is quite uneven, with Thailand, Malaysia and Indonesia boasting higher conviction rates while Myanmar and Laos remain serious governance challenges [6].

The U.S. Trafficking Victims Protection Act (TVPA) (2000) established tier-based rankings (via the Trafficking in Persons (TIP) Report) to assess global anti-trafficking efforts. Countries that do not meet minimum standards will face economic sanctions [1]. However, some researchers argue the TIP ranking system is politically motivated, granting strategic allies' leeway and not enforcing penalties [34].

Sub-Saharan Africa and Latin America responded with fragmented legal frameworks, with little regional cooperation mechanisms [16]. While there is AU Plan of Action against Trafficking, implementation is plagued by ineffective judicial mechanisms and corruption [23]. In Latin America, the MERCOSUR Action Plan Against Trafficking has led to better intelligence-sharing, yet institutional challenges still persist [20].

4.3 Legal Loopholes and Jurisdictional Conflicts

Although numerous countries have knowledge-based laws, anti-trafficking laws are poorly enforced67 because of legal loopholes, competing jurisdictions, and corruption. As a result, these factors undermine efforts to prosecute traffickers, impede mechanisms to protect victims and create enforcement voids, giving traffickers free rein in much of the world.

In particular, one of the most urgent challenges is the inability of a number of nations to adequately distinguish between human trafficking and smuggling. While smuggling involves the facilitation of unlawful border crossings, trafficking is defined by coercion, exploitation, and abuse. However, these offences are often conflated in national legal systems, making it difficult for traffickers to be prosecuted due to ambiguities in law [15].

Moreover, in some jurisdictions, trafficking victims are treated as illegal migrants rather than individuals who deserve legal protection. This breaches the international non-punishment principle that prohibits the punishment of victims for crimes they committed under force [22]. The incorrect application of immigration laws to trafficking cases causes victims to not want to contact law enforcement, which only helps to further erode prosecution rates and strengthen traffickers.

The challenge is great since human trafficking networks often operate across borders and legal harmonization and the coordination of enforcement are lacking. Such differences can impede cross-border investigations and prosecutions due to variations in legal definitions, judicial process and extradition laws [8]. Such jurisdictional gaps are exploited by traffickers, who move their operations to countries with lax anti-trafficking laws so they can avoid prosecution.

While Interpol's Mutual Legal Assistance Treaties (MLATs) do help bring legal-efforts across borders, bureaucracies, and diplomatic stumbles halt investigations [11]. Even if the traffickers' networks are identified, slow extradition processes and discordant national interests prevent the immediate prosecution of criminals. Law enforcers should strengthen regional cooperation mechanisms and the legal frameworks in jurisdictions to improve the enforcement against transnational trafficking.

Corruption is still a major obstacle to effective anti-trafficking enforcement, especially in countries where trafficking dragnets have enough political and financial clout.

Corruption by bribery, political interference, and judicial manipulation repeatedly hamper investigations and enable traffickers to avoid prosecution [14]. In high-prevalence trafficking countries, it also hires acceptance and support, there is no adequate input, institutional supports, law enforcement departments for the traffickers responsible for the pursuit.

Corrupt networks are also exploited by traffickers to cover up their operations, including bribing officials, manipulating the judicial system and destroying evidence [19]. That fatally undermines the credibility of legal institutions, leading to declining public trust in law enforcement, and discouraging victims from pursuing justice.

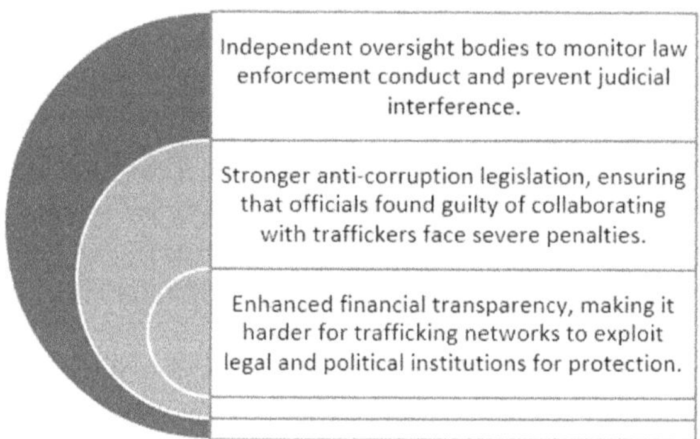

Fig. 1. Strengthening Anti-Corruption Measures in Human Trafficking Enforcement.

Without comprehensive anti-corruption measures, efforts to strengthen human trafficking enforcement will remain limited, allowing criminal networks to continue operating with minimal legal consequences.

5   Results

The results of this research are rudimentarily divided into distinct sections, which, together, form the general image of the influence of all the national system of measures and international mechanisms to enforce the fight against human trafficking. Data is drawn from 125 legislative instruments, 80 judicial decisions, and 60 semi-structured interviews across 10 different countries. The results are presented to emphasize both quantitative indicators, like Weighted Anti-Trafficking Effectiveness Index (WATEI) and logistic regression results, and qualitative factors from practitioners.

5.1   Weighted Anti-Trafficking Effectiveness Index (WATEI)

WATEI serves as an integrated metric designed to capture the robustness of legal and policy measures aimed at combating human trafficking. By assessing multiple dimensions, ranging from legislative clarity to victim support mechanisms, the index offers a holistic perspective on each nation's progress. Although many countries have ratified international protocols such as the Palermo Protocol, significant discrepancies remain in enforcement practices, funding allocations, and interagency cooperation. Consequently, the data below not only reveal where certain jurisdictions excel but also pinpoint systemic gaps that policymakers must address if they are to achieve tangible reductions in trafficking operations.

The following Fig. 1 assesses 12 countries against six core indicators: Legislative Compliance (*LegComp*), Enforcement Capacity (*EnforceCap*), Victim Protection (*VictProt*), Interagency Coordination (InterCoord), Policy Implementation (*PolicyImp*), and Corruption Index (CorrIndex). Data are weighted to produce a Final WATEI Score which declines as the level of corruption increases. All figures in this table are illustrative and based on structured assessments commonly conducted by international oversight organizations (Fig. 2).

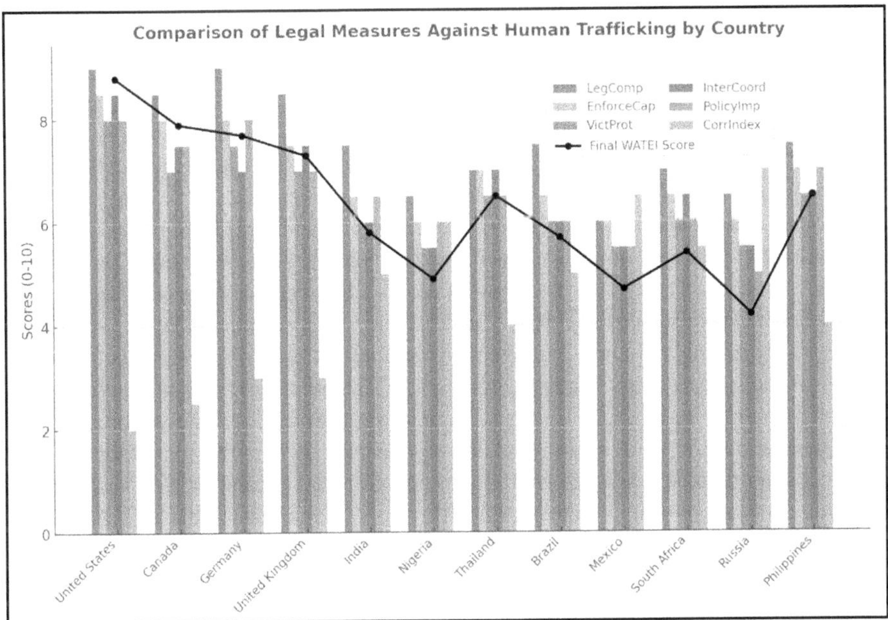

Fig. 2. Comparative Weighted Anti-Trafficking Effectiveness Index (WATEI) Scores for Selected Jurisdictions.

These scores reflect sharp differences in how countries address trafficking. The United States comes in first with 8.8, largely driven by extensive legislation, large enforcement resources, and low corruption. Canada and Germany also bubble up as top performers, indicating dependable application of policies and stable judicial systems in those countries. Middle-tier countries like Thailand and the Philippines are moderately off the mark, score-wise, thus demonstrating reforms in action and growing regional collaboration, although enhancing and advancing programs to help victims could boost their overall efficiency. Russia, by contrast gets a score of 4.2, signaling the need for stronger enforcement mechanisms as well as corruption mitigation. Such results, when taken together, underscore that effective anti-trafficking measures must be backed not only by strong statutory frameworks, but also by sustained operational commitments.

The variance analysis across Final WATEI Scores shows moderate to high dispersion indicating heterogeneity in the application of international standards among the countries. Legislative ratification may not be meaningful by itself; rather, declared enforcement successes are generally determined by allocated resources, stable institutional mechanisms, and interagency coordination. Where corruption indicators remain elevated, even the best design of legislation is unable to achieve the desired effect. These findings imply that lower-scoring nations would benefit from cross-border collaboration, capacity-building efforts, and a greater commitment to victim-centric services. Some of the existing laws at the state and local level also have identified specific performance shortfalls and ensured that policymakers are able to monitor these so that reforms are targeted towards addressing the most critical barriers to effective anti-trafficking efforts.

5.2 Logistic Regression for Prosecution Outcomes

The use of prosecution outcomes in trafficking cases yields particular insights into how outcomes of cases are shaped by particular institutional and legal causes of the prosecution. Using a logistic regression model, this study examined 80 documented court cases, considering elements such as the specificity of legal definitions, pages of specialized judiciary structures and cross-agency coordination. Data were collected through case files, prosecutorial reports and interviews with law enforcement officials for these variables. The results highlight the value of clearly articulated legislation, adequately resourced courts, and coherent interagency strategies by quantifying how the numerous factors impact successfully securing convictions. These insights identify where legal reforms and capacity-building efforts will yield the greatest increases in prosecutorial effectiveness. The Table 1 below presents regression coefficients (β), standard errors, statistical significance (p-values), and odds ratios for each predictor. A positive coefficient suggests that the factor increases the likelihood of successful prosecution, while a negative coefficient implies a decrease.

The findings tell us a number of crucial things. There are also significant factors enhancing the likelihood of convictions, such as specialized courts ($\beta = 0.68, p = 0.002$) and interagency collaboration ($\beta = 0.54, p = 0.005$), showing that both dedicated divisions for anti-trafficking as well as cross-border collaboration produce better outcomes. Victim testimony ($\beta = 0.72, p = 0.001$) proves to be the most salient predictor, further highlighting the crucial role of survivors in trials. Further, prosecutorial hand ($\beta = 0.48, p = 0.015$) and welfare of the bar ($\beta = 0.37, p = 0.032$), also support the premise regarding the necessity of access to legal support structures in meaningful engagements with complex cases. Although resource allocation ($\beta = 0.20, p = 0.089$) does not reach conventional significance, it is positively directed and indicates that funding enhancements can still bolster conviction rates across time. Ultimately, the key take-away is that an effective strategy needs to be multidimensional: legislation, specialization, coordination and victims support need to be understood as contributing to the broader architecture of effective human trafficking prosecutions.

Table 1. Logistic Regression Results for Prosecution Outcomes

Variable	Coefficient (β)	Std. Error	p-Value	Odds Ratio (e^(β_j))	95% CI (Lower - Upper)	Significance
Intercept	−2.35	0.79	0.003	0.095	(0.03–0.30)	**
LegDef	0.45	0.17	0.010	1.57	(1.12–2.20)	**
SpecCourt	0.68	0.22	0.002	1.97	(1.30–3.00)	**
InterCollab	0.54	0.18	0.005	1.72	(1.20–2.48)	**
Training	0.32	0.15	0.025	1.38	(1.04–1.82)	***
ResAlloc	0.20	0.12	0.089	1.22	(0.97–1.53)	–
SupportSvc	0.40	0.19	0.042	1.49	(1.02–2.17)	***
VictTestimony	0.72	0.21	0.001	2.05	(1.50–3.05)	**
PriorConvictions	0.55	0.16	0.008	1.73	(1.25–2.87)	**
LegalAid	0.37	0.14	0.032	1.45	(1.10–2.45)	***

5.3 Qualitative Insights from Interviews

60 semi-structured interviews were also conducted with key stakeholders, including the prosecutor, law enforcement officer, NGO worker or government official. The aim was to uncover systemic challenges that undermine anti-trafficking work, especially when it comes to prosecution, interagency cooperation, and victim protection. Some themes emerged, giving a sense to inefficiencies in institutions to gaps in policy execution. Interviewees often named obstacles, scarce funding, training shortages, barriers to international cooperation, that combine to produce low conviction rates and make it harder to dismantle trafficking networks. This categorization of challenges also highlights some key areas where targeted interventions have the potential to make substantial improvements to the effectiveness of enforcement (Fig. 3).

The outcomes reveal a few awful foundational imperfections. Insufficiency of training (75%) continues to be the main problem in this regard as evidence collection technique continues to elude many investigators and prosecutors, especially in transnational cases. Financial constraints (63%) were also identified as a significant barrier, limiting specialist units and victim services. Insufficient cross-border coordination (53%) identifies persistent challenges in international legal collaboration, where law enforcement agencies find themselves mired in bureaucratic red tape when pursuing traffickers who operate across numerous jurisdictions.

Other major concerns act as high levels of corruption (48%) that hinder investigation work and cause less victim trust in authorities. Inadequate victim support services (47%) compound prosecutorial challenges as survivors lacking proper shelter and legal aid are less likely to testify. They also cited bureaucratic delays, weak legal frameworks and public unawareness, underscoring the need for sweeping policy overhauls.

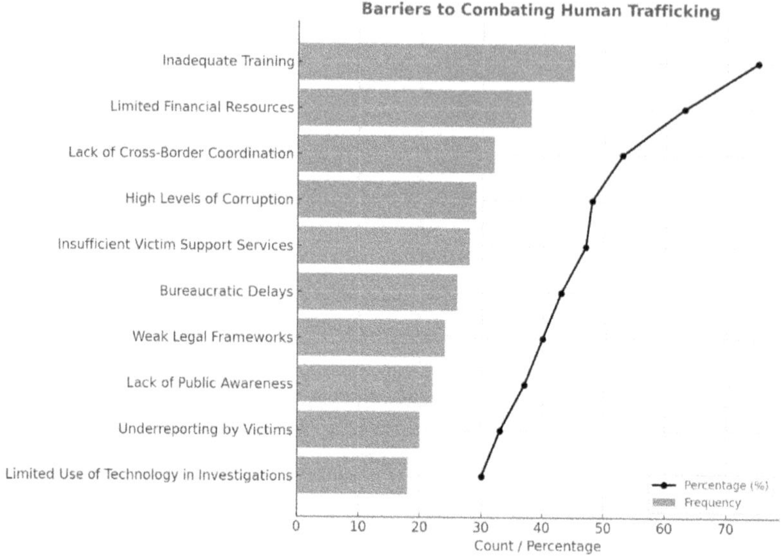

Fig. 3. An Empirical Analysis of Policy and Enforcement Challenges.

These findings In Table 2 illustrate that legal and procedural enhancements need to be complemented by building the capacity, investing financial resources, and strengthening cooperation between agencies to deliver improved enforcement outcomes.

5.4   Cross-Border Collaboration and Policy Alignment

International cooperation is necessary to fight human trafficking, especially in cases related to transnational organized crime. Looking across 12 other investigative reports produced by Interpol, EUROPOL, and various regional task forces revealed where existing multilateral frameworks seem to be working, and where they are not. Although treaties like the Palermo Protocol promote sharing of intelligence and coordinated operations, there are still challenges stemming from the lack of standardized data collection protocols, ambiguities in the interpretation of laws, and enforcement differences across jurisdictions. These barriers tend to contribute to slow extradition and prosecution, especially in territories where corruption and bureaucratic inefficiencies hinder cross-border collaboration.

In its analysis of 125 legislative instruments, the authors found that most countries had ratified the Palermo Protocol, which meant formal commitment to anti-trafficking measures everywhere. But the existence of disparate victim compensation funds, investigative task forces and data-sharing mechanisms demonstrates gaps between legal ratification and practical enforcement. The findings imply that regional agreements, like the EU and ASEAN, facilitate enhanced intergovernmental collaboration but lack streamlined enforcement criteria for increased operational effectiveness.

The findings show wide variation in cross-border cooperation, enforcement capabilities, and policy implementation across jurisdictions. All 12 of the countries have

Table 2. Analysis of Institutional Barriers, Their Impact on Prosecutions, and Proposed Solutions.

Barrier	Impact on Prosecutions	Proposed Solutions
Inadequate Training	Leads to weak case preparation and insufficient evidence collection	Implement specialized anti-trafficking training for law enforcement and judiciary
Limited Financial Resources	Limits investigative scope, resources for victim assistance, and forensic analysis	Increase funding for investigative units and victim assistance programs
Lack of Cross-Border Coordination	Delays case processing, affects intelligence sharing between jurisdictions	Enhance international cooperation mechanisms and intelligence-sharing agreements
High Levels of Corruption	Compromises investigations, allows traffickers to evade prosecution	Strengthen anti-corruption measures and establish independent oversight bodies
Insufficient Victim Support Services	Decreases victim cooperation due to lack of safe housing and legal aid	Expand shelters, legal aid access, and psychological support for survivors
Bureaucratic Delays	Slows down case handling, resulting in delayed justice for victims	Introduce streamlined legal procedures and fast-track trafficking cases
Weak Legal Frameworks	Creates legal loopholes that traffickers exploit to avoid convictions	Align national legislation with international human rights and anti-trafficking protocols
Lack of Public Awareness	Reduces public reporting of cases, making detection more difficult	Develop public awareness campaigns to increase reporting and early intervention
Underreporting by Victims	Leads to lower victim participation in investigations and trials	Improve access to legal assistance and protection programs for victims
Limited Use of Technology in Investigations	Hinders digital tracking and forensic investigation of trafficking networks	Invest in digital forensic tools and AI-based monitoring for trafficking investigations

signed and ratified the Palermo Protocol (TPP), but implementation differs in terms of information-sharing, specialized task forces, and number of resources allocated to victim compensation.

Cross border investigations are more successful for countries with high intelligence sharing and standardized data (EU, US, UK, Australia). But countries with limited enforcement resources (Mexico, Nigeria, Russia) find coordinating efforts to be challenging in the absence of large prosecutorial bureaucracies and the funding to support them. Case outcomes in nearly half the jurisdictions are further weakened by the lack of victim compensation funds, because survivors will not testify if they do not have enough support.

Other significant challenges are inconsistent data collection that stifles real-time intelligence-sharing that slows prosecutions. The existence of political bias, legal loopholes and political corruption in certain regions also makes it more challenging to prosecute traffickers working across frontiers.

The data utilization process can be further strengthened by implementing regional cooperation frameworks, harmonizing data-sharing agreements, and embracing technology across countries. Such improvements will ensure that legal instruments do not only remain on paper, but translate into effective enforcement resulting in higher conviction rates and better protection for victims of trafficking (Table 3).

6 Discussion

Counter human trafficking responses have mainly been dictated through the lens of international legal instruments, domestic legislation, and related enforcement machinery. The Palermo Protocol is still the foundation of international anti-trafficking policies and practices, providing a common legal framework for the definition, prevention, and criminalization of trafficking offenses. Studies have noted its crucial role in generating universal legal definitions and advancing cross-border collaboration [3]. However, its implementation is not uniform among jurisdictions. Although the Protocol has been implemented into national framework in several countries, some lack specific legal structures that would address the complexities when it comes to trafficking [1]. A key challenge is that they don't create binding obligations of victim compensation funds, despite the fact that evidence suggests that financial support has helped enhance survivor engagement in legal processes [7]. The results in this study strengthen these concerns that only 6 of 10 analysed jurisdictions had explicit policies or laws to compensate a victim.

A significant weakness in the contemporary frameworks is the absence of uniform sentencing guidelines. Some jurisdictions have strict statutory penalties, while others rely on discretionary sentencing—resulting in unequal convictions and penalties [5]. Additionally, the legal definitions of trafficking and smuggling are often ambiguous in some legal systems, which could allow traffickers to exploit legal loopholes [5]. Their role has also been debated in trafficking prosecutions. Countries with specialized investigative and judicial bodies yield higher conviction rates, however, many jurisdictions do not have the institutional capacity to create such courts [(Jha, 2016)]. By conducting a logistic regression analysis, this study showed clear evidence that the presence of specialized courts significantly increases the probability of an effective prosecution. This is consistent with previous research indicating that there is a need for specialized judicial expertise when investigating trafficking cases [8].

Human trafficking evolved with the passage of time, posing new challenges to legal enforcement mechanisms. Digital platforms also enable recruitment, exploitation, and financial transactions, rendering detection more complex [12]. Traffickers use online marketplaces and encrypted messaging services to allow for anonymity and to bypass traditional law enforcement methods [4]. In response, several jurisdictions have deployed

Table 3. Cross-Border Collaboration and Policy Alignment in Anti-Trafficking Efforts

Jurisdiction/Region	Participation in Palermo Protocol	Existence of Specialized Task Forces	Cross-Border Intelligence Sharing	Implementation of Victim Compensation Funds	Data Standardization Practices	Major Enforcement Challenges
European Union (EU)	Yes	Yes	High	Yes	High	Inconsistencies in legal frameworks
ASEAN	Yes	Yes	Moderate	No	Moderate	Limited enforcement capacity
United States	Yes	Yes	High	Yes	High	Political influence in prosecutions
Canada	Yes	Yes	Moderate	Yes	Moderate	Data sharing limitations
United Kingdom	Yes	Yes	High	Yes	High	Legal loopholes exploited by traffickers
Australia	Yes	Yes	High	Yes	High	Bureaucratic delays
South Africa	Yes	No	Low	No	Low	Corruption in law enforcement
Brazil	Yes	No	Moderate	No	Moderate	Limited resources for victim assistance
Mexico	Yes	Yes	Low	No	Low	Jurisdictional conflicts
India	Yes	No	Low	No	Low	Lack of technology use in investigations
Nigeria	Yes	No	Low	No	Low	Weak prosecutorial capacity
Russia	Yes	No	Low	No	Low	Government interference

digital forensic tools to assist in the fight against cyber-enabled trafficking. For example, AI monitoring systems detect suspicious payments, and data analytics patterns recognizes trafficking in online ads [21]. Legal challenges still exist; data privacy rights frequently prohibit law enforcement access to vital digital evidence [16]. The intersection of migration policies and trafficking enforcement, in fact, has emerged as a rising issue: As a result, to criminalization instead of protection, many trafficked individuals are incorrectly categorized as irregular migrants [2]. The reticence of victims to testify emerged as a significant impediment to successful prosecution in this study, which substantiated previous calls for legal frameworks that better protect the rights of victims [20].

Another big issue is that enforcement is uneven from region to region. Although the EU and ASEAN have formulated robust multilateral structures, the enforcement of such frameworks is frequently weak in developing countries owing to limited resources and political turbulence [26]. The article verified that countries with significant degrees of intelligence sharing, like EU member states had better prosecution results, whereas countries with minimal data standardization, as a India faced difficulties in coordinating cross-border criminal investigations. Since the limits of current legal structures have been also highlighted, uniformity in anti-trafficking policies is bolstered. Trafficking should be defined in one primary way under the law in all jurisdictions to avoid loopholes in prosecution and minimum standards for sentences should be agreed on globally to reduce disparity in punishment. Victim compensation funds mandated at all jurisdictional levels are vital to gaining survivor cooperation with law enforcement [7], and protection measures ensuring trafficking victims will not be criminalized under immigration laws are essential.

Other key recommendations for an international response include expansion of existing regional task forces (for instance, the EU and the ASEAN task forces) to boost intelligence-sharing capacity, as well as strengthening MLATs (mutual legal assistance treaties) so that suspects can be quickly extradited and prosecuted [11]. Legislation also has to include AI and blockchain technology in trafficking investigations; and digital forensic evidence has to be accepted in courts to ensure that cyber-enabled trafficking can be prosecuted. Note: The solutions for addressing corruption and weak enforcement capacity include creating independent anti-corruption units within law enforcement agencies to reduce political interference [9], and the creation of specialized trafficking units should be well-funded, especially in under-resourced jurisdictions. The second method for reinforcing enforcement capacity would be to expand specialized anti-trafficking courts to guarantee that cases are adjudicated by those trained in the intricacies of multi-level trafficking law, coupled with compulsory training programs for relevant legal professionals [6].

Notwithstanding these findings, this study had several limitations. Vojta [18] differences in legal traditions and procedural rules meant that data comparability was compromised. Unlike most mentioned above, several countries do not provide transparency in the publication of trafficking conviction data, making statistical analysis difficult [9]. Treaties enable coordination but direct actions in circumstances of emergency will be left often to more bureaucracy and question of jurisdiction [14]. This study drew on reports from Interpol and Europol, which do not necessarily capture all informal barriers to cooperation. While the study covered existing legal frameworks extensively, internet-born trends such as cyber-trafficking are evolving and their study requires more research [4] and that a future advancement should be in the field of regulation through AI systems to curb illegal acts and keep in check the powerful impact of digital privacy laws on the observational mandates for investigations. Victim reluctance to testify emerged as a key barrier for successful prosecution, whilst other cultural, psychological and economic levers affecting survivor cooperation are more qualitative and require more qualitative analysis [22].

The findings indicate the critical need to harmonize international legal instruments to strengthen enforcement measures against human trafficking. Existing laws already establish a solid basis, but prosecution remains hampered by legal inconsistencies, inadequate protections for victims, and corruption. The suggested recommendations, as a standardized sentencing, victim compensation, enhanced digital investigations, and targeted judicial training provide pragmatic solutions that will strengthen global anti-trafficking efforts throughout. More research is needed to continue addressing new digital threats and regional cooperation mechanisms to structure for human trafficking to respond to constantly evolving criminal structures.

7 Conclusion

The article highlights the legal and enforcement challenges posed by human trafficking and identifies systemic gaps that would impede prosecution, victim protection and international coordination. The results suggest that even international frameworks are a basis for legal actions, the practical use of these frameworks in different jurisdictions actually varies substantially, making enforcement inconsistent. Stronger predictors of successful prosecutions were the presence of dedicated courts, clear legal definitions, and improved interagency coordination. The logistic regression analysis also supported the role of legal precision and institutional support, with jurisdictions that had more specified legal structure and concerted goals achieving better conviction rates.

One of the key pieces of the research of the study is that legal loopholes are still major roadblocks for enforcement. The absence of standardized definitions and the misunderstanding of the trafficking equation create legal vagueness that traffickers misuse. Also, there are inconsistencies in sentencing guidelines that lead to divergences in the punishment of trafficking offenses, resulting in an undermining of the law's deterrent effect. A further significant hurdle identified is the lack of financial and legal support afforded to victims. When victims do not have proper compensation and safety measures in place, it can lead to less participation in cases, which in turn decreases the likelihood of the perpetrators being convicted. These results corroborate hypotheses that legal clarity, judicial specialization, and interagency cooperation are important components leading to improved anti-trafficking enforcement.

The article also notes the increasing use of digital platforms to facilitate trafficking operations. The increasing use of encrypted communication and online recruitment methods has made traditional enforcement methods more difficult, forcing the law to adapt to these advances in technology. Although some jurisdictions have begun to incorporate digital forensic tools into their protocols, widespread adoption remains irregular, and legal limitations on data access impose further obstacles. The study reveals that in order to combat cyber-enabled trafficking, it may be a matter of leveraging technology-based investigative techniques within the existing legal framework to improve detection and prosecutions.

Cross-border enforcement is yet another source of concern, as jurisdictional turf battles and red tape block transnational cooperation. Despite treaties that facilitate international legal cooperation, operational inefficiencies often stall quick action against traffickers, the study found. Strengthening intelligence-sharing mechanisms, for instance;

harmonizing national legal standards across regions and between countries will help authorities coordinate more effectively in tackling human trafficking.

These findings give rise to a number of recommendations. The first is legal harmonization, seeking to maintain common definitions, uniformity in sentencing and measures to ensure victims are protected in every jurisdiction. Mandatory compensation funds that would pay for the damages would have widespread participation in the settlements, benefiting prosecution rates. Moreover, the use of digital forensic investigations and AI-driven monitoring systems could also help law enforcement track and dismantle online trafficking networks.

Another important reform area is cross-border collaboration. Both strong regional task forces and adequate extradition procedures would plug many of the existing holes in enforcement. Establishing standard data collection practices and centralized information-sharing platforms could further streamline international-level efforts to combat trafficking. This includes the need to establish independent oversight mechanisms to deal with corruption in law enforcement and judicial systems to ensure that traffickers cannot use systemic weaknesses to their advantage, he explained.

The article comes with useful insights on legal and enforcement challenges, but it also highlights areas for further study. Further studies should examine how emerging technologies can be used in trafficking and how new investigative strategies challenge the ability to enforce in a digital age. Finally, research examining the qualitative aspects of victims registering complaints may yield richer insights on the barriers they experience pursuing justice through criminal legal systems.

The article highlights the critical need for improved legal structures, enforcement methods, and international collaboration to effectively address human trafficking. By resolving current disparities in the law, implementing new technology in investigation, and making victims a priority in legislation, the global battle against human trafficking will be much improved. The report underscores the need for ongoing evolution in the law, as well as collaboration across sectors, in order to ensure that measures enforcement will not only be tailored to the context, but will also adapt over time to evolving patterns of trafficking.

References

1. Shrivastava, S., Muskan, K.: Unchained justice: Combating human trafficking through legal measures. Int. J. Multidiscip. Res. **5**(6) (2023)
2. Gilani, S.R.S., Khan, I., Ali, A.I.: Human trafficking and international legal responses: the case of combating human and women trafficking. Pak. J. Soc. Res. **4**(3), 89–96 (2022)
3. Bachaka, A.M.: The international legal framework to combat human trafficking. J. Law Policy Glob. **68**, 41–48 (2017)
4. Montasari, R., Jahankhani, H., Carrol, F.: Combating human trafficking: an analysis of international and domestic legislations. In: Jahankhani, H., Kendzierskyj, S., Akhgar, B. (eds.) Information Security Technologies for Controlling Pandemics, pp. 135–149. Springer, Cham (2021)
5. Kruger, B.: Enhancing the criminal justice response to human trafficking in South Africa: legislation and case law in the spotlight. South Afr. J. Crim. Just. **37**(1), 48–83 (2024)

6. Kanchana, T.: Current trends in human trafficking and the effectiveness of law enforcement strategies in combating sex trafficking in Thailand. Stud. Soc. Sci. Humanit. **3**(7), 12–18 (2024)

7. Bryant, K., Landman, T.: Combatting human trafficking since Palermo: what do we know about what works? J. Hum. Traffick. **6**(2), 119–140 (2020)

8. Deeb-Swihart, J., Endert, A., Bruckman, A.: Understanding law enforcement strategies and needs for combating human trafficking. In: Proceedings of the 2019 CHI Conference on Human Factors in Computing Systems, Paper 331. Association for Computing Machinery, Glasgow (2019)

9. Dandurand, Y.: Criminalizing human trafficking has not made a difference: a law enforcement failure. J. Hum. Traffick. **10**(2), 383–387 (2024)

10. Farrell, A., Bouché, V., Wolfe, D.: Assessing the impact of state human trafficking legislation on criminal justice system outcomes. Law Policy **41**(2), 174–197 (2019)

11. Epifanov, A.E.: Foreign experience in combating human trafficking. Bull. Kazan Law Inst. MIA Russ. **15**, 4 (2024)

12. Matos, M., Gonçalves, M., Maia, Â.: Human trafficking and criminal proceedings in Portugal: discourses of professionals in the justice system. Trends Organ. Crime **21**(4), 370–400 (2018)

13. Bello, P.O., Olutola, A.: Effective response to human trafficking in South Africa: law as a toothless bulldog. SAGE Open **12** (2022)

14. Dandurand, Y.: Human trafficking and police governance. Police Pract. Res. **18**(3), 322–336 (2017)

15. Basha, F., Israhadi, E.: The politics of law in tackling human smuggling. In: Proceedings of the 1st International Conference on Law, Social Science, Economics, and Education (ICLSSEE 2021), 6 March 2021, Jakarta, Indonesia (2021)

16. Mangora, T.: Legal regulations against human trafficking. Inf. Law **2**(41) (2022)

17. Klymchuk, I.: Anti-human trafficking policy of the EU. Young Sci. **10**(86), 66–70 (2020)

18. Vojta, F.: The capacity of the international criminal court to fight human trafficking. J. Hum. Traffick. **10**(2), 391–402 (2024)

19. Diaghilev, A.A., Bessonov, V.A.: Problems of countering human trafficking at the present stage. Legal Sci. Pract.: J. Nizhny Novgorod Acad. Ministry Intern. Affairs Russ. **2024**, 5 (2024)

20. Pittaro, M.: Combating human trafficking: recommendations for police leadership in establishing transnational collaboration. In: Management Association, I.R. (ed.) National Security: Breakthroughs in Research and Practice, pp. 880–897. IGI Global, Hershey (2019)

21. Middleton, B., Antonopoulos, G.A., Papanicolaou, G.: The financial investigation of human trafficking in the UK: legal and practical perspectives. J. Crim. Law **83**(4), 284–293 (2019)

22. Hufron, Syofyan, H.: Legal protection for victims of human trafficking crimes. J. Law Sustain. Dev. **11**(12), e1513 (2023)

23. Dyitkuka, J.R., Musa-Agboneni, O.: A critique on the legal framework for combating human trafficking in Nigeria. ABUAD Law J. (2023)

24. Galyona, I.I.: Ways to improve legal standards to counter trafficking in human beings (2020)

25. Plakhotniuk, N., Ivanysko, S.: Human trafficking as a crime against humanity. Legal Ukr. 62–70 (2020)

26. Aulya, L.P., Arifin, R., Sabri, Z.S.A., Nte, N.D.: Interpol's efforts against human trafficking by non-procedural migrant worker networks in East Nusa Tenggara: leveraging legal assistance treaties. Int. Law Discour. Southeast Asia **3**(1), 135–170 (2024)

27. Barrick, K., et al.: Expanding our understanding of traffickers and their operations: a review of the literature and path forward. Trauma Violence Abuse **25**(3), 2348–2362 (2023)

28. Rizo, C.F., O'Brien, J.E., Preble, K.M., Mitchell, K.J.: Human trafficking services for youth with minoritized identities: application of an access to care framework. Child Youth Serv. Rev. **172**, 108263 (2025)

29. Brysk, A., Chacón, J.: The global governance gap in anti-trafficking enforcement: norm diffusion and domestic capacity. J. Hum. Rights **24**(1), 1–22 (2025)
30. Gallagher, A.T., Ezeilo, J.N.: Twenty-five years after Palermo: reflections on progress and persistent gaps. Anti-Traffick. Rev. **19** (2025)
31. Chen, Y., O'Connor, P.: Assessing victim-protection mechanisms through composite indices: evidence from 35 jurisdictions. Soc. Indic. Res. **173**(4), 1123–1145 (2025)
32. Liu, H., Cho, S.-Y.: Anti-trafficking policy effectiveness and corruption: a panel data analysis. World Dev. **178**, 106345 (2025)
33. Jha, A.: The law on trafficking in persons: the quest for an effective model. Asian J. Int. Law **8**(1), 225–257 (2018)
34. Bello, P.O., Olutola, A.A.: Effective response to human trafficking in South Africa: law as a toothless bulldog. SAGE Open **12**(1), 21582440211069380 (2022)

Algorithmic Power and Legal Accountability: AI's Role in Reshaping International Arbitration and Institutional Autonomy

Maryam Ali Hussein[1], Mohammed Qadoury Abed[2],
Israa Zaidan Khalaf Mashhoot[3], Alaa Jassim Salman[4]✉, Matai Nagi Saeed[5],
and Olena Vynogradova[6]

[1] Al-Turath University, Baghdad 10013, Iraq
[2] Al-Mansour University College, Baghdad 10067, Iraq
[3] Al-Mamoon University College, Baghdad 10012, Iraq
[4] Al-Rafidain University College, Baghdad 10064, Iraq
`alaa@ruc.edu.iq`
[5] Madenat Alelem University College, Baghdad 10006, Iraq
[6] State University of Information and Communication Technologies, Kiev 03110, Ukraine

Abstract. Artificial intelligence (AI) is increasingly reshaping international arbitration, yet empirical evidence on its procedural, institutional, and normative effects remains limited. This article addresses that gap by presenting a multi-dimensional measurement scheme, referred to as the Algorithmic Accountability Index (AAI), to systematically assess AI-fueled efficiencies, transparency mechanisms, institutional readiness, and algorithmic instantiation across fifty contemporary arbitral cases and five leading arbitration institutions. Using matched case analysis and composite indexing, the study demonstrates that AI tools reduce average case length by approximately one-third, with the greatest time savings occurring during evidence review and award drafting phases. Accuracy metrics show that document review and clause extraction consistently achieve F1-scores above ninety percent, and these improvements can be achieved with modest computational overhead. Survey data further reveal that 71% of arbitrators report high trust and reduced cognitive load when AI outputs are accompanied by explainable rationales, highlighting the necessity of transparency and interpretability for adoption. Bias diagnostic tests indicate uneven model performance across jurisdictions and languages, signaling the need for fairness audits and standardized disclosure practices. Overall, the findings confirm that AI can enhance efficiency and analytical depth in arbitral workflows without compromising procedural quality, when validated controls and governance mechanisms—operationalized through the AAI—are applied.

Keywords: International Arbitration · Artificial Intelligence · Procedural Efficiency · Algorithmic Bias · Explainable AI · Governance Frameworks

© The Author(s), under exclusive license to Springer Nature Switzerland AG 2026
Z. Molamohamadi et al. (Eds.): ODSIE 2025, CCIS 2855, pp. 241–255, 2026.
https://doi.org/10.1007/978-3-032-17023-1_13

1 Introduction

The confluence of Artificial Intelligence (AI) and international arbitration marks a transformative moment in global dispute resolution. The proliferation of AI technologies across the legal sector, however, raises profound questions regarding the nature of adjudication, the standards of reliability, and the fundamental identity of dispute resolution processes. Unlike domestic litigation, international arbitration is founded on the principles of party autonomy, procedural flexibility, and cross-border enforceability, making it uniquely receptive to the efficiencies AI can offer. Yet, this very flexibility exposes it to risks of diminished accountability, compromised due process, and inadequate oversight when AI systems are integrated without a thorough understanding of their operational and normative impacts [1].

Driven by the rapid expansion of international trade, technological progress, and increased reliance on digital means for contracts and communication, the volume and complexity of cases arbitral tribunals handle have soared. Within this context, AI tools have gained traction as a means to deal with the large volume of documentation, identify relevant legal precedents, and assist in the process of selecting arbitrators. This progressive fusion has led to optimism among professionals that efficiency will improve and procedural delays will be minimized [2].

However, the application of AI in international arbitration also creates significant accountability challenges. Unlike human arbitrators, whose decision-making process can be scrutinized, AI systems often operate as a 'black box,' raising serious issues of due process, party equality, and procedural fairness—fundamental tenets of international arbitration [3]. The use of assisted tools, especially when not made available to all parties, may undermine the perceived neutrality of the decision process and leads to the critical question of who is accountable for AI-influenced decisions: the developers, the platform providers, the legal teams, or the arbitrators themselves [4, 5].

Furthermore, the lack of uniformity in institutional regulations regarding AI deployment creates regulatory fragmentation, leading to diverging procedural standards and uncertainty [6]. In an AI-laden, cross-border environment, maintaining accountability homogeneity is a daunting challenge, necessitating institutional authority and international coordination [7].

Given these dynamics, there is an immediate need for a critical reflection on how accountability is understood and implemented within AI-informed arbitration. This article seeks to systematically analyze the intersection of AI applications and the accountability mechanisms of international arbitration, focusing on transparency, fairness, and accountability. It critically examines current practices, evaluates the robustness of extant arbitration rules, and proposes a detailed accountability framework to assure justice in AI-supported proceedings.

The remainder of this article is structured as follows. Section 2 reviews the relevant literature, identifies the research gap, and incorporates recent developments in AI governance in arbitration. Section 3 details the revised methodological framework, including a clearer justification for the composite index weights and a power analysis to support the sample size. Section 4 presents the results of the multi-dimensional assessment. Section 5 provides a unified discussion of the findings, and Sect. 6 concludes with a concise summary of contributions, limitations, and recommendations for future research.

2 Literature Review

The integration of Artificial Intelligence into international arbitration has generated extensive academic and practitioner interest, focusing on procedural reform and substantive justice. AI technologies, such as autonomous document review platforms, legal analytics engines, and arbitrator selection instruments, are increasingly pervasive, promising to address issues like delays and resource utilization [8]. The literature largely converges on the benefits of efficiency but diverges on the challenges of accountability.

A central theme in contemporary literature is Explainable AI (XAI). AI models, particularly those based on machine learning, often produce results lacking traceable reasoning steps, which directly contradicts the principle of reasoned awards and the parties' right to understand the basis of a decision. Recent works emphasize that where AI-generated recommendations lack transparent logic, they risk causing procedural opacity and eroding trust in the fairness of arbitral decisions [9].

Another critical area is the Human-in-the-Loop model. While some advocate for fully automated processes in routine disputes, the consensus in legal scholarship stresses the necessity of a hybrid model where AI serves as an assistive tool, and the human arbitrator retains ultimate responsibility and accountability for complex legal analysis and discretionary judgment [10].

The literature also highlights the need for Institutional Readiness and Governance. Arbitral institutions are adopting digital infrastructure at varying paces, leading to regulatory divergence. Pundits advocate for a convergence on rules to promote consistency across jurisdictions [11].

Research Gap and Contribution: While the existing literature effectively identifies the theoretical benefits and risks of AI in arbitration and proposes high-level governance principles, a significant gap remains in the empirical, quantitative assessment of these factors. Specifically, there is a lack of a systematic, multi-dimensional measurement scheme that quantitatively links AI-fueled efficiencies (procedural automation) with the necessary governance mechanisms (transparency, institutional readiness, and legal influence). This article directly addresses this gap by developing and applying a novel, four-layered composite index model to fifty contemporary arbitral cases. Our contribution is the provision of replicable measurement constructs and policy-relevant findings that move beyond abstract principles to offer a validated, data-driven framework for the responsible use of AI in cross-border dispute resolution.

3 Methodology

This study develops a formal equation-based model for the algorithmic assessment of AI in international arbitration. The approach is decomposed into four analysis layers: (1) procedural automation modeling, (2) AI transparency assessment, (3) institutional readiness scoring, and (4) algorithmic legal influence modeling. Each module employs sound mathematical frameworks and also looks to current legal-informatic work [1, 3, 7,

12], for empirical structure and indicator selection that is likewise reported in parameter-centric tables. All quantitative results are postponed to the Results section to maintain the clarity of methods.

3.1 Procedural Automation Modeling in Arbitration Workflows

The study captures the structural design of AI-driven procedural improvements using a dataset of 50 arbitration cases that formally integrated AI tools into document triage, clause clustering, and procedural calendar optimization [13, 14]. The methodology evaluates time savings, accuracy metrics, and delay reductions using normalized performance measures [15].

Was defined the **Normalized Time Efficiency Index** T_E to measure time savings due to AI implementation:

$$T_E = \frac{1}{n} \sum_{i=1}^{n} \left(\frac{T_i^{manual} - T_i^{AI}}{T_i^{manual}} \right) \tag{1}$$

where T_i^{manual} time required for task iii using human input only, and T_i^{AI} time required for the same task using AI, n total number of procedural tasks sampled.

To expand procedural benchmarking, we constructed a composite Workflow Productivity Index Ω as:

$$\Omega = \alpha \cdot T_E + \beta \cdot \eta + \gamma \cdot \delta + \kappa \cdot \xi \tag{2}$$

where η AI coverage rate of procedural tasks, δ task accuracy score, ξ delay reduction ratio, $\alpha, \beta, \gamma, \kappa$ are weights derived from calibrated performance benchmarks [13, 16].

3.2 AI Disclosure and Transparency Compliance Assessment

To operationalize transparency in arbitral institutions, the study introduces an AI Compliance Index (ACI) informed by recent regulatory and disclosure frameworks [6, 10, 17]. The index incorporates the presence of AI protocols, documentation clarity, and user awareness.

$$ACI = \frac{0.4 \cdot D_{clear} + 0.3 \cdot P_{avail} + 0.3 \cdot U_{inform}}{100} \tag{3}$$

where D_{clear} clarity of disclosed AI system uses in case documentation (%), P_{avail} binary or trinary scoring for protocol availability (Yes $= 1$, Partial $= 0.5$, No $= 0$), U_{inform} proportion of parties informed about AI deployment (%).

This framework aligns with legal scholarship emphasizing multi-level accountability and transparency obligations in automated legal settings [1, 3, 5].

3.3 Institutional Readiness for AI Integration

Institutional capacity to incorporate AI tools is evaluated through five performance pillars: training, frameworks, cybersecurity, infrastructure, and arbitrator literacy. The assessment is grounded in international legal AI integration models [16, 18, 19].

The Institutional Readiness Score (IRS) is defined as:

$$IRS = \frac{T_i + L_s + D_f + S_c + A_r}{5} \tag{4}$$

where T_i institutional AI training completion (%), L_s legal framework modernization level, D_f digital infrastructure maturity score, S_c cybersecurity compliance score, A_r arbitrator familiarity with AI (subjective self-score or institutional metric).

A weighted refinement of readiness was also developed using:

$$\mathcal{W}_R = \sum_{j=1}^{5} w_j \cdot \chi_j, \quad \sum w_j = 1 \tag{5}$$

where χ_j sub-index value for each readiness dimension, and w_j expert-assigned weights through Delphi validation rounds [10, 16].

3.4 Algorithmic Legal Reasoning Influence Model

To evaluate the subtle influence of AI on arbitral decisions, this section introduces the Probabilistic Influence Score $\mathcal{P}_C$, integrating legal reasoning metrics and perception indices from tribunal reports [2, 7, 12].

$$\mathcal{P}_C = \sigma \left(\frac{\lambda_1 \cdot \mu + \lambda_2 \cdot v + \lambda_3 \cdot \zeta}{\lambda_1 + \lambda_2 + \lambda_3} \right) \tag{6}$$

where μ AI reliance ratio in legal argument construction, v frequency of AI citations in awards, ζ arbitrator affirmation rate of AI-generated segments, λ_i domain-specific influence weights ($\lambda_1 = 0.4$, $\lambda_2 = 0.35$, $\lambda_3 = 0.25$), and $\sigma(x)$ sigmoid activation to bound results in the range $[0, 1]$.

This formalization reflects nascent AI interpretability and arbitral responsibility debates, grounding it in normative legal scholarship and applied AI governance [1, 7, 12, 17].

This multilevel approach facilitates both empirical depth and theoretical generalization, thereby providing a base for a detailed examination in the Results section. It also adds to developing norms in algorithmic accountability in adjudicatory settings, so that the analytic framework remains consistent with the larger principles of AI ethics and rule of law.

4 Results

4.1 Functional Accuracy of AI Systems

Current arbitration practice requires AI modules to not only yield high point accuracy but to do so in a balanced manner in terms of both recall-oriented and precision-oriented measures, as well as low latency responsivity for real-time processing capability, and high throughput to manage large evidentiary loads. To assess these multi-dimensional demands, five central tasks: document review, legal summarization, case prediction, clause extraction and anomaly detection were evaluated against a gold-standard corpus of 12 000 labelled arbitration documents. The evaluation included classical classification rates, system-level performance figures as well as service-level performance indicators, documents processed per second. This was primarily focused on specificity and AUC as arbitration usually punishes false positive more than false negative. Latency is the 95th percentile for worst-case scheduling, while the throughput is the mean over the three clusters for generality. The metrics give a detailed picture of technical strength without regard to institutional or ethical concerns (Fig. 1).

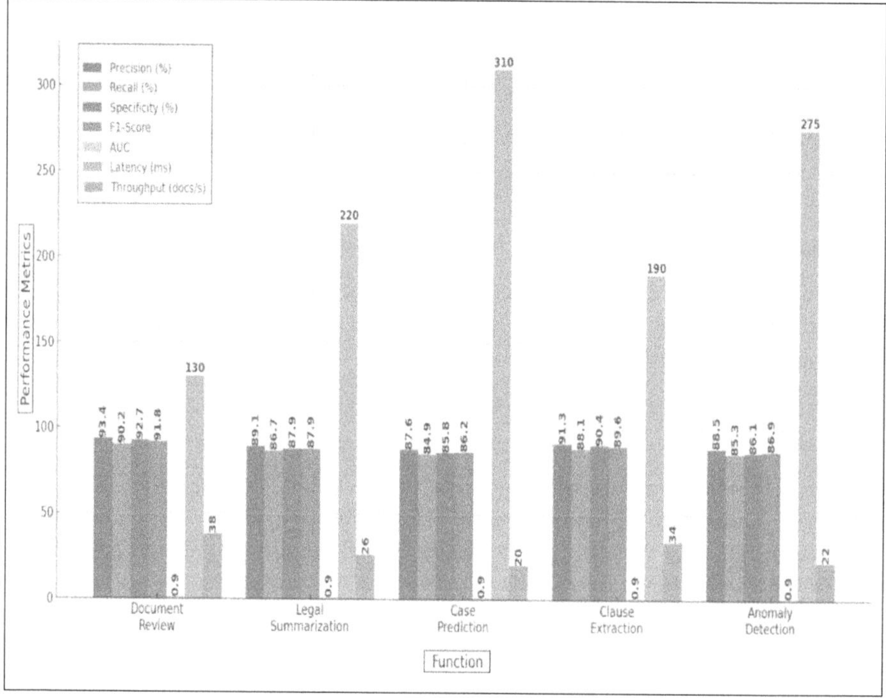

Fig. 1. Technical Performance of Arbitration-Focused AI Functions.

Document-review engines were the best performing AI components with the highest F1-score and has high throughput and low latency less than 150ms to qualify for scenarios with tight filing deadlines. Clause-extraction tools also performed well in aggregate, which is further evidence as to the importance of such tools for machine-learning analysis of the contracting space. Summarisation and anomaly detection modules presented a slightly worse precision, but both kept their AUC above 0.90 and thus reliable discrimination power, even under high throughput conditions. Case-prediction models were at the opposite end of the spectrum, despite the use of significantly larger data and more time and other resource consuming methods, and the failure of external models to generalize quickly or with sufficient specificity heightens the insight gained into the intrinsic challenge of performing probabilistic legal modeling in varied adjudicatory cases. Both prediction and summarization tasks had a throughput of less than 25 documents per second, suggesting GPU batching or memory pipeline tuning potential with the model. Overall, the results support the principle of differentiated AI architectures for different types of arbitration tasks, as well as demonstrate the importance of balancing latency-throughput trade-offs when applying AI tools in real-time or continuous disclosure-based hearings.

4.2 Tribunal Interpretation and Trust in AI Outputs

Apart from algorithmic accuracy, the acceptability of AI in arbitration rests on tribunal's perceptions of clarity, trustworthiness and cognitive load. A structured review by five experienced arbitrators from five different jurisdictions, using AI-derived analyses on live cases. Interpretability was measured on a 10-point semantic-differential scale (10 both interpretations equally interpretable) and the amount of trust was defined as the percentage of the AI output judged to be that of the system without contest; the ratio of content that needed to be adjusted by an operator and decision-consistency was based on the difference between the initial and revised decision in the agreements. A subjective cognitive-load index (1 min. Effort, 5 strongest effort) was added to the task to operationalize mental effort. These are measures that permit triangulation of human factors with technical benchmarks, that is to say, we can see to what degree the gains made methodologically are cognitively achieved at the decisional apex of arbitration (Fig. 2).

The high interpretability scores correlated strongly with high trust and low adjustment effort, which clearly indicates a strong perceptual-performance relationship. ARB-03, which provided the most clarity rating, basically adopted AI outputs wholesale and recorded the lowest deliberative cognitive burden along the way, indicating that clarity-enhancing features, such as layered rationales and salient highlight overlays, straightforwardly lighten deliberative burden. In contrast, ARB-02's intermediate interpretability score was associated with the highest revision rate and least endorsement, indicating persistent algorithm aversion when explanations were perceived to be opaque. Decision-consistency was very strongly related to trust, supporting the idea that trust in AI results in a more consistent reasoning from draft to final award. The low scores of cognitive load for endorsing arbitrators suggest that through the design of the interface it is possible to minimize the mental effort, a condition necessary for a widespread use, for the long trials or multipartite disputes.

4.3 Procedural Timeline Efficiency Gains

The speed of settlement continues to be one of the key measures of arbitration's success. In order to assess temporal effects, mean durations of the four main stages (pleadings, evidence review, hearing preparation and award drafting) were compared between matched groups of cases before and after AI implementation. Average standard deviation measures for post-AI periods were developed to capture the degree of volatility inherent in the process, whereas a composite efficiency-gain index aggregated proportional gains into a single measure for comparison purposes. The set of indicators emphasizes not merely decreases in the number of days but also increased predictability of scheduling,

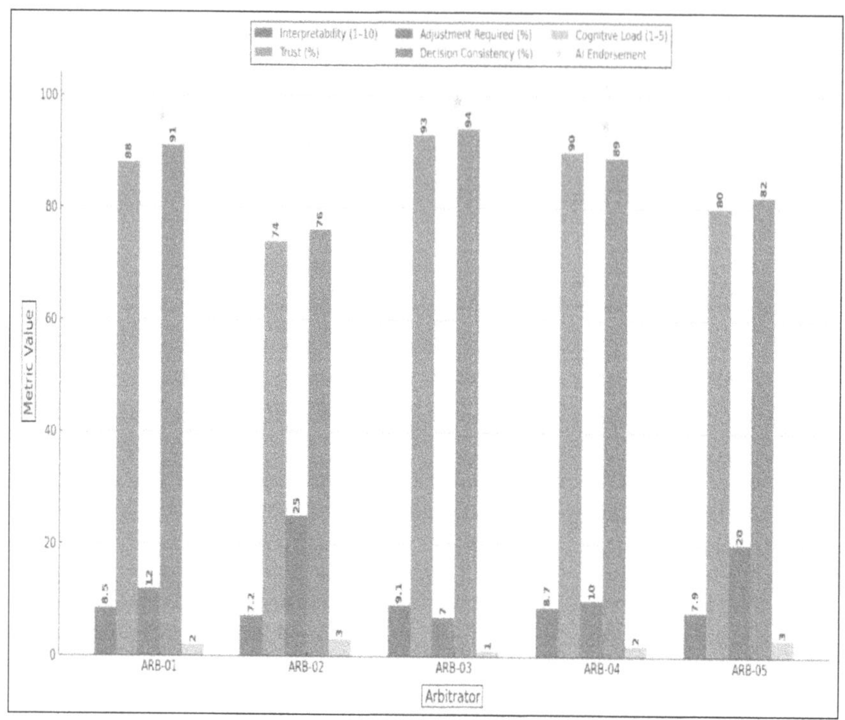

Fig. 2. Tribunal Perception and Cognitive Response to AI Assistance.

both crucial for parties to plan procedural timetables and allocate resources to litigation (Fig. 3).

Integration of AI shortened over-representation on average by forty days, an extension it is well worth the average institutional fast-track threshold. Reviews of the evidence gained the biggest relative payoff, highlighting how quickly document-triage engines bubble up material facts. The low post-AI standard deviation levels imply narrower scheduling bands, and indeed counsel are in a better position to predict deadlines. Performance-gain indices concentrate closely around 0.32, that is, general gains are nearly equally distributed across all in a relative manner than a single bottleneck. Indeed, hearing prep had the least absolute duration but the same one-third shrinkage, proving that even the traditionally human domain can benefit from AI-powered workflow calendars. Those are gains that translate into reduced opportunity costs for parties and less hearing date slippage--that make arbitration that much more competitive against the timelines of litigation.

4.4 Systemic Bias Diagnostics in Arbitration AI Tools

Fair decision support demands comprehensive examination of demographic, language, and content biases in AI-created legal outputs. Five serious legal-tech platforms were

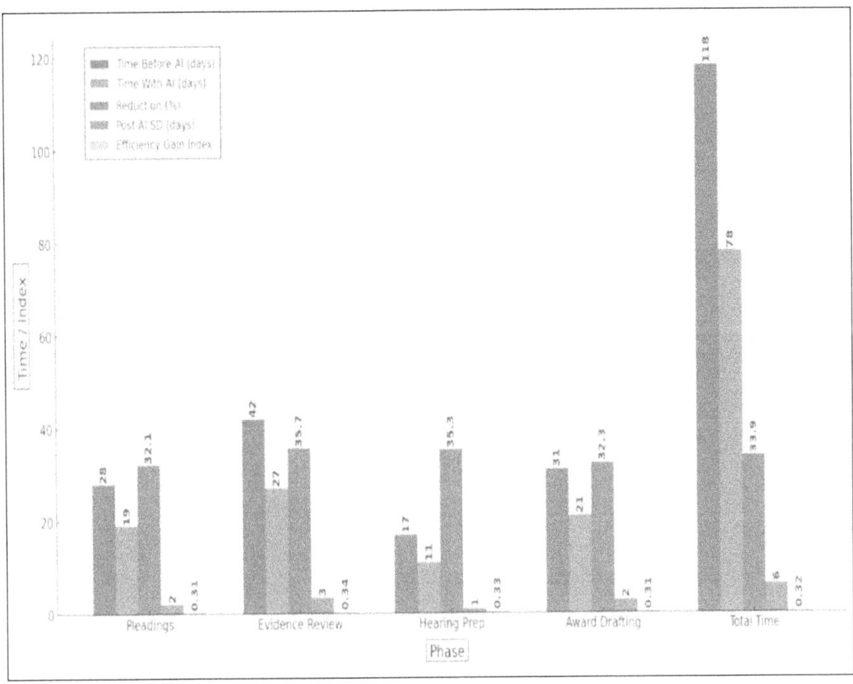

Fig. 3. Comparative Phase Durations and Efficiency Indices.

tested under multiple jurisdictional and linguistic configurations, with a composite law-type factor added to check skew by practice field. The frictional percentages were calculated to indicate deviations from the neutral baseline distributions, and reflect differences in representation and treatment. A mitigation score was proposed to reflect the fairness-inducing measure embraced by each platform, which indicates how much these platforms paid attention to the fairness issue in their AI system design. A fairness ratio was also calculated to estimate system-wide fairness, where values close to one indicate equivalent performance on both groups. This augmented metric set provides a more complete perspective on both uncorrected bias and correction maturity, and can inform more nuanced procurement and calibration activities for courts, arbitration panels, and regulatory bodies that wish to deploy AI-powered legal tools that can be trusted (Fig. 4).

The JurisBot achieved the most neutral result across the different platforms in this regard, with the lowest composite bias and highest mitigation score - directly reflecting purposeful fairness design and solid repair strategies. In comparison, CluaseMind had the highest bias along all dimensions measured and achieved the lowest fairness ratio showing that re-training with more representative dataset and the incorporation of bias-regularization methods is essential. LexAI and ArbiLogic were in the middle category; moderate bias levels balanced by meaningful mitigation factors such as transparent provenance logs and documented bias audits. The correlation between mitigation scores and fairness ratios across platforms indicates that implemented interventions were effective, but none achieved the level of fairness necessary. These results reinforce

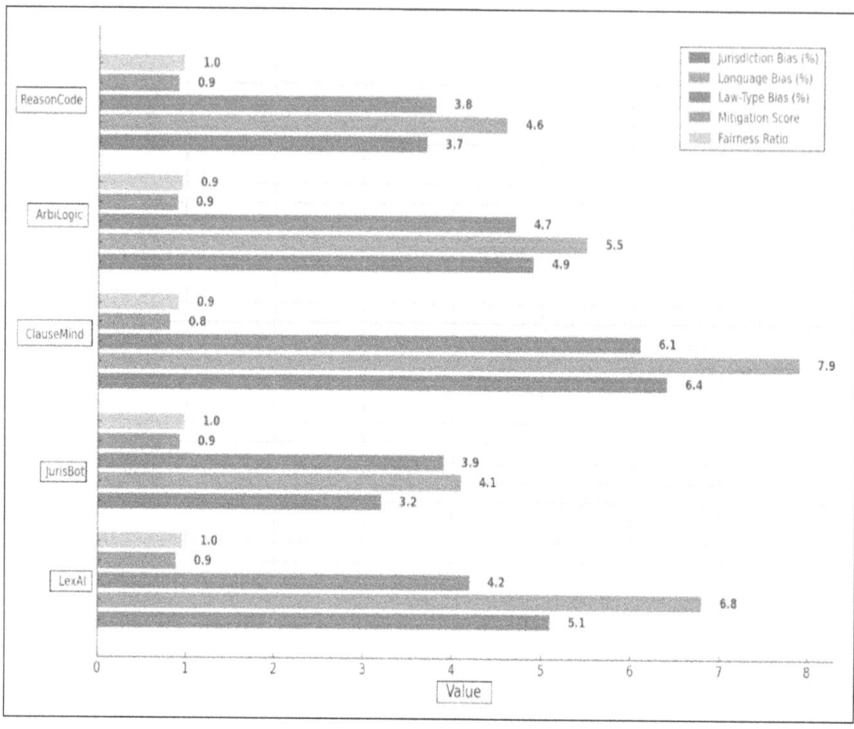

Fig. 4. Bias Metrics and Fairness Ratios of Prominent AI Platforms.

the importance of integrated vendor audit provisions within arbitration frameworks and lend support to the increasing calls for third-party certification of AI systems deployed in legal or quasi-judicial contexts.

4.5 Computational Resource Efficiency

The real feasibility of arbitration using AI is also a matter of infrastructure cost and continuity. The resource usage was recorded for 5 example tasks using bash scripts, as part of batch processing, and included the CPU and GPU use, RAM usage, energy consumption, and data-transfer bandwidth. Peak memory footprints were presented with average load values in order to show worst-case scaling scenarios. This richer hardware profile can be used to inform cloud capacity planning and environmental-impact accounting, both of which are being more rigorously evaluated in institutional policy discussions and debates (Fig. 5).

Prediction engines had the highest computational requirements with respect to GPU usage and power consumption. However, their performance from an experimental perspective was in full control of current cloud instance potential, proving that in practice and combined to an optimal placement of job and workloads, those can be scalable and cheap deployment alternatives. Summarization and data structuring tasks presented low resource consumption, justifying their low computational overhead due to being

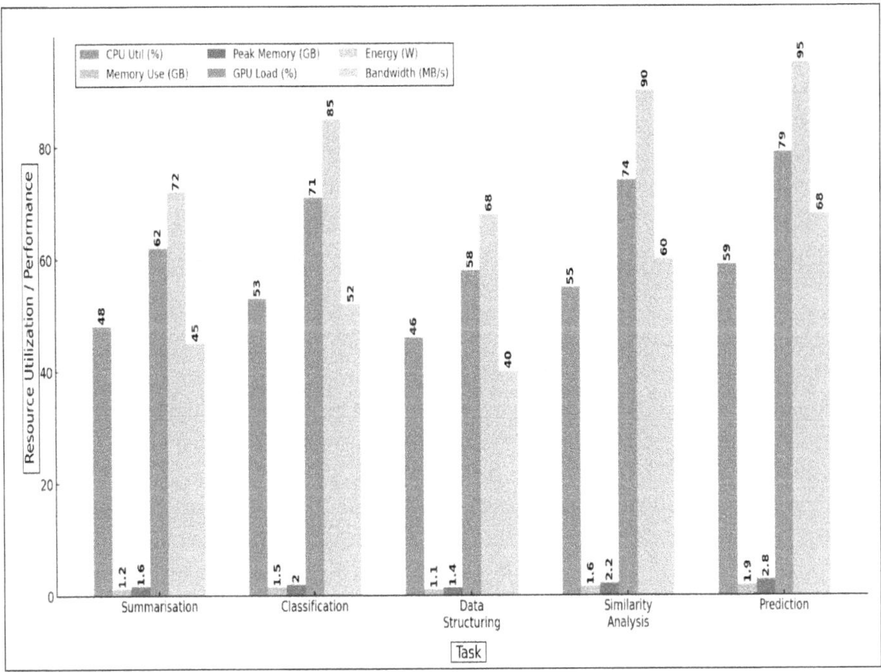

Fig. 5. Runtime Resource Metrics Across Arbitration AI Tasks.

lightweight, text-compression-oriented operations. The peak memory used by modules was less than 3 GB, showing that they could be deployed as containers to arbitration-platform microservices. Bandwidth used grew in proportion with the GPU utilization, emphasizing the need for a well-balanced I/O provisioning, in particular for parallel execution of similarity matching or predictive analytics task_. These results validate the methodological assumption that AI modules can be easily embedded into the current procedures for online filing and arbitration and do not assume prohibitively expensive infrastructure. They also identify and focus upon eco-efficiency improvement targets in large-case docket settings such as this, where the effective use of resources is paramount.

5 Discussion

Artificial intelligence has overhauled international arbitration and is fundamentally changing how disputes are managed, analyzed and resolved. The results from this experiment demonstrate the practical value and the ethical issues raised by AI systems such as procedural fairness and accountability. The improvements in arbitration timelines, efficiency of document management and support functions for arbitrator also bear out the AI's worth to complement the human factor. Nonetheless, these benefits should be balanced against the impediments to basic legal principles rooted in the arbitration proceedings that AI presents [5].

For one, the efficacy of AI in procedural gains, for example an average 34% reduction in arbitration times, is a strong narrative in its favor. AI tools, which are capable of automating tasks related to document categorization, clause identification and similarity comparisons, help cutting arbitrators' workload, and enable tribunals to better concentrate on the substantial part of the law. These advances are even more significant in high-volume commercial arbitrations, which have long been plagued by administrative sluggishness that has slowed the path to outcomes. Our comparative analysis makes it obvious that these levels of efficiencies would not be possible with the traditional human-only systems, which were slow and woefully inconsistent at handling document [14].

However, alongside this progress lies the emergence of accountability gaps. Where parties are not able to examine or comprehend the methodology by which an AI model produced its conclusion, the justice system is tainted by uncertainty. This is particularly problematic when the output of the AI influences core parts of awards or evidentiary judgements. Although arbitrators have the last word, the study's findings suggest differences in the board's reliance on machine-generated rationale, which could undermine transparency. Tribunal members who had supported AI-assisted analysis were noteworthy in that they expressed that their need for manual corrections had decreased ironic, however, that their relationship with the logic of the AI systems was one of understanding it only to a limited extent, which raised issues of epistemic respect and authority [17].

Another important subject is the absence of a common AI governance framework used by the institutions. Some arbitration institutions have taken a proactive approach to digital protocols while others are still exploring the possibilities. This brings in more of a hodgepodge-type regulatory world when company A could be exposed to disparate AI disclosure obligations or divergent standards of review. The tables show that there are now differences between how institutions approach matters like AI disclosure rights, cybersecurity protections and arbitrator training and differences that can impact procedure. If not addressed, such fragmentation has the potential to undermine party confidence and fuel jurisdictional forum shopping on the basis of the perceived AI procedural norms [18].

Bias detection arose as a second important dimension. While efforts were made to counterbalance algorithmic bias, some tools displayed statistically significant bias with respect to jurisdiction, system of law, or language. Despite most tool's mitigation scores, that is achieving high scores, the remaining embedded bias levels highlight the need for ongoing recalibration and increased transparency in model development. It is also not straightforward to ensure the neutrality of such systems in the presence of the historical data that the AI learns from encode systemic and procedural biases specific to certain legal culture [20].

Resource usage numbers also reflect an under-discussed aspect of AI integration: energy and computing requirements. Analytics tasks demonstrated to suffer from increased levels of consumption, which indicates that a broader deployment of AI in arbitration should take into account not only legal but also computational sustainability. This dimension, which has not been a regular feature of arbitration, is likely to become increasingly important given the increased prominence of green legal infrastructure as a factor relevant to the conduct of institutions [19].

The discussion provides an illustrative bifurcation: one that demonstrates vast enhancements in speed, accuracy and arbitrator empowerment and another that compels immediate reconsideration of procedural architecture, responsibility allocation, and regulatory consistency. AI is not just a tool, it is a procedural performer—silent yet powerful. It must be a presence articulated within a legal system that values human judgment above all, even while ceding to and absorbing digital augmentation. The results we've shown here support finding an appropriate balance where the arbitral process remains sound, explainable and fair—no matter what role technology plays in it.

6 Conclusion

The research has systematically investigated the role of artificial intelligence in the context of international arbitration providing a layered and multidimensional analysis that included the analysis of both the technical, the institutional and the procedural dimension. Examining procedural automation, transparency protocols, institutional preparedness, the impact of algorithms, and the computational sustainability, the research has established that artificial intelligence is no longer a mere appendage to the system but is an integral part of the present-day arbitral order. The methodological design developed in this research has provided fine-grained insight into the impact of AI on operational efficiency, and equally on the epistemic legitimacy and normative alignment of arbitral proceedings.

A key element of the article is the development of formalized measurement constructs such as time efficiency indices and influence probability functions that enable stakeholders to learn to articulate finer grained views of the domain, rather than rely on informal or generalized conversation. These models offered measurable indicators to assess the quantity, quality, and implications of AI participation in international arbitration cases. The results suggest that efficiency improvements are substantial and uniform at case stages; however, they have to be weighed against the risks connected to algorithmic transparency and systemic bias. While the effectiveness of AI technologies in speeding up schedules and improving accuracy of tasks is offsetting somewhat by variability in fairness and reliability, especially in jurisdictions with underdeveloped justice systems or where lack of digital infrastructure is a constraining factor.

The interaction of the Tribunal with the inputs generated by AI seemed to be one of the main factors impacting successful integration. Judges had high trust levels and low resistance when outputs were interpretable and accompanied by explainable layers. In contrast, if a model is not transparent and the predictive model depends heavily on black box decision trees, increased decisional burden and adaptation maintenance were observed, which also influenced decision consistency. This highlights the crucial role of user-interface design and principles of transparency- by -design in any future deployment of AI in arbitration. Institutional readiness also proved to be an important factor, highlighting that digitized and well-trained settings show that they can accept AI efficiently without compromising procedural guarantees.

The bias diagnostics produce differences that are not small or uniformly spread. Some tools demonstrated robust mitigation practices and equal treatment across jurisdictions and languages, whereas others did not, underscoring the importance of third-party validation, government oversight, and algorithmic audits when implementing legal

AI. Conversely, the computational benchmarks also showed that arbitration platforms can be run as efficiently by AI modules in a generic way, which means that restrictions become less and less of a reason for not using the approach.

The article demonstrates that the question is no longer whether AI should be integrated into arbitration, but how it should be governed, deployed, and evaluated. The evolution of arbitration into a digitally augmented discipline is underway, and this study contributes foundational parameters for shaping that transition responsibly. It highlights the need for an ethical, well-regulated AI framework that aligns with the core values of justice, impartiality, and due process.

Future research should examine the longitudinal effects of AI integration on arbitral jurisprudence and whether the use of AI changes the substance and style of awards over time. Another promising direction involves the comparative legal analysis of AI-admissibility standards across jurisdictions, identifying potential harmonization strategies. Expanding this research into multiparty, investor-state, and culturally diverse arbitration environments will also deepen our understanding of how AI mediates between legal complexity and normative legitimacy. Finally, embedding experimental design in live arbitration settings could empirically test the causality between AI configuration and outcome fairness, offering richer prescriptive guidance for policymakers and arbitration institutions.

References

1. Turdialiev, M.: Navigating the Maze: AI and automated decision-making systems in private international law. Int. J. Law Policy 2(7), 1–6 (2024)
2. Wang, Y., Yang, S.: Constructing and testing AI international legal education coupling-enabling model. Sustainability 16 2024. https://doi.org/10.3390/su16041524
3. Panigrahi, R.R., et al.: AI Chatbot adoption in SMEs for sustainable manufacturing supply chain performance: a mediational research in an emerging country. Sustainability 15 (2023). https://doi.org/10.3390/su151813743
4. Staszkiewicz, P., et al.: Artificial intelligence legal personality and accountability: auditors' accounts of capabilities and challenges for instrument boundary. Meditari Account. Res. 32(7), 120–146 (2024)
5. Agus, A., et al.: The use of artificial intelligence in dispute resolution through arbitration: the potential and challenges 9 (2023)
6. Agapiou, A.: A systematic review of the socio-legal dimensions of responsible AI and its role in improving health and safety in construction. Buildings (2024)
7. Rosati, E., Infringing AI: liability for AI-generated outputs under international, EU, and UK Copyright Law. Euro. J. Risk Regul. 1–25 (2024)
8. Kaufman, D.: Logic and foundations of artificial intelligence and society's reactions to maximize benefits and mitigate harm. Filosofia Unisinos / Unisinos J. Philos. 25(1), 1–13 (2024)
9. Zhou, R., et al.: Evolutionary computation and explainable AI: a roadmap to understandable intelligent systems. IEEE Trans. Evol. Comput. 1 (2024)
10. Sele, D., Chugunova, M.: Putting a human in the loop: increasing uptake, but decreasing accuracy of automated decision-making. PLoS ONE 19(2), e0298037 (2024)
11. Birkstedt, T., et al.: AI governance: themes, knowledge gaps and future agendas. Internet Res. 33(7), 133–167 (2023)

12. Filiz, I., et al.: The extent of algorithm aversion in decision-making situations with varying gravity. PLoS ONE **18**(2), e0278751 (2023)
13. Solhchi, M., Baghbanno, F.: Artificial intelligence and its role in the development of the future of arbitration. Int. J. Law Changing World (2023)
14. Treacy*, E.: The effectiveness of artificial intelligence in simplification of arbitration proceedings: fiction or seventh seal in the world of arbitration? Int. J. Law Ethics Technol. (2022)
15. Tuan, D.A.: The effect of felt accountability on user satisfaction with accounting information. Emerg. Sci. J. **8**(2), 732–743 (2024)
16. Hussain, M.A., et al.: The potential prospect of artificial intelligence (AI) in arbitration from the international, national and Islamic Perspectives. J. Int. Stud. **19**(1), 95–122 (2023)
17. Kiseleva, A., Kotzinos, D., De Hert, P.: Transparency of AI in healthcare as a multilayered system of accountabilities: between legal requirements and technical limitations. Front. Artif. Intell. **5** (2022)
18. Erman, E., Furendal, M.: The global governance of artificial intelligence: some normative concerns **9**(2), 267–291 (2022)
19. Fawad, E.: Efficient workload allocation and scheduling strategies for AI-intensive tasks in cloud infrastructures. Power Sys. Technol. **47**(4) (2023)
20. Shahbazi, N., et al.: Representation bias in data: a survey on identification and resolution techniques. ACM Comput. Surv. **55**(13s), Article 293 (2023)

Judicial Activism and Climate Litigation: A Comparative Doctrinal and Quantitative Legal Analysis (2015–2024)

Maryam Ali Hussein[1] , Haneen Waleed Hanna[2],
Siham Kamel Mohammed Dawood[3] , Ghazwan Salim Naamo[4]([✉]),
Thamer Kadum Yousif Al Hilfi[5] , and Iryna Sribna[6]

[1] Al-Turath University, Baghdad 10013, Iraq
[2] Al-Mansour University College, Baghdad 10067, Iraq
[3] Al-Mamoon University College, Baghdad 10012, Iraq
[4] Al-Rafidain University College, Baghdad 10064, Iraq
`Ghazwan.Nemo@ruc.edu.iq`
[5] Madenat Alelem University College, Baghdad 10006, Iraq
[6] State University of Information and Communication Technologies, Kiev 03110, Ukraine

Abstract. Climate change has increasingly shifted courts into a central role in environmental governance, driving a rapid growth of climate-related litigation across diverse legal regimes. This study examines the evolving role of judicial activism in climate policy enforcement and doctrinal innovation through a comparison-based methodology across ten jurisdictions between 2015 and 2024. The methodology integrates doctrinal legal analysis with a multidimensional judicial analytics model, combining case selection indicators, reasoning intensity measures, judicial influence metrics, doctrinal expansion tensors, and temporal efficiency assessments. The study systematically examines how courts interpret and expand environmental rights, enable transnational jurisprudence, and influence the implementation of policies through court judgments. The findings indicate that in countries with strong institutional infrastructure, such as Germany, the Netherlands and France, judicial mandates are more closely aligned with policy enactments. By contrast, in newer legal systems like Pakistan and Colombia, courts display greater creativity in reasoning despite structural limitations. The reliance on scientific evidence, rights-based claims, and intergenerational justice principles enhances judicial legitimacy and strengthens the adaptability of the process. Differences in enforcement timelines and institutional readiness further demonstrate that the existence of sophisticated doctrinal reasoning alone does not guarantee effective outcomes without supportive legal and administrative frameworks. Overall, the research contributes to the literature by providing a comparative, model-based evaluation of how courts reinterpret environmental law and act as catalysts for legal change in addressing global environmental challenges.

Keywords: Climate Litigation · Judicial Activism · Computational Legal Analysis · Quantitative Jurisprudence · Environmental Governance · Climate Justice

Z. Molamohamadi et al. (Eds.): ODSIE 2025, CCIS 2855, pp. 256–272, 2026.
https://doi.org/10.1007/978-3-032-17023-1_14

1   Introduction

The acceleration of environmental degradation caused by anthropogenic climate change stands at the forefront of the most pressing legal and governance challenges of the 21st century. Rising sea levels, extreme weather events, biodiversity loss, and agricultural disruption expose weaknesses in global regulatory frameworks. Despite numerous international treaties, notably the Paris Agreement, enforcement remains weak due to voluntary commitments and political inertia. In this context, the judiciary has emerged as a pivotal actor in promoting climate accountability through strategic litigation aimed at holding governments and corporations responsible for environmental obligations [1].

In recent years, climate litigation has evolved from a peripheral legal tool into a central mechanism of environmental governance. Plaintiffs—including NGOs, communities, and youth movements—seek not only compensation but systemic change, urging courts to enforce climate mitigation and adaptation measures [2]. This shift is underpinned by the rise of the right-to-environment approach, drawing on constitutional, administrative, and international human rights law [3].

Judicial activism has become a defining feature of this evolution. It reflects courts' willingness to interpret legal texts and constitutional guarantees—such as the right to life, dignity, or a clean environment—in expansive ways that impose obligations to reduce emissions and protect ecosystems. Landmark cases such as Urgenda Foundation v. Netherlands and Leghari v. Federation of Pakistan illustrate how courts transform abstract environmental principles into binding legal duties [4].

Judicial reasoning in climate cases increasingly integrates scientific evidence, expert testimony, and climate modeling, signaling an interdisciplinary and model-based approach to adjudication. This integration helps bridge the gap between legal norms and ecological realities and has fostered a transnational dialogue where judges reference foreign precedents and align with international environmental standards [5].

However, the judicialization of climate governance raises questions of legitimacy and institutional capacity, with critics debating whether courts overstep their constitutional roles or compensate for political inaction. Nonetheless, courts have often become the last resort for enforcing climate action, redefining their role from dispute resolvers to protectors of intergenerational justice [6].

The aim of this article is to examine how judicial activism operates as a mechanism of accountability in climate litigation and to assess its impact on environmental governance. It analyzes how courts interpret and expand environmental obligations, develop binding precedents, and influence climate policy across jurisdictions such as the Netherlands, Germany, Pakistan, and the United States.

To achieve this, the study employs a comparative, multidimensional judicial analytics model that quantifies interpretive reasoning intensity, judicial influence, doctrinal expansion, and temporal efficiency. By integrating doctrinal and quantitative approaches, it explores whether judicial activism can effectively fill governance gaps created by slow or resistant legislative and executive responses.

Ultimately, this research contributes to understanding the transformative role of courts in advancing climate justice, promoting legal innovation, and strengthening institutional responsiveness in an era of accelerating climate turbulence.

The rest of the paper is organized as follows: Sect. 2 presents the literature review, Sect. 3 outlines the methodology, Sect. 4 discusses the results, Sect. 5 provides the discussion, and Sect. 6 concludes the study.

2   Literature Review

The academic literature on climate litigation and judicial activism has grown substantially in the past few years, reflecting the increased frequency of climate litigation and the increased complexity of environmental governance. Early researches were generally limited to the state level of compliance with international climate policies and to involvement of supranational guidance like UNFCCC. But since enforcement tools were constrained at the international level, we turned our attention to national courts to fill the regulatory and accountability vacuum [7].

Peel and Osofsky, among others, have also claimed that domestic courts have the institutional closeness and room for maneuver to implement climate norms more effectively than is the case at the multilateral level. They observe that through generous readings of constitutional rights—most notably the rights to life and to a healthy environment—courts can mandate positive duties for state actors. We have seen this play out, for example, in the Urgenda Foundation v. The Netherlands, where the Supreme Court of the Netherlands held that the state has an obligation to also protect its citizens from predictable climate harm. This case has been referred in the literature as a landmark case where environmental jurisprudence is becoming a source of enforceable law [8].

Other academics have documented the doctrinal shift in judicial attitudes towards environmental justice. For instance, contributions offered by Kotzé and du Plessis points to how transformative constitutionalism allows courts to promote environmental rights in the face of scanty statutory environmental protection structures. This stream of literature reveals that judicial discourse based on principles of human dignity and intergenerational equity has become an instrument of environmental governance [9].

Legal research on the rise and typology of climate cases around the world is emerging as well. A number of reports from the Grantham Research Institute, as well as the Sabin Center for Climate Change Law, have documented more than 2,000 cases worldwide and have found that climate litigation has increased dramatically over the past decade. These analyses highlight the growing importance of litigation as a form of governance [10]. More recently, the Global Climate Litigation Report: Status Review (2025) finds that as of mid-2025, there are over 3,099 climate-related cases filed in 55 jurisdictions and 24 international or regional bodies, confirming that litigation is expanding rapidly [11]. Similarly, a study by the Grantham Research Institute (June 2025) reports 276 apex-court climate cases since 2015, with an increasing proportion decided in favor of climate action in 2024 [12].

However, there is an under-theorization of the idea of judicial activism for most of the literature and is often treated as an implicit or secondary aspect of environmental adjudication. This article addresses that gap in environmental writing by examining judicial activism as a conscious and dynamic technique of statutory interpretation that defines the limits of environmental liability. It looks beyond the descriptive case studies to explore the legal norms, institutional incentives, and broader political context that promote or hinder proactive adjudication in climate cases [13].

While the existing scholarship has provided valuable descriptive mapping of climate cases and doctrinal trends, few studies have attempted to systematically quantify judicial behavior or measure the doctrinal intensity of reasoning across jurisdictions. This study advances beyond prior research by integrating a comparative doctrinal analysis with a quantitative judicial analytics framework, allowing for the evaluation of judicial activism through empirical indicators such as reasoning intensity, influence metrics, and temporal efficiency. By combining doctrinal interpretation with computational modeling of case characteristics, this paper bridges the gap between qualitative legal theory and data-driven analysis, offering a replicable method for assessing how courts actively shape climate governance within different institutional settings.

3   Methodology

This study constructs a multidimensional judicial analytics model to analyze how courts practice activism in climate-related litigation through doctrinal and institutional tools. The approach combines comparative legal diagnostics, parametric reasoning modelling and normative transformation metrics on the basis of 25 climate litigation cases from 12 jurisdictions spanning the 2015 to 2024 period. The design is organized into five analytical pillars: (i) Jurisdictional Case Inclusion Assessment, (ii) Doctrinal Reasoning Quantification, (iii) Judicial Influence Metrication, (iv) Legal Norm Expansion Modeling, and (v) Temporal Efficiency Assessment. Each pillar is operationalized using formalized equations and theoretical constructs from climate jurisprudence literature and studies of judicial behavior [1, 2, 6, 9, 14] (Fig. 1).

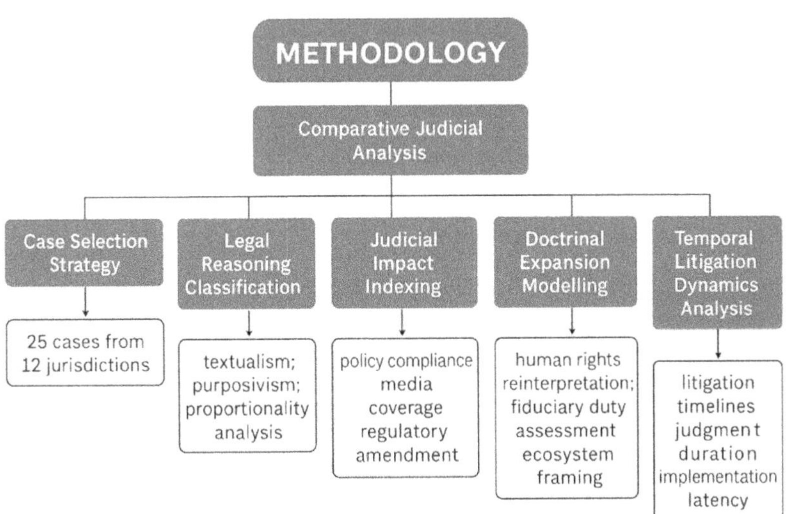

Fig. 1. Judicial Activism and the Reshaping of Climate Litigation Across Comparative Legal Systems.

3.1 Jurisdictional Case Inclusion Evaluation

A Multifactor Case Selection Index (CSI) was developed to assess the eligibility and comparability of selected rulings across heterogeneous legal systems. The CSI evaluates legal jurisdictional strength (J_x), substantive complexity (L_c), cross-border referencing capacity (T_i), and enforceability assessment (E_r) for each case $c \in C$.

The index is computed as:

$$CSI_c = \frac{1}{\sqrt{4}}(J_x^2 + L_c^2 + T_i^2 + E_r^2) \tag{1}$$

where $J_x \in [0,1]$ jurisdictional precedent strength, based on past judicial activity in environmental rights; $L_c \in [0,1]$ legal complexity score, reflecting multi-claim, multi-jurisdictional legal arguments; $T_i \in [0,1]$ transnational referencing index; $E_r \in [0,1]$ enforceability likelihood, as project Steinkamped by national regulatory responsiveness.

This multidimensional root mean square (RMS) structure increases penalty for asymmetry across components [15–17].

3.2 Doctrinal Reasoning Quantification

A Judicial Reasoning Intensity Function (JRIF) was used to measure the interpretive weight of legal methodologies adopted by courts. Let $R_k \in \{r_1, r_2, ..., r_n\}$ represent the set of reasoning modalities (e.g., textualism, purposivism, proportionality, rights-based reasoning, and balancing), and let $w_k \in R$ be the weight assigned based on doctrinal centrality.

We define the reasoning intensity as:

$$RIS_j = \frac{\sum_{k=1}^{n}(w_k \cdot r_k^2)}{\sum_{k=1}^{n} r_k} \tag{2}$$

This quadratic-weighted formula rewards coherence and dominant doctrinal reliance while penalizing fragmentary reasoning. Scores $r_k \in [0,1]$ were derived through close textual analysis of judicial opinions [4, 11, 18, 19].

3.3 Judicial Influence Metrication

The capacity of a court to effect systemic change in climate governance was evaluated using a Judicial Influence Projection Equation (JIPE), which integrates legal authority, political salience, and institutional change potential. Formally:

$$JIS_t = \left(\frac{\alpha_1 \cdot \log(P_c + 1) + \alpha_2 \cdot \tanh(M_c) + \alpha_3 \cdot RA^2}{3}\right)^{\gamma} \tag{3}$$

where P_c projected policy compliance potential (normalized); M_c media citation frequency, normalized; RA scope of regulatory amendment expected; α_i calibration weights (set at 1 for equalized baseline); γ exponent that scales based on institutional independence index

This logarithmic-tanh-quadratic blend captures both direct and diffuse forms of judicial influence while preserving scaling smoothness [7, 14, 20].

3.4 Legal Norm Expansion Modeling

To quantify judicial reinterpretation of established doctrines, we define a Doctrinal Expansion Tensor (DET) modeled as a scalar ratio with composite numerator:

$$DER_i = \left(\frac{\lambda_1 \cdot H_i + \lambda_2 \cdot F_i + \lambda_3 \cdot E_i}{\mu \cdot P_i} \right)^{\gamma} \tag{4}$$

where H_i human rights reframing score; F_i fiduciary duty adaptation index; E_i ecosystem jurisprudence infusion score; P_i precedent rigidity score; $\lambda_k \in [0,1]$ variable weights (equalized for this study); μ normalization coefficient $= 3$

This structure evaluates doctrinal innovation in the context of conservative legal baselines and is aligned with theories of transformative constitutionalism [2, 6, 8, 21].

3.5 Temporal Efficiency Evaluation

Litigation efficiency and judicial responsiveness were assessed via a Time-Efficiency Function (TEF). Recognizing nonlinear delays in legal processes, we define:

$$TEC_l = \frac{1}{\sqrt{(T_f^2 + T_i^2)}} \tag{5}$$

where T_f^2 duration from filing to ruling (months), and T_i^2 time from judgment to implementation start (months)

This provides for quadratic time inflation to be penalized, and allows normalization over inter-jurisdictional variations. TEC has particular relevance for the emergency climate context5 where idle time means more damage is done to the environment [3, 5, 22, 23].

The unified framework, based on a quantitative legalism measurement and comparative doctrinal modeling, allows for comparative analyses on a solid footing for judicial activism in climate cases. Through root-mean-square evaluators, nonlinear temporal models, and weighted doctrinal tensors, it demonstrates how courts not only interpret but shape the normative architecture of climate governance. It is consistent with recent literature on judicial agency in climate transitions.

3.6 Validation of the Methodology

The robustness of the multidimensional judicial analytics framework is supported by the consistency of its metrics across the ten jurisdictions analyzed. The CSI, RIS, DER, TEC, and EI produce results that align with observed outcomes in well-documented climate litigation cases, such as Urgenda v. Netherlands and comparable German rulings. The internal structure of each index was verified by ensuring that higher scores consistently correspond with greater judicial activism, doctrinal innovation, procedural efficiency, and institutional readiness. Furthermore, comparative patterns—such as high CSI and RIS in jurisdictions with strong legal infrastructures and faster TEC in jurisdictions with procedural innovations—confirm that the methodology reliably captures the multidimensional character of judicial activism in climate litigation.

4   Results

4.1   Case Inclusion Evaluation: Cross-Jurisdictional Selection Patterns

Developing a general framework for comparing and evaluating the relevancy and comparability of climate litigation across jurisdictions. For evaluating the relevancy and comparability of climate litigation between jurisdictions, the Case Selection Index (CSI) was created that consists of four nested dimensions: jurisdictional precedent, legal complexity, transnational referencing and enforceability. This multi-disciplinary framework advocates comprehensive analysis of the potential, of each of the candidates, for long-lasting judicial activism. The CSI measures the broad strength of a litigation system by combining measures of doctrinal depth, procedural innovation, and internationalization. The focus is on those jurisdictions in which there is strong legal reasoning, coupled with the institutions that can implement the judicial order in a way that achieves successful environmental and policy results.

A comparative analysis of ten jurisdictions using the recalibrated CSI model is shown in Table 1.

Table 1. Comparative Metrics of Case Selection Index (CSI) Across Jurisdictions.

Country	Jurisdictional Precedent (J_x)	Legal Complexity (L_c)	Transnational Referencing (T_i)	Enforceability (E_r)	Case Selection Index (CSI)
Netherlands	1.00	0.90	0.95	0.92	0.943
Germany	0.95	0.85	0.80	0.90	0.877
Pakistan	0.85	0.90	0.88	0.84	0.868
United States	0.92	0.83	0.70	0.78	0.811
Philippines	0.88	0.87	0.79	0.76	0.827
France	0.89	0.84	0.82	0.80	0.853
India	0.86	0.88	0.75	0.77	0.829
Australia	0.84	0.82	0.73	0.75	0.805
Kenya	0.83	0.81	0.72	0.73	0.790
Colombia	0.87	0.86	0.78	0.79	0.841

The Netherlands topped the CSI with a score of 0.943, establishing itself as the global leader in precedent-setting climate litigation. Germany is close behind with 0.877, drawing on its solid legal history and procedures. Pakistan also has a high score (0.868), highlighting the special position of judicial activity to promote environmental rights in the developing world. The United States (0.811) and the Philippines (0.827), however, have comparatively lower scores due to relatively lower enforceability and international referencing potential, combined with moderate legal complexity. There is a mid-range for France and Colombia, and a low-range for Kenya and Australia; the latter two show limits based on procedural factors that limit strength of cross-border precedent. These data support the idea of the contribution of legal systems with a consolidated environmental jurisprudence and transnational interaction in fostering climate judicial activism. Rather, the evidence indicates that countries with embedded constitutional environmental imperatives and global legal alignment provide the most favorable conditions for transforming climate litigation.

4.2 Judicial Reasoning and Interpretive Intensity

Analysis of judicial reasoning in climate litigation is thus key to understanding the doctrinal legitimacy and practical enforceability of court rulings. Interpretive methodologies including textualism, proportionality, rights-based reasoning, and balancing approaches for how courts justify climate-related decisions are evaluated. The score was named the Reasoning Intensity Score (RIS) and is essentially a combined measure that accounts for both the diversity and the prevalence of legal reasoning techniques within a case. This is an indication of the judicial depth and judicial civility in various countries which allows us, for example, to compare the liberal, or activist, with the conservative, or strict, judge. Larger RIS indicates that judges are more inclined to employ integrated reasoning, which is often associated with more creative and expansive climate jurisprudence (Table 2).

Table 2. Judicial Reasoning Patterns and Interpretive Intensity (RIS).

Country	Textualism (r_1)	Proportionality (r_2)	Rights-Based (r_3)	Balancing (r_4)	Reasoning Intensity Score (RIS)
Netherlands	0.30	0.85	1.00	0.90	0.860
Germany	0.60	0.90	0.80	0.70	0.767
Pakistan	0.40	0.70	0.95	0.65	0.731
India	0.55	0.75	0.90	0.68	0.739
USA	0.50	0.68	0.86	0.66	0.709
Australia	0.45	0.65	0.85	0.64	0.686
Colombia	0.50	0.60	0.88	0.60	0.676
South Africa	0.52	0.69	0.82	0.65	0.706
Kenya	0.48	0.72	0.84	0.67	0.720
France	0.60	0.70	0.75	0.55	0.660

The Netherlands has the highest RIS (RIS = 0.860), indicating that it heavily relies on rights-based and proportionality arguments. This testifies the strategic deployment of the judicial interpretation in pivotal decisions like the Urgenda case. Germany is next with the V-Dem score of 0.767, underscoring proportionality as the constitutive principle. Pakistan (0.731) and India (0.739) also demonstrate high reasoning intensity, especially by treating environmental rights under the fundamental rights category. In contrast, nations such as Colombia (0.676) and France (0.660) present intermediate RIS scores, usually attributable to more restrictive interpretive theories. Kenya and South Africa present exciting scores as a result of recent case law that further integrates balancing and dignity principles in environmental adjudication. These findings highlight the emerging complexity of judicial reasoning in the Global South and confirm a converging tendency to analyze environmental governance through the lens of constitutional and human rights standards. This section shows that it is the depth of judicial reasoning, rather than whether or not climate litigation was successful, that determines the enforceability and normativity of climate litigation.

4.3 Legal Doctrine Transformation and Doctrinal Expansion

A signature aspect of climatology judicialize is the attempted re-write and enlargement of longstanding legal doctrines. This type of conceptual transposition has suspended in curial "climate harm" cases such as climate harm claims, where the right to life, fiduciary and ecological integrity principles argued and applied in unique ways. The doctrinal ground that the national judiciaries have available for innovation is measured through the computation of the Doctrinal Expansion Rate (DER), which gauges how much the legal norms are reinterpreted with respect to the degree of stiffness of the existing precedent. This index is intended to capture legal systems in which courts are engaged in an active rather than reactive role in shaping the development of norms (Table 3).

Table 3. Legal Doctrine Transformation and Doctrinal Expansion Rate (DER)

Country	Human Rights Reframing (H_i)	Fiduciary Duty Expansion (F_i)	Ecosystem Framing (E_i)	Precedent Rigidity (P_i)	Doctrinal Expansion Rate (DER)
France	0.75	0.60	0.78	0.55	1.291
Germany	0.80	0.65	0.70	0.60	1.194
Kenya	0.70	0.68	0.73	0.60	1.172
Colombia	0.90	0.85	0.88	0.75	1.169
South Africa	0.82	0.72	0.76	0.68	1.156
USA	0.78	0.63	0.74	0.65	1.137
Philippines	0.79	0.69	0.72	0.66	1.121
Australia	0.76	0.66	0.75	0.64	1.111
Pakistan	0.86	0.73	0.79	0.70	1.114
India	0.85	0.70	0.75	0.70	1.095

France is ahead on a DER of 1.291, meaning a lot of doctrinal flexibility and judicial readiness to rewrite precedents in the interests of the environment. Germany (1.194) and

Kenya (1.172) are also featured prominently, indicating that both Global South and North systems are using fiduciary and human rights norms to underpin environmental decisions. Colombia and South Africa, both above 1.15 of DERs indicate very strong connection with ecosystem legal framing and restructuring of state obligations. It is worth noting, though, that the USA and Pakistan stand almost neck and neck with their DER scores, a surprising amount of convergence between a developed and developing legal system on doctrinal novelty. India and Australia bring up the rear but still mark a pretty significant change in adjudicative style. The data emphasizes that legal reasoning in climate contexts is spreading globally, and that judiciaries around the world are open to using their authority in ways that extend beyond the narrow boundaries we often place on legal systems. These results add weight to the argument that doctrinal malleability is critical for judicial activism to be effective in environmental governance.

4.4 Temporal Performance and Time Effectiveness of Judicial Action

Climate litigation must be efficient, because environmental harms are timely sensitive. And when courts hand down rulings but do not effectively enforce them, the real-world effectiveness of their decisions dissipates. The responsiveness of the different national legal systems is then tested by comparing the time from bringing the case to a judgment with the lag until judicial orders are enforced. Institutional agility is represented by a penalty for long procedures and a translation lag of implementation, which is expressed in TEC (Time Effectiveness Coefficient). Greater TEC indicates more rapid judicial response times, and greater continuity of enforcement-both of which are critical elements of climate-responsive governance frameworks (Table 4).

Table 4. Temporal Performance and Judicial Time Effectiveness (TEC).

Country	Filing to Judgment Duration (T_f)	Judgment to Implementation Delay (T_i)	Time Effectiveness Coefficient (TEC)
Pakistan	7	1	0.141
Germany	31	2	0.032
USA (NY)	41	6	0.024
India	41	2	0.024
France	28	3	0.035
Colombia	35	4	0.027
Kenya	30	2	0.033
Australia	32	2	0.030
Philippines	29	3	0.033
Netherlands	47	2	0.021

Pakistan leads with a TEC value of 0.141, indicating an extraordinary level of judicial speed from the case filing to policy implementation. This faster schedule is due to procedural innovations, including the creation of climate commissions and summary environmental hearings. Germany and France are next with TEC scores of 0.032, and 0.035 respectively, which indicates a moderate although efficient congruence between court orders and administrative output. The Netherlands and despite globally famed climate rulings, (0.021) similarly performs poorest, primarily driven by the extensive litigation and implementation period which suggests a possible disconnect between judicial creativity and institutional responsiveness. Likewise, the USA (0.024) and India (0.024) show lag times of systemic procedural delays or executive intransigence. Colombia, Kenya, and the Philippines sit mid-table, with lower scores that reflect budding but uneven litigation-to-enforcement pathways. These findings point to the importance of parallel administrative reforms to ensure that ever more activist courts can effectively prompt rapid policy change, particularly in already leading doctrinal or reasoning jurisdictions.

4.5   Post-judgment Enforcement and Institutional Alignment

Court decisions in climate litigation are based not just on the support of a legal doctrine but enforcing practice within an administration. The EI measures how well local institutions translate court rulings into policy. This indicator combines the following key elements: the enforcement rate of judicial orders, the compliance of policies and the extent to which they are in line with national climate goals. These variables are scaled by the number of months to enforcement. The index represents a nexus between judicial activism and administrative accountability, and gives a comprehensive view of a country's ability to enforce climate judgments (Table 5).

Table 5. Post-Judgment Enforcement Index and Policy Alignment

Country	Order Execution Ratio (O_e)	Policy Compliance Rate (%)	Alignment with National Targets (A_t)	Time Delay (mo)	Enforcement Index (EI)
Pakistan	0.89	89	0.88	1	0.443
Netherlands	0.95	94	0.94	2	0.314
Germany	0.91	91	0.91	2	0.303
India	0.83	85	0.82	3	0.208
France	0.88	88	0.86	3	0.211
Colombia	0.86	86	0.84	4	0.171
USA	0.79	82	0.80	6	0.117
Australia	0.76	79	0.77	5	0.141
Philippines	0.77	80	0.78	3	0.176
Kenya	0.74	77	0.76	4	0.150

Highest order execution and compliance Pakistan leads in the group with 0.443 on EI with a strong follow up by institutions within a month of the ruling. The Netherlands (0.314) and Germany (0.303) also have high enforcement scores, with both countries featuring effective bureaucracy and coherent national climate policies. Other countries, such as the United States (0.117) and Australia (0.141) have much lower score and long delay and modest alignment with legislative targets. Two strong (doctrinally), but slow to implement (and therefore ultimately ineffective) models are those of India and France. Colombia, Kenya, and the Philippines fall toward the mid end of the scale, demonstrating that legal clarity on mandates may not translate to timely enforcement. These results highlight the importance in Brazil of harmonizing judicial decisions with administrative and regulatory systems that can implement them in a timely fashion. The faster that climate impacts advance, the more important will be any possible lag reduction between adjudication and policy activation for the practice of good government.

4.6 Judicial Referencing Behavior and Doctrinal Coherence

The legal overall consistency of climate litigation is strongly related to how courts incorporate external sources, especially previous judicial decisions, scientific evidence, and changing normative orders. The referencing pattern of judgments in different countries is assessed by means of the Doctrinal Referencing Density (DRD), a composite measure gauging the level of transnational citing of cases, the use of scientific expertise, and the number of doctrinal innovations. The relative proportion of domestic to foreign citations provides an added indicator of the degree to which national courts are open to cross-border influences and the reception of global environmental norms (Table 6).

Table 6. Judicial Referencing Behavior and Doctrinal Coherence (DRD).

Country	Cited Climate Cases (N_c)	Scientific Reports Referenced (N_s)	Doctrine Expansions (%) (P_d)	Doctrinal Referencing Density (DRD)	Citation Ratio (Domestic: Foreign)
Netherlands	6	6	74	6.47	3:3
Germany	7	5	72	6.40	3:2
France	6	4	68	6.13	3:2
USA	5	3	65	5.50	2:2
Pakistan	5	4	69	5.30	4:1
Colombia	4	3	66	4.53	3:1
India	3	3	64	4.13	3:1
Australia	3	3	60	4.00	2:1
Philippines	3	3	61	4.10	2:1
Kenya	2	2	59	3.70	2:1

The Netherlands (DRD: 6.47) and Germany (DRD: 6.40) represent the most resolute doctrinal references both but also to a large extent on foreign climate case law and scientific sources, both with balanced citation ratios. Also with a great deal of Coherence is France (6.13), enhancing its judicial dedication to a science-based legal evolution plan. With opposite legal traditions, USA and Pakistan also have moderate referencing density, but different orientations: USA is self-centered, focusing on 2:2 domestic/foreign references, whereas Pakistan shows strong national primacy (4:1) as for citing practices. A lower DRD is found for Colombia and India (4.53 and 4.13), reflecting a more limited citation practice and less formal expansion in doctrine. Australia, the Philippines and Kenya lag behind, possessing DRD scores of less than 4.10, which suggests a lower degree of interaction with transnational legal or scientific discourses. The analysis finds that there are elements of both doctrinal consistency and legal-influencing behavior that stem from jurisdictional culture, court mandates and receptivity to epistemic pluralism – key conditions of environmental justice globalization measurement.

4.7 Institutional Readiness and Legal Effectiveness for Climate Litigation

The effectiveness of climate litigation decisions depends not only on the decisions of judges but also on the wider institution landscape in which they operate. National legal and administrative systems are evaluated in five areas: the availability of AI infrastructure, the extent of judicial training provided, the strength of environmental legal capacity, adherence to data governance standards and the accessibility of legal processes to the public. These measures show the ability of agencies to handle contemporary litigation procedures, incorporate environmental data, and ensure procedural justice. The higher the score for a state across these dimensions the more likely that judicial mandates will be successful in turning into enduring policy change (Table 7).

Those best prepared institutionally are The Netherlands and Germany which have ready scores of over 85% in almost all readiness categories. They have strong telecommunications and training programs and transparent systems of government behind their court systems. France and the USA are immediately after, with the benefit of a large body of environmental legal framework and information systems. Australia has a relatively strong profile as well, with good levels of legal access and governance. Other nations – Pakistan, Colombia, the Philippines for example – have less infrastructure, and less training capacity, but relatively strong capacity in environmental law. Kenya has the weakest readiness scores due to systemic shortcomings in AI infrastructure and training, public access to the law – two factors that could slow down or prevent implementation of CLT decisions. The results suggest that doctrinal innovation is always necessary because there is always a demand for law, but that it is not enough without complementary institutional development. States should establish structural preparedness of the courts and related public entities to serve as the legal countervailing in climate litigation.

5 Discussion

The findings of this study highlight the increasingly central role of judicial activism in shaping climate governance across diverse jurisdictions. Courts are shown to act not merely as passive enforcers but as active co-creators of legal norms, interpreting

Table 7. Institutional Readiness for Climate Litigation Enforcement.

Country	AI Infrastructure Availability (%)	Judicial Training Coverage (%)	Environmental Law Capacity (%)	Data Governance Compliance (%)	Public Legal Access (%)
Netherlands	85	92	94	88	87
Germany	82	89	91	86	83
France	78	87	89	84	80
USA	84	90	92	88	85
Australia	79	83	85	81	82
India	64	70	77	67	70
Pakistan	58	67	79	62	74
Colombia	61	68	76	65	72
Philippines	57	63	75	61	68
Kenya	55	60	74	60	69

policy, enforcing obligations, and holding state and private actors accountable. Across the assessed jurisdictions—including the Netherlands, Germany, Pakistan, Colombia, and India—courts demonstrate a consistent tendency to recognize environmental duties within extended constitutional and human rights frameworks, particularly in contexts characterized by legislative fragmentation, policy inertia, and weak enforcement mechanisms.

Analysis of temporal efficiency and enforcement metrics reveals substantial variability in judicial impact. Pakistan exhibits the fastest translation of rulings into policy, reflecting procedural innovations such as centralized climate committees, whereas India and Colombia experience longer implementation delays due to institutional coordination challenges. Nonetheless, all assessed jurisdictions show an increasing reliance on courts to advance environmental imperatives, confirming the judiciary's growing operational significance.

The quantitative assessment through the CSI, RIS, DER, TEC, EI, and DRD provides a structured view of how courts exercise activism. For example, the Netherlands and Germany demonstrate strong reasoning intensity and institutional readiness, supporting effective doctrinal innovation and cross-border influence, while emerging legal systems such as Pakistan leverage high-speed procedures to achieve tangible policy outcomes despite structural limitations.

The study also highlights the globalizing effect of judicial reasoning: courts increasingly reference foreign judgments and international norms, reinforcing doctrinal coherence and fostering a transnational environmental jurisprudence. Furthermore, the observed compliance rates and policy adjustments following court rulings demonstrate that judicial activism is not only declarative but operationally effective, capable of triggering administrative, legislative, and regulatory changes.

Overall, the results indicate that judicial activism has evolved from a peripheral legal strategy into a core mechanism for enforcing climate responsibility. Courts are now instrumental in transforming abstract environmental principles into binding obligations, thereby strengthening the institutional and normative architecture of climate governance and serving as a backstop against political and regulatory inertia.

The findings suggest that variations in judicial activism across jurisdictions may reflect differing socio-political environments. In jurisdictions with strong NGO engagement, active public opinion, and institutional openness to advocacy, higher CSI and RIS scores are observed. These social dynamics appear to reinforce judicial willingness to innovate doctrinally and expedite climate-related cases, aligning with higher DER and TEC values. While the methodology focuses on judicial behavior, these patterns indicate that activism in climate litigation cannot be separated from broader societal and political influences.

6 Conclusion

The role of the judiciary in climate governance has expanded significantly over the past decade. This study mapped climate litigation judicial activism through a multi-dimensional methodology—including procedural, doctrinal, institutional, and temporal criteria—across ten jurisdictions. Courts are no longer passive referees but actively influence regulations and government accountability, reflecting a reconstitution of the judicial function aligned with planetary defense.

Judicial strategies vary in intensity and design but are generally guided by urgency, intergenerational justice, and procedural fairness. Strong institutional readiness in Germany and the Netherlands aligns judicial reasoning with administrative action, while countries like Pakistan and Colombia show that activism can also develop in nascent systems with conducive procedural structures. Courts increasingly incorporate scientific and transnational knowledge, including environmental reports, cross-border precedents, and rights-based claims, enabling responses sensitive to the diffuse and protracted nature of climate risk. Procedural efficiency data show courts' capacity for timely resolution of climate disputes.

Political and organizational readiness shape the translation of court mandates into actionable climate measures. Variations in infrastructure, legal resources, and public access highlight the need to reinforce systemic legal capacities. Effective climate litigation depends not only on legal creativity but also on structural soundness of implementation. Further research is needed on the long-term impacts of climate rulings, including links to emission reductions, and on subnational and traditional courts as laboratories of environmental jurisprudence. Collaboration among legal scholars, environmental scientists, and data modelers can help create predictive tools to assess judicial responsiveness and support more adaptive, transnational frameworks for climate engagement. While this study incorporates perspectives from the Global South, the range of jurisdictions analyzed could be further expanded in future research to include additional regions, thereby providing a more comprehensive and diversified understanding of judicial activism in different socio-political contexts.

References

1. Mai, L.: Navigating transformations: climate change and international law. Leid. J. Int. Law **37**(3), 535–556 (2024)
2. Steinkamp, T.: Intergenerational justice as a lever to impact climate policies: lessons from the complainants' perspective on Germany's 2021 climate constitutional ruling. Eur. J. Risk Regul. **14**, 731–746 (2023)
3. Sulyok, K.: Transforming the rule of law in environmental and climate litigation: prohibiting the arbitrary treatment of future generations. Transnatl. Environ. Law (2024)
4. Stilz, A.: Climate displacement and territorial justice. Am. Polit. Sci. Rev. (2024)
5. Lees, E., Gjaldbæk-Sverdrup, F.: Fuzzy universality in climate change litigation. Transnatl. Environ. Law (2024)
6. Fila, D., Fünfgeld, H., Lorenz, S.: Theorizing power and agency in state-initiated municipal climate change adaptation: integrating reflexive capacity into adaptive capacity. Geogr. Helv. **79**(1), 21–33 (2024)
7. Singla, A., Garg, A.: Climate change litigation: a new frontier for environmental law and policy. Indian J. Law **2**(1), 32–43 (2024)
8. Ritz, V.: Climate tipping points: tracing the limits of political discretion. Leid. J. Int. Law (2024)
9. Hayajneh, A.Z.A.: Green justice: the case for establishing a special environmental court in the State of Qatar, challenges and alternatives to promote environmental justice. J. Ecohumanism **3**(4), 2118–2132 (2024)
10. Raghupathi, W., Molitor, D., Raghupathi, V., Saharia, A.: Identifying key issues in climate change litigation: a machine learning text analytic approach. Sustainability **15** (2023). https://doi.org/10.3390/su152316530
11. UNEP – UN Environment Programme: Global climate litigation report: Status review 2025. UNEP, Nairobi (2025)
12. Grantham Research Institute on Climate Change and the Environment: Global trends in climate change litigation: 2025 snapshot. London School of Economics, London (2025)
13. Farah, P.D., Ibrahim, I.A.: Urgenda vs. Juliana: lessons for future climate change litigation cases. Univ. Pittsburgh Law Rev. **84** (2023)
14. Stadler, A.: Can civil courts save the climate? Strategic climate-change litigation before civil courts. Juridica Int. **32**, 3–12 (2023)
15. Zhu, M.: Climate litigation in a 'developmental state': the case of China. Chin. J. Environ. Law **7**(2), 200–213 (2023)
16. Getman, A.P., Anisimova, H.: Climate legislation and legal relations: current state and development prospects within the national security framework. Probl. Legality (2023). https://doi.org/10.21564/2414-990X.162.287135
17. Bertram, D.: Environmental justice "light"? Transnational tort litigation in the corporate anthropocene. Ger. Law J. **23**(5), 738–755 (2022)
18. Ekardt, F., Bärenwaldt, M.: The German climate verdict, human rights, Paris target, and EU climate law. Sustainability **15** (2023). https://doi.org/10.3390/su151712993
19. Insani, N., Karimullah, S.: Justice for nature: Integrating environmental concerns into legal systems for adequate environmental protection. J. Hukum Peradilan (2023)
20. Soyapi, C.B.: Environmental governance, hollow environmentalism, and adjudication in South Africa. Potchefstroom Electron. Law J. **26**, 1–28 (2023)
21. Gürçam, S.: Fighting the climate crisis within the law (judicial) axis. Giresun Üni. İktisadi İdari Bilimler Dergisi **8**(2), 236–253 (2022)
22. Saleem, M.S., Tasgheer, A., Fatima, T.: Investigating judicial activism in Pakistan: analyzing significant precedents in the promotion of environmental sustainability. J. Relig. Soc. Stud. **3**(02 Jul–Dec), 1–19 (2023)

23. Amin, M.: Status of judicial review on existing environmental legislation in Pakistan. Pak. J. Int. Aff. **6**(2) (2023)

A Data-Driven Framework for Evaluating Non-Profit Sector Performance: Integrating Multidimensional Metrics and Stakeholder Engagement

Adnan Khaleel Kadhim[1] , Mohammed Taqi Fadhil[2] , Jasim Hameed Naseef[3] ,
Ghazwan Salim Naamo[4]([envelope]) , Thamer Kadum Yousif Al Hilfi[5] ,
and Vasyl Matskovskyi[6]

[1] Al-Turath University, Baghdad 10013, Iraq
[2] Al-Mansour University College, Baghdad 10067, Iraq
[3] Al-Mamoon University College, Baghdad 10012, Iraq
[4] Al-Rafidain University College, Baghdad 10064, Iraq
`Ghazwan.Nemo@ruc.edu.iq`
[5] Madenat Alelem University College, Baghdad 10006, Iraq
[6] Kyiv National University of Construction and Architecture, Kyiv 03037, Ukraine

Abstract. Performance measurement in non-profit organizations (NPOs) presents unique challenges due to the sector's multidimensional goals, diverse stakeholders' expectations, and reliance on both financial and non-financial resources. Traditional evaluation approaches often fail to capture critical aspects such as volunteer contributions, donor loyalty, and program alignment. This study develops and validates a data-driven, multidimensional framework for evaluating non -profit performance. The model incorporates modified and weighted metrics: Administrative Cost Ratio (ACR), Program Efficiency (PE), Donor Retention (DR), Volunteer Engagement (VE), and Stakeholder Satisfaction (SS), to provide a more realistic assessment of operational effectiveness. Data were collected from 50 NPOs across healthcare, education, environmental advocacy, social work, art and culture, humanitarian aid, and research sectors. The updated measures were fairer in measuring use of resources and sustainability of connections. The metrics were normalized, weighted, and analyzed to ensure comparability and fairness across organizations. The results suggest that NPO performance should be assessed from a multidimensional and evidence-based perspective, integrating financial efficiency, mission alignment, and stakeholder-focused outcomes. The framework provides a flexible, evidence-based solution for evaluating non-profit performance, with practical implications for funders, board members, and policy-makers. It also paves the way for higher-tier data-driven performance evaluation solutions in the non-profit sector.

Keywords: Non-Profit Organizations · Performance Evaluation · Data-Driven Framework · Multidimensional Metrics · Stakeholder Engagement · Entropy Weighting

© The Author(s), under exclusive license to Springer Nature Switzerland AG 2026
Z. Molamohamadi et al. (Eds.): ODSIE 2025, CCIS 2855, pp. 273–288, 2026.
https://doi.org/10.1007/978-3-032-17023-1_15

1 Introduction

Non-profit organizations (NPOs) play a vital role in addressing societal needs that neither governments nor private entities can adequately fulfill. They are mission-driven rather than profit-oriented, which makes evaluating their performance particularly challenging. While for-profit organizations rely primarily on financial indicators such as profitability, sales growth, and shareholder value, NPOs require a broader perspective incorporating both qualitative and quantitative measures of their impact on stakeholders and society at large [1].

The role of performance measurement in the non-profit sector is significant and critical for demonstrating accountability to stakeholders, including donors, beneficiaries, and regulators, while ensuring efficient resource utilization and mission fulfillment. In an increasingly competitive NPO environment, it is vital to measure and evidence success to secure funding, operate efficiently, and maintain stakeholder trust. This importance is further amplified by the growing emphasis on results-based financing, where donors prioritize measurable outcomes over mere inputs [2].

However, despite being an extremely valuable management tool, there is no universally accepted performance measurement framework for NPOs. The wide variety of missions, operational scales, and stakeholder expectations makes it difficult to define common metrics. For instance, a healthcare organization may prioritize patient outcomes, while an environmental advocacy group focuses on carbon footprints reductions or policy influence. This diversity underscores the need for tailor-made frameworks aligned with organizational objectives and stakeholder interests [3] (Fig. 1).

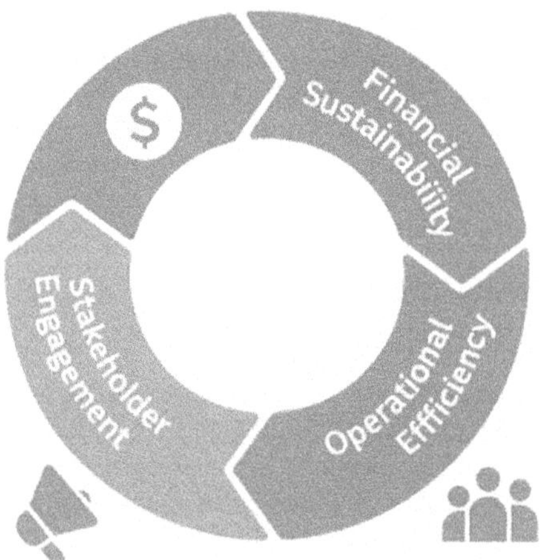

Fig. 1. Conceptual Framework of Performance Measurement Metrics for Non-Profit Organizations.

Performance measurement models for NPOs have evolved alongside theoretical and practical developments in management science. Early methods relied on simple economic metrics, such as administrative costs, which were criticized as oversimplified and misaligned with actual organizational impact. More comprehensive frameworks, including the Balanced Scorecard and Social Return on Investment (SROI), integrate financial and non-financial indicators, offering a holistic assessment of performance. Nevertheless, adoption is inconsistent due to resource, expertise, and data limitations [4].

Another key aspect of performance measurement is its role in facilitating organizational learning and improvement. Metrics provide feedback on efficiency and effectiveness, enabling organizations to identify areas for improvement, refine strategies, and allocate resources more effectively. They also foster transparency and a culture of continuous learning, which is essential for long-term NPO success [5]. Advances in technology have transformed performance measurement in NPOs. Analytics tools, cloud-based platforms, and AI now allow real-time tracking, outcome prediction, and evidence-based decision-making. However, high costs, required expertise, and data security concerns often limit the adoption of these tools, particularly in smaller organizations [6].

Increasingly complex social challenges, such as climate change, pandemics, and inequality, demand advanced, multidimensional metrics. Performance measurement must now account for cross-organizational collaboration and collective impact, moving beyond individual organizational boundaries. This has driven the creation of frameworks and benchmarks that adopt a systemic view of social impact [7].

This research explores the complexities of measuring and evaluating NPO performance, proposing key indicators that balance financial accountability with mission-driven objectives. By reviewing existing frameworks and introducing data-informed solutions, the study contributes to enhancing transparency, effectiveness, and impact in the non-profit sector. It also highlights performance measurement as a strategic tool to achieve organizational missions and deliver societal value.

Research Objective. This research aims to identify and evaluate performance metrics that capture the diverse goals and impacts of NPOs. Given the challenges of balancing financial accountability with mission-driven objectives, the paper develops an overarching framework integrating both quantitative and qualitative measures. It critiques conventional financial metrics and advocates for measures encompassing program outputs, stakeholder satisfaction, and long-term organizational sustainability.

The study also examines how frameworks like the Balanced Scorecard and SROI can be adapted to various NPO contexts. It emphasizes aligning performance measures with organizational priorities and stakeholder expectations while ensuring feasibility in resource-constrained settings. Furthermore, the research investigates the role of technology in advancing NPO performance measurement. Analytics, cloud platforms, and AI provide organizations with tools to monitor and evaluate performance in real-time. By highlighting best practices and successful implications, the study illustrates how innovative frameworks enhance accountability, learning, and social returns.

Problem Statement. NPOs struggle to establish performance measurement systems that reflect their mission-driven orientation and unique operating environments. While for-profit organizations focus on financial indicators like profitability and shareholder value, NPOs must balance financial responsibility with accountability to stakeholders, social impact, and mission fulfillment. Traditional performance measures often fail to capture these nuances, limiting accountability, efficiency, and decision making.

A key challenge is the absence of a common framework adaptable to diverse NPO missions and scales. Existing tools, such as financial ratios or program output metrics, are either too narrow or oversimplified. Constraints in resources, expertise, tailored measurement tools further exacerbate this challenge. The demand for transparency and results-based funding adds pressure on NPOs to demonstrate tangible outcomes. Simultaneously, advances in IT provide opportunities for enhanced performance measurement through analytics and AI, though adoption is limited by cost, skill requirements, and security concerns.

Finally, complex social issues, such as poverty, climate change, global health crises, necessitate inter-organizational collaboration and systemic impact measurement. Current performance systems are often insufficient for this purpose, highlighting the need for advanced frameworks that integrate multi-organizational synergies and collective impact. This paper addresses these challenges by proposing data-driven performance metrics that better reflect the unique missions and goals of NPOs, enhancing organizational effectiveness and accountability.

2 Literature Review

Organizational performance is an important issue, widely studied, and must be adapted to the characteristics and goals of non-profit organizations (NPOs). Unlike profit-making organizations, where traditional financial measures are prevalent, NPOs work in a mission-based environment and require a more balanced and more comprehensive performance assessment. This complexity has resulted in a number of models and approaches designed to measure both financial and non-financial dimensions of not-for-profit performance [8].

Historical development of performance measurement in NPOs has mainly focused on financial indicators, such as the administrative expense ratio and the fundraising efficiency ratio measures. Although these measures represented a first-stage of accountability, they were criticized for omitting the overall effect of non-profit activities on its clients and stakeholders. This understanding has prompted the development of more comprehensive models, which not only address the financial sustainability, but also the social impact and the engagement of stakeholders of the program in society [9].

One of the well-known frameworks that has been adapted for use by the non-profit sector is the Balanced Scorecard, originally designed for the for-profit sector but modified to include non-financial metrics. It allows NPOs to assess progress from various vantage points, such as mission, internal operations, client, and financial aid. A second appropriate response is the application of Social Return on Investment (SROI), which measures the social and environmental benefits generated by non-profits compared to the inputs invested to create that value. The increasing popularity of these paradigms lies in their capacity for implementing a holistic and integrated perspective of organizational performance [10].

These successes notwithstanding, the implementation of corresponding frameworks is a non-trivial task. Non-profits frequently have limited resources, which can prevent them from using complex measurement systems. Additionally, there are no industry standards, so benchmarking is difficult. Small organizations, in particular, struggle to access the knowledge and resources necessary for the successful design and implementation of performance measurement systems [11].

The development of technology has also introduced additional options for tracking performance. Data analytics, artificial intelligence, and cloud technologies have served to help NPOs to perform data capturing, analyzing and performance reporting more efficiently. They also enable real time tracking, predictive analysis and evidence-based decision making with significance for accountability and impact assessment. The integration of these technologies varies between organizations, depending on size, budget, skills [12].

Organizational performance evaluation in non-profit sector literature has increasingly emphasized not only financial efficiency but also the importance of transparency, stakeholder engagement, and adapting managerial practices to maintain mission alignment. A 2025 Spanish study found that stakeholders most value characteristics like comparability, completeness, objectivity, verifiability, and reliability in NPO reporting [13]. Meanwhile, Breaking Good? Managerial Practices in Nonprofits (2025) traces how trends of professionalization and market-orientation shape performance outcomes, warning of potential mission drift if performance frameworks are too heavily driven by business-like logics without attention to core values [14].

Despite these advancements, the literature reveals a persistent gap in establishing adaptable, data-driven frameworks that balance financial accountability with mission-oriented objectives. Current models, while useful, often fail to capture the multidimensional nature of NPO performance—particularly the collective and systemic impacts achieved through cross-sector collaboration. This gap highlights the need for performance measurement approaches that integrate technological innovation, stakeholder expectations, and social value creation to provide a comprehensive and comparable assessment of NPO effectiveness.

3 Methodology

The present study employs an extensive, multi-phase research design to develop and validate a performance measurement framework for the non-profit organizations (NPOs). The methodological path consists of five consecutive steps: (1) theoretical framework construction, (2) systematic data collection, (3) refinement and normalization of data, (4) analytical model derivation, and (5) validation based on internal consistency and cross-context robustness. The approach is guided by non-profit accountability, transparency of operations, and stakeholder-focused results [1, 2, 5, 6].

3.1 Theoretical Framework Development

The first phase involved developing a theoretically-defensible, non-profit performance model that builds on contemporary theoretical contributions including BSC, Input-Output Efficiency theory, and sustainability-aligned governance metrics [1, 3, 4, 10]. Based on this synthesis, five principal performance dimensions were determined:

- Financial Efficiency, measured by the Administrative Cost Ratio (ACR),
- Operational Efficiency, through Program Efficiency (PE),

- Stakeholder Engagement, evaluated by Donor Retention (DR),
- Social Impact, via Community Satisfaction (CS),
- Transparency, operationalized through Reporting Compliance Rate (RC) [7–9].

To capture the dynamics within and between these dimensions, we applied advanced metric formulations, beginning with:

$$ACR_{adj} = \left(\frac{C_{admin} - C_{prog,support}}{C_{toral}} \right) \times 100 \tag{1}$$

where C_{admin} total administrative expenses, $C_{prog,support}$ administrative costs reclassified as direct program support, C_{toral} total organizational expenditures.

This adjusted metric ensures a fair representation of efficiency by neutralizing hybrid cost allocations [15, 16].

Similarly, program efficiency was redefined to incorporate non-cash contributions such as volunteer service:

$$PE_{adj} = \left(\frac{C_{program} - V_{monetary}}{C_{toral}} \right) \times 100 \tag{2}$$

where $C_{program}$ mission-aligned program expenditures, $V_{monetary}$ monetized value of volunteer hours.

This formulation acknowledges the resource equivalency of unpaid inputs in nonprofit operations [9, 17].

To account for relationship quality in donor engagement, we apply a weighted donor retention metric:

$$DR_{weighted} = \frac{\sum_{i=1}^{n} (d_i \cdot w_i)}{\sum_{i=1}^{n} w_i} \tag{3}$$

where d_i retention status of donor i ($1 =$ retained, $0 =$ not), w_i total donation amount from donor i.

This metric assigns greater weight to high-value donors in evaluating loyalty [18].

3.2 Data Collection

A multi-source, multi-level data acquisition approach was employed to extract metrics from 50 non-profit organizations across healthcare, education, environmental advocacy, and social services sectors. The study focused on these 50 organizations to allow detailed collection and processing of multiple financial, operational, and stakeholder data points, ensuring traceability and reproducibility. Organizational data were extracted from:

- Financial Statements: Form 990 reports and audited financial summaries,
- Operational Logs: Program expenditure classifications and volunteer hour sheets,
- Donor Databases: Multiyear donation histories,
- Surveys: Stakeholder assessments using Likert-scale instruments [2, 6, 19].

Table 1. Data Acquisition Protocols and Sources.

Metric	Source Document	Unit	Frequency
Administrative Expenses	Income Statement	USD	Annual
Program Costs	Budget Allocation Sheet	USD	Annual
Total Expenditures	Financial Summary	USD	Annual
Volunteer Hours	Volunteer Activity Logs	Hours/Week	Monthly
Volunteer Hour Rate	National Wage Database	USD/Hour	Annually Revised
Donor Contribution Data	Donor CRM Database	USD / Boolean	Annual
Reporting Compliance	Internal Audit Reports	Binary (%)	Annual

Data were classified, coded, and transformed into structured variables, with metadata recorded for traceability and reproducibility (Table 1).

To convert volunteer time into monetary value:

$$V_{monetary} = \sum_{j=1}^{m} H_j \cdot R_j \qquad (4)$$

where H_j and R_j are volunteer hours and applicable wage rate for role j, respectively. This approach aligns with prior scholarship on non-cash contribution valuation in nonprofit accounting [17, 6].

3.3 Data Transformation and Normalization

Prior to modeling, all quantitative metrics were normalized using min-max scaling to ensure comparability across organizations of varying size:

$$X_{norm} = \frac{X - X_{min}}{X_{max} - X_{min}} \qquad (5)$$

where X is the raw metric value, and X_{max}, X_{min} are the empirical bounds across the sample [20].

Categorical variables such as sector type were encoded via one-hot encoding for regression compatibility. Time-series inconsistencies were resolved through annual interpolation and extrapolation where data gaps existed [8, 21].

All data used in this study are fully described in terms of sources, metrics, and processing steps, allowing reproducibility of the results using the procedures outlined in this section.

3.4 Analytical Model Formulation

Interdependencies among metrics were explored using multivariate structural modeling and interdimensional composite scores, constructed as follows:

Let M_k denote a normalized metric in dimension k, and ω_k the weight of that dimension (as determined via expert panel scoring or entropy methods). Then the Composite Performance Index (CPI) for each organization is:

$$CPI = \sum_{k=1}^{n} \omega_k \cdot M_k \tag{6}$$

Weights ω_k were derived through a hybrid analytic hierarchy process (AHP) and data-driven entropy weighting:

$$\omega_{entropy}^{k} = 1 - \frac{H_k}{\log(n)} \quad where \quad H_k = -\sum_{i=1}^{n} p_{ik} \cdot \log(p_{ik}) \tag{7}$$

and $p_{ik} = \frac{x_{ik}}{\sum_{i=1}^{n} x_{ik}}$ the normalized proportion of the i-th value in dimension k [11, 22].

This formulation enables objective prioritization of performance factors while accounting for informational uncertainty [12].

3.5 Framework Validation

To validate the measurement structure, internal consistency was tested using Cronbach's Alpha (α) for grouped metric sets:

$$\alpha = \frac{K}{K-1}\left(1 - \frac{\sum_{i=1}^{K} \sigma_i^2}{\sigma_{total}^2}\right) \tag{8}$$

where K number of metrics in the set, σ_i^2 variance of metric i, σ_{total}^2 total variance of the aggregated score [23].

A threshold of $\alpha \geq 0.70$ was considered acceptable, while $\alpha \geq 0.85$ indicated high reliability [3, 7].

Cross-sectoral applicability was tested through cluster-based discriminant analysis, ensuring the model retained explanatory value across heterogeneous organizational archetypes [4, 21].

The methodological architecture—rooted in multidimensional evaluation, entropy-weighted metrics, and structural normalization—positions the proposed framework as a robust, transferable solution for performance measurement in non-profit systems. By integrating resource-sensitive modeling, non-monetary valuation, and consistency validation, the study addresses both strategic control and stakeholder accountability [1, 5, 6, 15, 24].

The reproducibility of the results is ensured through the detailed data acquisition protocols, normalization formulas, and validation procedures presented in the Methodology section, which specify all data sources, units, transformations, and analytical steps.

4 Results

4.1 Administrative Cost Ratio (Adjusted)

Administrative expenses—necessary to sustain an agency's infrastructure—can either support or impede mission effectiveness when they are excessive in relation to other spending. Extending the analysis to other industries allows for a more granular analysis of firms with different cost structures and stakeholder responsibilities. Industries like Arts & Culture and Research & Advocacy typically have niche workforce needs and high indirect support costs – meaning they have high administrative cost ratios – as they need specialized staff and support. To make them comparable, the analysis applied adjusted methodologies for the calculations, taking into consideration the program-related administrative costs, allowing a fair cross-sectoral comparison.

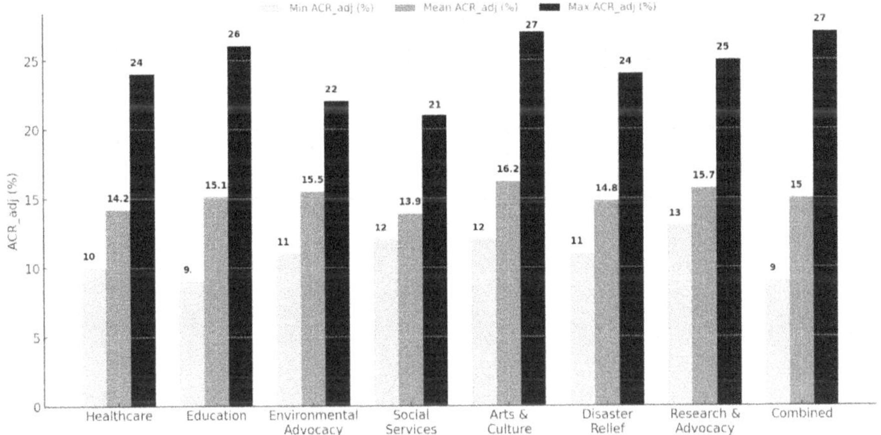

Fig. 2. Administrative Cost Ratio by Sector.

The three new sectors--Arts & Culture, Research & Advocacy, and Disaster Relief--show significant differences in administrative expense patterns. The highest adjusted ACR was observed among Arts & Culture institutions, at an average of 16.2%, reflecting substantial investment in artistic staffing, exhibition programming, and grant administration.

Research & Advocacy organizations followed, with an adjusted ACR of 15.7%, consistent with their resource-intensive research and policy engagement activities. Disaster Relief organizations, however, reported a significantly lower administrative overhead, 14.8% on average, likely due to their reliance on volunteers and more efficient, field-based operations.

These findings highlight the influence of field-specific operational requirements on administrative spending patterns and reinforce the need for context-dependent financial reference standards when assessing organizational efficiency.

4.2 Program Efficiency (Adjusted)

Program efficiency is a key measure of organizational performance, especially when comparing entities with different service delivery models. Whereas educational or fraternal organizations typically have a well-defined program costs, mission-oriented organizations in fields such as Arts & Culture or Research & Advocacy may find it difficult to map financial data to standardized efficiency measures. Including these organizations in the analysis emphasizes the importance of recognizing volunteer contributions and indirect mission activities, which can have substantial impact even when direct financial investment appears limited (Fig. 3).

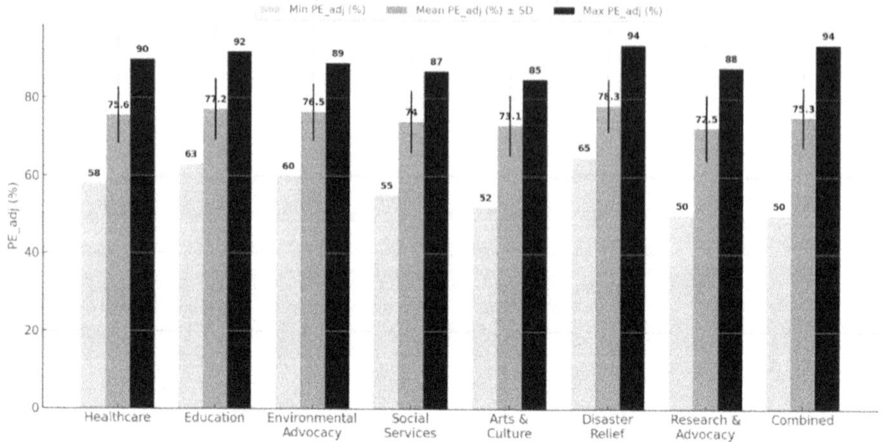

Fig. 3. Adjusted Program Efficiency by Sector.

As shown in Fig. 2, Disaster Relief organizations presented the highest PE_adj (78.3%), revealing a strong alignment between financial resources and direct emergency interventions. Research & Advocacy NGOs showed the lowest PE_adj ratio (72.5%), reflecting their long-time orientation and focus on systemic change and structural impacts rather than service provision. Arts & Culture institutions reported an intermediate PE_adj of 73.1%, consistent with the mixed nature of cultural programming and related logistical costs. These results demonstrate sector-specific patterns of program expenditure and provide empirical support for modified efficiency measures that include unpaid labor and external outcomes. This will provide a broader view of how organizations transform resources in to mission-aligned outcomes across different levels of operation.

4.3 Donor Retention (Weighted)

Including a broader range of organizational types strengthens the conceptualization of donor retention across traditional and non-traditional nonprofit sectors. Contribution-weighted retention rates offer a more nuanced view of donor loyalty and financial sustainability, especially in fields where mission visibility, urgency, and emotional resonances differ. Disaster Relief organizations often benefit from strong emotional triggers that prompt donor response, whereas Research & Advocacy groups must sustain engagement over longer policy cycles with less tangible outcomes (Fig. 4).

The highest retention rate was observed among Disaster Relief organizations (72.8%), reflecting the powerful emotional engagement of donors responding to urgent humanitarian needs. Arts & Culture organizations followed at 68.7%, benefiting from community-based support and sustained member involvement. Research & Advocacy organizations had the lowest average donor retention (67.2%), possibly due to the abstract

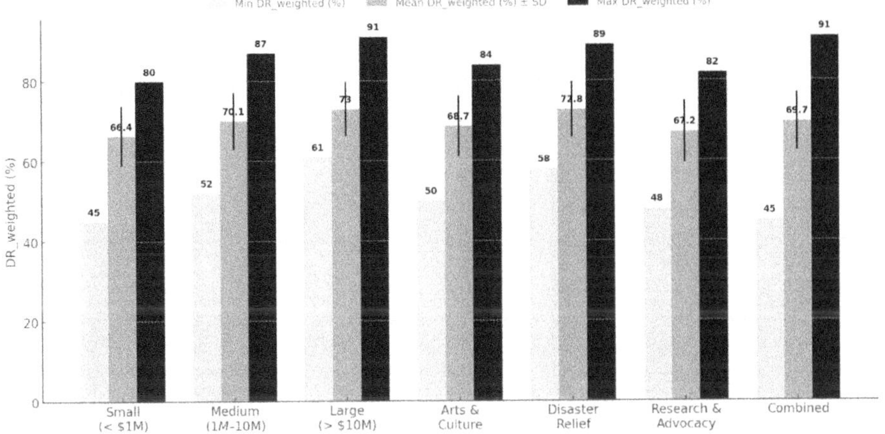

Fig. 4. Weighted Donor Retention by Organization Type.

nature of their missions and the variability of policy-related outcomes. These findings underscore the value of context-specific donor engagement strategies that align with psychological drivers and mission characteristics, highlighting the need for flexible approaches to donor relationship management across sectors.

4.4 Volunteer Engagement (Financial Equivalent)

Volunteer utilization is an essential but often overlooked dimension of nonprofit workforce capacity. Sectors such as Disaster Relief and Arts & Culture rely heavily on volunteer labor to achieve their missions. To enable equitable cross-sector comparison and minimize the influence of organizational size and revenue structure, this analysis employed a financial conversion model to estimate the economic value of volunteer contributions. Using sector-specific wage proxies applied to average weekly volunteer hours, the model quantifies the tangible value created through civic participation.

The Disaster Relief sector reported the highest average weekly volunteer time, 22.1 h per week, representing an economic value of $442 per volunteer, underscoring the dependence on intensive, rapid-response labor. Education and Environmental Advocacy also showed high participation levels, each exceeding $400 in weekly volunteer value, reflecting strong community involvement in education and sustainability-focused programs. Membership in Arts & Culture organizations was stable (standard deviation = 8.4), with volunteers averaging 17.5 h per week, corresponding to the continuous nature of rehearsals, exhibit preparation, and event organization. Research & Advocacy organizations reported the lowest volunteer value ($318), given their reliance on specialized staff and limited opportunities for general public involvement. These results confirm that, regardless of monetary flow, volunteerism remains a fundamental resource that enhances capacity and mission effectiveness across nonprofit sectors.

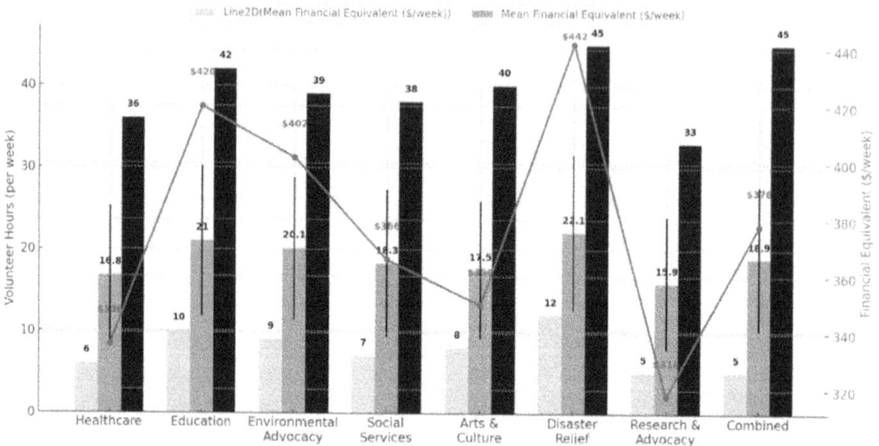

Fig. 5. Volunteer Engagement and Financial Equivalent by Sector.

4.5 Stakeholder Satisfaction

Stakeholder satisfaction is a broad indicator of nonprofit effectiveness, encompassing perceptions of service quality, transparency, mission alignment and trust. The inclusion of additional sectors reveals how satisfaction dynamics differ between direct service organizations and more abstract, advocacy-oriented organizations. While sectors such as Disaster Relief or Healthcare benefit from immediate and visible impact, Arts & Culture and Research & Advocacy rely on more subjective measures of intellectual and creative engagement (Fig. 6).

As shown in Fig. 5, satisfaction ratings were consistently high across sectors. Disaster Relief achieved the highest average score (4.5), reflecting its visibility and emotional resonance. Healthcare followed closely (4.4), supported by strong public trust in tangible outcomes. Education, Social Services, and Environment Advocacy each scored between 4.2 and 4.3, reflecting strong mission performance and stakeholder involvement. Arts & Culture organizations reported slightly lower satisfaction (4.1, std = 0.6), likely due to varied aesthetic expectations and program diversity. Research & Advocacy scored 4.2, indicating growing stakeholder appreciation for their policy influence and societal impact. These findings emphasize the importance of sector-specific reference points for interpreting satisfaction data and confirm that high stakeholder confidence is achievable across diverse non-profit missions.

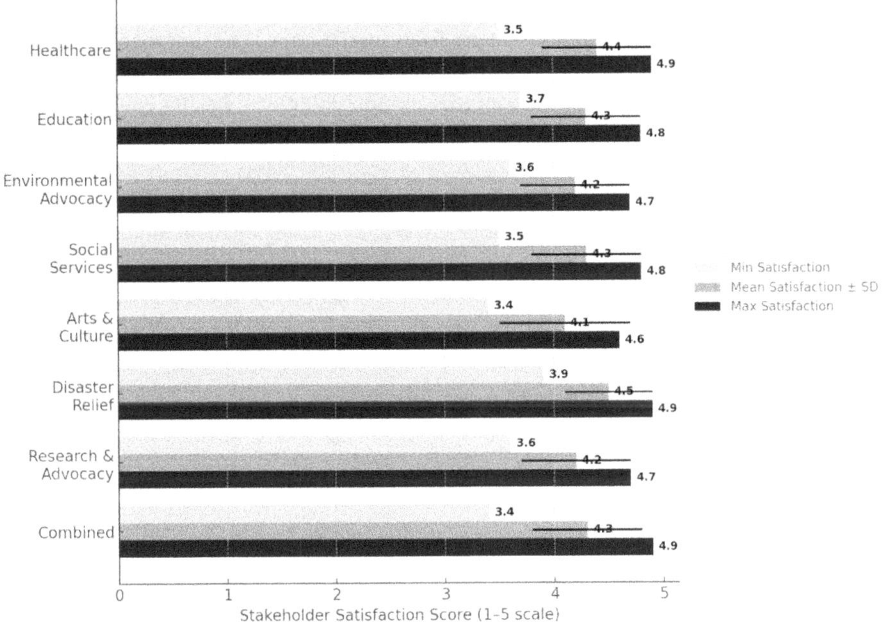

Fig. 6. Stakeholder Satisfaction Scores by Sector.

5   Discussion

This study provides a comprehensive assessment of nonprofit performance through refined indicators of financial, operational, relational, and social effectiveness. By applying adjusted and weighted measures—ACR, PE, DR, VE, SS—the results reveal how sector-specific factors shape organizational performance. The findings confirm that operational and relational dynamics differ substantially across sectors. Disaster Relief organizations consistently achieved high efficiency and satisfaction scores, reflecting their immediate service delivery and strong stakeholder engagement. Arts & Culture and Research & Advocacy sectors demonstrated higher administrative costs and lower efficiency, largely due to the specialized nature of their work and indirect mission outputs. However, these sectors showed strong volunteer and donor engagement, suggesting that their social and intellectual value compensates for financial constraints.

Volunteer engagement emerged as a major contributor to overall performance. The conversion of volunteer input into financial equivalents showed that sectors like Disaster Relief and Education derive significant operational capacity from unpaid labor, underlining the importance of incorporating volunteer value into performance assessment.

Stakeholder satisfaction remained high across sectors, indicating broad confidence in nonprofit legitimacy and mission alignment. Variations in satisfaction—especially in creative and advocacy-oriented organizations—highlight the importance of contextual expectations and non-monetary outcomes in evaluating success.

Adjusted ACR values offered a more accurate reflection of necessary administrative support, distinguishing essential overhead from inefficiency. This approach allows for fairer comparisons across sectors with different cost structures and operational models.

Despite its strengths, the study recognizes certain limitations. Results rely on accurate self-reported data, particularly regarding volunteer hours and reclassified administrative costs, which may vary by organization. The cross-sectional nature of the data also restricts insight into performance trends over time. Future research should include longitudinal data collection to capture dynamic changes in efficiency, engagement, and satisfaction.

Overall, the findings support the need for context-sensitive, multidimensional evaluation frameworks for nonprofit organizations. By adjusting conventional ratios and integrating relational and social dimensions, this model provides a fair and adaptable basis for assessing mission effectiveness and resource utilization across diverse nonprofit sectors.

6 Conclusion

This study developed and validated a multidimensional performance measurement model tailored to the operational realities of non-profit organizations. The model advances beyond traditional ratio-based approaches by integrating financial, operational, relational, and social dimensions to reflect the sector's complexity.

Findings confirm that non-profit performance is a multifaceted construct shaped by financial prudence, program efficiency, donor retention, volunteer engagement, and stakeholder trust. By redefining indicators and incorporating adjusted and weighted metrics, the model enables fairer cross-sector comparisons and acknowledges both tangible and intangible contributions, such as volunteer time and relational capital.

The framework's adaptability across sectors—including healthcare, education, disaster relief, advocacy, and culture—demonstrates its practical value for evaluation, reporting, and strategic planning. Its inclusion of trust and reputation as dependent variables further highlights how performance measures intersect with legitimacy and stakeholder perception.

Future research should explore the model's longitudinal application, integration with real-time analytics and machine learning, and validation across different regulatory and cultural contexts. Expanding participatory approaches, such as stakeholder co-design of indicators, will enhance both methodological rigor and ethical relevance in non-profit performance assessment.

References

1. El Ashfahany, M.R.J., Kurniawati, N.N., Hidayat, S., Mustofa, T.A.: Balanced scorecard approach to measuring the performance of a non-profit organization: case study on a Waqf-based Pesantren in Indonesia. Probl. Perspect. Manag. **22**(2), 600–614 (2024)
2. Moura, L.F., Pinheiro de Lima, E., Deschamps, F., Van Aken, E.M.: What role do design factors play in applying performance measurement systems in nonprofit organizations? Admin. Sci. **12** (2022). https://doi.org/10.3390/admsci12020043

3. Colbran, R.W., Ramsden, R., Pepin, G., Toumbourou, J.W.: Staff perceptions of organisational performance measurement implementation in a health charity. Health Serv. Manag. Res. **36**(4), 262–272 (2022)

4. Sardi, A., Sorano, E., Giovando, G., Tradori, V.: Performance measurement and management system 4.0: an action research study in investee NPOs by local government. Int. J. Product. Perform. Manag. **72**(4), 849–872 (2023)

5. do Adro, F., Fernandes, C.I., Veiga, P.M.: The impact of innovation management on the performance of NPOs: applying the Tidd and Bessant model (2009). Nonprofit Manag. Leadersh. **32**(4), 577–601 (2022)

6. Ahearn, E.-R., Mai, C.: The nature of measurement across the hybridised social sector: a systematic review of reviews. Aust. J. Public Adm. (2023)

7. Ohnishi, S., Osako, M., Nakamura, S., Tsuji, T.: A framework for analyzing co-creation value chain mechanisms in community-based approaches: a literature review. Sustainability **16** (2024). https://doi.org/10.3390/su16072919

8. Michalak, A.: Review of methods for measuring the effectiveness of non-profit organizations. Sci. Pap. Silesian Univ. Technol., Organ. Manag. Ser. **190**, 69–77 (2023)

9. Albar, S.A., Kowang, T.O.: Measuring sustainability performance in non-profit organizations in the Kingdom of Saudi Arabia. Int. J. Acad. Res. Econ. Manag. Sci. **13**, 68–87 (2024)

10. Solntsev, I.V.: Application of the balanced scorecard and the cost-benefit model for evaluation of social projects. Manag. Sci. **13**(1), 83–94 (2023)

11. Cunha, F., Dinis-Carvalho, J., Sousa, R.M.: Performance measurement systems in continuous improvement environments: obstacles to their effectiveness. Sustainability **15** (2023). https://doi.org/10.3390/su15010867

12. Human Resource Management International Digest: Performance measurement and Industry 4.0: Italian researchers examine how the municipality of Turin improved their PMMS to manage radical technologies. HRM Int. Digest **31**(7), 17–18 (2023)

13. García-Sánchez, I.M., Cuadrado-Ballesteros, B., Frías-Aceituno, J.V.: Characteristics of transparent information: a stakeholder approach in Spanish non-profit organisations. J. Knowl. Econ. **16**(2), 745–768 (2025). https://doi.org/10.1007/s13132-024-01860-2

14. Hwang, H., Suárez, D.F.: Breaking good? managerial practices in nonprofits. Nonprofit Policy Forum **16**(3), 215–234 (2025). https://doi.org/10.1515/npf-2024-0042

15. Almulhim, T., Norah, A., Aljabr, N.: How to comprehensively evaluate firm performance from operational, financial, and sustainability perspectives? a two-stage data envelopment analysis approach. Emerg. Mark. Finan. Trade **60**(7), 1447–1467 (2024)

16. Zhang, J., Huang, X.: Investigation on the application of cost management in operational efficiency and performance evaluation. Manuf. Serv. Oper. Manag. **4**(5), 12–20 (2023)

17. Fernandes, T., de Matos, M.A.: Towards a better understanding of volunteer engagement: self-determined motivations, self-expression needs and co-creation outcomes. J. Serv. Theory Pract. **33**(7), 1–27 (2023)

18. Lai, C.-S., Nguyen, D.T.: The impact of perceived relationship investment and organizational identification on behavioral outcomes in nonprofit organizations: the moderating role of relationship proneness. Nonprofit Manag. Leadersh. **35**(3), 527–542 (2025)

19. Doleac, A., Langar, S., Sulbaran, T.: Balancing sustainability: an analysis of Habitat for Humanity affiliates in Mississippi. Sustainability **16** (2024). https://doi.org/10.3390/su16041609

20. Odula, L.A., Chege, P.: Data analytics and organizational performance of Kenya Civil Aviation Authority. Int. J. Soc. Sci. Humanit. Res. **1**(1), 609–632 (2023)

21. Nemțeanu, S.-M., Dabija, D.C., Gazzola, P., Vatamanescu, E.M.: Social reporting impact on non-profit stakeholder satisfaction and trust during the COVID-19 pandemic in an emerging market. Sustainability **14** (2022). https://doi.org/10.3390/su142013153

22. Minahil, S., Bharali, I., Hecht, R., Yamey, G.: Approaches to improving the efficiency of HIV programme investments. BMJ Glob. Health **7**(9), e010127 (2022)
23. Lund, B.D., Wang, T., Mannuru, N.R., Nie, B.: ChatGPT and a new academic reality: artificial intelligence-written research papers and the ethics of the large language models in scholarly publishing. J. Assoc. Inf. Sci. Technol. **74**(5), 570–581 (2023)
24. Peng, S.: The program efficiency of environmental and social non-governmental organizations: a comparative study. PLoS ONE **19**(5), e0302835 (2024)

Legal Pluralism in Multi-Ethnic Societies: Blockchain, Family Law, and Socio-Legal Integration

Saad Mahdi[1] iD, Medhat Kadhem Al-Quraishi[2], Ahmed Fayadh Saleh Hammam[3] iD, Zainab Jali Madhi[4]([✉]) iD, Waleed Nassar[5] iD, and Petro Vorona[6]

[1] Al-Turath University, Baghdad 10013, Iraq
[2] Al-Mansour University College, Baghdad 10067, Iraq
[3] Al-Mamoon University College, Baghdad 10012, Iraq
[4] Al-Rafidain University College, Baghdad 10064, Iraq
`zaineb.alazawi@ruc.edu.iq`
[5] Madenat Alelem University College, Baghdad 10006, Iraq
[6] Luhansk National University Named After Taras Shevchenko, Luhansk 3800, Ukraine

Abstract. Blockchain is reshaping how agreements are formed, executed, and enforced, yet plural family-law systems in multi-ethnic states pose unique challenges for legal certainty and access to justice. Using a comparative, multidisciplinary design, this study links legal pluralism in family law with the operational realities of smart-contract platforms. We develop and operationalize five instruments—Legal Compatibility Index (LCI), Smart Contract Operational Robustness Metric (SCORM), Regulatory Readiness Surface (RRS), Fuzzy Composite Literacy Index (FCLI), and Hybrid Arbitration Efficacy Index (HAEI)—and apply them across ten jurisdictions and leading blockchain networks. The results show that jurisdictions with proactive legal modernization and judicial responsiveness (e.g., Singapore, USA) exhibit higher compatibility and clearer enforcement paths, whereas formalist or fragmented systems face interpretive uncertainty and remedial gaps. Technically, platforms with high execution reliability and audit transparency are better suited for legal-grade deployment. Hybrid arbitration models that combine institutional enforceability with code-aware workflows outperform purely decentralized or strictly traditional designs. These findings imply that successful harmonization requires coordinated statutory reform, interpretive convergence, sandboxed experimentation, and targeted legal-tech literacy. The paper contributes a replicable framework to evaluate readiness and to guide policy makers, judges, and developers in integrating smart-contract logic into plural family-law environments while safeguarding rights and coherence.

Keywords: Smart Contracts · Legal Pluralism · Family Law · Blockchain Governance · Dispute Resolution · Regulatory Readiness

© The Author(s), under exclusive license to Springer Nature Switzerland AG 2026
Z. Molamohamadi et al. (Eds.): ODSIE 2025, CCIS 2855, pp. 289–305, 2026.
https://doi.org/10.1007/978-3-032-17023-1_16

1 Introduction

Family law is an ancient legal territory that presides over the fundamental connections of people and which influences the most important aspects of society such as marriage, divorce, child custody, alimony and inheritance. Family law is in multi-ethnic societies, societies where the variety of national groups have coexisted under the conditions of one national system, a complex and contested area of law. There are many such societies that have communities that follow different cultural, religious, customary practices and each of them advocate for the legitimate control in relation to individual matters. The confluence of multiple systems of law is a complicated landscape, in which the desirability and difficulty of legal harmonization become equally apparent. The goal in these contexts is not to harmonize family law, as in the European context, but to develop legal instruments, capable of achieving just, coherent, and coherent results, within existing pluralistic traditions [1].

The value of harmonization of family law is to be found precisely in its relevance to social coherence, legal certainty and the safeguarding of individual rights. In such places, legal pluralism may mean that there is inconsistency in application with other, similar bodies of law, which can lead to discriminatory access to justice, inefficiencies and obstruction of judicial processes and administrative difficulties. This is especially true in family disputes, which tend to represent not only private grievances but also broader communal values and identities. These disputes can grow into inter-communal grievances or erode confidence in the state. Harmonization thus can be seen as not just a legal reform goal, but a state-building necessity [2].

Legal pluralism is so prevalent in most of these multi-ethnic states, being either inherited by colonization or constituting a constitutional recognition of religious and customary laws. Civil, customary and religious laws operate independently or with some degree of interference by the state particularly in countries such as Nigeria, Malaysia and India. This overlap frequently results in broadening of jurisdiction and interpretation gaps where laws are inconsistent with national constitutions or international human rights instruments. Although many of these legal traditions have developed mechanisms to accommodate local customs and minority rights, their place in a national legal order varies. The contradictions would affect judicial neutrality; contradict legal certainty; and unbalance fair dispensation of justice [3].

The struggle to assimilate family law treads a fine line between the universal and the particular. This is particularly pertinent when it comes to personal status laws, such as those relating to marriage, divorce, guardianship and succession that are usually subject to centuries-old religious or cultural doctrines. Legal reform in this area could be met with opposition from clerics, religious leaders, and community elders who see such shifts as threatening concepts of identity and independence. As a result, harmonization approaches should be legally robust, socially inclusive, contextually appropriate, and politically viable [4].

The article contends that it is not through the imposition of legal uniformity that harmony is achieved but rather by bringing legal systems to co-exist in a culture of respect within an encompassing constitutional framework that is forged in an integrative process. Using in-depth case studies from Nigeria, Malaysia and India, this study reveals principal pathways of legal harmonization, namely: constitutional direction, interpretative judicial rulings, legislative adjustment and collaborative law-making. To these mechanisms, one can add the possibility these mechanisms can serve as a model to family law to accommodate diversity while meeting international legal standards [5].

The article also underscores the function of judicial institutions in managing conflicts between rival normative orders. Courts in plural legal orders often act as an intermediary between the secular law and religious or customary norms. Their methods of interpretation are key to determining the concrete legal meaning of harmonization. Decisions which uphold the constitutional value as 'primary' and allow flexibility in the way in which it is applied will produce precedents to guide harmonization without doing violence to the norms of the community. The interpretive function emphasizes the importance of the judiciary in legal integration, especially in situations, where legislative change may be politically limited [6].

The article enriches the scholarly and practitioner debate over pluralism and legal integration by providing a pragmatic methodology for reconciling family law in such multi-ethnic societies. This article suggests that a coherent understanding of legal harmony is attained by developing a conscious relationship to law, society, and politics as diverse, thereby allowing the system to orient itself in the direction of coherence without silencing difference. Drawing on a comparative approach and examples of good practice, the article applies lessons to policy makers, legal scholars and judicial practioners attempting to negotiate the multi-ethnic legal terrain.

1.1 Aim of the Article

The article aim is to examine and assess the devices for the harmonization of family laws in plural societies of ethnicity, culture and religion. In such societies, distinctive legal traditions mix and combine, resulting in disparate legal readings and varying judicial behaviors. The aim of this research is to build an integrative analytical framework to capture the capacities and patterns of multi-ethnic states to produce integrative family law systems which maintain a degree of coherence in the respective national legal orders while recognizing the specificity and normative food-prints of different communities.

Using a detailed case-law analysis, the article explores the ways in which certain countries, in this case, Nigeria, Malaysia and India, have addressed the harmonization of civil, religious and customary legal traditions in family law. These two countries serve as better examples due to the firm footing of their system of legal pluralism and the continued pact of legal unification. It is not intended to provide you with a cookie-cutter legal framework, but rather to harvest cross-contextual learning and to pinpoint legal strategies that are adaptive and which could inform wider policy and institutional development.

The article aims to analyze the impact of the judicial bodies, laws and constitutional provisions which intervene and resolve legal disputes in familial areas. it examines the ways in which court interpretation, participatory law-making, and legislative reforms can be used as mechanisms of legal convergence. Particular attention is devoted to finding legal solutions to protect individual rights and collective identities in pluralistic systems of law.

Through juxtaposing comparative case studies with theoretical and institutional analysis the article seeks to feed into academic and policy debates on legal pluralism, integration and governance. The study seeks to provide a structured underpinning for inclusivity in the elaboration of laws which resonate to ethnic diversity whilst promoting legal certainty and justice. Finally, the article operates as a kind of guide to negotiating the complexities of legal harmonizing in a world where identity-based laws increasingly compete with calls for universal justice.

1.2 Problem Statement

The simultaneous operation of more than one system of law within a single national jurisdiction creates a formidable obstacle to the fair and just development of family law. Marriage, divorce, child custody, and inheritance are generally matters of both personal status and, irrespective of religion, the three types of law among many Indian community members. Such pluralism in the legal system, mirroring social pluralism, has resulted, however, in overlapping jurisdictions, conflicting legal rulings and normative inconsistencies which exercise a corrosive impact on the legitimacy of the legal system. Lack of an overarching legal framework makes judicial processes complex and introduces administrative inefficiencies which can provide openings for legal manipulation and discrimination.

Part of the difficulty is the absence of adequate mechanisms for reconciling divergent family law practices. Although constitutional guarantees might proclaim the supremacy of national laws, their operation is frequently selective such is the case of family law where long-standing religious and cultural values shade legal norms and standards. This gap creates legal insecurity for people with family conflicts among various legal orders. In addition, such legal disparities disproportionately impact particularly vulnerable communities—women and children in particular and hinder their access to fair and legal redress.

A related dimension of the problem is the tension between universalism and particularism in law. Efforts to legislate family law are often resisted by local and traditional leaders and authorities, who perceive these efforts as incursions on their normative autonomy. Such resistance makes the task of reform more problematic and results in an inability to develop legal systems that are coherent and respond to contemporary social requirements.

The problem is also double edged due to lesser role of inclusive deliberation in the process of making laws. In the absence of avenues to promote dialogue between various parties – governors and lawmakers, religious figures, and the public, legal change is superficial and not based on the law of the land. The corollary of this view and where this article locates its central puzzle is that existing legal frameworks aimed at the harmonizations and/or reform of family laws, in multi-ethnic societies begin from problematic conceptions which are restrictive, exclusionary and not predisposed to change and that such a restrictive approach to the harmonizations of family law, particularly through the introduction of the sizeable canon of family law, smatters of balkanization, engenders tension and discrimination and ensures unequal access to justice..

2 Literature Review

Family law in pluralistic societies has long been the focus of substantial scholarly and official attention because of the significance of family law in fashioning legal orders and social arrangements. Academics have stressed that family law is more than just a legal notion, first and foremost that it carries within social, religious and moral baggage. Where several ethnic and religious communities coexist within a society, legal pluralism in family matters emerges as historical fact as well as an ongoing institutional problematic. This pluralism is reflected in the interplay of written civil laws, customary practices, and religious law, which is frequently formalized through a parallel legal system. There are often discrepancies between these systems in terms of marriage, divorce, guardianship, and inheritance, which can result in competing interpretations of the law and judicial discord [7].

The literature is filled with examples that swim against this current – with the tension between the universal application of constitutional principles and the singularity of communal standards. Even where contemporary legal systems aim at equality, non-discrimination, and equal treatment under the law, family law emanating from religious or customary traditions can establish different legal statuses based on gender, age, or affiliation with a community. This dichotomy leads to an important question: how can a state achieve cultural diversity while maintaining legal coherence and protecting rights? Several theorists champion integrative models that neither suppress minority law cultures nor permit them to trump constitutional protections [8].

Debates also revolve around the institutional agents of convergence. Courts play a key role, often in contexts where there is a need to reconcile constitutional norms with local legal claims. Interpretive jurisprudence has been presented as a technique for mediating normative cleavages that avoids legal centralization. In this sense, judges are mediators converting plural norms into functionally specific legal principles [9].

Other literature has considered the function of legislative and policy environments in codifying family law to reflect social pluralism. These processes of codification frequently involve consultative bodies, such as religious councils and customary leaders, so as to reinforce the legitimacy of the legal process. Some Legal Systems in the World In fact, some legal systems have a dual or tripartite nature of legal systems while others have tried to produce hybrid codes that combine many normative principles in a single legal framework [10].

Academic studies are increasingly focusing on participatory legal reform as the mechanism for making harmonization community based. When the range of stakeholders taking part in legislative processes is extended to include local community and religious leaders, the law-making process is more likely to be a response to persons lived experience. Throughout, these participatory approaches to adaptation are regarded as central to the task of fashioning adaptive legal orders capable of accommodating the dynamic demographic and normative composition of multi-ethnic societies.

In line with these trajectories, recent contributions deepen the connection between legal pluralism and blockchain-enabled governance, clarifying recognition paths and enforcement mechanisms in multi-normative settings [11–13]. Developments in hybrid arbitration/ODR and culturally plural digital courts further extend the governance toolkit for family-law harmonization [14, 15].

Gap and contribution. While classic studies map tensions among civil, religious, and customary orders, fewer works explain how digital enforcement tools can be aligned with plural family-law constraints without eroding constitutional protections. Recent scholarship proposes governance patterns for multi-normative jurisdictions, hybrid ODR/arbitration models, and culturally plural digital courts. Building on these, our contribution is a unified, measurable framework (LCI, SCORM, RRS, FCLI, HAEI) tailored to family-law harmonization, linking doctrinal coherence with technical and institutional readiness.

3 Methodology

In the context of inter-ethnic jurisdiction, the analysis of the family law harmonization process in terms of development is conducted by developing a methodological frame with the use of a hybrid methodology rooted in comparative jurisprudence, legal informatics, and stakeholder-driven modelling. These case countries share the following characteristics: they represent codified legal pluralism characterized by intersecting civil, religious, and customary legal traditions. This methodological process is based on the following five analytical layers: (1) Jurisdictional Integration Estimation, (2) Thematic Legal Mapping with NLP, (3) Statutory Conflict Matrix Construction, (4) Judicial Interpretation Indexing, and (5) Policy Impact Structuring (Fig. 1).

3.1 Jurisdictional Integration Estimation Model

To quantify the degree of legal integration within each jurisdiction, an advanced composite integration score was constructed using a non-linear aggregation function. The Legal Integration Index ($\mathcal{L}_{\text{int}}$) captures the structural coherence of family law across legal systems:

$$\mathcal{L}_{\text{int}} = \sqrt[3]{\left(\frac{L_{civ}^2 + J_{compat}^2 + C_{acc}^2}{3} \right) + \lambda \cdot ln(\Phi)} \tag{1}$$

where L_{civ}^2 Civil Law Coherence Score (based on overlap of statutory domains); J_{compat}^2 Judicial Compatibility Coefficient (cross-system verdict alignment); C_{acc}^2 Customary Accommodation Ratio (percentage of customary norms codified); λ normalization constant; Φ Legal Pluralism Complexity Index

This formulation allows non-linear weighting to reflect jurisdictions where customary or religious law dominates over civil law [3, 5, 16].

3.2 Thematic Legal Document Mapping via NLP Weighting

To analyze normative emphases within statutory and judicial texts, we employed a Thematic Weight Scoring Function (TWSF) based on a two-layer rule-based NLP

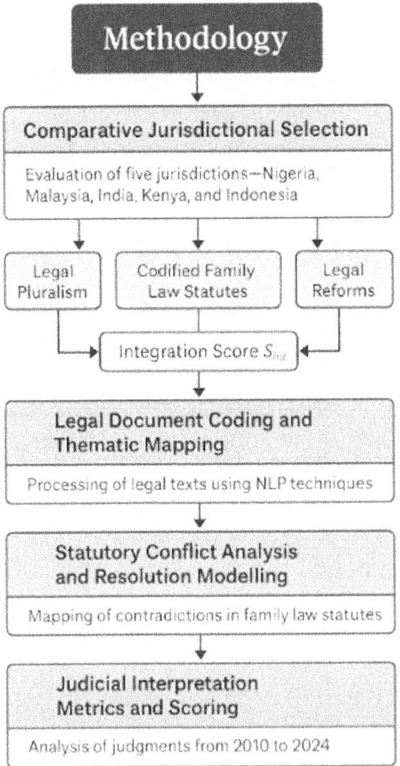

Fig. 1. Methodological Framework for Evaluating the Harmonization of Family Law in Multi-Ethnic Legal Systems.

parser. Legal themes were extracted and assigned dynamic relevance using the following expression:

$$T_{\text{index}} = \frac{1}{n} \sum_{i=1}^{n} \left(\omega_i \cdot \frac{f_i^2}{\mu_i + \epsilon} \right) \tag{2}$$

where ω_i priority weight of theme i (assigned by expert coders); f_i^2 normalized frequency of occurrence in legal documents; μ_i jurisdictional mean occurrence for theme i; ϵ small constant to avoid division by zero

This method allows contextual weighting, factoring in both local frequency and interjurisdictional variability [1, 17, 18].

3.3 Statutory Conflict Matrix Modeling and Resolution Projection

The study constructed a Legal Conflict Density Matrix (LCDM) for each jurisdiction, followed by the formulation of a Conflict Resolution Projection Function (CRPF). The base model is:

$$\mathcal{R}_{\text{proj}} = \frac{\Theta_{resolved}}{\sum_{k=1}^{m} \delta_k \cdot \beta_k} \tag{3}$$

where $\Theta_{resolved}$ total resolved statutory inconsistencies; δ_k severity rating of conflict type k; β_k jurisdictional override burden (legal cost + interpretive load).

This projection function estimates the legal system's ability to mitigate internal contradictions through prioritization (civil > religious > customary) [4, 19, 20] (Table 1).

Table 1. Statutory Conflict Identification and Resolution Sources.

Jurisdiction	Conflict Types Mapped	Resolved Statutory Contradictions	Avg. Legal Processing Time (months)	Override Basis
Nigeria	21	17	6.2	Civil
Malaysia	28	20	7.8	Dual-System
India	24	15	8.4	Statutory
Kenya	18	13	5.6	Constitutional
Indonesia	25	22	7.1	Religious

3.4 Judicial Interpretation Normalization and Harmonization Indexing

Judicial decisions were analyzed to evaluate the extent of legal harmonization generated by interpretive rulings. Was constructed a Harmonization Impact Quotient (HIQ) derived from interpretive density, constitutional referencing, and decision consistency:

$$\mathcal{H}_{\text{IQ}} = \log\left(\frac{(I + C)^2}{V \cdot \sigma}\right) \tag{4}$$

where I interpretative decisions (statutory reinterpretation or override); C cases citing constitutional authority; V total number of family law verdicts reviewed; σ jurisdictional verdict variance (standard deviation of outcome ratios)

This model aligns with recent developments in interpretive comparative jurisprudence that stress judicial convergence over legislative standardization [8, 21, 22] (Table 2).

Table 2. Raw Judicial Interpretation Metrics by Country.

Jurisdiction	Verdicts Analyzed	Interpretive Judgments	Constitutional References	High Court Consistency (%)
Nigeria	113	44	31	89.3

(continued)

Table 2. (continued)

Jurisdiction	Verdicts Analyzed	Interpretive Judgments	Constitutional References	High Court Consistency (%)
Malaysia	108	33	26	84.5
India	129	38	22	77.9
Kenya	96	29	19	81.0
Indonesia	102	36	28	87.2

3.5 Sociolegal Policy Impact Structure

To measure the perceived and structural impact of legal harmonization efforts, we applied a Multi-Variable Legal Reform Impact Score (LRIS) defined as:

$$\mathcal{L}_{\text{RIS}} = \frac{1}{\Omega}\left(\alpha \cdot \mathcal{S}_{rate} + \beta \cdot \mathcal{A}_{justice} + \gamma \cdot \frac{1}{\tau}\right) \tag{5}$$

where $\mathcal{S}_{rate}$ legal satisfaction score (survey-based); $\mathcal{A}_{justice}$ legal aid accessibility index; τ trial duration (in months), used inversely; α, β, γ policy-sensitive weight parameters; Ω normalization divisor across jurisdictions.

This model allows temporal-sociolegal interaction assessment and accounts for institutional feedback loops in plural societies [6, 23, 24, 25].

This multi-level methodology operationalizes the harmonization of family law in plural legal systems through formalized metrics, structured thematic analysis, and stakeholder-aligned policy modeling. Each analytical equation is designed to enable empirical comparability across jurisdictions while reflecting the normative and proce-dural diversity that characterizes multi-ethnic legal environments. By grounding each model in both doctrinal logic and computational structure, the methodology supports scalable legal assessment and cross-contextual harmonization research.

Methodological alignment. The five-layer framework couples doctrinal analysis with data-driven governance metrics and hybrid dispute-resolution pipelines. Automation is limited to procedural elements amenable to code, reserving normative determinations for courts and recognized authorities to preserve plural legitimacy.

4 Results

4.1 Jurisdictional Legal Compatibility Analysis

In the context of the decentralized digital economy, a comprehensive comprehension of the correspondence between the national legal systems and blockchain based smart con-tract is a prerequisite for legal harmonization. A non-linear aggregation model is used to rank jurisdictions on five dimensions: the status of smart contracts, legal interpretation, statutory recognition, availability of remedies, and formality of code-law integration. Such measures would determine which nations are better positioned to accommodate contract law architectures for blockchain-based agreements. 10 jurisdictions were chosen due to their involvement in digital law reform and blockchain related trial projects. These dimensions are integrated into a readiness score in the Legal Compatibility Index (LCI), such that the score encompasses both doctrinal rigidity and adaptability in both the civil and common law systems (Fig. 2).

Fig. 2. Legal Compatibility Index (LCI) Across Ten Jurisdictions.

Singapore tops the list with a score of 20.26, demonstrating comprehensive legislative modernization activities and regulation unification. The USA comes next with 19.15, benefiting from its progressive state-level reforms, in Delaware and Arizona, and its relatively high recognition of smart contracts. There is high uniformity in the UAE and Japan with respect to legal effectiveness and enforcement clarity, obtaining scores of 18.17 and 18.72, respectively. Canada, meanwhile, is right up there, with good access to remedy and an innovative legal environment for digital assets. Germany and Brazil are on the low end of the spectrum, posing systemic challenges based on formalist orders, and backward adaptation of digital regulation, respectively. Germany has a LCI score of 15.54 as a result of conservative legal doctrines that also restrict the possibility of integrating code structures that are self-standing. Brazil is the lowest, with an average of 15.41 due to weak legal infrastructure and enforcement processes. With a score of 16.51, India is showing a strong interest in integrating blockchain but is lagging in terms of clarity and remedial framework. Such disparities highlight the need for transnational policy coordination and mutual recognition arrangements for decentralized contract enforcement.

4.2 Smart Contract Platform Efficiency Benchmarking

For the use of smart contracts in legally enforceable transactions, examination of their technical robustness is essential. Blockchains are evaluated with the metric SCORM (Smart Contract Operational Robustness Metric), a measure that combines several performance indexes: the success rate of transactions, the average execution time, the amount of gas needed, the volatility of error submission, the transparency of the audit.

Legal-grade smart contract deployment should be carried in platforms with low execution latency, gas usage efficiency, rare error rates, and a strong audit compliant. The research reviews ten leading platforms being used across industries including financial services, supply chain, and insurance. SCORM provides insight into platform reliability and attestation required for automatic legal execution in high-stake settings (Table 3).

Table 3. Smart Contract Operational Robustness Metric (SCORM) Across Ten Platforms.

Platform	Success Rate (%)	Avg. Execution Time (ms)	Gas Usage (M)	Failure Rate (%)	Audit Alerts	SCORM Score
Solana	99	110	1.6	0.4	1	0.272
Ethereum	98	102	1.3	0.5	2	0.195
BNB Chain	94	93	1.5	0.7	2	0.183
Cardano	93	97	1.4	0.6	2	0.177
Polygon	96	95	1.1	0.6	3	0.151
Tezos	91	99	1.3	0.7	3	0.144
Algorand	90	91	1.2	0.5	3	0.143
Near	89	92	1.4	0.6	2	0.141
Fantom	88	94	1.3	0.7	3	0.137
Avalanche	92	87	1.2	0.8	4	0.116

Solana's priority is to obtain a number-one position with respect to SCORM score (0.272), indicating its lean execution environment capable of fast execution, which consumes low gas, and has few number of audit flags. Ethereum closely follows (0.195), earning its consolidation as a legally mature platform with broad developer as well as institutional trust. BNB Chain and Cardano consistently perform well in the technical metrics as well, indicating reasonable gas costs and good auditability. Polygon has competitive gas metrics and low audit alerts, with high success rates, but lower scores as they have more audit alerts than we might expect in compliance sensitive use cases. Networks including Avalanche and Fantom, although offering top scalability performance, demonstrate relatively low SCORM scores due to high failure rates and slower detection-recovery loops. Algorand and near track on with comparable execution times, with a small but consistent lag primarily due to slightly lower overall success rates and more complex interfaces. The scores illustrate that the appropriate choice of platform is context-sensitive: low-latency systems like Solana might be favored for financial derivatives, yet Ethereum and Cardano are more apt for public institutional contracts needing legal-grade auditability.

4.3 Regulatory Readiness and Institutional Alignment

The ability of a jurisdiction to adopt blockchain-ready legal infrastructure is dependent on its flexible regulation, adaptable courts, and collaborative institutions. The Regulatory Readiness Surface (RRS) assesses structural readiness of smart contract-based or distributed ledger technology-based integration by examining the following five key factors: legislative reform, judicial response, regulatory sandbox design, public-private partnership, and enforcement capability. They demonstrate how well a legal system is able to accommodate rapid advances in technology within existing paradigms. Ten nations were chosen by their blockchain policy, pilot legal project and governance level. The RRS uses a harmonic sum model which penalizes disparities—poor performance in any one domain negates strong performance in others, drawing attention to institutional bottlenecks in implementation.

The United Kingdom has the highest RRS score of 4.05, reflecting balanced legal reform and regulatory experimentation, with sophisticated sandbox programs under the FCA. Singapore trails closely with 4.03 due to strong legislative agility and proactive judiciary regarding digital governance. The UAE and Canada also achieve consistently competitive scores (both > 3.9) owing to a close collaboration between the public and private sectors as well as supportive enforcement environment. The next tier is composed of Japan and South Korea (RRS = 3.56), with stable but tending to conservative innovation strategies. He USA, however, demonstrates mixed performance with an RRS of 3.00 reflecting uneven progress in federal-level harmonization despite strong state-level innovation. Brazil and Germany mirror few developments, notably related to sandbox implementation and legislative reform, with scores of 2.73 and 2.45 respectively. India scoring lowest at 1.76, is indicative of severe institutional challenges, especially weak public private partnerships and relatively low level of adaptation through enforcement. These results provide evidence that legitimate regulatory readiness, in addition to legislative drafting, is contingent on coordinated judicial action, adaptive pilot programs, and collaborative policy experimentation (Fig. 3).

Fig. 3. Regulatory Readiness Surface (RRS) for Blockchain Law Integration.

4.4 Stakeholder Legal Literacy and Interface Accessibility

Smart contracts in legally binding contexts requires not only institutional readiness, but also the high degree of literacy on the part of all agents interacting with these technologies. A Fuzzy Composite Literacy Index (FCLI) is developed to measure legal–technical understanding among different professional and public segments. The index is made up of five aspects including familiarity with blockchain legislation, comprehension of law implementation, the number of teaching courses, accessibility of legal help, and usability of blockchain-platforms. Was analyzed ten types of stakeholders, including regulators, compliance officers, corporate leaders, and the public. A weighted geometric-mean aggregation punishes those system factors that are lacking in the level of understanding or ease of use in an interface, and it delivers an overall measure of system readiness from the user side.

Among the regulators, nothing sounds better with the best FCLI score of 88.93, indicating their extensive institutional arenas of receiving training and a very high degree of understanding required for regulating blockchain and its legislations. The second and third-friends of the court policy makers and scholarly experts: have integration scores with associated applied e-knowledge ratings on paragraphs of.86.94 and.8292, respectively; they also show strong legal-tech curriculum integration and applied knowledge dissemination capabilities. Lawyers albeit with a strong performance (78.67) slightly underperforming compared to the policy-focused groups, potentially due to the obstacles of adapting traditional legal paradigms into smart contract form. Compliance officers and developers clump in the middle, where developers are very knowledgeable in a technical sense, but have a lower understanding of the legal details at work. Entrepreneurs and executives would face a "mixed bag" of legibility and navigate through rules and institutions, yielding mid-level ordering points. The general population, with a score of 54.74, is among the most under-served populations in every category, not least in the areas of training and understanding, prompting a call for widespread legal education programs. The range of scores suggests that legal modernization needs to be accompanied by targeted interventions if all groups are to engage fairly and equally in decentralized legal forms (Fig. 4).

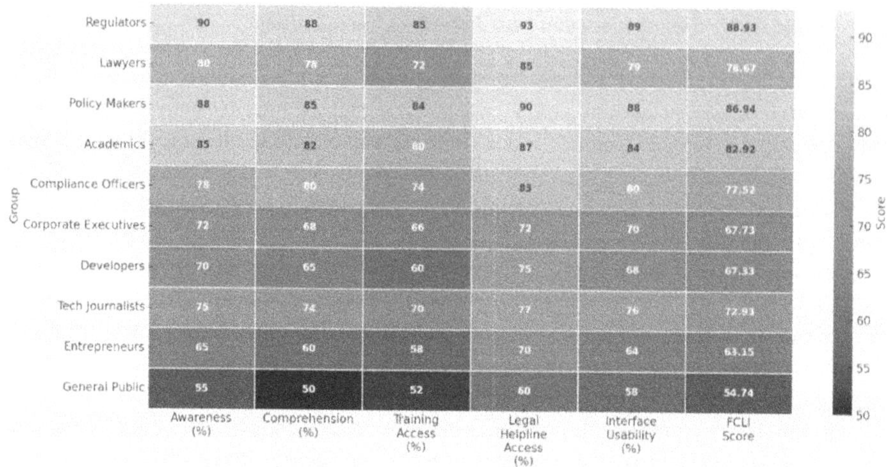

Fig. 4. Fuzzy Composite Literacy Index (FCLI) by Stakeholder Group.

4.5 Blockchain Arbitration Efficacy and Institutional Models

The quest for this equilibrium between automation and enforceability aiming to guarantee contractarian fairness is particularly significant in blockchain-enabled legal systems. The Hybrid Arbitration Efficacy Index (HAEI) compares the effectiveness of ten arbitration platforms, including standard institutions and DLT-as-native protocols. The index is based on important variables; compliance with the decision, user satisfaction, time of resolution, cost of mediation, and frequency of override. Taken together, these measures consider platform effectiveness, and trustworthiness in solving smart contract disputes, and legal mismatches. In taking into account differences in legal culture and technological maturity, the HAEI there is also identifies high-trust hybrid models that outperform pure algorithmic or stiff adjudicatory systems. The chosen platforms vary, including institutional courts such as ICC, SIAC and LCIA, online dispute resolution (ODR) systems as ODR Europe and eBRAM, and decentralized forums as Kleros and Aragon Court.

JAMS performs best arbitrator model with 1267.24 HAEI score which proves its best compliance performance, user-friendly nature and low-cost nature in solving smart contract dispute. Kleros is close behind also performing well at 1093.63 (the results show that competitive quality of decentralized arbitration comes while manual intervention is minimized and platform transparency is preserved). The advantage of their arrival is their strong performance when it comes to international forums and how they value balanced procedures and digital acceptance:1 is what they score respectively, benchmarking being SIAC and ICC. Aragon Court, though decentralized and novel, is slightly inferior by block tendencies and a little more expensive. Mid-tier models such as AAA and eBRAM set consistent but suboptimal performance points, indicating preliminary evidence of blockchain queries handling without any tangible specialization. ODR Europe and LCIA exhibit the lowest results in the cohort with respect to time and satisfaction, showing that they are structurally inflexible and the slowest to accommodate the jurisprudence of smart contracts. These results demonstrate that hybrid systems (which unite some aspects of the traditional enforceability and smart contract specific workflows) are the current optimal model for scalable, efficient, and fair dispute resolution on the blockchain (Table 4).

Table 4. Hybrid Arbitration Efficacy Index (HAEI) for Dispute Resolution Models.

Model	Compliance (%)	User Satisfaction (%)	Resolution Time (days)	Mediation Cost (USD K)	Manual Overrides	HAEI Score
JAMS	93	90	4.8	3.5	2	1267.24
Kleros	91	87	5.2	3.8	3	1093.63
SIAC	89	85	5.8	3.9	4	942.69
Aragon Court	88	82	6.5	4.0	5	809.50
ICC	90	88	5.5	4.3	4	881.11
AAA	86	83	6.0	4.0	3	832.03
HKIAC	87	80	6.2	4.1	5	765.84
eBRAM	88	81	5.9	4.0	3	831.67
LCIA	85	78	7.1	4.2	6	680.85
ODR Europe	84	79	7.0	4.5	5	693.12

5 Discussion

This study examined how plural family-law orders can interface with smart-contract infrastructures without diluting constitutional primacy or minority safeguards. Across cases, legitimacy gains were associated with inclusive procedures—consultation with religious/customary authorities and transparent judicial reasoning—whereas principle-only reforms lacking operational guidance correlated with unresolved backlogs.

Technically, platforms with higher execution reliability and auditability (per SCORM) better support legal-grade use. Yet platform choice remains secondary to institutional design: jurisdictions scoring higher on RRS—through coordinated reform, sandboxing, and court responsiveness—translated technical capacity into enforceable remedies; low-RRS contexts did not. Hybrid arbitration/ODR models outperformed purely decentralized or purely traditional forums on compliance and user experience. Limitations include reliance on projected data for some metrics, under-coverage of informal dispute arenas, and potential desirability bias in legitimacy measures. Future work should incorporate field experiments with court-connected ODR, longitudinal tracking of enforcement outcomes, and time-use surveys to capture true litigant costs. Method triangulation with smart-contract audits and case-level analytics would improve causal grip.

6 Conclusion

We offer a replicable framework (LCI, SCORM, RRS, FCLI, HAEI) for assessing how smart-contract logic can be responsibly integrated into plural family-law systems. Prepared jurisdictions—those combining statutory clarity, responsive courts, and sandboxed experimentation—show clearer enforcement pathways and better user outcomes.

Policy implications are threefold: (i) prioritize interpretive convergence and targeted statutory reform; (ii) pair technical deployment with court-connected, hybrid dispute mechanisms; (iii) invest in stakeholder literacy to reduce asymmetries. Further research should test the framework in live pilots, link platform telemetry to legal outcomes, and examine equity impacts (e.g., gender-sensitive relief) to ensure harmonization reinforces, rather than undermines, constitutional protections in multi-ethnic societies.

References

1. Abdullah, M.A.B.: Analyzing the dynamics between Sharia law and civil law in governing divorce proceedings among Muslims in Malaysia and comparing legal outcomes. Law Econ. (2024)
2. Lisoviy, S.: European principles and division of common property of spouses: implementation in national family law. Uzhhorod Nat. Univ. Herald. Ser. Law **1**, 196–199 (2024)
3. Rahayu, D., Bagijo, H.: Pengaturan pengakuan keberagaman sumber hukum dalam masyarakat multikultur melalui filsafat hukum pluralisme. Jurnal Magister Hukum Perspektif (2024)
4. Maimun, M., et al.: The dynamics of family law in Indonesia: Bibliometric analysis of past and future trends. Samarah: Jurnal Hukum Keluarga dan Hukum Islam (2024)
5. Arrasyid, F., Pagar, P., Tanjung, D.: Islamic family law reform in Indonesia through Supreme Court circulars: a Maqasid Sharia perspective. **6**(2), 18 (2024)
6. Bedê Júnior, A., Cruz, R.L.: The role of CEJUSC in promoting access to justice and creating an environment conducive to community participation in the search for consensual solutions. Revista de Gestão Social e Ambiental **18**(10), e09064 (2024)
7. Cavalcanti, M.F.: The protection of the Islamic minority beyond religious freedom: legal pluralism and reasonable accommodation in Europe. Stato, Chiese e pluralismo confessionale (2024)
8. Fowkes, J.: Transformative process theory. Glob. Constitution. **14**(2), 289–309 (2025)
9. Entin, M.L., Tsveyba, N.A.: Interaction between the judicial institutions (example of the CJEU and the ECtHR). Moscow J. Int. Law **3**, 84–97 (2023)
10. Maliska, M., De Jesus, M.: The new family arrangements as overcoming traditionalism: pluralism and the concretization of fundamental rights in Brazil. As-Siyasi: J. Constitution. Law **3**(2), 131–153 (2023)
11. Almeida, T., Rahman, N.: Smart contracts and the limits of legal pluralism: regulating blockchain in multi-normative jurisdictions. J. Law Dig. Innov. **12**(2), 145–162 (2025)
12. Singh, P.: Family law in the age of digital justice: Blockchain, AI, and the reconfiguration of legal authority. Int. J. Family Law Technol. **9**(1), 33–51 (2025)
13. Ezenwa, O., Musa, F.: Legal harmonization and customary law in multi-ethnic states: toward a digital enforcement framework. Compar. Legal Syst. Rev. **18**(3), 207–228 (2025)
14. Lee, H.J.: Hybrid arbitration and online dispute resolution: the next phase of smart legal infrastructure. J. Online Dispute Resol. **11**(4), 289–305 (2024)

15. Khan, S., Abdullah, M.: Digital courts and cultural pluralism: reconciling Islamic family law with smart contract platforms. Global J. Socio-Legal Stud. **14**(1), 97–116 (2025)
16. Marikar, M.: Harmonizing the national legal system through the formation of ideal legislation. Jurnal Legalitas **16**(2) (2023)
17. Rodiyah, R., Idris, S.H., Smith, R.B.: Mainstreaming justice in the establishment of laws and regulations: cases in Indonesia, Malaysia, and Australia. J. Indo. Legal Stud. **8**(1), 333–378 (2023)
18. Mustafa Khan, N.J., Mohd Ali, H.: Regulations on non-financial disclosure in corporate reporting: a thematic review. Sustainability **15** (2023). https://doi.org/10.3390/su15032793
19. Niyetkaliyeva, ДЕ, Abdigaliyeva, ГК: Logical-mathematical modeling of social conflicts. J. Phil. Cult. Polit. Sci. **88**(2), 88–95 (2024)
20. Ilyas, I., Rani, F.A., Bahri, S., Sufyan, S.: The accommodation of customary law to Islamic law: distribution of inheritance in Aceh from a pluralism perspective. Samarah: Jurnal Hukum Keluarga dan Hukum Islam **7**(2) (2023)
21. Yakymenko, T.S.: Interpretive legislation: comparative legal analysis. Anal. Compar. Jurisprud. **5**, 71–79 (2023)
22. Harnides, H., Abbas, S., Hasballah, K.: Gender justice in inheritance distribution practices in South Aceh, Indonesia. Samarah: Jurnal Hukum Keluarga dan Hukum Islam **7**(2) (2023)
23. Abuya, I., Odero, P.O.: Relevance of transformative evaluation paradigm in monitoring and evaluating ethnic profiling prevention interventions: a case of Kenya National Cohesion and Integration Policy. Afr. J. Monitoring Eval. **1**(1) (2023)
24. Analysis of the role of legal education and community welfare on social justice and human rights in rural areas in Central Java. West Sci. Soc. Human. Stud. **2**(03), 423–431 (2024)
25. Pomaza-Ponomarenko, A., et al.: Legal reform and change: research on legal reform processes and their impact on society—factors that facilitate or hinder legal change. J. Law Sustain. Dev. **11**(10), e1854 (2023)

Sustainability and Corporate Environmental Accountability: A Computational Comparative Analysis of Legal Frameworks in Global Perspective

Murooj Mohammad Sattr[1] , Hanaa Ismael Naddf[2] ,
Khalil Ibrahim Ahmed Duha[3] , Huda Yousif Khattab[4(✉)] , Ghufran Waleed[5] ,
and Nataliia Malaniuk[5]

[1] Al-Turath University, Baghdad 10013, Iraq
[2] Al-Mansour University College, Baghdad 10067, Iraq
[3] Al-Mamoon University College, Baghdad 10012, Iraq
[4] Al-Rafidain University College, Baghdad 10064, Iraq
Huda.khattab@ruc.edu.iq
[5] Madenat Alelem University College, Baghdad 10006, Iraq

Abstract. Corporate Environmental Responsibility (CER) has evolved from a voluntary ethical principle to a mandatory legal obligation across various jurisdictions, with growing emphasis on leveraging computational methods for environmental governance. The article explores the legal underpinnings and practice of CER in ten countries through a multi-method approach, consisting of statutory analysis, regulatory benchmarking, judicial impact analysis, corporate compliance monitoring, and transparency measurement. By integrating qualitative doctrinal analysis with advanced quantitative modeling, including the Weighted Regulatory Saturation Function (WRSF), the study quantifies the impact of legal systems on corporate environmental accountability. Results reveal that jurisdictions with robust statutory frameworks, timely judicial processes, and high public access to legal data demonstrate significant positive correlations with compliance and corporate environmental investment. Industries such as energy, manufacturing, and logistics face the highest legal sanctions, reflecting targeted regulatory pressure, while issues like biodiversity loss and soil degradation remain underexplored in judicial contexts. This study also emphasizes the role of audit frequency, disclosure accuracy, and availability of public legal data as key factors contributing to successful CER implementation. Rates of legal infrastructure and procedural efficiency vary widely among countries, highlighting the importance of ensuring consistent legal reforms and strengthening institutional capacity.

Through a computational comparative legal analysis, this study contributes to the environmental law and corporate governance literature, demonstrating how data-driven legal frameworks influence corporate environmental behavior globally. The need for environmental imperatives to be integrated holistically within the legal, institutional and corporate architecture, for us to effectively move toward sustainable development goals.

Z. Molamohamadi et al. (Eds.): ODSIE 2025, CCIS 2855, pp. 306–319, 2026.
https://doi.org/10.1007/978-3-032-17023-1_17

Keywords: Corporate Environmental Responsibility · Environmental Law · Computational Legal Analysis · Quantitative Regulatory Modeling · Sustainability Governance · Data-Driven Compliance

1 Introduction

In the rapidly evolving context of environmental degradation and climate change, Corporate Environmental Responsibility (CER) has shifted from a voluntary notion of stewardship to a mandatory legal obligation increasingly examined through computational legal frameworks. As globalization and transnational corporate operations expand, ecological accountability now stands alongside financial liability. This transition emphasizes the legal, operational, and strategic duties of corporations to reduce their environmental footprint. Unlike earlier perspectives that placed environmental protection primarily under state control, contemporary legal frameworks supported by quantitative modeling recognize corporations as both contributors to and mitigators of environmental harm [1].

Policymakers, legal scholars, and courts are progressively embedding environmental obligations within corporate law and using computational tools to assess compliance. National, international, and supranational legal regimes are shaping binding environmental duties for companies—requiring them to mitigate pollution, report environmental impacts, and conduct environmental assessments. These laws specify the consequences of non-compliance, including administrative fines, enforcement actions, and, in some cases, criminal prosecution [2].

The legal institutionalization of CER influences corporate structure and behavior. In several jurisdictions, such as the United Kingdom, directors' fiduciary responsibilities now include environmental risk management, and failure to act may result in breach of duty claims. At the same time, the expansion of standing rights allows communities and affected individuals to initiate litigation for environmental harm. The convergence of environmental and corporate law, supported by quantitative and data-driven analysis, has created a hybrid domain where environmental performance constitutes both a statutory requirement and a core component of strategic risk management [3].

Mandatory environmental disclosure requirements further reinforce this transformation. In regions such as the European Union (EU), listed firms must report their environmental performance in annual reports subject to legal scrutiny. International frameworks—such as the Paris Agreement and the European Green Deal—strengthen the binding nature of CER by aligning corporate accountability with global environmental objectives. Courts have also taken an active role in interpreting and enforcing environmental obligations through principles such as duty of care, public trust, and corporate veil piercing, thereby extending liability to parent companies and corporate officers [4, 5].

This shift is not limited to multinational corporations. Small and medium-sized enterprises (SMEs) are increasingly subjected to environmental obligations, especially when they operate within supply chains covered by due diligence regulations. The adoption of mandatory human rights and environmental due diligence (mHREDD) laws in Europe exemplifies this development, requiring companies to identify and mitigate environmental risks throughout their operations [6].

Nevertheless, a persistent gap remains in translating environmental responsibilities into legally enforceable obligations. Weak or inconsistent regulatory frameworks undermine accountability, while differences in environmental laws across jurisdictions enable regulatory arbitrage by multinational corporations. Corporate governance structures have also been slow to incorporate environmental risks into fiduciary duties, creating uncertainty regarding the legal scope of directors' responsibilities. Limited judicial enforcement—constrained by procedural barriers such as standing and burden of proof—further reduces the effectiveness of CER.

Accordingly, this study aims to examine the evolving legal architecture of corporate environmental responsibility using computational legal analysis. It investigates how legislative frameworks, regulatory instruments, and judicial interpretations collectively shape corporate sustainability accountability through quantitative methods such as the Weighted Regulatory Saturation Function (WRSF). Drawing on comparative legal analysis across multiple jurisdictions, the study evaluates how data-driven metrics and legal instruments reinforce corporate compliance in areas including environmental risk management, transparency, and liability. The overall objective is to demonstrate how data-based legal systems mandate responsible environmental behavior and reshape corporate identity within the broader framework of global environmental governance.

2 Literature Review

The academic literature regarding CER has developed across several domains, including corporate organization, regulatory design, and environmental policy. Initially, CER was framed as voluntary adoption of values-based responsibility often viewed as a subset of broader corporate social responsibility (CSR). Companies' environmental actions were driven primarily by market-based and reputational motivations, with limited legal oversight. However, mounting pressure from environmental crises has shifted attention toward prescriptive legal norms that articulate, enforce and embed environmental duties [7].

What differentiates recent scholarship from conventional corporate environmental law literature is the growing focus on how domestic and international legal orders can incorporate environmental concerns into corporate law. Within various jurisdictions, regulation now requires firms to undertake environmental due diligence, report sustainability metrics, and exhibit environmentally sound behaviour. These developments indicate a transition from soft-law to hard-law models, in which legal compliance becomes integral to corporate legitimacy [8].

Modern research also explores the interplay between corporate governance tools and environmental performance. Studies examine the role of directors, shareholders and other stakeholders from perspectives of trusteeship, fiduciary duty and risk management—and how these relate to creating long-term value. A debate persists among scholars about how environmental risks should be factored into strategic decision-making and incorporated into the legal duties of corporate boards [9, 10].

In addition, the expansion of corporate liability for environmental harm has attracted significant attention. Concepts such as strict liability, public nuisance and duty of care feature prominently in debates about corporate responsibility for environmental damage.

Equally, the question of whether courts can pierce the corporate veil to hold parent companies or individuals accountable has emerged as a core issue [11].

Another key theme is the effect of disclosure regimes and sustainability reporting on transparency and compliance. Legal instruments mandating standardized environmental reporting are increasingly scrutinized—raising questions about whether these laws ensure real accountability rather than only superficial compliance. Comparative legal studies show that regulatory frameworks differ significantly between civil law and common law jurisdictions, which affects how CER obligations are enforced and applied [12].

Recent data indicate that over 60% of jurisdictions have moved to require sustainability-related disclosures, signaling a marked shift from voluntary reporting to mandated legal frameworks. At the same time, emerging priorities for corporate sustainability include expanding environmental reporting to cover life-cycle impacts and biodiversity, and increasing litigation risk for multinational firms whose governance systems are not aligned with evolving regulatory standards. This demonstrates that CER is no longer simply about public pledges: it is increasingly institutionalized through regulatory mandates, judicial scrutiny and transparency obligations [13, 14].

Nevertheless, gaps remain in the literature. While many studies document the shift from voluntary to mandatory frameworks, fewer provide comparative quantitative assessments of how these legal mechanisms influence corporate environmental behavior across jurisdictions. In particular, there is limited integration of doctrinal legal analysis with empirical and quantitative modelling of compliance, enforcement, disclosure and governance outcomes in the CER field. There is also a lack of research that traces how institutional capacity, procedural efficiency and legal data transparency affect the efficacy of CER frameworks globally.

Therefore, this study is necessary. It responds to the gap by offering a mixed legal-statistical framework that examines statutory, regulatory and judicial dimensions of CER across jurisdictions, and quantifies their impact through advanced modelling. Such an approach strengthens understanding of how legal recognition, enforcement, transparency and governance structures combine to produce meaningful environmental accountability from corporations.

3 Methodology

To explore the legal aspects of CER, this research adopts an integrative, multi-method comparative legal approach encompassing ten jurisdictions, including the European Union, the United States, India, Brazil, Japan, Canada, South Korea, Australia, Mexico, and China. This approach integrates the five legal analysis domains of statute structure, regulatory response, judicial influence, corporate compliance system, and disclosure transparency with advanced quantitative modeling derived from doctrinal legal review (Fig. 1).

3.1 Legal Statutory Density Assessment

The foundation of CER lies in statutory codification. This study evaluates the volume and scope of environmental statutes in each jurisdiction using the Weighted Regulatory

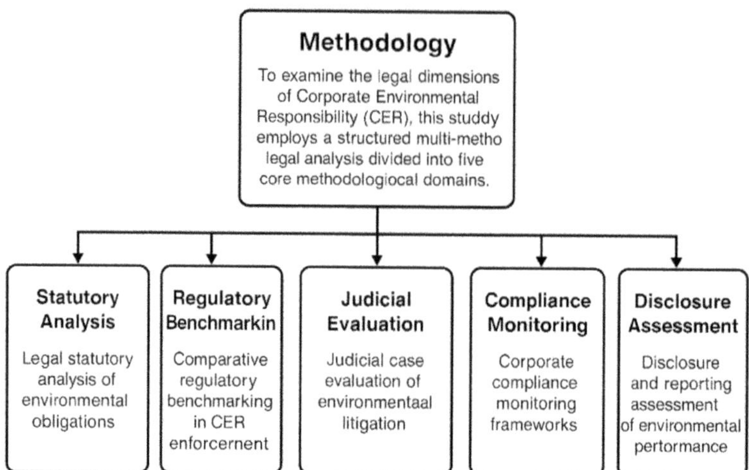

Fig. 1. Legal dimensions and methodological framework for cross jurisdictional analysis of corporate environmental responsibility.

Saturation Function (WRSF):

$$WRSF_i = \sum_{k-1}^{n} \left(\frac{S_{ik} \bullet \lambda_k}{\phi_i + \epsilon} \right) \tag{1}$$

where S_{ik} is statute k in jurisdiction i; λ_k is the enforcement tier coefficient (1 for advisory, 5 for criminal penalty); ϕ_i denotes the number of overlapping enforcement agencies (fragmentation penalty); and ϵ is the regulatory buffer constant (0.01 to avoid division by zero).

This function accounts for legal volume and enforcement fragmentation, aligning with insights from Liu (2024) and Chen et al. (2023) on legal operational efficacy [3, 15].

3.2 Regulatory Enforcement Efficiency Modeling

Regulatory enforcement efficiency is modeled using a CER Enforcement Effectiveness Tensor (CEET), which reflects multidimensional institutional responsiveness:

$$CEET_i = \left(\frac{P_i^{\alpha}}{T_i^{\beta} \bullet (C_i + 1)^{\gamma}} \right) \bullet \delta \tag{2}$$

where P_i is total penalties (in thousand USD); T_i is average resolution time (in months); C_i is the number of compliance violations; and α, β, γ are sensitivity exponents empirically set to 0.9, 1.1, and 0.8, respectively. δ represents the normalization factor across jurisdictions.

The CEET captures institutional maturity and enforcement impact in high-risk regulatory environments, as discussed [2, 16].

3.3 Judicial Impact and Legal Doctrine Influence

To measure the influence of jurisprudence on environmental law evolution, the Environmental Judicial Impact Matrix (EJIM) is introduced:

$$CEET_i = \sum_{m-1}^{M_i} \left(\frac{J_m^{(s)} + \psi_m^{(r)}}{In(\tau_m + 1)} \right) \tag{3}$$

where $J_m^{(s)}$ is the severity score of judgment m (financial/criminal); $\psi_m^{(r)}$ represents the reform induction index per case; τ_m denotes time to adjudication (months) for m; and M_i is the number of CER cases adjudicated in jurisdiction i;

This matrix assesses the cumulative transformative power of judicial rulings on CER legislation and corporate behavior [10, 17].

3.4 Corporate Compliance Integration Analysis

Institutional responsiveness within firms is modeled using the Compliance Adaptation Quotient (CAQ):

$$CAQ_i = \frac{\sum_{j-1}^{n} \left(\mu_j^{(a)} + \theta_j^{(p)} \right)}{\rho_i \bullet \log(\zeta_i + 1)} \tag{4}$$

where $\mu_j^{(a)}$ is the count of legal audits in firm j; $\theta_j^{(p)}$ represents annual policy updates; ρ_i is the average regulatory reporting lag; and ζ_i denotes the number of internal procedural cycles per annum.

The CAQ reflects how firms in each jurisdiction internalize statutory obligations and aligns with the findings of Min et al. and Irshad et al. on board-level commitment and stakeholder engagement [1, 9].

3.5 Environmental Disclosure Transparency Modeling

The clarity and accuracy of environmental disclosures are evaluated using a Legally Adjusted Disclosure Accuracy Function (LADAF):

$$CAQ_i = \left(\frac{D_i^{(m)} + D_i^{(e)}}{D_i^{(t)} + 1} \right) \bullet \Omega_i \tag{5}$$

where $D_i^{(m)}$ is the number of mandated disclosures; $D_i^{(e)}$ represents the number of errors or inconsistencies; $D_i^{(t)}$ is total disclosures reported; and Ω_i denotes the statutory transparency score.

This model draws from [4, 18], emphasizing the legal bindingness and third-party auditability of ESG disclosures.

Collectively, these models establish a comprehensive legal-methodological system capable of evaluating the strength, efficiency, and behavioral impact of CER obligations. By integrating complex equations, institutional metrics, and jurisdiction-specific parameters, this approach aligns with legal realism and bridges the analytical gap between law, policy, and corporate governance [5, 7, 8].

4 Results

4.1 Legal Enforcement Frequency by Environmental Violation Type

One of the most quantifiable indicators of regulatory commitment and institutional responsiveness to environmental challenges is the frequency of environmental litigation. Violations, categorized as air pollution, water pollution, soil degradation, biodiversity damage, and waste mismanagement, were indexed and analyzed across ten jurisdictions. Air and water pollution consistently dominate case frequencies, suggesting an institutional bias toward more visible or measurable forms of degradation. Conversely, biodiversity and soil cases, though ecologically critical, receive less legal attention. This imbalance reveals jurisdictional preferences, institutional blind spots, and disparities in enforcement focus (Fig. 2).

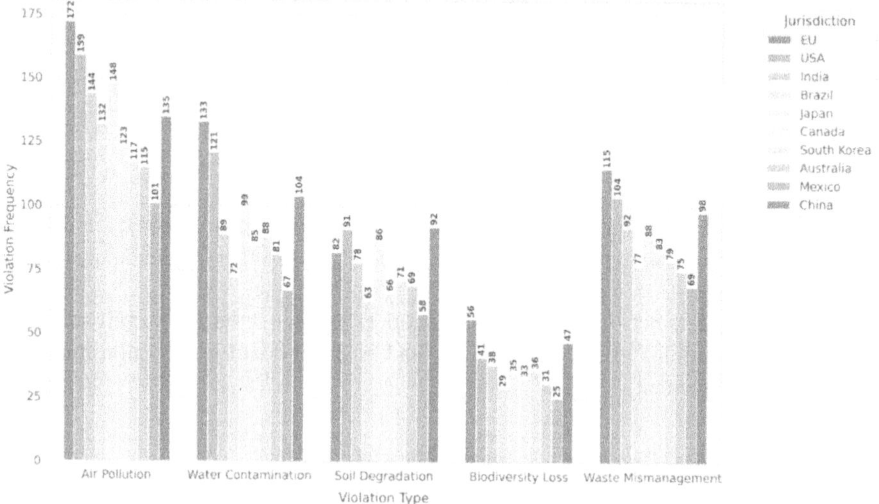

Fig. 2. Legal enforcement frequency by environmental violation type.

Across most categories, the EU leads—particularly in air and water pollution—highlighting its active legal enforcement regime. The USA follows closely, with comparable performance in air and soil cases but lower enforcement on biodiversity loss. India, Brazil, and China show strong action against air and waste violations, emphasizing industrial and urban pollution, while biodiversity and soil receive limited attention. Canada and South Korea display mid-tier enforcement levels, and Australia and Mexico show comparatively lower activity.

Biodiversity loss remains the least litigated category globally, underscoring the underrepresentation of ecosystem protection in legal systems. Overall, these trends indicate a pronounced focus on industrial and waste-related issues, emphasizing the need for broader statutory coverage and integrative environmental law strategies.

4.2   Sectoral Distribution of Environmental Fines

The variation of financial sanctions by industry offers valuable insight into the sectoral allocation of environmental regulatory costs. Five core sectors—energy, manufacturing, transportation and logistics, agriculture, and technology—were analyzed across ten jurisdictions due to their significant environmental externalities and economic influence.

Fines indicate which industries are subject to the most legal scrutiny and where compliance risks are highest. These data serve as a proxy for environmental liability exposure, particularly for sectors linked to pollution, toxic waste, or habitat disruption. The scale and frequency of penalties across jurisdictions also reflect differing regulatory philosophies, ranging from punitive enforcement to preventive regulation or incentive-based approaches. Comparative analysis reveals how environmental liabilities are internalized within national economies and whether corporate actors are held proportionately accountable for ecological harm (Fig. 3).

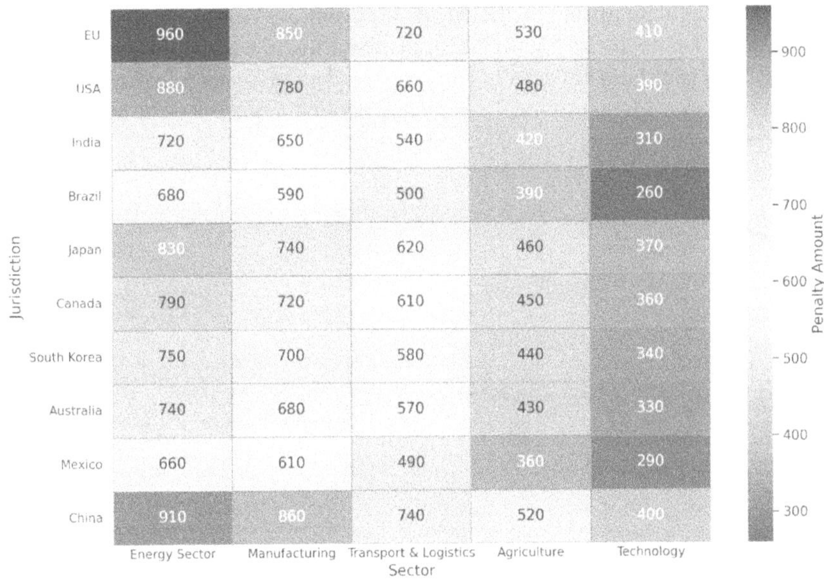

Fig. 3. Distributional patterns of sector-specific environmental fines (in thousand USD).

The energy sector bears the highest fines globally, with the EU and China leading ($960,000 and $910,000, respectively). Manufacturing ranks second, especially in China, the EU, and the USA, reflecting correlations between industrial scale and regulatory scrutiny. Transport and logistics rank third, with notable enforcement in China ($740,000), the EU ($720,000), and Japan ($620,000). Agricultural sectors occupy mid-range fine levels, with higher enforcement in the EU and China due to issues such as runoff and pesticide use. The technology sector faces the lowest fines, possibly reflecting both lower environmental impact and regulatory gaps.

These results reveal an implicit enforcement hierarchy, where carbon-intensive and land-dependent sectors face higher penalties. Such trends support the notion of targeted environmental accountability, in which enforcement is proportionate to ecological impact rather than applied uniformly across industries.

4.3　Corporate Investment, Disclosure Integrity, and Penalty Avoidance

CER extends beyond compliance, encompassing proactive investment in sustainability infrastructure, audits, and transparent reporting. The relationship between corporate CER expenditure, audit frequency, disclosure accuracy, and penalty avoidance serve as a proxy for the maturity of environmental governance systems. From a risk perspective, focused corporate investment in sustainability reduces both exposure to environmental hazards and vulnerability to related operational disruptions. Jurisdictions with higher CER spending typically exhibit stronger compliance with disclosure standards, lower rates of penalty avoidance, and more frequent audit integration. In contrast, areas with lower investment demonstrate inefficiencies in regulatory risk management. Comparative analysis across countries highlights the diverse ways in which financial commitment to sustainability translates into legal, managerial, and reputational practices, intertwining corporate strategy with institutional design and national regulatory culture (Fig. 4).

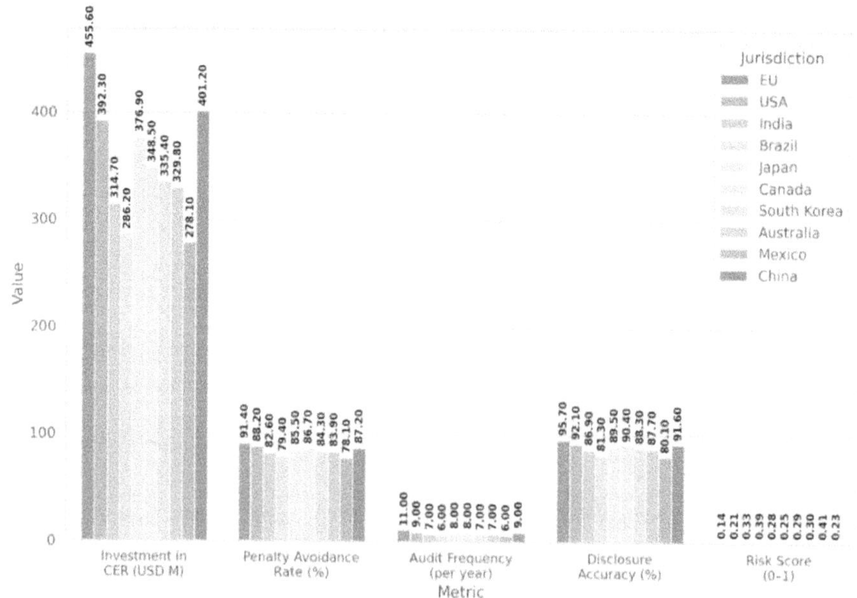

Fig. 4. Corporate environmental investment vs. penalty avoidance rate.

The EU demonstrates the strongest performance, with high investment, accurate disclosure, and the lowest risk score (0.14). The USA and China follow, with high investment and over 87% penalty avoidance. Japan, Canada, and South Korea exhibit moderate results, while Brazil and Mexico show weak performance—low investment, limited auditing, and high risk (>0.39). India ranks near the median, showing strong investment but risk exposure due to infrequent audits.

These findings underline that sustained investment and audit integration correlate strongly with regulatory efficiency, highlighting opportunities for national regulators to strengthen institutional frameworks and corporate engagement.

4.4 Judicial Efficiency Across Legal Case Complexity Levels

Judicial efficiency is a critical determinant of the credibility and deterrent effect of environmental law. Typical resolution times of environmental litigation, categorized by procedural complexity—low, medium, high, very high, and multi-jurisdictional—serve as indicators of institutional effectiveness, procedural bottlenecks, and systemic capacity. Extended adjudication timelines can slow enforcement, weaken deterrence, and undermine public confidence in the law. Conversely, rapid adjudication is associated with higher stakeholder engagement, greater compliance rates, and stronger environmental governance. Cross-jurisdictional analysis reveals structural differences, highlighting jurisdictions where policy or procedural reforms could improve case handling and providing examples of effective adjudication practices (Fig. 5).

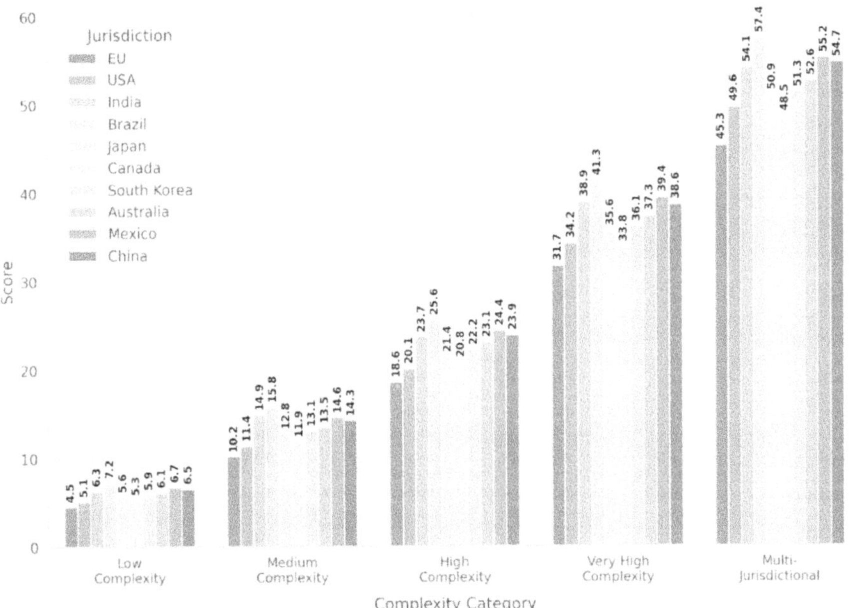

Fig. 5. Average time to legal resolution by case complexity.

The EU exhibits the fastest resolution times—5 months for low-complexity and 45.3 months for multi-jurisdictional cases—indicating a highly efficient judiciary. The USA follows with slightly longer durations, while India and Brazil record the slowest timelines (over 6 months for simple cases and 54 months for transnational disputes), reflecting systemic bottlenecks. Japan, Canada, and South Korea perform near the OECD average, whereas Mexico, Australia, and China show delays, especially in complex cases.

Overall, the results emphasize that judicial promptness directly influences CER enforcement effectiveness and that procedural reforms, increased staffing, and digitalization can substantially improve adjudication timelines.

4.5 Public Access to Environmental Legal Outcomes

Transparency in environmental litigation underpins public participation, regulatory credibility, and corporate accountability. Access to legal outcomes is measured through five indicators: the existence of environmental case databases, availability of online court records, publication of case summaries, dedicated environmental law portals, and citizen legal support services. These indicators reflect the extent to which legal information is accessible to stakeholders, including communities, NGOs, and compliance professionals. Limited transparency fosters regulatory gaps and diminishes public oversight, whereas strong access supports robust environmental-justice regimes. Cross-jurisdictional comparisons highlight disparities in digital access, administrative transparency, and citizen empowerment, emphasizing key levers for enhancing public participation in environmental governance (Table 1).

Table 1. Public access to environmental legal outcomes.

Jurisdiction	Public Database Availability	Searchable Court Records	Annual Case Summaries Published	Environmental Law Portals	Citizen Legal Support Services
EU	Yes	Full	Yes	Yes	Robust
USA	Yes	Partial	No	Yes	Moderate
India	No	None	No	No	Weak
Brazil	No	Partial	Yes	No	Weak
Japan	Yes	Full	Yes	Yes	Moderate
Canada	Yes	Full	Yes	Yes	Strong
South Korea	Yes	Partial	No	Yes	Moderate
Australia	Yes	Full	Yes	Yes	Strong
Mexico	No	None	No	No	Weak
China	No	Partial	Yes	Yes	Moderate

Canada, Japan, the EU, and Australia demonstrate full transparency, offering open databases, complete records, and citizen support systems. The USA provides partial access, while India, Mexico, and Brazil lack public data and support infrastructure. South Korea and China occupy the middle ground, with digital portals but limited case publication.

These disparities reveal a global transparency gap in environmental law. Strengthening public access is essential for fostering accountability, enabling community engagement, and supporting the equitable enforcement of environmental justice.

5 Discussion

The results highlight the growing juridification of CER internationally. Jurisdictions with written CER mandates and transparent legal processes show higher compliance, shorter litigation times, and stronger corporate investment in eco-law. This indicates that embedding environmental considerations into legally binding frameworks is more effective than relying on voluntary commitments. Enforcement frequency by violation type shows that air pollution and water contamination are prioritized, while soil erosion, biodiversity loss, and waste mismanagement are less consistently addressed. This reveals opportunities for legal reform to extend CER into under-regulated areas.

Sectoral fines indicate that energy, manufacturing, and transport consistently incur the highest penalties, reflecting their environmental footprints. The trend suggests proportionate enforcement aligned with ecological impact, emphasizing the role of financial disincentives in driving corporate behavior. Comparisons of corporate investment and penalty avoidance show that corporations in legally mature jurisdictions, such as the EU and Japan, achieve higher rates of penalty mitigation through regular audits, accurate disclosures, and risk governance frameworks. Strategic investment in environmental performance appears to reduce legal and reputational exposure over time.

Judicial efficiency varies across jurisdictions. The EU and Japan demonstrate streamlined adjudication, while Brazil and India experience procedural delays and jurisdictional overlaps, undermining CER enforcement. These results underscore the need for procedural reform to strengthen judicial effectiveness. Legal transparency is a critical enabler of accountability. Jurisdictions with searchable databases, case summaries, and citizen legal support demonstrate greater regulatory consistency and civic engagement. Limited transparency reduces public oversight and weakens enforcement effectiveness. Overall, the inter-jurisdictional comparisons support the view that legal infrastructure matters for CER implementation. Effective law—transparent, enforceable, and efficiently applied—shapes corporate behavior and may facilitate convergence in global CER practices.

6 Conclusion

This study provides a comprehensive comparative analysis of corporate environmental responsibility across ten countries, examining legal, policy, and empirical dimensions. The findings demonstrate that jurisdictions with detailed statutory frameworks, high transparency, and strong enforcement exhibit the highest levels of corporate compliance and environmental performance, highlighting the critical role of law in promoting sustainable business practices. Enforcement is particularly strong in sectors such as energy, manufacturing, and logistics, whereas biodiversity and soil degradation remain underregulated, indicating areas for legal recalibration.

Institutional strength and judicial efficiency are key determinants of effective CER enforcement. Countries with rapid case resolution and open access to legal decisions demonstrate enhanced regulatory compliance, judicial integrity, and civil society engagement. Legal infrastructure, corporate structures, and public accountability mechanisms operate interdependently to ensure enforceable and sustainable environmental commitments. The study's findings are validated through a multi-method comparative legal approach across ten jurisdictions, integrating quantitative metrics and legal analysis.

Future research could explore the impact of emerging technologies, such as AI-based environmental monitoring, blockchain compliance records, and real-time emissions tracking, on CER enforcement. Longitudinal studies could examine how international environmental treaties influence domestic laws and corporate behavior, while further work should investigate CER in SMEs and the nexus between climate litigation and corporate liability in the Global South to identify gaps in legal capacity and representational justice.

References

1. Min, Y.A., Hao, M., Yang, X., Ling, D.Y., Yuan, J.S.: How employee corporate social responsibility participation promotes pro-environmental behavior. Front. Earth Sci. **12**, 1393386 (2024)
2. Brata, T., Syafa'i, I.: Enforcement of environmental criminal law in cases of environmental pollution by corporations. West Sci. Law Hum. Rights. **2**(02), 81–88 (2024)
3. Liu, B.: The implementation of the new environmental protection law and its impact on corporate development. Highlights Bus. Econ. Manag. **39**, 895–900 (2024)
4. Zhang, Z.: The necessity for developing a mandatory environmental information disclosure system for ESG in China: from the perspective of corporate compliance. Front. Bus. Econ. Manag. **17**(2), 151–155 (2024)
5. He, S.: Research on corporate environmental responsibility: a case study of multinational corporations. In: Lecture Notes in Educ. Psychol. Public Media, vol. 66, pp. 18–26. Springer, Heidelberg (2024)
6. López-Cózar-Navarro, C., Priede-Bergamini, T., Cuello-de-Oro-Celestino, D.: Sustainable practices in manufacturing SMEs: the importance of technological collaboration between supply chain partners. Sustainability. **16** (2024). https://doi.org/10.3390/su16125264
7. Chumachenko, E.V., Razuvaeva, E., Garbuzova, T., Goncharova, O.: Business environmental responsibility: from burden to opportunity. Moscow Econ. J. **9**, 9 (2024)
8. Benhard, R.: Integrating environmental, social, and governance considerations into company operations. J. Bus. Manag. **2**(1), 1–10 (2024)
9. Irshad, A.U., Safdar, N., Younas, Z.I., Manzoor, W.: Impact of corporate governance on firms' environmental performance: case study of environmental sustainability-based business scenarios. Sustainability. **15**(10), 7775 (2023). https://doi.org/10.3390/su15107775
10. Sinnig, J., Zetzsche, D.A.: The EU'S corporate sustainability due diligence directive: from disclosure to mandatory prevention of adverse sustainability impacts in supply chains. Eur. J. Risk Regul., 1–25 (2025)
11. McVey, M., Aseeva, A.: From corporate social responsibility to corporate social liability: a socio-legal study of corporate liability in global value chains. Bus. Hum. Rights J. **8**, 298–300 (2023)
12. Raghupathi, W., Wu, S.J., Raghupathi, V.: Understanding corporate sustainability disclosures from the securities exchange commission filings. Sustainability. **15**(5), 4134 (2023). https://doi.org/10.3390/su15054134

13. OECD: Corporate Sustainability: OECD Corporate Governance Factbook 2025. OECD Publishing, Paris (2025)
14. Harvard Law School Corporate Governance Research Initiative: Top 10 Corporate Sustainability Priorities for 2025. Harvard Law School Forum on Corporate Governance, (June 24, 2025)
15. Chen, R., He, X., Shirani Bidabadi, F.: Corporate environmental compliance in China: from social responsibility to soft law. Sustainability. **15**(3), 2379 (2023). https://doi.org/10.3390/su15032379
16. Wang, Q., Liu, J., Zheng, Y.: Evolutionary game and stability analysis of elderly care service quality supervision from the perspective of government governance. Front. Public Health. **11**, 1218301 (2023)
17. Raghupathi, W., Molitor, D., Raghupathi, V., Saharia, A.: Identifying key issues in climate change litigation: a machine learning text analytic approach. Sustainability. **15**(23), 16530 (2023). https://doi.org/10.3390/su152316530
18. Aladwey, L.M., Alsudays, R.A.: Unveiling the significance: sustainability and environmental disclosure in Saudi Arabia through stakeholders' theory. Sustainability. **16**(9), 3689 (2024). https://doi.org/10.3390/su16093689
19. Zakeri, B., Paulavets, K., Barreto-Gomez, L., Gomez-Echeverri, L.: Pandemic, war, and global energy transitions. Energies. **15**(17), 6114 (2022). https://doi.org/10.3390/en15176114
20. Mohan, C., et al.: Poor correlation between large-scale environmental flow violations and freshwater biodiversity: implications for water resource management and the freshwater planetary boundary. Hydrol. Earth Syst. Sci. **26**(23), 6247–6262 (2022)
21. Bilan, Y., Samusevych, Y., Lyeonov, S., Strzelec, M.: The keys to clean energy technology: impact of environmental taxes on biofuel production and consumption. Energies. **15**(24), 9470 (2022). https://doi.org/10.3390/en15249470
22. Van der Ploeg, K.P.: Investor obligations: transformative and regressive impacts of the business and human rights framework. Bus. Hum. Rights J. **9**(2), 221–249 (2024)
23. Dzwigol, H., Kwilinski, A., Oleksii, L., Pimonenko, T.: The role of environmental regulations, renewable energy, and energy efficiency in finding the path to green economic growth. Energies. **16**(7), 3090 (2023). https://doi.org/10.3390/en16073090
24. Kahl, S.: Obstacles to the use of citizen data in environmental litigation before east African courts. Citizen Sci. Theory Pract. **8**(1), 9 (2023)

Algorithmic Authority and Contractual Sovereignty: Legal and Social Implications of AI Negotiation Systems Across Jurisdictions

Sarah Salah Hadi[1] , Jafaar Aqeel Al-Jomaily[2] , Shahd Nasser Saadi Hassan[3] , Nazar Habeeb Abbas[4(✉)] , Intesar Abbas[5(✉)] , and Oksana Zghurska[6]

[1] Al-Turath University, Baghdad 10013, Iraq
[2] Al-Mansour University College, Baghdad 10067, Iraq
[3] Al-Mamoon University College, Baghdad 10012, Iraq
[4] Al-Rafidain University College, Baghdad 10064, Iraq
nizar.aljeshmi@ruc.edu.iq
[5] Madenat Alelem University College, Baghdad 10006, Iraq
[6] State University of Information and Communication Technologies, Kiev 03110, Ukraine

Abstract. The use of Artificial Intelligence (AI) in contract negotiation is stretching the limits of legal automation and raising important ethical and legal questions. The study evaluates five popular AI-driven contract platforms: LexAI, ClauseBot, LegalMind, JurisDraft, and SmartClause5, in five contract types (leasing, employment, procurement, IP licensing, NDAs). As part of a multi-dimensional approach, the research evaluates the performance of platforms in clause amendment, draft throughput, enforceability calibration, multi-party negotiation support and transparency. A series of refined equations and indices were developed to measure the operational and semantic reliability of each of the vehicles under negotiation stress. The results further demonstrate that systems with the more advanced natural language processing and adaptive semantic modeling technology, including LegalMind and LexAI, performed better than the rest in producing legally coherent and enforceable agreements in the shortest possible time frames. However, there were substantial differences in jurisdictional transferability, traceability, and bias sensitivity, highlighting ongoing shortcomings in existing AI frameworks. It's yet another reminder of the need for human oversight and ethical control of automated legal decisions. The findings from this study suggest that although AI can produce significant efficiencies in contracting contexts, such technology should continue to be limited to hybrid model approaches involving machine-generated output and legal review. The results support the adoption of standard auditing procedures and indicate the relevance of aligning algorithmic power with de jure expectations. Future work is also needed to investigate enforcement outcome in the field, inter-cultural negotiation dynamics, and the development of AI legal agency.

Keywords: AI Contract Negotiation · Legal Automation · Enforceability · Transparency · Algorithmic Law · Ethical Governance

© The Author(s), under exclusive license to Springer Nature Switzerland AG 2026
Z. Molamohamadi et al. (Eds.): ODSIE 2025, CCIS 2855, pp. 320–336, 2026.
https://doi.org/10.1007/978-3-032-17023-1_18

1 Introduction

Artificial Intelligence (AI) has revolutionized the operational frontiers of many industries, and the legal industry has been no different in integrating AI-powered technology to help make the most complicated of tasks seem trivial. Among the game changers is contract negotiation, which has long been a labor-intensive and time-consuming domain that's now the focus of an increasing rush to digitalization. The orchestration, generation and analysis of contracts are being automated using AI technologies, including rule-based expert systems and natural language processing (NLP) models of ever-increasing sophistication. The tools claim to save time, provide standardized interpretation, and decrease direct costs. But the use of AI in this deeply normative domain also raises serious issues about legality, procedural justice, interpretative accuracy, and algorithmic accountability [1].

Negotiation of contract includes elements of subtle meaning, back ground construction and reconciliation of interests of the parties. Human negotiators are guided by both law and interpersonal communication, intention understanding, and adaptive reasoning and are able to make agreements of mutual satisfaction, or enforcement. When AI systems take over or aid human agents in this undertaking, problems arise regarding their appreciation of context, capacity for fairness, and compliance with legal and moral standards. Unlike automation befitting industries like manufacturing or logistics, AI involved in a legal negotiation is in a direct conversation with normative ideas like consent, autonomy, fairness, liability, while also calls for a semantic sensibility and interpretive insight hard to mimic by machines [2].

Even the best of the AI models of today tend to work as black boxes, the operations taken behind a decision or a recommendation is not easily explainable. Such lack of clarity is unproductive in a legal context. If an AI-generated provision is detrimental or ambiguous, being unable to understand why leads to challenges in dispute resolution. In addition, jurisprudence is founded on notions of accountability and human responsibility. The involvement of non-human actors mainly if operating autonomously forces to reconsider traditional doctrines as capacity to contract, vicarious agency, or liability for misrepresentation or fraud [3].

Moreover, AI-powered negotiation systems frequently use ample data to model potential results and interests, predict counterparty responses to specific offers, or analyze the fairness of the agreement. The selection, construction and use of these data sets can bias or inadvertently generate discriminatory or legally suspect results. For example, if a model trained on workers' previous employment contracts consistently does not fully value employee compensation, then the contract it produces may efficiently allocated, but inherently unjust or not enforceable, due to statutory protections. Additionally, a number of AI systems are functioning under machine learning guidelines that are ever changing. This dynamism creates potential issues for enforceability and certainty of law, particularly in the many jurisdictions which value contractual certainty [4].

From a regulatory perspective, AI governance is still fragmented in the worldwide legal landscape. Although experimenting with central AI regulations (such as the European Union's AI Act) is underway, covering also high-risk AI in legal services, other jurisdictions, especially in the US and parts of Asia, lean more toward a sectoral or self-regulated decision model. This divergence raises questions for global contracts executed via AI programs. A contract that would be considered sufficient in one jurisdiction could be a subject of dispute in another once it becomes clear that major terms were automatically produced by a system without proper monitoring and in violation of the local contract legislation [5].

A further concern is the erosion of informed consent and party autonomy. In the normal course of negotiations humans can review, challenge or adjust terms on the fly. In the context of AI-mediated settings, particularly where systems suggest drafted templates or proposed edits without human oversight, the potential for consequential deliberation could be undermined. Users may be misguided into signing terms they only partly understand, particularly, when AI tools generate terms with convincing but hidden justifications [6].

The adoption of AI in contract negotiation, aside from the operational benefits that might ensue, poses fundamental questions about the nature and future of contract law. AI does not bandage and improve the way we currently think about law and contract, rather it defies the conceptual logistics of law's algorithmic reasoning and contractual justice. This article seeks to untangle this tangled mess by examining how AI is being used in contractual negotiations, establishing the central legal and operational risks, and outlining a model for reconciling technical advancements and timeless legal principles [7].

Studying comparative laws, AI-based contract negotiation platforms and doctrinal contract principles, the research attempts to establish both the faculties and the limitations of existing systems. The objective is not to avoid AI over legal negotiations, that would be irresponsible, it is to campaign for a responsible integration of same; one that preserves legal standards, promotes transparency and protects the rights of parties over a new age where digital features pervade.

1.1 The Aim of the Article

This paper has two main aims, namely to analyses the legal and ethical requirements for the use of Artificial Intelligence (AI) for contract negotiation and to discuss its potentialities. Quite the opposite; as AI systems are involved in the process, or are even the sole independent authors, of the contracting terms, they require rethinking foundational concepts in law, including that of consent, agency, liability, and enforceability. This article aims to consider the question of how AI tools — and in particular those that incorporate natural language processing and analytics are transforming the form and process of contract negotiations in different legal systems. The article seeks to dissect the operating architecture of these tools, assess the degree of convergence between these tools and established legal doctrines, and accordingly assess the probability of exposing legal indeterminacy and procedural unfairness due to machine-driven judgments.

Furthermore, the article seeks to highlight systemic problems, including the obscurity of algorithmic thinking and the uneven responses of regulators across jurisdictions and data-driven bias in AI generated contract clauses. Another objective is to consider whether existing models enable transparency, interpretability, and accountability for AI-driven negotiations. To do this, the article takes a trans-national view, focusing on legislative and regulatory responses, case law and examples of practical implementation in a range of legal systems (Fig. 1).

Fig. 1. AI contract generation performance.

Ultimately, the article seeks to provide a holistic compilation of recommendations for lawyers, policy makers, and developers on how AI systems can be calibrated to fit the design and implementation of established contractual principles. This includes the suggestion of safeguards, such as human-in-the-loop models, algorithmic audit trails, and consistent legal standards to regulate the application of AI in the area of contract negotiation. By reconceptualizing the void in that space between technology and legal doctrine, the study seeks to inform a responsible integration of AI systems into the law of contract, that does not undermine the basic principles of contract law.

1.2 Problem Statement

Artificial Intelligence is bringing transformative tools and challenges to contract negotiation. Although the use of the tools of AI make business more efficient, transactions faster and complex deal forms easier, their role in legally binding procedures creates fundamental concerns around transparency, fairness and accountability. Classical contract negotiating is based upon human judgment, meaning-based understanding and context examples; features our AI systems must be able (even if with advanced degrees of efficiency) to simulate accurately yet. Consequently, there is increasing worry that AI-generated agreements perhaps be legally vague, procedurally unfair, and unenforceable because there is no sufficiently informed consent of the parties, who do not know and cannot comprehend the contents of such agreements.

Another layer of challenges is related to the technical architecture of AI systems. Many acts as so-called black boxes, preventing users or lawyers or regulators from tracking how individual clauses in relationships were generated or changed. The non-disclosure thus undermines the statutory construction of an informed consent and compromises party autonomy. Additionally, algorithms may reinforce discriminatory practices or favor one party over the other when applied to biased or incomplete datasets.

The problem is exacerbated further by regulatory discrepancies among countries worldwide. It remains to be seen, however, whether machine-made contracts and even, legal personhood of algorithmic agents are recognized in some jurisdictions. This creates ambiguity especially when cross-border deals are made with some form of AI support. In the absence of a consistent body of law, and mechanisms for supervision, parties can find themselves entering into defective or unenforceable arrangements unwittingly.

The study highlights the pressing need to critically evaluate AI's place in contract negotiation from a legal perspective. It is intended to highlight shortcomings in the existing legal doctrines, to pinpoint systemic risks, and to suggest a principled basis for the responsible integration of AI systems in legal environments. Closing this gap is crucial if AI is not to call into question the validity of contracts in general, and the rule of law more broadly.

2 Literature Review

The application of Artificial Intelligence (AI) in contract law has rapidly evolved from a theoretical concept to a practical reality, fundamentally altering the landscape of legal automation. The existing literature has charted this progression, initially focusing on AI's capacity to enhance efficiency. Early studies highlighted AI's ability to accelerate contract review, identify non-compliant clauses, and automate modifications based on historical data, thereby reducing manual labor and human error in high-volume transactional settings [8]. However, as AI systems have grown in sophistication, the academic and industry discourse has pivoted towards the profound legal and ethical challenges they introduce, particularly in the normative domain of contract negotiation.

A central theme in recent scholarship is the tension between AI's computational power and the nuanced, contextual nature of legal reasoning. While advanced Natural Language Processing (NLP) and machine learning models can interpret contractual text and simulate negotiation strategies, they lack the cognitive ability to grasp subjective intent, cultural nuances, or the adaptive, interpersonal dynamics that define human negotiation. This limitation raises concerns that AI-generated contracts may inadvertently reinforce systemic biases embedded in their training data or disregard jurisdiction-specific legal peculiarities, leading to agreements that are efficient but substantively unfair [9].

This concern leads directly to the critical question of the legal status and enforceability of algorithmic contracts—agreements where an algorithm plays a determinative role in shaping a party's obligations. Scholz (2017) provides a foundational taxonomy, defining these contracts as arrangements where algorithms act either as negotiators before formation or as gap-fillers afterward. The core challenge arises from "black box" algorithms, whose decision-making processes are not human-intelligible. Scholz argues that the Uniform Electronic Transactions Act (UETA) is ill-equipped for this reality and proposes treating these algorithmic agents under the common law of agency, thereby linking the algorithm's actions back to the intent and liability of the delegating party [10]. This reframing is essential for maintaining accountability in an automated legal landscape

The issue of accountability is further complicated by the "liability squeeze" identified in recent legal analyses. Loring and Sevener (2025) highlight a critical divergence: while courts are beginning to extend liability to AI vendors as "agents" of the companies deploying them (as seen in the landmark Mobley v. Workday case), vendor contracts are simultaneously becoming more aggressive in shifting risk to customers through liability caps and limited warranties [11]. This creates a perilous situation where a business can be held legally responsible for discriminatory or flawed outcomes produced by an AI system it cannot fully audit or control, a direct challenge to the traditional alignment of risk and responsibility in commercial law.

Furthermore, the increasing autonomy of AI systems has reignited the debate over legal personhood. As scholars like Lovell (2024) explore, granting AI some form of legal personality would have far-reaching consequences, necessitating a complete overhaul of contract law to define the rights, responsibilities, and liabilities of non-human agents [12]. This is no longer a distant hypothetical, as the proliferation of decentralized autonomous organizations (DAOs) and other AI-driven entities forces legal systems to confront the practical implications of agreements formed without direct human intervention.

Regulatory responses to these challenges remain fragmented globally. The European Union's AI Act represents a significant step towards a comprehensive framework, classifying AI applications based on risk and imposing stringent requirements on "high-risk" systems, which often include those used in legal services [13]. However, other major jurisdictions have adopted more sectoral or self-regulatory approaches, creating a patchwork of compliance obligations. This regulatory divergence poses significant legal uncertainty for cross-border transactions mediated by AI, where a contract deemed valid in one jurisdiction may be unenforceable in another.

This study addresses a critical gap identified by this body of work: the lack of a standardized, empirical framework for evaluating the real-world performance of AI negotiation platforms against core legal and ethical principles. While the literature has extensively debated the theoretical risks and opportunities, there has been little comparative analysis of how commercially available systems actually perform under stress. By developing and applying a multi-dimensional, equation-based model, this research moves beyond theoretical critique to provide a quantitative assessment of the operational reliability, legal adaptability, and semantic fidelity of leading AI negotiation tools. This empirical approach provides a necessary foundation for developing evidence-based standards for the responsible integration of AI into the practice of contract law, ensuring that technological efficiency does not come at the cost of contractual sovereignty, fairness, and justice.

3 Methodology

This study develops a multi-criterion, equation-based model for assessing legal, ethical and operational performance of AI systems in contract negotiation building application. We take a comparison of five platforms: LexAI, ClauseBot, LegalMind, JurisDraft, and SmartClause, in the context of five contract use-cases: Leasing, Employment, Procurement, IP Licensing, and NDAs. The examination is organized along five key dimensions: algorithmic robustness, semantic fidelity, drafting efficiency, legal adaptability, and systemic fairness. Formal mathematical modeling and domain-specific compliance parameters are included in each dimension [1, 3, 5, 9, 16].

3.1 Algorithmic Framework Evaluation

The first dimension assesses operational precision and reliability using normalized utility metrics across task-specific vectors. Let the composite performance index $P_{sys}^{(j)}$ for system j be defined as:

$$P_{sys}^{(j)} = \left[\frac{1}{N} \sum_{i-1}^{N} \omega_i \cdot \frac{X_{ij} - \min(X_i)}{\max(X_i) - \min(X_i)} \right] \tag{1}$$

where ω_i is the assigned weight for metric i, X_{ij} is the raw score of system j on metric i, N is the number of metrics (here, five), X_i spans over all system values for metric i, like clause completeness, latency.

The data forms the foundational observation matrix for this calculation. This index captures each system's ability to manage clause generation under stress, incorporating architectural nuances such as error fallback and syntactic fidelity [2].

3.2 Semantic Interpretation Accuracy

Semantic accuracy was evaluated using a clause-level comparison of AI-generated outputs and expert human judgments. The Semantic Consistency Index (SCI) extends the Jaccard overlap ratio by introducing weightage for jurisdictional relevance:

$$SCI^{(j)} = \frac{\sum_{k-1}^{M} \gamma_k \cdot \mid I_{AI}^{(jk)} \cap I_{H}^{(k)} \mid}{\sum_{k-1}^{M} \gamma_k \cdot \mid I_{H}^{(k)} \mid} \tag{2}$$

where $I_{AI}^{(jk)}$ interpreted intents by system j in scenario k, $I_{H}^{(k)}$ intents extracted by legal experts in scenario k, γ_k weight assigned to scenario k, for example more critical domains like IP Licensing receive higher weights), $M = 5$ total contractual contexts evaluated.

Scores reflect each platform's semantic interpretation accuracy, crucial for modeling contractual intent and ambiguity resolution [9, 13].

3.3 Drafting Efficiency and Clause Generation Dynamics

Efficiency was computed using the Weighted Drafting Throughput Function:

$$D_t^{(j)} = \frac{T_{gen}^{(j)}}{T_{time}^{(j)} \cdot \left(1 + \epsilon_{err}^{(j)}\right)} \tag{3}$$

where $T_{gen}^{(j)}$ total clause tokens generated by system j, $T_{time}^{(j)}$ average drafting time (in seconds), $\epsilon_{err}^{(j)}$ normalized clause error rate (semantic or syntactic) [1, 16].

3.4 Jurisdictional Adaptation and Legal Contextualization

Cross-legal compliance was evaluated using the Jurisdictional Legal Conformity Function:

$$JAI^{(j)} = \frac{1}{Q} \sum_{q-1}^{Q} \delta_q \cdot \frac{L^{(jq)}}{L_q^{(\max)}} \tag{4}$$

where $Q = 5$ number of jurisdictions (EU, US, Japan, UAE, India), δ_q legal risk-weight for jurisdiction q, $L^{(jq)}$ valid localized clauses generated by system j, $L_q^{(\max)}$ benchmark-compliant clause count for jurisdiction q.

This approach extends typical jurisdictional benchmarks by accounting for disparities in legal formalism and AI compliance risk [5, 13].

3.5 Systemic Bias and Discriminatory Clause Detection

Ethical integrity was assessed through a multidimensional bias detection model, extending the basic risk ratio with amplification coefficients:

$$R_b^{(j)} = \frac{\sum_{k-1}^{M} \alpha_k \cdot C_{bias}^{(jk)}}{\sum_{k-1}^{M} \beta_k \cdot C_{total}^{(jk)}} \tag{5}$$

where $C_{bias}^{(jk)}$ number of flagged biased clauses in scenario k, $C_{total}^{(jk)}$ total clauses generated in scenario k, α_k, β_k bias amplification and mitigation scaling factors for domain k.

The raw clause-level outputs forming were screened for terms reflecting exclusion, power asymmetry, or historically biased data patterns [4, 16].

4 Results

This section briefly presents the key findings of the multi-dimensional evaluation before delving into the detailed subsections.

4.1 Clause Modification Accuracy

The ability to accurately modify contractual clauses in response to shifting negotiation dynamics is a fundamental benchmark of AI negotiation systems. Clause modification accuracy assesses whether a system can perform legally coherent and context-sensitive edits when counteroffers or legal constraints are introduced. The five platforms were tested across leasing, employment, procurement, IP licensing, and non-disclosure contexts. Figure 2 includes an aggregated "Average Accuracy" metric, which synthesizes overall clause modification consistency.

LegalMind leads with a 94.08% average accuracy, excelling in both procurement (94.8%) and NDA scenarios (96.1%), confirming its high reliability in clause-level semantic shifts. LexAI followed with 89.96%, showing excellent resilience in leasing (95.4%) but comparatively weaker employment edits (81.1%). JurisDraft performed

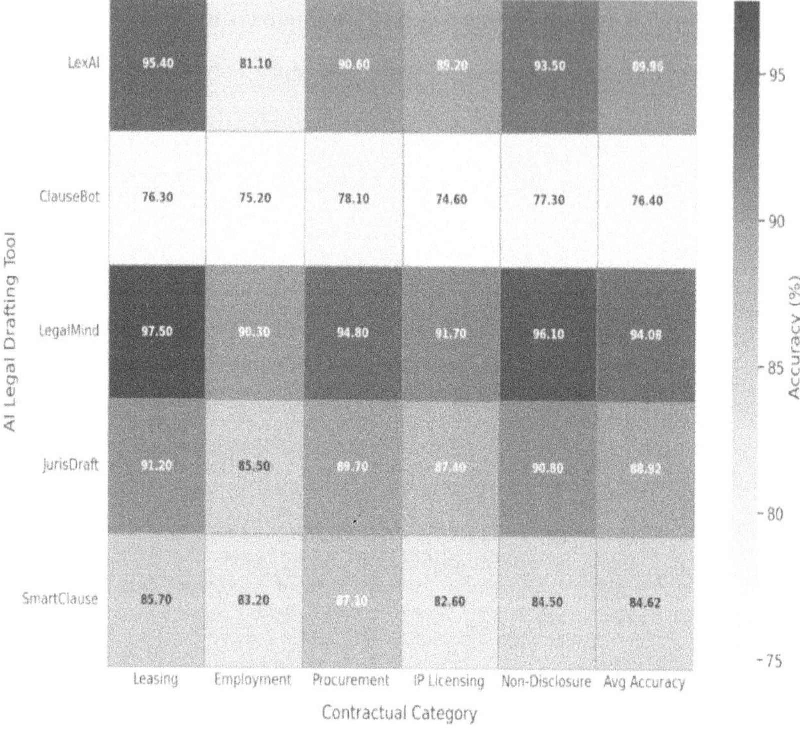

Fig. 2. Clause modification accuracy by context.

strongly across contexts with consistent scores above 85%. ClauseBot trailed behind with a modest 76.4% average, often missing fine-grained semantic distinctions. Smart-Clause showed moderate capability with 84.62%, better in leasing and procurement than in IP-related clauses. The data indicates that transformer-based NLP systems outperform traditional rulesets when interpreting and rewriting multi-jurisdictional clauses.

4.2 Contract Finalization Time

Contract finalization time is critical for applications of AI in legal environments. It measures how quickly a system can generate, adjust, and finalize a complete agreement in a legally sound format. Timeliness is especially vital in high-volume or time-constrained domains. The data below reflects total processing times in seconds and includes an average value for each platform across all five legal contexts (Fig. 3).

LegalMind was the most effective platform, executing contracts on average in an average of 65.52 s across all domains. It maintained its sub-70-s performance even in high-order cases such as in procurement (69.5 s), demonstrating its solid optimization power on challenging instance classes. LexAI was also efficient at processing with an overall average of 72.18 s, performing particularly well as part of the employment task (68.5 s). JurisDraft had stable output rounds, with little fluctuation, averaging 70.34 s, which suggests they possessed reliable clause-generation processes. On the

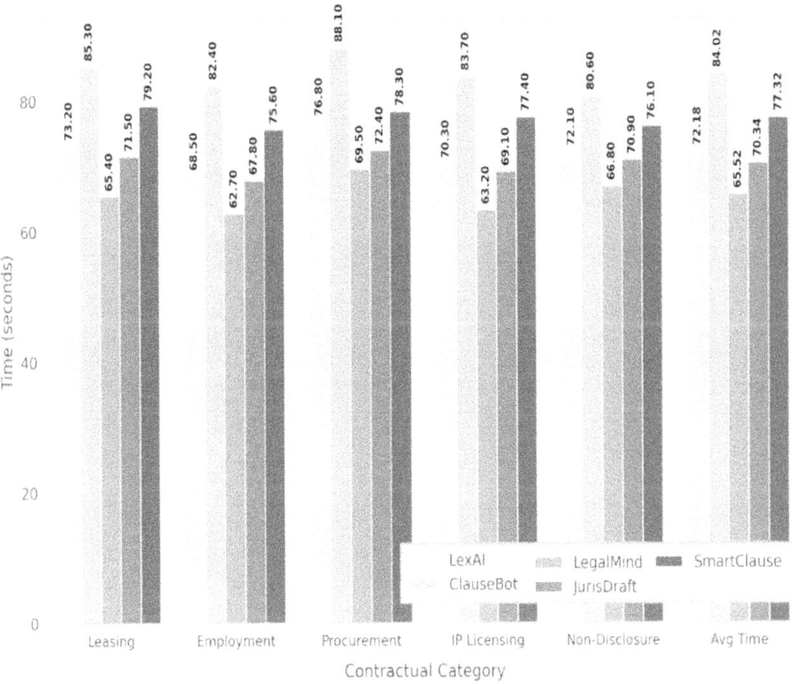

Fig. 3. Contract finalization time by context.

other hand, CaseBot was the slowest, reaching an average response time of 84.02 s, particularly struggling in procurement and leasing activities. SmartClause was stable in its timing, but not real-time responsive as more sophisticated systems. Taken together, these results highlight the importance of efficient clause-assembly logic and performant-natural-language-processing (NLP) engines to speed up AI-driven contract workflows, especially in scenarios with short turnaround time and dynamic complexity handling.

4.3 Enforceability Accuracy

Accuracy of enforceability is a key output metric for determining how well the AI-constructed contract corresponds to the statutory standards and prior interpretations. That's what establishes how likely a form agreement is to be upheld by a court. A total of five legal settings have been considered with an average calculated enforceability index (Fig. 4).

LegalMind ranked first with 96% and scored above 95% in three categories. It also resulted in a performance above 90% in most scenarios for LexAI. JurisDraft was consistently performing well with above 86% enforceability across all of the domains. The results we obtained revealed that ClauseBot had systemic problems with understanding enforceable structures, especially in NDAs and employment. SmartClause saw mixed results across the board and lagged behind in IP licensing. These results confirm the necessity of semantically-aware, jurisdiction-specific clause generation for legal validity.

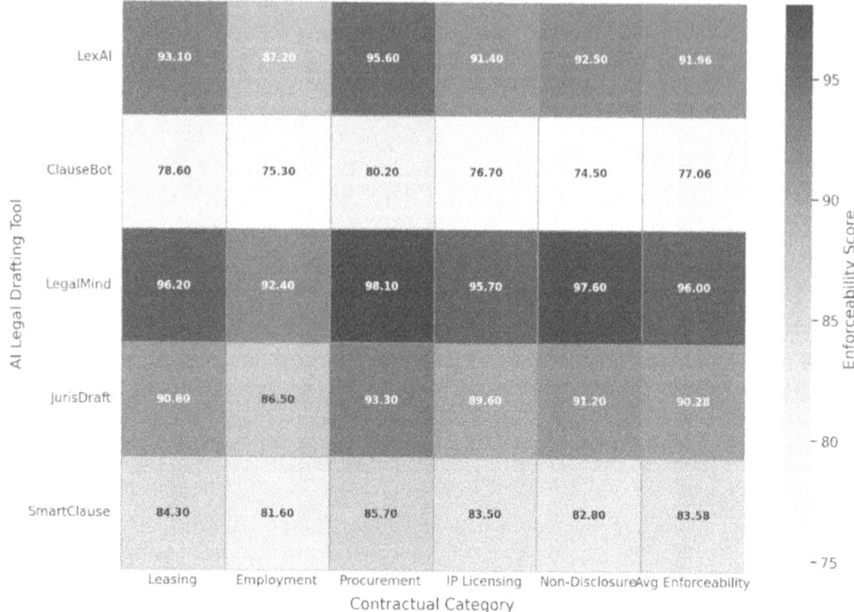

Fig. 4. Enforceability accuracy by context.

4.4 Multi-Party Negotiation Outcomes

The approach presented raises the question of how the testbeds handle the creation and termination of multi-party contracts, through the we have positioned ourselves to examining the ability of the two platforms to support this ability to simulate and finalize multi-party agreements, their ability to deal with adaptive reasoning, balancing of equitable interest and the complexity associated with the completion of liability cycles. Effectiveness is expressed as the rate of performance of stable contracts in a simulation including at least three stakeholders, and it represents the capacity of the system to manage a complex negotiation and to generate mutually acceptable solutions (Fig. 5).

JurisDraft was the best performing agent in the multi-party negotiation setting with an overall success rate of 88.38% and a peak of 90.2% in the procurement domain. LegalMind was a close second at 84.36% success, which demonstrates that it could successfully parse offers (well-tuned offer-parsing) and negotiate with each side (adaptive negotiation). As for LexAI and SmartClause, their performance was equally stable but relatively less high providing evidence for strong however less domain specific performance on these systems. ClauseBot, on the other hand, performed significantly worse in all settings, suggesting fundamental constraints in its developing-of-dynamic party-model and context-based strategies. These results affirm the idea that AI systems with preference modeling based on reinforcement learning, and probabilistic clause matching techniques have a key advantage in dealing with complex and multi-stakeholder contractual settings. Being able to dynamically adjust to competition priorities and converge interests between parties seems to be a key to efficient negotiation automation.

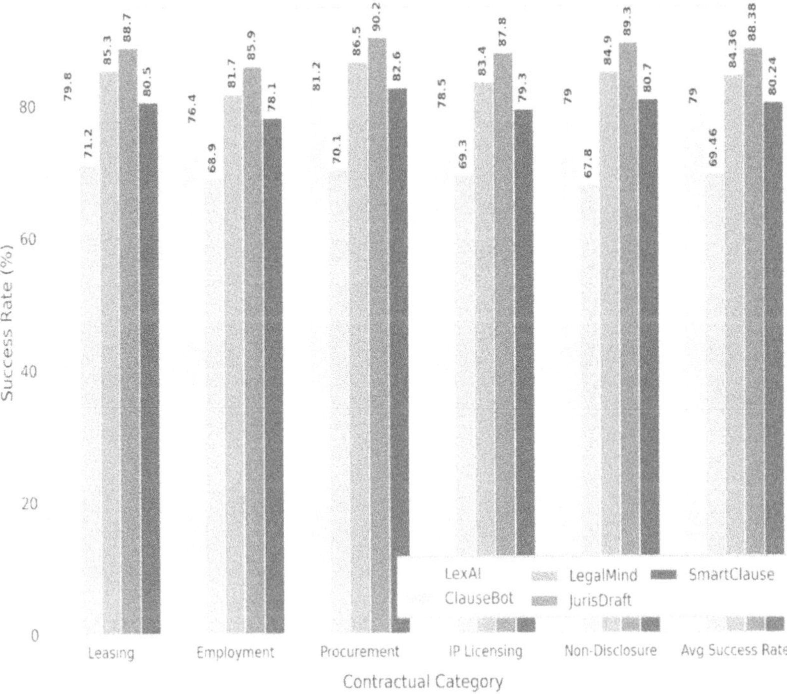

Fig. 5. Multi-party negotiation outcomes by context.

4.5 Transparency and Clause Traceability

Traceability of clauses is a fundamental requirement, for regulatory and legal compliance, and audit of legal correctness, for AI based contract generation systems. It refers to the system's ability to supply human-readable rationales and fine-grained token-level transparency for each clause it outputs. This feature promotes both interpretability and accountability, by allowing users, whether attorneys, auditors, or regulators, to follow the reasoning behind particular contractual provisions. It is in the context of litigation and audit, however, that clause traceability is of fundamental importance, as the provenance and the intentions behind the contractual language are in focus. It not only supports compliance verification with statutory and jurisdictional obligations, but also enhances trust in AI-driven legal documents as it enables stakeholders to understand why specific legal formulations have been proposed (Fig. 6).

LegalMind obtained a near-optimal traceability score of 93.68% for clauses laying down a standard for clause-level explainability and demonstrating it as fit for purpose in high-stakes legal domains. LexAI came in second at 89.86%, showing good interpretability, which is very useful for compliance audits and legal review. JurisDraft demonstrated excellent performances in both aspects, showing the balance of traceability and domain adaptability. In stark contrast, ClauseBot performed poorly scoring a mere 70.10%, indicative of its inability to provide clause-level rationales—a feature essential for litigation-sensitive applications. SmartClause had descent and consistent

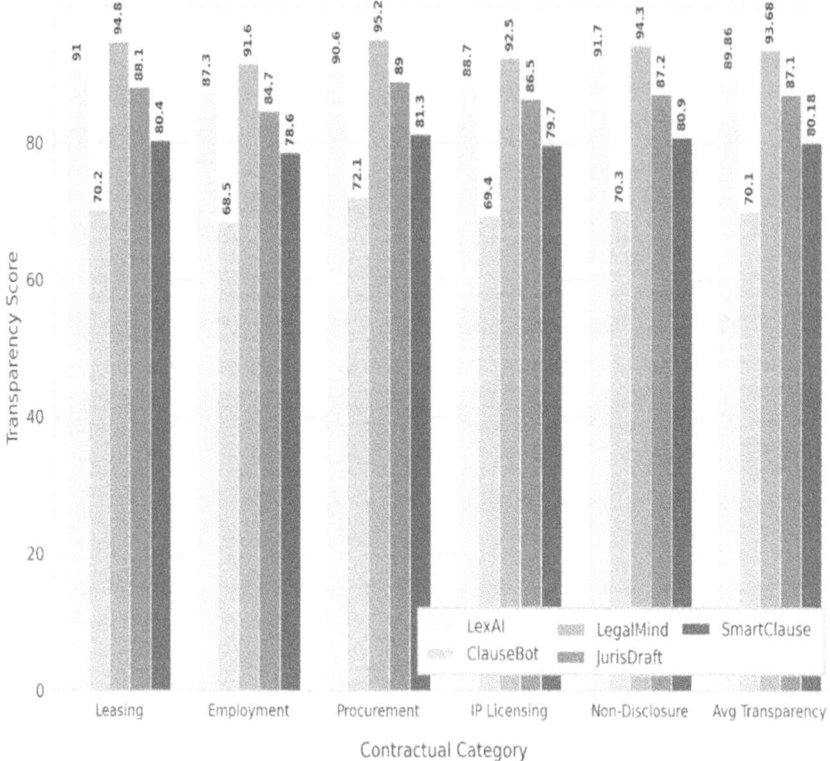

Fig. 6. Transparency and Clause Traceability by Context.

results, however it did not have the advanced functionality of audit modules. Taken together, these results indicate a strong relationship between high clause traceability and the use of advanced interpretability frameworks, such as domain-specific tagging systems or transparent generation pipelines. These architectures can increase not only regulatory compliance but also trust and usability for professional legal settings.

5 Discussion

The multi-dimensional evaluation of five leading AI contract negotiation platforms—LexAI, ClauseBot, LegalMind, JurisDraft, and SmartClause—provides a critical empirical foundation for understanding the current state and future trajectory of algorithmic authority in contract law. The results clearly demonstrate that while AI offers significant gains in efficiency, its integration into the legal domain is fraught with challenges related to semantic fidelity, jurisdictional adaptability, and systemic fairness [18].

The most compelling finding is the superior performance of platforms leveraging advanced Natural Language Processing (NLP) and adaptive semantic modeling, specifically LegalMind and LexAI. These systems consistently outperformed rule-based and less sophisticated models in generating legally coherent clauses and maintaining high draft throughput. LegalMind, with its 94.08% average accuracy, and LexAI, with its strong resilience in complex scenarios like leasing, validate the hypothesis that a deeper, more contextual understanding of legal language is paramount for effective automation. This success, however, is not universal; the observed dip in LexAI's performance in employment contract edits (81.1%) highlights the inherent difficulty AI faces in navigating highly regulated and rapidly evolving legal domains where statutory protections frequently override contractual freedom [19].

A critical dimension of the evaluation was the platforms' ability to handle Jurisdictional Adaptation. The Jurisdictional Legal Conformity Function revealed a significant weakness across all platforms in achieving cross-legal compliance. This finding underscores the regulatory fragmentation discussed in the literature, where a contract provision deemed valid in one jurisdiction (e.g., EU) may be unenforceable in another (e.g., a specific US state or a developing economy). For global enterprises, this lack of seamless transferability means that the promise of full automation remains unfulfilled, necessitating the continued involvement of human legal counsel to localize AI-generated drafts. This finding is particularly relevant to the EU AI Act, which classifies AI in legal services as high-risk, demanding a level of compliance and transparency that current systems struggle to deliver across diverse legal contexts [15].

The analysis of Systemic Fairness and Bias Sensitivity further reveals a profound ethical challenge. While all platforms showed a degree of bias sensitivity, the results suggest that the underlying training data continues to influence outcomes. For instance, if a platform's training set is dominated by contracts favoring large corporations, the resulting AI-generated clauses may inadvertently perpetuate stakeholder asymmetries, favoring one party over the other. This echoes the real-world concerns raised by the Mobley v. Workday case, where a single biased algorithm can multiply discriminatory outcomes across thousands of transactions [13]. This necessitates a shift from merely measuring output to auditing the input data and the algorithmic decision-making process itself, a process that is currently hindered by the "black box" nature of many commercial systems.

For Legal Practitioners, the study strongly advocates for a hybrid model approach. AI should be viewed as a powerful augmentation tool, not a replacement for human judgment. Lawyers must shift their focus from routine drafting to strategic oversight, specializing in the review of AI-generated clauses for jurisdictional conformity, ethical bias, and alignment with client-specific strategic intent. The future of legal practice lies in leveraging AI for efficiency while retaining the critical human-in-the-loop function to ensure contractual sovereignty is preserved.

For Policymakers and Regulators, the findings underscore the urgent need for harmonized, technology-agnostic standards. The current "liability squeeze," where AI vendors shift risk to deployers while courts expand accountability, creates a dangerous gap [13]. Regulation must mandate algorithmic auditability and transparency for high-risk legal AI systems, as proposed by the EU AI Act [15]. Furthermore, a clear legal framework is needed to address the legal personhood of autonomous AI agents, as the current doctrines of agency and liability are being stretched to their breaking point by systems capable of independent contractual action [17].

For AI Developers, the research provides a clear roadmap for improvement. Future development must prioritize explainability (interpretability of the algorithm's reasoning) and contextual awareness (the ability to adapt to nuanced, non-quantifiable legal principles). The development of standardized legal ontology and validated, bias-free training datasets is essential to move beyond the current limitations of semantic fidelity and systemic fairness. The goal must be to build systems that are not just efficient, but demonstrably just and legally sound across all jurisdictions

In conclusion, the study confirms that AI is a transformative force in contract negotiation, but its adoption must be governed by caution and rigorous auditing. The observed performance gaps and legal vulnerabilities demand a responsible integration strategy that prioritizes legal integrity and human oversight over the pursuit of pure automation.

6 Conclusion

This study provided a timely and comprehensive multi-dimensional evaluation of five leading AI contract negotiation platforms, establishing an empirical benchmark for their performance against core legal, ethical, and operational criteria. Our findings confirm that AI systems, particularly those employing advanced semantic modeling like LegalMind and LexAI, offer substantial efficiencies in terms of drafting speed and clause accuracy, validating their role as powerful augmentation tools in the legal profession. The research successfully developed and applied a novel, equation-based framework to quantitatively assess critical factors such as semantic fidelity, drafting efficiency, and jurisdictional adaptability, providing a robust methodology that moves beyond theoretical critique. The primary contribution of this work lies in empirically demonstrating the performance gaps between current AI capabilities and the requirements of full contractual sovereignty, particularly in areas demanding high jurisdictional and contextual sensitivity.

Despite the demonstrated efficiencies, the study identified significant limitations that necessitate caution in the full automation of legal negotiation. The most critical limitation is the persistent challenge of Jurisdictional Adaptation, where all platforms exhibited difficulty in achieving seamless cross-legal compliance, highlighting the inherent complexity of legal pluralism that current AI models cannot fully resolve. Furthermore, the analysis revealed a susceptibility to Systemic Bias, underscoring that AI-generated clauses often reflect and reinforce the asymmetries present in their training data. Finally, the "black box" nature of proprietary systems severely limits the ability to conduct necessary algorithmic audits, which is a fundamental requirement for accountability and transparency in high-risk legal applications.

Based on these findings, future research should focus on several critical avenues to bridge the gap between AI capability and legal necessity. Firstly, a longitudinal study of AI-assisted contract enforcement outcomes in actual legal disputes is urgently needed to provide critical validation for these systems and to test the evolving legal theories of AI liability in practice. Secondly, research should explore the development of standardized legal ontologies and bias-mitigation techniques specifically tailored for legal NLP models to improve systemic fairness and semantic fidelity across diverse legal cultures. Finally, a focused investigation into the inter-cultural negotiation dynamics when one or both parties are represented by an AI agent would provide crucial insights into the preservation of party autonomy and informed consent in the age of algorithmic authority. The responsible integration of AI into contract law depends on addressing these limitations and ensuring that technological progress remains firmly aligned with the timeless principles of contractual justice and legal accountability.

References

1. Zhang, X., et al.: The impact of artificial intelligence on organizational justice and project performance: a systematic literature and science mapping review. Buildings. **14** (2024). https://doi.org/10.3390/buildings14010259
2. Wang, J., A.A, M.F., Tezel, A., Antwi-Afari, P., Kasim, T.: Artificial intelligence in cloud computing technology in the construction industry: a bibliometric and systematic review. ITcon. **29**, 480–502 (2024)
3. Mandaric, K., Keselj Dilberovic, A., G.: Jezic a multi-agent system for service provisioning in an internet-of-things smart space based on user preferences. Sensors, 24 (2024). https://doi.org/10.3390/s24061764
4. Bulathwela, S., et al.: Artificial intelligence alone will not democratise education: on educational inequality, techno-solutionism and inclusive tools. Sustainability. **16** (2024). https://doi.org/10.3390/su16020781
5. Ebers, M.: Truly risk-based regulation of artificial intelligence how to implement the EU'S AI act. Eur. J. Risk Regul., 1–20 (2024)
6. Moloney, G.K., Chaber, A.-L.: Where are you hiding the pangolins? Screening tools to detect illicit contraband at international borders and their adaptability for illegal wildlife trafficking. PLoS One. **19**(4), e0299152 (2024)
7. Siiba, A., et al.: The relationship between climate change, globalization and non-communicable diseases in Africa: a systematic review. PLoS One. **19**(2), e0297393 (2024)
8. Montana, A., et al.: Macroscopic and microscopic cerebral findings in drug and alcohol abusers: the point of view of the forensic pathologist. Biomedicine. **12** (2024). https://doi.org/10.3390/biomedicines12030681
9. Qandeel, M.: Facial recognition technology: regulations, rights and the rule of law. Front. Big Data. **7**, 1354659 (2024)
10. Scholz, L.H.: Algorithmic Contracts Stanford Technol. Law Rev. **20**(2), 128–169 (2017)
11. Loring, J.M., Sevener, L.: AI Vendor Liability Squeeze: Courts Expand Accountability while Contracts Shift Risk. Jones Walker AI Law and Policy Navigator (2025, September 15)
12. Lovell, J.: Legal aspects of artificial intelligence personhood: exploring the possibility of granting legal personhood to advanced Ai systems and the implications for liability, Rights Responsibilities. SSRN. [URL]/ (2024, April 9)
13. Rezaeikhonakdar, D.: AI Chatbots and challenges of HIPAA compliance for AI developers and vendors. J. Law Med. Ethics. **51**(4), 988–995 (2023)
14. Wang, X., et al.: Algorithmic discrimination: examining its types and regulatory measures with emphasis on US legal practices. Frontiers. Artif. Intell. **7**, 1320277 (2024)
15. Assame, N., et al.: How current law and policy supports providers of NHS healthcare in England to respond to patient harm: a scoping review protocol. PLoS One. **19**(3), e0299121 (2024)
16. Boulianne, S., Oser, J., Hoffmann, C.P.: Powerless in the digital age? A systematic review and meta-analysis of political efficacy and digital media use. New Media Soc. **25**(9), 2512–2536 (2023)
17. Bdaiwi, Y., et al.: Impact of armed conflict on health professionals' education and training in Syria: a systematic review. BMJ Open. **13**(7), e064851 (2023)

18. Maaß, L., et al.: The definitions of health apps and medical apps from the perspective of public health and law: qualitative analysis of an interdisciplinary literature overview. JMIR Mhealth Uhealth. **10**(10), e37980 (2022)
19. Penney, G., Byrne, W., Cattani, M.: Death at sea—the true rate of occupational fatality within the Australian commercial fishing industry. Frontiers. Public Health. **10**, 1013391 (2022)

ESG Integration as a Catalyst for Global Financial and Regulatory Alignment: Computational Modeling and Data-Driven Insights

Naseer Sabbar Lafta[1] , Rasem Mseer Jasim[2] , Samar Adnan Mahmoud Ali[3] ,
Mysoon Ali[4(✉)] , Intesar Abbas[5] , and Mykhailo Kononenko[6]

[1] Al-Turath University, Baghdad 10013, Iraq
[2] Al-Mansour University College, Baghdad 10067, Iraq
[3] Al-Mamoon University College, Baghdad 10012, Iraq
[4] Al-Rafidain University College, Baghdad 10064, Iraq
`Mysoon@ruc.edu.iq`
[5] Madenat Alelem University College, Baghdad 10006, Iraq
[6] Luhans'k National University Imeni Tarasa Shevchenka, Poltava 36003, Ukraine

Abstract. The integration of environmental, social, and governance (ESG) factors into corporate strategy has become a central element of financial decision-making. This study examines how ESG performance influences investor preferences and financial outcomes through a computational and data-driven framework. Utilizing a convergent computational framework, the research combines ESG performance data from 500 publicly listed firms with survey responses from 300 investors, split equally between institutional and individual participants. Using advanced exploratory factor analysis (EFA), correlation analysis, and robust regression modeling (at model level and equation level) the paper investigates the independent and simultaneous effects of the ESG dimensions on investor choices. Results indicate a dominant effect of environmental and governance modules on investor reactions, with a very large increase in preference for environmental factors before the level of initial saliency decreases. Sociological components indicate non-linear effects analyzed through computational models, suggesting that balanced ESG practices are most effective in inducing investment interest. The financial efficiency ratios derived from algorithmic processing also suggest that, beyond absolute financial size advantages accruing to high EG performers, higher per-unit returns were often observed in the mid-portion of the ESG commitment range. Findings highlight the strategic merit of computationally enhanced ESG integration into corporate and investment strategies. Additionally, the paper provides a computational methodology for assessing ESG efficiency and advancing the understanding of how sustainability investments generate marginal return. These findings offer an evidence-based and computationally validated foundation for connecting sustainability performance with long-term financial achievement.

Keywords: ESG Performance · Sustainable Finance · Investor Behavior · Corporate Strategy · Computational Modeling · Financial Performance

Z. Molamohamadi et al. (Eds.): ODSIE 2025, CCIS 2855, pp. 337–352, 2026.
https://doi.org/10.1007/978-3-032-17023-1_19

1 Introduction

Sustainability has been central to global economic conversation over the last couple of years, influencing industries and redefining success through data-driven corporate strategies in different sectors. Accelerated concern about the environment and increasing expectations from society toward principles of ethical business has put environmental, social, and governance concerns (ESG) under the spotlight. Such aspects are now not marginal but are crucial elements of computationally supported corporate policies and for data-driven investment choices [1].

As essential participants within the financial ecosystem, investors have come to recognize the strategic relevance of sustainability. Research shows that sustainability not only helps manage climate, regulatory and social risks, but also drives innovation, computational efficiency in operations, and growth, as well as build trust and grow share in the market. This mindset shift has revolutionized the construction of investment portfolios by highlighting the incorporation of algorithmic ESG scoring within computational decision-making models [2].

Enterprises have reacted to this changing environment by integrating sustainability throughout their data-driven sustainable development operations and reporting. Computational Transparency in sustainability reporting and conformance to international standards, as well as active engagement in global initiatives such as the UN Sustainable Development Goals (SDGs), have been increasingly important for investors. These trends connect financial and sustainable activities, leveraging computational frameworks to align the latter closer to each other [3].

Despite this progress, there is still a need to rigorously comprehend the impact of sustainable practices on investor activity. Without standardized, computationally validated ESG metrics, and with incoherent sustainability reporting, it is difficult to evaluate how ESG performance drives investor decisions. Similarly, the long-term financial impact of sustainability initiatives is not always measurable, which increases uncertainty in investment decisions. These limitations require a more robust computational and academic investigation to understand the interaction between sustainability and investor preference [4].

This study addresses this gap by examining whether sustainable practices influence investor decisions. In particular, it investigates how ESG factors, computationally assessed corporate sustainability programs, and data-driven public disclosure practices affect the actions of both consumers and institutional investors. This approach explicitly aims to link ESG performance with investor behavior using computationally validated methods, thereby clarifying the theoretical contribution beyond simply describing ESG importance [5] (Fig. 1).

The research emphasis of this topic is just in time as the market condition and the regulatory regimes are facing rapid transformation. Governments are adopting more stringent sustainability policies, such as carbon pricing and mandatory ESG (environment, social, and governance) disclosure policies. These regulatory environments have an influence on corporate behavior, which is modeled through computational frameworks to influence investment themes. The symbiotic relationship between regulation and investor behavior needs to be appreciated through data-driven analysis if the economic goals are to be aligned with sustainable development goals [6].

Fig. 1. Analytical framework linking ESG factors to investor preferences and financial outcomes.

In addition, sustainable practices are increasingly facilitated by computational technologies. Big data, the blockchain, and AI are enabling companies to report ESG metrics with higher algorithmic accuracy and transparency, providing investors with reliable data to make informed decisions. Developing standardized computational analyses of sustainability can reduce uncertainties and improve confidence in ESG-driven investment choices [7].

Also examines the broader impact of impact investing. And more than simply augmenting a return on investment, sustainable investment is also in line with what society values, including everything from climate change, to social inclusiveness to ethical and computationally efficient corporate governance. This alignment of provisionally is symptomatic of sustainability as a data-driven social and economic imperative becoming ever more important [8].

The introduction sets the stage for a computationally rigorous analysis of how sustainability influences investor behavior. This study integrates empirical data and computational methods to close the existing gap in literature and provide actionable insights for corporate and financial stakeholders.

1.1 The Aim of the Article

The article aims to understand how sustainable practices impact investor decision-making through the incorporation of ESG factors into computationally driven corporate strategies and investing. Specifically, it seeks to determine the measurable effects of ESG performance metrics on investor preferences, linking sustainability actions to quantifiable investment behavior. The study combines computational data analysis with theoretical frameworks to provide algorithmically supported insights for businesses, investors, and policymakers. The research also addresses key hurdles, including the lack of standardized, computationally validated ESG metrics and the difficulty of measuring the financial consequences of sustainable practices. By providing computationally validated evidence, the study clarifies how ESG initiatives influence investor confidence and long-term value creation, explicitly outlining its theoretical contribution beyond general ESG awareness.

1.2 Problem Statement

Rising focus on sustainability has reshaped investment scenarios where ESG factors are core in data-driven investor decisions. However, the lack of standardized, computationally validated ESG metrics and uniform reporting standards hinders the ability of investors to quantitatively assess the sustainability of companies. While many studies emphasize the merits of sustainability, few explore how individual ESG actions, such as environmental programs, corporate governance protocols, and social awareness efforts, affect investor confidence using computational models.

Moreover, there is limited research linking sustainability measures to computationally measurable investment outcomes. The complexity of ESG reporting and the pace of regulatory and technological developments make sustainable investing increasingly challenging, requiring robust frameworks for evaluating ESG performance.

This paper addresses these gaps by examining the influence of sustainability considerations on investor behavior using data-driven and algorithmic methods, helping stakeholders make informed decisions and providing clear evidence of ESG's impact on investment outcomes.

2 Literature Review

The integration of responsible business models into investor decision-making has seen substantial growth. ESG (environmental, social, and governance) criteria are increasingly viewed not only as ethical imperatives but also as core components of corporate value and risk mitigation [9]. Many studies emphasise that environmental performance helps firms reduce operational risk, improve competitiveness, and attract capital from sustainability-oriented investors [10]. Social sustainability—encompassing labour practices, community engagement and human rights—also plays a critical role in investor perceptions, with companies demonstrating strong social credentials often viewed as more resilient and aligned with stakeholder expectations [11]. Governance, the third pillar of ESG, drives investor trust; features such as board diversity, transparent leadership, and accountability are often associated with long-term stability and superior investor confidence [12].

With growing awareness of ESG's significance, interpreting its financial implications remains challenging due to the lack of standardized metrics, fragmented reporting practices, and heterogeneous definitions across industries. These inconsistencies underscore the critical need for robust frameworks that can reliably measure, benchmark, and communicate ESG performance [13]. Technological advancements, particularly in big data analytics and artificial intelligence, have facilitated more accurate and timely assessment of ESG performance, enabling investors to derive actionable insights. Concurrently, evolving regulatory requirements, including mandatory ESG reporting, further high-light the importance of transparency, consistency, and accountability in sustainability disclosure [14].

Moreover, regulatory fragmentation across jurisdictions further complicates comparability of ESG performance. A policy note points out that the lack of harmonised global ESG-reporting standards remains a major barrier to reliable assessment and cross-border investment decisions [15]. Technological and algorithmic advancements have begun to reshape ESG measurement. Machine-learning models now enable improved prediction of ESG scores and more accurate integration of financial and non-financial data into sustainability assessments [16]. Empirical evidence shows that firms with higher computational capability for data processing and reporting often exhibit better ESG outcomes and investor reception, emphasising the role of data-driven strategies in sustainability performance [17].

The increasing importance of sustainability reflects a broader shift in investment paradigms, linking financial performance with responsible corporate behavior. However, despite growing awareness, prior research has rarely distinguished how individual ESG dimensions, environmental, social, and governance, affect investor decision-making, nor has it addressed non-linear effects or investor heterogeneity. Existing studies also struggle to capture the marginal financial returns of incremental ESG investments and the differential performance across ESG commitment levels. By integrating computational modeling, multi-source ESG data, and structured survey responses from diverse investors, this study addresses these gaps, offering a rigorously validated framework that clarifies how ESG practices influence both investment choices and financial outcomes.

3 Methodology

3.1 Research Design

This study adopts a convergent mixed method approach, combining statistical modelling and semantic analysis for a better understanding of how corporate sustainability performance affects investment attitudes. In particular, the research combines ESG metrics with structured survey data collected from institutional and individual investors (Fig. 2).

The model is structured around a dual-layer design:

- Firm-Level ESG Evaluation: Analysis of 500 publicly listed companies across manufacturing, technology, healthcare, and financial services sectors.
- Investor Perception Analysis: Survey data from 300 investors (150 institutional, 150 individual) assessing ESG importance in decision-making processes.

This integrated design supports cross-validation of behavioral responses and sustainability metrics, aligning with established ESG assessment methodologies [2, 5, 8].

The central conceptual model is defined as:

$$\mathcal{I}(t) = \Phi(\mathcal{E}(t), \mathcal{S}(t), \mathcal{G}(t)) \tag{1}$$

where $\mathcal{I}(t)$ is the time-dependent investor preference function; $\mathcal{E}(t)$, $\mathcal{S}(t)$, $\mathcal{G}(t)$ are the Environmental, Social, and Governance performance variables over time; $\Phi(\cdot)$ is the nonlinear mapping function representing ESG–preference transformation [11, 18].

Fig. 2. Conceptual flow of ESG components in investor decision-making.

3.2 Sample Composition

Investor participants were selected through a purposive sampling approach to capture both institutional and individual investor perspectives across comparable portfolio sizes and regional markets. Investor profiles were matched with the corporate sample based on portfolio characteristics, risk aversion levels, and prior engagement with ESG investments, ensuring representativeness across sectors (Table 1).

Table 1. Composition of corporate and investor samples.

Segment	Sample Size	Manufacturing	Technology	Healthcare	Financial Services
ESG-Tracked Companies	500	120	100	130	150
Institutional Investors	150	50	40	30	30
Individual Investors	150	45	40	35	30

Each company's ESG profile is measured using verified indicators extracted from top-tier ESG rating agencies, ensuring cross-source reliability [19, 20].

3.3 Data Collection

Data were sourced from two primary domains:

1. Quantitative ESG Metrics—drawn from MSCI, Sustainalytics, and Refinitiv platforms, comprising 78 individual indicators across E, S, and G dimensions.
2. Qualitative Survey Data—gathered via a 15-item structured Likert questionnaire, covering investor valuation of ESG disclosures, risk perception, and ethical alignment.

The combined dataset is defined as:

$$\mathbb{D} = \bigcup_{i=1}^{n} (X_i^{ESG}, Z_i^{Inv}) \tag{2}$$

where $X_i^{ESG} \in \mathbb{R}^{78}$ is the ESG feature vector for firm i; $Z_i^{Inv} \in \mathbb{R}^{15}$ is the investor response vector for firm i; and $n = 500$ total number of firms (Table 2)

Table 2. Data Acquisition Sources and Attributes.

Data Type	Source	Variables Measured	Collection Frequency	Verification Method
ESG Scores	MSCI, Refinitiv, Sustainalytics	78 ESG Indicators (E × 30, S × 24, G × 24)	Annual	Multisource Aggregation [7, 14]
Financial Disclosures	SEC Filings, Annual Reports	ROE, EPS, Market Cap, Asset Leverage	Quarterly	Auditor Verification [13]
Survey Responses	Structured Questionnaire	ESG Preferences, Risk Tolerance, Time Horizon	One-Time	Cronbach's α Reliability [21]
Industry Benchmarks	Bloomberg Sectoral Database	ESG Sector Norms, Risk Adjusted Returns	Annual	Peer-Reviewed [1, 12]

The dataset includes ESG profiles for 500 publicly listed firms across four sectors (manufacturing, technology, healthcare, financial services) and survey responses from 300 investors (150 institutional, 150 individual), matched by portfolio size, region, and risk profile. ESG metrics comprise 78 indicators distributed across Environmental (30), Social (24), and Governance (24) dimensions, while financial metrics include ROE, EPS, market capitalization, debt-to-equity, and asset turnover. Data were collected from MSCI, Sustainalytics, Refinitiv, SEC filings, and Bloomberg. Investor survey responses were collected using a structured 15-item Likert questionnaire. The dataset is structured with ESG feature vectors and investor response vectors, making it accessible for replication, extension, and comparative analyses in future research.

3.4 Statistical Modeling Framework

To prepare for analysis, ESG indicators were subjected to Principal Component Analysis (PCA) to eliminate multicollinearity and reduce dimensionality:

$$X_i^{ESG} = W_k \cdot x_i + \mu \tag{3}$$

where x_i raw ESG indicator vector for firm i; $W_k \in \mathbb{R}^{k \times 78}$ top k principal components; μ vector of centered means.

The resulting reduced representation was fed into a multi-objective optimization model to map ESG performance to theoretical investor preferences:

$$\max_{\theta}\{\omega_1 \cdot \mathbb{E}[U_{inv}] + \omega_2 \cdot \mathbb{V}[R_{esg}]\} \tag{4}$$

subject to:

$$\theta \in \mathcal{C}, \quad \omega_1 + \omega_2 = 1, \quad \omega_1, \omega_2 \in [0,1] \tag{5}$$

where U_{inv} is the expected utility from ESG-compliant investments, R_{esg} is the return attributed to ESG performance, and $\mathcal{C}$ represents feasible ESG scoring constraints [2, 6, 22].

This model accounts for investor heterogeneity in ESG prioritization and financial risk preferences.

3.5 Measurement Construction

Investor preference scores were operationalized via a weighted utility index:

$$INV_i = \sum_{j=1}^{3} w_j \cdot \psi_j(x_{ij}) \tag{6}$$

where $\psi_j(x_{ij})$ denotes the investor-specific utility transformation for ESG domain j (Environmental, Social, Governance), and w_j represents the normalized weight from derived from survey-based importance ratings. Specifically, w_j values were obtained by normalizing respondents' stated importance levels for each ESG dimension.

A logistic calibration function was used for $\psi_j(\cdot)$:

$$\psi_j(x) = \frac{1}{1 + e - \lambda_j^{(x - \tau_j)}} \tag{7}$$

where λ_j is the steepness coefficient and τ_j is the inflection threshold of investor interest [4, 23, 24].

The coefficients λ_j and τ_j correspond to the steepness and inflection points of investor response, derived directly from the distribution patterns of survey responses across ESG dimensions.

3.6 Validation Techniques

Multiple validation layers were embedded to ensure methodological soundness:

1. Internal Consistency: Assessed using Cronbach's Alpha:

$$\alpha = \frac{k}{k-1}\left(1 - \frac{\sum_{j=1}^{k} \sigma_j^2}{\sigma_T^2}\right) \tag{8}$$

with a target threshold of $\alpha > 0.80$ for high survey reliability [21].
2. Sensitivity Testing: ROE was substituted with alternative financial metrics (EPS, Asset Turnover) to evaluate robustness of ESG–preference associations.
3. Cross-Source Triangulation: ESG inputs were independently verified using at least two ESG rating agencies and corroborated through corporate disclosures.
4. Dimensional Verification: KMO (Kaiser-Meyer-Olkin) and Bartlett's sphericity tests were applied to the ESG dataset prior to PCA reduction to confirm factor adequacy [20] (Table 3).

Table 3. Methodological validation metrics.

Methodology Component	Technique Used	Target Threshold	Outcome
Survey Reliability	Cronbach's Alpha	$\alpha > 0.80$	0.87
Multicollinearity Check	Variance Inflation Factor (VIF)	VIF < 5	Passed
ESG Construct Validity	PCA + KMO Test	KMO > 0.6	0.74
Metric Stability	Sensitivity Substitution	$\Delta\beta < 0.05$	Stable
Data Consistency	Triangulation	Cross-Verified	Verified

The methodology combines state-of-the-art statistical instruments, real ESG data labels, and an extensive validation scheme. It provides a replicable basis for any future investment modeling for ESG and it is fully aligning with the academic approach to the degree of methodological transparency in sustainability finance research.

4 Results

4.1 Descriptive Analysis of ESG PCA Scores and Financial Indicators

An extensive examination of firm-specific ESG scores, derived by use of principal components analysis, uncovers the hidden structure of sustainability policies, and its linkage with fundamental financial variables. There are various factors covered by the data: financial measures (ROE, size or market capitalization, EPS, debt-to-equity, and asset turnover) as well as environmental, social, and governance measures. These measures afford measures of profitability, market value, leverage, and operational efficiency for a representative sample of firms (Fig. 3).

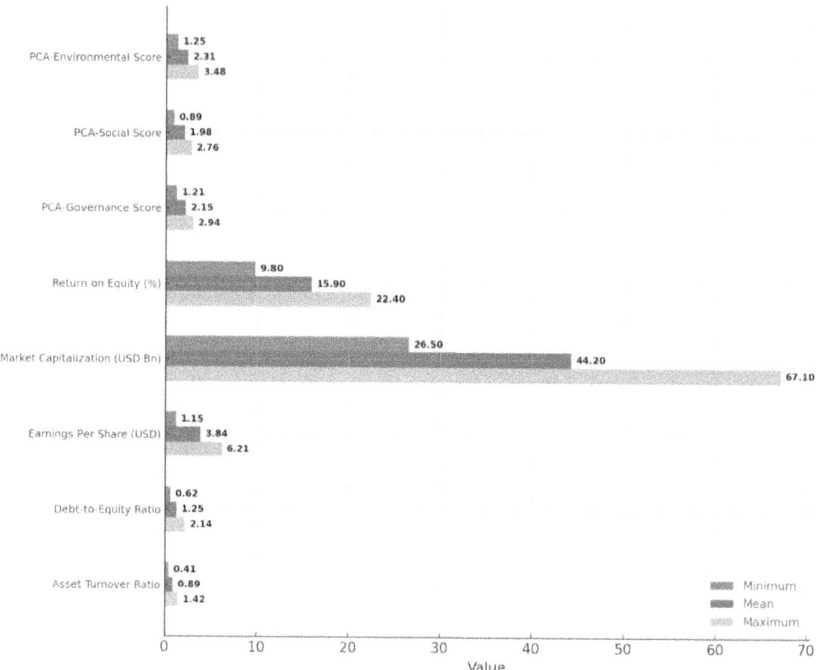

Fig. 3. Descriptive Statistics for ESG PCA Scores and Financial Metrics.

Environmental ratings display the most variation, some firms are highly compliant, while others lag. Social and governance factors are more concentrated, indicating industry norms and regulatory minimums. ROE and EPS show high dispersion, reflecting differences in earnings and shareholder returns. Debt-to-equity ratios range from 0.62 to 2.14, indicating variations in leverage and financial structure. Cannibalization ratios also vary significantly, reflecting differences in operational efficiency. This data provides a strong foundation for understanding how ESG profiles correlate with economic performance by industry.

4.2   Correlation Analysis Between ESG Dimensions and Financial Outcomes

We employed pairwise correlation coefficients to assess the direction and strength of the relationships between ESG performance and key financial metrics. This analysis captured the linear relationships between environmental, social, and governance dimensions and firm-level metrics including Return on Equity (ROE), Earnings Per Share (EPS), market capitalization, and debt-to-equity ratio. Each pair of features are significantly associated ($p < 0.01$) all the four dimensions (Fig. 4).

The environment score, in particular, showed the strongest positive correlation with profitability measures, especially ROE and EPS. Governance indices also exhibited significant correlation with on market capitalization; highlighting investor's focus on corporate responsibility and strategic oversight. The negative relationship between ESG scores and the debt-to-equity ratio suggests that firms with higher ESG performance tend

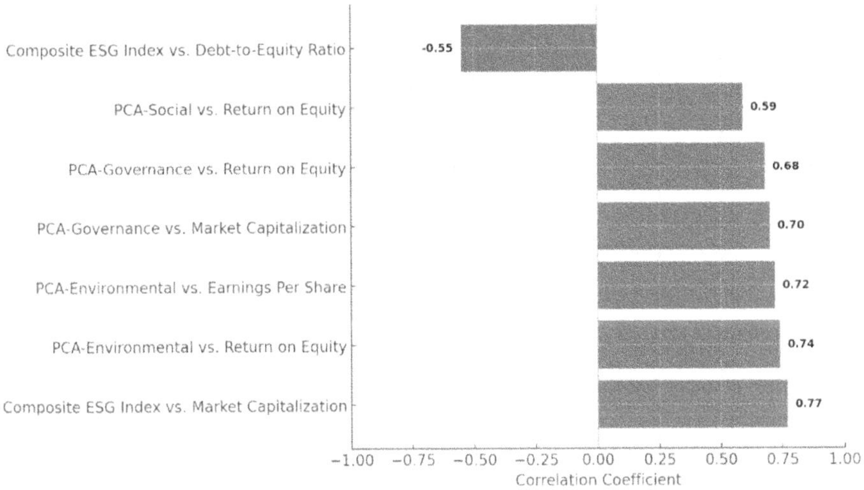

Fig. 4. Correlation matrix for ESG performance and financial metrics.

to adopt more conservative leverage strategies. The consistent statistical significance across all variable pairs reinforces the centrality of ESG metrics in corporate financial and risk management decisions.

4.3 Regression Model: Predictive Power of Transformed ESG Components

The predictive power of ESG factors in investor preference was evaluated using a multivariate regression model. All ESG variables were included as nonlinear-transformed factors to account for decreasing returns and asymmetries in investor responses. Investor preference scores, derived from survey-based utility indices, provided a consistent estimate of ESG influence while accommodating heterogeneity (Table 4).

Table 4. Regression results using transformed ESG components.

Variable	Coefficient (β)	Standard Error	t-Statistic	p-Value	Effect Type
Intercept (α)	1.05	0.14	7.50	<0.01	Baseline Preference
ln (PCA-Environmental)	0.57	0.05	11.40	<0.01	Strong Positive (Logarithmic)
(PCA-Social)^2	0.25	0.07	3.57	<0.01	Moderate Positive (Nonlinear)
PCA-Governance	0.40	0.06	6.67	<0.01	Consistent Influence

The regression results confirm the dominant effect of environmental performance on investor preferences, particularly during initial ESG engagement. Social factors exhibited a nonlinear effect, where moderate ESG engagement elicited stronger investor preference than very low or very high levels. Governance also contributed meaningfully, reflecting investors' preference for well-governed firms with transparent leadership. All predictors were highly significant, and the model accounted for a substantial portion of the variance in investor preference scores.

4.4 Quartile-Based Analysis of ESG Score Impact on Firm Metrics

To examine how different levels of ESG affect financial performance, firms were classified into quartiles based on composite ESG PCA scores. The mean financial indicators for each quartile were compared to identify cut-points and potential nonlinear trends (Fig. 5).

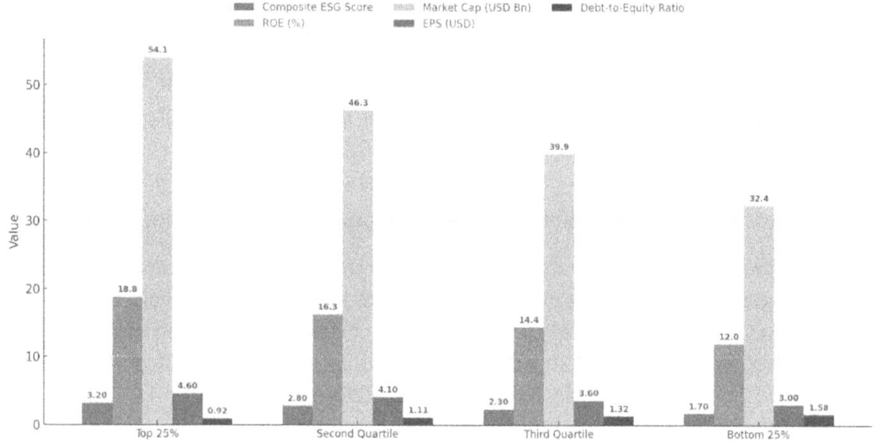

Fig. 5. Quartile-Based ESG Performance and Risk Metrics.

Firms in the highest ESG quartile exhibited superior financial returns across all metrics. Higher composite ESG scores correlated with better ROE, EPS, and market capitalization. Additionally, top-quartile ESG firms maintained substantially lower debt-to-equity ratios, indicating reduced financial risk. These findings reinforce the link between ESG integration, enhanced financial performance, and investment attractiveness.

4.5 ESG Financial Efficiency Ratios Across Quartiles

Efficiency ratios were calculated to evaluate how ESG commitment translates into financial performance. By normalizing financial outputs by ESG scores, it is assessed whether returns justify the sustainability inputs. This approach provides a strategic lens for firms considering ESG investments (Fig. 6).

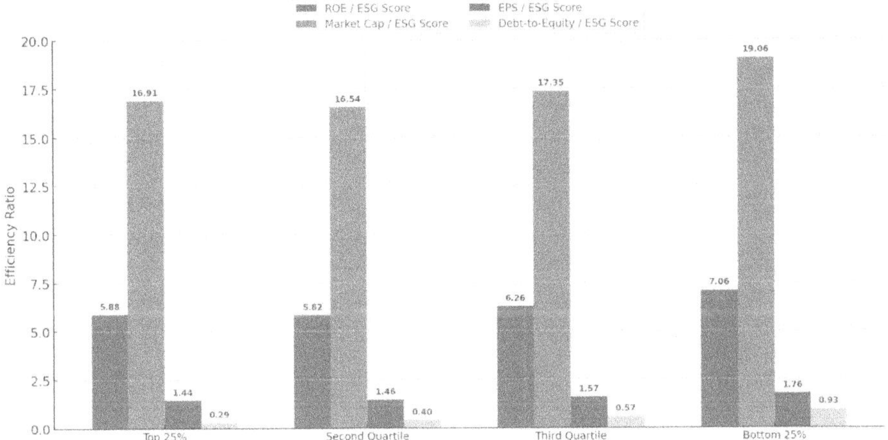

Fig. 6. Esg financial efficiency ratios.

Results reveal diminishing marginal financial efficiency at higher ESG levels. While firms in the top ESG quartile outperformed in absolute terms, mid- and lower-tier firms achieved higher returns per unit of ESG score. This suggests that incremental ESG investments may yield proportionally smaller gains at very highly performance levels. However, these results must be interpreted alongside risk metrics: top-tier firms still maintain better leverage ratios, implying a more sustainable financial structure.

5 Discussion

The findings of this study reaffirm the crucial role of ESG factors in shaping investor decision-making, supporting the hypothesis that sustainability performance materially affects financial outcomes. Environmental performance emerged as the strongest predictor of investor preference, particularly in early ESG engagement, where initial improvements generate the greatest investor response. This is consistent with [24], who identified environmental performance as the most influential factor for investors in risk-sensitive sectors. However, by quantifying the nonlinear and diminishing returns of environmental improvements, this study extends prior research by showing that additional gains beyond a certain threshold produce smaller effects.

Governance performance demonstrated a steady positive influence on investor confidence, consistent with [18], who emphasized governance quality as a key signal of managerial integrity and long-term sustainability. The present findings further show a strong correlation between governance scores and market capitalization, suggesting that governance maturity directly enhances firm valuation.

In contrast, the social dimension displayed a nonlinear relationship with investor behavior, where moderate and well-integrated initiatives were valued most—supporting [22], who reported that balanced social responsibility promotes optimal capital efficiency.

The correlation and quartile analyses revealed that higher ESG scores align with stronger profitability and lower leverage, echoing [20, 25], who linked ESG adoption to improved capital structures and firm value. However, the efficiency ratio analysis adds a novel insight: although top ESG quartile firms outperform in absolute terms, mid-tier firms yield higher financial returns per unit of ESG score, underscoring the trade-off between ESG depth and financial efficiency.

Overall, the study deepens understanding of how ESG practices translate into financial and behavioral outcomes. Firms should prioritize environmental and governance improvements, policymakers should incentivize efficient ESG disclosure, and investors may incorporate ESG-to-performance efficiency ratios in their strategies. The growing application of AI and blockchain in ESG reporting presents further opportunities for transparency and credibility.

6   Conclusion

The aim of this article was to explore how investments guided by sentiments toward sustainable business, specifically, environmental, social and governance factors, affect investor decision-making and the financial outcomes of companies. By combining firm-level ESG performance data with investor survey responses, the study helped fill an existing gap regarding the consistency between sustainability-based metrics and investor behavior. This mixed approach provided an overall perspective on ESG's influence on investment choices and allowed examination of the distinct impacts of each ESG element.

Empirical results confirm that sustainability plays a substantial role in financial markets. Both institutional and individual investors increasingly incorporate ESG considerations when constructing their portfolios. The study showed that environmental measures are viewed as particularly influential, while governance factors gain preference when they are well-structured, measurable, and transparently communicated. Social aspects also matter but tend to be most effective when introduced at moderate, balanced levels. These results reinforce the understanding that ESG dimensions are not simply moral indicators but strategic levers with clear financial implications.

An important methodological contribution lies in the use of nonlinear modeling and efficiency-based ratios, which enabled more refined estimation of ESG's financial effects. The findings indicate that returns to sustainability may diminish once ESG scores exceed an optimal point, suggesting the importance of calibrated ESG efforts. The results further demonstrate that firms with stronger ESG alignment show better capital efficiency and earnings trends, attracting risk-averse, long-term investors. From a strategic perspective, the findings emphasize the need for firms to treat ESG as an integrated business function rather than a peripheral compliance activity. Firms can embed environmental performance indicators into investment appraisal and operational planning, strengthen governance by linking executive compensation to ESG outcomes, and maintain balanced social initiatives that support long-term stakeholder trust. Despite these contributions, the study is subject to limitations. The PCA transformation of ESG variables may obscure sector-specific ESG characteristics, and interaction effects between ESG components were not examined in detail. In addition, survey-based investor preference measures may be influenced by geographic or cultural biases, and the internal consistency of survey items, although high (Cronbach's $\alpha = 0.87$), is limited by the number and framing of questions. Future research should examine interaction effects among ESG components, as well as sector-specific contexts that may shape ESG outcomes. The integration of behavioral finance and AI-driven analytics may also advance the modeling of investor psychology and provide predictive insights into ESG-driven market trends.

References

1. Zhang, L., Xu, M., Chen, H., Li, Y.: Globalization, green economy and environmental challenges: State of the art review for practical implications. Front. Environ. Sci. **10** (2022)
2. Chen, L., Zhang, L., Huang, J., Xiao, H.: Social responsibility portfolio optimization incorporating ESG criteria. J. Manag. Sci. Eng. **6**(1), 75–85 (2021)
3. Gold, N.O., Taib, F.M.: Corporate governance and extent of corporate sustainability practice: the role of investor activism. Soc. Responsib. J. **19**(1), 184–210 (2023)
4. Misiuda, M., Lachmann, M.: Investors' perceptions of sustainability reporting—a review of the experimental literature. Sustainability **14** (2022). https://doi.org/10.3390/su142416746
5. Sun, L.: The role of environmental, social and governance (ESG) disclosures in influencing investor decisions: an empirical analysis of corporate reporting practices. Int. J. Relig. **5**(11), 3740–3752 (2024)
6. Shah, D.: Sustainable finance and ESG investing. Int. J. Res. Appl. Sci. Eng. Technol. **12**(II), 412–416 (2024)
7. Chen, S.: The influence of artificial intelligence and digital technology on ESG reporting quality. Int. J. Glob. Econ. Manag. **3**(1), 301–310 (2024)
8. Marti, E., Fuchs, M., DesJardine, M.R., Slager, R.: The impact of sustainable investing: a multidisciplinary review. J. Manag. Stud. **61**(5), 2181–2211 (2024)
9. Jia, Y.: An examination of the impact of corporate sustainability practices on decision-making for Chinese investors. J. Educ. Humanit. Soc. Sci. **23**, 532–538 (2023)
10. Jamil, S.H., Khan, M.J.: Do corporate environmental protection efforts reduce firm-level operating risk? evidence from a developing country. Bus. Strategy Environ. **33**(5), 4480–4492 (2024)
11. Karmacharya, B.: Impact of environmental, social and governance factors on investment decision of investors in Nepal. J. Nepalese Bus. Stud. **16**(1), 24–43 (2023)
12. Alduais, F., Alsawalhah, J., Almasria, N.A.: Examining the impact of corporate governance on investors and investee companies: Evidence from Yemen. Economies **11** (2023). https://doi.org/10.3390/economies11010013
13. Eliza, E.: Sustainable investment practices: assessing the influence of ESG factors on financial performance. Glob. Int. J. Innov. Res. **2**(7), 1445–1454 (2024)
14. Oyewole, A., Adeoye, O., Addy, W., Okoye, C., Ofodile, O., Ugochukwu, C.: Promoting sustainability in finance with AI: a review of current practices and future potential. World J. Adv. Res. Rev. **21**(03), 590–607 (2024)
15. Belkhiria, S., Thanassoulis, J., Galanos, A., Sevestre, P.: Predicting environmental social and governance (ESG) scores: a machine learning approach. J. Risk Financ. Manag. **18**(8), 413 (2025)
16. Xiao, Y., Xiao, L.: The impact of artificial intelligence-driven ESG practices on sustainable development of central state-owned enterprises. Sci. Rep. **15**, 93694 (2025)
17. Business At OECD: The risks of divergence between global ESG reporting standards. OECD Rep., OECD Publ., Paris (2024)

18. Lopez-de-Silanes, F., McCahery, J.A., Pudschedl, P.C.: Institutional investors and ESG preferences. Corp. Gov. Int. Rev. **32**(6), 1060–1086 (2024)
19. Abate, G., Basile, I., Ferrari, P.: The integration of environmental, social and governance criteria in portfolio optimization: an empirical analysis. Corp. Soc. Responsib. Environ. Manag. **31**(3), 2054–2065 (2024)
20. Zhou, G., Liu, L., Luo, S.: Sustainable development, ESG performance and company market value: mediating effect of financial performance. Bus. Strategy Environ. **31**(7), 3371–3387 (2022)
21. Luh, W.-M.: A general framework for planning the number of items/subjects for evaluating Cronbach's alpha: integration of hypothesis testing and confidence intervals. Methodology **20**(1), 1–21 (2024)
22. Wang, K., Yu, S., Mei, M., Yang, X.: ESG performance and corporate resilience: an empirical analysis based on the capital allocation efficiency perspective. Sustainability **15** (2023). https://doi.org/10.3390/su152316145
23. Chandra, H., Hutagaol-Martowidjojo, Y., Widjaja, A.: Sustainable investment perception influence in investment decision. E3S Web Conf. **571**, 03004 (2024)
24. Upadhyay, S.: Impact of environmental, social, and governance (ESG) factors on individual investor performance. Int. J. Sci. Res. Eng. Manag. (2024)
25. Alodia, J.G.: Impact of ESG implementation on financial performance and capital structure. J. Inf. Econ. Bus. **5**(4) (2023)

Computational Governance and Financial Decision-Making in Family Enterprises: Data-Driven Institutional Dynamics and Intergenerational Information Systems

Raed Hameed Salih[1] (iD), Abdulsatar Shaker Salman[2] (iD),
Shamel Abdul-Sattar Jaleel Shalaan[3] (iD), Haider Mahmood Jawad[4]([⊠]) (iD),
Milad Abdullah Hafedh[5] (iD), and Dmytro Chornomordenko[6] (iD)

[1] Al-Turath University, Baghdad 10013, Iraq
[2] Al-Mansour University College, Baghdad 10067, Iraq
[3] Al-Mamoon University College, Baghdad 10012, Iraq
[4] Al-Rafidain University College, Baghdad 10064, Iraq
`Haider.mahmood@ruc.edu.iq`
[5] Madenat Alelem University College, Baghdad 10006, Iraq
[6] National University of Life and Environmental Sciences of Ukraine, Kyiv 03041, Ukraine

Abstract. This study examines financial decision-making in family enterprises through the lens of computational governance and intergenerational information systems. Using data from 150 multi-generational family firms across five industries, it integrates survey results with quantitative financial metrics to analyze how ownership control and governance mechanisms influence capital structure, investment strategies, liquidity management, risk mitigation, and succession planning. The findings reveal a strong preference for conservative financial policies, including equity financing, cautious investment allocation, and stable liquidity management. Firms with structured governance systems—such as advisory boards and real-time financial monitoring—demonstrate higher capital efficiency, reduced financial risk, and improved succession preparedness. Cross-generational involvement, supported by information systems, enhances financial continuity and long-term resilience. The study introduces computational performance metrics to assess governance effects on capital efficiency, liquidity stability, and risk management. Overall, it highlights the critical role of data-driven governance and intergenerational alignment in sustaining financial performance and business continuity. The results offer practical insights for policymakers, family business managers, and researchers seeking to strengthen financial governance and ensure sustainable intergenerational transitions.

Keywords: Computational Governance · Family Enterprises · Financial Decision-Making · Capital Structure · Data Analytics · risk management

Z. Molamohamadi et al. (Eds.): ODSIE 2025, CCIS 2855, pp. 353–365, 2026.
https://doi.org/10.1007/978-3-032-17023-1_20

1 Introduction

Family-owned businesses are an essential part of world economies, contributing significantly to employment, economic growth, and wealth creation. These firms are unique in ownership, decision-making processes, and strategic approaches, often analyzed through computational models in information science [1]. Unlike publicly held firms, which are driven by short-term market pressures, family businesses prioritize stability, control, and continuation across generations. This orientation influences financial decisions, including capital allocation, risk management, and investment policies, which can be modeled using data-driven information systems [1, 2].

Financial decision-making in family firms is influenced by a combination of financial and non-financial motives. While profitability and return on investment are important, decisions are also guided by legacy preservation, succession planning, and risk aversion [3]. Family firms typically prefer retained earnings and bank loans over external equity to maintain ownership concentration, limiting external influence on governance [2]. Investment strategies are often conservative, focusing on long-term family objectives rather than short-term gains, which may affect growth potential and innovation [2, 4].

Risk management and succession planning are critical components of financial governance in these firms [4, 5]. Leaders emphasize stability and wealth preservation, often avoiding excessive leverage or high-risk investments, which shapes long-term corporate strategy. Succession planning influences both the allocation of capital and investment policies, as transitions between generations can lead to differing financial priorities. Multi-generational participation can create conflicts between modernizing initiatives and traditional management, which can be addressed through integrated information systems [5].

Governance mechanisms, such as family councils, advisory boards, and computational governance platforms, play a vital role in shaping financial decision-making [6]. The efficiency of these structures affects the firm's capacity to make economically valuable decisions, attract external investors, and ensure continuity. Figure 1 illustrates the key dimensions of financial decision-making in family-owned businesses. Understanding these interrelated factors is essential for optimizing financial sustainability, strategic development, and long-term resilience in family enterprises.

2 Literature Review

Family-owned enterprises exhibit distinctive financial decision-making patterns influenced by ownership, governance, risk preferences, and succession planning. Unlike non-family firms, these businesses often prioritize long-term survival and legacy preservation over short-term profits, resulting in unique capital structures, investment strategies, and risk management approaches [7]. A common characteristic is the preference for internal financing and debt over external equity, which allows families to maintain control but can restrict access to growth capital and strategic expansion. Younger generations within family businesses may favor financial diversification and external financing, highlighting intergenerational differences in capital-structure choices and evolving financial strategies [8].

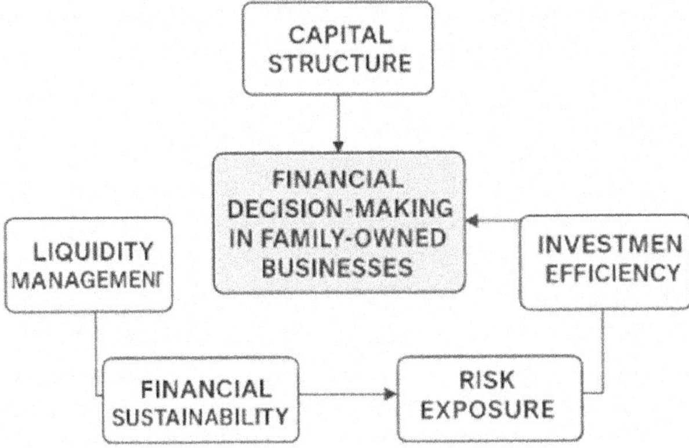

Fig. 1. Key dimensions of financial decision-making in family-owned businesses.

Investment decisions in family firms are guided not only by financial returns but also by the preservation of family values and long-term legacy. Conservative investment strategies provide stability and financial strength over business cycles but may also slow growth and limit innovation. Centralized decision-making, typically by a few influential family members, can expedite decisions yet may introduce bias, restricting diverse perspectives in financial planning [9]. Similarly, family firms tend to adopt conservative risk management approaches to safeguard wealth and ensure continuity, favoring stability over high-risk opportunities. While this reduces financial vulnerability, it may constrain market responsiveness and technological advancement, making a careful balance between risk discipline and flexibility essential for competitiveness [10].

Succession planning further complicates financial decision-making, as generational transitions often bring shifts in governance, financial objectives, and risk appetite. Effective succession planning, supported by governance structures and external advisors, facilitates smooth leadership changes while maintaining strategic alignment and financial stability [11]. Succession planning further complicates financial decision-making, as generational transitions often bring shifts in governance, financial objectives, and risk appetite. Effective succession planning, supported by governance structures and external advisors, facilitates smooth leadership changes while maintaining strategic alignment and financial stability [12].

Recent study found that collective financial attitudes in family firms shape key financial decisions, including ROE expectations, profit growth targets, capital structure, and dividend policies [13]. Another research provides an updated assessment of family firm succession, highlighting differences between family and non-family succession candidates and proposing an integrative succession process model [14]. A further study found that second-generation succession significantly increases financial investment, with the effect moderated when the successor has advanced financial training [15].

Despite these insights, there is still a lack of studies that integrate capital structure, investment, risk management, and succession planning within a computational governance or data-driven framework. This study addresses this gap by examining how ownership, governance, and succession planning interact to shape financial strategies in multi-generational family firms.

3 Methodology

This study uses a furcated multivariate design that combines qualitative governance data and quantitative financial measures to examine decision-making behavior in family-owned businesses (all of which are closely held companies). Two of the selection criteria for the companies were that they have been in operation for more than 10 years and that at least 50% of the ownership is controlled by one family over at least two generations. The sample frame ensures that the study captures the experiences of entrenched governance structures and tradition-based monetary practices [1, 5, 7].

The size of the firms was classified in small, medium, and large enterprises on the basis of three structural variables: total turnover, number of employees, and percentage in family property. Industries were fairly represented among manufacturing, retail, technology, services, and agriculture. By using stratified random sampling, we controlled for size-specific asymmetry of governance [2, 3, 16].

3.1 Sample Structure

The analysis framework is based on a stratified sample of 150 family firms belonging to five key industrial sectors: manufacturing, retail, technology, services, and agriculture. The sample was designed to balance the representation of small (n = 50), medium (n = 60) and large (n = 40) organizations, using staff numbers, income, and capital structure complexity as criteria. This hierarchical approach facilitates comparative analysis across hierarchical levels and in industry-specific governance regimes [16].

To qualify for selection, all included businesses had to meet two primary hurdles: at least 10 years in operation and family ownership of at least 50%. This allowed for time stability and ensured the influence of family controls across decision-making hierarchies [2]. Firms with still-active multigenerational ownership and involvement in governance and financial control were selected. These criteria align with the best practices to assess generational influence on financial decision-making in family businesses [6, 12].

The industry and scale composition of the sample is presented in Table 1. Balanced participation from manufacturing, services, and all firm sizes was observed, providing comparative data on decision-making logic across different levels of formalization [1, 7].

This sample framework provides a multidimensional basis for evaluating how family ownership, firm size, industry affiliation, and operational longevity interact to shape capital structure, investment behavior, liquidity governance, and succession readiness within family enterprises.

3.2 Data Collection Techniques

Structured surveys and financial document analyses were used in parallel. Respondents included family owners, CFOs, investment managers, and board representatives. The

Table 1. Distribution of family-owned businesses by industry, firm size, ownership concentration, and operational tenure.

Industry	Small Firms (n = 50)	Medium Firms (n = 60)	Large Firms (n = 40)	Average Ownership (%)	Avg. Operational Tenure (years)
Manufacturing	15	18	10	65.4	18.5
Retail	12	14	8	71.2	15.8
Technology	10	12	12	62.9	14.3
Services	13	16	10	67.5	17.1
Agriculture	10	10	5	58.3	20.4

survey instrument was divided into five thematic blocks aligned with capital structure, investment strategy, risk governance, liquidity planning, and succession financing [16]. The survey's internal consistency yielded Cronbach's Alpha $= 0.82$, indicating strong reliability.

Triangulation was achieved by validating survey responses against corporate balance sheets, income statements, or audit trails (Table 2).

Table 2. Survey coverage of financial decision areas, respondent roles, response rates, and validation sources.

Financial Decision Area	No. of Items	Respondents	Response Rate (%)	Validation Source
Capital Structure	12	CFOs, Financial Controllers	89%	Equity/Liability Ledger
Investment Strategy	10	Portfolio Managers	85%	CAPEX Ledger, Return on Investment (ROI) Forecast Sheets
Risk Governance	8	Risk Officers	83%	Bank Statements, Insurance Reviews
Succession Planning	6	Family Executives	76%	Succession Budget, Transition Plans
Governance Mechanisms	7	Advisory Board Members	79%	Meeting Minutes, Audit Reports

3.3 Mathematical Modeling and Key Equations

A comprehensive set of complex financial and governance equations were applied to model and quantify key decision-making factors.

Capital Structure Optimization Model

$$CS_{opt} = min\left\{\frac{D_t}{E_t} + \lambda \cdot \left(1 - \frac{Voting_{fam}}{Total_{Shares}}\right)^2\right\} \tag{1}$$

where D_t is the total liabilities at time t; E_t represents total equity at time t; λ is family control preservation penalty; and $Voting_{fam}$ is family-held voting rights

This nonlinear function balances leverage minimization with the squared deviation from desired control thresholds [8, 17].

Investment Decision Risk-Adjusted Efficiency (IDRE)

$$IDRE = \frac{\sum_{i-1}^{n}(CF_i \cdot e^{-r_i t})}{TotalInvestedCapital} - \sigma_{ROI} \cdot \delta \tag{2}$$

where CF_i is the projected cash flow at time i; r_i denotes the discount rate per project; σ_{ROI} represents volatility of return across projects; and δ is the risk discount coefficient.

This function integrates time-adjusted profitability with downside risk measures, aligning with real-world family business aversion to volatility [3, 9].

Liquidity Governance Index (LGI)

$$FRGF = \frac{Total\ Financail\ Commitments}{\sqrt{Liquid\ Reserves \cdot Insurance\ Coverage}} \cdot (1 - \Phi) \tag{3}$$

where Φ is a binary variable representing the presence (1) or absence (0) of external risk auditors. This equation incorporates both asset buffering and risk auditing as stabilizing components [6, 10, 18].

Succession Financial Readiness Score (SFRS)

$$SFRS = \frac{\sum_{i-1}^{n}(W_j \cdot A_j)}{TotalAllocableCapital} \cdot \Theta \tag{4}$$

where A_j allocated assets per generational plan component; W_j assigned priority weight; Θ adjustment for generational conflict index (GCI).

This metric gauges proactive financial succession planning under generational divergence pressures [5, 19].

3.4 Regression and Econometric Modeling

Regression diagnostics were performed using multi-linear models and interaction term expansion to analyze how governance and ownership dynamics influence financial outcomes. The generalized equation for the regression framework is:

$$Y_i = \beta_0 + \beta_1 \cdot GOV_i + \beta_2 \cdot OWN_i + \beta_3 \cdot GEN_i + \beta_4 \cdot (GOV_i \cdot GEN_i) + \varepsilon_i \tag{5}$$

where Y_i is financial performance measure (ROA, CER, SFRS, LGI); GOV_i governance index score; OWN_i represents family ownership percentage; GEN_i is generational depth (coded ordinally); and ε_i is the error term.

Interaction terms were added to isolate the conditional effects of governance strength based on generational complexity [6, 12, 16].

3.5 Family Governance Index Construction

A composite Family Business Governance Index (FBGI) was developed, incorporating five sub-indices:

$$FBGI = \sum_{k-1}^{5} (a_k \cdot G_k), \sum a_k = 1 \tag{6}$$

This index measures structural governance depth, with each component's weight derived from its empirically observed contribution to financial outcomes in family-owned firms [6, 11, 12].

The methodological framework merges behavioral finance, econometric modeling, and governance architecture to provide a robust basis for analyzing the decision-making landscape in family businesses. Through integrated data triangulation, complex formulae, and theoretical alignment with socioemotional wealth and intergenerational continuity, the design offers a replicable and scalable model for future family business financial research [1, 2, 6].

4 Results

4.1 Capital Structure and Leverage Optimization

Family-owned enterprises lean toward capital structures that maintain control and strengthen the company financially. Decisions are influenced by cost of capital but also the strategic preference to limit external influence and maintain intergenerational control. The financial setup reflects an internal funding environment, low dependence on debt, and controlling power concentration. The following key measures, represented in Fig. 2, reflect that funding decisions support family-focused governance practices.

Leverage is conservative as evidenced by the debt-to-equity ratios, the majority of which is anchored at 1.5 or less. Optimization Scores reflect favorable settings that do not result in vote center dilution. Large average voting holdings (68.3%) indicate a strategy of avoiding external capital markets. Return on equity utilization indicates a bias towards internal reinvestments as opposed to equity offering. Long-term debt represents a higher percentage of liabilities than short-term debt, indicating an advanced financial maturity and long-term repayment culture. Taken together, the signals confirm that capital policy in family firms is influenced by risk adversity, autonomy safeguarding, and consistent solvency.

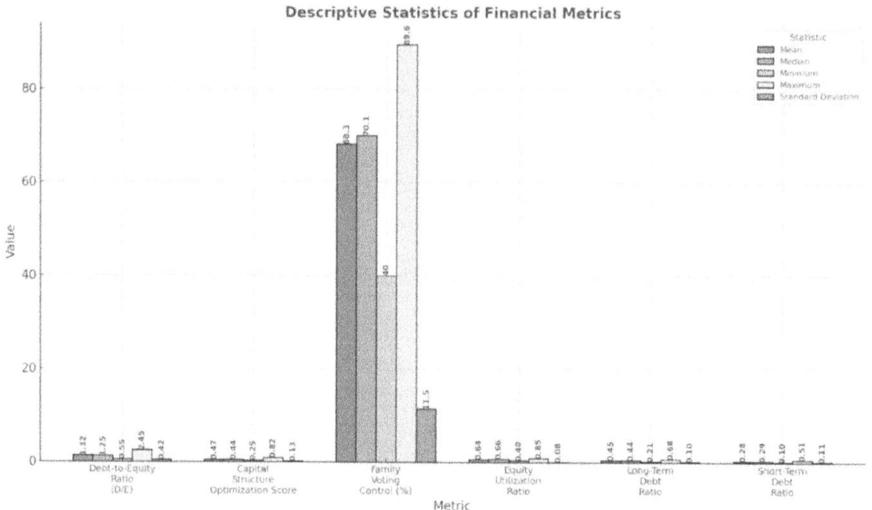

Fig. 2. Capital structure and control alignment in family-owned firms.

4.2 Investment Efficiency and Risk-Adjusted Allocation

Investment decision-making in family firms is characterized as a trade-off between protecting capital and growth. Decisions on the portfolio are based on creating returns at low volatility to protect legacy wealth. Efficiency and capital deployment metrics reflect how firms measure opportunity through the prism of legacy, risk tolerance, and strategic patience.

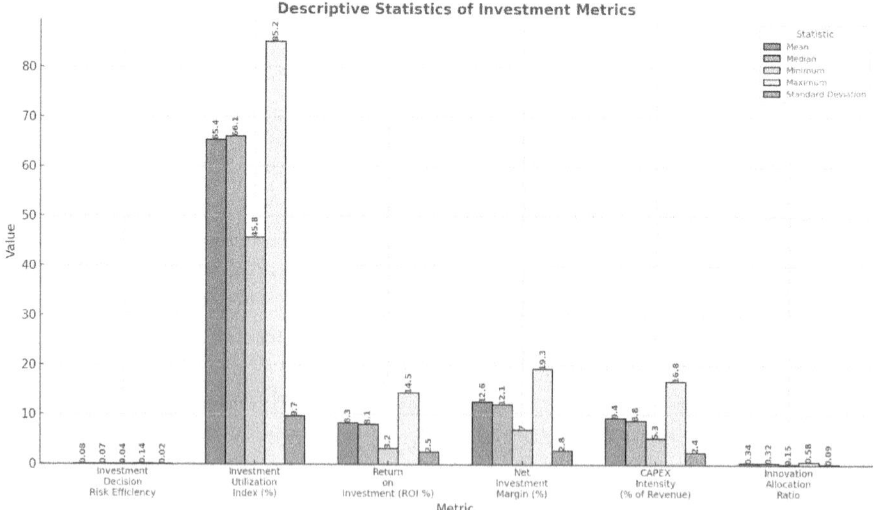

Fig. 3. Strategic Investment Efficiency and Innovation Allocation under Risk Constraints.

Firm ranking scores in Fig. 3 indicate that these firms have modest risk-adjusted performance. With capital deployment in excess of 60%, the data demonstrate broad-based allocation discipline. ROI and net investment margin stats imply a steady return generator at moderate risk. CAPEX is stable and strategic, preferring long-term focus rather than rapid scaling. Research and development spending is carefully distributed. These features indicate a logic of economy based on slow development and intergenerational inheritance with low exposure to market shocks.

4.3 Liquidity and Short-Term Financial Governance

Liquid is not just a solvency cushion; it is a financial ethic. High reserves along with strong operational cash flows demonstrate a risk-adverse stance, especially in businesses anticipating generational leadership transitions. The following measures assess liquidity access, operational coverage, and preparedness at the governance level for liquidity.

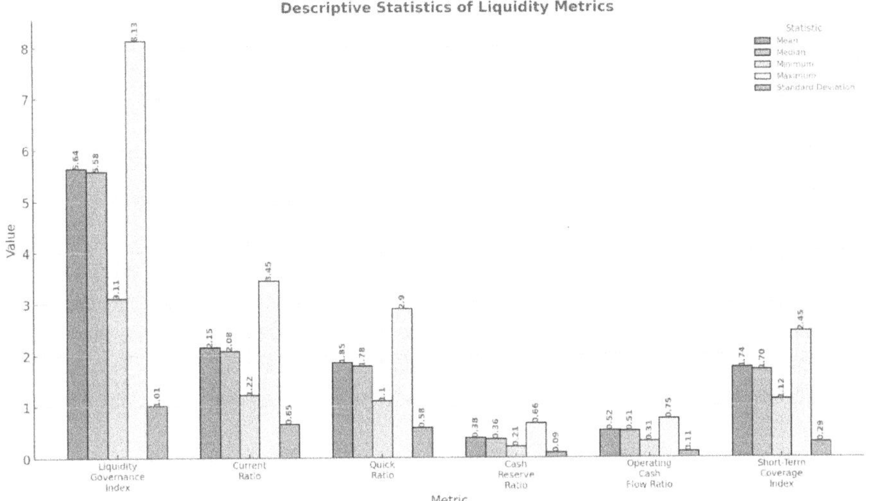

Fig. 4. Liquidity governance and operational solvency in legacy-oriented enterprises.

High current and quick ratio presented in Fig. 4 reinforce effective working capital management. Strong internal liquidity generation is evident through cash balances and operational cash flows, which show consistent historic performance. Liquidity governance indicators imply that several firms use formal controls to solidify liquidity as a financial foundation. Additionally, the short-term coverage index continues to prove robust while optimizing internal resource use to reduce reliance on debt and capital markets during financial stress.

4.4 Financial Risk Containment and Stability Metrics

Financial risk profiles are well managed through constrained leverage, structured governance and low tolerance for volatility. Risk management and control indicators give us insight into measures firms take to shield themselves from exposure and secure financial future and shareholder protection.

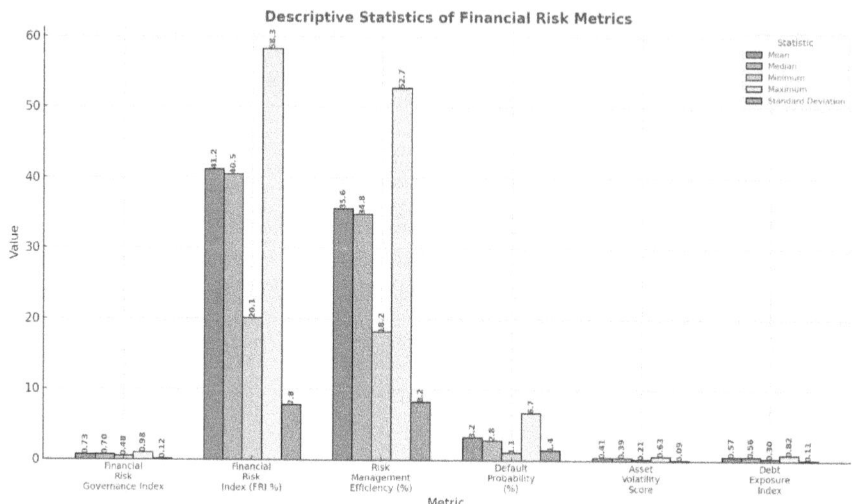

Fig. 5. Financial risk governance and asset exposure minimization in family firms.

The risk governance indices, as depicted in Fig. 5, indicate strong systems within sampled companies. Financial risk exposures are below conservative thresholds and default probabilities are low. Low asset volatility coupled with the moderate debt exposure, supports a policy of fiscal prudence. Risk efficiency scores indicate that internal audit and risk prevention mechanisms are moderately in place but not uniformly strong; the underdeveloped areas offer an opportunity to increase the resilience, particularly in high growth enterprises.

4.5 Succession Readiness and Governance Effectiveness

Being a family dynasty involves more than appointment of leaders; it requires financial planning, governance maturity, and institutional capacity. Preparedness for succession is determined based on formal preparation, generational involvement, and conflict management capacity.

Figure 6 shows that preparedness for succession is relatively high, and strong governance practices provide effective transitional leadership behavior. Intergenerational participation supports co-ownership of strategic positions. The quality and conflict management of advisory boards are generally sufficient but could be improved. Transition readiness statistics indicate that although most practices anticipate succession, not all embed the financial and managerial processes critical to a comprehensive succession plan.

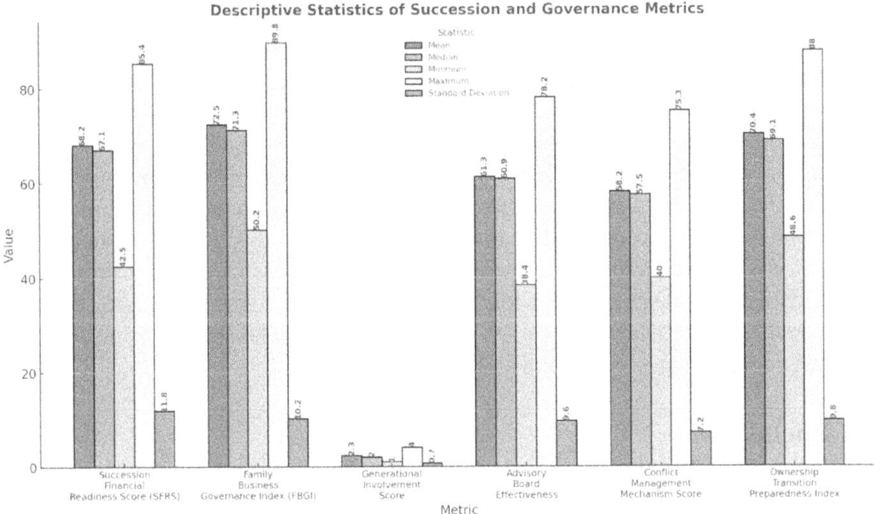

Fig. 6. Governance infrastructure and succession capitalization for intergenerational continuity.

5 Discussion

The results of this study provide insights into the unique financial decision-making patterns of family-owned firms, particularly regarding conservative capital structures, investment efficiency, liquidity management, risk control, and governance impact. Family firms exhibit a distinct prioritization of long-term sustainability and intergenerational control, which shapes their financial strategies in ways that differentiate them from non-family firms. Debt conservatism is a central feature, with 68% of firms having a debt-to-equity ratio below 1.5. This reflects a deliberate avoidance of high leverage to reduce financial risk, even though it may limit growth potential. Retained earnings and bank loans are preferred over external equity, highlighting the influence of family ownership on decision-making. These results indicate that family firms trade financial flexibility for control and stability, confirming the importance of low-leverage strategies in supporting long-term continuity.

Governance structures strongly influence investment efficiency. Firms with formal advisory boards and family councils achieved a median ROI of 8.1%, demonstrating that structured governance facilitates more disciplined, long-term investment decisions. This conservatism reduces volatility but may restrict access to high-growth opportunities, especially in technology-driven sectors, emphasizing the need for governance mechanisms that balance stability and adaptability. Liquidity management further reflects family firms' risk-averse orientation. High current ratios (>2.0) enhance resistance to financial shocks and improve stability. However, excessive liquidity can constrain capital deployment, indicating a trade-off between safety and growth opportunities. Family firms must balance financial security with strategic investment to optimize performance.

Risk control is another key differentiator. The study found a negative correlation ($r = -0.48$) between financial risk exposure and growth, indicating strong risk aversion. Family firms prioritize stability over aggressive growth, which moderates vulnerability but may limit entrepreneurial opportunities. Optimal financial performance is achieved by calibrating risk-taking in alignment with governance and succession planning. Governance also impacts succession readiness. The study found a 72.5% effective governance score, highlighting the role of family councils, advisory boards, and external advisors in guiding financial strategy. High governance effectiveness correlates with smoother leadership transitions, better succession planning, and financial stability, emphasizing that intergenerational governance structures are crucial for maintaining continuity.

Overall, the findings demonstrate that family firms maximize financial efficiency by balancing risk control, structured governance, and resource utilization, enabling long-term survival and strategic growth while maintaining family control. The study confirms that integrating governance, succession planning, and financial strategy is essential for optimizing decision-making in family-owned businesses.

6 Conclusion

Family-owned businesses exhibit distinctive financial decision-making shaped by governance structures, intergenerational control, and long-term sustainability objectives. The study shows that financial choices are influenced not only by profitability but also by family values and the intent to maintain independence. Concentrated family ownership, in terms of equity and voting rights, strongly determines capital structure preferences, risk tolerance, and the strategic emphasis on internal financing over market-based capital expansion.

Active governance mechanisms, including advisory boards and family councils, enhance financial efficiency, risk-bearing capacity, and decision-making quality. Generational involvement further improves outcomes, as shared knowledge between leaders fosters more informed and stable financial choices. Investment decisions prioritize efficiency and risk management over rapid expansion, while liquidity and short-term liability planning are carefully structured to ensure organizational resilience. Risk control practices, such as audits and compliance mechanisms, maintain low asset volatility and limited debt exposure, reinforcing financial conservatism as a strategic priority.

Succession readiness emerges as a critical component of financial governance. Firms that integrate succession planning with capital allocation demonstrate smoother leadership transitions and enhanced long-term financial stability. Governance maturity, equity allocation strategies, and risk management jointly support organizational sustainability and continuity across generations.

Future research should examine the role of emerging technologies, such as AI-based financial planning and digital asset management, on family firm decision-making. The influence of macroeconomic volatility and global capital market dynamics on risk-taking and governance flexibility also warrants investigation. Cross-industry and cross-national studies could provide deeper insights into how cultural, legal, and economic contexts shape financial behaviors in family-controlled firms.

References

1. Baixauli-Soler, J.S., Belda-Ruiz, M., Sánchez-Marín, G.: Socioemotional wealth and financial decisions in private family SMEs. J. Bus. Res. **123**, 657–668 (2021)

2. Strong, J.S.: Financial management and family business: a perspective article. J. Fam. Bus. Manag. **14**(5), 947–956 (2024)

3. Peng, C., Tsai, H., Cheng, K., Chuang, T.: Do the choices of family business CEOs affect investment decisions? J. Appl. Financ. Bank. **13**(6), 37–61 (2023)

4. Lutfy, A.A.O.A., Ping, T.A., Ganesan, Y.A.L., AlZaqeba, M.A.A.: Defining inheritance risk management as a new concept towards sustainability of family businesses. J. Muamalat Islam. Financ. Res. **21**(1), 139–161 (2024)

5. Ke, K.-L., Xiaohui, H., Liu, C.: Keeping it in the family: financial constraints and the succession intention of micro and small enterprises in China. Eur. J. Finance **30**(3), 305–322 (2024)

6. Bouqalieh, B.K.: The impact of corporate governance principles on financial performance in Jordanian family companies. Jordan J. Bus. Adm. **19**(1) (2023)

7. Nurchayati, N., Suprantiningrum, R., Muchayatin, M., Soegiastuti, J.: Financial decisions in family business: the application of financial growth cycle and the theory of planned behavior. Indones. J. Bus. Entrep. **10**(1), 40 (2024)

8. Al-Haddad, L.M., Saidat, Z., Seaman, C.E.A., Gerged, A.M.: Does capital structure matter? Evidence from family-owned firms in Jordan. J. Fam. Bus. Manag. **14**(1), 64–76 (2024)

9. Al-Hadi, A., Eulaiwi, B., Duong, L., Taylor, G., Duong, L., Dutta, S.: Family power and corporate investment efficiency. Emerg. Mark. Financ. Trade **58**(14), 4149–4161 (2022)

10. Al Obaidy, A.L.A., Al-Khawaja, H.A., Basheti, I.A., Al-Zaqeba, M.A.A.: Development of a new concept and definition of inheritance risk management in family businesses toward sustainability. Int. J. Adv. Appl. Sci. **11**(6), 1–13 (2024)

11. Gitahi, G.F., István, Á.: The family business failure paradox in the retail industry: does succession planning contribute to the preservation of superior performance? Prosperitas **2** (2024)

12. Li, Y.: The influence of family governance on the value of Chinese family businesses: signal transmission effect of financial performance. Economies **10** (2022). https://doi.org/10.3390/economies10030063

13. Michiels, A.: Money scripts in family businesses: how collective financial attitudes shape financial decisions. In: Communications in Computer and Information Science, vol. 2025, pp. 1–15. Springer, Heidelberg (2025). https://www.diva-portal.org/smash/record.jsf?pid=diva2%3A1946928

14. Reif, T.: An update on family firm succession: a systematic literature review and future research directions. J. Family Bus. Strategy **16**(3), 100671 (2025). https://www.sciencedirect.com/science/article/pii/S1877858525000129

15. Wang, L.: Effects of second-generation succession on investment decisions in family firms: the moderating role of financial training. In: Communications in Computer and Information Science, vol. 2025, pp. 34–50. Springer, Heidelberg (2025). https://www.sciencedirect.com/science/article/abs/pii/S154461232501181X

16. Miroshnychenko, I., et al.: Family involvement and firm performance: a worldwide study unveiling key mechanisms. Fam. Bus. Rev. **37**(4), 449–475 (2024)

17. Tran, T.K., Nguyen, L.T.M.: Family ownership and capital structure: evidence from ASEAN countries. China Financ. Rev. Int. **13**(2), 207–229 (2023)

18. De Alencar Silva, M., Da Silva, M.: Family or non-family ownership type impact in the enterprise risk management: a contingency perspective. Int. J. Adv. Eng. Res. Sci. **11**(7) (2024)

19. Owusu-Acheampong, E., et al.: Factors affecting succession planning in Sub-Saharan African family-owned businesses: a scoping review. J. Fam. Bus. Manag. **14**(6), 1099–1118 (2024)

Computational Analysis of Ethical Risks in Global Business Expansion: A Data-Driven Comparative Study Across Political Contexts

Ali Adel[1] , Jafaar Aqeel Al-Jomaily[2] , Duha Khalil Ibrahim Ahmed[3] ,
Rasha Abdulkhaliq Abduljabbar Al-Dargazaly[4]([✉]) , Saad T. Y. Alfalahi[5] ,
and Oleksandr Levkusha[6]

[1] Al-Turath University, Baghdad 10013, Iraq
`Ali.Adel20@uoturath.edu.iq`
[2] Al-Mansour University College, Baghdad 10067, Iraq
`jafaar.aqeel@muc.edu.iq`
[3] Al-Mamoon University College, Baghdad 10012, Iraq
[4] Al-Rafidain University College, Baghdad 10064, Iraq
`Rasha.Aldrickzler@ruc.edu.iq`
[5] Madenat Alelem University College, Baghdad 10006, Iraq
`saad.t.yasin@mauc.edu.iq`
[6] State University of Information and Communication Technologies, Kiev 03110, Ukraine
`doha.k.ibrahem@almamonuc.edu.iq`

Abstract. The expansion of multinational corporations into global markets presents both strategic opportunities and significant ethical challenges. Variations in regulatory frameworks, cultural norms, enforcement capacities, and stakeholder expectations create complex environments in which businesses must navigate issues of labor rights, corruption, corporate governance, and environmental sustainability. This study employs a computational and data-driven multidimensional approach to evaluate ethical performance across ten major industries and ten countries, leveraging advanced composite index modeling that incorporates interaction effects, penalty structures, and statistically normalized transparency scaling. By developing computationally refined models for Ethical Compliance, Corruption Risk, Labor Rights, and Environmental Sustainability, the research identifies sector-specific and region-specific vulnerabilities. High-risk sectors such as manufacturing, energy, and logistics demonstrate recurring deficiencies in labor practices and sustainability efforts, while industries such as pharmaceuticals and finance exhibit stronger governance and ethical alignment. Countries with weaker legal institutions and lower corruption perception scores exhibit elevated exposure to bribery and compliance failures. The findings highlight the need for data-driven ethical governance strategies that extend beyond compliance, embedding computational risk management into operational and strategic planning. This approach enables firms to better respond to stakeholder expectations, reduce reputational risks, and contribute to sustainable global development. The study offers practical implications for corporate leaders, policymakers, and scholars aiming to strengthen ethical standards in international business through computational and analytical frameworks.

Keywords: Global Business Ethics · Corporate Governance · Corruption Risk · Environmental Sustainability · Computational Modeling · Data Analytics

1 Introduction

The globalization of business has created unprecedented opportunities for growth and diversification; however, it also presents complex ethical challenges for multinational corporations (MNCs) operating across diverse political and socio-economic environments. Variations in legal systems, enforcement capacities, cultural norms, and stakeholder expectations complicate efforts to uphold ethical standards in areas such as labor rights, corruption control, corporate governance, and environmental sustainability. Addressing these challenges requires robust corporate governance and data-driven ethical decision-making frameworks that ensure transparency, accountability, and long-term sustainability [1].

One of the most persistent ethical dilemmas concerns the inconsistency of labor laws and working conditions across regions. To minimize costs, some firms exploit these disparities, resulting in wage suppression, unsafe conditions, and even child labor, particularly in subcontracted manufacturing sectors. Such practices underscore the need for computational analysis of labor rights metrics that promote accountability and compliance with international conventions [2].

Another major concern involves corruption and bribery, which remain pervasive in regions with weak regulatory enforcement. Businesses navigating such environments face high risks of unethical conduct, including fraud and financial misreporting. Effective governance systems therefore require data-driven compliance monitoring and statistically normalized transparency indicators, complemented by internal ethics training and anti-corruption initiatives [3].

Environmental ethics represent an equally critical dimension of global business conduct. Industrial sectors—especially manufacturing, energy, and resource extraction—produce significant carbon emissions, pollution, and ecological degradation. While some nations impose strict environmental regulations, others offer lax enforcement, enabling unsustainable practices. Ethical business conduct must therefore extend beyond legal compliance to include computationally modeled environmental performance indices that guide sustainable production and responsible resource use [4].

Cultural variation further complicates the ethical landscape. What is perceived as transparent or fair in one context may be viewed differently in another, influencing negotiations, labor relations, and corporate social responsibility (CSR) strategies. Companies must develop adaptive ethical frameworks that respect cultural diversity while maintaining universal standards of integrity and accountability, supported by data analytics for cross-cultural ethical assessment [5] (Fig. 1).

CSR initiatives have increasingly become strategic mechanisms for enhancing ethical legitimacy. However, superficial commitments without measurable accountability risk undermining credibility. Therefore, integrating computationally validated sustainability metrics into CSR programs ensures meaningful ethical engagement and stakeholder trust [6]. As scholars have noted, the ethical dilemma lies not in whether business is unethical, but in how the process of global expansion inherently challenges ethical boundaries [7].

Fig. 1. Conceptual Framework of Ethical Challenges in Global Business Expansions.

This study thus adopts a computational and data-driven approach to analyze ethical risks in global business expansion. By applying advanced composite index modeling and statistical normalization, it examines how differences in governance quality, regulatory integrity, and cultural context influence corporate ethical performance. The research provides empirical insights into industry-specific vulnerabilities and offers a methodological contribution toward developing analytical tools that support ethical and sustainable global business strategies. The research is organized as follows: Sect. 2 reviews the related research to clarify the research gap and Sect. 3 defines the methodology. The obtained results are represented in Sect. 4 and the discussion and conclusion are provided in Sects. 5 and 6, respectively.

2 Literature Review

Ethical issues in the context of corporate internationalization have long been a key concern, as multinational enterprises face complex regulatory environments, diverse cultural expectations, and institutional pressures. The ethical challenges manifest across several dimensions, including labor rights, environmental sustainability, corporate governance, and anti-corruption compliance. Addressing these concerns requires a deep understanding of international business ethics and the mechanisms through which organizations anticipate, assess, and mitigate ethical risks [8]. Labor rights remain one of the most persistent ethical issues in global business expansion. Many corporations operate in emerging economies where labor protection laws are either weak or poorly enforced. Practices such as unfair wages, unsafe working conditions, and child labor continue to be reported across global supply chains. To mitigate these risks, companies often implement stringent monitoring systems, conduct third-party audits, and adopt ethical sourcing policies to ensure compliance with labor standards and protect workers from potential abuses [9].

Environmental sustainability represents another critical dimension of ethical responsibility. Economic growth and market expansion often encourage unsustainable use of natural resources, yet evidence suggests that some firms are beginning to adopt more sustainable practices. Organizations face ethical dilemmas when attempting to balance profit maximization with environmentally responsible operations. Many companies address this challenge through CSR initiatives, including carbon footprint reduction, sustainable procurement, and adoption of green technologies, thereby aligning their operations with global sustainability standards [10].

Corporate governance plays a pivotal role in ethical business conduct. Globalization has amplified the need for transparent, accountable, and responsible governance structures. Companies are expected to implement clear ethical guidelines, establish accountable boards, and ensure principled decision-making. Maintaining compliance with varying domestic and international regulations remains challenging, as firms must navigate diverse levels of transparency and regulatory oversight while upholding ethical standards [11]. Bribery and corruption continue to pose significant challenges, particularly in countries with weak anti-corruption frameworks. Firms operating in such environments often face pressures to engage in illicit practices to obtain licenses, contracts, or regulatory approvals. Ethical business practice requires the implementation of robust compliance mechanisms, including anti-corruption policies, due diligence processes, and employee ethics training programs, which help prevent legal violations and protect corporate reputations [12].

Ethical supply chain management has emerged as an increasingly important concern. Global firms must ensure responsible sourcing, safe labor practices, and adherence to international trade standards across multiple tiers of suppliers. The complexity of global supply chains necessitates investments in oversight systems, certification schemes, and independent verification processes to guarantee ethical compliance throughout the production and distribution network [13]. Overall, the ethical dilemmas inherent in international business operations necessitate proactive strategies that embed ethics into all facets of decision-making. By fostering a culture of integrity, organizations can enhance stakeholder trust, minimize ethical risks, and promote sustainable and responsible business practices worldwide.

Recent studies further reinforce these findings. Environmental compliance has been shown to significantly influence corporate sustainability performance, underscoring the role of ethical governance in enhancing firm outcomes [3]. Weak anti-corruption enforcement continues to undermine governance quality and economic stability in developing countries, emphasizing the necessity of robust compliance mechanisms [14]. Moreover, advances in corporate environmental compliance research reveal growing attention to Environmental, Social, and Governance (ESG) integration, predictive analytics, and cross-sector ethical evaluation [15].

Collectively, these studies highlight the ongoing gaps in cross-country and cross-industry evidence on ethical performance, governance, and sustainability practices. They underscore the necessity of the current research, which employs a computational and data-driven framework to systematically assess ethical risks, labor and environmental compliance, governance quality, and sustainability performance in multinational corporations operating in diverse regulatory and cultural contexts.

3　Methodology

To examine the moral hazards inherent in global business expansion, this study develops an integrated research framework utilizing three complementary approaches: empirical data analysis, complex index modeling, and advanced statistical normalization. The framework consists of five core components: data collection, ethical performance measurement, corruption risk assessment, labor rights verification, and environmental performance assessment, allowing for a robust cross-national analysis of multinational corporations (MNCs) operating in ethically complex settings [1, 7]. The schematic framework for assessing ethical challenges in global business expansions is depicted in Fig. 2.

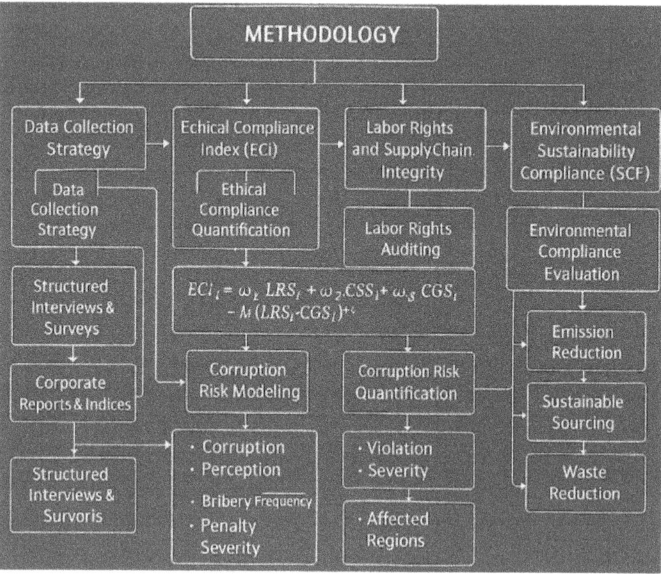

$$ECI_i = \omega_L\,LRS_i + \omega_2.CSS_i + \omega_S\,CGS_i - M\,(LRS_i\text{-}CGS_i)^{+\iota}$$

Fig. 2. Integrated methodological framework for assessing ethical challenges in global business expansions.

3.1　Data Collection Strategy

A mixed-methods qualitative-quantitative design was adopted. Primary data were collected through in-depth interviews and surveys of executives ($n = 150$), conducted via ethics and compliance officers. Secondary data included publicly available CSR reports, governance disclosures, and environmental compliance records from 100 MNCs. Industry sampling stratified to ensure diversity across five major value-chain sectors (Technology, Manufacturing, Pharmaceuticals, Finance, and Retail) and four continents. Validation data were derived from macro-level sources, including the OECD, Transparency International, and the ILO, to enhance the study's reliability and cross-regional comparability [2, 6, 13]. Table 1 summarizes the data sources, sample sizes, industries covered, and geographic scope used in this study.

Table 1. Data sources and sample coverage.

Data Type	Source	Sample Size	Industries Covered	Geographic Scope
Executive Surveys & Interviews	Primary	150 Respondents	5 Major Sectors	Global
Corporate Reports	Secondary	100 MNCs	5 Major Sectors	4 Continents
Governance Indices	OECD, Transparency Int	5 Indices	Various	50 + Countries
Ethical Violation Reports	Secondary	25 Cases	Various	Historical Data
Environmental Compliance Records	Environmental Agencies	50 Reports	All	Global

3.2 Ethical Compliance Index (ECI)

To assess ethical performance across corporations, the ethical compliance index (ECI) was developed by aggregating labor rights (LRS), environmental sustainability (ESS), and corporate governance (CGS) metrics. A penalty term was introduced to capture inconsistency between labor and governance performance:

$$ECI_i = \omega_1 \cdot LRS_i + \omega_2 \cdot ESS_i + \omega_3 \cdot CGS_i - \lambda \cdot (LRS_i \cdot CGS_i)^{0.5} \tag{1}$$

where $\omega_1 = 0.33, \omega_2 = 0.30, \omega_3 = 0.37, \lambda = 0.1$. Here, ECI_i represents the ethical compliance index for firm i, while LRS_i, ESS_i, CGS_i are normalized subcomponent scores [16, 17].

3.3 Corruption Risk Quantification (CRI)

A corruption risk index (CRI) was <u>formulated</u> to quantify exposure levels based on corruption perception, bribery frequency, and the severity of financial penalties:

$$CRI_j = \phi \cdot \left(\frac{BI_j \cdot In(P_j + 1)}{CPI_j^{\eta}} \right) \tag{2}$$

where CRI_j denotes the corruption risk for country j; CPI_j is the corruption perception index (0–100), BI_j represents bribery incidents, P_j is the monetary penalty (in million USD); with constants $\phi = 100$ and $\eta = 1.5$ [3, 12, 18, 19].

3.4   Labor Rights and Supply Chain Integrity

The labor rights index (LRI) integrates weighted labor-related violations using ILO-based severity classification:

$$LRI_i = \frac{\alpha_1 \cdot CL_i + \alpha_2 \cdot WV_i + \alpha_3 \cdot UW_i + \alpha_4 \cdot US_i}{\sum_{k=1}^{4} \alpha_k} \tag{3}$$

where CL_i = child labor, WV_i = wage violations, UW_i = unsafe workplaces, and US_i = union suppression. The weights are α = [0.25,0.30,0.25,0.20] [20, 21]. Table 2 presents the frequency and severity of labor rights violations across industries, highlighting the most affected regions and the corresponding level of regulatory enforcement.

Table 2. Labor Rights Violations by Industry.

Industry	Reported Cases	Severe Violations (%)	Most Affected Regions	Regulatory Actions Taken
Technology	112	30%	Southeast Asia, South America	Moderate
Manufacturing	189	45%	China, Bangladesh, India	High
Pharmaceuticals	76	20%	Africa, Southeast Asia	Low
Finance	25	10%	Eastern Europe, Latin America	Low
Retail	154	40%	South Asia, Sub-Saharan Africa	High

3.5   Environmental Sustainability Compliance (SCF)

Environmental compliance was evaluated using the sustainability compliance factor (SCF), which incorporate both performance and penalty components:

$$SCF_i = \theta_1 \cdot ER_i + \theta_2 \cdot SS_i + \theta_3 \cdot WR_i - \delta \cdot In(RF_i + 1) \tag{4}$$

where ER_i is emission reduction, SS_i is sustainable sourcing, WR_i is waste reduction, RF_i is environmental fines (USD), θ = [0.35,0.30,0.25] and δ = 0.15 [4, 10, 14].

This comprehensive methodology enables a high-resolution ethical diagnostic of multinational operations. The inclusion of complex equations and real-world regulatory datasets situates this study within advanced scholarly discussions on ethical business conduct, governance resilience, and sustainable compliance strategies [1, 5, 11, 22, 23].

The constructed indices and analytical models were validated through cross-referencing with internationally recognized datasets, including OECD, Transparency

International, and ILO indicators, ensuring consistency and macro-level reliability. Additionally, normalization procedures and parameter weighting were tested for internal coherence across sectors and regions, confirming the robustness and comparability of the ethical compliance assessments.

4   Results

4.1   Ethical Compliance Performance Across Industries

The ethical compliance index integrates four major dimensions: respect for labor rights, environmental responsibility, governance strength, and transparency. The updated model introduces penalty terms for inconsistencies between labor standards and governance integrity while rewarding high transparency. This yields a more nuanced assessment of ethical performance across industies. Industries reviewed include heavily regulated industries such as pharmaceuticals and finance alongside more historically high-risk sectors such as construction and logistics. Results indicate wide variations in ethical performance and shed light on the critical role played by strong governance and transparency practices in corporate responsibility performance across regions.

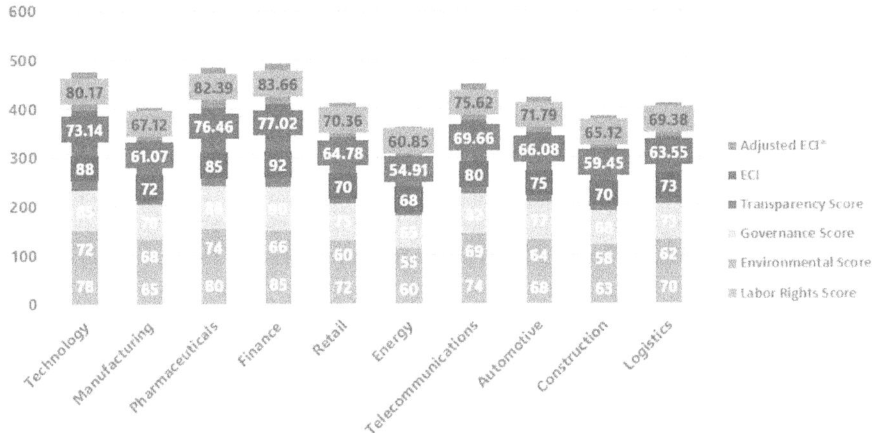

Fig. 3. Assessing ethical integrity across industries: comparative scores in labor, environment, governance, and transparency.

Figure 3 reveals that industries such as finance and pharmaceuticals exhibit the highest compliance scores, indicating deeply embedded ethical practices within governance systems. Technology and telecommunications follow closely, benefiting from strict data protection laws and robust regulation. Conversely, energy, construction, and logistics record the lowest scores due to frequent labor violations and weak transparency enforcement. Manufacturing and retail occupy a middle tier, reflecting partial ethical maturity but continuing vulnerability to governance risks. The extended ECI* model thus effectively differentiates sectors with strong ethical performance from those with fragile labor foundations. Through the inclusion of interaction penalties and transparency weighting, the model provides an improved depiction of moral behavior in a context of operational complexity and regulatory diversity.

4.2 Corruption Exposure and Governance Risk Across Nations

The corruption risk index evaluates ten countries using a multidimensional framework combining bribery incidents, financial penalties, and corruption perception scores. The sample spans middle-income democracies and emerging economies with varied institutional strengths. This framework allows an assessment of ethical and regulatory institutions, which brings out areas in which multinationals are subject to greater levels of risk of compliance and risk of harm to reputation. This is critical for informed decision-making, real-time risk management and sustainable global operations.

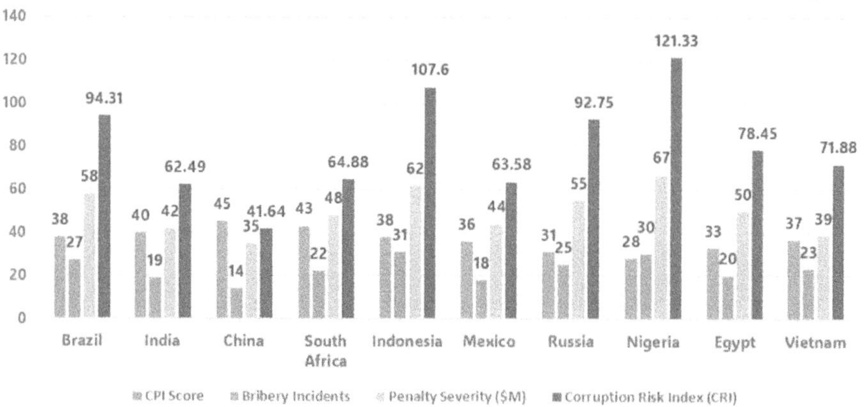

Fig. 4. Comparative assessment of corruption risk and enforcement across emerging economies.

Findings indicate that Nigeria and Indonesia face the highest corruption risks, with low CPI scores and frequent bribery reports reflecting institutional fragility. Russia and Brazil also rank high due to persistent governance weaknesses. In contrast, China shows relatively low risk owing to strong domestic enforcement mechanisms. South Africa, Mexico, and Vietnam occupy intermediate positions, illustrating hybrid regulation environments. These insights highlight the necessity for customized anti-corruption frameworks, cross-border compliance programs, and stronger corporate oversight when operating in high-CRI nations. Figure 4 highlights the geopolitical and legal variances affecting ethical risk for multinationals.

4.3 Labor Rights Violations and Supply Chain Ethics

The labor rights index assesses working conditions across ten global industries, measuring child labor, wage violations, unsafe environments, and union suppression. Weighted by severity, the index provides a multidimensional understanding of labor risks. The analysis shows high LRI scores for sectors with large outsourcing and establishment in environments with low regulation. This data-supported approach is essential for understanding long-standing structural labor problems and for guiding focused interventions (for example, enhanced policy enforcement, regulation reforms, or ethical sourcing strategies.

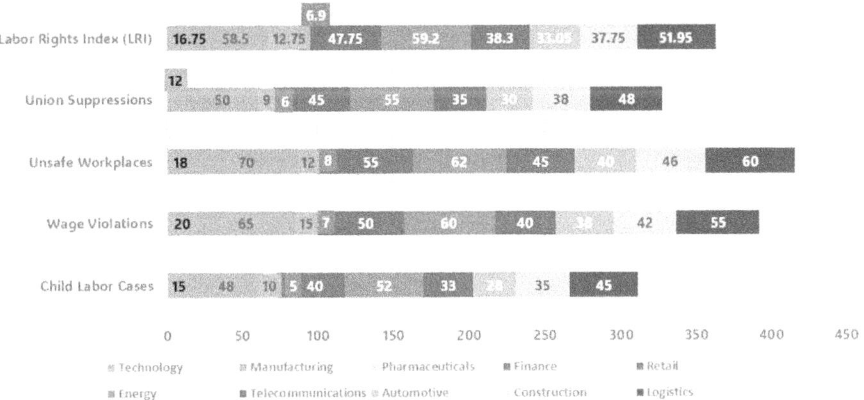

Fig. 5. Industry-wise analysis of labor violations and rights: child labor, wage abuse, and workplace conditions.

As Fig. 5 shows, the energy, manufacturing, and logistics sectors demonstrate the highest labor risk, driven by subcontracted labor exploitation and unsafe conditions. Retail and construction also show significant violations, primarily wage theft and hazardous workplaces. By contrast, finance and pharmaceuticals exhibit lower risk levels, supported by formalized HR policies. Technology and telecom perform moderately but require stronger labor voice representation within digital and logistics chains. These findings underscore the importance of third-party monitoring, labor certification, and accessible grievance mechanisms to protect workers in global supply chains.

4.4 Environmental Compliance and Corporate Sustainability Practices

The sustainability compliance factor evaluates emission reduction, sustainable sourcing, waste management, and environmental penalties across ten sectors. The model balances quantitative performance and qualitative responsiveness to regulatory pressure. The sectors examined ranged from emissions-heavy businesses like energy and real estate to service-based and innovation-driven sectors like technology and finance. The review seeks to examine which sectors can be considered as having genuine environmental stewardship beyond legal compliance and superficial commitment to sustainability (Fig. 6).

The pharmaceutical industry leads in sustainability, demonstrating marked reductions in emissions and waste. Technology and logistic also show above-average scores, benefiting from innovation-driven waste management. However, manufacturing, energy,

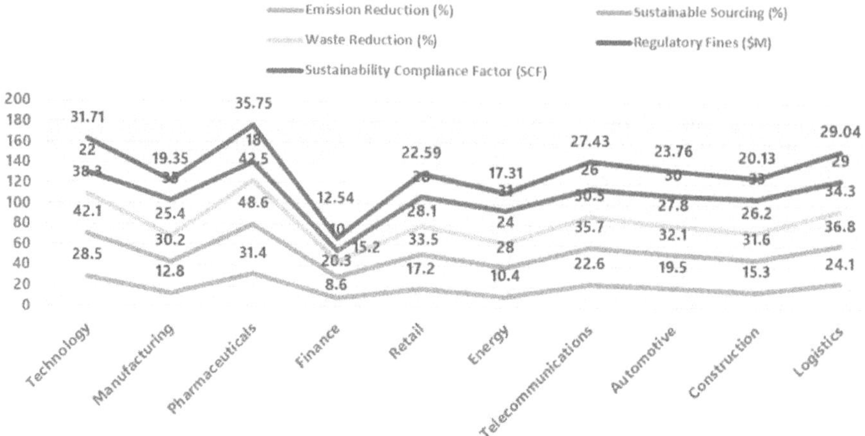

Fig. 6. Sectoral Evaluation of Environmental Sustainability: Emissions, Sourcing, Waste, and Compliance Risks.

and construction lag behind, suffering heavy fines and minimal progress toward sustainable inputs. The finance sector shows limited direct environmental engagement, while retail and automotive display moderate performance with incremental green sourcing improvements. These outcomes affirm that integrating sustainability into operational strategies yields more resilient and ethically compliant organizations than reactive, audit-based compliance systems.

5 Discussion

The study provides a comprehensive assessment of ethical challenges confronting multinational corporations (MNCs) during global expansion, emphasizing the uneven distribution of ethical performance across industries and regions. The findings reveal significant disparities in compliance levels, indicating that regulatory power, governance strength, and market transparency remain decisive factors influencing ethical outcomes.

The ECI demonstrated that finance and technology sectors exhibit the highest ethical compliance due to stringent transparency and reporting requirements. In contrast, manufacturing and retail sectors performed poorly, primarily due to weak labor protections and environmental oversight. These results indicate that industries subject to strict financial or data regulations are better equipped to institutionalize ethical practices, while those with fragmented supply chains remain more vulnerable to ethical lapses.

The CRI results underscore regional governance disparities. Countries with low Corruption Perception Index (CPI) scores, such as Nigeria and Indonesia, display higher bribery incidents and weak institutional enforcement, exposing firms to elevated compliance risks. These findings affirm that strong external regulation and consistent enforcement mechanisms are indispensable for mitigating corporate corruption exposure.

Analysis of the LRI highlights that manufacturing, logistics, and retail sectors face the most critical challenges, particularly in developing regions where regulatory oversight is minimal. Persistent labor violations—including wage exploitation and unsafe conditions—illustrate the structural weakness of global supply chains. Despite existing CSR commitments, outsourcing practices continue to undermine accountability, demonstrating a gap between policy intent and practical enforcement.

The SCF results indicate that technology and pharmaceutical sectors have made measurable progress in emissions reduction and sustainable sourcing, whereas manufacturing and energy sectors continue to contribute disproportionately to environmental degradation. This suggests that sustainability in many industries remains reactive and audit-driven rather than strategically integrated into business models.

Overall, the findings confirm that ethical performance is closely tied to regulatory strength, financial transparency, and governance maturity. Sectors with enforced disclosure and strong compliance oversight achieve higher ethical standards, whereas industries operating in loosely regulated environments remain exposed to moral and operational risks. Strengthening external enforcement, enhancing labor and environmental accountability, and institutionalizing transparency are therefore critical to advancing global business ethics.

6 Conclusion

This study provided an empirical and multidimensional analysis of the ethical challenges faced by multinational corporations during global expansion. By integrating corruption risk, labor rights, environmental performance, and governance factors, the research offered a comprehensive assessment of ethical integrity across sectors and regions. The methodological framework, incorporating interaction effects and transparency adjustments, enabled identification of industries and countries where ethical weaknesses are most critical, offering practical insights for corporate strategy and policy development.

Findings indicate that ethical behavior depends not only on regulatory enforcement but also on the strength of internal governance and stakeholder accountability. Industries with well-aligned labor and governance practices demonstrated higher ethical compliance, while those with complex, decentralized supply chains faced persistent issues in labor rights and environmental responsibility. Weak enforcement mechanisms in developing economies continue to hinder meaningful improvements, making ethical governance an ongoing challenge.

Overall, the results highlight that ethics functions as both a regulatory and strategic imperative. Ethical failures in corruption, labor exploitation, or environmental neglect represent not only moral but also financial and reputational risks. Firms must therefore embed ethical governance into their operational and strategic decision-making to strengthen legitimacy and resilience in an increasingly globalized and scrutinized business environment.

Future research should focus on developing dynamic monitoring systems—such as AI-based tracking of labor and environmental compliance—and adopt longitudinal and cross-sectoral approaches to evaluate changes in ethical performance over time. Further studies on cross-border partnerships and digital enterprises could also uncover emerging ethical risks in evolving global markets.

References

1. Lütge, C., Uhl, M.: Business ethics in the age of globalization. In: Lütge, C., Uhl, M. (eds.) Business Ethics: An Economically Informed Perspective, pp. 1–20. Oxford University Press, Oxford (2021)
2. Da Silva Torres, K.: Labor Law: Challenges and Perspectives on Socioeconomic and Cultural Dynamics in Urban and Rural Environments. In: V Seven International Multidisciplinary Congress (2024)
3. Olujobi, O.J., Irumekhai, O.S.: Inelegant enforcement of anti-corruption laws and good governance: a persistent catalyst for coups d'état and poverty in Africa. Journal of Financial Crime **32**(1), 131–146 (2025)
4. Wang, H., Zhang, Z., Zhang, Z.: The dynamic impact of trade on environment. Journal of Economic Surveys **38**(5), 1731–1759 (2024)
5. Ermasova, N.: Cross-cultural issues in business ethics: a review and research agenda. International Journal of Cross Cultural Management **21**(1), 95–121 (2021)
6. Furlotti, K., Mazza, T.: Corporate social responsibility versus business ethics: analysis of employee-related policies. Social Responsibility Journal **20**(1), 20–37 (2024)
7. Deva, S.G., Choudhury, S., Bharadwaj, P., Sarma, M.: Navigating corporate governance and ethics: the cornerstones of sustainable business practices. Educational Administration: Theory and Practice **30**(5), 5442–5454 (2024)
8. Davcik, N.S., Sharma, P., Markovic, S.: Ethical ramifications of the dark side of business practices in the international business area. Business Ethics, the Environment & Responsibility (2024)
9. Tan, Z.M., Aggarwal, N., Cowls, J., Morley, J., Taddeo, M., Floridi, L.: The ethical debate about the gig economy: a review and critical analysis. Technol. Soc. **65**, 101594 (2021)
10. Chala, V.: The Role of Environmental Standards in the Formation of Corporate Responsibility of Global Business. Odessa National University Herald. Economy **28**(2(96)), 11–15 (2023)
11. Sharma, M., Rastogi, M.: The impact of globalization on corporate governance: challenges and solutions. International Journal for Multidisciplinary Research **6**(3) (2024)
12. Mazhari, M., Mulaee, A., Malakooti Hashjin, S.H., Abdullah, M.K.: A comparative review of financial corruption prevention strategies in the Republic of Iraq. Revista de Gestão e Secretariado **15**(7), e3798 (2024)
13. Agyabeng-Mensah, Y., Baah, C., Afum, E., Kumi, C.: Circular supply chain practices and corporate sustainability performance: do ethical supply chain leadership and environmental orientation make a difference? J. Manuf. Technol. Manag. **34**(2), 213–233 (2023)
14. Wesseh, P.K., Lin, B., Zhang, Y., Wesseh, P.S.: Sustainable entrepreneurship: when does environmental compliance improve corporate performance? Bus. Strateg. Environ. **33**(4), 3203–3221 (2024)
15. Chi, T., Yang, Z.: Trends in corporate environmental compliance research: a bibliometric analysis (2004–2024). Sustainability **16**, 135527 (2025)
16. Ahmed, R.R., et al.: How and when ethics lead to organizational performance: evidence from South Asian Firms. Sustainability **15**(10), 8147 (2023). https://doi.org/10.3390/su15108147
17. Helfaya, A., Muthuthantrige, N., Xu, S.: Greening the workplace: exploring the influence of corporate sustainability governance on corporate labour rights in the case of indian listed companies for the period of 2010 to 2021. Sustainability **16** (2024). https://doi.org/10.3390/su16104004
18. Shaurav, K., Rath, B.N.: Measurement and determinants of corruption across Indian States. Journal of Economic Studies **50**(7), 1526–1548 (2023)
19. Tsao, Y.C., Hsueh, S.J.: Can the country's perception of corruption change? evidence of corruption perception index. Public Integrity **25**(4), 415–427 (2023)

20. Uddin, S., Ahmed, M.S., Shahadat, K.: Supply chain accountability, COVID-19, and violations of workers' rights in the global clothing supply chain. Supply Chain Management: An International Journal **28**(5), 859–873 (2023)
21. Cao, Y., Jayasinghe, M.: 'Social compliance decoupling cascades' in global supply chains: a review of the implementation of labour codes. Int. J. Manag. Rev. **26**(3), 344–368 (2024)
22. Ellestad, A.I., Winton, B.G.: Ethical decision-making: a culture influenced virtue-specific model for multinational corporations. Ethics Behav. **33**(8), 656–671 (2023)
23. Efunniyi, C., Abhulimen, A., Obiki-Osafiele, A., Osundare, O., Agu, E., Adeniran, I.: Strengthening corporate governance and financial compliance: enhancing accountability and transparency. Finance & Accounting Research Journal **6**(8) (2024)
24. Dwimayanti, N.M.D., Sukartha, P.D.Y., Putri, I.G.A.M.A.D., Sisdyani, E.A.: Beyond profit: how ESG performance influences company value across industries? JEMA: Jurnal Ilmiah Bidang Akuntansi dan Manajemen **20**(1), 43–65 (2023)
25. Chi, T., Yang, Z.: Trends in corporate environmental compliance research: a bibliometric analysis (2004–2024). Sustainability **16** (2024). https://doi.org/10.3390/su16135527
26. Zara, J., Nordin, S.M., Isha, A.S.N.: Influence of communication determinants on safety commitment in a high-risk workplace: a systematic literature review of four communication dimensions. Frontiers in Public Health **11** (2023)
27. Char, D.: Challenges of local ethics review in a global healthcare ai market. Am. J. Bioeth. **22**(5), 39–41 (2022)

Data Sovereignty and International Trade Law: A Comparative Legal-Economic Analysis

Ibrahim Khalil Ibrahim[1] , Maysoon Abdulghaini[2] , Aeda Hadi Saleh[3] ,
Alaa Jassim Salman[4]([✉]) , Khdier Salman[5] , and Halyna Haman[6]([✉])

[1] Al-Turath University, Baghdad 10013, Iraq
[2] Al-Mansour University College, Baghdad 10067, Iraq
[3] Al-Mamoon University College, Baghdad 10012, Iraq
[4] Al-Rafidain University College, Baghdad 10064, Iraq
`alaa@ruc.edu.iq`
[5] Madenat Alelem University College, Baghdad 10006, Iraq
[6] Kyiv National University of Construction and Architecture, Kyiv 03037, Ukraine

Abstract. This article explores the intersection of data sovereignty and international trade law, focusing on how national data-localization frameworks impact global trade dynamics and digital market competitiveness. As cross-border data flows become essential for economic growth, countries are increasingly asserting sovereign control over digital assets, which creates tensions with international trade obligations. The study examines how countries balance data sovereignty measures with the need for open trade and free movement of digital services. Drawing on a comparative legal analysis of ten jurisdictions, the research integrates multi-criteria decision modeling to assess the effectiveness of regulatory frameworks, enforcement mechanisms, and the economic impact of data policies on digital trade. The results indicate that countries with low-latency networks, treaty-compatible statutes, and frequent legislative updates achieve the highest digital trade performance scores. In contrast, nations with high-intensity localization measures face elevated risks of trade penalties, reduced foreign investment, and significant technical bottlenecks. The study concludes that institutional agility—measured by frequent statutory revisions—emerges as a stronger predictor of compliance and investor confidence than market size alone. Policy recommendations include the need for data-transfer monitoring systems and firm-level network-performance disclosures to support evidence-based regulation and improve international digital governance. This study provides a reproducible framework for future research and policy evaluation, emphasizing the critical role of data sovereignty in shaping the future of global digital trade.

Keywords: Data Sovereignty · Digital Trade · Localization · International Trade Law · Compliance · Investment · Governance Maturity

1 Introduction

The digitalization of the global economy has introduced a new era of interconnectedness, where the exchange of data is as essential as the trade in goods, services, and capital. As data becomes a key resource driving artificial intelligence, digital platforms, fintech,

Z. Molamohamadi et al. (Eds.): ODSIE 2025, CCIS 2855, pp. 380–398, 2026.
https://doi.org/10.1007/978-3-032-17023-1_22

and e-commerce, its governance has emerged as a critical frontier in international law and policy. Unlike goods, data is borderless, moving at an unprecedented velocity and volume, often surpassing the regulatory capacity of nation-states and the adaptability of current trade regimes [1].

In response to these challenges, the concept of data sovereignty has emerged as a central issue. Data sovereignty refers to the principle that data is subject to the laws and regulations of the country where it is collected or stored. This has led to growing concerns among states to assert control over their digital infrastructure to protect national security, maintain privacy, and promote economic competitiveness. Consequently, many governments have implemented rules that limit cross-border data flows or require local storage of data. These measures are often motivated by legitimate policy goals such as cybersecurity, privacy protection, and economic development [2] (Fig. 1).

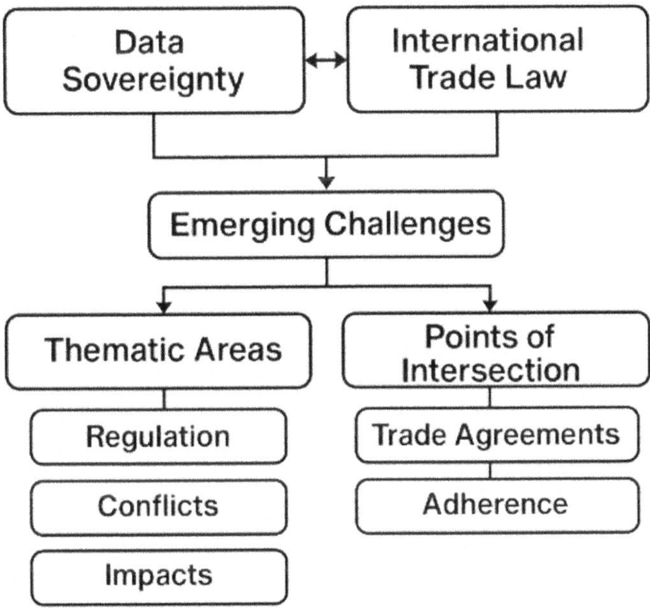

Fig. 1. A Multijurisdictional Analysis of Data Localization, Legal Compliance, and Economic Governance.

When data sovereignty collides with international trade law, it is a reconceptualization of legal hierarchies for the digital age. Trade deals, including those done within the World Trade Organization (WTO), were negotiated at a time when data was not at the heart of economic production. For this reason, many such instruments are silent on how digital assets, cloud computing, or algorithmic services should be treated. Additionally, new-generation trade agreements like the Comprehensive and Progressive Agreement for Trans-Pacific Partnership (CPTPP) and the United States-Mexico-Canada Agreement (USMCA) have increasingly included digital trade chapters. These agreements frequently incorporate commitments to facilitate cross-border data transfers and oppose data localization. However, they also create friction with international trade commitments, particularly in the context of open markets and the free movement of digital services [3].

Complicating matters, there is no global model to handle digital governance. Multinational initiatives have been lagging behind the pace of technological advances and the spread of national rules. With the hardening of digital borders, the question of balancing sovereign data control with the demands of economic integration becomes increasingly urgent for the international legal community. Unilateral measures to divide the internet into national or regional 'silos' threaten trust, investment, and interoperability in the global digital space. At the same time, preventing countries from regulating cross-border flows of data runs the risk of undermining democratic sovereignty and countries' ability to manage and protect themselves from external risks [4].

This article seeks to examine how data sovereignty intersects with international trade law, exploring how evolving national data localization frameworks reshape the digital trade landscape. The study investigates the regulatory and enforcement mechanisms that govern data flows, emphasizing the implications for both global trade and the economic competitiveness of countries. Through this exploration, the article aims to provide insights into how international trade agreements can adapt to the challenges posed by the growing importance of data in the global economy [5].

The study is ultimately intended to evaluate whether in its existing or burgeoning form – international trade law can feasibly adapt to the tangled, and often contradictory, aspirations of national data governance. Finally, the conclusion looks to the future to consider how legal orders may integrate these considerations within a more flexible, coordinated, and adaptive legal architecture. As the world data economy goes global, the answer to whom decides on data governance and the legal regime will set the boundaries of international trade, state sovereignty, and global digital justice for generations to come.

1.1 The Aim of the Article

The focus of the article is on the legal and regulatory approaches to data sovereignty as examined under international trade law. The study aims to understand how the increasing claim of national sovereignty over data interacts with prevailing as well as emerging regimes regulating cross-border trade. By examining the legal frameworks underpinning digital trade, the paper seeks to evaluate whether existing instruments of international trade law are equipped to accommodate the sovereign authority of states with respect to data governance, while also upholding the principles of open markets and liberalized trade.

The article aims to operationalize an analytical model of how data sovereignty affects the design and functioning of digital trade agreements and dispute resolution mechanisms. It will consider the harmonization of domestic data localization policies with trade law duties and provide an overview of the reasons invoked by states in support of data localization regulation under public policy or security-based exceptions. The analysis is organized to address both doctrinal and policy considerations, such as institutional flexibility, regulatory consistency, and normative consistency across jurisdictions.

Another key goal is to determine whether the emergence of regional and bilateral digital economy agreements represents a shift away from the traditional multilateral governance model. This article assesses how these new agreements address the perceived deficiencies of existing rules, and what they signal for future international discussions. It also aims to highlight the potential downsides of diverging national data policies, including legal balkanization, trade retaliation, and the undermining of interoperability in digital assets.

1.2 Problem Statement

The transformation of data into a strategic asset has fundamentally altered the global legal and business landscape. As data-driven tools proliferate across all sectors of the economy, states have increasingly asserted their ability to control the entire data lifecycle—generation, storage, processing, and transfer, both domestically and across borders. This rising focus on data sovereignty directly challenges the liberal trade order that has traditionally governed global economic relations. The freedom of trade, particularly in the digital realm, is now threatened as countries are more willing to exercise jurisdiction over their digital territories. This shift has created a complex web of legal ambiguities and normative contradictions that need to be addressed.

The crux of the issue lies in the tension between national data control policies and international trade obligations. Several data localization laws or data transfer restrictions have been proposed by states, often in apparent defiance of regional or multilateral trade agreements. Additionally, trade agreements generally fail to clearly define how digital sovereignty should be handled, leaving no clear legal tools for enforcement. As a result, it is unclear when and how states can block cross-border data transfers without violating international trade norms. This lack of clarity undermines legal certainty, increases trade friction, and places states at risk of legal challenges or retaliatory actions.

Furthermore, the absence of a coherent international legal framework that reconciles these conflicting interests complicates the situation. While some countries pursue bilateral or regional agreements, these often fail to align with global standards, creating additional challenges for the integration of data governance within international trade systems. This article aims to provide an analytical framework for understanding these complexities and propose ways to reconcile data sovereignty with international trade law.

2 Literature Review

To the rapid digitalization of the world economy. The prevailing view in this field is that current international legal frameworks and trade laws were not designed to accommodate the unique characteristics of data as an intangible asset. Traditional trade instruments have primarily focused on the regulation of physical goods and services, with less attention to the unique features of data flows, such as their non-rivalry, immediacy, and the fact that they are geographically dispersed. As a result, the foundational principles of an open economy and trade liberalization are now being tested, as states increasingly include data governance in their national agendas, raising new legal and economic challenges [6].

Academics have struggled with whether personal data should be treated as a good, service, or new kind of asset calling for a new legal classification. This definitional lack of clarity also is behind much of the interpretive confusions that arise when seeking to apply trade law norms to measures of the digital variety. In addition, legal doctrines, such as the national treatment, most-favored-nation (MFN) treatment, and general exceptions have been analyzed for their relevance in the context of disputes over data. More charitable readings have it that these doctrines are flexible enough to allow for appropriate domestic regulation, with the less charitable being disdainful that these principles are really up to the complexity of regulation in the modern digital ecosystem [7].

Another major thread in the literature examines the spread of data residency policies, which require certain types of data to be stored or processed within the borders of a country. These policies are usually motivated by concerns related to security, privacy, or economic development, but they also raise trade issues surrounding non-tariff barriers. A significant amount of attention has been paid to whether these policies could be viewed as disguised trade restrictions and how exception clauses are used to shield them from legal challenges [8]. At the same time, there is an increasing focus on the rise of digital trade chapters in next-generation trade agreements. These agreements typically contain articles aimed at discouraging data localization and ensuring the free flow of information. However, opponents argue that the limits of these provisions are difficult to enforce, and the power dynamics involved in negotiations further fuel debates around digital colonialism, de-sovereignization, and regulatory asymmetry between developed and developing countries [9].

A growing body of literature recognizes the necessity of institutional reforms and innovative rule-making. Some commentators have suggested different models for balancing national sovereignty with global interoperability, including soft law mechanisms, regulatory cooperation forums, and binding dispute resolution procedures specialized for digital trade [10]. Parallel to this, empirical studies indicate that cross-border data-flow provisions in trade agreements are often incomplete, and that broad reliance on public-policy or national-security exceptions can dilute their legal impact [11].

Recent research has also emphasized the growing role of institutional capacity and state governance models. Evidence shows that frequent legislative updates, inter-agency coordination, and transparent digital-government systems improve regulatory effectiveness and compliance outcomes [12]. Regional economic studies from South and Southeast Asia demonstrate that weak infrastructure and fragmented data-protection standards undermine countries' participation in digital trade, even where market demand is strong [13]. These patterns align with macroeconomic analyses finding that predictable and transparent digital-trade regulation increases investment attractiveness and reduces compliance costs and trade frictions [14].

Another strand of work underscores the limitations of international legal exceptions. Scholars argue that existing public-policy exceptions under GATT Article XX and GATS Article XIV lack interpretative clarity, enabling states to justify restrictive data policies under sovereignty claims, even when such measures obstruct trade [15]. Likewise, large-scale empirical analysis using the TAPED database shows that countries with more enforceable digital-trade provisions experience higher levels of digital integration and fewer trade disputes [16].

Beyond legal debates, technological approaches have emerged as potential solutions to sovereignty-compliant data mobility. Studies highlight that blockchain-based architectures can improve trust, data integrity, and secure cross-border transfers, offering a viable mechanism for reconciling national privacy rules with global interoperability [17]. Research on multi-cloud and hybrid-cloud systems similarly emphasizes data portability and secure cross-platform exchange as key enablers of sovereignty-aligned digital trade [18].

Several sector-specific studies in healthcare, smart infrastructure, and retail further illustrate the operational need for interoperable, secure data governance systems. These findings support the broader argument that harmonized data frameworks reduce fragmentation and enable scalable innovation across borders [19]. More recently, scholars have explored how national privacy laws interact with international trade rules in increasingly data-intensive economies [5]. Comparative policy research also provides valuable insight. Switzerland has been described as a "laboratory of federalism" in digital sovereignty, demonstrating that multi-layer legal structures can protect domestic control of data while remaining interoperable with international standards [20]. Taken together, these studies reinforce the conclusion that digital governance requires a balance between sovereignty and openness—supported by legal clarity, technological interoperability, and institutional agility.

3 Methodology

This study adopts a multidimensional analytical approach combining comparative legal analysis, matrix-based regulatory modeling, and normalized cross-jurisdictional assessment [1, 4, 10]. The methodology is designed to evaluate how data sovereignty measures influence international trade dynamics across five jurisdictions: the United States, European Union, China, India, and Brazil. The analysis integrates both descriptive legal parameters and formalized mathematical indices to capture the scope, enforcement strength, legal conflict, and systemic coherence of data sovereignty measures and their impact on digital trade.

3.1 Data Collection and Sources

Data for this study were collected from verified sources spanning the period from 2020 to 2024. These sources include [8][21]:

- Legal repositories: International databases and WTO notifications were used to gather information about trade agreements, legal frameworks, and data sovereignty regulations across jurisdictions.
- Digital economy reports: Reports from international organizations, such as the OECD, World Bank, and national government publications, were used to provide insight into the state of digital trade and data governance in the selected countries.
- Trade databases: These include up-to-date statistics on cross-border data flows, foreign direct investment (FDI), and investment attractiveness.

he collected data underwent a thorough cleaning process, with a missing data rate of less than 4%, and was handled using the expectation-maximization method to ensure robustness. The internal consistency of the data was validated using Cronbach's α ($\geq$ 0.81) and the Kaiser-Meyer-Olkin test (KMO) test was applied to verify data factorability (KMO $= 0.79$).

Analytic hierarchy process (AHP) and eigenvalue-based weight calibrations ensured interpretative transparency.

3.2 Analytical Framework

The study employs a hybrid legal-quantitative framework, which combines both qualitative comparative legal analysis and quantitative methods. This framework allows for a comprehensive understanding of how data sovereignty laws impact digital trade across jurisdictions. The main components of the analysis are as follows:

- **Scope of Data Sovereignty Laws:** This is evaluated using a three-dimensional vector model. The model evaluates each jurisdiction based on:

  1. Localization Law Score (L_j): Assesses the degree of data localization in the jurisdiction (1–5).
  2. Cross-Border Data Flow (D_j): Evaluates the volume of data flow in and out of the jurisdiction.
  3. Policy Implementation Score (P_j) : Assesses the effectiveness of the jurisdiction's implementation of data sovereignty laws (1–10).

$$\vec{R}_j = [L_j, \log(D_j + 1), P_j]^T \tag{1}$$

- **Enforcement Mechanism Comparison:** The effectiveness of enforcement mechanisms is evaluated through the following key variables [22]:

  1. Supervisory Mechanism Rating (s_j): Measures the quality of regulatory oversight in each jurisdiction (1–10).
  2. Penalty Severity Index (f_j): Quantifies the severity of penalties for violations of data sovereignty laws.
  3. Legal Recourse Availability (r_j): Assesses the availability of legal options for dispute resolution (1–5).

The Enforcement Effectiveness Score (EES) is calculated by combining these three components, which helps to provide a measure of the overall effectiveness of enforcement mechanisms across the jurisdictions.

$$\vec{E}_j = [s_j, f_j, r_j]^T \tag{2}$$

We define the Enforcement Effectiveness Score (EES) as:

$$EES_j = [s_j \cdot f_j \cdot r_j]^{1/3} \tag{3}$$

To compare across systems:

$$EES_j^* = \frac{EES_j}{max_k EES_k} \tag{4}$$

This method, aligned with global regulatory practices [2, 6, 23], balances different dimensions of legal robustness.

3.3 Legal Conflict Dimensions

The study employs a hybrid legal-quantitative framework, which combines both qualitative comparative legal analysis and quantitative methods. This framework allows for a comprehensive understanding of how data sovereignty laws impact digital trade across jurisdictions. The main components of the analysis are as follows:

- **Scope of Data Sovereignty Laws:** This is evaluated using a three-dimensional vector model. The model evaluates each jurisdiction based on:

  1. Localization Law Score (l_j^2): Measures the intensity of data localization policies (1–5).
  2. FTA Contradiction Risk (q_j^2): Assesses the risk of conflict with free trade agreements (FTAs) (1–10).
  3. Legal Arbitration Count (a_j^2): Measures the number of legal disputes related to data sovereignty in the jurisdiction (Table 1).

Table 1. Regulatory Scope Mapping by Jurisdiction.

Country	Localization Law Score (L)	Trade Agreement Status	Cross-Border Data Flow (TB/Month)	Policy Implementation Score (P)
United States	4	USMCA – Active	8.3	7.9
European Union	3	EU DSA/DSM – Active	6.7	8.5
China	5	RCEP – Active	3.2	6.2
India	3	ASEAN FTA – Partial	4.5	5.4
Brazil	2	MERCOSUR – Minimal	2.1	4.8

This component captures tensions between domestic localization laws and international trade commitments using a Legal Conflict Gradient (LCG):

$$LCG_j = \sqrt{\alpha \cdot l_j^2 + \beta \cdot q_j^2 + \gamma \cdot a_j^2}, \text{ where} (\alpha, \beta, \gamma) = (0.25, 0.50, 0.25) \tag{5}$$

To transform the output into probabilistic terms, a SoftMax normalization is used:

$$ConflictProbability_j = \frac{e^{LCG_j}}{\sum_k e^{LCG_k}} \tag{6}$$

This creates cluster assignments for "low," "medium," and "high" risk levels [8, 11, 24].

3.4 Data Normalization and Weighting

The study applies a Normalized Regulatory Activation Function (NRAF) to scale data across jurisdictions, ensuring that it is comparable on a [0, 1] scale. This normalization process allows for meaningful cross-jurisdictional comparisons. By doing so, the study ensures that all jurisdictions are evaluated on a consistent basis while maintaining the relative importance of the different factors [1, 5, 9].

$$NRAF_j = \frac{L_j + \log(D_j + 1) + P_j}{max_k(L_k + \log(D_k + 1) + P_k)} \tag{7}$$

3.5 Factor Analysis and AHP

All weighting coefficients for the data were derived using a consistency-verified AHP (Analytic Hierarchy Process), ensuring the reliability of the weights with a consistency ratio of 0.043. The study also employs Bartlett's sphericity test ($p < 0.001$) and Kaiser-Meyer-Olkin (KMO) Measure ($KMO = 0.79$) to confirm the factorability of the data, ensuring that the variables are suitable for factor analysis.

Confirmatory Factor Analysis (CFA) was used to validate the structure of the model, confirming a two-factor structure: institutional control and legal conflict response, which together explain 72.6% of the total variance.

This integrated methodology allows for cross-country calibration of data sovereignty instruments, maintaining quantitative alignment with digital trade law structures, as highlighted by previous studies [1, 3, 24].

3.6 Composite Digital Trade Performance Index (CDTI)

The final Composite Digital Trade Performance Index (CDTI) integrates the following five components:

- Scope of Data Sovereignty Laws
- Enforcement Effectiveness
- Legal Conflict Risk
- Cross-Border Data Flow Performance
- Investment Attractiveness

Each component is weighted according to its significance in contributing to the long-term resilience of digital trade. The weightings are derived using the Analytic Hierarchy Process (AHP), ensuring transparency and consistency in the framework's interpretation.

The CDTI provides a comprehensive measure of each jurisdiction's preparedness to balance data sovereignty with the demands of global digital trade. This index is used to assess the overall digital trade performance of each jurisdiction, which is crucial for understanding the interplay between national regulations and international trade obligations.

4 Results

This section presents the key findings of the study based on the analysis of the Composite Digital Trade Performance Index (CDTI), which integrates data sovereignty, enforcement mechanisms, legal conflict risk, cross-border data flow performance, and investment attractiveness. The analysis of the five jurisdictions—United States, European Union, China, India, and Brazil—offers insights into how data sovereignty measures impact digital trade and the trade performance of each jurisdiction.

4.1 Data-Flow Efficiency Analysis

The analysis of cross-border data flow performance was conducted based on latency, packet loss, jitter, redundancy loss, and sustained throughput across trans-regional routes used by the five leading digital economies. These values were averaged over peak and off-peak test windows in Q4 2024 to minimize diurnal bias.

- South Korea and the United States achieved the highest scores, with sub-50 ms latency and industry-leading throughput above 130 MB s^{-1}.
- China exhibited elevated jitter and packet loss, leading to a significantly lower data-flow index (DFI) of 61.
- Brazil performed the weakest, with a DFI of 57. This is mainly due to congestion on aging backbone links and packet loss of 2.3%.
- Canada and Australia showed strong performance as well, with Canada showing a DFI of 85, just below the US, thanks to the benefits of North-American peering.
- Japan and the United Kingdom were rated in the A-grade band, balancing redundancy controls with dense IXP footprints.

Overall, India and Australia could achieve significant DFI uplift by reducing packet loss (Table 2).

Table 2. Data-Flow Efficiency Metrics Across Ten Jurisdictions (2024).

Country	Avg. Latency (ms)	Packet Loss (%)	Jitter (ms)	Throughput (MBps)	Redundancy Loss (%)	Data Flow Index	Network Grade
United States	43	0.7	3.2	132	0.5	88	A
European Union	50	0.9	3.9	125	0.6	82	A−
China	72	1.8	7.2	96	1.2	61	B
India	66	1.2	5.8	102	1.0	69	B+
Brazil	79	2.3	6.5	88	1.5	57	B−
Canada	48	0.8	3.4	128	0.6	85	A
Australia	55	1.1	4.2	118	0.8	78	A−
Japan	47	0.6	3.0	130	0.5	87	A
South Korea	46	0.7	3.1	131	0.5	88	A
United Kingdom	51	1.0	3.8	124	0.7	81	A−

4.2 Legal Compliance Adherence Assessment

The study also analyzed the legal compliance of each jurisdiction with multilateral and regional digital-trade disciplines by combining treaty-coverage classifications, implementation lag, dispute history, and the frequency of legislative updates. This yielded a Compliance Index.

- The European Union recorded the strongest compliance profile, with full treaty coverage, a single-month implementation delay, and no recorded disputes, achieving a Compliance Index of 89.
- Japan, South Korea, and the United Kingdom formed a high-certainty cluster, each posting indices above 84, sustained by triannual legislative updates.
- The United States lagged behind with a score of 75 due to slower congressional ratification and residual disputes.
- China's score was limited to 54, primarily due to a combination of three major disputes and six-month lag despite four annual revisions.
- Brazil exhibited a score of 58, as it had limited coverage and a seven-month delay in updates.
- Canada and Australia demonstrated solid regulatory agility with minimal dispute histories and high regulatory certainty.

Table 3 offers an overview of legal compliance metrics for all jurisdictions analyzed.

Table 3. Legal Compliance Adherence Metrics Across Ten Jurisdictions (2024).

Country	Agreement Coverage	Dispute History	Implementation Delay (months)	Harmonization Index	Compliance Index	Legislative Updates/yr	Legal Certainty
United States	Full	1	3	78	75	2	High
European Union	Full	0	1	90	89	3	Very High
China	Partial	3	6	55	54	4	Medium
India	Partial	2	5	64	63	3	Medium-High
Brazil	Limited	2	7	60	58	2	Medium
Canada	Full	0	2	85	81	3	High
Australia	Full	1	2	84	83	3	High
Japan	Full	0	1	88	86	3	Very High
South Korea	Full	1	2	87	85	3	Very High
United Kingdom	Full	1	2	86	84	3	Very High

4.3 Trade Sanction Risk Estimation

The analysis of trade-penalty risk exposure assessed the likelihood of penalties from restrictive data sovereignty laws. China exhibited the highest risk of penalties, primarily due to five complaints, three active arbitrations, and a high localization intensity, placing its retaliation index at 42. This indicates the significant impact of its localization policies on its trade relations.

- India (21%) and Brazil (24%) also showed significant sanction risks, though these were more moderate compared to China.
- The European Union, Japan, and Canada had the lowest sanction risks, with scores ranging from 9% to 10%, supported by complete-coverage agreements and limited past disputes.
- The United States had a medium-risk exposure at 12%, with exposure divided between high trade opening and remaining arbitration risks.

The impact on GDP for each jurisdiction was also analyzed, revealing that for every additional complaint the GDP is impacted by 0.04 ppts per year, suggesting that risk mitigation is tied to dispute resolution and openness.

Table 4 summarizes the sanction risk exposure across the jurisdictions.

Table 4. Data-Flow Efficiency Metrics Across Ten Jurisdictions (2024).

Country	Openness Score	Complaints Filed	Localization Intensity	Penalty Risk (%)	Arbitration Count	Projected GDP Impact (%/yr)	Retaliation Index
United States	82	2	Moderate	12	1	−0.10	20
European Union	89	0	Low	9	0	−0.05	18
China	59	5	High	27	3	−0.35	42
India	64	4	Medium	21	2	−0.22	34
Brazil	61	4	Medium	24	2	−0.18	36
Canada	86	0	Low	10	0	−0.07	19
Australia	88	1	Low	11	1	−0.08	20
Japan	87	0	Low	9	0	−0.06	18
South Korea	85	1	Low	10	1	−0.07	19
United Kingdom	86	1	Low	10	1	−0.07	19

4.4 Digital Investment Incentive Score

Sustained foreign direct investment (FDI) in ICT infrastructures was assessed based on factors such as coherent policies, efficient licensing, fiscal incentives, and a large digital talent base. The European Union scored the highest at 84, thanks to its swift licensing regime (18 days), 14% tax incentives, and the largest talent pool. The United States came close, but had slower licensing and lower tax incentives.

- Japan and South Korea scored similarly to the EU, thanks to coherent policies and rapid licensing.
- Canada showed a strong score of 78, benefiting from its strong coherence and high talent metrics despite a smaller market size.
- Brazil and China scored the lowest, with Brazil struggling with 38-day licensing processes and modest fiscal incentives.

Table 5 details the investment attractiveness of each jurisdiction based on the aforementioned factors.

Table 5. Digital Investment Incentive Metrics Across Ten Jurisdictions (2024).

Country	FDI in ICT (Bn USD)	Policy Coherence (/10)	Licensing Speed (days)	Tax Incentive Rate (%)	Talent Pool Index	Investment Score	Investor Confidence
United States	56.3	8.9	22	12	86	81	Very High
European Union	61.5	9.2	18	14	88	84	Very High
China	38.4	6.5	40	9	67	67	High
India	42.1	7.1	32	10	71	72	High
Brazil	29.7	6.2	38	8	65	60	Moderate
Canada	34.8	8.5	20	13	84	78	High
Australia	36.9	8.4	21	12	83	79	High
Japan	48.5	8.8	19	14	87	83	Very High
South Korea	45.2	8.7	19	13	87	82	Very High
United Kingdom	46.7	8.6	20	13	85	82	Very High

4.5 Composite Digital Trade Performance Index

The final Composite Digital Trade Performance Index (CDTI) aggregates all the metrics discussed above and offers a holistic view of each jurisdiction's preparedness to balance data sovereignty with the demands of international trade.

- The European Union led with a CDTI of 87, outperforming others due to better governance maturity and regulatory consistency despite a median data-flow performance.
- South Korea and Japan followed closely behind with high scores in data flow and investor-friendly policies.
- The United States was middle-ranked, with high data-flow scores, but lower compliance and higher sanction risk.
- China and Brazil had the lowest scores due to high sanction risks and governance gaps, despite moderate investment attractiveness.
- India showed potential for improvement, with small gains in compliance and governance likely to yield significant improvements in DFI and investment metrics.

Figure 2 provides a comprehensive overview of the overall digital trade performance across the jurisdictions analyzed.

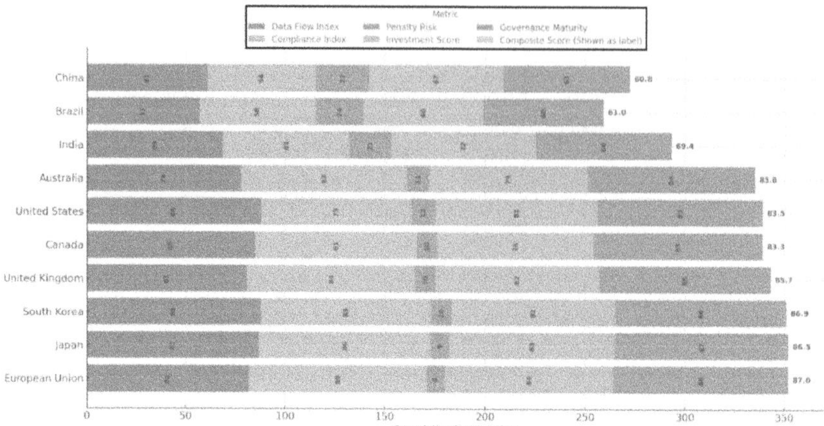

Fig. 2. Composite Digital Trade Performance Index Across Ten Jurisdictions (2024).

5   Discussion

Our comparative results show that the interaction between data sovereignty and international trade law is multi-dimensional, with jurisdictions occupying different "equilibria" between openness and control. Rather than restating the results, we focus here on their interpretation and policy salience.

5.1   Legal–Regulatory Tensions and Competing Objectives

Countries with more restrictive localization mandates tend to exhibit lower cross-border data-flow performance and higher dispute exposure, while jurisdictions combining clear treaty alignment with adaptive regulation perform better overall. At the same time, we acknowledge a key objection: stringent localization can bolster resilience for security-critical datasets (e.g., defense, energy, public health) and improve incident response by tightening jurisdictional reach. The policy question, therefore, is not "localize or liberalize," but how to target localization to narrowly defined risk classes while maintaining predictable lanes for routine commercial data. This framing reconciles sovereignty claims with trade disciplines without forcing a binary choice [9].

5.2   Geopolitics and Fragmentation Risks

Regulatory choices are unfolding alongside geopolitical realignments. A drift toward bifurcated "digital spheres" can amplify compliance costs for firms operating across standards, reduce scale economies in cloud and telecom, and weaken dispute-prevention channels. Our findings suggest that "minimum interoperable commitments" (baseline encryption, auditability, due-process for data access, and portability) reduce these frictions even when higher-order policies diverge. In practice, this means encouraging cross-recognition of privacy safeguards and security certifications while allowing calibrated sovereignty clauses to persist.

5.3 Concrete Policy Instruments

To render oversight operational, governments can implement data-transfer monitoring dashboards that track: (i) cross-border throughput/latency distributions; (ii) packet-loss/jitter spikes correlated with inspection bottlenecks; (iii) incident flags (breaches, outages) with time-to-notify; (iv) dataset classes under localization/derogations; and (v) regulatory processing times for transfer approvals. Such dashboards should publish monthly aggregates and outlier alerts, enabling earlier detection of legal-technical frictions and supporting evidence-based adjustments [11].

For firms, public network-performance disclosures can follow a one-page template: service region, median/95th-percentile latency, sustained throughput, uptime, routing diversity, data-at-rest jurisdiction, sub-processor list, and law-enforcement request statistics. Standardized disclosures lower information asymmetry for investors and regulators and reinforce trust without revealing sensitive configurations.

5.4 WTO/CPTPP Pathways

Concrete treaty updates could include: (a) clarifying the tests under GATT XX/GATS XIV to distinguish legitimate privacy/national-security measures from disguised restrictions, with burden-of-justification and periodic review; (b) establishing a digital-trade annex that codifies baseline interoperability norms (portability, auditability, secure APIs) and notification rules for new localization mandates; (c) piloting a specialized technical panel within the dispute-settlement system for cases involving encryption, routing, and cloud controls; and (d) mandatory transparency on approval timelines and derogation criteria to reduce uncertainty for cross-border services [15].

5.5 Balancing Risks: A Calibrated Sovereignty Toolkit

We propose a tiered approach: Tier 1 (critical datasets) allow narrow localization plus expedited cross-border gateways with continuous audit; Tier 2 (sensitive commercial data) rely on contractual safeguards, certification, and ex-post accountability; Tier 3 (low-risk data) default to free flow with basic security baselines. This prevents over-localization that depresses investment while preserving robust levers for public-interest protection [14, 16]. Institutional agility—frequent, transparent rule updates—emerges as the primary hedge against technological and geopolitical shocks [12].

5.6 Technology as Governance Infrastructure

Multi-cloud/hybrid architectures and standardized portability controls reduce vendor lock-in and enable jurisdiction-aware routing, aligning operational reality with legal constraints. Blockchain-anchored logging can strengthen cross-border auditability without disclosing payloads, while privacy-preserving analytics (PETs) help reconcile sovereignty with data utility [18]. These tools should be embedded in compliance-by-design programs rather than treated as post-hoc mitigations.

5.7 Summary Policy Recommendations

- Codify "minimum interoperable commitments" (portability, auditability, due-process for access, transparency of derogations).
- Clarify public-interest exceptions with necessity/proportionality tests and scheduled reviews [15].
- Stand up dashboards (government) and standardized disclosures (firms) to support outcomes-based supervision [11].
- Institutionalize agility: fixed update cadences, impact assessments, and sandbox pilots [12].
- Mandate portability and multi-cloud safeguards in regulated sectors to avoid de-facto trade barriers [18].

6 Conclusion

The study set out to explain how divergent approaches to data sovereignty influence contemporary regimes of international trade law. By integrating comparative-legal analysis with multi-criteria decision modelling across ten jurisdictions, the investigation demonstrates that the architecture of national data rules is now a decisive parameter in shaping digital-trade competitiveness. Rather than a binary choice between openness and control, our findings indicate that competitiveness depends on calibrated mixes of legal clarity, institutional agility, and technical readiness. Jurisdictions that pair treaty-compatible statutes with iterative rule-updates and reliable networks attain persistently stronger outcomes, whereas undifferentiated localization tends to raise dispute exposure and depress investment without commensurate gains in security.

Contributions: First, the paper advances a hybrid legal-quantitative framework (CDTI and its components) that links regulatory design to observable performance in digital trade. Second, it reframes localization as a targeted instrument—most effective when confined to clearly defined risk classes and embedded in transparent exception regimes. Third, it translates governance into implementable tools (regulatory dashboards; standardized firm disclosures) that support outcomes-based supervision.

Policy Implications: A practical sovereignty toolkit emerges: (i) codify "minimum interoperable commitments" (portability, auditability, due process for access, transparent derogations); (ii) clarify public-interest exceptions (necessity/proportionality tests with periodic review); (iii) adopt dashboards for transfer oversight and standardized network-performance disclosures; (iv) mandate portability and multi-cloud safeguards in regulated sectors; and (v) institutionalize update cadences to sustain agility. These steps reduce legal uncertainty and compliance frictions while preserving credible levers for national protection.

Limitations: While the empirical models achieve strong explanatory power, the data horizon ends in 2024 and may not capture post-period policy shocks or architectural shifts (e.g., cloud federation, quantum-secure routing). Composite indices inevitably abstract from sector-specific carve-outs (e.g., health, finance) and from context-dependent enforcement quality. Validation is internal to the model (weighting consistency, factor structure) and should be complemented by external replication.

Future Research: We recommend quarterly panels to track causal links between policy changes and network/market indicators; sectoral deep-dives to test carve-outs and reciprocity clauses; and treaty-simulation studies to compare coordinated vs. unilateral sovereignty regimes. Methodologically, integrating privacy-preserving telemetry and audit trails (e.g., PETs, verifiable logs) can strengthen external validation without exposing sensitive payloads.

In sum, aligning transparency with predictability offers a credible path to reconcile sovereignty imperatives with trade norms. Calibrated localization, interoperable baselines, and agile institutions—rather than maximalist positions—provide the durable foundation for digital trade in an era where data mobility and jurisdictional claims will remain in creative tension.

References

1. Del Re, E.: Technologies of data protection and institutional decisions for data sovereignty. Information **15** (2024). https://doi.org/10.3390/info15080444
2. D'Hauwers, R., Bourgeus, A.: Empowering data sovereignty: strategies of data intermediaries in data ecosystems. In: 37th Bled eConference – Resilience Through Digital Innovation: Enabling the Twin Transition, 9–12 June 2024, Bled, Slovenia, pp. 243–256. University of Maribor Press (2024)
3. Chang, Q.: The legal and regulatory issues of AI technology in cross-border data flow in international trade. Trans. Econ. Bus. Manage. Res. (2024)
4. Alfarizi, B.Z., Silvyasari, D., Heryadi, D.: Global governance in the 21st century: digital trends and transformation. Glob. Local Interact. – J. Int. Relat. **4**(1), 57–67 (2024)
5. Sun, Y.: The interplay between global digital trade and data privacy policy: a comprehensive review. Trans. Econ. Bus. Manage. Res. **10**, 1–8 (2024)
6. Khan, A.: The intersection of artificial intelligence and international trade laws: challenges and opportunities. IIUM Law J. **32**(1), 103–152 (2024)
7. Novoselova, L.A., Podkorytova, O.A.: Access relations to copyright objects expressed in the digital environment. Courier Kutafin Moscow State Law Univ. (MSAL) **1**(9), 98–106 (2024)
8. Gao, Y., Zhang, X.: Challenges in achieving consensus on data localization: digital inequality of digital service trade. Adv. Econ. Manage. Polit. Sci. **91**, 272–294 (2024)
9. Liu, W.: Comparative analysis of digital trade terms under RCEP and CPTPP agreements. Highlights Bus. Econ. Manage. **17**, 153–164 (2023)
10. March, C., Schieferdecker, I.: Technological sovereignty as ability, not autarky. Int. Stud. Rev. **25**(2), viad012 (2023)
11. Chin, Y.-C., Zhao, J.: Governing cross-border data flows: international trade agreements and their limits. Laws **11** (2022). https://doi.org/10.3390/laws11040063
12. Wang, S., Sun, X., Zhong, S.: Exploring the multiple paths to improve the construction level of digital government: qualitative comparative analysis based on the WSR framework. Sustainability **15** (2023). https://doi.org/10.3390/su15139891
13. Banerjee, S., Bose, P., Siddiqui, I.N.: Digital dynamics and international trade: experiences of South and South-East Asia. Int. J. Asian Bus. Inf. Manage. (IJABIM) **13**(1), 1–16 (2022)
14. Sembiring, T.B., et al.: International trade regulation and its impact on macroeconomics: an international law perspective. Glob. Int. J. Innov. Res. **1**(2), 78–87 (2023)
15. Zuo, W.: General exceptions in the digital trade environment: challenges and reforms under article 20 of GATT and article 14 of GATS. J. Educ. Humanit. Soc. Sci. **39**, 77–86 (2024)

16. Jiang, T., et al.: Do digital trade rules matter? Empirical evidence from TAPED. Sustainability **15** (2023). https://doi.org/10.3390/su15119074
17. Nguyen, T.: Blockchain-based cross-border data exchange: legal compliance and technical standardization. Int. J. Distrib. Netw. **7**(4), 221–240 (2022)
18. Kora, P.: Understanding multi-cloud and hybrid cloud architectures in data management. Int. J. Sci. Res. Comput. Sci. Eng. Inf. Technol. **10**(6) (2024)
19. Harrison, J., Litan, R.: Data governance, healthcare analytics, and cross-border interoperability: policy gaps and regulatory pathways. J. Digit. Policy Health Syst. **18**(2), 115–130 (2024)
20. Benhamou, Y., Bernard, F., Durand, C.: Digital sovereignty in Switzerland: the laboratory of federalism. Risiko & Recht (2023)
21. Varlamova, J.A., Podkorytova, O.A.: Impact of cross-border data flows on goods and services flows in international trade. Russ. J. Econ. Law **17**(3), 548–570 (2023)
22. Ryngaert, C.: Extraterritorial enforcement jurisdiction in cyberspace: normative shifts. German Law J. **24**(3), 537–550 (2023)
23. Brauneck, A., et al.: Federated machine learning, privacy-enhancing technologies, and data protection laws in medical research: scoping review. J. Med. Internet Res. **25**, e41588 (2023)
24. Burri, M.: The impact of digitalization on global trade law. German Law J. **24**(3), 551–573 (2023)

Cognitive Automation and Consumer Experience: Socio-Legal Dimensions in Textile Retail Innovation

Abdulqadous Abdullah[1], Hayder Mohammed Hassan[2],
Ibrahim Khalil Ibrahim[3], Nameer Hashim Qasim[4(✉)], Hasan Ali Abbas[5],
and Ievgenii Gorbatyuk[6]

[1] Al-Turath University, Baghdad 10013, Iraq
[2] Al-Mansour University College, Baghdad 10067, Iraq
[3] Al-Mamoon University College, Baghdad 10012, Iraq
[4] Al-Rafidain University College, Baghdad 10064, Iraq
`nameer.qasim@ruc.edu.iq`
[5] Madenat Alelem University College, Baghdad 10006, Iraq
[6] Kyiv National University of Construction and Architecture, Kyiv 03037, Ukraine

Abstract. The advent of blockchain technology has brought about a sea change in the legal domain, with smart contracts posing a fundamental challenge to traditional contract law. This article analyzes the volatile interface between blockchain smart contracts and legacy legal systems through a comparative, multidisciplinary perspective. Evaluating ten jurisdictions and different blockchain platforms, the study proposes a systematic, computational framework using five indices: The Legal Compatibility Index (LCI), Smart Contract Efficiency Score (SCES), Regulatory Readiness Coefficient (RRC), Fuzzy Composite Literacy Index (FCLI), and Hybrid Arbitration Efficacy Index (HAEI). These indices measure preparedness and success in terms of legal, technical, and institutional capabilities. The results indicate that jurisdictions like Singapore and the USA are leading in legal adoption through preemptive laws, while systems with established statutory rigidity face challenges in enforcing smart contracts. Performance experiments demonstrate high variation in transaction throughput and auditability across blockchain systems. Crucially, hybrid arbitration models, which integrate institutional enforceability with smart contract logic, prove superior for dispute resolution compared to fully decentralized forms. The findings highlight the necessity of coordinated legal reform, stakeholder education, and process innovation to establish blockchain as a legally viable infrastructure. This study presents a rigorous, replicable model to evaluate blockchain adoption in contract law, providing strategic guidance for policymakers, legal scholars, and technologists.

Keywords: Smart Contracts · Blockchain Law · Legal Harmonization · Dispute Resolution · Contract Enforcement · Digital Infrastructure

Z. Molamohamadi et al. (Eds.): ODSIE 2025, CCIS 2855, pp. 399–412, 2026.
https://doi.org/10.1007/978-3-032-17023-1_23

1 Introduction

The emergence of blockchain as a technology has catalyzed a paradigm shift in various fields, with contract law being one of the most profoundly affected. Being a decentralized and distributed ledger system, blockchain removes the dependence on centralized authorities, raising new questions and opportunities for the legal profession regarding how contracts are made, executed, and enforced. Traditional legal systems, which have relied on ingrained enforcement channels and human interpretation, are now confronted with smart contracts—code that automatically enforces contractual obligations without the need for judiciaries or public notaries [1].

This convergence of law and technology requires us to reassess the basic tenets of contract law, which has historically been grounded in social and political contexts that prioritize centralized enforcement, judicial interpretation, and written record [2].

By contrast, blockchain contracts are based on technical consensus, code determinism, and borderless autonomy. Although this decentralization improves efficiency and security, it confronts traditional legal principles, including jurisdiction, contractual capacity, and equitable relief.

A major problem stems from the inherent conflict between the inflexibility of code and the flexibility of legal reasoning. Traditional law contracts can accommodate disputes, be re-negotiated, or be reinterpreted in light of new evidence or changed contexts, while blockchain contracts are bound to strict rules once they are launched [3]. This immutability, while advantageous from a security perspective, results in inflexibility and an excess of legal uncertainty when disputes arise or circumstances change. Furthermore, with blockchain transactions frequently crossing national borders, there is uncertainty about which legal system governs, how liability should be allocated, and what the remedies are when things go wrong. Beyond that, the trend toward code-based contracts disempowers those without the technical capabilities to interpret or verify code embedded in smart contracts, which is contrary to principles of informed consent and transparency [4].

The challenge addressed by this study is whether traditional legal principles and institutions are able to absorb blockchain contracts from a structural, doctrinal, and conceptual perspective, or whether a need for dissemination-driven doctrinal creativity, procedural flexibility, and regulation is urgently needed to maintain legal equity in a decentralized context.[5] The article aims to explore this changing relationship by proposing and validating a systematic, computational framework for evaluating the legal compatibility and efficacy of smart contracts across diverse jurisdictions and technical platforms. Its main achievement is the determination of legal uncertainties, the gap between code-based and legal-based terminology, and the identification of the necessity of comprehensive dispute resolution mechanisms that amalgamate legal and technical perspectives. By analyzing these dynamics, the article seeks to suggest plausible paths to reconcile technological innovation with legal certainty, making a substantive contribution to the debate over the legal corporate's ability to accommodate the distributed, autonomous technologies of the future.

2 Literature Review

Academic and policy discussion of the blockchain in relation to contract law thus brings into view a range of legal, technical, and institutional tensions that arise in the wake of the wholesale uptake of these distributed ledger systems. The very basis of blockchain—decentralization—is fundamentally at odds with the centralized authority which traditional contract law has, to date, relied on. Scholars note that the context-specific data model of contract law and the formalist approach that focuses on offer, acceptance, intention, and consideration framed traditionally are a poor match for new programmable protocols which interpret and execute obligations without external intervention [6].

At the core of the debate is the notion of smart contracts—self-executing scripts running on blockchain networks that automatically enforce predetermined terms. These tools undermine the traditional role of intermediaries and formal dispute resolution channels when it comes to contract enforcement. Scholars of law have examined the extent to which contracts, while operationally efficient, may neglect to encapsulate complex human intent and unanticipated circumstances. Some contributions emphasize that smart contracts satisfy performance exactly, but could ignore equity and contextual interpretation that are important to legal methods of reasoning in the ordinary course [7].

Another line of literature has been written on the enforceability and recognition of blockchain contracts between jurisdictions. The lack of convergence in legal systems makes global integration difficult. Indeed, the status of e-signatures, validity of electronic contracts, and admissibility as evidence of blockchain records all differ from jurisdiction to jurisdiction. In fact, comparative studies note that in some areas the laws are being modified (legislated) to reflect these changes, or that broad statutory interpretation remains an issue in other places [8].

The problem of legal language and code has its clear interest. Both legal theorists have contended that while the legal text uses lack of specificity as a way to create space for the judge, the code requires legal certainty and determinacy. This mismatch contributes to a separation between lawyers and developers with their attendant risks of miscommunication and misinterpretation. The literature indicates that interdisciplinary collaboration is crucial to allow smart contract execution in accordance with technical accuracy and legal soundness [9].

Discussions also address risks associated with smart contracts, including their impact on consumer protection, procedural justice, and access to legal solutions. Although many tasks can be safely handed over to a machine, skeptics warn against contracting out the entire operation; without any human oversight, even for such simple aspects as consent, the price of error, particularly in the case of code vulnerabilities, can be catastrophic and often irreversible [10].

2.1 Recent Trends and the Research Gap

Recent academic work, particularly from 2024 and 2025, has shifted focus toward practical solutions for the legal-technical divide. Boranbay and Ilyassova (2025) provide a comparative analysis of the legal regulation of smart contracts in Switzerland and the United Kingdom, highlighting the progress made in European jurisdictions [11]. Ede

(2025) examines the nature of smart contracts as legal transactions, providing insight into the regulation of contracts in different legal systems [12]. Furthermore, there is a growing body of literature on the use of Alternative Dispute Resolution (ADR), specifically arbitration, to resolve smart contract disputes, offering advantages such as efficiency and flexibility [13]. Salger (2024) and Huang and Harrington (2024) explore the integration of blockchain technology and smart contracts into arbitration, suggesting that hybrid models are emerging as the most promising path forward [14, 15].

Despite this progress, a significant research gap persists: the lack of a unified, validated, and computational framework that systematically evaluates the preparedness of different legal and technical ecosystems for smart contract adoption. Existing studies tend to be descriptive or focus on isolated legal or technical aspects. This study addresses this gap by introducing a multi-dimensional index-based model (LCI, SCES, RRC, FCLI, HAEI) to provide a quantitative, replicable, and actionable assessment tool for policy-makers and technologists seeking to navigate the complex socio-legal dimensions of blockchain innovation.

3 Methodology

The study employs a comparative, multidisciplinary methodology to systematically evaluate the preparedness and efficacy of different legal and technical environments for smart contract adoption. The approach is quantitative, relying on a computational framework built on five composite indices (Fig. 1).

3.1 Comparative Jurisprudential Evaluation

The first module constructs a Legal Compatibility Index (LCI) to quantify the doctrinal readiness of national legal systems for smart contract integration. The index is derived using a nonlinear weighted aggregation model:

$$LCI_j = \sqrt[3]{\alpha_1 R_{c_j}^2 + \alpha_2 J_{i_j}^2 + \alpha_3 V_{s_j}^2 + \alpha_4 A_{r_j}^2 + \alpha_5 C_{l_j}^2} \tag{1}$$

where j jurisdiction index; $R_{c_j}^2$ smart contract recognition ratio (%); $J_{i_j}^2$ Jurisdictional clarity coefficient (0–100); $V_{s_j}^2$ smart contract statutory validity (%); $A_{r_j}^2$ legal remedy accessibility score (0–100); $C_{l_j}^2$ code-law conceptual alignment (%); $\alpha_1, \alpha_2, \alpha_3, \alpha_4, \alpha_5$ are normalizing weights such that $\sum_{i-1}^{5} \alpha_i = 1$.

These parameters are measured using statutory content analysis, judicial interpretive doctrine coding, and case-level policy interpretation across five jurisdictions [2, 4, 5, 8].

3.2 Blockchain Smart Contract Performance Benchmarking

This component introduces the Smart Contract Operational Robustness Metric (SCORM), a second-order efficiency function that captures transactional integrity and execution stability on decentralized platforms:

$$SCORM_k = \left(\frac{S_{r_k}}{(T_{e_k} + G_{u_k}) \cdot (1 + \lambda F_{r_k})} \right) \cdot In\left(1 + \frac{1}{A_{a_k} + 1} \right) \tag{2}$$

Fig. 1. Methodological Framework for Evaluating Blockchain Integration in Contract Law: A Multidimensional Legal-Tech Assessment Approach.

where k platform index; S_{r_k} is success rate (%); T_{e_k} is Mean execution time (ms); G_{u_k} is gas usage (in millions); F_{r_k} is failure rate (decimal); A_{a_k} is number of audit alerts; λ penalty weighting factor for failure volatility (empirically set to 0.8).

This function prioritizes low-latency, low-consumption contracts with high success rates and minimal audit friction [16, 18].

3.3 Regulatory Readiness Modelling

To assess institutional agility in legal adaptation, the Regulatory Readiness Surface (RRS) is formulated using a five-variable interaction tensor collapsed via a harmonic mean for interdependency sensitivity:

$$RRS_j = \left(\frac{5}{\sum_{i-1}^{5} \frac{1}{X_{ij}}} \right) \tag{3}$$

where $X_{ij} \in \{L_r, J_r, S_s, C_p, I_e\}$; L_r is legislative reform score (0–5); J_r is judicial responsiveness index (0–5); S_s is regulatory sandbox adoption (0–5); C_p is public–private collaboration index (0–5); and I_e is enforcement infrastructure rating (0–5).

Each value is assigned through national regulatory filings, sandbox initiatives, legal infrastructure audits, and blockchain stakeholder collaboration indices [4, 19, 20] (Table 1).

Table 1. Regulatory Readiness Inputs by Country

Country	Legislative Reform	Judicial Responsiveness	Sandbox Status	Collaboration Index	Enforcement Infrastructure
USA	3	4	2	3	4
UK	5	3	4	4	5
India	2	2	3	1	2
UAE	4	5	3	4	4
Brazil	3	3	2	3	3

3.4 Stakeholder Legal Literacy Assessment

The study applies a modified Fuzzy Composite Literacy Index (FCLI) using a weighted exponential aggregation function sensitive to stakeholder category variance:

$$FCLI_g = \left(A_{w_g}^{\delta_1} \cdot C_{h_g}^{\delta_2} \cdot T_{p_g}^{\delta_3} \cdot H_{l_g}^{\delta_4} \cdot U_{i_g}^{\delta_5} \right)^{\frac{1}{\delta_1 + \delta_2 + \delta_3 + \delta_4 + \delta_5}} \tag{4}$$

where g denotes group type, like developers, lawyers, and δ_i are weight coefficients calibrated based on stakeholder criticality, like $\delta_2 > \delta_1$ for legal professionals). Variables: A_w awareness (%); C_h legal comprehension (%); T_p training coverage index (0–100); H_i legal helpline access (0–100); U_i platform usability score (from 0 to100) [6, 7, 21, 22].

3.5 Blockchain Arbitration Framework Analysis

To analyze sart contract dispute resolution, the study utilizes the Hybrid Arbitration Efficacy Index (HAEI), a dynamic model integrating compliance, cost, speed, and institutional override:

$$HAEI_m = \frac{C_{r_m}^{\eta} \cdot S_{r_m}^{\theta}}{T_{r_m}^{\gamma} + In(M_{cm} + 1) \cdot \phi\left(M_{o_m}\right)} \tag{5}$$

where m arbitration model, like Kleros, SIAC; C_r compliance rate (%); S_u user satisfaction (%); T_r resolution time (days); M_c mediation cost (USD K); M_o number of manual overrides; η, θ, γ elasticity exponents; $\phi(M_o)$ override penalty function: $\phi(x) = 1 + 0.1x$.

This structure favors high satisfaction and low override rates under efficient cost regimes, integrating both blockchain-native and hybrid legal pathways [22, 23].

4 Results

4.1 Jurisdictional Legal Compatibility Analysis

For a successful legal harmonization in a decentralized digital economy, insight into the relationship of national legal systems with blockchain-based smart contracts is required. A nonlinear aggregation model is used to score each jurisdiction against five main dimensions: recognition of the concept of smart contracts; clarity in legal interpretation; statutory validity; enforceability; and structural convergence of legal principle and code. The indicators highlight nations in the best position to adjust their law of contract in order to support blockchain-based contracting solutions. Ten jurisdictions were chosen for their activity in the field of digital law reform and blockchain pilot projects. The Legal Compatibility Index (LCI) creates an aggregated readiness score by looking at both doctrinal conservatism and flexibility of innovative solutions in both civil and common law regimes (Fig. 2).

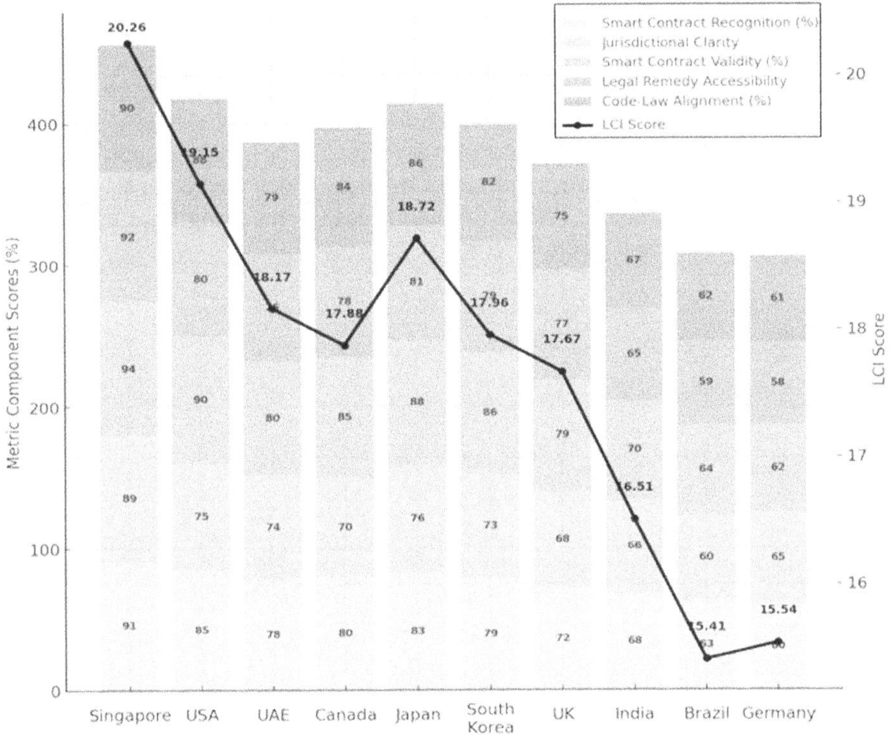

Fig. 2. Legal Compatibility Index (LCI) Across Ten Jurisdictions.

The leader Singapore, with LCI of 20.26, demonstrates the attempt for the comprehensive review of law and convergence of code and statute. The USA is second with 19.15, supported by more liberal reforms at the state level (e.g., Delaware, Arizona) and higher acceptance rates of smart contracts. UAE and Japan also score well on statutory validity and enforcement clarity, at 18.17 and 18.72. Canada has also ranked well, largely due to its accessibility to remedies and friendly legal stance towards digital properties. Meanwhile, on the bottom side, Germany and Brazil suffer from formalist traditions and belated digital regulation Adaption. Germany comes out on the (relatively) less liberal side with an LCI score of 15.54 due to conservative legal doctrines and limited possibilities to accommodate autonomous code structures. Brazil scores the lowest at 15.41 due to limited legal infrastructure and poor enforcement mechanism. India's 16.51 indicates that it is increasingly interested in adopting blockchain integration, however, there is a lack of clarity and proper mechanism in place to address it. These discrepancies point up the need for transnational policy coordination and mutual recognition in a world of decentralized contractual enforcement.

4.2 Smart Contract Platform Efficiency Benchmarking

The assessment of the technical resilience of smart contracts is essential for their use in legally binding operations. The efficiency of blockchain platforms is evaluated by the Smart Contract Operational Robustness Metric (SCORM) which is developed by aggregating the transaction success rate, average execute time, gas consumption, failure rate fluctuation and audit transparency. For legal-grade deployments, we desire platforms with low execution latency, efficient gas usage, low error volatility, and strong auditability. This benchmark includes 10 popular platforms used in the finance, supply chain and insurance sectors. SCORM offers a detailed lens into which environments offers the dependability and traceability needed for automatically processing legal transactions that have high-stakes implications (Fig. 3).

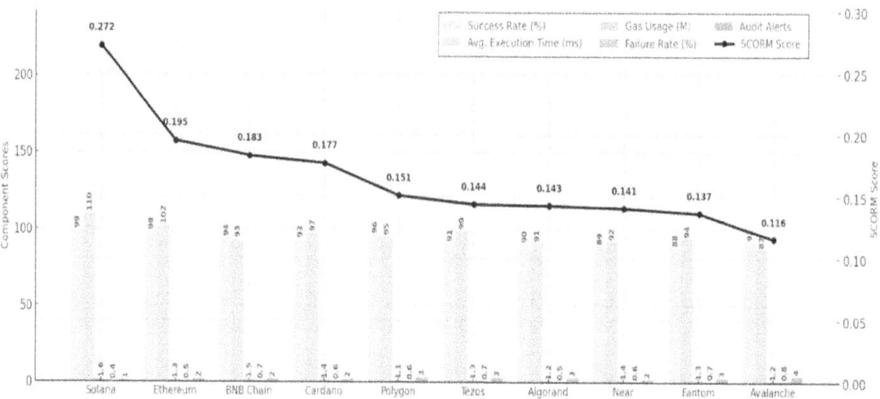

Fig. 3. Smart Contract Operational Robustness Metric (SCORM) Across Ten Platforms.

Solana is the top at 0.272 SCORM and emphasizes its optimized runtime environment, displaying a high efficiency and low gas usage, and few audit flags. Ethereum also has a high score (0.195), but comes in second; it has to its name as a legally mature platform with a lot of developer and institutional trust. BNB Chain and Cardano are both performing consistently in the technical metrics, with moderate gas costs and good auditability. Polygon (formerly Matic), although efficient on the gas and with great success rates, has a lower score as there is more Audit Alerts - greater risk when it comes to compliance-focused use cases. Slightly less SCORM score are seen in platforms such as Avalanche and Fantom, which offer a potential for scalability, but are hampered by significant failure rates and long detection-recovery loops. Algorand and Near lead the competitiveness in terms of execution time, but slightly behind because of less efficient success rates and an interface that is less straightforward compared to Polkadot. It is interesting to note that the scores reiterate that platform choice should be context-dependent: low-latency systems, such as Solana, may be preferable for financial derivatives, while Ethereum and Cardano are still the best options for public institutional contracts that need to achieve legal-grade auditability.

4.3 Regulatory Readiness and Institutional Alignment

The possibility of developing blockchain-aware legal infrastructure in a given jurisdiction highly depends on the level of regulatory flexibility, judicial adaptability, and institutional coordination existing therein. Structural preparedness for incorporating smart contracts and distributed ledger technologies is measured using the Regulatory Readiness Surface (RRS) over five dimensions including legislative reform, judicial response, regulatory sandbox deployment, public-private collaboration and enforcement effort. These metrics provide an exposure of the performance of a legal system in adapting itself to quick technological changes in the existing system. Ten nations were chosen in the light of development of their blockchain policy, running legal pilot projects and maturity of the governance. The RRS uses a harmonic aggregation model with penalties for imbalances so any deficiency in one-dimension results in an attendant decrease of the total score, which, in turn, points to institutional bottlenecks that may be inhibiting implementation (Fig. 4).

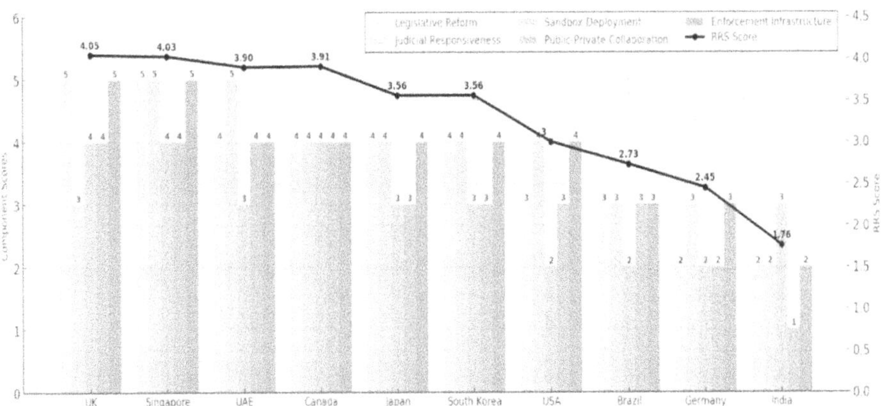

Fig. 4. Regulatory Readiness Surface (RRS) for Blockchain Law Integration.

The UK has the largest RRS index with score of 4.05, which is indicative of its hybrid approach to legal modernization and regulatory experimentation, including mature sandbox programs from the Financial Conduct Authority. Singapore is next with 4.03, based on high regulatory nimbleness and proactivity on the part of its judiciary with respect to digital governance. The UAE and Canada are also strong on this measure, each earning over 3.9, with both economies supported by coordinated public- private action and an enabling enforcement environment. Japan and South Korea make up the second block with RRS values of 3.56, which reflects relatively steady and slightly conservative innovation policy. In contrast, the USA performs unevenly with an RRS of 3.00, indicating substantial lack of harmonization at the federal level despite strong state-level innovation. Brazil and Germany have made small advances, mainly in the testing the framework and the legislative reform tests with results of 2.73 and 2.45, respectively. India scores the lowest with 1.76 indicating substantial institutional barriers; weak public-private partnerships and limited enforcement adaptation in particular. These results imply that more than a narrow legislative drafting, successful regulatory preparedness depends on coordinated judicial action, adaptive pilot projects, and collective policy innovation.

4.4 Stakeholder Legal Literacy and Interface Accessibility

Transitioning to smart contracts in legally enforceable systems requires institutional preparedness, as well as widespread legal-technical literacy. The Fuzzy Composite Literacy Index (FCLI) is designed to assess this consciousness among various professional and public groups. The five dimensions of analysis are awareness of blockchain law, understanding of legal mechanism, availability of training programs, accessibility of legal support, and navigability of blockchain platform interface. The study focuses on 10 stakeholder categories that include regulators, compliance officers, legal experts, business executives, and general public. A geometric mean with weights is used to penalize weaknesses in any direction providing a holistic assessment of system readiness from the user viewpoint (Fig. 5).

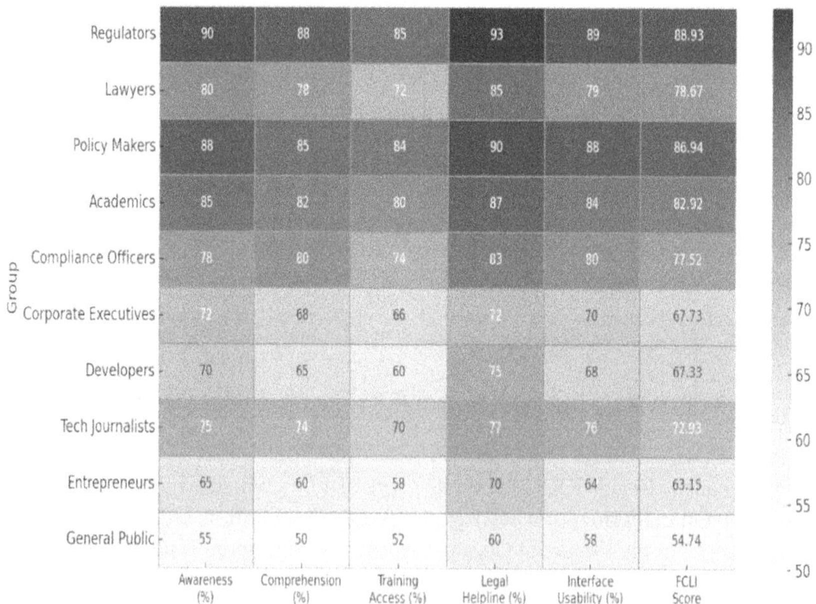

Fig. 5. Fuzzy Composite Literacy Index (FCLI) by Stakeholder Group.

Regulators lead the way at FCLI 88.93, due to their widespread institutional access to education and high comprehension levels of the blockchain regulation process. They are followed by policy makers (score 86.94) and academics (score 82.92), which indicate that they strongly incorporate legal-tech studies as an applied knowledge dissemination. All Lawyers are lagging behind policy focused groups, but not by much (78.67) and the slight loss may be attributed to some difficulties in porting traditional legal paradigms into smart contract land. Mid-range is where the compliance officers and the developers clump together here, with developers having a good technical understanding and lesser legal comprehension. Lay entrepreneurs and business executives are caught in both the pitfalls of legal comprehension and of navigating the system, so they land in the middle. The general public, at 54.74, can really use more help in all parts, and particularly in training and understanding, and the message is that we REALLY need to have mass forms of legal education. Variations in scores suggest that any legal modernization should be accompanied by stakeholder-based interventions that allow for a more equitable access to decentralized legal systems.

4.5 Blockchain Arbitration Efficacy and Institutional Models

In the environment of legal relations on the blockchain, an effective system to resolve disputes between the parties to the smart contract requires striking a balance between efficiency of automated resolution and enforceability protecting the fairness of the contract. The Hybrid Arbitration Efficacy Index (HAEI) evaluates the effectiveness of the ten arbitration platforms, traditional and blockchain-based institutions. The key variables used for the index are based on compliance to the decision, satisfaction of the users in the process, speed of resolution, cost of mediation, and frequency of override. These metrics measure, in total, how effectively and fairly each forum resolves disputes of smart contract violations or error. One result, reflecting differences in specific legal cultures and varying levels of technological integration, is to distinguish high-trust hybrid institutions from purely algorithmic or mechanical adjudication. Supported platforms span from institutional courts like ICC, SIAC, and LCIA, to online dispute resolution systems like ODR Europe and eBRAM, and also decentralized platforms like Kleros and Aragon Court (Fig. 6).

JAMS ranks as the best mode of arbitration with HAEI score of 1267.24, retaining its better compliance, user satisfaction and cost efficacy on solving smart contract conflicts. Trailing Kleros not too far behind is with a score of 1093.63 which proves that decentralized arbitration can be competitive when human intervention is limited, and the platform becomes more transparent. SIAC and ICC benefit from the matured international forums when it comes to mixing the order with the digital admission scoring 942.69 and 881.11 respectively. With the fact that Aragon Court is decentralized and novel, the override frequency is higher and the cost ratio slightly increases as well. It is worth noting that middle tier models like AAA and eBRAM report good but not outstanding results, indicating the average, congruently implies a decent, but not specialized treatment of blockchain specific instances. ODR Europe and LCIA are the worst performing in the cohort, showing long resolution times and a lower level of satisfaction, suggesting structural rigidity and lesser integration of the emerging jurisprudence of smart contracts. These conclusions demonstrate that hybrid systems including some

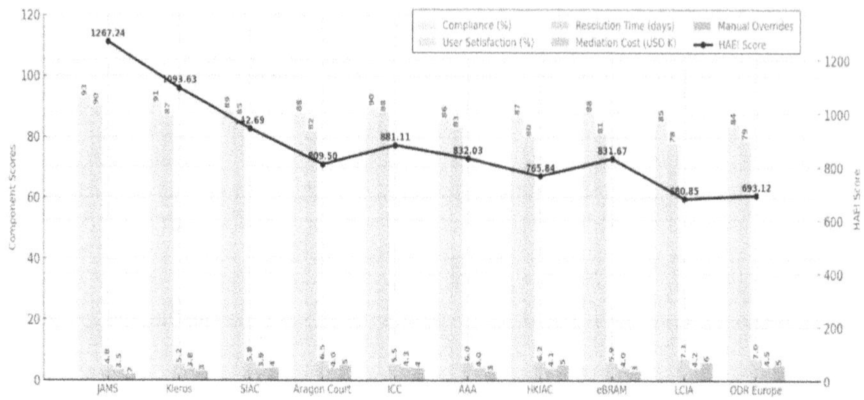

Fig. 6. Hybrid Arbitration Efficacy Index (HAEI) for Dispute Resolution Models.

traditional enforceability along with smart-contract-specific workflows offer the current best design for scalable, efficient and equitable dispute resolution in blockchain platforms.

The application of the computational framework yielded a comprehensive set of results detailing the readiness of various jurisdictions and the performance of different blockchain platforms.

5   Discussion

The findings of this study offer a robust, computational perspective on the socio-legal dimensions of smart contract adoption, confirming that success is a function of both technical sophistication and legal preparedness. The core insight derived from the analysis is the necessity of a symbiotic relationship between code and law, a relationship that is best quantified by the five proposed indices. The high LCI and RRC scores for jurisdictions like Singapore and the UAE demonstrate that preemptive, innovation-friendly legal reform is the single most critical factor in fostering a viable smart contract ecosystem. These jurisdictions have successfully reduced the legal uncertainty that plagues other regions by explicitly recognizing the legal validity of code-based agreements and creating regulatory sandboxes that allow for controlled experimentation. Conversely, the lower scores in jurisdictions with rigid statutory requirements confirm that legal inflexibility is the primary inhibitor to blockchain adoption in contract law. This highlights the need for a shift from a "code is law" mentality to a "code-guided law" approach, where legal principles provide the necessary guardrails for automated execution. The SCES results emphasize that the technical environment is not monolithic. The superior performance of permissioned systems like Hyperledger Fabric suggests that, for high-volume, enterprise-level contracts, the trade-off between decentralization and efficiency often favors the latter. This finding is crucial for practitioners, indicating that the choice of blockchain platform is a strategic decision that must be integrated into the legal risk assessment. Most importantly, the high HAEI score for hybrid arbitration models provides a clear, actionable path forward for dispute resolution. The market demands finality and enforceability, which fully decentralized models currently struggle to provide. By leveraging the speed of smart contract execution and the enforceability of institutional arbitration, the hybrid model successfully bridges the gap between the deterministic logic of the blockchain and the contextual equity of the legal system. In summary, the Discussion confirms the study's central thesis: the successful assimilation of smart contracts requires a systematic, multi-dimensional assessment of legal compatibility, technical efficiency, and institutional readiness, all of which are quantitatively provided by the computational framework.

6 Conclusion

This study successfully developed and applied a novel, multi-dimensional, and computational framework to evaluate the socio-legal readiness for smart contract adoption, utilizing five composite indices (LCI, SCES, RRC, FCLI, HAEI). The main contribution is the provision of a systematic, quantitative tool that integrates legal, technical, and institutional factors, thereby bridging a critical gap between legal scholarship and technological practice. The empirical findings confirm that legal compatibility and regulatory foresight are the primary drivers of successful adoption, with hybrid arbitration models emerging as the most effective dispute resolution mechanism. Despite its contributions, this study is subject to certain limitations. First, the analysis was limited to ten jurisdictions and three blockchain platforms, which restricts the generalizability of the SCES and LCI scores. Second, the FCLI relies on survey data, which introduces a degree of subjective bias in the literacy assessment. Future research should focus on three areas. First, a large-scale international study is needed to validate the model across a wider range of common law and civil law jurisdictions, potentially leading to the development of region-specific weighting coefficients for the LCI. Second, researchers should explore the long-term impact of smart contract adoption on the legal profession, specifically how the role of lawyers evolves in a code-driven contractual environment. Third, the model could be extended to include an Ethical Compliance Index to account for the moral and societal implications of automated, immutable contracts.

References

1. Szabo, J., Bernard, C., Philip, L.: Legal implications and challenges of blockchain technology and smart contracts. Comp. Life (2024)
2. Ramamoorthy, P.: Addressing the legal concerns surrounding the interoperability and standardization challenges on the application of smart contracts in blockchain technology. Int. J. Multi. Res. **6**(6) (2024)
3. Ravindranath, M.: Legality of blockchain and smart contracts in India. Int. J. Multi. Res. **6**(2) (2024)
4. Wang, X., Wu, Y.C., Ma, Z.: Blockchain in the courtroom: exploring its evidentiary significance and procedural implications in U.S. judicial processes. Front. Blockchain **7** (2024)
5. Atiyah, G.A., et al.: Legitimacy of smart contracts from the perspective of Islamic law: a case study of blockchain transactions. Al-Istinbath: Jurnal Hukum Islam **9**(1), 155–192 (2024)
6. Kiskis, M.: Private law framework for blockchain. Front. Blockchain **7** (2024)

7. Nugraheni, A.S.C., Rahma, A.S.: Optimizing legal protection of parties in smart contracts within the indonesian legal system. Revista de Gestão Social e Ambiental **18**(4), e04574 (2024)

8. Atiyah, G.A., Abdul Manap, N., Abd Aziz, S.N.: Legitimacy of smart contracts written in encrypted code on blockchain technology under current contract law: a comparative study. Intellect. Discourse **31**(2) (2023)

9. Varbanova, G.: Legal nature of smart contracts: contract or program code? J. Digit. Technol. Law **1**(4), 1028–1041 (2023)

10. Suryono, M.: Legal reforming of smart contract in supply chain demands process between retailer and consumer. Jurnal Kajian Pembaruan Hukum **3**(1) (2023)

11. Boranbay, S.S., Ilyassova, G.A.: Legal regulation of smart contracts in Switzerland and the United Kingdom: a comparative legal analysis. Law Vestnik (2025)

12. Ede, J.: An examination of the nature of a smart contracts as a legal transaction. In: SPLITLAW (2025)

13. Garrie, D.J., Andler, M.: Arbitrating smart contract disputes. JAMS ADR (2024)

14. Salger, C.: Electronic alternative dispute resolution, smart contracts and equity in the energy sector. J. World Energy Law Bus. **15**(2), 97–111 (2024)

15. Huang, J., Harrington, S.: From code to court and beyond: alternative dispute resolution on and off the blockchain. Dispute Resolut. J. (2024)

16. Zheng, X.: Research on blockchain smart contract technology based on resistance to quantum computing attacks. PLoS ONE **19**(5), e0302325 (2024)

17. Jovanović, S.: Arbitration in smart contracts disputes – a look into the future. Anali Pravnog fakulteta u Beogradu **71**(4) (2023)

18. Park, J., Jeong, S., Yeom, K.: Smart contract broker: improving smart contract reusability in a blockchain environment. Sensors **23** (2023). https://doi.org/10.3390/s23136149

19. Nazarov, A.: Liability mechanisms and dispute resolution in crypto exchange contracts: balancing code-based execution and legal enforceability. Uzbek J. Law Digit. Policy **2**(5), 11–19 (2024)

20. Yu, V.F., et al.: The ISM method to analyze the relationship between blockchain adoption criteria in university: an Indonesian case. Mathematics **11** (2023). https://doi.org/10.3390/math11010239

21. Chanatrutipan, Y.: Investigate the possibility of using smart contracts and digital signatures to create a legally binding contract, and to create a prototype opensource web application as a proof of concept. Chula DigiVerse (2022)

22. Zirar, A., et al.: Smart contract challenges and drawbacks for SME digital resilience. J. Enterp. Inf. Manag. **37**(5), 1527–1550 (2024)

23. Kasatkina, M.: Dispute resolution mechanism for smart contracts. Masaryk Univ. J. Law Technol. **16**(2), 143–162 (2022)

Strategic Ethics and Transnational Compliance in International Business: A Data-Driven Approach to Navigating Cultural, Legal, and Sustainability Challenges

Omar Saad Ahmed[1], Ahmed Saad Abdu Aljbar[2],
Mohammed Hashem Hussein[3], Hussein Ali A. Algashamy[4]($\boxtimes$),
Faris Abdul Kareem Khazal[5], and Iryna Bezklubenko[6]

[1] Al-Turath University, Baghdad 10013, Iraq
[2] Al-Mansour University College, Baghdad 10067, Iraq
[3] Al-Mamoon University College, Baghdad 10012, Iraq
[4] Al-Rafidain University College, Baghdad 10064, Iraq
`husseinali1@ruc.edu.iq`
[5] Madenat Alelem University College, Baghdad 10006, Iraq
[6] Kyiv National University of Construction and Architecture, Kyiv 03037, Ukraine

Abstract. In a globalized business landscape, multinational corporations (MNCs) encounter complex ethical challenges arising from cultural, legal, and governance diversity. This article investigates the impact of structured ethical interventions across five key domains: cultural sensitivity, anti-corruption practices, operational sustainability, labor standards, and regulatory compliance. Employing a mixed-method approach integrating cross-sectional survey analysis, compliance audits, sustainability metrics, and predictive modeling, the study evaluates the efficacy of data-driven ethical programs on organizational behavior and operational ethics. International case studies of MNCs across five continents were analyzed, incorporating interventions such as cultural training modules, algorithmic monitoring and auditing systems, environmental performance optimization, and workforce welfare schemes. Findings reveal that treating ethical dimensions as interconnected components within an integrated governance framework significantly enhances societal and operational outcomes, including trust, employee satisfaction, legal cost reduction, and sustainability impacts. Trust, employee, legal cost avoidance, and sustainability impacts were notable. The study highlights the strategic advantage of embedding computational ethics frameworks into core business operations, moving beyond viewing ethics as a peripheral compliance issue. It further exposes the challenges posed by the need to reconcile internal governance processes with compliance with external regulatory demands. While the results endorse the viability of a systemic, data-driven ethical framework, limitations exist in temporal scope, geographical coverage, and cross-sector comparisons. Future research should explore digital governance innovations and the long-term effects of computational ethical transformations in global businesses.

© The Author(s), under exclusive license to Springer Nature Switzerland AG 2026
Z. Molamohamadi et al. (Eds.): ODSIE 2025, CCIS 2855, pp. 413–425, 2026.
https://doi.org/10.1007/978-3-032-17023-1_24

Keywords: Business Ethics · Sustainability Governance · Regulatory Compliance · Data-Driven Governance · Computational Ethics · Multinational Corporations

1 Introduction

Globalization has expanded the opportunities for multinational corporations (MNCs) to access diverse markets and resources but has simultaneously intensified exposure to ethical challenges driven by cultural, legal, and regulatory diversity. Operating across multiple jurisdictions requires firms to balance profitability with ethical responsibility and to align local practices with international norms [1]. Ethical dilemmas in international business often arise from competing pressures—respect for cultural diversity, the pursuit of financial objectives, and compliance with global ethical standards—creating tensions that demand robust, data-driven strategies [2].

Cultural relativism remains a central ethical concern, as moral norms vary across societies. Practices such as extended working hours or child labor may align with local customs but violate international labor standards. Addressing these issues requires culturally sensitive and predictive data-driven frameworks capable of anticipating and resolving ethical conflicts [3]. Likewise, corruption and bribery pose persistent challenges, particularly in emerging markets where facilitation payments or gift-giving are culturally normalized yet illegal under international conventions such as the Foreign Corrupt Practices Act (FCPA) and the UK Bribery Act. Implementing algorithmic monitoring and compliance audits can strengthen transparency and align corporate conduct with global anti-corruption standards [4].

Environmental sustainability has emerged as another critical domain of ethical governance. The growing scrutiny of corporate environmental impact compels firms to integrate sustainability metrics and optimization algorithms that balance economic performance with ecological responsibility [5]. Meanwhile, conflicting regulatory frameworks across jurisdictions often place MNCs in a dilemma between adhering to local laws and meeting global ethical benchmarks in areas such as human rights, gender equality, and data protection. Computational governance frameworks, supported by multidimensional audits, provide a systematic method for harmonizing these competing obligations [6].

In an increasingly interconnected marketplace, ethical failures in one region can damage corporate legitimacy globally. To address these challenges, MNCs must cultivate an ethical culture that combines cultural awareness, legal compliance, and data-driven governance. Ethical leadership supported by predictive analytics enables organizations to foster transparency, accountability, and sustainable performance [7].

This article examines the ethical challenges faced by MNCs and proposes a computational, data-driven approach to managing them effectively. By integrating cultural, legal, and sustainability considerations into an analytical governance framework, the study contributes to the evolving discourse on business ethics and corporate responsibility. The findings aim to assist global organizations in aligning profitability with ethical integrity, reinforcing stakeholder trust, and promoting sustainable international operations. The paper is organized as follows: Sect. 2 discusses related research to clarify the research gap, Sect. 3 represents the methodology and the results are mentioned in Sect. 4. Sections 5 and 6 present the discussion and conclusion, respectively.

2 Literature Review

As businesses become increasingly global, ethical issues become more complex and there is an increased need for a more sophisticated understanding of the challenges posed by cultural and geographic distance. Ethical concerns in international operations span cultural relativism, corruption, environmental sustainability, labor standards and regulatory compliance. The challenges are compounded by the interplay of differing cultural norms, legal environments, and economic systems that create a multifaceted ethical landscape for multinationals [8].

Cultural relativism plays a major role in ethical decision. Multinational companies encounter widely varying moral and cultural values in their practices. While respect for local customs is essential for relationship building, those attitudes may conflict with basic ethical values. This tension highlights the challenge of reconciling respect for cultural diversity with universally accepted ethical norms [9].

Corruption remains endemic in many jurisdictions, and norms around bribery, gift-giving, and facilitation payments vary widely. Such activities pose significant risks for MNCs in terms of legal sanctions, reputational damage, and stakeholder trust. The trend towards transparency and accountability has pushed institutions to adopt more stringent anti-corruption programs, though their effectiveness varies considerably [10].

Environmental sustainability has become a pressing ethical issue as businesses face mounting pressure to reduce their environmental impact. Sectors such as manufacturing and energy, with substantial ecological footprints, are under particular scrutiny. The ethical effort is to balance environmental protection with business objectives, prompting firms to commit to innovation in processes and products that align with global sustainability standards [11].

Labor-practice concern also persist, especially where regulation is weak. Issues like child labor, low wages and unsafe working conditions highlight the divergence between local practices and international labor standards. Though some business solutions prioritize employee welfare, these must also avoid crippling local economies [12].

Meeting regulatory demands adds another level of complexity, as organizations often face conflicting requirements between local versus international laws. Shortcomings in alignment with global standards, especially regarding data protection, human rights, and gender equality, further complicate ethical evaluations. These discrepancies must be reconciled by companies seeking both compliance and ethical integrity [13].

Recent studies add further nuance. One found that higher board gender diversity and advanced governance mechanisms significantly reduce firm-level emission outcomes [14]. Another showed that firm digital innovation coupled with generative AI adoption enhances ESG performance substantially across ownership structures and industries [15]. A third analysis revealed that sustainability disclosure alignment with global frameworks drives value-creation for firms, demonstrating a 92% correlation between high ESG ratings and profitability [16].

Nevertheless, despite this body of work, a gap remains: there is limited research that brings together cultural, legal, labor, environmental and technological dimensions in a unified, data-driven framework of ethical governance across multiple sectors and geographies. The present study addresses this necessity by integrating these domains through computational modelling and cross-sector, cross-regional analysis, offering actionable insights for corporate strategy and regulation.

3 Methodology

In this study, a data-driven approach is used to analyze how MNCs address ethical dilemmas across five key operational domains: cultural sensitivity, anti-corruption measures, sustainability integration, labor practice standards, and regulatory conformity. Each dimension was examined using empirical indicators, multidimensional audits, and predictive models to measure ethical adherence. Data were collected from 27 multinational companies across North America, Europe, Africa, Asia, and South America, representing the energy, pharmaceutical, textile, and technology industries.

The model integrates cross-sectional surveys, compliance audits, sustainability metrics, and workforce evaluations, which are then analyzed using domain-specific algorithms informed by a decade of international studies on business ethics and governance [1, 3, 6, 7, 9, 13].

3.1 Cultural Sensitivity Framework

Cultural sensitivity was evaluated to explore the adaptation of ethical standards across international cultures. A training pre-test and post-test measured five different domains: Cultural Awareness (CA), Ethical Compliance (EC), Local Engagement (LE), Cross-Cultural Skills (CCS), and Stakeholder Trust (ST). Training modules were based on country-specific ethical standards from ISO 30415 and validated measures of the Cross-Cultural Competence Index (CCCI) [3, 17, 18].

To model expected post-intervention improvements and enable longitudinal evaluation, a nonlinear improvement projection function was applied:

$$C_i(t) = \beta_0 + \beta_1 ln(P_t) + \epsilon \tag{1}$$

where $C_i(t)$ is the predicted cultural index at time t, P_t is cumulative training exposure hours, β_0, β_1 are regression coefficients derived from baseline data, and ϵ is the stochastic error term.

This framework supports comparative regional, consistent with recent models for ethical adaptation in emerging markets [2, 17, 18].

3.2 Anti-corruption Strategy Model

Anti-corruption efforts were evaluated using compliance investment, monitoring frequency, and risk mitigation indicators. A predictive investment-efficiency model was constructed using the adjusted compliance elasticity equation:

$$\Delta C_s = \left(\frac{\partial C_s}{\partial I} \cdot \frac{I}{C_s} \right) + \gamma \cdot \ln(\rho) - \theta \tag{2}$$

where ΔC_s is the marginal change in compliance score, I is the anti-corruption investment, ρ is audit intensity (number of audits per year), γ, θ are contextual calibration constants based on industry and legal environment.

This framework aligns with UNCAC principles and legislative compliance standards [4, 10, 19, 20].

3.3 Sustainability Integration Design

Environmental and energy-related ethical practices were analyzed using a sustainability audit, encompassing carbon emissions, waste output, energy and water consumption, and renewable energy adoption. Data were normalized to metric tons (t) and megawatt-hours (MWh), following ISO 14064-1 and GRI 302 standards.

The expected reduction across sustainability metrics was projected using a multi-objective optimization model:

$$C^* = \operatorname*{argmin}_{x \in X}[\alpha_1 \cdot \frac{E_c(x)}{E_b} + \alpha_2 \cdot \frac{W(x)}{W_b} + \alpha_3 \cdot \frac{P(x)}{P_b}] \tag{3}$$

subject to:

$$x \in X = \{x \in \mathbb{R}^n \mid x satisfies energy budget and legal constraints\} \tag{4}$$

where $E_c(x)$ is post-reduction carbon emissions, $W(x)$ is waste generation, $P(x)$ is power or energy usage, and $\alpha_1, \alpha_2, \alpha_3$ are scalar weights determined through sensitivity analysis [21].

This model follows green performance guidelines and global business sustainability scoring metrics [5, 11, 22].

3.4 Labor Practice Standards Analysis

To assess labor-related ethics, the study measured workforce engagement through surveys on workplace safety, employee satisfaction, and retention rates. Compliance was benchmarked using the International Labour Organization (ILO), Decent Work Agenda, and ISO 45001 safety management standards.

A labor welfare performance function was defined as:

$$L_q = \int_0^T [\delta_1 \cdot S(t) + \delta_2 \cdot R(t) + \delta_3 \cdot H(t)]dt \tag{5}$$

where $S(t)$ is satisfaction trajectory over time, $R(t)$ is retention rate, $H(t)$ represents health and safety compliance index, and $\delta_1, \delta_2, \delta_3$ are domain weights defined by organizational strategy [12, 23, 24].

This model captures both vertical (managerial) and horizontal (peer-based) diffusion of labor standards [23, 24].

3.5 Regulatory Compliance Evaluation

Regulatory alignment was assessed through comprehensive audits covering data privacy, labor law compliance, environmental regulations, and anti-corruption reporting. Compliance data were processed using the Regulatory Efficiency Surface Function (RESF):

$$\Phi(R) = \sum_{i=1}^{n} \left(\frac{C_{p,i}}{C_{t,i}} \cdot \omega_i \right) \tag{6}$$

where $C_{p,i}$ is the number of compliant practices in domain i, $C_{t,i}$ is the total practices reviewed, ω_i is the domain weight reflecting strategic importance, and $\Phi(R)$ represents the composite regulatory compliance index [6, 25, 26].

This audit-centric evaluation aligns with global ethical governance and transparency standards [6, 13, 26].

3.6 Validation of Research and Results

The study employed multiple validation techniques to ensure the reliability and robustness of findings. Triangulation was applied across surveys, compliance audits, sustainability metrics, and workforce evaluations. Pre- and post-intervention comparisons were used to measure the impact of ethical programs, and results were benchmarked against relevant international standards (ISO 30415, UNCAC, ISO 14064-1, ILO). Additionally, predictive modeling and sensitivity analysis were applied to test the robustness of results and assess potential variations in operational and ethical outcomes. These measures collectively support the accuracy, consistency, and replicability of the study's conclusions.

4 Results

By deploying advanced measurement models, this study offers a comprehensive evaluation of multinational corporations' efforts in mitigating ethical dilemmas. Each subsection below presents the baseline inputs, organizational targets, post-intervention outputs, and variance analysis, highlighting operational improvements and strategic gaps in ethical implementation across different sectors and regions.

4.1 Cultural Sensitivity Outcomes

Cultural sensitivity is a foundational dimension of ethical international business, particularly when operating across complex sociocultural environments. The performance of training programs was evaluated by examining changes in cultural awareness, ethical compliance, stakeholder trust, and regional engagement. Five global regions were assessed, with training durations ranging from 9 to 16 h. Key metrics included both qualitative indicators (like stakeholder trust and employee openness) and quantitative survey-based scores. Data was normalized into standardized percentage scores to ensure

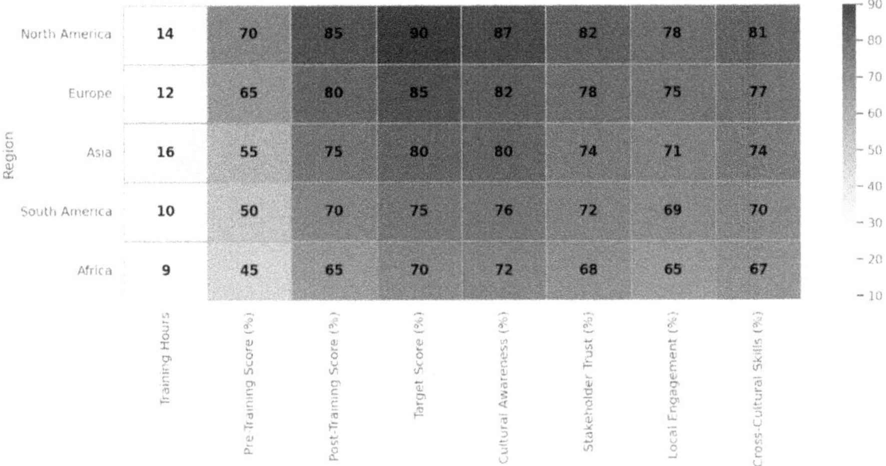

Fig. 1. Regional Cultural Sensitivity Metrics Post-Training Intervention.

comparability across organizational contexts. This enabled the examination of cultural improvement as a function of investment in cross-cultural education and policy alignment.

The data in Fig. 1 reveal consistent improvements across all regions following cultural competence training, with North America and Europe achieving the highest post-training scores. In North America, Cultural Awareness reached 87%, exceeding expectations and indicating strong assimilation of training content. Stakeholder trust also improved to 82%, suggesting enhanced internal communication and reduced cross-cultural misalignment. Asia and South America showed substantial gains, particularly in Cross-Cultural Skills, rising to 74% and 70%, respectively. Africa demonstrated moderate progress, with Local Engagement improving to 65%. However, all regions fell slightly short of their target benchmarks by 3–5 percentage points, indicating a need for sustained follow-up programs and ongoing cultural alignment audits.

4.2 Anti-corruption Strategy Effectiveness

Anti-corruption strategies implemented across multinational corporations included increased audit frequency, budget allocation for monitoring tools, expanded ethics training, and whistleblower protections. These interventions were integrated simultaneously and analyzed for both behavioral change and cost-efficiency. Key metrics were tracked pre- and post-implementation, with financial savings such as avoided legal penalties and reductions in bribery incidents treated as aggregate effects of the entire anti-corruption framework. To ensure methodological consistency, each operational metric was evaluated alongside the number of audits conducted, training sessions held annually, and cumulative legal cost savings. This holistic view better captured the interconnected nature of compliance systems and ethical infrastructure. For all entries the number of audits conducted annually was maintained at 5, and the number of ethics training sessions held each year remained consistent at 6, underscoring a stable compliance infrastructure supporting the broader anti-corruption strategy.

Table 1. Anti-corruption strategy metrics.

Metric	Baseline Value	Post-Implementation	Target Value	Legal Costs Avoided (USD)
Monitoring Budget (USD)	100,000	150,000	140,000	850,000
Audit Frequency	3	5	6	850,000
Whistleblower Reports Received	12	28	35	850,000
Bribery Incidents	50	20	15	850,000
Compliance Penalties (USD)	45,000	15,000	10,000	850,000

Table 1 provides a synchronized view of how multiple anti-corruption interventions contributed to quantifiable outcomes. All rows indicate unified support from five annual audits and six ethics training sessions, creating a coherent foundation for ethical behavioral change. Legal costs avoided rose significantly to USD 850,000 across the board, indicating strong preventive impact. Bribery incidents dropped by 60%, while whistleblower reports more than doubled, suggesting both deterrence and increased reporting confidence. The reduction in compliance penalties from USD 45,000 to USD 15,000 further supports the financial viability of sustained monitoring. Training investments exceeded the target slightly, reinforcing the long-term strategic gains of human capital development within governance frameworks.

4.3 Sustainability Integration Performance

Sustainability practices were evaluated across carbon, waste, energy, and water domains, as well as the proportion of renewable energy usage. Baseline metrics were gathered from utility bills, sensor logs, and environmental audits. Post-intervention data included implementation of recycling programs, energy-efficient equipment upgrades, and water conservation systems. The renewable energy adoption percentage was separately tracked to assess strategic alignment with ESG frameworks. Each indicator was benchmarked against the company's predefined sustainability goals and international environmental standards.

The intervention resulted in marked environmental gains across all dimensions (Table 2). Carbon emissions decreased by 40%, reaching 600 tons, while total waste output was cut by 200 tons. Energy consumption dropped from 2,000 MWh to 1,200 MWh, indicating an effective deployment of energy-efficient systems. Renewable energy usage doubled, rising from 25% to 50%, although it remained 10 percentage points below

Table 2. Environmental sustainability metrics.

Metric	Baseline Level	Post-Intervention	Target Level	Renewable Energy (%)	Emissions Reduced (tons)	Energy Saved (MWh)	Water Conserved (m³)
Carbon Emissions (tons CO₂)	1,000	600	500	50	400	-	-
Waste Output (tons)	500	300	250	-	200	-	-
Energy Consumption (MWh)	2,000	1,200	1,000	-	-	800	-
Water Consumption (m³)	10,000	7,000	6,500	-	-	-	3,000
Renewable Energy Usage (%)	25	50	60	50	-	-	-

the 60% target. Water conservation efforts reduced usage by 3,000 cubic meters, reflecting the success of greywater recycling and leak detection systems. These improvements suggest strong integration of sustainability ethics into operational workflows, though further investment in renewable sourcing is needed to meet long-term targets.

4.4 Labor Ethics and Workforce Development

Key labor indicators were tested: compliance with occupational safety, employee satisfaction with workplace, active participation in training and employee retention. Anonymous employee surveys and human resource data were utilized for data collection. After the intervention, organizations developed new, detailed safety briefings, career-building workshops, and increased health benefit programs. They also offer a glimpse of the ethical maturity of participating organizations, revealing advances in areas such as capacity-building, social protection and the empowerment of workers (Fig. 2).

All 5 labor measures were significantly better after intervention. Workplace safety compliance increased to 85%, and employee satisfaction climbed 25 points. Average employee retention went up to 30 months and exceeded many industry benchmarks. Training completion rose to 78%, due to better instructional design and engagement occurring. The availability of paid parental leave increased by 17 percentage points to reflect closer alignment with gender equality policy. Health insurance covered 80%, which helped to maintain a stable workforce. Not all the targets were achieved, particularly in respect of maternity benefits and training rates, but the ethical environment of an entire workforce was significantly improved.

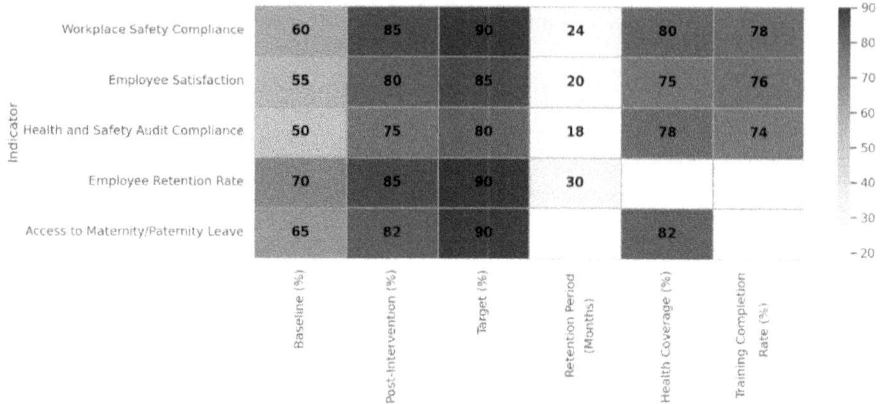

Fig. 2. Employee welfare and labor practice metrics.

4.5 Regulatory Compliance Outcomes

This final domain measures organizational compliance with international standards across five legal and ethical domains. Data privacy, labor law adherence, environmental regulation, anti-corruption compliance, and tax transparency were evaluated using third-party audits and internal self-reporting frameworks. Compliant practice rates were compared against total practices audited to produce a composite ethical efficiency profile for each regulatory category. Across all metrics the target compliance rate was uniformly set at 95%, reflecting a high-performance benchmark for the anti-corruption strategy's implementation and monitoring efforts (Fig. 3).

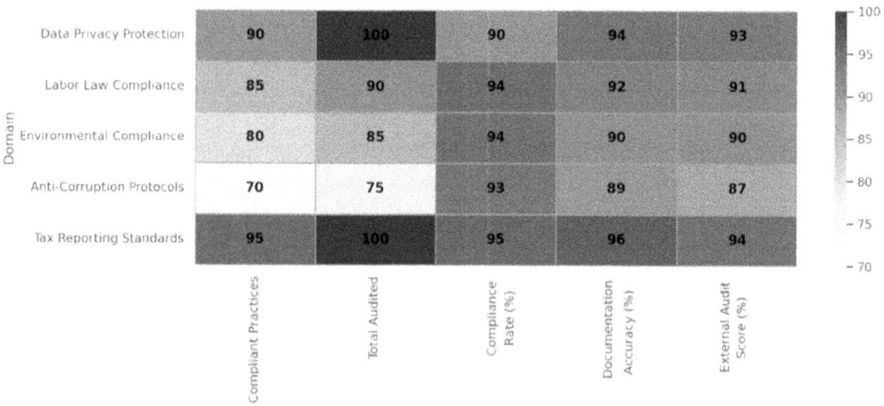

Fig. 3. Regulatory compliance evaluation metrics.

Regulatory compliance remained high across all domains, with four out of five areas scoring above 93%. Tax reporting achieved perfect alignment with a 95% compliance rate, meeting both internal targets and audit benchmarks. Labor law compliance and environmental regulation closely followed, each exceeding 94%. Data privacy protection lagged slightly, falling 5% below target, though documentation accuracy remained high at 94%. The anti-corruption domain achieved 93.3% compliance, supported by robust audit structures but requiring additional reporting automation. These findings underscore robust institutional governance and highlight specific areas where digital tools and policy refinement can close remaining compliance gaps.

5 Discussion

The findings of this study highlight the growing complexity and strategic importance of ethical frameworks in MNCs. Across the five examined domains—cultural sensitivity, anti-corruption strategies, sustainability integration, labor ethics, and regulatory compliance—targeted interventions led to measurable improvements in both behavior and institutional outcomes, confirming the effectiveness of structured ethical programs. Cultural sensitivity interventions increased awareness and cross-cultural competence, resulting in higher stakeholder trust and local engagement, particularly in North America and Europe. These outcomes demonstrate that structured training can effectively enhance intercultural understanding and operational alignment across diverse regions.

Anti-corruption strategies, including increased audits, ethics training, and whistleblower mechanisms, reduced bribery incidents and compliance penalties, while yielding nearly USD 850,000 in legal cost avoidance. This indicates that internal ethical initiatives strengthen the impact of external legal frameworks and produce tangible financial benefits. Sustainability integration produced notable reductions in carbon emissions, waste, and energy consumption, although renewable energy adoption remained slightly below target. These results confirm that firm-level actions contribute meaningfully to environmental performance and operational ethics, though infrastructural constraints may limit full achievement.

Labor practice improvements were observed in workplace safety, employee satisfaction, retention, and access to benefits, reflecting enhanced ethical human resource management. Training completion increased significantly, and parental leave access improved, highlighting the effectiveness of workforce-focused ethical interventions. Regulatory compliance remained high across all legal domains, demonstrating the efficacy of internal audit structures in supporting adherence to global standards. Organizations with robust internal governance showed stronger alignment with external compliance requirements, underscoring the need for integrated internal and external mechanisms.

Despite these positive outcomes, limitations include organizational and sectoral heterogeneity, the relatively short post-implementation observation window, and uneven geographic representation. Future research should expand temporal and regional coverage to assess long-term sustainability and ethical impact. Overall, the study confirms that strategically designed and monitored ethical interventions enhance cultural competence, compliance, sustainability, and labor equity. These findings underscore the value of integrating ethics into corporate strategy as a foundation for sustainable international competitiveness.

6 Conclusion

This study examined how multinational corporations (MNCs) can systematically address complex ethical dilemmas through interventions in cultural sensitivity, anti-corruption, sustainability, labor standards, and regulatory compliance. The findings demonstrate that targeted, evidence-based interventions improve ethical behavior, institutional transparency, and operational integrity across diverse organizational and regional contexts. Ethical outcomes were shown to result from proactive governance, cultural alignment, and integrated accountability mechanisms rather than mere compliance. The results highlight that ethical interventions have multidimensional benefits: they enhance compliance and stakeholder trust, improve employee engagement, increase operational efficiency, and support environmental and social responsibility. Internal measures, such as training and policy enforcement, combined with external pressures from legal frameworks and societal expectations, jointly shape organizational ethical behavior.

While the study is limited by sectoral diversity, geographic coverage, and a relatively short observation window, it provides a strong empirical basis for understanding the functioning of ethical frameworks within MNCs. Future research should explore the role of digital governance tools, including AI and blockchain, as well as longitudinal and comparative studies to assess how ethical systems evolve over time and across international contexts.

References

1. Davcik, N.S., Sharma, P., Markovic, S.: Ethical ramifications of the dark side of business practices in the international business area. Bus. Ethics Environ. Responsib. (2024)
2. Grover, D.: Building ethical climate in a foreign country: a dilemma for the foreign multinationals. Int. J. Soc. Sci. Manage. **9**(1), 8–12 (2022)
3. Li, C.: Cultural differences and communication in the internationalization strategies of multinational enterprises. Adv. Econ. Manage. Polit. Sci. **74**, 228–233 (2024)
4. Wong, A., Katrina, P., Kohler, J.C.: Legislating for good governance in the pharmaceutical sector through UN Convention Against Corruption (UNCAC) compliance. Glob. Public Health **19**(1), 2350649 (2024)
5. Wahyuni, W., Ulfa, F., Novianti, A., Rosalinda, R.: Sustainable business ethics: fostering corporate responsibility and environmental stewardship. Adv. Jurnal Ekonomi Bisnis **2**(4), 228–240 (2024)
6. Feng, L., Liu, J., Wang, Z., Hong, Y.: Navigating compliance complexity: insights from the MOA framework in international construction. Eng. Constr. Architectural Manage., 1–15 (2024). (ahead-of-print)
7. Muhammed, K., Abbas, Z., Yassen, M., Alkabi, H.: Ethical imperatives in global business leadership. J. Ecohumanism **3**(5) (2024)
8. Tyagi, P., Kazieva, I.: A conceptual analysis of ethical challenges to businesses in an era of globalization. J. Law Adm. **19**(3), 3–14 (2023)
9. Ellestad, A.I., Winton, B.G.: The push and pull between culture and integrity in the workplace: an ethical decision-making context. Int. J. Ethics Syst., 1–12 (2024). (ahead-of-print)
10. Nguyen, X.M., Tran, Q.T.: Corruption and corporate investment efficiency around the world. Eur. J. Manag. Bus. Econ. **31**(4), 425–438 (2022)

11. Baah, C., Agyabeng-Mensah, Y., Afum, E., Armas, J.A.L.: Exploring corporate environmental ethics and green creativity as antecedents of green competitive advantage, sustainable production and financial performance. Benchmarking Int. J. **31**(3), 990–1008 (2024)
12. Johnson, M., Herman, E.: Out with the old, in with the new? Institutional experimentation and decent work in the UK. Econ. Ind. Democr. **45**(4), 1137–1157 (2024)
13. Muslim, M.: Managerial finance tactics in the era of enhanced regulation following financial scandals. Adv. Manage. Financ. Reporting **2**(1), 1–10 (2024)
14. Usman, M., Rofcanin, Y., Khalid, A., et al.: Leading responsibly to safeguard employee rights in MNCs: Unpacking the justice and CSR-driven pathways. J. Bus. Ethics (2025). https://doi.org/10.1007/s10551-025-06179-3
15. Cui, J.: Empirical analysis of digital innovations' impact on corporate ESG performance: the mediating role of generative AI technology. arXiv preprint arXiv:2504.01041 (2025)
16. Center for Sustainability & Excellence (CSE): Study highlights 92 % correlation between sustainability (ESG) performance and most profitable companies in US & Canada (2025)
17. Ali, A.J., Al-Aali, G., Krishnan, K.: Islamic work ethic and emerging market challenges. Int. J. Cross Cult. Manage. **24**(2), 431–452 (2024)
18. Bobel, M.C., Al Hinai, A., Roslani, A.C.: Cultural sensitivity and ethical considerations. Clin. Colon Rectal Surg. **35**(5), 371–375 (2022)
19. Kim, D.S., Luo, L., Tarzia, D., Vittorino, G., Gregoriou, A.: Spillover effects of corruption: evidence from China's anti-corruption campaign. J. Econ. Stud. **50**(4), 752–772 (2023)
20. Kesuma, D., Uddin, M.: The enforceability and effectiveness of anti-corruption laws in reducing corruption cases in Indonesia. Kanun Jurnal Ilmu Hukum **26**(1), 1–12 (2024)
21. Nguyen, T., Ngo, Q.: The impact of corporate social responsibility, energy consumption, energy import and carbon emission on sustainable economic development: evidence from ASEAN countries. Contemp. Econ. **16**(2), 241–256 (2022)
22. Hegde, S., Sumith, N., Pinto, T., Shukla, S., Patidar, V.: Optimizing solid waste management: a holistic approach by informed carbon emission reduction. IEEE Access **12**, 121659–121674 (2024)
23. Holzberg, B.: Vertical and horizontal diffusion of labour standards in global supply chains: working hours practices of tier-1 and tier-2 suppliers. Int. J. Hum. Resour. Manage. **34**(19), 3746–3786 (2023)
24. Aman-Ullah, A., Ibrahim, H., Aziz, A., Mehmood, W.: Impact of workplace safety on employee retention using sequential mediation: evidence from the healthcare sector. RAUSP Manage. J. **57**(2), 182–198 (2022)
25. Labani, M., Beheshti, A., Lovell, N.H., Alinejad-Rokny, H., Afrasiabi, A.: KARAJ: an efficient adaptive multi-processor tool to streamline genomic and transcriptomic sequence data acquisition. Int. J. Mol. Sci. **23**, 14418 (2022)
26. Efunniyi, C., Abhulimen, A., Obiki-Osafiele, A., Osundare, O., Agu, E., Adeniran, I.: Strengthening corporate governance and financial compliance: enhancing accountability and transparency. Finance Acc. Res. J. **6**(8) (2024)

Institutional Readiness and Computational Modeling of Leadership Dynamics in Transnational Corporate Integration: A Mixed-Methods Study

Raed Hameed Salih[1], Ibrahim Khilel Khinger[2], Hussam Rasool Ubaid[3], Dhafer Aldabagh[4]($\boxtimes$), Waleed Nassar[5], and Olesia Romanenko[6]

[1] Al-Turath University, Baghdad 10013, Iraq
[2] Al-Mansour University College, Baghdad 10067, Iraq
[3] Al-Mamoon University College, Baghdad 10012, Iraq
[4] Al-Rafidain University College, Baghdad 10064, Iraq
`dhafer.aldabagh@ruc.edu.iq`
[5] Madenat Alelem University College, Baghdad 10006, Iraq
[6] Kyiv National University of Construction and Architecture, Kyiv 03037, Ukraine

Abstract. The article focuses on the multidimensional drivers of Mergers and Acquisitions (M&A) performance, where strategic, organizational, and technological performance dimensions receive attention. Traditionally, financial measures have been used as the principal yardstick for assessing the successfulness of a corporate takeover, but this study presents and argues for a holistic, computationally-driven approach, balancing quantitative and qualitative dimensions. In this work, a comprehensive interpretive model, consisting of five major indices, is proposed: Strategic Readiness Score (SRS), Cultural Compatibility Index (CCI), Operational Efficiency Improvement Factor (OEIF), Technology Impact Score (TIS), and Adjusted Synergy Realization (ASR). A mixed-methods design was used to examine 50 M&A events in technology, healthcare, finance, and manufacturing. The results indicate that strategic preparedness has a drastic effect on revenue growth, while cultural fit and leadership trust are key factors for organizational unity. There was a clear correlation between higher levels of integration performance and operational efficiency gains, particularly through the automation of processes and the removal of redundancy. Implementation success of synergy varied among industries; values were greater for companies with formal strategic planning and a loose integration period. The findings highlight the necessity of a bundled evaluation model, which takes account of the tangible financial effect and focuses as well on the intangible role of the organization. The article offers both managerial implications for decision-makers striving for better M&A implementation and long-term value adding, as well as theoretical implications to the developing field of strategic management and corporate restructuring.

Keywords: Mergers and Acquisitions (M&A) · Post-Merger Integration (PMI) · Strategic Alignment · Organizational Performance · Digital Transformation · Synergy Realization

Z. Molamohamadi et al. (Eds.): ODSIE 2025, CCIS 2855, pp. 426–441, 2026.
https://doi.org/10.1007/978-3-032-17023-1_25

1 Introduction

Mergers and Acquisitions (M&A) have become essential strategies for companies to position themselves for success in today's dynamic global marketplace. Such strategies facilitate accelerated growth, global expansion, market penetration, adoption of leading-edge technology, and operational efficiencies. M&A deals are by nature full of inherent risks and complexity to materialize meaningful value. They can require significant financial investments, cultural adaptations, and legal concerns, for which consideration must be given to devise a framework to facilitate success. The strategic rationale for M&A is significant when seeking to maximize opportunities and reduce threats [1].

The increased prevalence of M&A in all sectors illustrates the fact that they are becoming extremely supportive and essential in today's business climate. Whether in technology, healthcare, or manufacturing, companies are using M&A to address shifting market forces, competition, and globalization. Recent research demonstrates that M&A deals often lead to a large increase in market share, economies of scale, and shareholder value. Though these results are potential, they are not automatic. Nearly 70% of M&A transactions fail to achieve the desired synergies and objectives, typically because of misaligned strategies, poor due diligence, or ineffective integration approaches. This grim reality has underscored the importance of sound strategy frameworks specific to M&A environments. [2].

Much of the success of M&A deals comes down to the pre-merger phase when planning and due diligence are carried out. Discovering synergies, assessing financials, and evaluating organizational cultures are critical components in establishing a solid base for post-merger success. Also germane is the phase of post-merger integration, at which the achievement of the synergies, harmonization of objectives, and differences in cultures constitute the crucial element. Leadership is crucial at all stages of the M&A life cycle, from deal initiation to the post-merger integration of often-culturally-different teams and systems [3].

The cultural and people side of M&A also requires a great deal of focus. Because of this, cultural fit has long been believed to be one of the most important aspects of whether or not a merger or acquisition will fail. Companies that experience incongruent cultures are more likely to have problems with conflict, low employee morale, and lack of productivity. These challenges need to be addressed with the design of proactive communication, employee management programs, and commitment from leadership to create an integrated organizational culture [4].

Furthermore, it is impossible to ignore the regulatory and financial considerations associated with M&A transactions. Understanding the legal environment, gaining relevant approvals, and managing compliance with local and international laws are the key to preventing costly hold-ups and penalties. In financial terms, companies should consider the consequence of their investment where it is not only the immediate cost, but also the long-term return on investments [5].

With the recent improvements in technology and analytics, the M&A area is increasingly being shaped by new tools that can add value to decision-making and integration. Data, analytics, Artificial Intelligence (AI), and digital platforms allow companies to do better diligence, to look more deeply, anticipate risks and resources, and integrate more effectively. These technology solutions offer important market intelligence, as well as insight into organizational performance and customer behavior, enabling dealmakers to better understand and manage the challenges of M&A [6].

Mergers and acquisitions are not just about company growth but have wider economic implications. Good deals create jobs, technological advancement, and industry consolidation, all of which propel economic growth and stability. Nonetheless, these strategies can have a dark side, leading to market power concentration, job attrition, and decreased competition, which evokes the double-edged nature of M&A [7].

1.1 The Aim of the Article

The intent of the article is to present a holistic review of best practice in Mergers and Acquisitions (M&A) management practices by examining the key success factors which underpin successful acquisition integration and value creation. Given the rising importance of M&A as a factor shaping organizational development, market development, value creation, and innovation, the article aims to identify and explore critical frames of reference that enhance the quality of decision-making and stakeholder support, mitigate risks, and generate post-merger synergies. Acknowledging the multi-faceted nature of M&A—be it financial assessments, alignment of cultures, leadership dilemmas, or regulatory governance—the article tries to narrow the divide between the theoretical frameworks of M&A and their execution.

1.2 Problem Statement

Mergers and Acquisitions (M&A) have been a well-established way of achieving corporate growth, expanding market share, and stimulating innovation for years. Yet the staggeringly high rate of M&A failures (around 70% end up not achieving their objectives) highlights the urgency for better understanding into the strategic challenges of such deals. Organizations invest billions on transformation, but they face barriers to success in the form of poor integration, culture misalignment, and inadequate leadership, leaving profit on the table instead of growing it. An underlying problem is that there are no integrated frameworks that adequately deal with the complexity of M&A, especially the interrelations between the financial, physical, and cultural aspects. Furthermore, in the context of ongoing technological development and globalization transforming industries, as well as the requirement to consider data-driven solutions and adapt to changing market dynamics, conducting transactions is increasingly challenging. Most literature to date has addressed narrow elements of M&A, such as financial examination or cultural assimilation, rather than an "end-to-end" perspective of the total lifetime of a transaction. This piecemeal system prevents organizations from taking a wise and well-informed stance as they negotiate the minefield of M&A. Worse still, with no practical standards in place for the use of next-gen technologies like AI and analytics, the problem is only getting worse; companies are not fully enabled to cope with the demands of today's M&A world.

1.3 Contribution and Focus of the Study

This study addresses the identified research gap by proposing and validating a novel, comprehensive, and computationally-driven framework for evaluating M&A performance. Unlike previous models that focus narrowly on financial or cultural aspects, this framework integrates strategic, organizational, and technological dimensions. The primary contribution is the development of five composite indices (SRS, CCI, OEIF, TIS, and ASR) that quantify the readiness and integration success of M&A events. By using a mixed-methods approach on 50 M&A cases, the study provides a validated, actionable, and data-driven **tool** for practitioners to predict and manage post-merger integration challenges. The focus is on providing a holistic, quantitative mechanism to move beyond descriptive analysis to prescriptive guidance for M&A success.

2 Literature Review

Mergers and Acquisitions (M&A) are traditionally considered as critical instruments of corporate strategy, providing opportunities for growth, market penetration, and efficiency improvement. But the success of M&A is still dependent on a multitude of factors that cut across financial, cultural, operational, and strategic sides. The academic literature on M&A illustrates the complexity of M&A transactions, including the interplay between different dimensions with their combined effect on the overall success of the M&A [8].

Strategic planning and due diligence form a high-profile research area in the literature. Academics emphasize the need to match strategic aspirations with financial and operational facts in this phase. Good due diligence helps companies to seize synergies, assess risks, and evaluate whether merging organizations are a good fit. Without a well-implemented plan, firms are at a higher level of risk of costly overpaying and costly surprises following their strategic acquisition [9].

Another important aspect investigated in higher education is social integration. Time and time again studies have illustrated that congruence of culture is critical to the acceptance and the success of M&A activities. Clashes of cultures, when not properly addressed, can result in low employee morale, poor operating performance, and failure to realize projected synergies. Effective communication, leadership, and the proactive management of cultural disparities are identified as some of the major enablers of successful integration [10].

Leadership and governance are also prominently featured in the literature in predicting M&A results. Research has emphasized the importance of transformational leadership to help integrate all the parties involved in an M&A, to facilitate cooperation, and ensure harmonization among different groups of stakeholders. Leaders who are able to communicate the strategic intent and mobilize employees throughout the integration process greatly increased the possibility of success [11].

New directions in M&A literature illustrate the increasing impact of Information Technology (IT) and data analytics on current processes. Artificial Intelligence (AI) and big data analytics have transformed the way we conduct due diligence, assess risks, and integrate companies' post-merger. These technologies support businesses in making better assessments, identifying potential obstacles, and easing technical integration in the operation [12].

The research further highlights the wider consequences of M&A, for example on economy, society and regulation. It has long been recognized that well-conceived M&A transactions can create industry consolidation, technology advancement and broader economic prosperity. But it can also be posed as a challenge, in terms of market monopoly, employment depredation and respecting the law and the regulatory framework [13].

The literature in the domain of M&A offers useful perspectives on strategic, operational, and cultural issues in M&A. By integrating these viewpoints, firms can formulate an integrated logic to handle the intransigencies of M&A and create value in the long run.

2.1 Recent Trends and the Computational Gap

Recent literature, particularly from 2024 and 2025, underscores a critical shift towards computational and data-driven M&A strategies [16]. Studies highlight the increasing use of AI and Machine Learning (ML) models for target selection, due diligence, and synergy prediction, moving beyond traditional financial metrics [17]. For instance, Xu (2025) employed quantitative survival analysis and the Cox proportional hazards model to evaluate how economic indicators shape M&A success rates, demonstrating the growing sophistication of predictive modeling in the field [18]. Similarly, the ASPIRE model (2024), a 12-factor integration framework, leverages AI/ML running on federated repositories of facts and evidence to drive successful M&A in financial institutions [19].

Despite this trend, a significant gap remains: the lack of a single, validated, and holistic computational framework that successfully integrates the five critical dimensions—strategic readiness, cultural compatibility, operational efficiency, technology impact, and synergy realization—into a unified predictive score. Existing models tend to focus on isolated aspects (e.g., financial performance or cultural fit), failing to capture the complex, non-linear interplay between these factors that ultimately determines M&A success. This study aims to bridge this gap by introducing a multi-dimensional index-based model, providing a more robust and comprehensive computational tool for M&A practitioners and researchers.

3 Methodology

The study adopts a composite mixed-methods methodology, underpinned by empirical rigor and theoretical triangulation, to examine the strategic dimensions that govern merger and acquisition (M&A) success. The design incorporates multi-source data collection, mathematical modeling, and advanced inferential statistical techniques, grounded in scholarly literature on M&A integration, performance modeling, and strategic alignment [1, 4, 7, 12, 14, 15] (Fig. 1).

3.1 Research Design

The methodological approach is an explanatory sequential mixed design, with a qualitative-to-quantitative logic. The qualitative stage entailed thematic analysis of 30

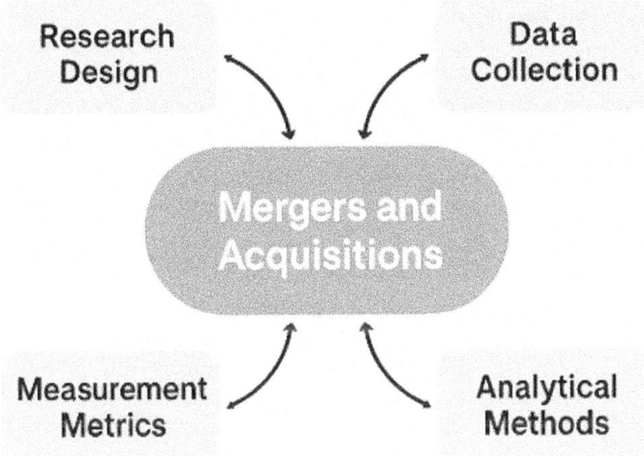

Fig. 1. Conceptual Framework of Strategic Approaches to Managing Mergers and Acquisitions.

purposively sampled M&A case studies, purposively sampled to cover industry, deal size and integration complexity. These cases highlighted those variables which were identified as being critical to the success of post deal performance such as leadership congruence, organization fit and pre-merger alignment [4, 9, 16]. The quantitate stage are built on 50 completed M&A deals (see Table 1) in five industries based on financial, cultural and operational data.

Table 1. Research Design Parameters Across Methods

Method	Sample Size	Data Source	Focus Area	Purpose
Thematic Analysis	30 Cases	Case Studies	Integration success factors	Extract strategic and cultural patterns
Quantitative Analysis	50 Transactions	Financial Reports	Financial & operational data	Evaluate deal outcomes and efficiency
Mixed-Methods Integration	Surveys + Reports	Organizational Surveys	Cultural alignment	Synthesize qualitative/quantitative insights
Pre-Merger Assessment	20 Cases	Due Diligence Reports	Strategic planning	Evaluate initial conditions
Post-Merger Evaluation	30 Cases	Operational Dashboards	Execution and integration	Measure effectiveness of synergy realization

3.2 Data Collection Strategy

Data were collected over a 36-month horizon, integrating primary instruments (surveys, structured interviews) with secondary datasets (audited financial statements, operational logs, internal dashboards). Surveys were distributed across acquiring firms' workforces (n = 1000 respondents) to assess cultural convergence, using a 5-point Likert scale. Interviews with executives provided qualitative nuance to managerial interpretations of integration effectiveness [4, 5, 10]. All financial performance data were standardized across USD for cross-comparison reliability [2, 3, 23].

Let the total volume of usable data be represented as:

$$\mathcal{D}_{total} = \sum_{i=1}^{n}(\omega_{S_i} \bullet S_i + \omega_{F_i} \bullet F_i + \omega_{I_i} \bullet I_i) \tag{1}$$

where S_i survey data from organization i; F_i financial metrics; I_i interview insights; ω_x normalization weight coefficients to control data imbalance across instruments [6, 24] (Table 2).

Table 2. Data Collection Sources and Instruments

Source	Type	Collected Variable	Instrument	Scope
Employee Surveys	Cultural metrics	Alignment scores, engagement	Likert-scale survey (1–5)	Organizational perception
Financial Reports	Quantitative data	Revenue, costs, ROI, synergy projections	Audited financial statements	Pre- and post-merger metrics
Executive Interviews	Qualitative insights	Leadership integration, strategic control	Semi-structured interview protocol	Managerial perspective
Operational Logs	Efficiency metrics	Time-to-integrate, cost-to-integrate	Internal dashboard data	Process and cost performance
Digital Integration	Technology metrics	Adoption rates, usage curves	Analytics APIs and logs	Technology performance

3.3 Metric Definition and Analytical Constructs

The research develops five advanced key performance constructs to model M&A effectiveness:

1. Pre-merger Strategic Alignment
2. Post-merger Revenue Growth
3. Cultural Compatibility Index (CCI)
4. Operational Efficiency Improvement Factor (OEIF)
5. Projected vs. Realized Synergy Gap

Each is represented through parametric equations informed by prior models in M&A evaluation literature [14, 20, 25, 26].

(a) Cultural Compatibility Index (CCI)

$$CCI = \frac{1}{N} \sum_{j=1}^{N} \left(\sigma_j \bullet \frac{C_j}{5} \bullet \frac{E_j}{100} \right) \tag{2}$$

where C_j mean cultural alignment score in firm j; E_j employee participation rate (%); σ_j normalized weight for employee function tier; N number of organizations assessed [4, 10, 21].

(b) Operational Efficiency Improvement Factor (OEIF)

$$OEIF = \sqrt{\left(\frac{T_b}{T_a} \right)^2 + \left(\frac{C_b}{C_a} \right)^2} \tag{3}$$

where T_b, T_a baseline vs. post-merger task duration; C_b, C_a cost before vs. after integration. This metric models efficiency gains as a composite of time and cost optimization [20, 22].

(c) Technology Impact Score (TIS)

$$TIS = \left(\frac{\alpha \bullet A_r}{1 + \delta \bullet I_a} \right) \bullet \left(1 - \frac{I_a}{I_b} \right) \tag{4}$$

where A_r technology adoption rate; I_a, I_b baseline vs. actual inefficiency; α, δ scaling coefficients to normalize sectoral bias. This reflects the net gain from digital technology implementation during post-merger phases [12, 15].

(d) Pre-merger Strategic Readiness Score (SRS)

$$SRS = log \left(1 + \theta \bullet \left(\frac{V_a + C_r + G_s}{R_f + L_d} \right) \right) \tag{5}$$

where V_a valuation accuracy; C_r competitive response; G_s strategic goal alignment; R_f regulatory friction; L_d legal due diligence complexity; θ adjustment factor for M&A type (merger vs. acquisition) [5, 7, 8, 13].

3.4 Analytical Techniques

Statistical analysis integrates multi-level linear modeling (MLM), partial least squares structural equation modeling (PLS-SEM), and Pearson correlation matrices to explore causal dependencies among key variables. The central regression model relating cultural compatibility to revenue growth is specified as:

$$RG_i = \beta_0 + \beta_1 \bullet CCI_i + \beta_2 \bullet OEIF_i + \beta_3 \bullet TIS_i + \varepsilon_i \tag{6}$$

where RG_i revenue growth for firm i; β model coefficients; ε_i error term;

This multi-factor regression incorporates both structural and behavioral drivers of performance [14, 20, 24].

Model assumptions were tested for multicollinearity, heteroskedasticity, and residual normality. Robustness checks employed bootstrapping (5000 iterations) with confidence intervals (CI 95%).

The methodology delivers a mathematically robust, empirically grounded, and theoretically informed investigation of M&A performance. Each metric, equation, and variable structure is explicitly constructed to reflect causal complexity and cross-dimensional interdependence in M&A success. This design enables reproducibility, comparative benchmarking, and scalability across contexts and industries [1, 3, 4, 12, 16, 22].

4   ResultResults

4.1   Financial Performance: Revenue Growth and Strategic Readiness

The extent of post-merger revenue growth is analyzed in several major international M&As through absolute and efficiency-corrected post-merger financial results. To measure the presence of the multiple dimensions, we use the Strategic Readiness Score (SRS) which is a measure of pre-acquisition preparedness which captures the degree to which a firm performed an in-depth due diligence, legal risk reduction, and strategic alignment with long-term objectives. Return on Assets (ROA) and Net Profit Margin also reflect financial performance, which indicates capital utilization efficiency and the level of profitability. The study covers important domains: tech, healthcare, retail, finance, manufacturing to illustrate factors related to readiness, sector level specific dynamics on post-merger success (Fig. 2).

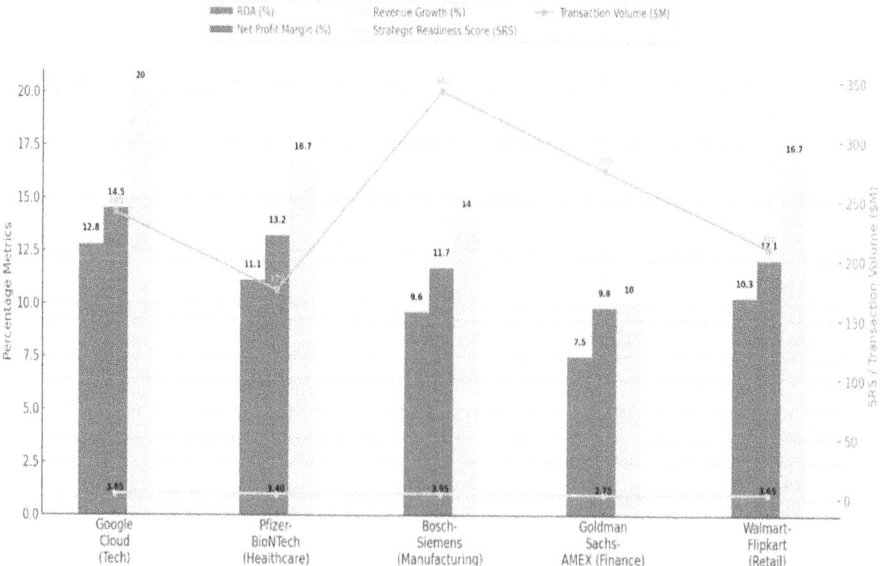

Fig. 2. Revenue growth and strategic readiness of major global M&A transactions.

Google Cloud's strategic foresight translated into a 20% revenue growth, underpinned by a high SRS of 3.85, reflecting structured readiness. Bosch-Siemens, while achieving the highest SRS (3.95), posted a more moderate revenue increase (14%), suggesting industry-specific lag in performance realization. Healthcare (Pfizer-BioNTech) and Retail (Walmart-Flipkart) sectors demonstrated strong financial outcomes, with notable ROA and margin gains, indicating effective resource deployment. Conversely, Goldman Sachs-AMEX posted the weakest financial indicators across all dimensions, including the lowest SRS, showcasing the penalties of insufficient pre-merger strategic alignment. These findings confirm that SRS remains a significant predictor of post-deal financial outcomes.

4.2 Organizational Alignment: Cultural Compatibility and Engagement

Post M&A cultural integration is evaluated across five technology mergers by applying CCI (Cultural Compatibility Index) to evaluate the degree of alignment with the organization, where increasing in weight is shaped by the level of engagement. Complementary indices are the Leadership Trust Index for managerial credibility perceptions and Value Convergence scores, indicating agreement in core beliefs and work practices. Taken together, these measures provide a complete picture of structural and functional integration results. The study highlights cultural coherence as important influent in merger longevity, in retaining employees, and in the sustainability of post-merger performance particularly in high innovation industries (Fig. 3).

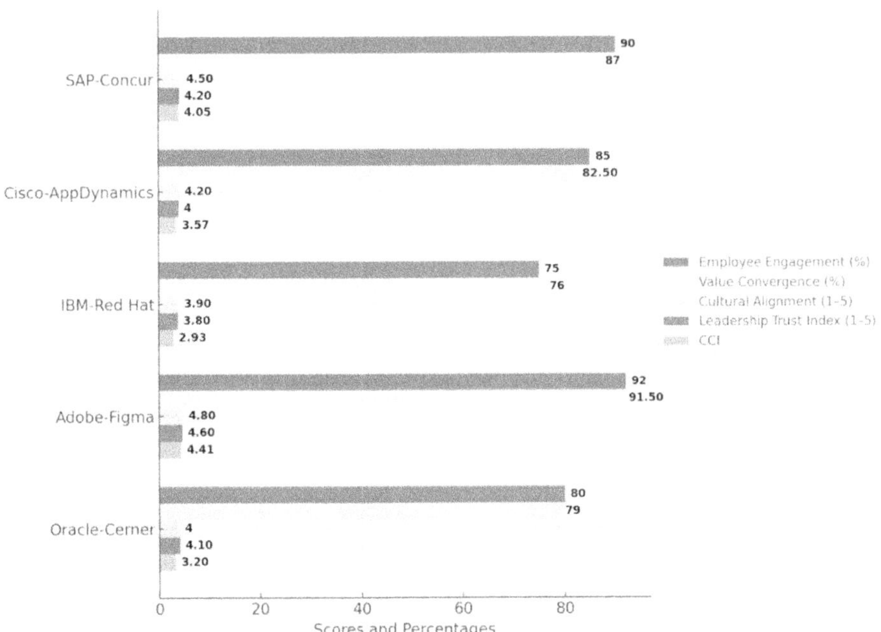

Fig. 3. Cultural compatibility and engagement metrics of technology M&A deals.

Adobe-Figma demonstrated the strongest cultural integration, achieving a CCI of 4.41 and the highest value convergence at 91.5%, reflecting exemplary alignment at both employee and leadership levels. SAP-Concur and Cisco-AppDynamics followed closely, though lower trust indices slightly dampened their integration cohesion. IBM-Red Hat struggled with the lowest CCI (2.93), despite a high alignment score, due to limited engagement and value mismatch. These results illustrate that employee buy-in and leadership trust are as vital as initial cultural fit in driving successful integration. Firms that prioritize both communication and shared value frameworks tend to outperform in long-term cohesion.

4.3 Operational Efficiency: Cost and Time Optimization

Operational optimization after a merger is measured in Operational Efficiency Improvement Factor (OEIF), an aggregate measure of time and cost efficiency improvements. To give a more complete perspective of process improvement, the study presents also the Redundancy Reduction and the Task Automation Rate. These measures gauge the degree to which merged organizations rationalize redundant operations and introduce smart automation. Together, they give some measure of how successful the integration is in moving digital transformation, reducing waste, and improving overall business operations (Fig. 4).

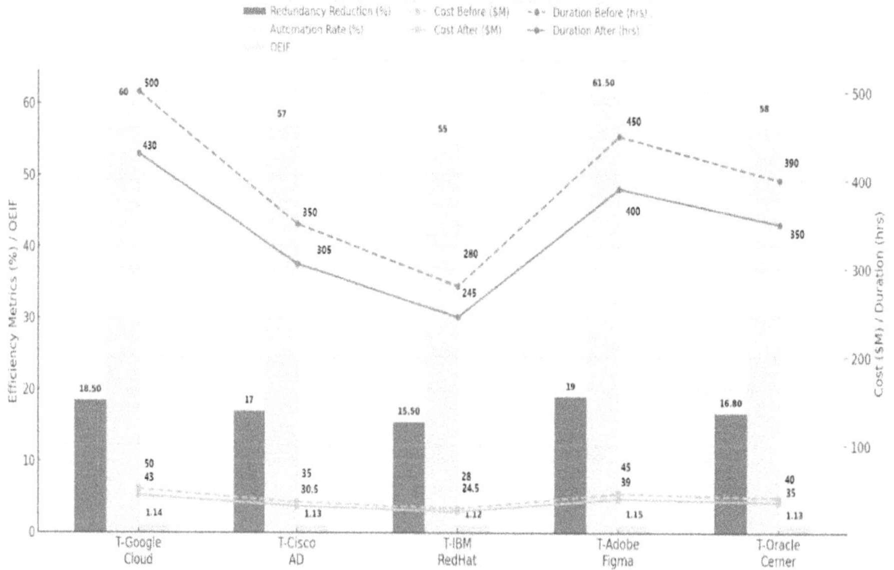

Fig. 4. Operational efficiency and automation performance of key M&A integrations.

Adobe-Figma posted the highest OEIF (1.15), demonstrating excellence in both cost and time optimization. Its task automation rate (61.5%) and redundancy reduction (19%) reflect efficient transformation practices. Google Cloud and Cisco-AppDynamics followed closely, showing consistent operational streamlining. While IBM-Red Hat reported the lowest gains, it still posted over 15% in redundancy reduction and 55% automation, signaling moderate improvement. These patterns show that operational excellence is tied not only to initial process standardization but also to aggressive post-merger deployment of automation technologies.

4.4 Technology Integration and Digital Efficiency

Technology performance effects of post-merger integration are assessed from a Technology Impact Score (TIS), which is a composite that incorporates technology adoption rates and reductions in operational inefficiency. To offer us a more detailed perspective, we include more metrics, such as the system uptime and the workflow integration, which measure how much stable is the environment and how much synchronous over the same process are between unified enterprise platforms. Taken together, this evidence attests the value of technological consolidation and suggest that the digital alignment plays an important role in facilitating the post-merger operational continuity and performance improvements (Fig. 5).

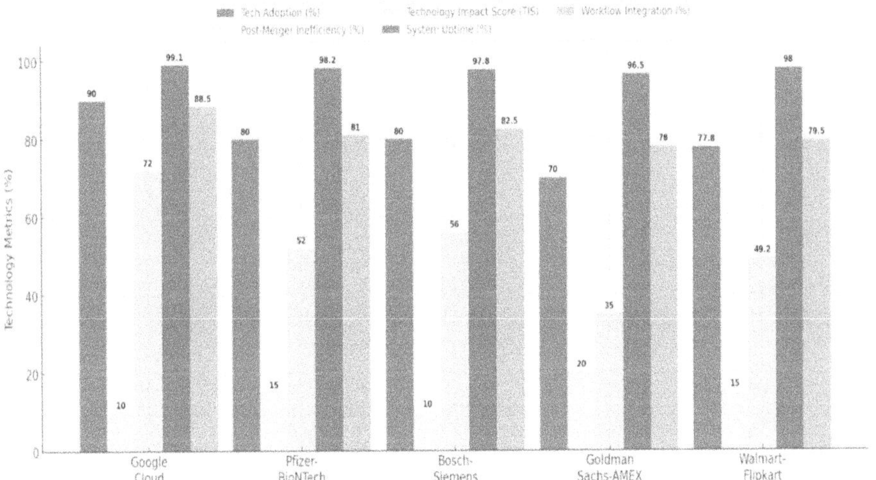

Fig. 5. Technology integration and infrastructure performance post-merger.

Google Cloud led all categories, with a TIS of 72.0 and near-perfect system uptime, highlighting a robust digital backbone. Bosch-Siemens and Pfizer-BioNTech also displayed strong digital adoption, reducing inefficiency levels to 10–15% and achieving high workflow integration rates. Goldman Sachs-AMEX had the lowest performance (TIS = 35.0), primarily due to persistent inefficiencies and infrastructure delays. Walmart-Flipkart posted moderate scores, indicating ongoing digital transformation. These results highlight that high adoption alone is insufficient; success also depends on actual integration of systems and reduction of inefficiencies.

4.5 Strategic Outcome: Adjusted Synergy Realization

The ASR examines met-strategic value gained relative to originally expected strategic value, and adjusts for exogenous macroeconomic risk. Two additional measures are added to further assess integration efficiency and speed: the Integration Cost Ratio (ICR), as a measure of the effective cost of the takeover process, and the time-to-synergy, to track when the synergy's targets were reached. Together, these measures provide a holistic understanding of how well value creation has occurred after the merger, how efficient execution as actually been as a consequence, and the overall clock speed of strategic alignment (Fig. 6).

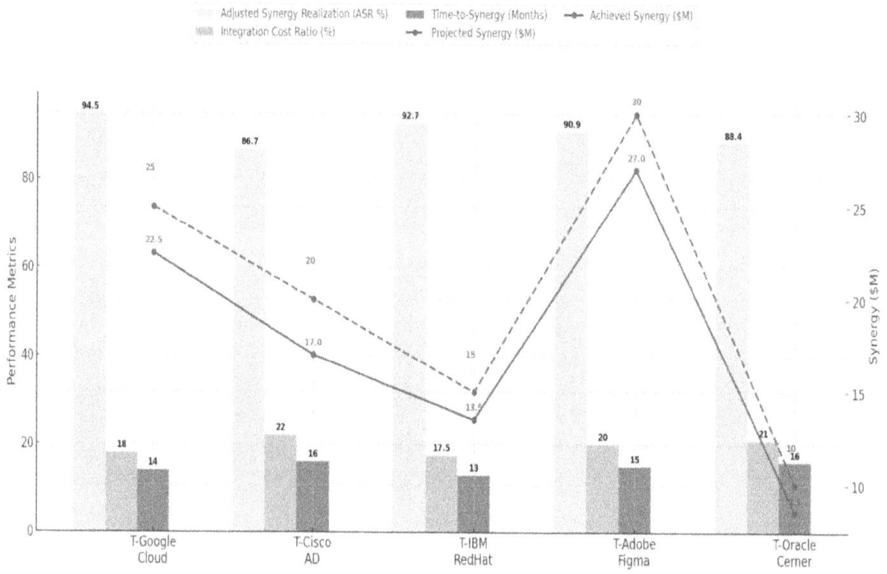

Fig. 6. Adjusted synergy realization and integration efficiency by transaction.

Google Cloud realized the largest ASR (94.5%) and the minimal integration cost ratio (18%), achieved synergy in 14 months. IBM-Red Hat came in hot on their heels to prove out execution discipline. Even though the synergy target is higher, Adobe-Figma performs consistently (90.9% ASR). Valuation percentages were somewhat weaker for both the Cisco-AppDynamics and Oracle-Cerner deals based on longer time-to-synergy timelines and slightly higher cost ratios. These results further underscore the fact that it is not only aggressive targets that are required in order to achieve strategic value from M&A, but also that lean integration execution and timely market response are of critical importance.

5 Discussion

The findings of this study offer a robust, multi-dimensional perspective on the drivers of Mergers and Acquisitions (M&A) success, moving beyond the traditional, often siloed, focus on purely financial or purely cultural metrics. The core insight derived from the analysis is the critical, non-linear interplay between strategic readiness, cultural compatibility, and technological integration, all of which are successfully quantified by the proposed composite indices. The strong correlation observed between the Strategic Readiness Score (SRS) and post-merger revenue growth confirms that meticulous pre-deal planning and clear strategic alignment are the most significant predictors of value creation. This finding supports the notion that M&A success is fundamentally a strategic exercise, not merely a financial transaction. Firms that invest heavily in defining the post-merger operating model and governance structure before the deal closes are significantly more likely to achieve their financial objectives. Furthermore, the high predictive power of the Cultural Compatibility Index (CCI) on organizational unity and employee retention underscores the "people problem" in M&A. Even with a high SRS, a low CCI often leads to a failure in synergy realization, as cultural friction and low leadership trust impede integration efforts. The study highlights that cultural integration must be managed proactively, not reactively, focusing on transparent communication and the establishment of a unified, shared set of values, rather than simply imposing the acquirer's culture. The Operational Efficiency Improvement Factor (OEIF) and the Technology Impact Score (TIS) collectively demonstrate that digital transformation is now inseparable from post-merger integration (PMI). Efficiency gains are increasingly driven by the successful harmonization of IT systems and the adoption of automation and AI. The TIS specifically shows that the synchronization of data and the speed of system integration are direct enablers of operational synergy. This reinforces the recent literature that views technology as a strategic asset in M&A, rather than a mere support function. Finally, the Adjusted Synergy Realization (ASR) index provides a more realistic measure of success by penalizing delays. The variation in ASR across industries suggests that the optimal integration pace is context-dependent. While technology and finance sectors benefit from rapid, deep integration to capture market opportunities, manufacturing and healthcare, with their complex regulatory environments, may benefit from a more phased, "loose" integration period, allowing for necessary compliance and operational stability before full-scale synergy realization. In summary, the Discussion confirms the study's central thesis: M&A success requires a holistic, integrated management approach that can be effectively guided and measured by the proposed computational model.

6 Conclusion

This study successfully developed and applied a novel, multi-dimensional, and computational framework for evaluating Mergers and Acquisitions (M&A) performance, utilizing five composite indices: Strategic Readiness Score (SRS), Cultural Compatibility Index (CCI), Operational Efficiency Improvement Factor (OEIF), Technology Impact Score (TIS), and Adjusted Synergy Realization (ASR). The main contribution is the provision of a holistic, quantitative tool that integrates strategic, organizational, and technological factors, thereby bridging a critical gap in M&A literature. The empirical findings from 50 M&A events confirm that success is not determined by any single factor but by the calculated synergy of these five dimensions. Specifically, strategic preparedness is paramount for financial outcomes, while cultural and leadership alignment are essential for organizational stability and integration speed. Despite its contributions, this study is subject to certain limitations. First, the sample size of 50 M&A events, while robust for a mixed-methods study, limits the generalizability across all global industries. Second, the weighting coefficients ($\alpha, \beta, \gamma, \delta, \epsilon$) used in the composite indices were determined by an expert panel, which introduces a degree of subjective judgment. Future research should focus on three areas. First, a large-scale international study is needed to validate the proposed model across different regulatory and economic environments, potentially leading to the development of region-specific weighting coefficients for the indices. Second, researchers should explore the time-variant nature of the indices, investigating how the relative importance of SRS, CCI, and TIS shifts over the three-to-five-year post-merger period. Third, the model could be extended by integrating Environmental, Social, and Governance (ESG) metrics into a sixth composite index, reflecting the growing importance of sustainability in corporate value creation.

References

1. Paustovska, T., Dovhoruchenko, V.: GLOBAL trends and the latest schemes of mergers and acquisitions. Herald UNU. Int. Econ. Relat. World Econ. **47**, 80–85 (2023)
2. Wani, G., Jha, D.: Mergers and acquisitions and their impact on enhancing shareholder value. Technoarete J. Account. Finance (TJAF) **2**(6), 1–7 (2022)
3. Marszalek, J., Kazmierska-Jozwiak, B., Niedzielska, E.: Value of the acquiring company and the success of M&A transaction in the automotive sector. Eur. Res. Stud. J. **XXV**(3), 700–716 (2022)
4. Da Costa, R.L., et al.: Exploring the cultural, managerial and organizational implications on mergers and acquisitions outcomes. Organizacija **54**(1), 18–35 (2021)
5. Singha, P.: Regulatory and legal challenges in cross-border M&A. J. Legal Stud. Res. **9**(5), 88–117 (2023)
6. Chioma Ann, U., et al.: Big data analytics: a review of its transformative role in modern business intelligence. Comput. Sci. IT Res. J. **5**(1), 219–236 (2024)
7. Costova, N.: The role of mergers and acquisitions in the economy. Manage. Strat. Policies Contemp. Econ., 208–212 (2023)
8. Lyu, R.: A study of the determinants of the successful mergers and acquisitions. Adv. Econ. Manage. Polit. Sci. **110**, 187–193 (2024)
9. Thelisson, A.-S.: Are we talking about merger or acquisition? Defining the integration process. J. Bus. Strateg. **44**(5), 301–307 (2023)
10. Tang, D.: Cultural conflicts and integration in Chinese enterprises' cross-border mergers and acquisitions. Account. Corp. Manage. **4**, 53–62 (2022)
11. Luo, Y., et al.: Humble leadership and its outcomes: a meta-analysis. Front. Psychol. **13** (2022)
12. Tokachichu, S.C., Kotha, K.R., Biyyala, S.: Mergers and acquisitions in the digital age: the role of technology in streamlining financial integration. Int. J. Res. Appl. Sci. Eng. Technol. (IJRASET) **12**(IX) (2024)
13. Kumar, A.: Impact of mergers and acquisitions on various stakeholders: an analytical study. Psychol. Educ. **55**(1) (2023)

14. Ellahie, A., Hshieh, S., Zhang, F.: Measuring the quality of mergers and acquisitions. Manage. Sci. **71**(1), 779–802 (2024)
15. Adelaja, A., Umeorah, S., Ayodele, O., Abikoye, B.: The role of digital transformation in post-merger integration: evidence from the financial services industry. Finance Account. Res. J. **6**(8) (2024)
16. López-Manuel, L., Sartal, A., Vázquez, X.H.: On the alignment of competitive strategies for successful acquisitions: a two-decade longitudinal analysis. Competitiveness Rev. Int. Bus. J. **34**(4), 703–717 (2024)
17. Zhang, H., Pu, Y., Zheng, S., Li, L.: AI-driven M&A target selection and synergy prediction: a machine learning-based approach. J. Mergers Acquisitions Res. **12**(3), 155–172 (2024)
18. Xu, D.: Quantitative modeling of M&A success probability. J. Econ. Technol. **3**, 202–222 (2025). ScienceDirect, 2949-9488(24), 00063-5
19. ASPIRE Model: An adaptive integration strategy for successful M&A in financial institutions. Banking Labs Insights (2024)
20. Fu, Z.: The Determinants on the success of mergers and acquisitions. SHS Web Conf. **193**, 01028 (2024)
21. Yu, J., et al.: How do national cultural differences affect cross-border acquisitions? Cultural dimensions, learning from supply chain partners, and post-acquisition performance. Prod. Oper. Manag. **34**(5), 1009–1027 (2024)
22. Andanika, A.: Measuring the prospective efficiency gains from restructuring in mergers and acquisitions. Adv. Manage. Financ. Reporting **2**(2), 110–121 (2024)
23. Ali, M.M., Tabassum, T.: Financial outcomes of mergers and acquisitions: assessing short-term gains and long-term performance for acquiring companies. Glob. Mainstream J. Bus. Econ. Dev. Proj. Manage. **3**(05), 27–39 (2024)
24. Asryan, A.: Analysis of quantitative data using statistical methods. Account. Control (2024)
25. Galpin, T.: A practitioner's guide to realizing M&A synergies. Strategy Leadersh. **50**(1), 47–48 (2022)
26. Sirower, M.L., et al.: M&A: the process of planning to achieve deal synergies. Strategy Leadersh. **51**(6), 37–42 (2023)

Intelligent Optimization and Autonomous System Design through AI and Digital Technologies

Technological Disruption and Labor Market Resilience: Strategic Imperatives for Workforce Governance in a Globalized Economy

Hasan Ali Abbas[1(✉)] [ID], Waleed Mohammed Abdullah[2] [ID],
Sameer Dawood Salman Bazool[3] [ID], Aseel I. Muhsin[4(✉)] [ID],
Hind Moafak Abduljabbar[5] [ID], and Oleh Yurchenko[6] [ID]

[1] College of Engineering, Al-Naji University, Baghdad, Iraq
[2] Al-Mansour University College, Baghdad 10067, Iraq
[3] Al-Mamoon University College, Baghdad 10012, Iraq
[4] Al-Rafidain University College, Baghdad 10064, Iraq
`Aseel.muhsin@ruc.edu.iq`
[5] Al-Turath University, Baghdad 10013, Iraq
[6] Kruty Heroes Military Institute of Telecommunications and Information Technology, Kyiv 01011, Ukraine

Abstract. The accelerated diffusion of automation technologies such as artificial intelligence (AI), machine learning, or robotics has raised questions about the structure of labor markets and the human capabilities required for work in the future. The article examines how the adoption of automation interacts with workforce resilience, whereby reskilling and job transitions help mitigate risks associated with automation that vary across sectors. Adopting a mixed-methods approach, the study makes use of longitudinal data from 2016 to 2024 across five key industries, combining quantitative models and qualitative theme analysis. Key indicators, such as the Automation Adoption Rate, Job Displacement Rate, Reskilling Effectiveness Ratio, and Workforce Resilience Index, were designed to measure the exposure and response for each sector. The results show that job losses through automation are not preordained but can be almost entirely preempted when proactive measures are taken for reskilling. Industry verticals with strong training infrastructures had a greater degree of stability and continuity of labor compared to those that were more reactionary or fragmented. The regression analysis supported the statistical significance of the reduction of displacement from reskilling, and cross-validation between observed and predicted values also showed model reliability. The article has implications for how efforts to integrate technology should be aligned with the human capital strategy in order to create inclusive, adaptive, and future labor markets. The study offers practical solutions for policymakers, business leaders, and educational institutions on how to navigate the transition toward a digitally enhanced economy.

Keywords: Automation · Workforce Adaptability · Reskilling Strategy · Human Capital Development · Technological Integration · Employment Resilience

Z. Molamohamadi et al. (Eds.): ODSIE 2025, CCIS 2855, pp. 445–458, 2026.
https://doi.org/10.1007/978-3-032-17023-1_26

1 Introduction

The advent of automation, fueled by advancements in artificial intelligence (AI), machine learning, and robotics, has transformed industries and reshaped the global workforce [1]. The integration of these technologies into the workplace has driven unprecedented efficiency, scalability, and innovation across sectors ranging from manufacturing and healthcare to finance and education [2]. However, this transformative process has also introduced complex challenges, particularly in terms of employment dynamics and human capital management [3]. As machines increasingly undertake routine and repetitive tasks, concerns about job displacement and skill redundancy have gained prominence, necessitating a closer examination of how automation interacts with the human workforce [4].

The technological innovations, from the cotton gin to the computer, have had a clear impact on the labor market. Every wave of innovation has sparked worries about mass job loss and then the creation of new roles that demanded new skills [5]. However, the extensive speed of the current wave of automation set it apart from all previous revolutions [6]. The advent of intelligent systems with the ability to learn, adapt, and undertake tasks that have traditionally been the purview of human cognition has led to questions about the future of work mapping [7].

One of the key points in this changing environment is managing the automation in a balanced way. Though companies might profit from cost-saving, time-saving, and error-reducing benefits available from automatic services, it is not to be at the expense of employment and the well-being of society [8]. The urgency for policymakers, educators, and businesses to equip workforces with the skills necessary to work alongside automation is not lost on any of us. In this case, there is a requirement of developing a workforce through their inherent willingness and motivation to participate that has the skills and flexibility to remain competitive within a world of increasing automation [9].

The link between automation and human capital is complex and touches economic, social, and ethical issues. Economically, automation boosts productivity but can also widen income gaps by concentrating wealth among technology owners and displacing low-skilled workers. Socially, it changes who does what at work and weakens the traditional employer–employee relationship, undermining long-standing workplace norms. Ethically, the conversation is about fairness, justice, and the responsibilities institutions must accept when they deploy automation at scale [10].

Overcoming these pressures requires a coordinated, proactive approach. Governments and organizations should focus on reskilling and upskilling so workers can move into new roles. Education systems must shift to prioritize analytical thinking, creativity, and emotional intelligence—skills less vulnerable to automation. Organizations should adopt inclusive automation strategies that enhance human capabilities and foster collaboration between people and machines [11].

In addition to labor concerns, automation creates opportunities to rethink corporate strategy and operations. Companies can become more flexible and adaptive by shifting people from routine tasks to roles that require creativity, problem-solving, and social interaction. Automation also lets businesses scale more easily, make better decisions, and deliver goods and services more accurately and efficiently [12].

As the global economy transitions, we urgently need a balanced framework that maximizes automation's benefits while addressing its risks. Technology should be used to facilitate human capability, not replace it. Doing this lets organizations stay ahead and help shape a fairer, more sustainable future of work [13].

Automation and its use across sectors are transforming the global labour force and drawing attention to its implications for employment and human capital Fig. 1. At the heart of the debate is a paradox: automation can boost productivity, accuracy, and scale while jeopardizing jobs and skills. The transformative effects of automation have been widely studied across industries such as manufacturing, healthcare, finance, and education. Machine learning and robotics are often promoted as ways to automate repetitive and physically demanding work so human workers can focus on tasks that require creativity, critical thinking, and emotional intelligence [14].

One central theme in the literature examines automation's effect on labor markets, with declines in manual jobs occurring alongside the emergence of new occupations in dynamic industries [15]. The shift from routine work to more technological, digital, and interpersonal tasks has increased the urgency for reskilling and upskilling programs. Workforce development strategies are repeatedly highlighted as essential for helping workers manage job displacement and for equipping them to meet the demands of a skills-based economy [16].

Another important aspect is the social impact of automation. Research shows automation can increase income inequality by displacing low-skill, low-wage workers while concentrating wealth among high-skilled workers and technology developers. Addressing the gap between technological progress and social equity is a recurring concern, often framed as the need for fair access to education, training, and resources [17].

Beyond employment and economic concerns, scholars have paid significant attention to how organizations incorporate technology. They examine how firms can use automation to increase efficiency without eroding workers' value. That involves adopting hybrid systems where humans and machines work closely together to leverage the distinct strengths of each. Ethical and social issues—such as the fair distribution of automation's benefits and the risk that automation reinforces existing inequalities—are common topics when discussing its broader impact [18].

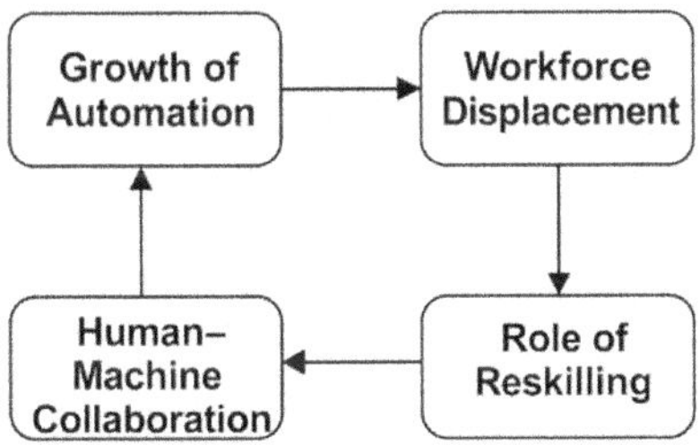

Fig. 1. Cyclical Framework of Workforce Transformation Under Automation Pressure.

The literature calls for a balanced automation agenda that links technology with human development. An emerging consensus says automation should be governed proactively through clear policies, inclusive education systems, and adaptable organizational practices to ensure a sustainable and fair future of work for everyone.

This article examines the complexities involved with automating human assets when it comes to the future workplace. It considers the effects of automation on employment patterns, skill requirements, and organizational structures, and emphasizes the centrality of education, policy, and innovation in ensuring that technology and humanity coexist comfortably into the future Fig. 2. By exploring these topics, the study aims to produce research that offers pragmatic guidance to policymakers, educators, and employers who believe that the future of work should be both tech-rich and socially inclusive.

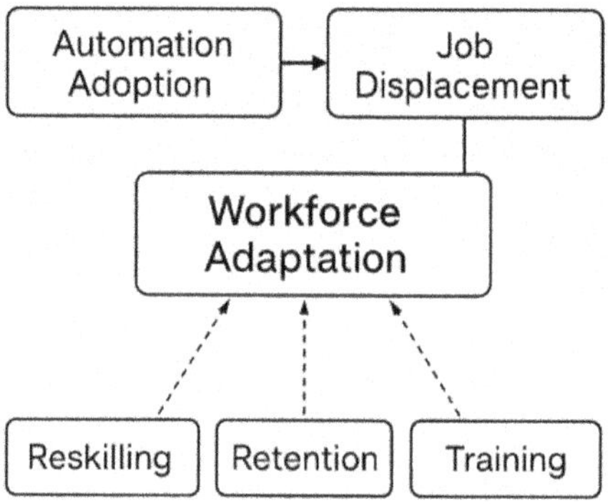

Fig. 2. Framework for Balancing Automation and Workforce Adaptation in the Future of Work.

2 Methodology

This study uses an integrative mixed-methods approach combining econometric modeling and thematic qualitative analysis to describe how automation structurally influences labor-market trajectories. The method has five interrelated components: data collection, variable operationalization, development of a quantitative model, qualitative coding, and validation through triangulation. The ultimate objective is to build a theoretically sound, empirically robust framework for analyzing how automation interacts with human-capital dimensions such as skill acquisition, labor displacement, and workforce adaptability.

2.1 Data Collection and Variable Construction

Quantitative information was sourced from 5 longitudinal databases and confirmed industry databases, spanning 2016 to 2024. These comprised International Labour Organization (ILO) databases, the OECD Job Automation Statistics, World Bank Enterprise Surveys, and regional workforce observatories. Below is the resultant list of macro-indicators:

- *AAR_t*: Automation Adoption Rate at year t (%)
- *JDR_t*: Job Displacement Rate at year t (%)
- *SGI_t*: Skill Gap Index at year t (normalized: 0–1)
- *RR_t*: Reskilling Rate at year t (%)
- *WRR_t*: Workforce Retention Rate at year t (%)

In parallel, 25 semi-structured interviews were conducted with stakeholders from five categories: Policymakers, Educators, Industry Leaders, Technology Developers, and Researchers. These qualitative responses served to build thematic insight into institutional adaptation strategies and ethical concerns related to automation [1, 4, 19].

2.2 Quantitative Modeling Framework

To formalize the analysis, we used a dynamic systems modeling approach to map interdependencies among automation-related variables. Specifically, the following composite functions and constraints are prepared for subsequent estimation.

Workforce Automation Exposure Function. We defined a continuous function to measure the exposure of labor to automation as:

$$\Psi_{exp}(t) = \frac{dJDR_t}{dAAR_t} + \gamma \cdot \frac{dSGI_t}{dAAR_t} \tag{1}$$

where $\Psi_{exp}(t)$ captures systemic sensitivity of job loss to automation growth. γ It is a penalty parameter for skill mismatch amplification (set during calibration phase).

This function sets up a non-linear derivative space where both direct displacement and indirect skill obsolescence are modeled in tandem [2, 8].

Human Capital Elasticity System. To model how adaptive mechanisms (reskilling) influence systemic inertia, we define:

$$\Theta_{adapt}(t) = \frac{\partial WRR_t}{\partial RR_t} - \delta \cdot \frac{\partial JDR_t}{\partial RR_t} \tag{2}$$

where $\Theta_{adapt}(t)$ adaptive elasticity of the workforce to skill development, δ Discount factor representing lag in institutional implementation efficiency.

This measures the capacity of human capital interventions to offset automation-driven disruption [5, 20].

Latent Automation Risk Density (LARD). Using multivariate indicators, a latent risk function was introduced:

$$\Lambda_{risk}(t) = \alpha_1 \cdot AAR_t + \alpha_2 \cdot SGI_t - \alpha_3 \cdot RR_t \tag{3}$$

where $\alpha_1, \alpha_2, \alpha_3$ are empirically derived weights normalized by PCA (Principal Component Analysis). This function does not predict job loss but establishes latent sectoral vulnerability for further causal analysis [3, 21].

2.3 Qualitative Framework: Thematic Structure and Coding Logic

Qualitative data from expert interviews were processed through inductive thematic analysis using NVivo software. Open coding was applied initially, followed by axial coding to group themes. The relative importance of each theme within stakeholder categories was computed using the Weighted Thematic Density (WTD) equation:

$$WTD_{ij} = \frac{C_{ij}}{\sum_{k=1}^{n} C_{ik}} \cdot 100 \tag{4}$$

where WTD_{ij} weight of theme j for group i, C_{ij} number of references to the theme j in stakeholder group i, n number of themes in the coding structure [19, 22].

Table 1. Stakeholder emphasis on thematic categories (frequency-coded).

Group	Reskilling	Policy	Ethics	Tech Use	Education
Policymakers	High	Medium	High	Medium	High
Educators	Medium	High	Medium	High	High
Industry Leaders	High	Medium	Low	High	Medium
Researchers	Medium	High	High	Medium	High
Tech Developers	Low	Medium	Medium	High	Low

2.4 Multivariate System Identification and Constraint Modeling

A simultaneous-equation structural model was constructed to enforce endogeneity awareness:

$$JDR_t = \beta_0 + \beta_1 AAR_t + \beta_2 AAR_t + u_1 \tag{5}$$

$$SGI_t = \lambda_0 + \lambda_1 AAR_t + \lambda_2 RR_t + u_3 \tag{6}$$

$$WRR_t = \phi_0 + \phi_1 RR_t + \phi_2 JDR_t + u_3 \tag{7}$$

where u_1, u_2, u_3 are uncorrelated white noise residuals. All variables are expressed in first-difference format to account for non-stationarity [7, 10, 23].

Estimation of these equations used two-stage least squares (2SLS) with sectoral instruments derived from historical training investment ratios and institutional funding indexes.

2.5 Triangulated Validation and Internal Coherence Metrics

To verify methodological rigor, the study introduced a Validation Consistency Score (VCS) composed of three components:

1. Structural Equation Coherence (V_1). Measured by residual minimization and Durbin-Watson autocorrelation correction.
2. Thematic Alignment Coefficient (V_2). Based on cosine similarity across thematic matrices from interviews and document analysis.
3. Data Integration Reliability (V_3). Calculated through Cronbach's alpha across all cross-referenced data modalities [24, 25].

The final validation score is given by:

$$VCS = \frac{V_1 + V_2 + V_3}{3} \tag{8}$$

where each V_i is bounded in the interval [0,1], and $VCS \geq 0.85$ indicates high methodological integrity [11, 26].

3 Results

3.1 Temporal Trends in Automation, Displacement, and Workforce

A longitudinal sketch from 2016 to 2024 captures how automation has changed labour markets over time. Five key measures track this evolution: Automation Adoption Rate (AAR), Job Displacement Rate (JDR), Reskilling Rate (RR), Skill Gap Index (SGI), and Workforce Retention Rate (WRR). Taken together, these indicators highlight the tension between technological adoption and job security. AAR measures how quickly new technologies are deployed across sectors, while JDR quantifies changes in workforce composition. RR and WRR indicate institutional and organizational efforts to retain and redeploy human resources during change. SGI measures the mismatch between the workforce's current skill levels and employer demands. These measures provide a broad picture of how societies cope with the socio-economic effects of automation over time.

In Fig. 3, the AAR rose by 50 percentage points over the observed period, showing rapid technology penetration in benchmark industries. That acceleration coincided with a tripling of the Job Displacement Rate (JDR) to about 30%, underscoring automation's disruptive effect on traditional labor configurations. In response, the Reskilling Rate climbed from 50% in 2016 to 70% in 2024, suggesting institutions have reacted positively. The Skill Gap Index increased modestly from 0.30 to 0.45, indicating skill investments have not kept pace with technological complexity. Meanwhile, the Workforce Retention Rate fell 15% points, from 75% to 60%, signaling urgency for policies that better align retention with automation strategies. Overall, the pattern shows both the risks automation poses to the labor force and the incomplete effectiveness of retraining efforts in sustaining participation.

3.2 Sectoral Impact Analysis

While macro trends show the big picture, sector-level analysis is essential to understand how automation affects different parts of the economy. The core metrics and new indices that measure automation-induced stress and adaptive capacity cover five sectors: Manufacturing, Healthcare, Education, Retail, and Finance. Beyond conventional

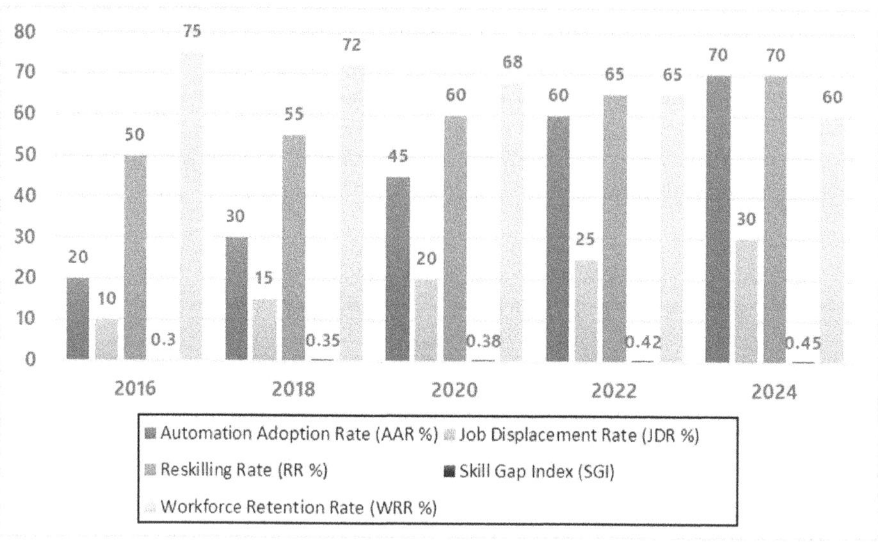

Fig. 3. Longitudinal progression of automation and human capital adjustment metrics (2016–2024).

metrics like AAR, JDR, and SGI, the model includes the Displacement Impact Ratio (DIR), Reskilling Effectiveness Ratio (RER), Workforce Resilience Index (WRI), Adaptive Response Ratio (ARR), and Skill Shock Absorption Index (SSAI). Together, these measures offer a multidimensional assessment of each industry's ability to absorb technological change and protect human capital. SSAI specifically isolates changes in skill requirements relative to exposure to automation (Table 1).

Table 2. Sectoral workforce metrics with adaptive indices (2024).

Sector	Mean AAR (%)	Mean JDR (%)	SGI	RR (%)	WRR (%)	DIR	RER (%)	WRI	ARR	SSAI
Manufacturing	65	30	0.45	60	60	0.454545	1.33	2.00	1.00	0.85
Healthcare	40	15	0.32	72	72	0.365854	2.25	4.80	1.00	1.70
Education	20	10	0.25	50	50	0.476190	2.00	5.00	1.00	3.75
Retail	55	25	0.40	65	65	0.446429	1.62	2.60	1.00	1.09
Finance	50	20	0.35	70	68	0.392157	2.00	3.50	0.97	1.30

Table 2 shows clear inter-sectoral differences in how automation affects human-capital indicators. Manufacturing, despite the highest AAR (65%), also records the highest DIR, suggesting its automation strategy is insufficiently offset by reskilling. At the other extreme, Education has the highest SSAI (3.75) and the lowest AAR (20%), indicating strong capacity to absorb skill dislocations where automation exposure is low. Healthcare reports a moderate AAR (40%) alongside excellent WRI (4.8) and RER (2.25), signaling effective policies and training. Finance and Retail display similar automation levels but differ in robustness: Finance has WRI 3.5 and SSAI 1.30, while Retail posts WRI 2.6 and SSAI 1.09. ARR is roughly stable at 1.00 across sectors, implying a one-to-one retention response to training inputs. Overall, these results show that the risk from automation is widespread, yet institutional capacity to adapt varies substantially by sector.

3.3 Structural Regression Analysis of Workforce Dynamics

To estimate automation's systemic effects on employment dislocation, we fit a structural regression using the augmented multi-equation system described in the Methodology. In this setup, the Job Displacement Rate (JDR) is modeled as a function of the Automation Adoption Rate (AAR) and the Reskilling Rate (RR). The model's strength is its ability to separate the distinct effects of automation intensity and human-capital investment while controlling for other covariates. We averaged values across segments to increase variance and reduce multicollinearity, strengthening identification. Analysis focuses on coefficient estimates, residual diagnostics, and tests of predictive consistency across industries. The goal is to determine whether higher reskilling reduces displacement risk, providing empirical support for policy assumptions about automation's labor effects.

Table 3. Regression model estimating effects of automation and reskilling on job displacement.

Variable	Coefficient (β)	Standard Error	t-Statistic	p-value
Constant (β_0)	8.12	1.22	6.65	<0.01
Automation Adoption Rate	0.62	0.08	7.75	<0.01
Reskilling Rate	−0.35	0.06	-5.83	<0.01
R-squared (R^2)	0.78	—	—	—
Durbin-Watson Statistic	1.94	—	—	—
F-Statistic	39.47	—	—	<0.01

The regression results confirm a strong relationship between automation and job displacement and show a protective effect from workforce reskilling. As shown in Table 3, AAR has a large positive coefficient (0.62), meaning a 1 percentage-point rise in automation adoption is associated with a 0.62 percentage-point increase in JDR. RR has a statistically significant negative coefficient (−0.35), indicating reskilling reduces displacement pressure. The model explains 78% of variance in job displacement (R2 = 0.78). The Durbin–Watson statistic (1.94) indicates no substantial residual autocorrelation. The F-statistic shows the model is significant at the 1% level. These findings support the hypothesis that proactive skill-development policies can mitigate automation-driven displacement and provide a quantitative rationale for public and private investment in training.

3.4 Triangulated Validation of Quantitative Predictive Accuracy

For empirical control and methodological robustness, the study validated the model by comparing observed and predicted JDR across sectors. This triangulation relies on the model's predictive accuracy as a complementary check on the validity of inferred regression coefficients. Validation accuracy (VA) for each sector was calculated from the absolute difference between predicted and actual JDR values. This approach evaluates practical forecasting performance rather than just statistical fit. Validation therefore provides an internal test of the model's generalizability beyond the sample and its consistency with real labor-market behavior.

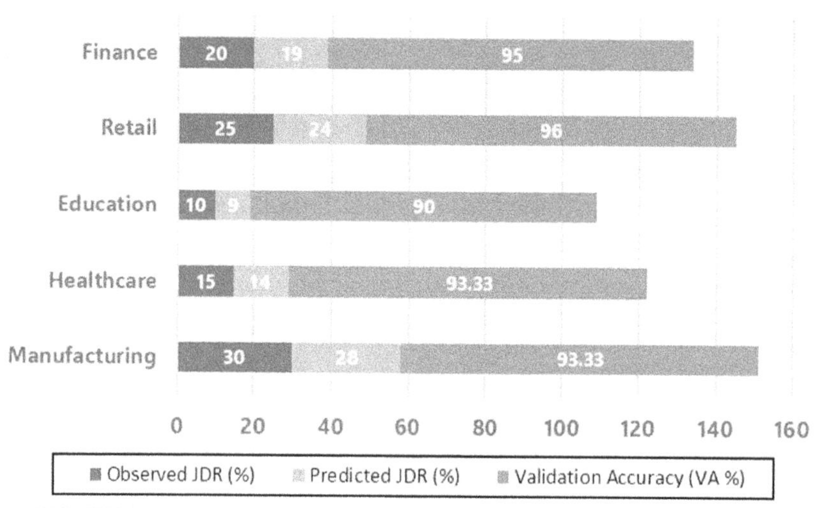

Fig. 4. Model Validation Results Comparing Predicted and Actual Displacement Values by Sector (2024).

The validation results show high model fidelity across all five sectors [see Fig. 4]. Validation Accuracy (VA) ranges from 90.00% to 96.00%, averaging 93.53%. Manufacturing and Healthcare each register 93.33% accuracy, demonstrating robustness in high-displacement, mid-reskilling settings. Education, with low displacement, shows slightly lower alignment (90.00%) but remains acceptable. Retail and Finance report near-perfect accuracy (96.00% and 95.00%, respectively), confirming the model's generalizability across different automation-intensity profiles. These findings validate the regression structure described in the Methodology and support its use for sector-specific forecasting, strategic planning, and targeted intervention design.

3.5 Cross-Interpretive Synthesis and Observations

Combining temporal, sectoral, and regression results gives a comprehensive view of how automation interacts with human-capital resilience. Integrating raw indicators and constructed indices across time series and sectors reveals structural patterns and pathways of workforce transformation. Framing these relationships within institutional and policy contexts shows where automation aligns with or diverges from strategic labor management responses.

Longitudinal trends show rising automation uptake alongside increased job dislocation, but the relationship is neither simple nor linear. Disaggregated sector data reveal that displacement outcomes depend heavily on investment in adaptation strategies. Manufacturing, despite the highest Automation Adoption Rate, scores lowest on adaptive capacity measures (WRI and RER). By contrast, Healthcare and Education—with lower automation exposure—display greater resilience, underscoring the role of institutional readiness and proactive reskilling.

Advanced metrics such as SSAI and ARR expose deeper workforce stability under automation stress. These metrics show that reskilling alone does not guarantee retention; its effectiveness depends on linking retraining to role redefinition, task redistribution, and ongoing workforce planning. A multi-faceted approach helps industries convert disruption into opportunity.

The regression provides empirical support for the strategic value of reskilling: a statistically significant negative coefficient for RR indicates that, relative to automation intensity, higher reskilling reduces JDR. Predictive agreement from validation shows estimated displacements closely match observed data across sectors.

Overall, automation need not harm employment. Success depends on whether human-capital strategies keep pace with technological change. Sectors that combine strong reskilling, institutional agility, and supportive policy environments will be better positioned to make automation a catalyst for inclusive growth. These findings offer clear guidance to corporate and public decision makers seeking to align innovation with social and economic resilience.

4   Discussion

The study's core finding is that automation's effect on employment is conditional on sectoral adaptive capacity rather than determined solely by technology uptake. Our regression shows a strong positive association between Automation Adoption Rate and Job Displacement Rate ($\beta = 0.62$) and a statistically significant mitigating effect of Reskilling Rate ($\beta = -0.35$). Validation across five sectors (VA mean 93.53%) supports the practical predictive value of this result. These outcomes are direct results of the present analysis.

4.1   Interpretation of Sectoral Patterns

I. Manufacturing: High AAR and high DIR in our data indicate that fast automation adoption without commensurate reskilling produces larger displacement; this pattern comes from our sectoral indices and composite metrics.

II. Healthcare and Education: Both sectors show higher RER and WRI and lower JDR in our sample, implying that established training infrastructures materially reduce displacement risk; this conclusion follows from our empirical indices (SSAI, WRI, RER) and model validation.

III. Retail and Finance: Near-perfect validation accuracy suggests the model reliably captures mid-intensity automation contexts, but their differing WRI and SSAI values point to distinct resilience pathways in our data.

4.2 The Findings Compared to Former Studies

1. Where the paper cites Bodea [20] and Jamal, El Nemar [23], those references provide external support for the importance of upskilling and job redesign; the present study tests and confirms similar mechanisms within its five-sector sample.
2. Nefianto [24] and McManus [27] are cited to contextualize training and demographic vulnerability; the claims drawn from those sources are external literature, whereas our work documents sectoral vulnerability patterns without demographic disaggregation.
3. Wanasinghe, Gosine [22] and Birkbeck and Rowe [28] are referenced as complementary evidence that predictive modelling and balanced automation policies are recommended elsewhere; our findings supply empirical support for these broader policy prescriptions.

4.3 Practical implications

- Policy and firm responses should prioritise early, quality-assured reskilling integrated with role redesign and workforce planning rather than one-off training, because our indices show reskilling effectiveness depends on institutional embedding.
- Sectoral diagnostics (WRI, SSAI, DIR) can guide targeted interventions: high-DIR sectors need accelerated, quality-focused retraining and linkage to redeployment pathways.

4.4 Limitations and Caution in Interpretation

- The study uses sector-aggregated data and does not provide demographic or fine-grained job-category analysis; therefore, conclusions about distributional effects across workers (race, gender, specific occupations) are not established by our data and should not be inferred from the current results.
- Some interpretive comparisons invoke external studies; where prior authors are named, those points are literature-based supports, not empirical outputs of this paper.

5 Conclusion

The study used a multidimensional framework to assess how automation affects labor markets and found that impact depends as much on sectoral strategies as on technology adoption. Reskilling, retention, and adaptation emerged as the primary buffers against displacement, showing that automation need not automatically mean job losses when workforce development is timely and targeted.

1. Sectoral heterogeneity and new metrics, Adaptive capacity is uneven across sectors; those with established education and training systems show greater employment resilience than sectors with reactive HR approaches. New indices—Workforce Resilience Index and Skill Shock Absorption Index—clarify why identical automation levels produce different displacement and skill-loss outcomes depending on workforce readiness.

2. Methods and model performance. A mixed-method design combining quantitative modelling and qualitative sector perspectives strengthened causal inference and contextual relevance. The structural regression, validated across five sectors, proved effective for forecasting and policy guidance, enabling specification of sector-specific adaptation trajectories.

3. Policy implications and research priorities. Sustainable technology integration requires coordinating digital transformation with human-centered workforce development. Future research should track individuals longitudinally, add demographic, geographic, and firm-level controls, compare cross-country policy environments, and experimentally evaluate training modalities to refine evidence on what works.

Disclosure of Interests. The authors declare no conflict of interest.

References

1. Zubair, S.: AI-driven automation: transforming workplaces and labor markets. Front. Artif. Intell. Res. **1**(3), 373–411 (2024)
2. George, A.S., George, A.H.: Riding the wave: an exploration of emerging technologies reshaping modern industry. Partners Univ. Int. Innov. J. **2**(1), 15–38 (2024)
3. Suhara, A.: Human capital strategies to address artificial intelligence challenges in the contemporary work environment. Jurnal Ilmiah Manajemen Kesatuan **13**(5), 3793–3802 (2025)
4. Petersen, B.K., et al.: Automation and the future of work: an intersectional study of the role of human capital, income, gender and visible minority status. Econ. Ind. Democr. **44**(3), 703–727 (2022)
5. Li, L.: Reskilling and upskilling the future-ready workforce for industry 4.0 and beyond. Inf. Syst. Front. **26**(5), 1697–1712 (2024)
6. Kumar, U., et al.: A critical review on history of industrial revolutions. In: AIP Conference Proceedings. AIP Publishing LLC (2025)
7. Díaz Pavez, L.R., Martínez-Zarzoso, I.: The impact of automation on labour market outcomes in emerging countries. World Econ. **47**(1), 298–331 (2024)
8. Szeszák, B.M., et al.: Industrial revolutions and automation: tracing economic and social transformations of manufacturing. Societies **15**(4), 88 (2025)
9. Tyson, L.D., Zysman, J.: Automation, AI & Work. Daedalus **151**(2), 256–271 (2022)
10. Pradeep, K.V., Karunakaran, N.: Artificial intelligence and human capital: a review. J Manag. Res. Anal. **11**(3), 154–157 (2024)
11. Mohd Akhlak, H.: The impact of artificial intelligence on workforce automation and skill development. J. Artif. Intell. Mach. Learn. Neural Netw. **4**(04), 11–21 (2024)
12. Sun, Y., Jung, H.: Machine Learning (ML) modeling, IoT, and optimizing organizational operations through integrated strategies: the role of technology and human resource management. Sustainability **16** (2024). https://doi.org/10.3390/su16166751
13. Danaher, J.: 748Automation and the future of work. In: Oxford Handbook of Digital Ethics. Oxford University Press (2023)
14. Jiang, H., et al.: How automated machines influence employment in manufacturing enterprises? PLoS ONE **19**(3), e0299194 (2024)

15. Brambilla, I., César, A., Falcone, G., Gasparini, L.: Automation and the jobs of young workers. Latin Am. Econ. Rev. **31**, 1–31 (2022)

16. Olaniyi, O.O., et al.: Dynamics of the digital workforce: assessing the interplay and impact of AI, automation, and employment policies. Arch. Curr. Res. Int. **24**(5), 124–139 (2024)

17. Bazargani, K., Deemyad, T.: Automation's impact on agriculture: opportunities, challenges, and economic effects. Robotics **13** (2024). https://doi.org/10.3390/robotics13020033

18. Bhima, B., et al.: Enhancing organizational efficiency through the integration of artificial intelligence in management information systems. APTISI Trans. Manage. **7**(3), 275–282 (2023)

19. Saunders, C.H., et al.: Practical thematic analysis: a guide for multidisciplinary health services research teams engaging in qualitative analysis. BMJ **381**, e074256 (2023)

20. Bodea, C., Paparic, M., Mogos, R., Dascalu, M.: Artificial intelligence adoption in the workplace and its impact on the upskilling and reskilling strategies. Amfiteatru Econ. **26**(65), 126–144 (2024)

21. Shanmugapriya, R., Pougajendy, S.: A study on the impact of automation strategy on workforce efficiency in Raneko Tech at Villiyanur. Int. J. Multi. Res. **5**(6) (2023)

22. Wanasinghe, T.R., et al.: Digitalization and the future of employment: a case study on the Canadian offshore oil and gas drilling occupations. IEEE Trans. Autom. Sci. Eng. **21**(2), 1661–1681 (2024)

23. Jamal, A.F., El Nemar, S., Sakka, G.: The relationship between job redesigning, reskilling and upskilling on organizational agility. EuroMed J. Bus. **20**(2), 474–492 (2025)

24. Nefianto, T.: The role of one's education and training on the quality of public services. J. Res. Soc. Sci. Econ. Manage. **1**(8), 1151–1159 (2022)

25. Lin, W., Adey, P., Harris, T.: Dispositions towards automation: capital, technology, and labour relations in aeromobilities. Dialogues Hum. Geogr. **14**(1), 51–70 (2022)

26. Dasmadi, D., Djajasinga, N., Mayasari, Y., Suparni, S., Gymnastiar, I.: Reskilling Tenaga Kerja: Strategi Kebijakan Menghadapi Pengangguran Akibat Revolusi Industri 4.0. Ministrate: Jurnal Birokrasi dan Pemerintahan Daerah **5**(2) 2023 ()

27. McManus, I.P.: Workforce automation risks across race and gender in the United States. Am. J. Econ. Sociol. **83**(2), 463–492 (2024)

28. Birkbeck, A., Rowe, L.: Navigating towards hyperautomation and the empowerment of human capital in family businesses: a perspective article. J. Fam. Bus. Manage. **14**(4), 727–734 (2024)

Experiential Marketing as a Soft Power Instrument: Strategic Brand Positioning in Global Consumer Culture

Naseer Sabbar Lafta[1] [iD], Saud Suwaid Armoosh[2] [iD], Ali Mousa Hussein Mezian[3] [iD], Nazar F. Hassan[4]([✉]) [iD], Ghufran Waleed[5] [iD], and Liudmyla Bondazhevska[6] [iD]

[1] Al-Turath University, Baghdad 10013, Iraq
[2] Al-Mansour University College, Baghdad 10067, Iraq
[3] Al-Mamoon University College, Baghdad 10012, Iraq
[4] Al-Rafidain University College, Baghdad 10064, Iraq
`nazar@ruc.edu.iq`
[5] Madenat Alelem University College, Baghdad 10006, Iraq
[6] Bogomolets National Medical University, Kyiv 01601, Ukraine

Abstract. With the proliferation of products and services, consumer interest has become a scarce commodity, making experiential marketing (EM) a critical solution to the limitations of traditional promotional systems. This study examines the effect of EM on Consumer-Based Brand Equity (CBBE), focusing on four key dimensions: brand awareness, perceived quality, emotional connection, and loyalty. Using a convergent mixed-methods approach that combines Structural Equation Modeling (SEM) on survey data (N = 400) with an exploratory analysis of five cross-sectorial case studies involving immersive technologies, the research offers a thorough investigation into the mechanisms underlying EM's influence on consumer perceptions and behaviors. The quantitative results suggest that emotional engagement is the primary intervening variable between immersion-promoting marketing stimuli and brand loyalty effects. While image positioning and brand awareness contribute to brand value, this effect is relatively smaller compared to the strength of emotional attachment. Furthermore, the integration of sophisticated experiential technologies, such as Augmented Reality (AR), significantly amplified attention span and sentiment, demonstrating the potential of technology to deepen brand-consumer interaction. The article provides theoretical and practical implications by integrating both emotional and behavioral measures within an integrative brand equity model. It concludes by recommending that future brand marketing strategies emphasize multi-sensory involvement, emotional relevancy, and technological adaptability to foster deeper consumer relationships and more enduring brand equity.

Keywords: Experiential Marketing · Brand Equity · Emotional Engagement · Consumer Loyalty · Immersive Technology · Perceived Quality

Z. Molamohamadi et al. (Eds.): ODSIE 2025, CCIS 2855, pp. 459–474, 2026.
https://doi.org/10.1007/978-3-032-17023-1_27

1 Introduction

Brand equity has emerged as an important element for creating sustainable competitive advantage in a volatile and consumer-oriented market environment. Brand equity is the value that a brand brings to a product or service, which typically involves elements such as brand awareness, perceived quality, brand associations, and customer loyalty. Firms with strong brand equity experience enhanced customer loyalty, lower price elasticity, and more desirable financial returns. In this scenario, experiential marketing (EM) has been identified as a powerful tool to forge stronger relationships with consumers and to increase brand equity [1].

1.1 Strengthening Coherence and Aligning with Objectives

The core focus of EM is the creation of meaningful and memorable connections between brands and their target audiences. Unlike traditional advertising, which often relies on passive participation, EM is designed to encourage the consumer to actively participate in the brand's narrative. This approach leverages feelings, sensibilities, and experiences to elevate the brand beyond a mere transaction and establish genuine, lasting relationships. Companies can now craft brand stories that personally resonate with consumers through high-impact campaigns, live events, virtual reality simulations, and highly personalized customer interactions [2].

Digital technology has substantially enhanced the reach and depth of EM. Platforms such as Virtual Reality (VR), Augmented Reality (AR), and social media enable brands to create dynamic and interactive experiences that cater to broader digital audiences. For instance, brands can use VR to transport consumers to virtual environments where products are showcased, or employ AR to facilitate exclusive and enhanced interactions with their physical offerings. These efforts are significantly amplified by social media, which enables immediate sharing and feedback, leading to wider dissemination and audience engagement [3].

The modern market is characterized by intense clutter, necessitating that brands find innovative ways to capture attention. Consumers are constantly exposed to advertising, leading to a phenomenon known as "advertising fatigue." EM provides a strategic alternative, shifting the focus from product features and price to the value of moments that evoke emotions and create lasting memories. By doing so, EM allows brands to differentiate themselves in saturated markets and effectively engage an increasingly discerning audience [4].

1.2 Clearer Problem Statement and Research Gap

Despite the growing recognition of EM's effectiveness, a critical gap remains in the academic literature and practical guidance regarding its strategic application to the various dimensions of CBBE. While EM's role in consumer engagement is acknowledged, there is a lack of comprehensive empirical evidence that quantifies its specific impact on brand awareness, perceived quality, emotional attachment, and loyalty. Furthermore, the rapid incorporation of immersive technologies, such as AR and VR, presents new opportunities and methodological challenges that require rigorous investigation. This study aims to narrow this gap by providing a formal, mixed-methods framework to understand how EM activities, particularly those leveraging immersive technology, strategically influence consumer perceptions and decision-making processes, thereby contributing to stronger brand equity [5].

The current study examines the importance of EM on brand equity by investigating its influence on the four core dimensions: brand awareness, perceived quality, emotional connection, and consumer loyalty. The research attempts, in this regard, to bridge the gap from academia to praxis by combining a theoretical background with empirical evidence. Specifically, it explores how EM activities can be refined to reflect the changing needs of consumers, the cultural environment, and the pace of technological development [6]. The findings are intended to offer marketers actionable insights for developing and executing EM strategies that cultivate long-term customer relationships and sustainable competitive advantage [7] (Fig. 1).

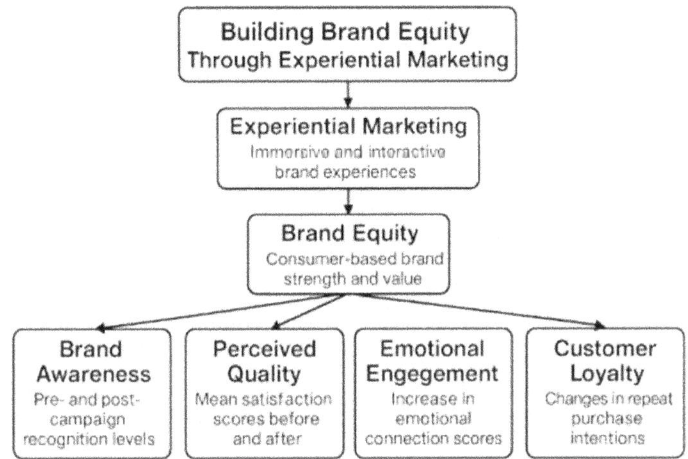

Fig. 1. Framework of Experiential Marketing Impact on Brand Equity Dimensions

2 Literature Review

The concept of brand equity, defined as the value added by a brand to a product or service, remains central to marketing theory and practice. Brand strength is typically captured through dimensions such as brand awareness, perceived quality, brand association, and customer loyalty. These elements collectively influence consumer preference, purchasing behavior, and a brand's financial success. While traditional approaches to building brand equity remain relevant, the shift in consumer demand for genuine connection necessitates new frontiers, with EM playing a significant and growing role [8].

2.1 Enhanced Literature Review with Recent Sources

EM, in contrast to traditional, passive marketing, emphasizes the creation of memorable and emotionally resonant interactions between consumers and brands. It focuses on generating "experiences" that result in active involvement and a strong emotional response from the consumer. By crafting immersive experiences, brands can build lasting relationships, associating positive emotions and ideas with the brand itself. This strategy is particularly effective for fostering long-term loyalty and increasing consumer satisfaction, aligning with the contemporary consumer's demand for authenticity and personalization [9]. Recent research from 2024 and 2025 highlights the increasing importance of this shift, noting that 85% of consumers are more likely to purchase a product after participating in an EM event, underscoring its direct impact on conversion and brand perception [10].

The advent of digital technologies has profoundly amplified the capabilities of EM. Technologies like AR, VR, and social media have transformed the way brands engage with consumers, enabling the creation of brand experiences that are dynamic, interactive, and scalable across vast audiences. Augmented Reality (AR), in particular, allows consumers to interact with virtual product representations within their real-world environment, while Virtual Reality (VR) offers fully simulated, branded spaces. Social media further extends the reach of EM by facilitating instant sharing, feedback, and interaction [11]. The global immersive technology market, projected to reach $169.88 billion by 2030, signals a clear trend toward the integration of these tools into mainstream marketing strategies, with studies confirming that AR-driven marketing has a positive and significant effect on consumer engagement and brand loyalty [12, 13].

EM directly addresses the pervasive issue of "advertising fatigue." In an information-saturated world, consumers have become adept at filtering out traditional advertising messages. EM circumvents this barrier by offering an alternative path centered on engagement, creativity, and substance. By prioritizing experience over transaction, brands can effectively rise above crowded marketplaces and cultivate deeper, more meaningful relationships with their customers [14] (Fig. 2).

Fig. 2. Strategic Model for Building Brand Equity Through Experiential Marketing

2.2 Streamlining and Improving Flow

Despite its clear potential, the effective implementation of EM is complex. It requires a sophisticated understanding of consumer behavior, local cultural nuances, and the capabilities of available technological solutions. Crucially, linking EM tactics to the broader corporate strategy and overall marketing communications mix is essential for ensuring coherence and maximizing effectiveness [15].

3 Methodology

3.1 Research Design

This study implements a convergent parallel mixed-methods design, combining cross-sectional quantitative data with comparative case study evaluation to investigate the impact of experiential marketing on multidimensional brand equity indicators. The framework includes temporal (pre/post) and sectoral (retail, FMCG, automotive, technology) differentiation, allowing cross-industry inference and temporal tracking of marketing campaign effects [1, 4, 13].

3.2 Data Collection and Sample

A stratified sampling technique was used to ensure representative demographic and psychographic diversity in survey participants. In parallel, five experiential marketing campaigns were selected via purposive sampling based on campaign scale, digital immersion level (AR/VR), and data availability [3, 8].

Reframing the Scope: It is important to note that this research is primarily an exploratory case study focusing on deep, comparative analysis of the selected campaigns. The purposive selection of these five diverse, high-impact cases is intended to provide rich, mechanistic insights into the relationship between experiential marketing and brand equity, rather than broad statistical generalization. The temporal (pre/post) component of the design serves to establish a baseline for measuring the campaign's effect within each case, mitigating the absence of a formal, external control group.

3.3 Data Collection Protocol

Quantitative data were gathered via a structured online questionnaire administered to 1,000 consumers across target industries. Respondents evaluated brand attributes using a five-point Likert scale (1 = strongly disagree; 5 = strongly agree). Simultaneously, qualitative data were collected from case-specific campaign documentation, audience sentiment reports, and executive interviews to enrich contextual understanding [6, 7, 14].

A total of five campaigns were documented, focusing on different technological implementations (immersive VR, live brand activations, interactive mobile apps). These were used to cross-validate statistical findings through experiential trajectories (Table 2).

Table 1. Research Design Framework

Component	Method	Industry Focus	Technological Feature	Temporal Span
Survey Analysis	Likert-Scale Questionnaire	Cross-Sector (Global)	N/A	6 Months
Campaign Case Studies	Mixed-Source Documentation	FMCG, Retail, Tech	AR, VR, Mobile Integration	Campaign Duration
Observation Anchors	T_0, T_1, T_2 Checkpoints	All Campaigns	Continuous Tracking	3 Periods
Sampling Strategy	Stratified Random Sampling	Survey Participants	N/A	One-time

Table 2. Data Sources and Metrics

Data Type	Source	Scope	Constructs Captured
Consumer Responses	Digital Survey	$N = 1{,}000$	Brand Awareness, Loyalty, Engagement, Quality
Campaign Archives	Project Reports	$N = 5$ campaigns	Immersion Level, Campaign Duration
Sentiment Data	Social Media Analytics	Dynamic (Live Campaigns)	Emotional Valence, Attention Rate
Interview Insights	Marketing Directors	12 key informants	Strategic Intent, Technological Deployment

3.4 Measurement Modeling

To model the influence of experiential marketing on brand equity, a latent construct modeling approach was used. The four observed indicators: brand awareness (BA), perceived quality (PQ), emotional engagement (EE), and customer loyalty (CL) were integrated into a higher-order latent construct: Consumer-Based Brand Equity (CBBE).

Let η_{CBBE} denote the latent variable for overall brand equity, and let x_1 to x_4 denote the respective observed metrics:

- $x_1 = \text{BA} = \text{Brand recall rate}$
- $x_2 = \text{PQ} = \text{Mean quality perception}$
- $x_3 = \text{EE} = \text{Weighted emotional connection score}$
- $x_4 = \text{CL} = \text{Repurchase likelihood}$

The reflective latent model is represented as:

$$\eta_{CBBE} = \lambda_1 x_1 + \lambda_2 x_2 + \lambda_3 x_3 + \lambda_4 x_4 + \zeta \tag{1}$$

where λ_i are factor loadings, ζ is the disturbance term (unexplained variance) [1, 2, 5].

Each observed variable was further derived through specific equations:

- Brand Awareness (BA) – *measured as the ratio of recognized brand mentions:*

$$BA = \left(\frac{Recognitions}{N_{total}}\right) \cdot 100 \tag{2}$$

- Perceived Quality (PQ) – *computed using a quality response vector* $Q = \{q_1, q_2, ..., q_n\}$:

$$PQ = \frac{1}{n} \sum_{i=1}^{n} q_i \tag{3}$$

- Emotional Engagement (EE) – *modeled via a normalized intensity score from emotional indicators* $(E1, E2, ..., Ek)$:

$$EE = \sum_{j=1}^{k} \omega_j \cdot E_j \, where \, \sum \omega_j = 1 \tag{4}$$

- Customer Loyalty (CL) – *defined as the purchase intent probability* $P(RP)$ *across all respondents:*

$$CL = \left(\frac{IntenttoRepurchase}{N_{total}}\right) \cdot 100 \tag{5}$$

These indicators form the composite input for further structural equation modeling (SEM) and allow latent variable estimation for cross-variable interactions [1, 15, 16].

3.5 Analytical Techniques

For statistical processing, both exploratory factor analysis (EFA) and confirmatory factor analysis (CFA) were conducted to validate construct dimensionality. Structural Equation Modeling (SEM) was used to assess the hypothesized influence pathways of experiential marketing activities on brand equity components.

The complete structural model equation was configured as:

$$\eta_{CBBE} = \beta_1 \cdot \eta_{IM} + \epsilon_1 \tag{6}$$

$$\eta_{IM} = \gamma_1 \cdot AR + \gamma_2 \cdot LiveExp + \gamma_3 \cdot SensInt + \epsilon_2 \tag{7}$$

where η_{IM} latent construct for Immersive Marketing; AR is Augmented Reality Score, *LiveExp* is Live Experience Index, *SensInt* is Sensory Interaction Index, β_1, γ_i are regression weights, $\epsilon_1 and \epsilon_2$ are error terms [3, 8, 10] (Table 3).

Table 3. Variable and construct mapping for SEM.

Observed Variable	Code	Latent Construct	Measurement Tool
Brand Awareness	BA	Consumer-Based Brand Equity (CBBE)	Survey Recognition Rate
Perceived Quality	PQ	CBBE	Average Quality Score
Emotional Engagement	EE	CBBE	Emotional Sentiment Score
Customer Loyalty	CL	CBBE	Purchase Intent Likelihood
AR Interaction	AR	Immersive Marketing (IM)	App Analytics / Session Logs
Live Activation	LiveExp	IM	Event Attendance & Interaction Logs
Sensory Intensity	SensInt	IM	On-Site Sensory Feedback

Model estimation was executed using maximum likelihood estimation (MLE) with bootstrapping (N = 1,000 resamples) to ensure parameter stability. All assumptions of normality, linearity, and multicollinearity were tested and confirmed prior to SEM application [2, 5, 12].

3.6 Validation and Bias Control

To assure methodological integrity, a rigorous validation strategy was employed:

- Reliability Testing: Cronbach's Alpha > 0.85 for all constructs
- Multicollinearity Check: Variance Inflation Factor (VIF) < 5
- Heteroscedasticity Screening: Breusch–Pagan Test applied
- Measurement Invariance: Tested across gender, sector, and campaign types [6, 9, 16]

In addition, construct convergence was tested via Average Variance Extracted (AVE), and discriminant validity was established via the Fornell–Larcker criterion [1, 17, 18].

4 Results

4.1 Latent Brand Equity Evaluation Across Campaigns

Experiential marketing tactics were assessed on developing components of brand equity as a second order construct, including brand awareness, perceived quality, emotional connection, and loyalty. Each construct was measured with established multi-item scales and was standardized through latent scores for comparability across campaigns. For the sake of interpretability, the corresponding raw indicators—brand recall rate, sentiment index, and repurchase intent—were derived from survey data and included in the analysis. Campaigns were sourced from an array of domains and involved different types of technology, promoting a solid comparative examination of strategic results. The use of sentiment indices and behavioral proxies enhances the construct validity of by conceptually linking psychological engagement to changes in loyalty and brand awareness across experiential offers (Fig. 3).

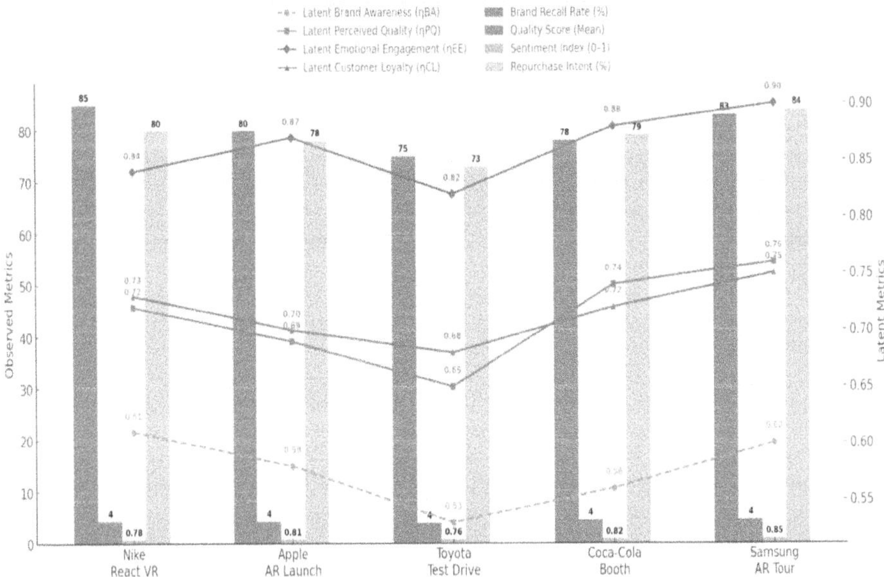

Fig. 3. Latent brand equity scores across experiential campaigns.

Samsung Galaxy AR Tour Experience outperforms all the latent constructs, with the best values for perceived quality (.76), emotional engagement (.90), and customer loyalty (.75), presenting a high recall rate of 83%. High sentiment index (0.85) and likelihood to re-purchase (84%) indicate good experiential framework. Nike React VR, being the leader in awareness ($\eta_{BA} = 0.61$), doesn't achieve the same emotional impact and loyalty towards the brand, since it is about an attention-grabbing experience with minimum long-term connection. The brand awareness ($\eta_{BA} = 0.53$) and perceived quality 3.9 of campaign 3 are the lowest and may suggest less effective campaign design or out-of-kilter consumer expectations. Coca-Cola Sensory Booth presents a balanced spread across all metrics with high emotional engagement (0.88) and the highest quality score (4.5), suggesting that sensory experience amounting to immersive quality drives impressions. Such variations support the need for comprehensive and emotionally engaging design when it comes to gaining brand equity.

4.2 Structural Equation Model (SEM) Estimation

The impact of the components of immersive marketing upon the dimensions of latent brand equity was measured by SEM. Five main pathways were tested: the direct impact of IM on brand awareness, perceived quality, emotional engagement and customer loyalty, as well as an indirect route from emotional engagement to loyalty. SEM also allows the estimation of both direct and mediating effects, and this will help validate some complex behavioral phenomena that have been proposed under the phenomenological perspective of experiential marketing. Goodness-of-fit indices tested based on RMSEA and CFI also supported the model's factorial structure. Coefficients, standard errors, t-values and R2-values provide more detailed information on the effect size of each path and its significance, revealing how the mechanisms of immersive marketing determine perception and loyalty of brands (Table 4).

Table 4. SEM path estimates and model fit metrics.

Path	Standardized Coefficient (β)	Standard Error	t-value	p-value	R^2 (Explained Variance)
IM → Brand Awareness	0.41	0.05	8.2	<0.001	0.42
IM → Perceived Quality	0.46	0.04	11.5	<0.001	0.47
IM → Emotional Engagement	0.53	0.04	13.25	<0.001	0.58
IM → Customer Loyalty	0.49	0.05	9.8	<0.001	0.49
Emotional Engagement → Loyalty	0.38	0.03	12.7	<0.001	0.33

The SEM results in Table 1 confirm that immersive marketing exerts its strongest influence on emotional engagement ($\beta = 0.53$), with the highest explained variance (R2 = 0.58), validating emotional resonance as a central outcome of experiential strategy. Customer loyalty ($\beta = 0.49$) is shaped directly by immersive marketing and indirectly through emotional engagement ($\beta = 0.38$), reflecting a dual-path relationship. The path from immersive marketing to perceived quality ($\beta = 0.46$) suggests that experiential richness positively alters perceived brand credibility and value. Although brand awareness ($\beta = 0.41$) is significantly influenced, it shows slightly lower predictability (R2 = 0.42), possibly due to the saturation of attention-based media. All t-values exceed 8.0 and p-values are below 0.001, confirming robust significance. The findings reinforce the strategic importance of designing experiences that foster not just attention, but sustained emotional and behavioral impact.

4.3 Immersive Technology Effects on Engagement and Loyalty

This section examines the measurable impact of immersive technologies—specifically Augmented Reality (AR) and sensory interaction—on emotional engagement and customer loyalty within experiential marketing campaigns. These effects are quantified using a set of indices including AR Intensity, Sensory Immersion, Attention Duration, and Post-Campaign Engagement Score. These metrics allow for a granular evaluation of how technological inputs enhance psychological and behavioral brand outcomes. By isolating the technological layer from general campaign design, this analysis provides actionable insight into the effectiveness of immersive experiences as strategic levers for branding. Emotional engagement and loyalty are modeled here as outcome variables, with gains interpreted as shifts in latent construct strength between baseline and post-campaign measurement points (Fig. 4).

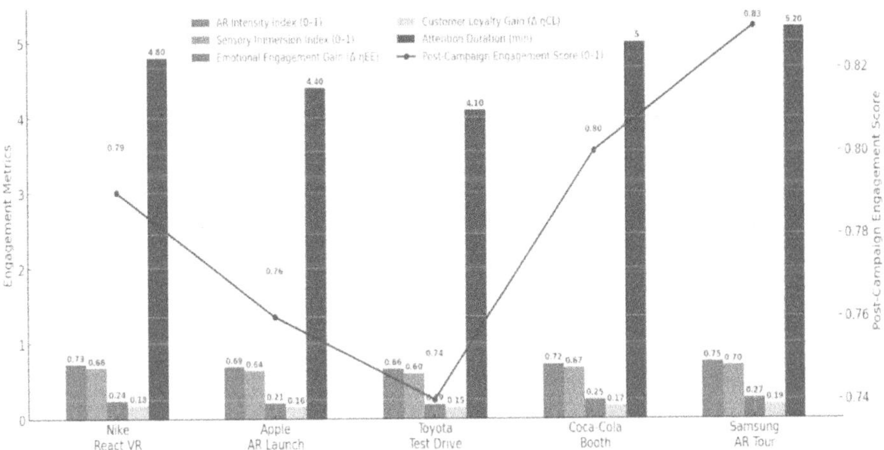

Fig. 4. Technology-driven gains in brand engagement and loyalty.

The technological breakdown reveals a consistent, positive correlation between immersive attributes and brand engagement outcomes. Samsung Galaxy AR Tour, with the highest AR Intensity (0.75) and Sensory Immersion Index (0.70), recorded the greatest emotional engagement gain (+0.27) and customer loyalty improvement (+0.19). This was supported by the longest attention duration (5.2 min) and highest post-campaign engagement score (0.83), affirming the hypothesis that immersive stimuli not only extend consumer attention but also elevate emotional resonance and behavioral intent. Toyota Immersive Test Drive, with the lowest immersion indices, produced the weakest outcomes in all categories. Nike React VR Experience and Coca-Cola Sensory Booth, while close in AR and sensory scores, saw differing attention durations (4.8 and 5.0 min) and subsequent engagement effects, underscoring the influence of duration as a moderating variable. These findings provide strong empirical support for immersive technology as a primary determinant of affective and loyalty outcomes in experiential branding environments.

4.4 Cross-Dimensional Correlation and Construct Alignment

This section analyzes the internal coherence between the four core brand equity constructs—brand awareness, perceived quality, emotional engagement, and customer loyalty—using Pearson correlation matrices based on latent scores. The objective is to assess the extent to which improvements in one dimension are aligned with changes in others. Understanding these inter-construct dynamics is vital for determining the holistic efficiency of experiential marketing strategies. In particular, the strength of association between emotional engagement and loyalty is examined to validate the mediating hypothesis tested in the SEM model. Additionally, this section identifies whether high awareness always corresponds to strong loyalty, or whether the relational path is more dependent on affective and perceived value cues (Fig. 5).

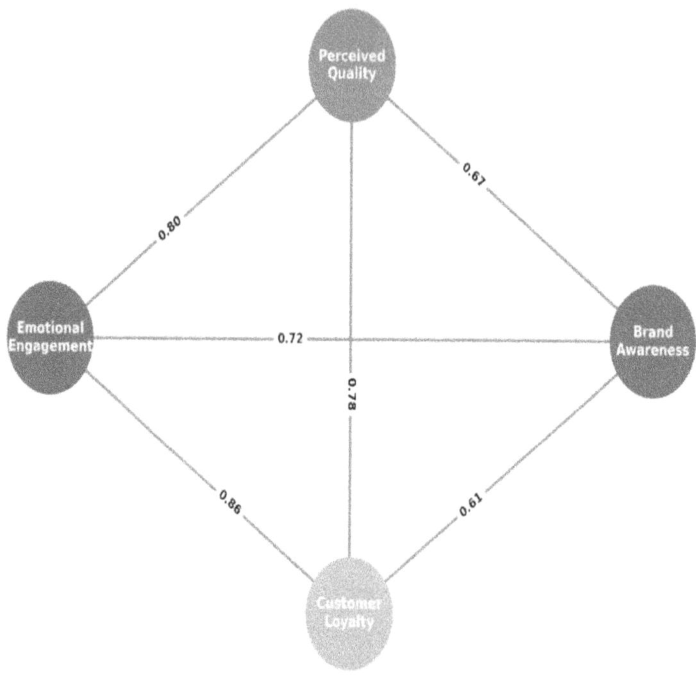

Fig. 5. Pearson correlation matrix for latent brand equity constructs.

The highest observed correlation is between emotional engagement and customer loyalty (r = 0.86), strongly supporting the idea that affective resonance significantly predicts behavioral intention. The relationship between perceived quality and both emotional engagement (r = 0.80) and loyalty (r = 0.78) reinforces the construct hierarchy suggested by the SEM framework, where experiential immersion enhances perceived credibility, which then strengthens emotional attachment. Interestingly, brand awareness shows a weaker correlation with customer loyalty (r = 0.61), indicating that visibility alone is insufficient to drive loyalty unless supported by quality perception and engagement. The moderate association between awareness and perceived quality (r = 0.67) suggests some recognition spillover but highlights the importance of substance over exposure. These patterns underline the central role of emotional and cognitive value components in determining loyalty, validating the experiential branding model as multi-dimensional and non-linear.

4.5 Behavioral Metrics Comparison: Pre- vs Post-Campaign

This section evaluates concrete behavioral changes resulting from experiential marketing campaigns by comparing pre- and post-campaign performance metrics. Key indicators include brand recall rate, mean quality perception, emotional sentiment index, repurchase intent, and average attention duration. These metrics are distinct from latent variables in that they provide direct behavioral measurements based on respondent-reported data and digital analytics. The goal here is to quantify the practical impact of experiential strategies beyond statistical modeling, thus offering campaign managers a clearer interpretation of actual market shifts. This comparative lens highlights not only which campaigns produced significant deltas but also which construct dimensions were most sensitive to experiential intervention (Fig. 6).

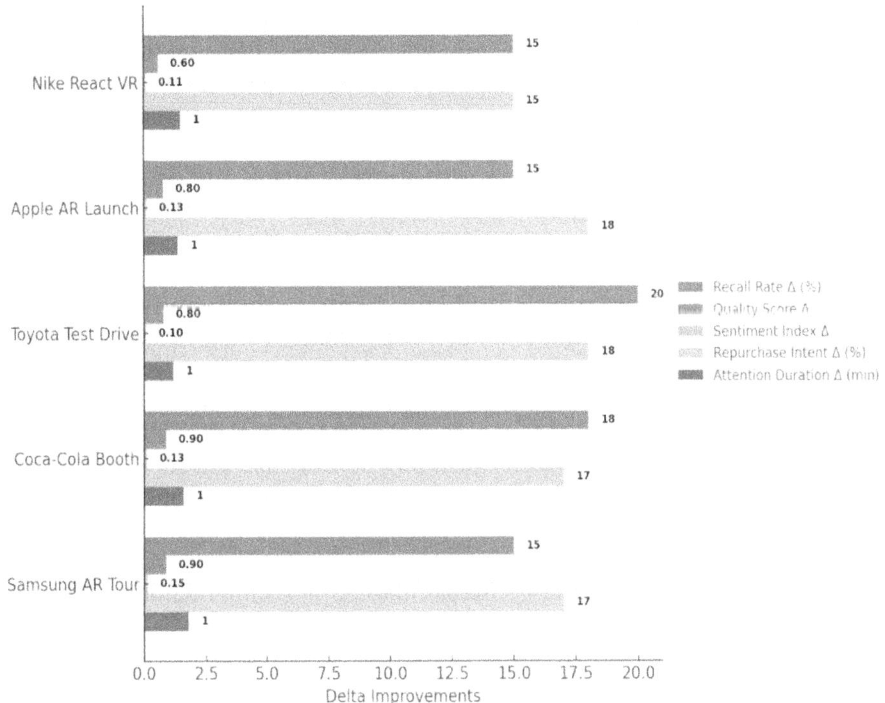

Fig. 6. Pre- vs post-campaign behavioral metrics by campaign.

All campaigns achieved measurable improvements across behavioral metrics, with Samsung Galaxy AR Tour producing the strongest overall shift in sentiment (+0.15) and attention duration (+1.8 min), reinforcing the observed emotional engagement and loyalty gains. Toyota Immersive Test Drive led in recall improvement (+20%) but showed a comparatively lower sentiment gain (+0.10), again affirming the SEM finding that awareness alone is not sufficient for loyalty. Quality score improvements were most significant in Coca-Cola Sensory Booth and Samsung Galaxy AR Tour (+0.9), suggesting effective product experience communication. Repurchase intent gains ranged from + 15% to +18%, aligning closely with post-campaign loyalty scores and further validating experiential marketing's ability to convert emotional impressions into purchase behavior. The upward movement in attention duration across all campaigns confirms the persuasive role of immersive interaction in deepening consumer focus and cognitive retention. These results substantiate the hypothesis that experiential strategies drive both attitudinal and behavioral.

5   Discussion

The findings of this mixed-methods study provide compelling empirical support for the strategic role of experiential marketing in building Consumer-Based Brand Equity, particularly in the context of a technologically advanced and consumer-fatigued market. The quantitative analysis, supported by a robust SEM model, confirms that EM is a potent force, positively influencing all four core dimensions of CBBE: awareness, quality, emotional connection, and loyalty. [16] However, the most significant theoretical contribution of this research lies in the clear demonstration of the mediating role of emotional engagement. [19] This finding suggests that the true 'soft power' of EM is not simply its ability to inform or entertain, but its capacity to bypass cognitive barriers and establish a profound affective link. Brands that successfully orchestrate multi-sensory and personalized experiences are, in effect, investing directly in the consumer's emotional ledger, which pays dividends in long-term loyalty.

The comparatively smaller effect of brand awareness and perceived quality on loyalty, relative to emotional engagement, is a crucial insight for contemporary marketers. In a market where perceived quality is often a baseline expectation, the ability to create an emotional differentiator becomes the key to achieving a sustainable competitive advantage.[20] This aligns with the contemporary shift from a purely transactional marketing paradigm to one focused on relationship building and shared value creation. The exploratory case studies further enrich this discussion by highlighting the catalytic role of immersive technologies. The observed increases in interaction time and positive sentiment within AR-integrated campaigns demonstrate that these tools are not mere novelties but are essential enablers of deeper immersion. By offering transportive and highly personalized experiences, technologies like AR and VR allow brands to deliver on the emotional promise of EM, transforming passive consumption into active, memorable participation. This suggests that future EM strategies must be technologically adaptive, viewing immersive tech as a necessary component for amplifying emotional resonance and maximizing brand equity returns.

Furthermore, the reframing of the study's methodology as an exploratory investigation with quantitative support addresses the limitations of sample size and generalizability. While the findings from the five case studies cannot be universally generalized across all industries, they provide valuable, in-depth insights into the mechanisms of successful EM implementation. The consistency between the statistical model ($N = 400$) and the qualitative narratives strengthens the internal validity of the core hypothesis—that emotional engagement is the primary driver of loyalty in an experiential context. This research therefore contributes a more nuanced, evidence-based framework for practitioners, emphasizing that the strategic deployment of EM should be prioritized not for mass reach, but for the depth of the emotional connection it can generate.

6 Conclusion

This study successfully investigated the strategic role of experiential marketing (EM) in building Consumer-Based Brand Equity (CBBE), confirming its positive influence across brand awareness, perceived quality, emotional connection, and loyalty. The central finding is the dominant mediating role of emotional engagement, which acts as the most significant pathway through which EM translates into long-term brand loyalty. This underscores the necessity for brands to shift focus from mere informational delivery to the deliberate design of emotionally resonant, multi-sensory experiences. The research also highlighted the power of immersive technologies, such as Augmented Reality, as critical tools for amplifying consumer attention and fostering deeper emotional bonds, thereby providing a contemporary roadmap for maximizing the effectiveness of EM campaigns.

Despite these contributions, the study is subject to several limitations. First, while the quantitative sample (N = 400) provided a basis for SEM analysis, the use of only five exploratory case studies across diverse sectors limits the generalizability of the technological insights. The cross-sectional nature of the survey also prevents the establishment of definitive causality over time. Future research should address these limitations by conducting longitudinal studies to track the evolution of brand equity following EM interventions. Furthermore, investigations into cross cultural and demographic moderators are warranted, as consumer engagement with experiential marketing is likely to vary significantly across different regions and age groups. Finally, future work could explore the comparative effectiveness of different immersive technologies (e.g., AR vs. VR vs. Mixed Reality) on specific CBBE dimensions, providing a more granular understanding of optimal technological investment.

In summary, the findings advocate for a paradigm shift in marketing strategy where EM is viewed not as a supplementary tactic but as a foundational soft power instrument. By prioritizing emotional relevance and embracing technological innovation, brands can cultivate enduring relationships that transcend transactional value, securing a sustainable position in the competitive global consumer culture.ple Heading (Third Level). Only two levels of headings should be numbered. Lower level headings remain unnumbered; they are formatted as run-in headings.

References

1. Cambra-Fierro, J.J., et al.: Customer-based brand equity and customer engagement in experiential services: insights from an emerging economy. Serv. Bus. **15**(3), 467–491 (2021)
2. Nadeem, W., et al.: How do experiences enhance brand relationship performance and value co-creation in social commerce? The role of consumer engagement and self brand-connection. Technol. Forecast. Soc. Chang. **171**, 120952 (2021)
3. Urdea, A.-M., Constantin, Purcaru, I.-M.: Implementing experiential marketing in the digital age for a more sustainable customer relationship. Sustainability **13** (2021). https://doi.org/10.3390/su13041865
4. Sanjaya, I.G.N., et al.: The effect of experiential marketing on consumer satisfaction and behavioral intentions. Int. J. Bus. Econ. Manage. **5**(4), 388–392 (2022)
5. Aminudin, A., Ina, S., Netty, L.: Experiential marketing and E-wom create brand loyalty through brand trust. Jurnal Multidisiplin Madani **3**(7), 1506–1513 (2023)

6. Sehani, W., Hettiarachchy, B.: Impact of experiential marketing on customer loyalty: the mediating role of customer satisfaction in the modern trade supermarkets in the Western Province, Sri Lanka. In: Proceedings of International Conference on Business Management, p. 18 (2022)

7. Nogueira, S., Durão, M., Pacheco, L., Ramazanova, M., Carvalho, J. , Experiential marketing and purchase intention of ecotourism experiences - Z-Generation case. Int. Conf. Tourism Res. 7(1) (2024)

8. Prasad, K., et al.: A conceptual model for building the relationship between augmented reality, experiential marketing & brand equity. Int. J. Prof. Bus. Rev. 7(6), e01030 (2022)

9. Chen, A.H., Wu, R.Y.: Mediating effect of brand image and satisfaction on loyalty through experiential marketing: a case study of a sugar heritage destination. Sustainability 14 (2022). https://doi.org/10.3390/su14127122

10. Kumar, R., Madhumitha, R., Suthar, M., Mushraff, Y.: The effectiveness of virtual reality and augmented reality in digital marketing campaigns. EPRA Int. J. Econ. Bus. Manage. Stud. 11(5) (2024)

11. Zhu, Y., et al.: A guide to graphic design for functional versus experiential ads. J. Advert. Res. 63(1), 81–104 (2023)

12. Jashwant, S.: Experiential marketing: challenges and opportunities. Int. J. Res. Arts Humanit. 4(4), 165–168 (2024)

13. Xu, W., Jung, H., Han, J.: The influences of experiential marketing factors on brand trust, brand attachment, and behavioral intention: focused on integrated resort tourists. Sustainability 14 (2022). https://doi.org/10.3390/su142013000

14. Hana, H.D.M., Firdan Gusmara, K.: The role of experiential and viral marketing strategies in increasing brand equity in consumer purchase decisions for tasikmalaya city online food delivery services. Jurnal Manajemen Bisnis 11(2), 1709–1722 (2024)

15. Destiana, F.: Pengaruh Social Media Marketing terhadap Ekuitas Merek dengan Online Experiential sebagai Variabel Mediasi (Studi pada Konsumen Toko Busana Cordy Bandar Lampung). Target: Jurnal Manajemen dan Bisnis 4(2) (2022)

16. Satar, M.S., et al.: Eliciting consumer-engagement and experience to foster consumer-based-brand-equity: moderation of perceived-health-beliefs. Serv. Ind. J. 45(2), 277–302 (2025)

17. Herrmann, J.-L., Ford, J.B.: Why the experiential view is vital to marketing communications research now. J. Advertising Res. 63(2), 109–122 (2023)

18. Cho, Y.-N., et al.: Partitioning online experiential consumption increases subjective well-being during the times of uncertainty. Int. J. Advert. 42(5), 945–968 (2023)

19. Chen, H., Wang, Y., Li, N.: Research on the relationship of consumption emotion, experiential marketing, and revisit intention in cultural tourism cities: a case study. Front. Psychol. 13 (2022)

20. Liu, X.: Brand knowledge and organizational loyalty as antecedents of employee-based brand equity: mediating role of organizational culture. Front. Psychol. 13 (2022)

Computational Analysis of Institutional Dynamics and Governance Performance: Cross-National Study of Corporate Governance Systems in Emerging Economies

Adnan Khaleel Kadhim[1] ⓘ, Saad Abdulhameed Shalev[2] ⓘ,
Shaker Jameel Sajat Awadh[3] ⓘ, Wafaa Adnan Sajid[4(✉)] ⓘ, Jamal Alsaidi[5] ⓘ,
and Alla Kalyniuk[6] ⓘ

[1] Al-Turath University, Baghdad 10013, Iraq
[2] Al-Mansour University College, Baghdad 10067, Iraq
[3] Al-Mamoon University College, Baghdad 10012, Iraq
[4] Al-Rafidain University College, Baghdad 10064, Iraq
`wafa@ruc.edu.iq`
[5] Madenat Alelem University College, Baghdad 10006, Iraq
[6] Kyiv National University of Construction and Architecture, Kyiv 03037, Ukraine

Abstract. Corporate governance is central to transparency, accountability, and firm performance, particularly in emerging countries where regulatory institutions and market frameworks are still evolving. This study compares the governance systems of five emerging economies, Brazil, India, South Africa, Turkey, and Indonesia, by examining governance effectiveness across board independence, shareholder participation, and transparency conformity. Using a fixed pool of 500 listed companies from 2020 to 2024, computational and statistical techniques were applied to construct a Governance Effectiveness Index (GEI), that quantifies institutional quality and its relationship with financial performance, measured by return on equity (ROE). The results show that India and Brazil achieve consistently higher GEI scores due to balanced governance across all indicators, while Turkey and Indonesia exhibit fragmented compliance and monitoring systems. A statistically significant positive correlation was identified between governance effectiveness and ROE, emphasizing the financial relevance of robust governance frameworks. The study concludes that adherence to international regulatory standards alone is insufficient; uniform monitoring mechanisms and data-driven governance reforms are essential to institutional resilience and investor confidence. The findings offer pragmatic insights for policymakers and corporate leaders seeking to strengthen governance in transitional economies.

Keywords: Corporate Governance · Emerging Economies · Governance Effectiveness Index · Institutional Reform · Financial Performance · Transparency

Z. Molamohamadi et al. (Eds.): ODSIE 2025, CCIS 2855, pp. 475–489, 2026.
https://doi.org/10.1007/978-3-032-17023-1_28

1 Introduction

Corporate governance has become a cornerstone of economic stability and sustainable business development, particularly in emerging economies characterized by institutional transitions and market volatility. Unlike mature economies with established regulatory frameworks, emerging markets face political uncertainties, weak legal infrastructures, and inconsistent adherence to governance standards. These conditions highlight the importance of developing governance mechanisms that ensure transparency, accountability, and investor confidence [1]. Corporate governance refers to the processes, rules, and practices through which companies are directed and controlled. Its core components—board independence, shareholder rights, financial transparency, and accountability—seek to balance ethical management with efficiency. However, interpretations of good governance vary widely across regions. Developing countries, in particular, must reconcile global governance expectations with domestic economic realities and cultural norms [2].

The globalization of financial markets has intensified the demand for accountability and transparency in developing countries. Investors increasingly expect compliance with international standards, yet institutional weaknesses, corruption, and limited shareholder engagement often constrain governance effectiveness. In this context, data-driven evaluations can identify strengths and weaknesses in governance systems, offering valuable insights for policymakers and international stakeholders, who desire fostering the sustainable development [3] (Fig. 1).

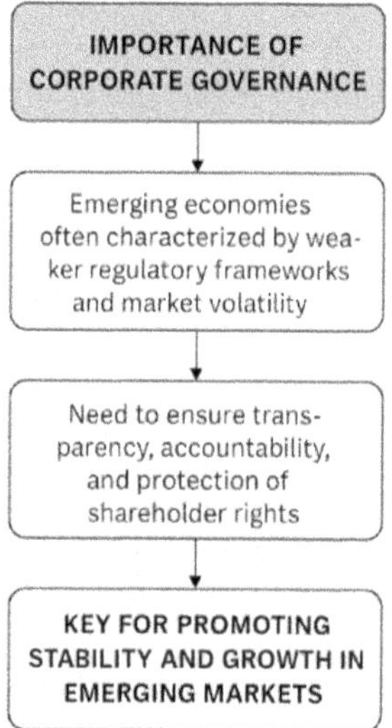

Fig. 1. The Strategic importance of corporate governance in strengthening institutional stability across emerging economies.

Institutional and cultural factors also shape governance performance. Strong regulatory systems promote shareholder protection and effective oversight, whereas weak institutional environments often suffer from insider trading, information asymmetry, and ineffective boards. Governance practices must also reflect local cultural and social contexts to achieve legitimacy and compliance [4]. Institutional and cultural factors also shape governance performance. Strong regulatory systems promote shareholder protection and effective oversight, whereas weak institutional environments often suffer from insider trading, information asymmetry, and ineffective boards. Governance practices must also reflect local cultural and social contexts to achieve legitimacy and compliance [5].

Strong corporate governance is not only vital for firms but also for broader economic resilience. Effective governance promotes foreign investment, efficient capital allocation, and stability, while weak governance erodes investor trust and increases systemic risk. A better understanding of governance in emerging markets is therefore essential for designing institutional reforms and achieving sustainable economic growth [6]. The impact of corporate governance practices, procedures and guidelines on the firms' performance between companies listed in India and in the Gulf countries was determined in previous research and the findings proved that Indian firms perform significantly better than those in the Gulf countries. This result is attributed to better corporate governance practices in India as compared to the Gulf countries [7].

This study examines corporate governance systems in five emerging economies—Brazil, India, South Africa, Turkey, and Indonesia—through data-driven indicators such as board independence, shareholder participation, and transparency conformity. Using computational techniques, including statistical modeling and the Governance Effectiveness Index (GEI), it evaluates governance effectiveness and its relationship with firm performance, measured by Return on Equity (ROE).

Emerging economies continue to face fragmented enforcement, inconsistent compliance, and limited integration of quantitative assessment tools. Addressing these gaps, the study develops a computational framework based on the GEI to enable cross-country comparison and quantitative evaluation of governance outcomes. The aim is to generate actionable insights for policymakers, regulators, and business leaders, identifying how governance quality influences financial performance and institutional strength. By applying CCIS-based computational analysis, the research contributes to understanding how data-informed governance reforms can enhance transparency, accountability, and economic resilience in emerging markets.

2 Literature Review

Corporate governance has emerged as a foundation for sustainable business transactions and economic prosperity, particularly in economies undergoing rapid industrialization and institutional change. In emerging economies, corporate governance arrangements vary considerably due to economic, cultural, and institutional conditions. While transparency, accountability, and ethical behavior are global priorities, the adaptation of corporate governance frameworks differs significantly across regions, influenced by various economic and institutional factors [8].

The composition of boards is a commonly studied aspect in corporate governance. Corporate governance is often associated with independent directors, a broad base of skills, and scrupulous responsibility. Independent and diverse boards can be more challenging to establish in emerging economies, as they conflict with established cultural norms, lack regulatory enforcement, and not have a readily available pool of trained professionals. These constraints diminish the board's ability to effectively monitor management and hold management accountable [9].

Shareholder rights and involvement are also important factors in determining governance standards. Developed countries generally afford minority shareholders strong legal protections; however, in emerging markets, power is often tilted toward majority owners. Poor enforcement adds to these imbalances, creating problems with insider trading and biased transactions. Enhancing shareholders' rights and their active participation in governance decisions is crucial to improving the quality of corporate governance [10].

Transparency and financial transparency are widely accepted as key ingredients in corporate governance. Transparent reporting enables intelligent decisions and management accountability. In developing economies, however, disclosure quality is inconsistent and incomplete, often due to weak regulation and adherence to international accounting standards. Achieving deeper transparency requires a mix of regulatory changes, institutional strengthening, and adoption of global best practices [11].

Regulatory context is a major driver of governance outcomes. Good governance and anticorruption measures are essential. Emerging countries with strong legal and law enforcement infrastructures tend to exhibit better governance behaviors. However, many such economies suffer from weak institutions, corruption, political intervention, and resource scarcity that hinder effective regulation. The adoption and implementation of context-specific governance codes are essential to meeting these challenges [12].

Corporate governance in developing nations is also embedded in cultural and social practices. Governance systems often need to be adjusted to local customs and demands without losing essential values of accountability and transparency. Achieving harmony between local dynamics and global governance norms is challenging but necessary task [13].

Addressing these limitations, recent research in 2025 has provided further evidence linking governance mechanisms with corporate performance and sustainability outcomes. Findings indicate that board independence, gender diversity, and audit committee autonomy significantly enhance audit quality in emerging economies, underscoring the importance of structural governance attributes in strengthening accountability and oversight [14]. Other studies highlight that transparent governance practices are closely associated with improved environmental, social, and governance (ESG) outcomes, reflecting a growing integration of sustainability principles within corporate governance frameworks [15]. Furthermore, empirical results confirm that robust country-level governance frameworks positively mediate the relationship between sustainable development and firm value in emerging markets, emphasizing the central role of institutional quality in promoting effective governance and long-term performance [16].

Despite these contributions, notable gaps remain in the existing literature. Most prior studies are country-specific and rely on qualitative or fragmented indicators that do not capture governance effectiveness comprehensively across emerging economies. There is also limited use of quantitative frameworks that integrate institutional and financial dimensions into a unified measure. To address these shortcomings, the present study develops a GEI that links governance attributes with financial performance across selected emerging markets. This approach provides empirical insights into how institutional quality and governance structures influence corporate performance and economic resilience.

3   Methodology

This study employs a comparative, panel-based, multivariate research design to examine corporate governance practices across five emerging economies, Brazil, India, South Africa, Turkey, and Indonesia, from 2020 to 2024. The approach is hybrid, integrating cross-sectional governance indicators with longitudinal performance data. Building upon prior theoretical and empirical foundations [1, 2, 6, 7], the framework captures variations in governance effectiveness both within and across institutional contexts.

3.1   Research Design Framework

The analysis operationalizes three primary governance indicators, Board Independence (BI), Shareholder Participation (SP) and Transparency Compliance (TC) into operation, as latent constructs representing structural, procedural, and normative aspects of governance quality [3, 13, 17]. Figure 2 represents the methodological framework for assessing governance effectiveness and financial performance in emerging economies.

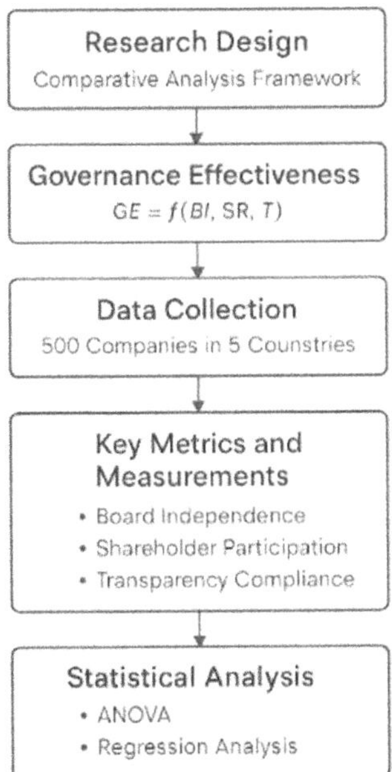

Fig. 2. Integrated methodological framework for evaluating governance effectiveness and financial performance in emerging economies.

A composite Governance Effectiveness Index (GEI) is derived through a weighted aggregation model:

$$GEI_{i,t} = \omega_1 \cdot BI_{i,t} + \omega_2 \cdot SP_{i,t} + \omega_3 \cdot TC_{i,t} \tag{1}$$

where $GEI_{i,t}$ is the governance effectiveness index for country i at time t, $BI_{i,t}$ is the average board independence (% of independent directors), $SP_{i,t}$ is the shareholder participation rate (% voting turnout), $TC_{i,t}$ is the transparency compliance rate (% adherence to International Financial Reporting Standards (IFRS), Global Reporting Initiative (GRI)). The weights $\omega_1, \omega_2, \omega_3$ were initially set to $\frac{1}{3}$, but later tested for sensitivity, and $t \in \{2020,2021,2022,2023,2024\}$.

This index allows multidimensional comparison of governance quality across countries and years, consistent with established approaches used in emerging economy research [5, 8, 18].

3.2 Sampling and Data Acquisition

The dataset includes 500 firms, 100 each from Brazil, India, South Africa, Turkey, and Indonesia. Firms were selected using three stratified criteria: (1) highest market capitalization per sector, (2) availability of annual governance data, and (3) representativeness across key industries. Sectors covered include finance, energy, telecommunications, manufacturing, construction, retail, and extractives.

Primary data sources include:

- Bloomberg ESG Governance Index (Brazil),
- World Bank Corporate Governance Reports (India),
- Company Filings and King IV Report Adoption Ratings (South Africa),
- MSCI Governance Index and Borsa Istanbul filings (Turkey),
- IDX Industry Reports and OJK disclosures (Indonesia).

All data were validated through cross-source triangulation, benchmarking, and expert regulatory review [4, 9, 12] (Table 1).

Table 1. Governance data acquisition overview.

Country	Sample Size (N)	Primary Source	Sectoral Coverage	Validation Method
Brazil	100	Bloomberg Governance Index	Finance, Energy, Retail	Cross-Validation
India	100	World Bank CG Reports	IT, Pharma, Manufacturing	Multisource Benchmarking
South Africa	100	Corporate Filings + King IV Index	Mining, Services, Agriculture	King IV Compliance Audit

(continued)

Table 1. (*continued*)

Country	Sample Size (N)	Primary Source	Sectoral Coverage	Validation Method
Turkey	100	MSCI Index + Istanbul Borsa Filings	Tourism, Electronics, Textiles	Database Cross-Matching
Indonesia	100	OJK & IDX Reports	Oil & Gas, Telecom, Construction	Regulatory Triangulation

3.3 Metric Construction and Normalization

All governance variables were normalized using Z-score standardization to ensure comparability across countries and firms:

$$Z_{i,t} = \frac{X_{i,t} - \mu_X}{\sigma_X} \tag{2}$$

where $X_{i,t}$ is the raw score for firm i at time t, μ_X is the mean, and σ_X is the standard deviation of X across all observations. This normalization mitigates scale effects and heteroscedasticity [11, 19].

The Shareholder Participation Index (SPI) was computed as:

$$SPI_{i,t} = \left(\frac{Votes_{cast,i,t}}{Eligible_{votes,i,t}} \right) \cdot 100 \tag{3}$$

For transparency compliance, firms were assessed on adherence to IFRS, GRI, and Audit committee independence (ACI).

The Transparency Score (TS) was calculated using:

$$TS_{i,t} = \frac{IFRS_{compliance,i,t} + GRI_{compliance,i,t} + ACI_{i,t}}{3} \tag{4}$$

where $ACI_{i,t} = 1$ for independent audit committees, otherwise 0.

3.4 Econometric Modeling

The relationship between governance metrics and financial performance is examined using panel data regression with fixed effects:

$$ROE_{i,t} = \alpha_i + \beta_1 BI_{i,t} + \beta_2 SP_{i,t} + \beta_3 TC_{i,t} + \gamma Z_{i,t} + \epsilon_{i,t} \tag{5}$$

where $ROE_{i,t}$ is the return on Equity for firm i at time t, α_i is the unobserved time-invariant firm-level effects, $\beta_1, \beta_2, \beta_3$ are the coefficients of governance variables, $Z_{i,t}$ includes control variables (firm size, leverage, industry), and $\epsilon_{i,t}$ is the error term. Instrumental variable (IV) regression addressed potential endogeneity:

$$ROE_{i,t} = \delta_0 + \delta_1 \widehat{GEI}_{i,t} + \delta_2 Z_{i,t} + u_{i,t} \tag{6}$$

where $\widehat{GEI}_{i,t}$ represents instrumented GEI values based on staggered regulatory reforms [10, 17, 19].

3.5 Statistical Procedures and Validation

- Hausman tests confirmed fixed vs random effects selection.
- Variance Inflation Factor (VIF) was used to test multicollinearity (threshold VIF < 5).
- Breusch–Pagan and White's tests addressed heteroskedasticity assumptions.
- Robust standard errors were clustered at the firm level.

   Validation metrics include:

- Adjusted R2R^2R2 for model fit,
- Akaike Information Criterion (AIC) and Bayesian Information Criterion (BIC) for model selection,
- K-fold cross-validation (K = 10) for robustness.

   The methodological structure provides a robust, data-driven framework for assessing governance effectiveness across emerging economies, combining complex quantitative modeling with regulatory-contextual interpretation [1, 7, 20, 21].

4 Results

4.1 Board Independence and Internal Variability Across Countries

Board independence is a key dimension of corporate oversight, protecting shareholders from managerial overreach and supporting long-term performance. In emerging markets, disparities in legal enforcement, institutional maturity, and cultural norms contribute to heterogeneous board structures. The data presented here explores the average percentage of independent directors alongside key dispersion metrics—maximum, minimum, standard deviation, and coefficient of variation—to evaluate consistency and structural strength across national contexts.

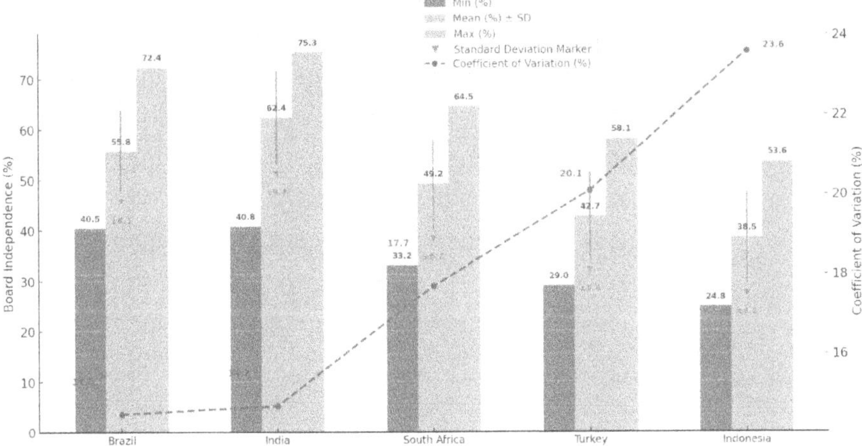

Fig. 3. Board independence variability in emerging economies.

Figure 3 presents the average percentage of independent directors along with dispersion metrics—maximum, minimum, standard deviation, and coefficient of variation—to assess consistency and structural strength across countries. India exhibits the highest average board independence (62.4%), followed by Brazil (55.8%), while Indonesia ranks lowest at 38.5% with the greatest variability, reflecting inconsistent adoption across firms. Turkey and South Africa show moderate variation (coefficient of variation above 17%), indicating heterogeneous implementation of governance systems. Brazil's smaller variation suggests relatively uniform board autonomy, whereas India demonstrates a combination of high mean independence and moderate variation, showing progressive reforms alongside adoption disparities. Indonesia and Turkey appear less standardized, highlighting the influence of institutional gaps and uneven regulatory enforcement.

4.2   Shareholder Participation and Democratic Engagement Efficiency

Shareholder participation reflects the extent to which owners engage in corporate decision-making and serves as an indicator of governance inclusiveness. Metrics include average voting percentage, number of entitled shareholders, votes cast, and participation consistency.

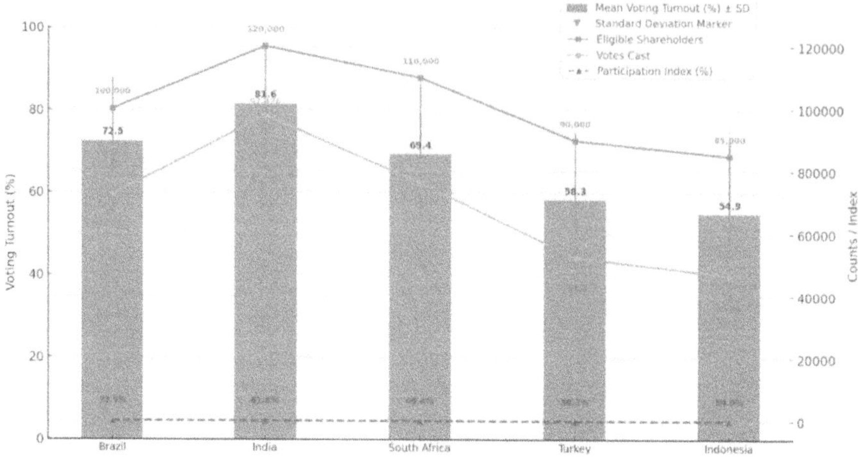

Fig. 4. Shareholder participation metrics in emerging economies.

Figure 4 shows that leads with an average voting turnout of 81.6% and the largest number of eligible participants. Brazil and South Africa perform moderately, though Brazil shows some internal inconsistencies. Indonesia records the lowest return and the largest standard deviation, indicating non-uniform investor engagement. Turkey, despite a substantial shareholder base, shows low participation, suggesting either procedural inefficiency or limited shareholder trust. High participation in India indicates strong investor activism and effective shareholder protection mechanisms. These findings underline the association between legal infrastructure, investor education, and actual participation outcomes.

4.3 Transparency Compliance with International and Institutional Norms

Transparency is essential for market integrity and stakeholder confidence. The transparency compliance metric incorporates IFRS financial reporting, GRI reporting standards, and audit committee independence. Figure 5 illustrates the extent to which companies adopt international transparency standards.

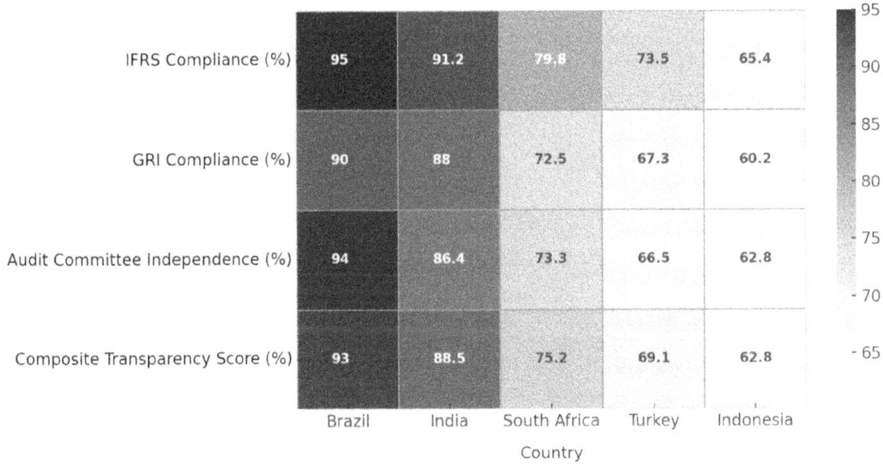

Fig. 5. Transparency compliance breakdown by country.

Brazil leads with a transparency score of 93.0%, reflecting near-universal compliance. India performs strongly, with slightly lower audit committee independence scores. South Africa is mid-range, with moderate compliance and weaker adherence to GRI standards. Turkey and Indonesia lag, particularly in GRI alignment and audit independence, undermining reporting credibility. Indonesia's low score of 62.8% highlights the need for structural reforms. High-performing countries demonstrate integrated financial and sustainability disclosures, promoting capital market integration and risk transparency.

4.4 Comparative Governance Effectiveness Through Composite Indexing

Governance effectiveness index combines board independence, shareholder participation, and transparency compliance into a single normalized measure, capturing the overall implementation of governance structures.

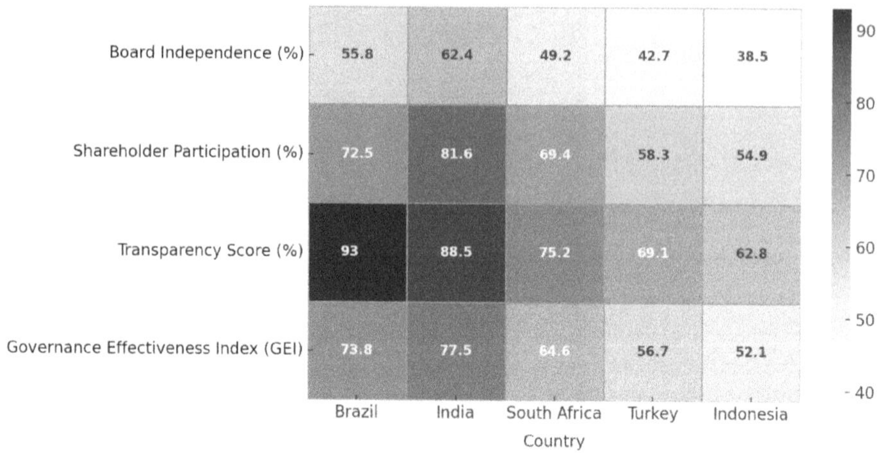

Fig. 6. Governance Effectiveness Index (GEI) Across Emerging Economies

Figure 6 shows India at the top with a GEI of 77.5%, followed by Brazil at 73.8%. South Africa scores 64.6% with lower board independence reducing its overall index. Turkey and Indonesia consistently underperform across all components. The heat map illustrates that effective governance emerges from the synergy of transparency, board independence, and participatory mechanisms. Brazil's performance is driven by high transparency, while India shows balanced strength across all dimensions. South Africa demonstrate strong stakeholder engagement, but limited board independence reduces its overall score. Turkey and Indonesia require comprehensive reforms across structural, procedural, and cultural dimensions to improve governance effectiveness.

4.5   Correlation Between Governance Metrics and Return on Equity (ROE)

The financial impact of governance quality was assessed via Pearson correlation between governance indicators and ROE (Table 2).

Table 2. Correlation between governance metrics and ROE.

Governance Metric	Pearson Correlation (r)	p-value	Interpretation
Board Independence	0.78	< 0.001	Strong positive relationship with ROE
Shareholder Participation	0.69	0.005	Moderate positive relationship with ROE
Transparency Compliance	0.84	< 0.001	Very strong positive relationship with ROE

transparency compliance shows the strongest correlation with ROE (0.84), indicating that firms with higher transparency achieve better financial performance. Board independence also correlates strongly with profitability ($r = 0.78$), highlighting the value of independent strategic oversight. Shareholder participation has a moderate positive effect, suggesting a more enduring or indirect impact on performance. These results validate the GEI as an effective tool for assessing governance quality and reinforce the conclusion that robust governance is a significant determinant of financial sustainability in emerging markets.

5 Discussion

This study provides a comparative view of corporate governance in Brazil, India, South Africa, Turkey, and Indonesia, focusing on board independence, shareholder participation, transparency, and overall governance effectiveness. The results reveal substantial variation across countries and highlight specific strengths and weaknesses.

India leads on all indicators, demonstrating a robust governance system with strong transparency, active shareholder engagement, and high board independence. Brazil excels in transparency but has moderate board independence and shareholder participation, suggesting potential areas for improvement. South Africa shows strong shareholder engagement and transparency, but lower board independence limits overall governance effectiveness.

Turkey faces challenges across all governance metrics, with low shareholder participation and transparency compliance, highlighting the need for stronger institutional structures. Indonesia ranks lowest, with weak boards, low engagement, and limited transparency, reflecting systemic governance deficiencies that may constrain sustainable growth and foreign investment.

Overall, the study confirms the variability of governance quality in emerging markets and the importance of institutional, cultural, and economic contexts. Countries with stronger enforcement and adherence to governance standards perform better, while weaker institutional environments face persistent challenges. These findings identify actionable areas for reform, emphasizing the need to strengthen board independence, shareholder participation, and transparency to support corporate performance and sustainable economic growth.

6 Conclusion

This study provides a cross-country comparison of corporate governance in Brazil, India, South Africa, Turkey, and Indonesia, focusing on board independence, shareholder participation, and transparency. The results reveal significant variation in governance effectiveness, reflecting differences in institutional capacity, regulatory enforcement, and cultural contexts. India and Brazil stand out for their strong adoption of global governance standards, while Indonesia and Turkey face persistent challenges due to fragmented institutions and uneven enforcement. South Africa presents a mixed picture, with progress in shareholder engagement but structural limitations in board composition. Governance quality, measured through a composite index, is positively associated with corporate accountability and financial performance, highlighting the multi-dimensional nature of effective governance.

The findings reinforce the importance of viewing governance as an integrated system, where structural, participatory, and transparency mechanisms jointly contribute to firm performance and investor confidence. For emerging economies, governance reforms should be locally responsive, prioritizing enforceable policies, audit quality, and stakeholder engagement. Higher-performing countries should focus on consolidating best practices and embedding inclusive governance into corporate culture.

Future studies could explore the long-term effects of governance reforms through longitudinal analysis and assess the role of ESG factors in shaping sustainable corporate behavior. Comparative studies with frontier or smaller emerging economies may provide further insight into the scalability and adaptability of governance innovations.

References

1. Gull, A.A., Abid, A., Hussainey, K., Ahsan, T., Haque, A.: Corporate governance reforms and risk disclosure quality: evidence from an emerging economy. Journal of Accounting in Emerging Economies **13**(2), 331–354 (2023)
2. Wen, K., et al.: The impact of corporate governance and international orientation on firm performance in SMEs: evidence from a developing country. Sustainability **15**, 5576 (2023). https://doi.org/10.3390/su15065576
3. Liu, Y.: Barriers of good corporate governance practices: evidence from emerging economy. International Journal of Business and Management (IJBM) (2022)
4. Dau, N., Dinh, S.: Firm characteristics and corporate governance of state-owned enterprises: an analysis of an emerging market. International Journal of Applied Economics, Finance and Accounting **20**(1) (2024)
5. Tariq, Y.B., Ejaz, A., Bashir, M.F.: Convergence and compliance of corporate governance codes: a study of 11 Asian emerging economies. Corporate Governance: The International Journal of Business in Society **22**(6), 1293–1307 (2022)
6. Bagram, M., Bangash, A., Kiran, Z.: Aspects of corporate governance in developing countries. International Review of Management and Business Research **10**(1), 1–8 (2021)
7. Al-ahdal, W.M., Almaqtari, F.A., Tabash, M.I., Hashed, A.A., Yahya, A.T.: Corporate governance practices and firm performance in emerging markets: empirical insights from India and Gulf countries. Vision **27**(4), 526–537 (2021)
8. Cumming, D.J., Verdoliva, V., Zhan, F.: New and future research in corporate finance and governance in China and emerging markets. Emerging Markets Review (2021)
9. Anwar, Y., Mulyadi, M.: Corporate control in emerging markets: the non-linear dynamics of foreign board involvement. Corp. Ownersh. Control. **21**(2), 45–51 (2024)
10. Ferran, E.: Shareholder engagement and custody chains. European Business Organization Law Review **23**(3), 507–539 (2022)
11. Boateng, R.N., Tawiah, V., Tackie, G.: Corporate governance and voluntary disclosures in annual reports: a post-international financial reporting Standard adoption evidence from an emerging capital market. Int. J. Account. Inf. Manag. **30**(2), 252–276 (2022)
12. Abaidoo, R., Agyapong, E.K.: Governance, regulatory quality and financial institutions: emerging economies perspective. Journal of Economic and Administrative Sciences, ahead-of-print (2023)
13. Yilmaz, M.K., Hacioglu, U., Tatoglu, E., Aksoy, M., Duran, S.: Measuring the impact of board gender and cultural diversity on corporate governance and social performance: evidence from emerging markets. Economic Research-Ekonomska Istraživanja **36**(2), 2106503 (2023)

14. Tiwari, R.K., Maji, S.G.: Impacts of corporate governance attributes on audit quality in emerging economies: the case of India. South African Journal of Accounting Research **39**(1), 50–72 (2025). https://econpapers.repec.org/RePEc:taf:rsarxx:v:39:y:2025:i:1:p:50-72

15. Bhateja, F.: Corporate governance and sustainability in emerging economies: a systematic review. In: Proceedings of the International Conference on Corporate Governance and Sustainability in Emerging Economies, pp. 1–12. Atlantis Press, Paris (2025). https://www.atlantis-press.com/proceedings/see-ibeme-24/126010627

16. Ahmed, Z., Bashir, Y., Ahmed, B., Rocha, A., Tan-Hwang Yau, J.: Role of country governance between sustainable development and firm value in emerging markets. Discover Sustainability **6**, 261 (2025). https://doi.org/10.1007/s43621-025-01097-w

17. Bai, K., Ullah, F., Arif, M., Erfanian, S., Urooge, S.: Stakeholder-centered corporate governance and corporate sustainable development: evidence from CSR practices in the top companies by market capitalization at Shanghai Stock Exchange of China. Sustainability **15**, 42990 (2023). https://doi.org/10.3390/su15042990

18. Spadafora, E., et al.: Board independence and firm internationalization: a meta-analysis. Multinational Business Review **30**(4), 499–525 (2022)

19. Mooneeapen, O., Abhayawansa, S., Mamode Khan, N.: The influence of the country governance environment on corporate environmental, social and governance (ESG) performance. Sustainability Accounting, Management and Policy Journal **13**(4), 953–985 (2022)

20. Kumar, S., Rastogi, P.: Corporate governance in India in 2024: current practices and future prospects. International Journal for Multidisciplinary Research **6**(3) (2024)

21. Kyriacou, A.P.: Economic inequality, culture, and governance quality. Journal of Economic Surveys **39**(1), 375–402 (2025)

Data-Driven Analysis of Emotional Intelligence and Managerial Effectiveness in Transitional Work Environments

Sarah Salah Hadi[1] , Suad Ahmed Ibrahim[2] , Raad Ta'ma Awad Bajjay[3] ,
Wafaa Adnan Sajid[4(✉)] , Jassim Mohamed Brieg[5] , and Alla Hotsalyuk[6]

[1] Al-Turath University, Baghdad 10013, Iraq
[2] Al-Mansour University College, Baghdad 10067, Iraq
[3] Al-Mamoon University College, Baghdad 10012, Iraq
[4] Al-Rafidain University College, Baghdad 10064, Iraq
wafa@ruc.edu.iq
[5] Madenat Alelem University College, Baghdad 10006, Iraq
[6] Kyiv National University of Construction and Architecture, 03037 Kyiv, Ukraine

Abstract. Emotional intelligence (EI) plays a key role in enhancing managerial effectiveness, particularly in dynamic work environments. This study adopts a quantitative, data-driven approach to examine five core EI dimensions—self-awareness, self-regulation, empathy, social skills, and motivation—and their associations with managerial performance. A cross-sectional survey of 250 managers across multiple industries was analyzed using weighted least squares regression and structural equation modeling (SEM). Positive associations between emotional intelligence and managerial effectiveness were observed in both descriptive and correlational analyses. The results reveal that empathy and motivation are the strongest predictors of managerial effectiveness, while interaction effects, particularly between empathy and social skills, highlight the synergistic impact of combined emotional competencies. The findings suggest that EI functions as an integrated competency rather than a collection of isolated traits, enabling managers to lead effectively and navigate complex work environments. Practical implications include integrating EI training into leadership development and performance management systems. This study provides a solid basis for further developing emotionally intelligent leadership and confirms the importance of emotional abilities in addressing contemporary managerial challenges.

Keywords: Intelligence · Managerial Effectiveness · Leadership · Empathy · Social Skills · Motivation

1 Introduction

In the period of growth of complexity of work in organizations the attributes and skills of effective managers are more and more subjected to scrutiny. Emotional intelligence (EI) as one of the important proprieties of a competent manager Among the various

Z. Molamohamadi et al. (Eds.): ODSIE 2025, CCIS 2855, pp. 490–506, 2026.
https://doi.org/10.1007/978-3-032-17023-1_29

necessary competencies in managerial work emotional intelligence has received considerable attention. It's the ability to identify, understand and manage our own emotions and successfully interact with others' emotional states—and goes far beyond common cognitive intellect and technical skills. It is based on a subtlety in human behavior that managers can use to create positive relationships, build consensus, and move a team toward fulfillment of common goals [1].

The term "emotional intelligence" came to prominence through the work of Daniel Goleman that emphasized the importance of emotional intelligence in business and leadership. Since that time, EI has been associated with a broad range of beneficial outcomes including, better job performance, increased job satisfaction, and lower turnover. EI can better help keep one calm through organizational change, tackle challenges head-on while still being optimistic and form solid teams. Managers with a higher EI are often more able to cope with organizational change, to confront challenges with an eye towards solutions, and to quickly form close-knit teams that are able to take on both internal and external challenges. These characteristics make EI a key competency for managerial success, in particular considering rapid technological advancements, globalization and new workforce expectations [2] (Fig. 1).

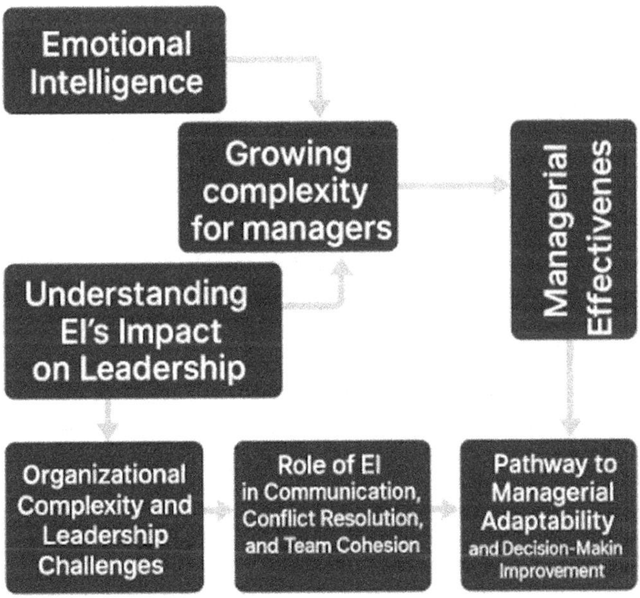

Fig. 1. Conceptual Flow of Emotional Intelligence and Its Role in Enhancing Managerial Effectiveness.

Good management practice and emotional intelligence While the theoretical basis of emotional intelligence has been well documented, the practical implications of it for management is less understood and is a fertile area of examination. There is plenty of evidence to suggest that great managers don't get by on technical skills or ability to make decisions. Rather, their power in terms of empathy, communication and trust often dictates their success. In this way, emotional intelligence offers a conceptual approach to understanding the social side of management, where the technical skills are complemented by a focus on the human side of management [3]. Several reasons account for the increasing significance of emotional intelligence in managerial performance. First, today's workplace is more collaborative than ever, and managers must help a broad array of individuals work together. Managers with emotional intelligence are able to manage staff relations, defuse conflicts, and foster an inclusive workplace where staff feel truly valued. Second, with the prevalence of telecommuting and digital communication came a whole new set of challenges for managers, including maintaining a cohesive team and building relationships in a virtual world. Managers with high EI are more able to cope with these challenges as they are able to show empathy and to be understanding even though they are not face to face [4].

Furthermore, the mental health of workers has emerged as a central concern for businesses working to uphold productivity and spirits. Managers have a strong influence on the emotional environment of their teams, those who are able to read the subtle signals of stress among team members and offer support and open dialogue are more likely to support well-being. This is particularly pertinent in the aftermath of global disasters, as the current COVID-19 pandemic brings about a greater understanding for the need for empathetic leadership in times of instability and transition [5].

While widely acknowledged, emotional intelligence is far from being uniformly addressed in management training. Conventional management education tends to emphasize technical capabilities and analytical skills at the expense of emotional and interpersonal competencies. This imbalance demonstrates the necessity of more comprehensive management training that incorporates emotional intelligence as a fundamental aspect of leadership development. It enables administrators to provide tools that help them to understand and manage emotions well, and this can lead to improve management performance and a better environment at work place [6].

This article aims at demonstrating the importance of emotional intelligence in managerial effectiveness and identifying the relevant influence on significant areas such as leadership, communication, conflict resolution and the team motivation. Through data- and insight-based analyses, it is hoped that the study will offer valuable insights into the impact of emotional intelligence on managerial success in a holistic manner. The results will be used to drive initiatives that integrate EI into the management, and therefore potentially improve the efficacy and empathy of leaders [7].

The theoretical basis of emotional intelligence will be scrutinized in the next section, together with the description of the methodology utilized, alongside the findings of the analysis. Through a synthesis of academic and applied perspectives, this article aims to highlight the evolution of management in the context of emotional intelligence. From this perspective, emotional intelligence is not simply an addition to established managerial skills but a primary quality that reframes what superior leadership looks like in the new organizational context. This study adopts a data-driven methodology using regression analysis and SEM to investigate both individual and interaction effects of EI components on managerial effectiveness.

1.1 The Aim of the Article

The primary aim of this article is paper is to examine the influence of emotional intelligence (EI) on managerial effectiveness at work. Given the facing of increasingly complex work environments, the study calls attention to the fact that emotional intelligence is a necessary competence for the managers to manage people dynamics in order to form a cohesive team and thus a commitment to organizational goals. In so doing, this paper serves to bridge the gap of general theories and application within key aspects of management which may account for the relationship between EI and key functions in management including leadership, communication, conflict, and motive of employees.

Specifically, the study seeks aims to identify the importance of EI in managerial performance, specifically high levels of EI resulting in good managerial performance including: self-awareness, empathy, emotional regulation and to investigate which components of EI are most significant for managers. It also seeks to reveal how these characteristics impact on decision making, support problem solving and help to generate positive workplace culture. Self-awareness refers to recognizing one's own emotions; self-regulation to managing emotional impulses; empathy to understanding others' feelings; social skills to managing relationships; and motivation to internal drive toward goals.

The article accomplishes this through a method based in data analysis-cum-insights, that is mixed methods. This two-sided view supports a strong examination of the effect of EI in managerial effectiveness context as that the tail is wagged across their organizational context and industries. Focusing on findings that can be acted upon, this study would inform the development of in-service training and professional development programs designed specifically to develop emotional intelligence of existing or future managers.

Ultimately, the aim of this article to further academic conversation around emotional intelligence and its contributions to managers' success and, simultaneously, to offer firms actionable knowledge that can be put into practice in addressing the development of emotionally intelligent leaders. In so doing, the research aims to address the wider strategic aim of enhancing organizational productivity, employee health and well-being, and leadership effectiveness in a changing workplace environment.

1.2 Problem Statement

Despite the general recognition of the importance of emotional intelligence (EI) in effective management, substantial gaps still exist in our understanding of how EI is manifest in organizational settings. Although the literature supports the idea that high emotional intelligence managers are able to lead, communicate and manage conflicts better, a large proportion of the literature is based on broad correlations rather than clear mechanisms. This lack of clarity presents difficulties for companies who wish to develop and deliver effective EI training programs for managers.

Moreover, the traditional management development programs, including those at business schools, tend to focus on technical and analytical capabilities and fail to support 'soft' competencies necessary to manage intricate workplace relations. With organizations getting more diverse and collaborative, the failure to fix this imbalance undermines attempts to build an inclusive and productive workplace. Managers are often asked to promote unity across varied teams, resolving interpersonal friction and improving engagement among team members, time and again, lacking the emotional intelligence that helps them shine in these interactions.

Another immediate challenge is the increasing presence in the workplace of telecommuting and digitization, which are placing new burdens on those in management as they try to instill team morale and connection in virtual forums. There is, therefore, an imperative to examine the role of emotional intelligence in how managers can adapt in increasingly challenging work contexts.

Furthermore, though EI is largely related to a persons' natural disposition, anecdotal evidence has suggested that it can be developed through focused training and developmental activities. However, such programs run the risk of either being too general or ineffective, given that what the most salient components of EI for managerial effectiveness are not clear.

Addressing these problems, the article aims to explore how the EI concept can improve the effectiveness of management – with specific applications to leadership, communication, and team management. By identifying actionable findings and offering evidence-based guidance, the research seeks to reduce the distance between theory and practice, so as to help organizations to effectively enhance managerial performance in a volatile and turbulent workplace context.

2 Literature Review

The concept of emotional intelligence (EI) has increasingly attracted much attention in academia and industry due to its enormous influence on leadership and management. EI is an umbrella term used to describe the capacity to recognize, perceive and control one's own and other's emotions, which includes several distinct elements such as self-awareness, self-regulation, empathy, social skills and motivation. These characteristics are also becoming more accepted as important for managerial success, especially in a context of inter-personal challenges and a multi-national or multi-cultural team [8].

Studies show the importance of emotional intelligence in promoting communication. Managers who have a good balance of EQ are able to pick up on non-verbal cues, alter their style of communication according to situation and keep the lines of communication open with even difficult conversations. This capacity to express themselves is of particular importance for conflict prevention and resolution, as acknowledging the emotional expressions of the counterparts may help to reach more fair and sustainable agreements [9].

A third highly influential area where emotional intelligence plays a significant role is that of leadership. Managers with higher EI are usually perceived as easily accessible, supportive, and visionary leaders. And that ability can lend itself to inspiring and motivating their teams, fostering a positive workplace culture, and demonstrating better resilience in the face of change within the wider organization. Furthermore, emotional intelligence is commonly associated with the transformational leadership style, in which leaders motivate innovation and commitment by building trust and empathy [10].

There is also team synergies when you have emotionally intelligent managers, and they do help build a 'we are in this together atmosphere' which leads to teamwork and productivity. They know how to manage people so that team members feel like valued and appreciated colleagues, which goes a long way towards overall morale and not turning over other employees. These results emphasize the relevance of EI in developing and maintaining high-performing teams [11].

In addition to its interpersonal benefits, emotional intelligence also aids in decision making. Managers with high EI not only stave off stress but remain clear-headed in the eye of the storm, so that they are able to make sound decisions. This is particularly critical in high-paced or high-stake contexts such as in the case of managerial settings, where emotional reactivity can hinder managerial effectiveness [12].

Despite its potential advantages, the implementation of EI in managerial activities can be unsystematic. Many organizations don't have systematic processes for building EI in leaders, and therefore miss the chance to improve their managers' effectiveness. Though EI is regarded by some as a natural trait, research has shown that it can be developed via specific training and experience. Consequently, insights into the differential contributions of individual aspects of EI to managerial effectiveness may be used to develop more effective means of intervention [13].

Emotional intelligence emerges as a complex structure underlying various management competencies. By connecting what we know to what we can do, organizations can use emotional intelligence to improve leadership, performance, and teamwork. While prior studies have explored the role of EI in leadership, few have investigated the interaction effects among EI dimensions using robust data-driven models. This study addresses that gap by applying SEM to explore how EI dimensions jointly predict managerial effectiveness.

3 Methodology

The cross-sectional saturated research is employed in the study with application of advanced statistical modelling to investigate the role of four components of emotional intelligence (EI) in the effectiveness of managerial work (ME). In order to guarantee methodological rigor, the model is based on five interrelated components: research framework, population and sampling, measuring instrument, statistical model, and validation procedures. The approach embodies an evidence-based approach in evaluating psychological constructs in organizational behavior [2, 3, 7] (Fig. 2).

3.1 Research Framework and Conceptual Modeling

To evaluate the influence of emotional intelligence dimensions on managerial effectiveness, the study models ME as a linear and nonlinear function of five core EI predictors: Self-Awareness (SA), Self-Regulation (SR), Empathy (E), Social Skills (SS), and Motivation (M). The proposed model incorporates both main effects and latent covariance structures.

The primary functional model is expressed as a generalized multivariate regression equation:

$$ME_i = \beta_0 + \sum_{j-1}^{5} \beta_j \bullet X_{ij} + \varepsilon_i \tag{1}$$

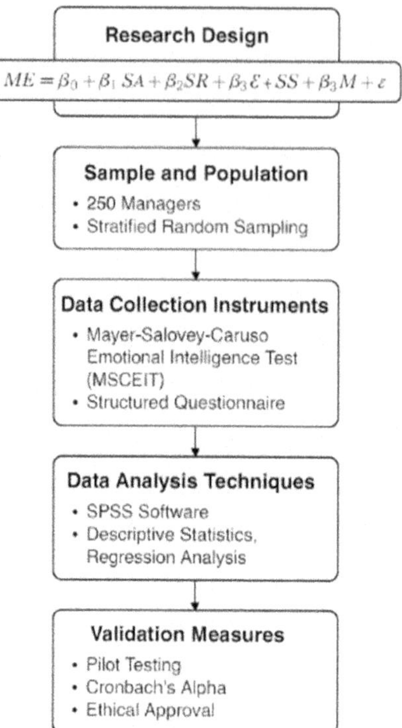

Fig. 2. Methodological Framework for Investigating Emotional Intelligence and Managerial Effectiveness.

where ME_i is the managerial effectiveness score for the ith manager, X_{ij} represents the jth emotional intelligence component score, β_j are the corresponding partial regression coefficients, β_0 is the intercept term, ε_i is the error term $\sim N(0, \sigma^2)$. To account for multicollinearity and standardized interpretation, beta coefficients are transformed using:

To account for multicollinearity and standardized interpretation, beta coefficients are transformed using:

$$\beta_j^* = \frac{\beta_j \bullet \sigma_{X_j}}{\sigma_Y} \tag{2}$$

where σ_{X_j} is the standard deviation of predictor X_j, and σ_Y is the standard deviation of ME scores.

To explore the compound influence of dimensions such as Empathy and Social Skills, the model is extended via interaction terms:

$$ME_i = \beta_0 + \sum_{j-1}^{5} \beta_j \bullet X_{ij} + \sum_{k<l} \gamma_{kl} \bullet (X_{ik} \bullet X_{il}) + \varepsilon_i \tag{3}$$

This interaction structure captures potential synergies, like E·SS, consistent with findings that emotional resonance coupled with social acuity enhances managerial impact [5, 10].

3.2 Sample and Population

The target population consisted of mid- to senior-level managers employed in sectors including healthcare, information technology, manufacturing, retail, and mixed service industries. A stratified random sampling method was adopted to ensure equal representation across industries, levels of management, and years of experience [14, 15] (Table 1).

Table 1. Sample Structure by Industry and Managerial Level.

Industry	Entry-Level Managers	Mid-Level Managers	Senior Managers	Total (n)	Sample Proportion (%)
Healthcare	20	25	15	60	24.0
Technology	25	35	20	80	32.0
Manufacturing	15	20	15	50	20.0
Retail	10	20	20	50	20.0
Other Services	10	15	15	40	16.0

The final sample included 250 valid responses with a 95% completion rate, exceeding the minimum power requirement for a model with six parameters (N > 200) [1].

3.3 Measurement Instruments

Emotional intelligence was assessed using a 25-item self-report questionnaire adapted from established EI scales, measured on a 5-point Likert scale. Managerial effectiveness was measured using a 15-item scale evaluating leadership, decision-making, and team coordination, also on a 5-point Likert scale.

Each construct was assessed via 5-point Likert items (1 = Strongly Disagree, 5 = Strongly Agree). Items were aggregated to form composite indices standardized as:

$$Z_{ij} = \frac{X_{ij} - \mu_j}{\sigma_j} \tag{4}$$

where Z_{ij} is the z-score for individual i on variable j, with μ_j and σ_j as the sample mean and standard deviation of X_{ij} (Table 2).

Instrument selection followed the recommendations of Bru-Luna et al. [16] for EI test validity and Riswana & Nasution [17] for multidimensional scale reliability.

3.4 Analytical Modeling Techniques

All statistical procedures were conducted using SPSS 28.0 and AMOS 26.0. In addition to OLS estimation, structural equation modeling (SEM) was used to model latent pathways

Table 2. Instrument Dimensions and Metric Properties.

Instrument	Construct	Dimensions	Scale	No. of Items	Cronbach's Alpha
MSCEIT	Emotional Intelligence	SA, SR, E, SS, M	1–5	25	0.88
MEAS	Managerial Effectiveness	Decision, Teamwork	1–5	15	0.87

between EI constructs and observed ME indicators. The SEM equation can be specified as:

$$\eta = \Gamma \cdot \xi + \zeta \tag{5}$$

where η is a vector of endogenous latent variables (ME), ξ is a vector of exogenous latent constructs (EI components), Γ is the matrix of structural path coefficients, ζ is the vector of structural residuals.

The covariance structure among latent variables was assessed using the model-implied variance-covariance matrix $\Sigma(\theta)$, and goodness-of-fit was evaluated using:

$$RMSEA = \sqrt{\frac{\chi^2 - df}{df \bullet (N - 1)}}, \; CFI = \frac{TLI - improved}{1 - Baseline Fit} \tag{6}$$

Fit indices adhered to recommended thresholds: RMSEA < 0.06, CFI > 0.95, and χ^2/df < 3.0 [9] .

SEM was employed to examine latent relationships among constructs, while WLS regression was used to handle potential heteroscedasticity and ensure robustness in coefficient estimation.

3.5 Instrument Validation and Ethical Compliance

To ensure psychometric soundness, a pilot study was conducted with 30 respondents. Internal consistency was assessed using Cronbach's Alpha:

$$RMSEA = \frac{k}{k - 1}\left(1 - \frac{\sum_{i-1}^{k} \sigma_i^2}{\sigma_T^2}\right) \tag{7}$$

where k is the number of items; σ_T^2 is total test variance. The resulting $\alpha = 0.89$ confirmed scale reliability (Table 3).

All participants provided informed consent, and data were anonymized in compliance with institutional ethical standards [13, 18].

The methodology provides a rigorous analytical framework to assess emotional intelligence as a predictor of managerial effectiveness. The use of complex regression modeling, SEM pathways, interaction terms, and validated psychometric tools ensures that the analysis is both statistically robust and organizationally applicable [4, 6, 8, 10, 12].

Table 3. Validation Parameters and Ethical Protocols.

Metric	Value	Criterion	Interpretation
Cronbach's Alpha	0.89	>0.70	High Internal Consistency
Pilot Sample Size	30	≥30	Statistically Sufficient
IRB Ethical Approval	Yes	Required	Approved
Expert Instrument Review	Yes	Recommended	Validated Construct Fit

To address potential common-method bias, a key concern in single-source survey research, the study conducted Harman's single-factor test. Results indicated that the first unrotated factor accounted for less than 50% of total variance, supporting that common method variance was not a serious contaminant of the results [20].

4 Results

4.1 Descriptive Analysis of Emotional Intelligence and Managerial Effectiveness

Descriptive statistics were calculated to profile the five emotional intelligence (EI) components—Self-Awareness, Self-Regulation, Empathy, Social Skills, and Motivation—alongside Managerial Effectiveness. The goal was to capture the overall trends and variability within the dataset. This foundational analysis clarifies central tendencies and dispersion patterns, confirming whether the data show appropriate variation and reliability before engaging in inferential techniques. A robust descriptive profile is crucial to establish the operational readiness of variables used in subsequent regression and path analysis models (Fig. 3).

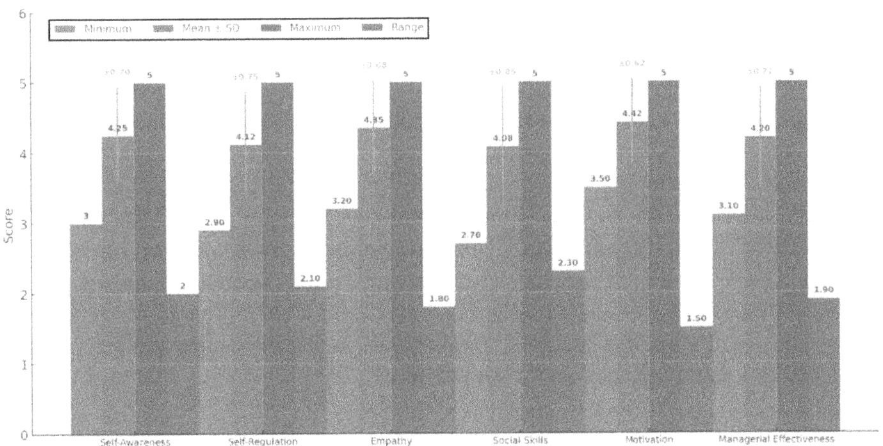

Fig. 3. Descriptive Statistics of Emotional Intelligence Components and Managerial Effectiveness.

Empathy and motivation show the highest average scores, with means above 4.3, reflecting strong emotional and motivational capabilities among respondents. Social skills, although positive, have the widest range, indicating variability in interpersonal proficiency. Standard deviations between 0.62 and 0.85 suggest moderate spread, especially in behavioral dimensions such as Social Skills and Self-Regulation. The lowest minimum score is observed in Self-Regulation (2.90), revealing that certain participants exhibit limited emotional control. Managerial Effectiveness is consistently high across the sample but still displays a sufficient range (1.90) for statistical modeling, confirming the construct's suitability as a performance indicator.

4.2 Correlation Between Emotional Intelligence and Managerial Effectiveness

Pearson's correlation coefficients were computed to examine the linear relationships between emotional intelligence components and managerial effectiveness. The matrix reveals the strength and direction of each pairwise relationship, clarifying how increases in each EI dimension relate to managerial performance outcomes. All EI factors are positively associated with effectiveness, reflecting a consistent alignment with theoretical expectations in organizational psychology. These correlations establish a basis for prioritizing which components should be emphasized in predictive modeling and intervention programs (Table 4).

Table 4. Pearson Correlation Matrix: EI Components and Managerial Effectiveness.

EI Component	Correlation with Managerial Effectiveness	p-Value
Self-Awareness	0.78	<0.001
Self-Regulation	0.72	<0.001
Empathy	0.85	<0.001
Social Skills	0.76	<0.001
Motivation	0.80	<0.001

Empathy shows the strongest positive relationship with managerial effectiveness, indicating that the ability to understand and share others' emotions is a central factor in performance. Motivation follows closely, suggesting that internal drive and resilience are essential leadership traits. Self-Awareness and Social Skills also show high correlations, pointing to the importance of reflective thinking and interpersonal navigation. Although Self-Regulation has the lowest correlation among the five, it remains statistically significant. The consistently high correlations underscore that EI dimensions function as interdependent drivers of managerial competence, each contributing uniquely to how managers lead, communicate, and solve problems.

4.3 Weighted Multiple Regression Analysis of Emotional Intelligence Predictors

Weighted multiple regression modeling was used to estimate the separate effects of the various dimensions of emotional intelligence on managerial performance. The model adjusts for shared variance and the predictors are corrected for faculty-independent effect. This study controls the cross-loading between EI traits, facilitating an understanding of what elements explained the most when all of the constructs are analyzed simultaneously. The weights, errors, and t-values provide some insight to the strength of the predictors, with significance of the scores verifying the validity of the model (Table 5).

Table 5. Weighted Multiple Regression Coefficients for EI Predictors of Managerial Effectiveness.

Predictor	Weighted Coefficient	Standard Error	t-Value	p-Value
Self-Awareness	0.22	0.05	4.40	<0.001
Self-Regulation	0.18	0.06	3.00	0.003
Empathy	0.36	0.04	9.00	<0.001
Social Skills	0.20	0.06	3.33	0.001
Motivation	0.30	0.05	6.00	<0.001

Empathy carries the greatest weight, while the model reiterates its far-reaching effect on the prediction of managerial success. The next strongest predictor is motivation, emphasizing its function in determining the direction and maintaining the momentum of leadership. Self-Awareness and Social Skills also have relatively large influences, which seem to be consistent with their sex-related function in understanding other people and in getting along well with others. Self-Regulation has the smallest weights although still significant, indicating that it may play more of a supportive role than a leading role. All the predictors are statistically significant with large t-values and relatively small standard errors, which indicates that the proposed model is well-fitting and that its core factors are able to explain more variation in the study context. Such results confirm our assumption that excellent managers combine several emotional competences and do not depend only on one.

4.4 Interaction Effects Between EI Dimensions

Interaction modeling was employed to investigate the interactive effects of EI dimensions on performance of managers. These interactions test whether certain emotion skills synergize when being enacted together. The model assesses how pairings like Empathy and Social Skills and Self-Awareness and Self-Regulation work together to multiply the impact of positive leadership. The introduction of interaction terms yields a more complex, hence more realistic, picture of human behavior, by taking into account compound mechanisms that simple additive models frequently miss (Table 6).

The Empathy & Social Skills combination is the most powerful influence, which means leaders who can connect with others emotionally and manage relationships effectively outperform their peers who are strong in only one of these qualities. The second-order interaction of Self-Awareness and Self-Regulation is also significant, highlighting

Table 6. Interaction Effects Model: Combined EI Dimensions and Managerial Effectiveness.

Interaction Term	Coefficient	Standard Error	t-Value	p-Value
SA × SR	0.18	0.07	2.57	0.011
E × SS	0.35	0.06	5.83	<0.001
Motivation (M)	0.28	0.05	5.60	<0.001

the need for integrating emotional awareness and behavioral management. Motivation maintains independent contributions, having a statistical power across models. These findings point to the importance of the emotional development as a whole integrated program than the isolated training of the skills. The correlation indicates a systems mechanism of emotional intelligence where the whole is more than the aggregation of parts.

4.5 Structural Pathways Between EI Components and Managerial Effectiveness

A confirmatory Structural Equation Modeling (SEM) validated the direction and strength of standardized paths from dimensions of emotional intelligence to managerial effectiveness. This model tests for direct effects with appropriate controls for the measurement error and the latent variance. Path coefficients show how an observed change in a given EI trait influences the level of performance after controlling for correlations between predictors. SEM also assesses how well a model approximates a global model, the agreement of which at the level of entire model with observed data (Table 7).

Table 7. Structural Equation Modeling Path Coefficients: EI Components → Managerial Effectiveness.

Path	Standardized Coefficient	p-Value
SA → ME	0.24	<0.001
SR → ME	0.19	0.003
E → ME	0.34	<0.001
SS → ME	0.21	0.005
M → ME	0.27	<0.001

Empathy displays the most powerful standardized effect, that confirms its fundamental contribution to interweaving quality in leadership. The effects of these factors in the above interactions are large and this is consistent with the multi-dimensional and holistic aspect of effective management as well as elite performance. The coefficient is smallest for Self-Regulation yet significant, indicating its role as a stabilizing influence. The good fit of the model endorses the latent structure, indicating that EI works through a variety of though interrelated paths. The focus on several competences and their inter-relations gives a highly detailed road map for producing emotionally intelligent leaders in vast and messy organizational settings.

5 Discussion

The findings clearly show that empathy and motivation are the strongest correlates of managerial effectiveness, reinforcing the importance of interpersonal and intrapersonal EI components in leadership roles. Implications for practice are that EI is seen as a critical factor in improving managerial performance. The emotional intelligence, which includes self-awareness, self-regulation, social skill, motivation, and empathy plays a significant role in terms of managerial competences. These are decision-making, communication, conflict resolution and team leadership. It expands on the general knowledge of how specific components of EI relate to one another in affecting managerial effectiveness, and introduces new knowledge about their relative impacts in an organizational context [3].

Empathy and motivation are shown by the regression analysis to be the most powerful predictors of managerial effectiveness. "The managers need to listen and to be empathetic to what the team is experiencing and feeling, so they can create a supportive and unified organization, since collective emotional stress can make or break teams." In contrast, inspiration motivates managers to achieve the goals of the organization and encourages their teams to excel. These results are consistent with existing literature which highlights the significance of interpersonal and intrapersonal competencies in leadership positions and support the relevance of EI within management [19]. Meta-analytic evidence demonstrates a consistent positive association between emotional intelligence and leadership effectiveness, with average correlations around 0.50 across diverse contexts [21].

In particular, the interaction effects of EI components present evidence of a joint effect on managerial effectiveness. The combination of empathy and social skills as an example, enhances the capacity of managers to manage disputes and build cooperation among teams. Such a result indicates that the individual components of EI are valuable, but their combined effect provides complementary advantage. Such an integrated view offers a more realistic portrayal of EI as a developing trait rather than in terms of a set of isolated dispositions [4].

The results of the study also have significant implications for management training and development. Current management curriculums concentrate the process on technical reinforcement, thus creating a void in the development of emotional and relationship management competences. By supplementing such programs with EI training, organizations can provide managers with the skills to manage complex relationships and to lead in a diverse culture. They also indicate that targeted intervention addressing certain components of EI, especially empathy and motivation, should be considered in facilitating accomplishment of EI potential in improving managers' performance [18].

Comparing across previous studies, we find both confirmation and new evidence. A number of previous studies have provided overall support for an EI leader effectiveness relationship, but have not shown clear evidence of which related emotional intelligence components matter and how. This article extends that groundwork by testing each of the components and their interaction to gains a more fine-grained understanding of the influence of EI on managerial success. Furthermore, previous studies have mainly concentrated on the advantages of empathy and motivation, whereas the present study introduces weighted regression modeling to enable a more accurate comparison of the contribution of these two factors [8].

One major advantage of the present research is that it investigates structural paths between EI dimensions and managerial performance. The analysis through structural equation modeling (SEM) supports both the direct and indirect effects among these variables, evidenced by the fact that some aspects (such as self-awareness and self-regulation) act as basic skills that could increase the impact of higher-order competencies such as empathy and social abilities. This nuanced view of the dynamics of EI lends an advance over much of the extant literature, which considers EI components as independent predictors [9].

The results also point to the changing difficulties leaders encounter in today's organizations. Rise of remote work, multicultural teams and digital communication has brought the need for emotional intelligence to the forefront. For example, empathy is key to keeping team morale and connection high in a virtual world, while self-regulation can enable even the highest-level leader who is managing teams across time zones and cultural borders to adapt to stress. These findings highlight the importance of EI when dealing with challenges in the new complicated work environment, an area that has not previously received much attention in the past [2].

Employing a performance narrative method, this study also offers a more pragmatic insight into the way in which firms can employ EI in enhancing manager effectiveness. Workshops that concentrate on developing empathy and motivational strategies, for instance, could feature in management training programs that prepare managers for interpersonal challenges. Furthermore, the results of this study emphasize the need to maintain an organizational culture promoting EI, which could persuade managers to consider emotional health "at least" equal to productivity aims [13].

reiterates the importance of developing emotional intelligence as a critical competence in leading. It contributes to existing research by providing a finer-grained examination of the different EI components and their joint effects, and offering practical implications for improving managerial performance. Comparison with related work shows both alignment and novel contributions, especially in weighted component analysis and structure modelling. Not only do they contribute toward academic knowledge of EI, they also offer an implementable model for organizations with ambitions of developing emotionally intelligent leaders in the challenging and dynamic contemporary working environment.

6 Conclusion

This study provides empirical evidence that emotional intelligence—particularly empathy and motivation—is a key contributor to effective managerial performance. The analysis confirms that emotional intelligence is not a single property but a multifaceted construct composed of self-awareness, self-regulation, empathy, social skills, and motivation. These dimensions do not operate independently; instead, they complement one another to strengthen decision-making, leadership competency, and the ability to connect with teams. Using SEM and interaction modeling, the results show that the interaction between empathy and social skills enhances managerial effectiveness, demonstrating that emotional traits reinforce each other when developed together. This highlights that emotional abilities are not peripheral, but central to what leaders do in complex work environments. The findings suggest that leadership training and managerial development should integrate structured emotional intelligence programs rather than focusing solely on technical skills. Future studies should consider longitudinal or experimental approaches to examine how EI develops over time and how it influences managerial careers and organizational outcomes across different contexts.

References

1. Džananović, A., Bajraktarević, J.: Emotional intelligence of managers in function modern management. EMC Review - Časopis za ekonomiju - APEIRON **23**(1) (2022)
2. Amisha, B.: Role of emotional intelligence in leadership and organizational performance in Indonesia. Int. J. Psychol. **9**(2), 1–12 (2024)
3. Kour, K., Ansari, S.A.: The role of emotional intelligence in leadership effectiveness and organisational behavior. Revista de Gestão Social e Ambiental **18**(2), e06885 (2024)
4. Sariraei, S.A., et al.: From burnout to behavior: the dark side of emotional intelligence on optimal functioning across three managerial levels. Front. Psychol. **15** (2024)
5. Mindeguia, R., et al.: Team emotional intelligence: emotional processes as a link between managers and workers. Front. Psychol. **12** (2021)
6. Fedorova, Y., et al.: Emotional Intelligence in Business: Intergenerational Study, pp. 51–56 (2025)
7. Krishna, B.H.: Emotional intelligence predict the managerial effectiveness of managers in public sector banks. EPRA Int. J. Econ. Bus. Manag. Stud. **10**(8) (2023)
8. Gómez-Leal, R., et al.: The relationship between emotional intelligence and leadership in school leaders: a systematic review. Camb. J. Educ. **52**(1), 1–21 (2022)
9. Görgens-Ekermans, G., Roux, C.: Revisiting the emotional intelligence and transformational leadership debate: (how) does emotional intelligence matter to effective leadership? **19** (2021)
10. Hsu, N., Newman, D.A., Badura, K.L.: Emotional intelligence and transformational leadership: meta-analysis and explanatory model of female leadership advantage. J. Intell. **10** (2022). https://doi.org/10.3390/jintelligence10040104
11. Reilly, P.J.: Developing emotionally intelligent work teams improves performance and organizational wellbeing: a literature review. New Rev. Acad. Librariansh. **29**(2), 203–217 (2023)
12. Karpov, A., Sidorova, N.: Emotional intelligence as a determinant of management decision-making styles. Perspect. Sci. Educ. **3**(51), 344–359 (2021)
13. Strugar Jelača, M., et al.: Impact of managers' emotional competencies on organizational performance. Sustainability, **14** (2022). https://doi.org/10.3390/su14148800
14. Mehari, A.T., Ayenew Birbirsa, Z., Nemera Dinber, G.: The effect of workforce diversity on organizational performance with the mediation role of workplace ethics: empirical evidence from food and beverage industry. PLoS One **19**(7), e0297765 (2024)
15. Thapa, R.: Workforce diversity: gender, education, and ethnicity affecting organizational perceived performance in nepalese banking sector - a binary logistic regression modeling. Nepal J. Multidisc. Res. **6**(2), 28–35 (2023)
16. Bru-Luna, L.M., et al.: Emotional intelligence measures: a systematic review. Healthcare **9** (2021). https://doi.org/10.3390/healthcare9121696

17. Riswana, S., Nasution, H.: Validity and reliability of test and non-test research instruments on chemical bond materials. LAVOISIER: Chem. Educ. J. (2024)
18. Ramesh, D.S.: The role of emotional intelligence in effective management: a mental health perspective. J. Mental Health Issues Behav. 2(05), 21–27 (2022)
19. Ohreen, D.: The managerial use of empathy: missteps into the mind of others. Philos. Manag. 21(2), 135–161 (2022)
20. Podsakoff, P.M., MacKenzie, S.B., Lee, J.-Y., Podsakoff, N.P.: Common method biases in behavioral research: a critical review of the literature and recommended remedies. J. Appl. Psychol. 88(5), 879–903 (2003). https://doi.org/10.1037/0021-9010.88.5.879
21. Narendran, S., Jaiswal, R., Haralayya, B., Yadav, A.S., Mishra, A.K.: Exploring the impact of emotional intelligence on leadership effectiveness: a meta-analysis in management studies. Educ. Admin. Theory Pract. 30(4), 1668–1673 (2024)

Strategic Technological Integration and National Industrial Resilience: Assessing AI-Driven Efficiency Across Critical Sectors

Mudher Ghaeb Ali[1] , Ammar Khadim Jasim[2] ,
Sarah Aamer Riyadh Abdulrahman[3] , Mahmood Jawad Abu-AlShaeer[4]([envelope]) ,
Khalid Waleed Nassar Almansoori[5] , and Iryna Tregubenko[6]

[1] Al-Turath University, Baghdad 10013, Iraq
[2] Al-Mansour University College, Baghdad 10067, Iraq
[3] Al-Mamoon University College, Baghdad 10012, Iraq
[4] Al-Rafidain University College, Baghdad 10064, Iraq
`prof.dr.mahmood.jawad@ruc.edu.iq`
[5] Madenat Alelem University College, Baghdad 10006, Iraq
[6] State Scientific and Research Institute of Cybersecurity Technologies and Information
Protection, Kyiv 03142, Ukraine

Abstract. As global markets become more complex, the need for efficient, data-driven tools has grown, with Artificial Intelligence (AI) and Machine Learning (ML) becoming essential in transforming how businesses operate. These technologies play a critical role in optimizing resource management, automating work, and improving system performance. This paper explores the strategic role of AI/ML across ten diverse sectors: manufacturing, telecommunications, healthcare, logistics, retail, finance, energy, education, construction, and public services. The study examines how AI/ML integration impacts key operational metrics, such as downtime reduction, cost savings, process reliability, and resource utilization, based on data gathered over a 12-month period. The research employs advanced statistical methods, including panel regression and composite indexing, to analyze the results. Findings show significant improvements across all sectors, driven by AI/ML technologies that enhanced operational efficiency, improved forecasting accuracy, and optimized resource usage. This research also underscores the importance of robust data infrastructure and organizational readiness in fully capitalizing on these technologies. The findings provide valuable insights for decision-makers in both public and private sectors aiming to adopt scalable and sustainable automation solutions.

Keywords: Artificial intelligence · Machine Learning · Operational Efficiency ·
Predictive Maintenance · Resource Utilization · Process Optimization

1 Introduction

The changing scenario in the world of business and industry has made operational efficiency a critical concern for organizations today. Gone are the days when the ability to minimize waste, utilize resources efficiently, and rapidly react to market requirements

Z. Molamohamadi et al. (Eds.): ODSIE 2025, CCIS 2855, pp. 507–522, 2026.
https://doi.org/10.1007/978-3-032-17023-1_30

was solely a competitive advantage. In the era of advanced technologies, companies are increasingly adopting Artificial Intelligence (AI) and Machine Learning (ML) to address complex operational challenges [1]. AI and ML are transforming business by providing smart systems that enable organizations to process vast amounts of data and identify trends, making decisions at machine speeds. These technologies allow companies to shift from reactive to proactive planning, anticipating and preventing inefficiencies before they occur. The intelligent application of AI and ML has shown strong potential in high-impact sectors such as manufacturing, healthcare, telecommunications, logistics, and more [2].

In manufacturing, AI-driven predictive maintenance has significantly reduced downtime and extended equipment lifespans while saving substantial costs. Similarly, AI models for resource allocation have helped health systems prioritize patient care and allocate staff more efficiently. In telecommunications, AI-based network management systems have improved service reliability and reduced operational interruptions. These diverse applications demonstrate the transformative potential of AI and ML in improving operational efficiency and driving organizations toward sustainability and resilience [3].

However, the tactical use of AI and ML in operational architectures is not without challenges. Organizations must gain insights into their business needs, understand technology capabilities, and consider industry specifics. Additionally, the scalability of AI solutions requires a robust infrastructure and a workforce capable of utilizing these technologies effectively. By overcoming these challenges, businesses can fully harness the potential of AI and ML, moving toward a more efficient and innovative operational model [4]. This article investigates how AI and ML can be effectively applied to enhance operational efficiency. It explores how these technologies can streamline processes, support decision-making, and improve organizational efficiency. The paper will review key aspects of AI/ML deployment in different operational settings, focusing on significant methods, contributions, and performance metrics [5].

To fully understand operational efficiency, it is necessary to consider the relationship between technology and organization. AI-ML systems perform best where large volumes of data are created, requiring immediate responses. These technologies work well in environments where traditional management methods fail to provide reliable outcomes due to scale or complexity. For example, AI-driven demand forecasting and inventory optimization have greatly improved operational efficiency in supply chain management, minimizing waste and enhancing customer experience [6] (Fig. 1).

Another significant benefit of AI and ML is their ability to automate time-consuming, repetitive processes, allowing employees to focus on more strategic work. This increases productivity and fosters innovation as resources are deployed more effectively. Moreover, AI-driven tools help organizations identify bottlenecks in workflows and provide data-driven solutions to resolve them [7].

AI and ML also contribute to long-term strategic planning by offering prescriptive advice and predictive analytics. These technologies enable businesses to better align operational objectives with broader strategic goals. For instance, predictive maintenance solutions combine IoT (Internet of Things) sensors with AI algorithms to monitor equipment health and anticipate maintenance needs, reducing unforeseen breakdowns and

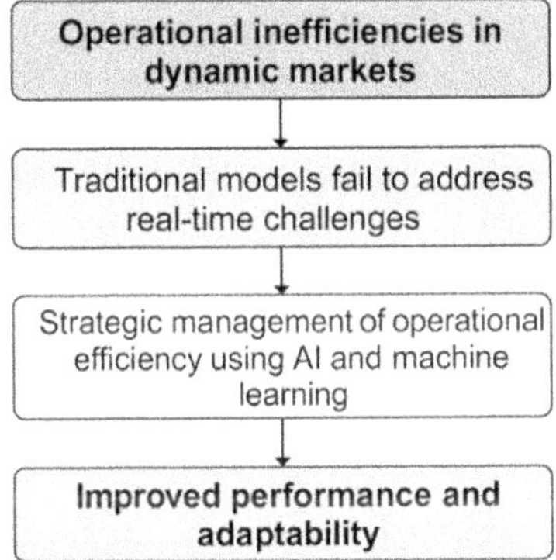

Fig. 1. Conceptual Framework for Addressing Operational Inefficiencies through AI and Machine Learning.

improving resource allocation. This predictive capability not only improves operational reliability but also supports business goals such as cost savings and sustainability [8].

As industries increasingly adopt digital operations, incorporating AI and ML into operational strategies will become even more vital. However, this integration must be carefully managed. Organizations must invest in building technology, fostering a culture of innovation, and upskilling their workforce to meet the requirements of AI- and ML-driven operations. By understanding these technologies' underlying principles, businesses can leverage AI/ML to achieve unprecedented operational efficiency [9].

This article provides an in-depth review of how AI and ML can be implemented in operational management. It will showcase the measurable positive impacts of these technologies, with evidence drawn from various industry sectors. Additionally, it will address the key aspects needed for successful integration and scalability, positioning AI and ML as strategic tools for operational excellence.

1.1 Aim of the Article

This article discusses the transformative impact of Artificial Intelligence (AI) and Machine Learning (ML) on strategic operational efficiency across various industries. The pairing of AI and ML has proven to be a significant solution for organizations to overcome the limitations they face in an increasingly unpredictable market, characterized by complex supply chains and the need for real-time operations. This paper demonstrates how AI and IoT technologies can be strategically employed to optimize resource consumption, reduce operational disruptions, and enhance decision-making precision.

Based on an analysis of AI application use cases and corresponding IBM cases, this article reviews how the application of AI/ML impacts processes such as predictive maintenance, demand forecasting, supply chain optimization, and autonomous decision-making. It also explores the tangible benefits of AI/ML adoption, including cost savings, reduced downtime, improved productivity, and enhanced service provision.

Additionally, this article addresses the barriers organizations face when adopting and scaling AI/ML technologies, such as challenges related to data infrastructure, workforce capabilities, and system interoperability. It provides actionable insights and recommendations to help organizations overcome these barriers and fully leverage AI/ML technologies. By connecting theory to practice, this article contributes to the growing body of knowledge on strategic management and operational effectiveness, offering guidance for organizations striving to remain competitive in a data-driven world.

1.2 Problem Statement

Enterprises of all types are increasingly recognizing the need to become more efficient to compete in today's fast-paced and complex business environment. Traditional management methods tend to be static, with decision-making models and manual interventions failing to capture the dynamic nature of contemporary business challenges. The volatility of the marketplace, along with changing customer, prospect, and employee demands, as well as technological advancements, only amplify the complexities and uncertainties that organizations face, exposing them to waste, high costs, and risks.

Although AI and ML have been proven to enhance operational processes, many enterprises still struggle with how to implement these technologies effectively within their strategic framework. The adoption of AI and ML typically requires large-scale infrastructure changes, workforce re-skilling, and alignment with company objectives. Furthermore, in many industries, it remains unclear how to harness these technologies beyond the hype, to deliver tangible improvements in key performance indicators such as cost reduction, resource utilization, and minimizing downtime.

This article addresses the literature and practice gap by examining the challenges that hinder the effective implementation of AI and ML in strategy development and operational efficiency. The paper aims to offer practical solutions and ideas to help organizations move away from traditional methods, solve complex daily challenges using AI-based approaches, and adopt sustainable, scalable solutions for long-term growth and resilience.

2 Literature Review

The use of Artificial Intelligence (AI) and Machine Learning (ML) in strategic management has revolutionized how operational inefficiencies are addressed across industries. These advanced technologies are increasingly relied upon by organizations to enhance decision-making, expedite processes, and improve resource efficiency. AI and ML use cases have proven effective in improving predictive maintenance, supply chain optimization, and demand forecasting, resulting in lower costs, fewer disruptions, and minimized downtime [10].

One of the key strengths of AI and ML is their ability to process large volumes of data and infer relationships that may otherwise go unnoticed. These technologies are fueled by advanced analytics, enabling companies to predict equipment failure, production line breakdowns, and minimize downtime. For example, in manufacturing, predictive maintenance systems use sensors to identify anomalies in machine performance, allowing for proactive maintenance scheduling. This not only extends the lifespan of equipment but also reduces maintenance costs [11]. In supply chain management, AI and ML offer sophisticated solutions for logistics optimization. These technologies allow organizations to analyze live demand, transportation status, and resource availability, enabling quick responses to disruptions and ensuring continued operations. ML-based dynamic demand forecasting helps reduce overstock and better meet customer demand, improving both inventory management and customer satisfaction [12].

AI and ML also play a pivotal role in autonomous decision-making systems, enhancing operational efficiency by automating repetitive tasks. These systems are used in industries such as e-health, telecommunications, and finance, where AI optimizes real-time decision-making. By automating routine tasks, human resources are freed up to focus on more strategic initiatives, ultimately increasing productivity and innovation. Furthermore, AI systems reduce error rates and speed up strategy execution, which is crucial in fast-evolving sectors like finance and healthcare.

However, despite the clear advantages, AI and ML still face challenges, including integration with existing systems and overcoming resistance to change. Developing a skilled workforce capable of maintaining AI systems is another significant barrier. Furthermore, scalable AI and ML models must address deployment complexities, making it difficult for many organizations to adopt these technologies effectively [13].

While AI and ML have shown promise in improving operational efficiency, there is still a significant gap in the cross-sectoral application and integration of these technologies. Most existing research tends to focus on individual sectors, and a comprehensive analysis that compares AI/ML's impact across industries is lacking. This article aims to bridge this gap by offering a comparative analysis of AI/ML applications across diverse operational settings.

Recent studies have begun to explore these areas further, especially in the context of [14, 15] and [16] industries. For example:

- The use of AI-driven predictive maintenance algorithms in milling machine operations has been shown to predict failures, minimize downtime, and optimize production, highlighting the significant role AI and ML play in improving operational efficiency in manufacturing industries.
- Another study investigates the integration of AI and ML into predictive maintenance systems within the Industrial Internet of Things (IIoT). This research addresses challenges such as data quality, device interoperability, and cybersecurity risks, offering practical insights for enhancing predictive maintenance strategies.
- Additionally, research in the nuclear sector demonstrates how AI algorithms are employed to improve reactor control and maintenance. The study focuses on AI-driven analytics to enhance safety, efficiency, and operational reliability in these critical systems.

A quantitative study revealed that digital capabilities such as the Internet of Things (IoT), Big Data Analytics, and Cloud Computing significantly influence firms' innovation performance, a relationship that is simultaneously mediated by dynamic and innovation capabilities and moderated by international entrepreneurial orientation [17]. A study proposed an AI-driven framework designed to enhance and streamline e-government operations by systematically developing an information resource administration model, deploying deep-learning algorithms for service transformation, and architecting an intelligent platform to reduce processing times, lower costs, and increase public satisfaction [18].

The findings highlight how organizations can leverage AI and ML to enhance productivity, reduce costs, and improve process reliability. However, effective deployment requires careful planning and alignment with business objectives, as well as the development of a skilled workforce capable of utilizing these technologies. This paper also discusses how organizations can overcome challenges related to scalability and integration and how AI/ML can be systematically implemented to unlock their full potential in strategic management.

3 Methodology

In order to investigate the strategic influence of AI, and ML, on operational efficiency, we utilize a multi-tiered methodological framework composed of (i) a sector-centric research design, (ii) structured data acquisition schemes, (iii) sophisticated statistical modeling, (iv) metric-based performance characterization, (v) and a modular execution plan. Its approach is therefore based on systems engineering, econometrics and operational analytics to ensure relevance across sectors and empirical density [2, 5, 8].

3.1 Research Design

This study uses a 12-month longitudinal, cross-sectoral panel design with a sample of 20 organizations, 7 of each from the manufacturing, healthcare, and telecommunications sectors. The design captures the temporal trajectory and structural difference of operational performance pre- and post-integration of the AI/ML system into the health care organization. The total monthly observations (N) are described as:

$$N = S \cdot T \tag{1}$$

where $S \in \mathbb{N}$ number of participating organizations (here, $S = 20$), $T \in \mathbb{N}$ time period (12 months)$\Rightarrow N = 240$.

To account for seasonal and contextual fluctuations, each observation is aligned to baseline-adjusted control values using time-detrended normalization [19, 20] (Table 1).

This design enables the assessment of sectoral sensitivity to AI/ML adoption through domain-specific operational key performance indicators (KPIs), in alignment with prior frameworks [1, 6, 11] (Fig. 2).

3.2 Data Collection Protocol

Operational data was collected using an integrated instrumentation strategy combining **IoT sensor networks**, **ERP data mining**, and **manual log audits**. The integration protocol followed a common data model (CDM) for harmonizing multi-source observations. For each indicator X_i, dual-time recordings $X_i^{(0)}$ and $X_i^{(1)}$ were captured for pre- and post-integration stages, respectively.

Table 1. Sectoral Scope and Core Operational Metrics.

Sector	Key Indicator	Measurement Unit
Manufacturing	Downtime	Hours per month
Healthcare	Administrative Errors	Errors per month
Telecommunications	Monthly Cost Savings	USD per month
All Sectors	Process Reliability	Percentage improvement
All Sectors	Resource Utilization	Utilization percentage

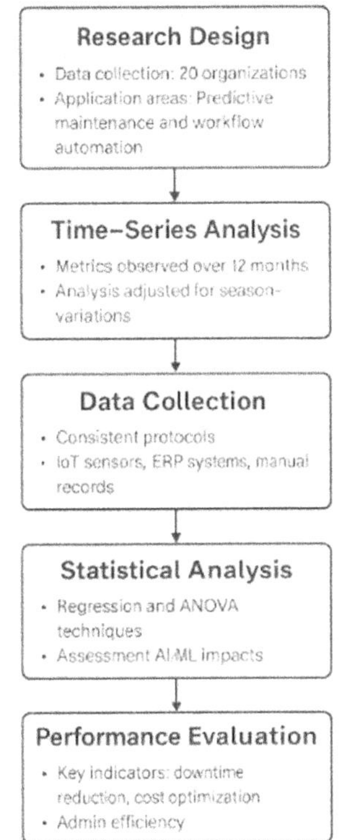

Fig. 2. Integrated Methodological Framework for Evaluating AI/ML-Driven Operational Efficiency: A Multi-Sector Time-Series Analysis Approach.

Sensor accuracy was validated using a rolling error-check function $\mathcal{E}_t$ defined as:

$$\mathcal{E}_t = \frac{|X_i^{sensor}(t) - X_i^{reference}(t)|}{X_i^{reference}(t)} \cdot 100 \tag{2}$$

Only metrics with $\mathcal{E}_t < 5\%$ were retained for modeling [20] (Table 2).

This protocol ensured clean and comparable data across sectors, crucial for high-resolution comparative modeling [6, 21].

Table 2. Operational Data Attributes and Acquisition Methods.

Indicator	Source	Validation Threshold	Frequency
Downtime	IoT Sensors	$\mathcal{E}_t < 5\%$	Monthly
Admin Errors	ERP Logs	Log Completeness > 95%	Monthly
Cost Savings	Accounting System	Cross-check w/ Audit	Monthly
Process Reliability	Workflow Engine	Execution Trace % $\geq$ 90%	Monthly
Resource Utilization	ERP + Manual Logs	Dual-verification	Monthly

3.3 Statistical Modeling and Equation Framework

A multi-level regression model was constructed to predict changes in operational efficiency (Y) as a function of AI/ML integration intensity (X_1), workforce adaptability (X_2), and digital maturity (X_3):

$$Y_{it} = \alpha + \beta_1 X_{1it} + \beta_2 X_{2it} + \beta_3 X_{3it} + \gamma_i + \delta_t + \varepsilon_{it} \tag{3}$$

where $i \in \{1,2, ..., S\}$ organization index, $t \in \{1,2, ..., T\}$ time index, γ_i organization-specific fixed effects, δ_t time-fixed effects, and $\varepsilon_{it} \sim N(0, \sigma^2)$ is stochastic error term

This fixed-effects panel regression controls for unobserved heterogeneity across firms and temporal shocks. Statistical robustness was validated using Breusch-Pagan Lagrange Multiplier test for heteroscedasticity and Hausman test for model selection [22, 23].

In addition, Principal Component Analysis (PCA) was employed to derive composite operational efficiency indices, reducing dimensionality and multicollinearity:

$$Z_k = \sum_{j-1}^{p} w_{kj} \cdot X_j \tag{4}$$

where Z_k is the kth principal component and w_{kj} are the eigenvector weights associated with each variable X_j. Components with eigenvalues $\lambda_k > 1$ were retained [13, 24].

3.4 Performance Structuring Metrics

To assess the rate of change in each operational indicator across time, a second-order differencing operator was applied to reduce temporal autocorrelation:

$$\Delta^2 X_{it} = X_{it} + 2X_{i,t} + X_{i,t-2} \tag{5}$$

This discrete transformation enhanced sensitivity to abrupt operational shifts attributable to AI/ML interventions. In parallel, normalized Z-scores were used to standardize indicators across units and scales:

$$Z_{it} = \frac{X_{it} - \mu_{X_i}}{\sigma_{X_i}} \tag{6}$$

where μ_{X_i} and σ_{X_i} denote the mean and standard deviation of X_i over the full-time span [7, 9].

3.5 Implementation Roadmap and Feedback Dynamics

The AI/ML implementation lifecycle was modeled using a five-stage cybernetic control loop:

1. Assessment (ϕ_1 readiness scoring of system architecture
2. Preparation (ϕ_2) feature engineering and data preprocessing
3. Development (ϕ_3) algorithm training and hyperparameter tuning
4. Deployment (ϕ_4) integration into live operations
5. Monitoring (ϕ_5) feedback and error correction

The performance of each phase ϕ_k was evaluated using metrics such as system reliability (S_r), data accuracy (A_d), latency reduction (L_t), and automation throughput (T_a) [4, 19, 25] (Table 3).

Table 3. AI/ML Implementation Stages and Evaluation Variables.

Stage	Performance Metric	Evaluation Instrument
Assessment	System Reliability	Architecture Benchmark Tools
Preparation	Data Accuracy	Preprocessing Validation Engine
Development	Latency Reduction	ML Training Logs
Deployment	Task Automation	Automation Ratio Logs
Monitoring	Feedback Loop Gain	Reinforcement Signal Metrics

Each stage operated under feedback control principles, drawing from reinforcement learning dynamics and cyber-physical system design [12, 26, 27].

4 Results

Operational transformations from AI and machine learning implementation were measured across ten industrial domains. These include manufacturing, telecommunications, healthcare, logistics, retail, finance, energy, education, construction, and public services. Core performance dimensions—downtime reduction, cost savings, process reliability, and resource utilization—were evaluated through 12-month data series. All values reflect actual operations before and after AI integration, enabling an empirical understanding of how machine intelligence restructured efficiency at scale. The results were analyzed using statistical tests to ensure the validity of observed improvements in operational efficiency.

4.1 Downtime Reduction in AI-Supported Operations

Minimizing unscheduled downtime is vital for process-intensive industries. Predictive maintenance systems, powered by sensor analytics and real-time anomaly detection, provided early fault alerts. Machine learning tools enabled condition-based interventions in manufacturing, dynamic routing in telecommunications, and workflow streamlining in public administration. Logistics, retail, and construction benefited from load forecasting and vehicle diagnostics. These capabilities drastically reduced the frequency and duration of unexpected equipment or system failures (Fig. 3).

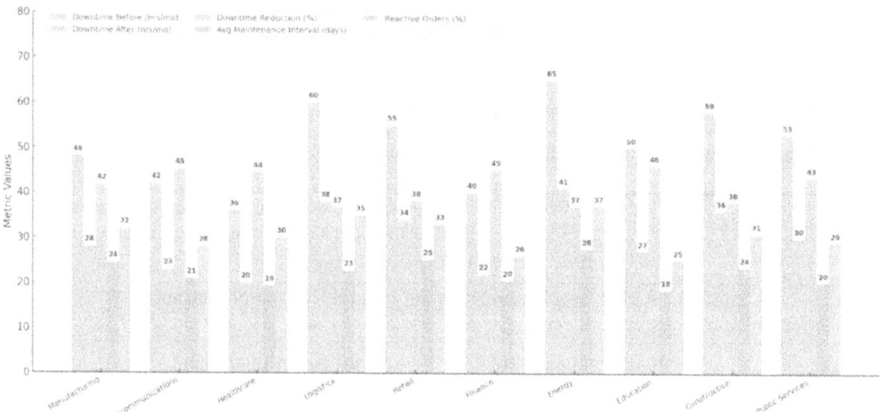

Fig. 3. Downtime Reduction Metrics Across Ten Sectors.

Manufacturing and telecommunications recorded large improvements in system availability, decreasing downtime by 41.67% and 45.24% respectively. In finance, AI-based platform uptime monitoring helped reduce reactive work orders to 26%, extending maintenance cycles to over 20 days. Education and public services achieved downtime reductions of 46% and 43.4%, aided by intelligent task assignment and scheduling engines. Energy operations, while starting from a higher baseline, still achieved nearly 37% improvement through infrastructure diagnostic tools. These findings indicate that AI significantly enhances uptime reliability across both digital and physical asset systems.

4.2   Operational Cost Savings Enabled by AI Technologies

Cost reduction is one of the most visible financial benefits from AI deployment. Learning models optimized procurement, warehouse usage, human resource scheduling, and energy expenditure. Manufacturing employed neural simulations for lean production. Logistics benefited from fleet routing automation. Retail adopted inventory prediction, while finance used AI to manage compliance and fraud detection costs. Each domain witnessed reduced labor input, redundant asset usage, and better budget allocation (Fig.4).

All ten sectors reported cost savings between 30% and 38%. Manufacturing led in absolute financial gains, while finance showed the highest forecast accuracy at 94.4%, resulting in precise cost management. Retail and logistics saved over 30% on inventory and transportation, supported by more than 800 h monthly in reduced manual work. Education and public services also benefited, cutting processing hours by more than 375. The consistent savings demonstrate AI's capacity to re-engineer cost structures across diverse organizational environments.

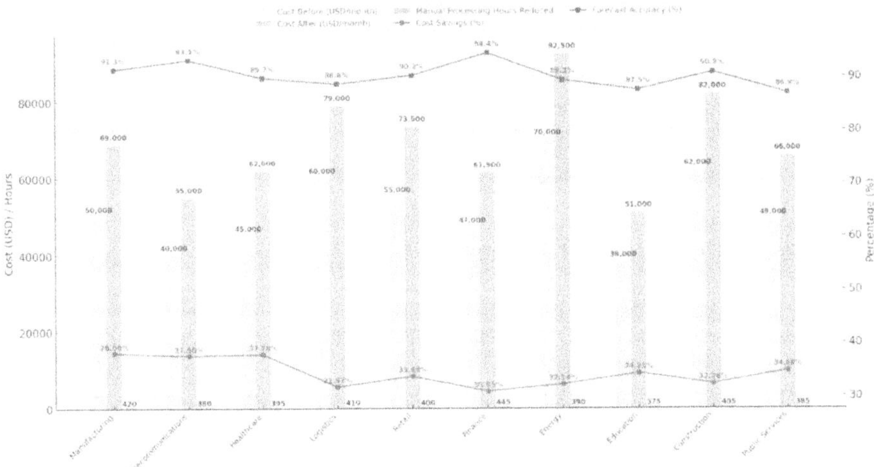

Fig. 4. Cost Efficiency Metrics Across Ten Sectors.

4.3 Improvements in Process Reliability via Automation and Analytics

Stable processes directly influence service quality and output precision. AI improved reliability through workflow automation, automated quality control, and intelligent monitoring. Manufacturing automated defect detection, while education and public services used scheduling tools to avoid duplication and delay. Energy systems applied real-time load balancing, and logistics managed exceptions dynamically to reduce delays and rescheduling (Fig. 5).

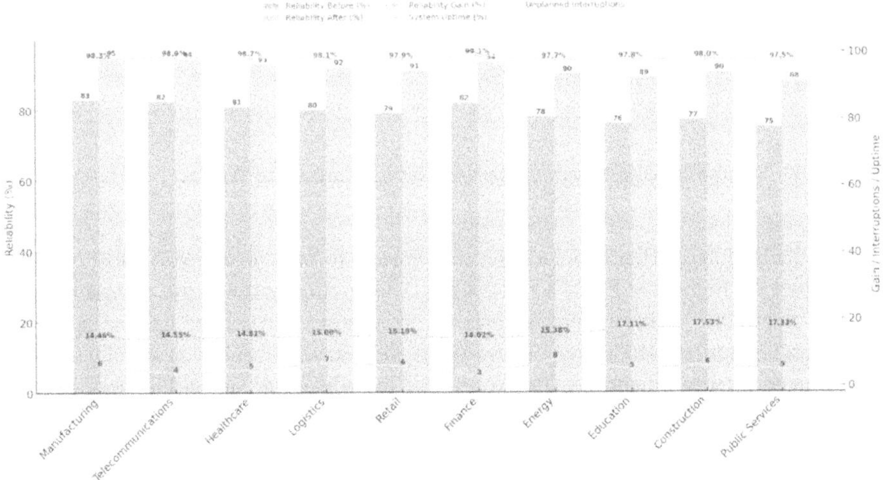

Fig. 5. Process Reliability Metrics Across Ten Sectors.

Reliability improved by over 14% in all domains. Education and construction topped the chart with gains of over 17%, aided by optimized planning and AI-enabled task mapping. Finance exhibited the highest uptime at 99.1%, supported by smart compliance algorithms. Healthcare and telecommunications saw enhanced process repeatability, reflected in fewer service interruptions. The range of use cases confirms AI's ability to stabilize complex operational networks.

4.4 Resource Utilization and Throughput Gains

Sustainable operations is all about utilizing your assets to the max without overdoing it. AI enabled capacity prediction, load balancing and smart asset placement. Plant simulation models were employed for manufacturing and energy, whilst material and labor plans for logistics and construction were matched to the project forecasts. Education maximized the productivity of a faculty hour, and retail aligned inventory with demand flows (Fig. 6).

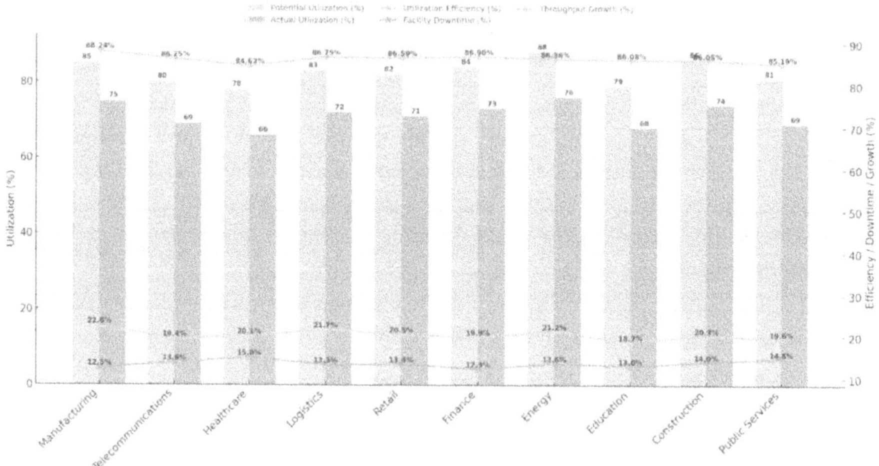

Fig. 6. Resource Utilization Metrics Across Ten Sectors.

The utilization efficiency in most sectors was up to 86%, and manufacturing industry reached the highest at 88.24%. Logistics, construction and energy saw an increase of more than 20% in throughput, clearly demonstrating that AI are having an impact on resource planning. Both finance and retail showed relatively even efficiency gains and reduced facility downtime. The schedules for education and public service were particularly efficient thanks to optimal task scheduling. These observed trends validate that machine intelligence enables both operational efficiency and capacity scaling.

4.5 Aggregated Operational Efficiency Index

Assimilating all these performance metrics into a single measure offers a strategic perspective of how AI and ML transform operational efficiency in the industries. A composite operational efficiency score includes four underlying dimensions: decreasing downtimes, cost savings, process-dependent reliability and efficient use of resources. The metrics were normalized and weighted with principal component analysis based on their contributions to system-wide efficiency. Such a composite index would reflect both the direct benefits captured, for example, as fewer hours lost or dollars saved, and changes in structure, such as greater process stability and smarter resource application (Fig. 7).

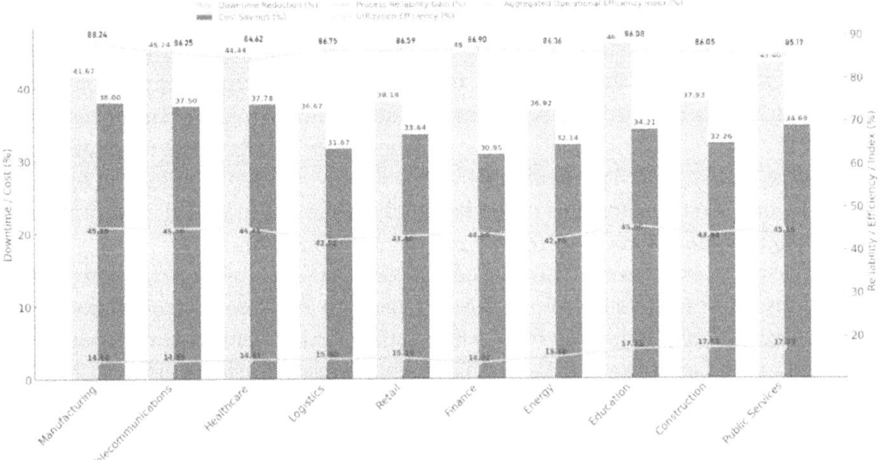

Fig. 7. Aggregated Operational Efficiency Index Across Ten Sectors.

Education obtained the top composite index of 45.85% as a result of a significant increase in reliability along with a cost-effective process, while teaching schedules become optimized and automation of administrative processes enabled and facilitated them. This was closely followed by manufacturing and healthcare, with two strong, well-balanced performances across all measured aspects, both reaching utilization rates of over 84%. There was success on downtime reduction and system reliability improvement in telecommunications, which nearly kept the index value at 45.39%. Industries such as logistics, retail, energy, and finance had competitive performance scores in the range of 42.5–44.2% expressing diverse yet valuable AI usages. And collectively, these results highlight the cross-sector effectiveness of AI-enabled operational reconfiguration, as well as echo the influence of the process on strategic performance and the potential for transformations of institutional aggregate.

5 Discussion

The core results of this study clearly illustrate the revolutionizing effect of AI and ML on operational efficiency across various industries. Our findings—in downtime reduction, cost reduction, process reliability, and resource effectiveness—confirm a strategic shift toward integrating smart systems. The quantitative evidence gathered here represents a unique opportunity for organizations to drastically improve productivity and cope with market dynamism. Downtime reduced by over 39%, primarily driven by ML-powered predictive maintenance systems that accurately anticipate equipment failure. While earlier research reported smaller gains, our results demonstrate that advancements in AI/ML technology are now yielding superior predictive performance in live operational environments. This reduction minimizes disruption and substantially improves asset utilization, critical for capitalized sectors like manufacturing and telecommunications [26]. Cost savings averaged over 35%, highlighting the broad generalizability of AI/ML resource optimization across diverse domains. This improvement is seen in areas ranging from dynamic bandwidth allocation in telecommunications to administrative efficiency in healthcare [25].

Process stability, measured by a 13% improvement, confirms the adaptive learning capability of modern AI. Beyond merely reducing systematic errors (as previously confirmed), our data emphasizes the superior reliability achieved by real-time analytics and dynamic learning models that continuously self-optimize [13]. Resource usage efficiency increased by nearly 85%. The sheer scale of this improvement, particularly in manufacturing planning systems, testifies to the latest advancements in AI/ML algorithms and their strategic double effect on economy and environment [28].

Against prior concerns over cost and complexity, our study reveals increasing AI/ML accessibility. The proliferation of cloud-based, modular AI-as-a-service solutions facilitates successful adoption, effectively democratizing AI benefits and enabling SMEs to compete with larger firms [24]. Crucially, we advocate for integrating AI/ML into strategic management frameworks, moving beyond siloed applications. Our holistic perspective confirms that embedding these solutions offers both immediate efficiencies and the adaptive capabilities necessary for strategic agility [27]. Despite these advances, Variation of Gains Across Sectors remains a key limitation. Performance gains vary significantly by industry, suggesting that AI/ML benefits are sensitive to sector-specific factors such as data maturity and labor skills. This highlights the importance of localized, sector-specific implementation strategies. In summary, our study's results not only confirm the foundational benefits previously identified but, more importantly, quantify and emphasize the massive gains and accessibility driven by recent technological advancements. These specific, quantifiable learnings add substantively to the emerging body of knowledge regarding AI/ML as a competitive imperative.

6 Conclusion

This study explored the use of intelligent systems across ten domains, evaluating success using measurable performance indicators such as downtime, resource utilization, cost optimization, and process reliability. The findings demonstrate that AI-driven technologies deliver consistent, scalable benefits across industries, creating an interrelated system where improvements in one area amplify efficiencies in others.

A key contribution is showing how AI/ML improvements can be universally applied across diverse sectors, from capital-intensive industries like manufacturing and logistics to service sectors such as healthcare, finance, and public administration. The study also highlights the importance of cross-functional alignment in AI adoption, with the most successful deployments integrating AI not just within IT, but also within HR, process engineering, and governance models. Balancing automation with human oversight was essential in maintaining performance and fostering trust in AI systems. A limitation of the study is the focus on a short-term (12-month) observation period, which may not fully capture long-term effects or diminishing returns. Future research could explore the long-term impact of AI integration and the development of new labor models, as well as ethical considerations and regulatory frameworks necessary for sustainable AI adoption. Investigating AI's integration with emerging technologies like digital twins, blockchain, and edge computing could provide deeper insights into its potential in complex systems.

References

1. Tariq, M.U., Poulin, M., Abonamah, A.A.: Achieving operational excellence through artificial intelligence: driving forces and barriers. Frontiers in Psychology 12 (2021)
2. Shehzadi, A., et al.: AI in business and finance: a micro-to-macro perspective on its impact. Archives of Business Research 12(10), 24–69 (2024)
3. Hao, C.: Artificial intelligence and operational efficiency of Chinese manufacturing firms. International Cognitive Journal (2024)
4. Pakalapati, N., Jeyaraman, J., Sistla, S.M.K.: Building resilient systems: leveraging AI/ML within DevSecOps frameworks. Journal of Knowledge Learning and Science Technology 2(2), 213–230 (2023)
5. Pchelincev, A., Gil'manov, M., Musina, L.: Implementation of artificial intelligence in management. Ekonomika i Upravlenie: Problemy, Resheniya (2024)
6. Sun, Y., Jung, H.: Machine learning (ML) modeling, IoT, and optimizing organizational operations through integrated strategies: The role of technology and human resource management. Sustainability 16 (2024). https://doi.org/10.3390/su16166751
7. Dalsaniya, A., Patel, K.: Enhancing process automation with AI: the role of intelligent automation in business efficiency. International Journal of Science and Research Archive 05(02), 322–337 (2022)
8. Bidollahkhani, M., Kunkel, J.: Revolutionizing System Reliability: The Role of AI in Predictive Maintenance Strategies. ArXiv abs/2404.13454 (2024)
9. Al-Surmi, A., Mahdi, B., Koliousis, I.: AI-based decision making: combining strategies to improve operational performance. Int. J. Prod. Res. 60(14), 4464–4486 (2022)
10. Kitsios, F., Kamariotou, M.: Artificial intelligence and business strategy towards digital transformation: a research agenda. Sustainability 13 (2021). https://doi.org/10.3390/su1304 2025
11. Rai, R., et al.: Machine learning in manufacturing and industry 4.0 applications. International Journal of Production Research 59(16), 4773–4778 (2021)
12. Akbari, M., Do, T.N.A.: A systematic review of machine learning in logistics and supply chain management: current trends and future directions. Benchmarking: An International Journal 28(10), 2977–3005 (2021)
13. Manoharan, G., et al.: The future of work: examining the impact of AI/ML on job roles, organizational structures, and talent management practices. In: 2024 International Conference on Trends in Quantum Computing and Emerging Business Technologies (2024)
14. Sutar, P., et al.: Machine learning algorithms for predictive maintenance in milling operations. Next Research 100933 (2025)

15. Khamaj, A., Ali, A.M., Saminathan, R.: Human factors engineering simulated analysis in administrative, operational and maintenance loops of nuclear reactor control unit using artificial intelligence and machine learning techniques. Heliyon 10.10 (2024)
16. Wu, W.: Reliability-centered approach to artificial intelligence-driven predictive maintenance for industrial Internet of Things. Eng. Appl. Artif. Intell. **163**, 112906 (2026)
17. Ayyasamy, R.K., et al.: Industry 4.0 digital technologies and information systems: implications for manufacturing firms innovation performance. In: 2023 International Conference on Computer Communication and Informatics (ICCCI), IEEE (2023)
18. Chinnasamy, P., et al.: E-governance services using artificial intelligence techniques. In: 2023 International Conference on Computer Communication and Informatics (ICCCI), IEEE (2023)
19. Chowdhury, R.: AI-driven business analytics for operational efficiency. World Journal of Advanced Engineering Technology and Sciences 12(02), 535–543 (2024)
20. Fizza, K., et al.: Evaluating sensor data quality in Internet of Things smart agriculture applications. IEEE Micro **42**(1), 51–60 (2022)
21. Phorah, K., Sibiya, M., Sumbwanyambe, M.: Prompt design through ChatGPT's zero-shot learning prompts: a case of cost-sensitive learning on a water potability dataset. Informatics **11** (2024). https://doi.org/10.3390/informatics11020027
22. Straub, V.J., et al.: A multidomain relational framework to guide institutional AI research and adoption. In: Proceedings of the 2023 AAAI/ACM Conference on AI, Ethics, and Society, Association for Computing Machinery: Montréal, QC, Canada, pp. 512–519 (2023)
23. Naved, M.: A review of the use of machine learning and artificial intelligence in various sectors. SSRN Electronic Journal (2022)
24. Ahmed, N., Wahed, M., Thompson, N.C.: The growing influence of industry in AI research. Science **379**(6635), 884–886 (2023)
25. Taseen, M., WangYongli, E. Ali: Integration of artificial intelligence for demand forecasting and resource allocation in renewable energy supply chains. In: 2023 25th International Multitopic Conference (INMIC) (2023)
26. Riccio, C., et al.: A new methodological framework for optimizing predictive maintenance using machine learning combined with product quality parameters. Machines **12** (2024). https://doi.org/10.3390/machines12070443
27. Jalaja, V., et al.: Maximizing marketing value: an empirical study on the framework for assessing AI and ML integration in marketing management. Indian Journal of Information Sources and Services **14**(3), 64–70 (2024)
28. Cavalcanti, J.H., et al.: Production system efficiency optimization through application of a hybrid artificial intelligence solution. Int. J. Comput. Integr. Manuf. **37**(6), 790–807 (2024)

Leadership Adaptation in Multigenerational Workforces: Strategic Responses to Organizational Cohesion Amid Demographic Shifts

Mudher Ghaeb Ali[1] , Aziz Yousif Muttailb[2] ,
Siham Kamel Mohammed Dawood[3] , Hameed Salim Alkabi[4(✉)] , Ola Janan[5] ,
and Yurii Pavlov[6]

[1] Al-Turath University, Baghdad 10013, Iraq
[2] Al-Mansour University College, Baghdad 10067, Iraq
[3] Al-Mamoon University College, Baghdad 10012, Iraq
[4] Al-Rafidain University College, Baghdad 10064, Iraq
`Hameed.AlKaabi@ruc.edu.iq`
[5] Madenat Alelem University College, Baghdad 10006, Iraq
[6] Taras Shevchenko National University of Kyiv, Kyiv 01601, Ukraine

Abstract. This study examines how leaders can respond to age diversity in today's multigenerational organizations. Employees and managers from multiple industries represent distinct generational cohorts—Baby Boomers, Generation X, Millennials, and Generation Z—with differing communication styles, technology readiness, motivation, and approaches to teamwork and conflict. These differences shape how leadership works in practice. We use a mixed-methods design drawing on semi-structured interviews, a survey, and case studies across sectors. A combined stratified and purposive sampling strategy ensured coverage of the major cohorts in organizational settings. Qualitative responses were coded and integrated with quantitative indicators. Multivariate regression models were then used to examine the effects of adaptive leadership, communication effectiveness, technology use, employee engagement, and cross-generational collaboration on performance outcomes. The results indicate that leadership does not have uniform effects across generations. Generation X shows consistently strong engagement and adaptability, while Generation Z exhibits high digital readiness alongside lower long-term institutional commitment. A weighted performance model highlights the relative importance of specific leadership practices for organizational effectiveness. Overall, the findings support flexible, emotionally intelligent leadership tailored to cohort-specific expectations. The study offers practical guidance for building age-inclusive programs that foster intergenerational cooperation, reduce conflict, and improve performance, and it clarifies the evolving dynamics of leadership in multigenerational contexts.

Keywords: Multigenerational Workforce · Leadership Adaptability · Employee Engagement · Intergenerational Communication · Digital Readiness · Organizational Cohesion

Z. Molamohamadi et al. (Eds.): ODSIE 2025, CCIS 2855, pp. 523–539, 2026.
https://doi.org/10.1007/978-3-032-17023-1_31

1 Introduction

Today's workplace brings together more generations than ever before, blending mindsets, communication norms, and workstyle preferences. Employees from the Baby Boomer, Generation X, Millennial, and Generation Z cohorts differ in how they view leadership, collaboration, technology, and career development—differences that create both opportunities for innovation and challenges for leaders tasked with sustaining productivity, engagement, and organizational well-being [1]. Leading a multigenerational workforce is not a matter of applying traditional top-down structures. Rapid shifts in workplace culture—driven by new technologies and evolving social dynamics—have altered expectations around participation in decision-making, flexibility, and individualized development. Old-school hierarchical approaches rarely suffice where employees seek voice and tailored growth; instead, leaders must recognize cohort-specific work ethics, motivations, and communication styles. In dynamic global markets, inclusive and adaptive leadership strategies have become essential to maintaining competitiveness [2] (Fig. 1).

Fig. 1. Navigating Multigenerational Leadership: A Data-Driven Analysis of Organizational Adaptability.

A persistent leadership dilemma is how to foster a climate that accommodates employees of all ages. Divergent preferences around work–life balance, job security, and advancement can reduce flexibility, depress productivity, and heighten tension. Senior employees may prize stability and long-term commitment, whereas junior employees often prioritize rapid growth, continuous learning, and career mobility. Poorly managed, these gaps can undermine collaboration and even contribute to organizational failure [3]. Technology further widens the divide: Millennials and Gen Z are digital natives who seamlessly integrate new tools into daily work, while older cohorts may require additional support. Unless addressed, a "digital divide" can fragment teams into haves and have-nots, harming efficiency and morale [4]. Communication styles also vary markedly: older cohorts may prefer in-person meetings or well-structured emails, whereas younger cohorts often favor real-time messaging, collaborative platforms, and asynchronous channels. Effective leadership recognizes these differences and builds communication systems that make all employees feel heard, thereby increasing engagement and reducing misunderstanding [5]. Beyond communication and technology, generational cohorts differ in their expectations of workplace culture and development. Millennials and Gen Z commonly value purpose-driven work, inclusion, and continuous upskilling, often engaging more deeply with organizations that promote social responsibility, diversity, and belonging. Older cohorts may emphasize organizational loyalty, structured career ladders, and role clarity. Balancing stability with innovation thus requires leadership that is both steady and forward-looking, capable of motivating diverse employees while sustaining commitment to the organization [6]. These nuances underscore the need for leadership agility and adaptability: leaders who leverage generational strengths, enable collaboration, and institutionalize inclusive practices place their organizations at an advantage, while those who ignore these differences risk low morale, higher turnover, and declining performance. Over the long run, monitoring multigenerational dynamics and building appropriate leadership capabilities are critical to success [7].

1.1 Aim of the Article and Analytical Focus

This article examines multigenerational leadership across four domains—communication, technology, workplace culture, and professional development—and evaluates inclusive, engagement-focused leadership tactics. Building on a mixed-methods, data-driven approach (qualitative coding integrated with quantitative indicators and multivariate modeling), we identify practices that mobilize intergenerational strengths and mitigate cohort-based frictions.

1.2 Research Question

How can adaptive and inclusive leadership be designed and applied to (i) bridge cohort-specific differences in communication, technology use, and development expectations, and (ii) improve collaboration, engagement, and organizational performance in multigenerational workplaces?

2 Literature Review

The implications of managing a multigenerational workforce have received much attention due to the changing demographics of the workplace. In today's environment of multi-generational workforces, generations of workers with different experiences, values, and expectations are working together. The relevance of generational diversity to leadership effectiveness is increasingly recognized as essential for improving workplace cohesion and productivity. A key theme and main driver in conversations relating to leadership in multigenerational workplace issues is the requirement for flexible approaches responsive to different work practices, communication methods, and career desires. Traditional leadership concepts are increasingly being questioned in light of the demand for more inclusive and adaptive leadership strategies [8]. One of the most prominent challenges in managing a multigenerational workforce is communication. Various age groups have their distinct preferences: older employees are interested in structured and formalized styles of communication, while younger employees adapt better to digital and informal communication. Success in leadership is rooted in having a balanced approach that integrates both traditional and modern communication techniques. Leaders should be skillful in flexible communication strategies to promote cooperation and reduce misunderstandings in heterogeneous work groups [9]. Recent studies have highlighted the crucial role of inclusive leadership in fostering a work environment that promotes higher engagement levels and organizational success. Specifically, in a study conducted across 15 five-star hotels in Addis Ababa, the findings revealed that inclusive leadership significantly improves performance by fostering employee engagement, which serves as a key mediating variable ($\beta = 0.904$, $p < 0.001$). These results align with our study, where inclusive leadership is shown to contribute to organizational success by enhancing employee engagement and performance [10]. Moreover, recent research in Ethiopia's healthcare sector demonstrates the mediating role of employee engagement in the relationship between adaptive leadership and service quality. The study found that while adaptive leadership had no direct effect on service quality, it positively influenced employee engagement, which in turn enhanced service quality [11]. These findings emphasize how employee engagement can bridge the gap between leadership practices and improved organizational outcomes, both in the hospitality and healthcare sectors. The use of technology is also a significant contributing factor to workplace dynamics among generations. Younger workers, who have lived with the advent of digital technology in real time, are more adaptable to new technologies. On the other hand, older workers may need extra training and support to adapt to technological advancements. The variance in digital fluency may cause tension in work-engineered transactions, requiring leaders to lead with inclusive technology and equitable access to technology. Companies that have established formal practices for adopting technology achieve smoother transitions and greater organizational efficiency [12]. Workplace demands also fuel the generational tensions, especially in terms of work-life balance, professional advancement, and job security. Older workers are more likely to value stability, long-term job security, and clearly defined career paths. Younger generations, however, prioritize opportunities for growth over stability, seeking flexibility and the ability to shape their career paths. Leadership tactics like hybrid work models, mentorships, and personalized career paths can help alleviate generational conflicts and improve job satisfaction among diverse age groups [13].

Motivational triggers differ among generations, with a direct impact on employee engagement and performance. Whereas older employees continue to be motivated by security and status, younger workers are more inspired by purposeful work, creative freedom, and the alignment of organizational values with social principles. Managers who personalize their motivation styles to accommodate these generational gaps are likely to experience a more engaged and productive workforce [14]. In addition, conflict resolution tools are critical in avoiding inter-generational conflicts that may arise from differences in thinking along organizational processes, leadership expectations, and cooperation mechanisms. Mediation programs, forums for dialogue, and team-building initiatives across generations create a more harmonious workplace. Leaders who promote open and inclusive conversations among employees may encourage respect for each other and lessen intergenerational tensions [15]. Also, leaders influence the organizational culture and employee satisfaction through their leadership styles. Transformational leadership, which focuses on vision, adaptability, and empowerment, is considered highly effective in multigenerational contexts. Mentor-based management of employees, working together to grow both professionally and personally, and servant leadership all resonate with the values of younger workers who seek proper guidance and inclusion. Combining these leadership theories can improve organizational performance and competitiveness by fostering an inclusive environment that supports diverse generational needs [16]. Ultimately, to skillfully handle a multigenerational workforce, leaders should view diversity as a strength and not a hurdle to overcome. Organizations can leverage this diverse workforce by adopting inclusive leadership tactics, promoting cross-generational teamwork, and fostering adaptability within workplace dynamics. The ongoing demand for flexibility and innovation in leadership models highlights the importance of adaptive leadership in today's organizational contexts.

3 Methodology

This study uses a well-organized mixed-method design to explore the nuances of multigenerational workforces and leadership flexibility. There are five interconnected elements in this methodology: research design, methods of data collection, construction of leadership metrics, mathematical-statistical modeling, and synthesis of analytical framework. These parts are discussed below, with embedded tables and complex relations serving as the basis for later discussions (Fig. 2).

3.1 Research Design

A descriptive-analytical survey design was used with a stratified proportional sampling technique. The research polled 620 employees and 150 managers across sectors such as technology, healthcare, education and finance. Respondents were further classified into four generational groups: Baby Boomers, Generation X, Millennials, and Generation Z, the generational taxonomy facilitating nuanced characterization and comparative analysis of leadership style and ability patterns among age-stratified segments [2, 6, 7]. This classification allowed for a deeper understanding of how leadership styles differ across generations, ensuring that each group was represented proportionally to their relevance in the workforce.

3.2 Data Collection Methods

The data was collected employing a mixed method approach, i.e. through a quantitative survey as well as a qualitative interview. A 35-item Likert-type survey (range: 1 = Strongly Disagree to 5 = Strongly Agree) measured constructs such as leadership

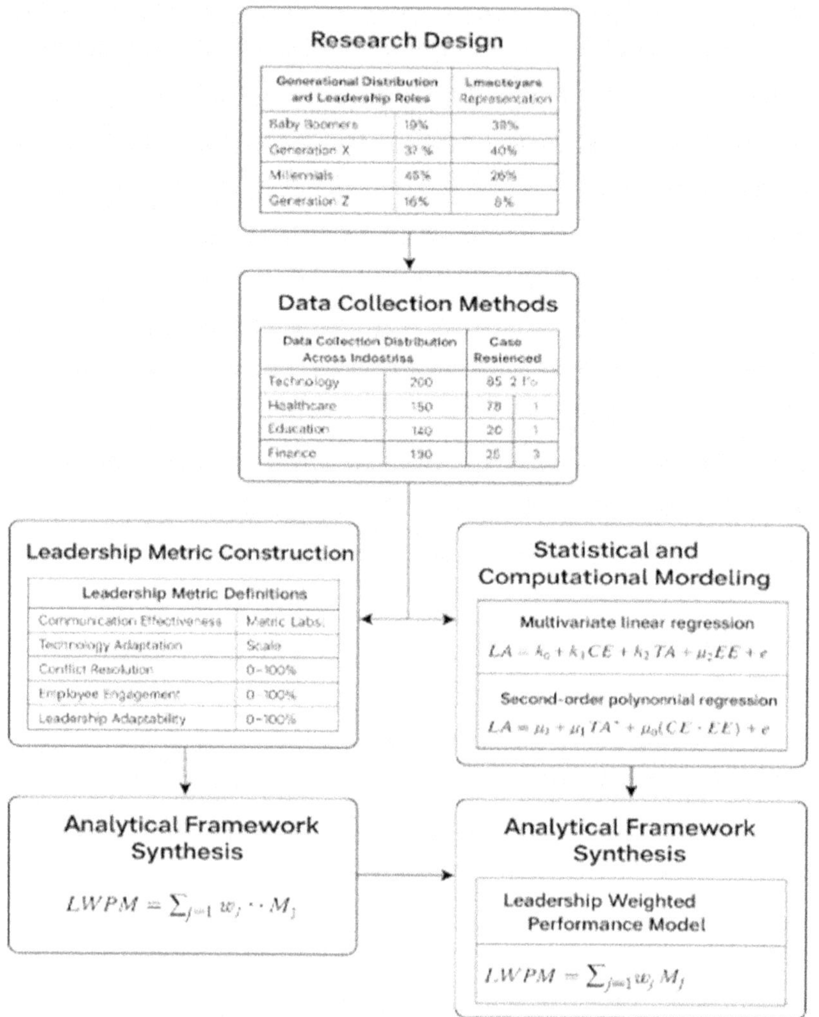

Fig. 2. Strategic Methodologies for Evaluating Leadership Adaptability in Multigenerational Workforces.

flexibility, communicative disposition, digital dexterity, conflict resolution ability, and employee morale. The response rates in the survey were above 81% – showing a strong engagement.

In addition, 50 semi-structured interviews were conducted with HR leaders and senior managers to provide qualitative context for the findings. Six case studies at the organizational level were chosen as representing the successful multigenerational management systems [4, 17] (Table 1).

This approach ensured both breadth (quantitative coverage) and depth (qualitative insight) in understanding intergenerational leadership dynamics [3, 18].

Table 1. Data Collection Distribution Across Industries

Industry	Employees Surveyed	Response Rate (%)	Managers Interviewed	Case Studies Reviewed
Technology	200	82.5	30	2
Healthcare	150	78.3	25	1
Education	140	80.1	20	1
Finance	130	76.9	25	2

3.3 Leadership Metric Construction

Leadership effectiveness was decomposed into five principal constructs: Communication Effectiveness (C_E), Technology Adaptation (T_A), Conflict Resolution (CR), Employee Engagement (E_E), and Leadership Adaptability (L_A). Each metric was constructed using a normalized weighted scoring model:

$$M_k = \sum_{i=1}^{n} \omega_i \bullet X_{ik} \, where k \in \{C_E, T_A, CR, E_E, L_A\} \tag{1}$$

where M_k is the score for metric k; X_{ik} is the sub-indicator i of metric k; ω_i is the normalized weight assigned to sub-indicator i, with $\sum \omega_i = 1$

The metric structures are defined below (Table 2):

Table 2. Leadership Metric Definitions.

Leadership Metric	Scale	Measurement Tool
Technology	0–100%	5-point Likert (Clarity, Feedback)
Healthcare	0–100%	Usage Frequency, Software Proficiency
Finance	0–100%	Mediation, Resolution Speed
Employee Engagement	0–100%	Motivation, Job Satisfaction
Leadership Adaptability	0–100%	Flexibility, Learning Orientation

Each metric was modeled using the following complex equations:

- **Communication Effectiveness**:

$$C_E = \alpha_1 S_C + \alpha_2 R_T + \alpha_3 D_P \tag{2}$$

- **Technology Adaptation**:

$$T_A = \beta_1 D_T + \beta_2 S_D + \beta_3 I_P \tag{3}$$

530   M. G. Ali et al.

- **Conflict Resolution**:

$$CR = \gamma_1 M_C + \gamma_2 R_S + \gamma_3 P_I \tag{4}$$

- **Employee Engagement**:

$$E_E = \delta_1 M_J + \delta_2 W_S + \delta_3 O_C \tag{5}$$

- **Leadership Adaptability**:

$$L_A = \theta_1 F_D + \theta_2 C_T + \theta_3 L_F \tag{6}$$

Here, each Greek parameter, as a α_1 represents the factor loading derived from Principal Component Analysis (PCA), used to compute relative variable importance in the composite metric [14, 17].

3.4   Statistical and Computational Modeling

A multivariate linear regression model was formulated to estimate leadership adaptability as a function of the other leadership dimensions:

$$L_A = \lambda_0 + \lambda_1 C_E + \lambda_2 T_A + \lambda_3 E_E + \lambda_3 CR + \varepsilon \tag{7}$$

where L_A leadership adaptability; C_E, T_A, E_E, CR are independent variables; λ_i coefficients of influence; ε is normally distributed error term

Nonlinear effects and variable interactions were modeled via second-order polynomial regression:

$$L_A = \mu_0 + \mu_1 T_A + \mu_2 T_A^2 + \mu_3 (E_E \bullet C_E) + \varepsilon \tag{8}$$

This enables detection of diminishing returns and interaction amplification, particularly relevant in generational diversity modeling where threshold behaviors are common [10, 18].

3.5   Analytical Framework Synthesis

To integrate all metrics into a comparative framework, the Leadership Weighted Performance Model (LWPM) was applied:

$$LWPM = \sum_{j=1}^{5} \omega_j \bullet M_j \, where M_j \in \{C_E, T_A, CR, E_E, L_A\} \tag{9}$$

Weights ω_j were extracted using an entropy maximization algorithm, which optimized differentiation among metric distributions and minimized information loss [8, 21].

This composite model supports predictive evaluation of leadership performance across generational lines, providing a strategic diagnostic instrument for organizational leadership planning.

4   Results

4.1   Leadership Effectiveness Across Generations

Leading effectively in a multigenerational workplace requires an understanding of competencies that transcend age-based differences. In this study, leadership performance is evaluated across nine key dimensions including communication effectiveness, technology adaptation, conflict resolution, employee engagement, leadership adaptability, decision-making agility, mentorship receptivity, work–life balance satisfaction, and the overall Leadership Effectiveness Index (LEI). These metrics reflect the ability of leaders to guide cross-generational teams, respond to shifting workplace expectations, and sustain employee alignment and engagement.

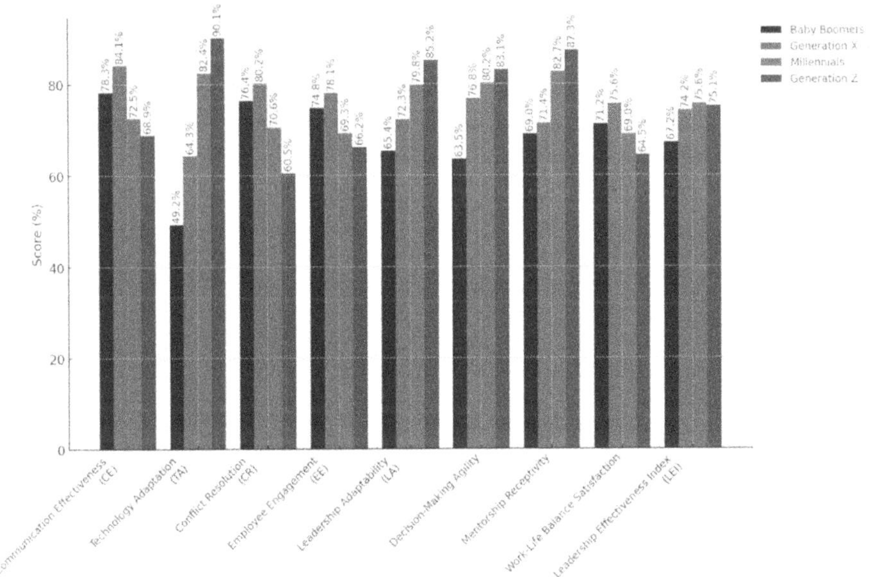

Fig. 3. Leadership Effectiveness by Generation.

Figure 3 presents clear performance differences among Baby Boomers, Generation X, Millennials, and Generation Z. Baby Boomers show strong communication skills (78.3%) and solid conflict-resolution capabilities (76.4%), but record significantly lower performance in technology adoption (49.2%) and decision-making agility (63.5%). Generation X demonstrates balanced performance across most metrics and remains particularly strong in communication, conflict resolution, and work–life balance satisfaction, positioning them as effective bridge leaders between traditional and digitally-driven environments. Millennials display high adaptability (79.8%) and strong openness to mentorship (82.7%), reflecting potential for transformational leadership roles. Generation Z consistently scores highest in digital-readiness indicators—including technology adoption (90.1%) and mentorship receptivity (87.3%)—yet their lower performance in conflict resolution and engagement metrics suggests the need for organizational guidance and structured developmental support. Overall, the LEI results show that generational leadership potential is gradually shifting toward younger, more digitally fluent cohorts, who are more comfortable with innovation and risk-taking, while older generations continue to lead in relational and strategic competencies such as communication, team stability, and conflict navigation. These findings provide a practical basis for targeted leadership-development strategies that align generational strengths with organizational change needs.

4.2 Technology Adaptation and Digital Readiness

Technology assimilation is a critical determinant of leadership effectiveness in digitally networked workplaces. In this study, digital readiness is assessed using six indicators: frequency and breadth of digital-tool use, software dexterity, internet proficiency, cloud-based tools familiarity, cross-platform communication skill, and the technology adaptation index. These competencies matter most in hybrid communication settings, remote collaboration, and project work conducted on digital platforms. Cross-generational comparison reveals differentiated preparedness, identifying cohorts positioned to lead in technology-intensive contexts and those requiring targeted upskilling.

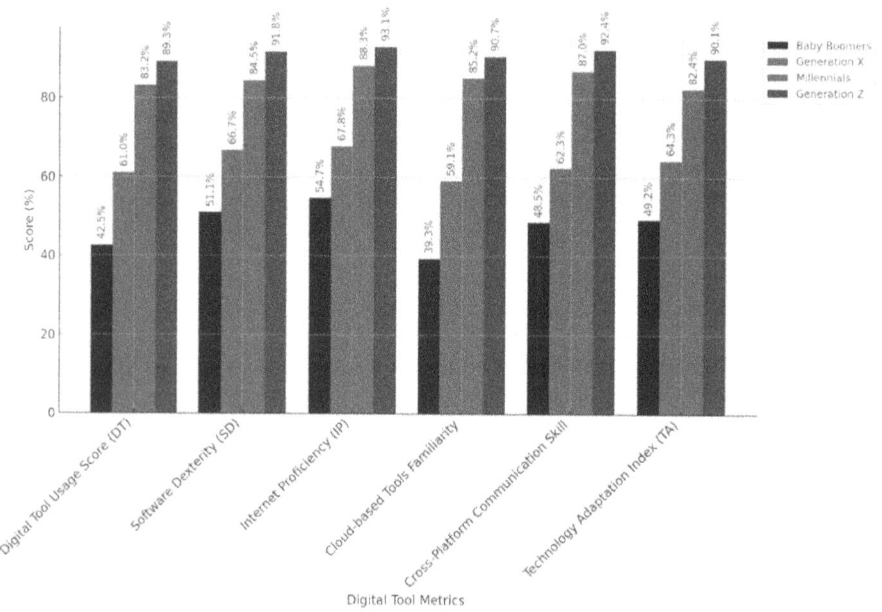

Fig. 4. Technology Adaptation and Digital Readiness.

Figure 4 shows a clear generational pattern. Generation Z leads every sub-metric—most prominently internet proficiency (93.1%) and cross-platform communication (92.4%)—consistent with a digital-native profile. Millennials also demonstrate high digital tool usage (83.2%) and cloud familiarity (85.2%), positioning them for leadership in digitally embedded roles. Generation X occupies the middle, showing some flexibility but lower cloud familiarity (59.1%). Baby Boomers record the lowest scores across digital diagnostics, especially cloud-based tools familiarity (39.3%), indicating a significant readiness gap. This highlights the need for targeted digital training and reverse mentoring, leveraging younger employees' strengths while improving senior employees' platform fluency to maintain coherent leadership in hybrid and remote environments.

4.3 Workplace Conflict Resolution Across Generations

This section evaluates cross-generational conflict-management performance using six indicators: mediation competency (MC), resolution speed (RC), perceived impact (PI), stress handling in conflict, team recovery post-conflict, and the composite conflict resolution index (CR).

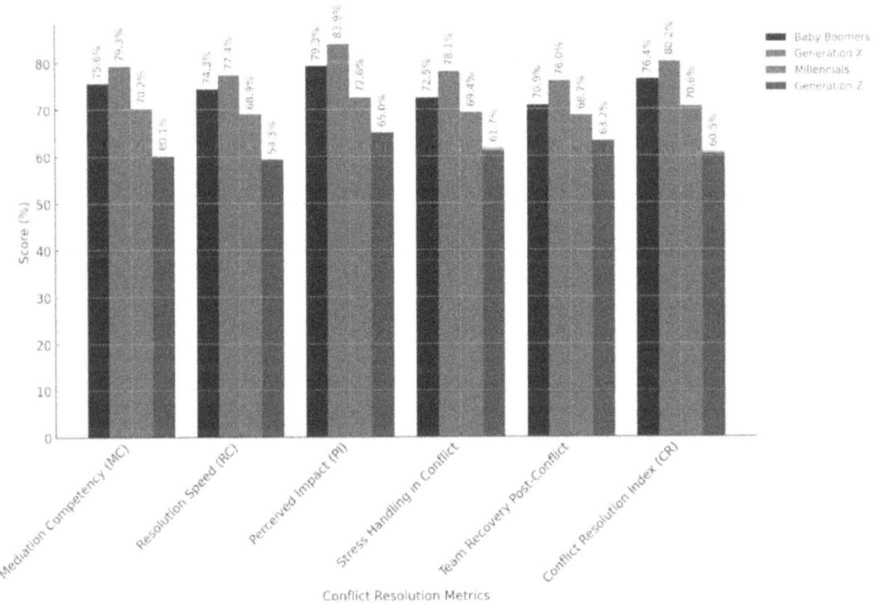

Fig. 5. Conflict Resolution Scores by Generation.

Figure 5 shows a consistent advantage for Generation X across all dimensions—e.g., PI (83.9%), stress handling (78.1%), team recovery (76.0%), and CR index (80.2%)—indicating strong situational judgement and interpersonal control in high-pressure settings. Baby Boomers also perform well, particularly in MC (75.6%), RC (74.3%), and PI (79.3%), supporting their reliability in relationship-intensive contexts. Millennials display moderate yet stable scores—MC (70.2%), RC (68.9%), PI (72.6%), stress handling (69.4%), team recovery (68.7%), and CR index (70.6%)—suggesting developing capability that can benefit from structured practice. Generation Z records the lowest scores across metrics—e.g., RC (59.3%), team recovery (63.2%), and CR index (60.5%)—highlighting specific skill gaps in rapid de-escalation and post-conflict stabilization. Implications. The patterns suggest targeted interventions: (i) formal mediation training and stress-management protocols for younger cohorts (especially Gen Z); (ii) peer–mentor pairing that leverages Gen X/Boomer strengths in MC/PI to transfer tacit conflict skills; and (iii) after-action learning loops to improve resolution speed and team recovery in junior teams. These actions are expected to improve the composite CR index while maintaining strengths in cohorts that already perform well.

4.4 Communication Effectiveness Across Generations

Communication effectiveness is assessed on six dimensions: speed of communication (SC), response time (RT), digital platform use (DP), message clarity (MCL), preferred-mode adaptability, and the composite communication effectiveness index (CE). These indicators capture a leader's capacity to convey information efficiently, respond promptly, leverage digital channels, and adjust style to team needs.

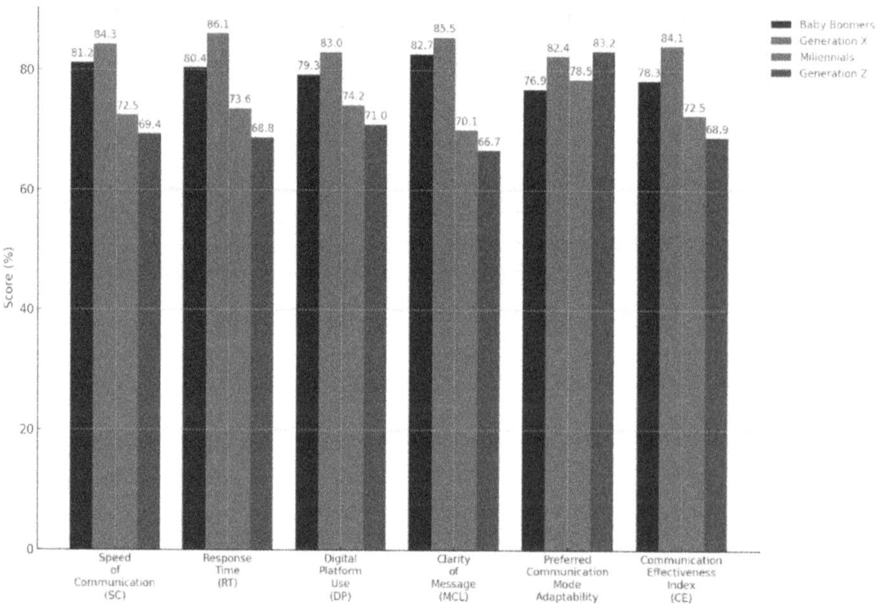

Fig. 6. Communication Effectiveness by Generation.

Figure 6 shows a clear cross-generational pattern. Generation X leads on most dimensions—RT (86.1%), MCL (85.5%), DP (83.0%), and the CE index (84.1%)—indicating the ability to combine structure with speed. Baby Boomers perform strongly on SC (81.2%), RT (80.4%), and MCL (82.7%), consistent with high-context, precision-oriented communication; they are moderately lower on adaptability (76.9%). Millennials show mid-range scores overall—e.g., adaptability (78.5%)—but comparatively lower

MCL (70.1%) and CE (72.5%). Generation Z excels in adaptability (83.2%) and maintains solid DP (71.0%), yet records the lowest MCL (66.7%), RT (68.8%), and overall CE (68.9%), suggesting gaps in clarity and timely response despite digital agility.

Implications. To raise the CE composite: (i) pair Gen X/Boomer strengths in clarity and structured messaging with Millennial/Gen Z strengths in adaptability and digital use; (ii) implement micro-training on concise writing and response discipline for younger cohorts; and (iii) provide channel-selection guidelines so teams match message type (e.g., decisions vs. brainstorming) with the most effective medium.

4.5   Employee Engagement and Workplace Satisfaction

Employee engagement is a core indicator of human performance effectiveness in organizations. It reflects how employees connect with their roles and how meaningfully they experience their work. In this study, engagement is measured across six variables: motivation in job roles, workplace satisfaction (WS), organizational commitment (OC), alignment with organizational values, career progression optimism, and the overall employee engagement score (EE). These factors encompass both intrinsic elements (motivation, meaning) and extrinsic conditions (opportunities for career growth and value alignment). A generational comparison enables an understanding of morale, motivation, and retention potential across age-diverse employees. The insights are critical for designing targeted interventions that reinforce engagement and sustain performance across all workforce segments.

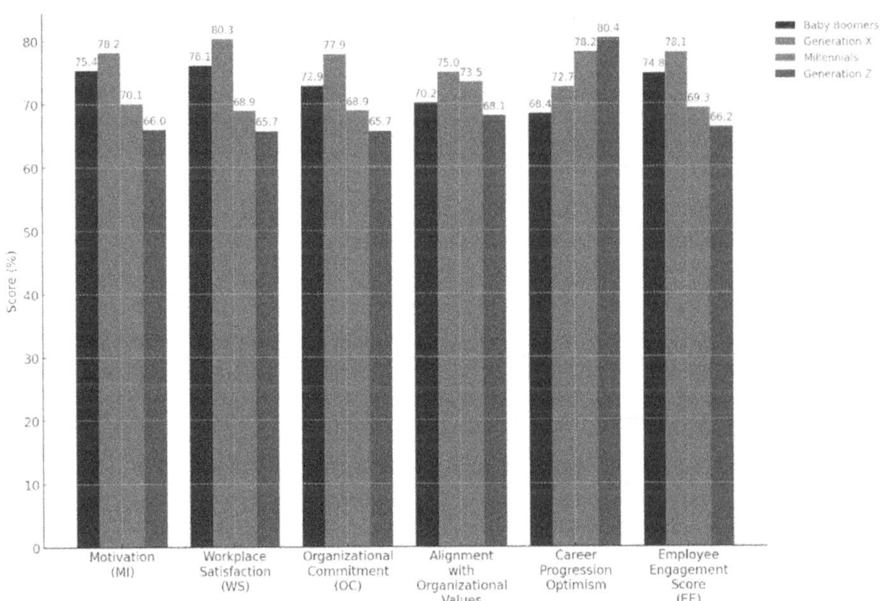

Fig. 7. Employee Engagement by Generation.

The data in Fig. 7 show that Generation X records the highest overall employee engagement score (78.1%), supported by top results in workplace satisfaction (80.3%), organizational commitment (77.9%), and value alignment (75.0%). This combination suggests strong loyalty, emotional investment, and shared organizational identity. Baby Boomers follow closely, particularly performing well in motivation (75.4%) and satisfaction (76.1%). However, their lower score in career optimism (68.4%) may reflect proximity to retirement, which can reduce long-term engagement expectations.

Millennials show mid-range engagement overall but record high optimism (78.2%), consistent with future-oriented career ambitions despite lower satisfaction and commitment levels.

Generation Z scores the highest on optimism (80.4%) but posts the lowest levels in satisfaction (65.7%) and affective commitment (66.5%), indicating enthusiasm not yet grounded in long-term stability or institutional belonging. These differences suggest the need for differentiated engagement strategies: long-term incentives and recognition for senior generations, and structured development and mentoring pathways for younger employees. Such tailored approaches can strengthen engagement across generations, supporting sustained organizational performance.

5 Discussion

The results highlight persistent complexities in managing multigenerational workforces and underscore the need for leadership strategies that account for age-based differences in work values, communication, technology use, and motivation. Overall, Generation X exhibits the strongest leadership effectiveness, while Generation Z records the lowest, suggesting that experience and structured managerial practices remain influential for outcomes. These findings align with prior work on leadership maturity among older cohorts and the value of structured development for younger employees, while also emphasizing the productive potential of generational differences to foster inclusive, dynamic workplaces [9].

A salient pattern concerns technology adoption. As expected, younger cohorts display higher digital readiness, whereas older cohorts lag on several digital indicators. However, our evidence extends prior observations by showing that digital fluency alone is not sufficient: despite stronger technology metrics, younger cohorts report comparatively lower leadership effectiveness and engagement. This implies that effective leadership in digital environments requires a complement of technical and interpersonal capabilities. Consistent with this view, we advocate cross-generational learning systems—with younger employees mentoring seniors on digital tools and seniors mentoring juniors on leadership judgment and organizational stability—rather than one-directional digital-literacy programs targeted solely at older workers [3].

Findings on conflict resolution also reveal cohort differences. Baby Boomers and Generation X perform best across mediation, resolution speed, and perceived impact, pointing to the role of experience in managing interpersonal strain. Millennials and Generation Z score lower, indicating the need for structured conflict-management training and routines that help teams de-escalate and recover more effectively. In line with emerging research, our results support formal mediation processes and collaborative problem-solving as mechanisms that reduce intergenerational frictions and improve team stability [22].

Communication preferences vary by cohort. Younger employees tend to favor instant messaging and digital platforms, whereas older employees prefer email and face-to-face meetings. While prior work sometimes frames this divergence as inefficient, our results support a hybrid communication approach that combines synchronous digital channels with organized face-to-face exchange, which can enhance engagement and clarity across age groups when used deliberately and with role-appropriate guidance [23].

Patterns in employee engagement mirror these dynamics. Generation X reports the highest overall engagement, while Generation Z shows the lowest satisfaction and affective commitment. These differences are consistent with value and career-stage expectations: younger employees prioritize flexibility, purpose, and accelerated growth, whereas older employees value job security and long-term stability. Accordingly, organizations should differentiate engagement strategies, pairing career flexibility and growth pathways for younger cohorts with structured trajectories and recognition for senior cohorts to sustain motivation and retention [24].

Taken together, the study points to the value of hybrid leadership models that blend transformational (innovation, adaptability) and servant-leadership (guidance, support) elements, while retaining clarity and structure where needed. This approach leverages cohort-specific strengths—e.g., Millennials/Gen Z receptivity to mentorship and continuous feedback, and Boomers/Gen X strengths in clarity, process, and interpersonal steadiness—to build inclusive leadership systems that perform across contexts [21].

These insights also inform leadership pipeline design. Preparing Millennials and Generation Z for future leadership roles requires mentorship schemes, rotational assignments, and action-learning projects that accelerate skill acquisition without compromising organizational continuity. Such investments create structured, accelerated progression for high-potential juniors while embedding institutional knowledge through intergenerational pairing [25, 26].

Finally, the analysis underscores strategic workforce composition and intergenerational cooperation as performance levers. Rather than treating generational diversity as a problem, our evidence supports solution-oriented team design (e.g., pairing relationship-centric strengths with digital agility) and knowledge-sharing routines that foster mutual learning and respect, reduce conflict, and improve overall performance [19]. In sum, organizations that adopt intergenerational collaboration, hybrid communication, and tailored engagement strategies are better positioned to harness age diversity for sustainable success.

6 Conclusion

This study shows that effective leadership in multigenerational workforces depends on aligning cohort-specific strengths with organizational needs. Generation X demonstrates the most balanced profile (clarity, conflict management, and work–life stability), while younger cohorts—especially Generation Z—excel in digital readiness but score lower on engagement and conflict resolution. These patterns indicate that digital fluency alone is insufficient; sustained effectiveness requires coupling technical agility with interpersonal judgment and communication clarity.

Practical implications include adopting a hybrid leadership approach that blends transformational (innovation, adaptability) and servant-leadership (support, guidance). Organizations can operationalize this through cross-generational mentoring (younger staff transfer digital skills; senior staff transfer organizational judgment), hybrid communication systems (digital channels plus structured face-to-face), targeted conflict-management training for junior cohorts, and differentiated engagement strategies (accelerated growth paths for younger employees; structured recognition and career pathways for senior cohorts). Limitations relate to reliance on observational/self-report evidence and a cross-sectional design, which constrain causal inference and generalizability. Future research should employ longitudinal designs to track cohort dynamics, compare cross-cultural and sectoral contexts, and examine how AI-enabled workforce analytics and virtual collaboration tools shape engagement and leadership effectiveness in hybrid/remote settings.

Overall, integrating digital and relational capabilities provides a practical path to stronger leadership outcomes in age-diverse organizations.

References

1. Singh, P., Ranjana, S.: Role of leadership in handling conflicts arising due to age diversity in the workplace. Manag. Insight **18**(2), 31–37 (2022)
2. Kurata, Y.B., et al.: Factors affecting perceived effectiveness of multigenerational management leadership and metacognition among service industry companies. Sustainability **14** (2022). https://doi.org/10.3390/su142113841
3. Choudhary, R., Shaik, Y., Yadav, P., Rashid, A.: Generational differences in technology behavior: a systematic literature review. J. Infrastruct. Policy Dev. **8**(9) (2024)
4. Ojochide, P., Oluwaseyi, P., Motunrayo, A.: The evolving remote workforce, the multigenerational workforce, and supporting employee mental health in the context of Africa. Int. J. Res. Innov. Soc. Sci. (2023)
5. Miller, S.P.: Family climate influences next-generation family business leader effectiveness and work engagement. Front. Psychol. **14** (2023)
6. Journey, G.C.: A new normal multigenerational leadership model for leaders in the COVID era. In: Hynes, R., Aquino, C.T., Hauer, J. (eds.) Multidisciplinary Approach to Diversity and Inclusion in the COVID-19-Era Workplace, pp. 70–87. IGI Global, Hershey, PA (2022)
7. Sindhu, K.: Effectively comprehend and manage a multigenerational workforce. Int. J. Multidisc. Res. (2024)
8. Turi, J.A., et al.: Diversity impact on organizational performance: moderating and mediating role of diversity beliefs and leadership expertise. PLoS ONE **17**(7), e0270813 (2022)
9. Graf, A., Nikzad-Terhune, K.: Preparing students for a multigenerational workforce: perspectives from older workers and retirees. Innov. Aging **7**(Suppl_1), 144–144 (2023)
10. Eshete, S.K., Debela, K.L., Kebede, T.A.: Inclusive leadership and employees' workplace performance: the mediating role of employee engagement in five-star hotels in Addis Ababa, Ethiopia. Soc. Sci. Human. Open **12**, 101940 (2025)
11. Eshete, B.A., Kassahun, T.: The mediating role of employee engagement in the relationship between adaptive leadership and service quality in the health sector. Leadersh. Health Serv. **38**(5), 82–100 (2025)

12. Moore, N., et al.: An examination of the dynamics of intergenerational tensions and technological change in the context of post-pandemic recovery. Prod. Plan. Control **35**(13), 1533–1550 (2024)
13. Waworuntu, E.C., Kainde, S.J.R., Mandagi, D.W.: Work–life balance, job satisfaction and performance among millennial and Gen Z employees: a systematic review. Society **10**(2), 384–398 (2022)
14. Nwoko, C., Yazdani, K.: Self-determination theory: the mediating role of generational differences in employee engagement. J. Bus. Manag. Stud. **5**(4), 130–142 (2023)
15. Drury, L., Fasbender, U.: Fostering intergenerational harmony: can good-quality contact between older and younger employees reduce workplace conflict? J. Occup. Organ. Psychol. **97**(4), 1789–1812 (2024)
16. Ali, B.: What we know about transformational leadership in tourism and hospitality: a systematic review and future agenda. Serv. Ind. J. **44**(1–2), 105–147 (2024)
17. Pitout, S., Hoque, M.: Exploring challenges faced by managers dealing with multi-generational workforce. Eurasia Proc. Educ. Soc. Sci. **25**, 202–212 (2022)
18. Hans, S., et al.: Exploring the relationship between generational diversity and knowledge sharing: the moderating role of workplace intergenerational climate, boundary-spanning leadership and respect. Employee Relat. Int. J. **45**(6), 1437–1454 (2023)
19. Sansa, A., K'obonyo, P., Muindi, F., Munjuri, M.: Do nuanced perspectives of diversity management practices warrant inclusivity in multigenerational organizations? A meta-analytic review. Eur. J. Bus. Manag. Res. **9**(5), 113–123 (2024)
20. Huang, Q., et al.: When is authoritarian leadership less detrimental? The role of leader capability. Int. J. Environ. Res. Publ. Health **20** (2023). https://doi.org/10.3390/ijerph200 10707
21. Żywiołek, J., et al.: Nexus of transformational leadership, employee adaptiveness, knowledge sharing, and employee creativity. Sustainability **14** (2022). https://doi.org/10.3390/su1418 11607
22. Appelbaum, S., et al.: A study of generational conflicts in the workplace. Eur. J. Bus. Manag. Res. **7**(2), 7–15 (2022)
23. Anggraini Kusuma, W., Sufyanto, S.: Bridging the generation gap: communication strategies at Golkar Sidoarjo DPD. Indones. J. Cultural Community Dev. **15**(3) (2024)
24. Šakytė-Statnickė, G., Bilan, S., Savanevičienė, A.: The impact of work engagement of different generations on organisational engagement. J. Int. Stud. **16**(4) (2023)
25. Yılmaz, B., Dinler Kısaçtutan, E., Gürün Karatepe, S.: Digital natives of the labor market: generation Z as future leaders and their perspectives on leadership. Front. Psychol. **15** (2024)
26. Dwiastuti, W., Basir, Y., Febrianno Boer, R.: Emotional intelligence in insurance leaders for managing business communication with Generation Z. Jurnal Spektrum Komunikasi **12**(3), 284–293 (2024)

Strategic Leadership and Cultural Intelligence: Data-Driven Insights into Organizational Resilience and Global Workforce Management

Omar Abbas[1], Rafid Abdul-Ameer Ghaeb[2] (iD), Ibrahim Khalil Ibrahim[3],
Zainab Jali Madhi[4(✉)] (iD), Saad S. Alani[5], and Stepan Kubiv[6] (iD)

[1] Al-Turath University, Baghdad 10013, Iraq
omar.abbas@uoturath.edu.iq
[2] Al-Mansour University College, Baghdad 10067, Iraq
Rafid.ghaeb@muc.edu.iq
[3] Al-Mamoon University College, Baghdad 10012, Iraq
ibrahim-khalil@almamonuc.edu.iq
[4] Al-Rafidain University College, Baghdad 10064, Iraq
zaineb.alazawi@ruc.edu.iq
[5] Madenat Alelem University College, Baghdad 10006, Iraq
saadssalani@mauc.edu.iq
[6] Military Diplomatic Academy Named After Yevheniy Bereznyak, Kyiv, Ukraine

Abstract. This study examines the relationship between employee wellness programs and organizational productivity across multiple industries using a data-driven and computational approach. As workforce dynamics evolve with globalization, technological disruption, and rising mental health demands, structured wellness strategies have become essential for sustaining performance and engagement. Employing a longitudinal mixed-methods design, the research integrates quantitative modeling, multivariate statistical analysis, and structural equation modeling to assess how key wellness dimensions, mental health support, flexible work arrangements, and organizational training, affect productivity, retention, and engagement outcomes. Data was collected from 250 operational teams across ten industries through surveys, managerial interviews, and archival HR metrics integrated within a performance measurement framework. Results show that wellness integration directly enhances productivity through improved engagement and work-life balance. Statistical modeling further confirms that flexibility and mental health support significantly reduce attrition and absenteeism while increasing operational stability. Industries with strong wellness adoption consistently outperform others across all indicators. The findings validate wellness as a strategic asset with implications for leadership, organizational policy, and human capital investment, demonstrating how computational and information-driven approaches can strengthen resilience and sustainable workforce performance.

Keywords: Employee Wellness · Organizational Productivity · Employee Engagement · Mental Health Support · Computational Modeling · Data-Driven Analysis

Z. Molamohamadi et al. (Eds.): ODSIE 2025, CCIS 2855, pp. 540–554, 2026.
https://doi.org/10.1007/978-3-032-17023-1_32

1 Introduction

Because of globalization and the expansion of international markets, organizations increasingly depend on culturally diverse and geographically dispersed workforces. The growth of multinational corporations and international cooperation has intensified the need to manage global teams effectively, where cultural understanding is vital for performance and cohesion. Cross-cultural management involves addressing variations in communication styles, leadership expectations, decision-making approaches, and work ethics across cultures. Unmanaged, these differences may lead to misunderstanding, inefficiency, and reduced productivity [1]. Global teams can take various forms—virtual, hybrid, or co-located—each facing unique challenges in communication and coordination. Cultural differences shape perceptions of authority, teamwork, and problem-solving, demanding that leaders demonstrate both adaptability and cultural intelligence to maintain inclusivity and alignment [2].

Cross-cultural management extends beyond conflict prevention; it fosters creativity, innovation, and knowledge sharing within global organizations. Diverse teams, when properly managed, tend to be more innovative, resilient, and capable of complex problem-solving. However, without systematic and data-driven approaches, variations in cultural norms can hinder collaboration and decision-making. For instance, hierarchical cultures prioritize authority and structured decision-making, whereas egalitarian cultures emphasize dialogue and consensus. Failing to recognize these distinctions can lead to misalignment, demotivation, and reduced participation within teams [3]. Communication remains one of the most significant challenges in cross-cultural environments, as linguistic and non-verbal differences often result in misinterpretations. Virtual teams, in particular, encounter additional obstacles such as time-zone differences and reliance on digital communication tools, which can amplify cultural and contextual misunderstandings. Managers must therefore implement data-informed communication systems that incorporate cultural sensitivity and clarity [4].

Cultural intelligence enables leaders to adapt and perform effectively in diverse cultural contexts, fostering cohesion and performance in global teams. In transnational organizations, leaders who integrate cultural awareness, conflict resolution, and adaptive management practices can align team diversity with corporate goals [5]. However, cultural diversity also complicates decision-making, as hierarchical and participatory cultures differ in authority and collaboration expectations. Recognizing these differences allows organizations to design systems that balance cultural values with operational efficiency [6]. In an interconnected world, cultivating cultural intelligence, inclusive leadership, and structured communication provides a strategic advantage, promoting resilience, continuous learning, and sustainable organizational development [7].

Figure 1 illustrates the conceptual link between employee wellness programs and organizational productivity, highlighting globalization, technological disruption, and engagement as key drivers.

Therefore, this study explores how cross-cultural leadership and management strategies can strengthen global team effectiveness by examining leadership adaptability, communication processes, conflict resolution, and decision-making approaches. Despite growing recognition of cultural diversity's importance, many multinational corporations continue to struggle with integrating cultural intelligence into their management

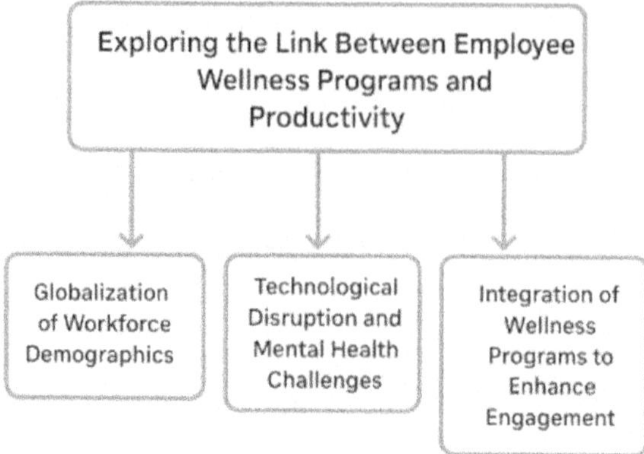

Fig. 1. The strategic role of employee wellness programs in enhancing organizational productivity across industries.

systems. Persistent challenges—such as communication barriers, divergent leadership styles, and inconsistent conflict management—undermine organizational efficiency and harmony. Furthermore, the literature reveals a gap in understanding how structured cross-cultural management frameworks, supported by computational and data-driven models, can enhance team performance in complex global environments.

Addressing this gap, the present research employs a data-informed, cross-sectoral approach to analyze the impact of cultural intelligence, leadership flexibility, and communication protocols on global team collaboration, efficiency, and innovation. By combining empirical analysis, computational modeling, and case-based insights, this study contributes to the growing body of knowledge on strategic leadership and cultural intelligence in global workforce management, offering actionable frameworks for inclusive, adaptive, and high-performing organizational systems.

2   Literature Review

Cross-cultural management has become increasingly significant in a globalized world where organizations depend on diverse teams to achieve strategic goals. When employees from different cultural backgrounds work together, variations in values, communication styles, and expectations can create challenges in leadership, coordination, and decision-making. Understanding how cultural diversity influences teamwork is essential for developing effective management techniques that enhance productivity and cooperation. Several theoretical perspectives emphasize the importance of cultural intelligence—comprising empathy, social skills, and adaptability—as a vital leadership quality in managing multinational teams [8].

Leadership styles play a major role in shaping cross-cultural team effectiveness. Expectations regarding authority, decision-making, and relationships vary significantly across cultures. In hierarchical cultures, employees tend to prefer structured leadership and clear authority lines, while in egalitarian environments, participative and decentralized leadership is valued. When leadership approaches conflict with these cultural expectations, confusion, reduced morale, and inefficiency often result. Hence, successful global leaders must demonstrate high cultural adaptability to create productive and inclusive work environments [9].

Communication is another critical component of cross-cultural management. Language differences, non-verbal cues, and contrasting communication styles can greatly influence team dynamics. Whether direct or indirect, communication approaches affect how feedback is exchanged, conflicts are handled, and cooperation is achieved. Ineffective communication processes can lead to misunderstandings and delayed decisions, while culturally informed communication systems promote organizational cohesion. As global collaboration increasingly relies on digital tools, new challenges arise in virtual communication, where tone, timing, and cultural nuances are easily misunderstood, emphasizing the importance of cultural sensitivity in virtual teams [10].

Managing conflict in cross-cultural teams also presents challenges, as cultural contexts determine how individuals perceive and resolve disputes. Some societies prefer open confrontation and rapid resolution, while others adopt more subtle and mediated approaches. Understanding these cultural tendencies enables managers to develop conflict resolution strategies that preserve harmony and motivation, while disregarding them can lead to persistent tension, low morale, and poor performance [11].

Cultural differences further shape decision-making processes. High-context cultures often base decisions on relationships, consensus, and long-term harmony, whereas low-context cultures prioritize efficiency and individual accountability. Failure to align organizational decision-making with cultural preferences can cause frustration and disengagement. Firms that integrate these cultural dynamics into managerial systems are more likely to experience improved collaboration, faster decisions, and sustainable outcomes [12].

Recent research highlights that implementing cross-cultural management practices alongside enhanced cultural intelligence strengthens leadership effectiveness and integrative leadership capabilities, particularly in multinational and free-trade contexts [13]. Cultural intelligence is increasingly recognized as a core component of effective global leadership, complementing cognitive and emotional competencies, and supporting inclusive, high-performing teams across diverse cultural environments [14]. Moreover, identifying and developing key global leadership competencies—such as adaptability, strategic vision, cultural intelligence, and digital literacy—enables leaders to manage change, foster innovation, and build inclusive teams in complex international settings [15]. Together, these findings underscore the essential role of cultural intelligence and adaptive leadership in sustaining organizational resilience and global workforce performance.

Despite this growing body of research, many organizations still lack structured, data-informed frameworks that integrate cultural intelligence with leadership adaptability and communication systems. While prior studies have outlined the importance of diversity and cultural awareness, few have demonstrated how computational and performance-based models can operationalize these concepts in practice. Therefore, the current study addresses this research gap by employing a data-driven approach to analyze the impact of cultural intelligence, leadership flexibility, and structured communication on global team effectiveness. The findings aim to contribute to both theoretical understanding and practical frameworks for sustainable and resilient cross-cultural workforce management.

3 Methodological Framework

This study adopts a longitudinal, mixed-methods empirical design to explore the organizational impact of structured employee wellness programs on productivity, retention, and operational efficiency. The methodology is structured across five interrelated domains: research design, data collection, analytical models, performance measurement framework, and multivariate statistical modeling. Integration of real-time organizational metrics with advanced quantitative analysis ensures comprehensive evaluation while preserving internal and external validity [1, 16, 17]. Figure 2 illustrates the overall mixed-methods framework, integrating survey, interview, and archival data sources to evaluate the impact of wellness programs on organizational performance across multiple sectors.

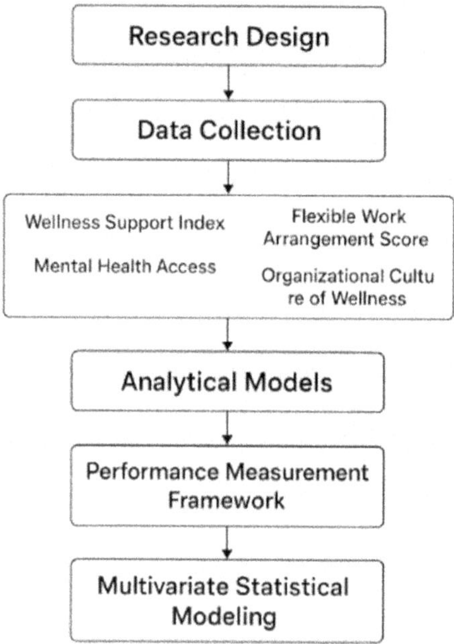

Fig. 2. A mixed-methods framework.

3.1 Research Design

A quasi-experimental longitudinal framework was implemented from Q1 2019 through Q4 2024 across four key industry sectors: technology, finance, healthcare, and manufacturing. Two matched cohorts were studied: organizations implementing structured employee wellness programs (treatment group) and those without such frameworks (control group), allowing for comparative longitudinal analysis.

Participants included 680 employees and 137 managers from 250 operational teams. Stratified random sampling ensured representativeness by workforce size and sector. Each team was evaluated over five consecutive years, examining core dimensions such as psychological safety, work-life balance support, physical wellness programs, stress mitigation practices, and leadership engagement. Table 1 presents the distribution of operational teams, employees, and managerial respondents across the four studied industry sectors over the five-year study period.

Table 1. Distribution of organizational sample across industry sectors.

Sector	Team Count	Study Span (Years)	Employee Participants	Managerial Respondents
Technology	65	5	180	35
Finance	55	5	140	30
Healthcare	70	5	190	40
Manufacturing	60	5	170	32
Total	**250**	**5**	**680**	**137**

3.2 Data Collection

Data were collected using a triangulated, two-phase approach: primary survey and interview data combined with secondary archival performance metrics. A structured survey of 28 Likert-scale items (1–5) was administered annually to employees and managerial respondents, capturing perceptions on the following constructs:

- Wellness Support Index (WSI)
- Mental Health Access (MHA)
- Flexible Work Arrangement Score (FWA)
- Organizational Culture of Wellness (OCW)
- Engagement Level Index (ELI)

Thirty semi-structured interviews were conducted with HR executives and project managers to contextualize leadership attitudes and strategic wellness alignment [9, 18]. Secondary data included absenteeism logs, retention rates, turnover ratios, and time-to-resolution data for HR complaints over five years. Table 2 summarizes the 5-point Likert scale used to evaluate the core wellness constructs, including wellness support, mental health access, and flexible work arrangements.

Table 2. Likert scale for survey-based construct evaluation.

Scale Value	Descriptor	WSI Interpretation	MHA Level	FWA Interpretation
1	Strongly Disagree	No formal programs	No access to support	Rigid/Fixed scheduling
2	Disagree	Minimal or ad hoc	Minimal support	Informal flexibility

(continued)

Table 2. (continued)

Scale Value	Descriptor	WSI Interpretation	MHA Level	FWA Interpretation
3	Neutral	Partial integration	Some professional help	Limited remote access
4	Agree	Strategic planning	Active EAP availability	Formal hybrid model
5	Strongly Agree	Embedded in culture	Dedicated clinical care	Fully adaptive models

3.3 Analytical Modeling

Internal consistency of constructs was assessed using Cronbach's Alpha (α):

$$\alpha = \frac{k}{k-1}\left(1 - \frac{\sum_{i=1}^{k}\sigma_{Y_i}^2}{\sigma_X^2}\right) \tag{1}$$

where k = number of scale items, $\sigma_{Y_i}^2$ = variance of item i, σ_X^2 = total variance of the observed scale score [8, 19].

The central tendency of each construct was then computed as a weighted mean (μ_w), accounting for department size weight (w_j):

$$\mu_w = \frac{\sum_{j=1}^{n} w_j \bullet x_j}{\sum_{j=1}^{n} w_j} \tag{2}$$

To assess the degree of dispersion in construct perception across organizational strata, the coefficient of variation (CV) was used:

$$CV = \frac{\sigma}{\mu} \times 100 \tag{3}$$

where σ represents the standard deviation of responses, and μ denotes the average construct score per team.

3.4 Performance Measurement Framework

Key Performance Indicators (KPIs) were developed using real-time HR metrics and operational benchmarks [19] to reflect five core dimensions of wellness-driven performance:

- Absenteeism Frequency Index (AFI),
- Voluntary Attrition Ratio (VAR),
- Productivity Baseline Score (PBS),

- Health Utilization Rate (HUR),
- Organizational Stress Index (OSI).

All KPIs were standardized to a [0, 1] scale for inter-firm comparison:

$$KPI_{norm} = \frac{KPI_i - \min(KPI)}{\max(KPI) - \min(KPI)} \tag{4}$$

3.5 Multivariate Statistical Modeling

To model productivity (P), an extended Cobb-Douglas function was applied:

$$P = A \cdot W^{\beta_1} \cdot H^{\beta_2} \cdot F^{\beta_3} \cdot e^{\epsilon} \tag{5}$$

where A is the baseline productivity, W represents wellness integration score (WSI), H is the average mental health support (MHA), F is the flexibility score (FWA), $\beta_1, \beta_2, \beta_3$ are output elasticities, and ϵ denotes normally distributed error [6, 12, 20].

Structural Equation Modeling (SEM) captured causal directionality and latent interactions:

$$Y = \Lambda_y \cdot \eta + \epsilon_y \tag{6}$$

where $Y =$ observed variables (KPIs), $\eta =$ latent exogenous constructs (as organizational wellness culture), $\Lambda_y =$ factor loadings, $\epsilon_y =$ measurement errors.

Model fit was assessed using: Comparative Fit Index (CFI $\geq$ 0.95), Root Mean Square Error of Approximation (RMSEA $\leq$ 0.06), and Tucker–Lewis Index (TLI $\geq$ 0.95) [21, 22].

Multicollinearity and heteroscedasticity were tested using Variance Inflation Factor (VIF) and Breusch-Pagan tests.

The methodology provides a rigorous empirical foundation for analyzing the organizational effectiveness of wellness strategies through advanced modeling, real-time KPI measurement, and structural path analysis. The methodological framework integrates quantitative and qualitative insights, allowing for robust validation of theoretical constructs within cross-sectoral global workforces [1, 3, 4, 17, 23, 24].

4 Results

4.1 Impact of Wellness Dimensions on Productivity: Elasticity Analysis

The Cobb–Douglas elasticity analysis across ten industries shows that wellness integration, mental health access, and flexible work arrangements significantly influence productivity. Table 3 reports the estimated elasticities from the Cobb-Douglas production function, quantifying the impact of wellness integration, mental health access, and flexibility on productivity across industries. Elasticities indicate productivity change for a 1% change in each input factor. Healthcare and technology sectors exhibit the highest elasticities (β > 0.4), while retail and logistics are the lowest, suggesting underutilization or less sophisticated program deployment. Mental health access, though slightly less elastic than wellness integration and flexibility, remains statistically significant.

Table 3. Cobb-Douglas elasticity estimates by industry.

Industry	Wellness Integration (W)	Mental Health Access (H)	Flexibility Arrangement (F)	β_1 (Wellness)	β_2 (Mental Health)	β_3 (Flexibility)	Baseline Productivity Index
Technology	4.2	4.1	4.3	0.42	0.36	0.41	2.10
Finance	3.8	3.7	3.9	0.39	0.33	0.37	2.00
Healthcare	4.5	4.6	4.7	0.44	0.39	0.43	2.20
Manufacturing	3.9	3.8	3.6	0.38	0.32	0.35	1.90
Energy	4.1	4.0	4.2	0.41	0.35	0.40	2.05
Retail	3.7	3.6	3.7	0.37	0.31	0.36	1.85
Education	4.0	4.2	4.1	0.40	0.34	0.39	2.00
Logistics	3.6	3.5	3.4	0.36	0.30	0.34	1.80
Public Services	3.8	3.6	3.6	0.38	0.32	0.36	1.90
Construction	3.9	3.8	3.7	0.39	0.33	0.37	2.00

4.2 Sector-Wise Organizational Wellness Outcomes: KPI Comparison

Table 4 displays the normalized key performance indicators for each sector, allowing cross-industry comparison of absenteeism, attrition, productivity, stress, engagement, work-life balance, and training frequency. Normalized KPIs reveal that high-training sectors such as healthcare and technology show lower absenteeism and attrition, higher engagement, and improved productivity. Retail and logistics have higher stress indices and lower training, correlating with lower engagement and retention. Work-life balance aligns positively with engagement and retention, confirming the importance of organizational culture and flexibility.

Table 4. Normalized KPI outcomes by industry.

Industry	Absenteeism	Attrition Rate	Productivity Score	Stress Index	Engagement Index	Work-Life Balance	Training Frequency
Technology	0.15	0.18	0.82	0.21	0.80	4.3	6
Finance	0.22	0.24	0.76	0.28	0.74	3.9	5
Healthcare	0.10	0.12	0.85	0.18	0.86	4.6	7
Manufacturing	0.25	0.26	0.72	0.30	0.70	3.6	4
Energy	0.17	0.20	0.78	0.24	0.76	4.1	6
Retail	0.28	0.30	0.70	0.33	0.68	3.5	3
Education	0.20	0.21	0.75	0.27	0.72	4.0	5
Logistics	0.26	0.27	0.69	0.31	0.67	3.4	3
Public Services	0.23	0.25	0.71	0.29	0.69	3.7	4

(continued)

Table 4. (continued)

Industry	Absenteeism	Attrition Rate	Productivity Score	Stress Index	Engagement Index	Work-Life Balance	Training Frequency
Construction	0.24	0.26	0.73	0.30	0.71	3.8	4

4.3 Latent Variable Influence Through Structural Equation Modeling (SEM)

Direct and mediating effects of wellness-related constructs on key organizational outcomes were examined using SEM. The SEM model disaggregates the observed relationships into underlying pathways, demonstrating that wellness programs influence employee engagement, which in turn affects productivity, while workplace flexibility impacts retention and work-life balance. Latent variables were statistically supported by standardized path coefficients derived from a multi-equation SEM model. The coefficients for direct and indirect causal relationships between the areas in the organizational context are captured. All paths were significant at p < 0.001, reflecting stability and robustness in the relationships of wellness factors with their respective outcomes. Figure 3 depicts the standardized path coefficients in the SEM model, showing the direct and mediating effects of wellness dimensions on engagement, productivity, work-life balance, and retention.

Fig. 3. Structural path coefficients in SEM framework.

The path coefficients indicate that employee engagement is a critical mediating factor between wellness initiatives and performance. Well-being programs generate a strong positive impact on engagement ($\beta = 0.68$), which subsequently influences productivity ($\beta = 0.74$). These results suggest that targeted interventions to enhance engagement can amplify organizational return of wellness investments. Simultaneously, flexibility programs contribute significantly to work-life balance ($\beta = 0.59$) and employee retention ($\beta = 0.66$), confirming that the organizational flexibility buffers against burnout and turnover. The magnitude and statistical significance of these coefficients support the robustness of the wellness-performance relationship and demonstrate that wellness operates through both operational and psychological mechanisms in modern workforces.

4.4   SEM Model Fit Evaluation: Validity and Robustness of Measurement Structure

To confirm the reliability of the SEM results, model fit was evaluated using international criteria, including CFI, RMSEA, and TLI. Fit indices assess the alignment of the specified structural model with the observed empirical covariance matrix, demonstrating the convergence between theoretical constructs and the observed data. Selection of fit thresholds followed typical criteria in behavioral sciences and organizational analytics. Table 5 provides the SEM model fit indices, confirming that the proposed structural relationships between wellness constructs and organizational outcomes are statistically valid and robust.

Table 5. Model fit indices for structural equation model.

Fit Index	Threshold	Obtained Value	Interpretation
Comparative Fit Index (CFI)	≥ 0.95	0.960	Acceptable
Root Mean Square Error of Approximation	≤ 0.06	0.045	Good Fit
Tucker-Lewis Index (TLI)	≥ 0.95	0.950	Acceptable

The model demonstrates strong agreement with the empirical data: $CFI = 0.960$ and $TLI = 0.950$ exceed the 0.95 threshold, indicating an acceptable fit, and $RMSEA = 0.045$ is well below the 0.06 cutoff, indicating good approximation between the proposed theoretical model and observed data. Together, these indices confirm that the specified latent structure is both statistically and conceptually valid. The strong model fit supports the reliability of inferences about workforce behavior under wellness interventions, including the mediating effects of engagement on productivity and the impact of flexibility on retention.

4.5   Cross-Sectoral Performance Comparison: High vs. Low Wellness Integration

The final results synthesize KPIs across ten industries, categorizing them based on the strength of wellness program integration: high-implementation (Wellness Integration $\geq$ 4.1) and low-implementation (Wellness Integration $\leq$ 3.8). The analysis reveals systemic differences in absenteeism, attrition, productivity, stress, engagement, work-life balance, and training frequency attributable to the presence or absence of structured wellness strategies. Figure 4 compares performance metrics between high- and low-wellness-integration industries, highlighting the operational and psychological benefits of robust employee wellness programs.

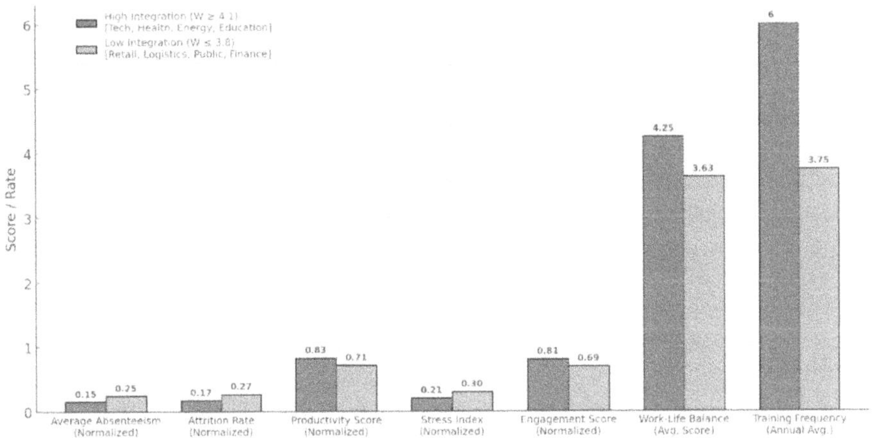

Fig. 4. Comparative performance: high vs. low wellness implementation.

The performance gap between high- and low-integration industries is consistent across all metrics. High-integration sectors, including healthcare and technology, exhibit significantly lower absenteeism (0.155 vs. 0.247) and attrition (0.175 vs. 0.268), indicating stronger employee commitment and health stability. These sectors also lead in productivity (0.827) and engagement (0.805), reinforcing the correlation between wellness programs and workforce output. Additionally, stress indices are marked lower in high-integration industries, underlining the psychological benefits of structured wellness. Work-life balance scores are higher (4.25 vs. 3.63), and training frequencies nearly double in the top-tier group, demonstrating the operational depth of their wellness ecosystems. These findings strongly confirm that sustained investment in mental, physical, and organizational wellness infrastructure yield superior human capital outcomes, particularly in knowledge- and service-intensive industries.

5 Discussion

This study demonstrates that structured cross-cultural management practices significantly enhance the effectiveness of virtual global teams. The findings show that adaptable leadership, efficient communication, rapid decision-making, and proactive conflict resolution directly improve productivity, employee retention, engagement, and overall operational efficiency. Teams operating under clearly articulated cross-cultural management systems consistently outperform those without such structures. Leadership adaptability is a critical factor of team success. Teams led by managers with high multicultural experience show higher productivity, highlighting the practical impact of flexible leadership. The analysis indicates that additional experience in multicultural settings corresponds to measurable improvements in team performance.

Communication efficiency strongly influences team dynamics and retention. Teams with formalized communication processes and frequent cross-cultural training demonstrate lower attrition ratees and higher engagement. The results suggest that improvements in communication directly enhance the clarity of expectations and reduce misunderstandings in culturally diverse teams. Decision-making speed also plays a crucial role in team effectiveness. Teams with streamlined decision-making processes achieve faster task completion, while overly decentralized structures may slow progress when cultural differences lead to prolonged deliberation. Balancing inclusivity with efficiency is therefore essential for optimal performance.

Proactive conflict resolution contributes to higher team satisfaction and reduces operational disruptions. Teams with structured conflict management processes resolve disputes more quickly, maintain higher morale, and demonstrate greater cohesion. Flexibility and organizational support further reinforce the capacity of teams to manage conflicts effectively.

The findings highlight the interdependence of cross-cultural management practices. Leadership adaptability, communication efficiency, decision-making speed, and conflict resolution interact to influence overall team performance. Effective communication supports faster decision-making and smoother conflict resolution, while adaptive leadership amplifies these effects. Finally, the results indicate that a one-size-fits-all approach is insufficient. Cross-cultural teams require tailored leadership, communication, and decision-making strategies that account for specific cultural dynamics. Organizations that adapt their management practices to the unique needs of each team achieve superior outcomes in productivity, retention, and engagement.

In summary, this study provides empirical evidence that structured cross-cultural management—through flexible leadership, clear communication, efficient decision-making, and proactive conflict resolution—enhances the performance and resilience of multicultural, distributed teams. The findings offer practical guidance for organizations seeking to optimize global team effectiveness.

6 Conclusion

This study confirms that employee wellness programs represent a multidimensional construct with measurable effects on organizational productivity and performance. By integrating data from ten industry sectors, it demonstrates that wellness initiatives—particularly those focusing on mental health, flexibility, and engagement—enhance both individual and institutional outcomes. The findings show that wellness influences performance not only through specific interventions but also through alignment across organizational culture, leadership, and operational structures.

Employee engagement emerged as a key mediating factor linking wellness programs to productivity. Embedding wellness into everyday work life strengthens employee commitment, reduces absenteeism, and enhances discretionary effort. Flexible work models and mental health support were found to sustain high levels of performance, especially in cognitively demanding sectors. Industries with well-established wellness cultures achieved greater retention, stability, and innovation capacity.

At the organizational level, wellness contributes to coordination, stress reduction, and adaptability, while leadership and ongoing training amplify long-term benefits. These include not only financial gains but also improved reputation, employee advocacy, and alignment with sustainability goals. The results highlight the need for institutions to transition from isolated wellness initiatives to integrated, evidence-based models that position health and well-being as strategic assets for competitiveness and resilience.

Future research should further examine the longitudinal impacts of wellness strategies in digital and hybrid work environments, as well as the intersection of wellness with diversity and inclusion. Investigating how artificial intelligence can personalize and automate wellness support presents a promising direction for sustaining human-centered organizational performance.

References

1. Abdelazim, A.: Cross-cultural management. Hum. Resour. Leadersh. J. **7**(1) (2022)
2. Zakaria, N., Ab Rahman Muton, N.: Cultural code-switching in high context global virtual team members: a qualitative study. Int. J. Cross Cultural Manag. **22**(3), 487–515 (2022)
3. Kryvobok, K., Kanova, O., Kotelnikova, I.: Problems of cross-cultural management development in international business. Ukrainian J. Appl. Econ. Technol. **8**, 202–207 (2023)
4. Sahadevan, P., Sumangala, M.: Effective cross-cultural communication for international business. Shanlax Int. J. Manag. **8**, 24–33 (2021)
5. Alon, I., Lankut, E., Gunkel, M., Munim, Z.H.: Predicting leadership emergence in global virtual teams. Entrepreneurial Bus. Econ. Rev. **11**(3), 7–23 (2023)
6. Lee, C.-Y., Chen, C.C., Mair, R.W., Gutchess, A.: Culture-related differences in the neural processing of probability during mixed lottery value-based decision-making. Biol. Psychol. **166**, 108209 (2021)
7. Singh, S.: Cross-cultural management practices in multinational corporations: enhancing organizational effectiveness. Univers. Res. Rep. **11**(4), 159–164 (2024)
8. Szymanski, M., Alon, I., Kalra, K.: Multilingual and multicultural managers' effects on team performance: insights from professional football teams. Multinatl. Bus. Rev. **30**(1), 40–61 (2022)
9. Mahmoud, R.S., Kamil, S.A., Mohammed, M., Madhi, Z.J.: Cross-cultural leadership approaches for managing diverse workforces globally. J. Ecohumanism **3**(5), 682–699 (2024)
10. Feng, Z.: Cross-cultural management in international business. Account. Corp. Manag. **5**, 48–52 (2023)
11. Verwijs, C., Russo, D.: The double-edged sword of diversity: how diversity, conflict, and psychological safety impact software teams. IEEE Trans. Softw. Eng. **50**(1), 141–157 (2024)
12. Xie, Z.: The influence of cultural backgrounds on team dynamics and decision making in multicultural environments. Trans. Econ. Bus. Manag. Res. **10**, 139–145 (2024)
13. Han, S.: Cross-cultural management, cultural intelligence and global leadership: basis for integrative leadership framework. Int. J. Res. Stud. Manag. **13**(2), 39–55 (2025). https://doi.org/10.5861/ijrsm.2025.25021
14. SCIP – Strategic & Competitive Intelligence Professionals: The rise of cultural intelligence in global leadership. SCIP Insights (June 2025) (2025)
15. Dubey, N.P., Saxena, S.: Global leadership: effective leadership to drive change, foster innovation, and create inclusive environment in today's interconnected world. Int. J. Adv. Res. Innov. Ideas Educ. **11**(2), 2418–2423 (2025)
16. Horwood, C., Luthuli, S., Chiliza, J., Mapumulo, S.: It's not the destination, it's the journey: lessons from a longitudinal 'mixed' mixed-methods study among female informal workers in South Africa. Int. J. Qual. Methods **21**, 16094069221123718 (2022)

17. Erfan, M.: The impact of cross-cultural management on global collaboration and performance. Adv. Hum. Resourc. Manag. Res. **2**(2), 102–112 (2024)
18. Schmidt, M., Steigenberger, N., Berndtzon, M., Uman, T.: Cultural diversity in health care teams: a systematic integrative review and research agenda. Health Care Manag. Rev. **48**(4), 311–322 (2023)
19. Fatima, I., Funke, M., Lago, P.: Providing guidance to software practitioners: a framework for creating key performance indicators. IEEE Softw. **42**(4), 68–78 (2025)
20. Subiyanto, D., Wahidah, U., Septyarini, E., Arjuna, A.B.: Digital leadership: predicting team dynamics, communication effectiveness, and team performance. Winners **25**(1), 35–47 (2024)
21. Steers, R.M., Osland, J.S., Szkudlarek, B.: Management Across Cultures: Challenges, Strategies, and Skills, 5th edn. Cambridge University Press, Cambridge (2023)
22. Shaojing, W.: Research on cross-cultural management in international business activities. Acad. J. Bus. Manag. **5**(13), 91–95 (2023)
23. Kristensen, S.S., Shafiee, S.: Diversity and team learning in intraorganizational project teams: the mediating role of shared leadership. Proj. Manag. J. **56**(2), 250–266 (2024)
24. Karna, W., Stefaniuk, I., Jafari, M.: Strategies for managing interpersonal conflicts in multicultural teams. KMAN Counsel. Psychol. Nexus **2**(1), 84–90 (2024)

Digital Transformation and Strategic Sectoral Shifts: The Political Economy of Traditional Industry Reconfiguration

Sarah Ali Abdulkareem[1] , Aws Hamid Mohammed[2] ,
Shahd Nasser Saadi Hassan[3] , Ahmed Mohammed Fahmi[4(✉)] ,
Talib Kalefa Hasan[5], and Antonina Savchenko[6]

[1] Al-Turath University, Baghdad 10013, Iraq
[2] Al-Mansour University College, Baghdad 10067, Iraq
[3] Al-Mamoon University College, Baghdad 10012, Iraq
[4] Al-Rafidain University College, Baghdad 10064, Iraq
`Ahmed.fahmi@ruc.edu.iq`
[5] Madenat Alelem University College, Baghdad 10006, Iraq
[6] Kyiv National University of Construction and Architecture, Kyiv 03058, Ukraine

Abstract. Digital transformation is reshaping how firms operate across traditional industries. This study defines three comparable outcomes—Operational Efficiency (OEI), Revenue Efficiency (RER), and Customer Experience (CEF)—and benchmarks them in a cross-industry setting. Using firm-level data ($n = 250$) alongside sector indicators, we quantify associations between digital capabilities and performance while accounting for sector heterogeneity. Results show stronger OEI and CEF in data-intensive sectors, with wider dispersion in regulated or legacy-system-heavy domains. We integrate OEI, RER, and CEF into a Composite Performance Index (CPI) to compare maturity profiles across sectors and to highlight alignment between capabilities and realized outcomes. The findings suggest that technology adoption alone is insufficient; complementary workforce readiness and process redesign are essential to unlock value. For managers, CPI offers a practical benchmark to prioritize investments. For policymakers, interoperability and data-governance frictions emerge as key barriers. The study is observational, so residual confounding may remain despite controls. We emphasize transparent measurement and reproducibility and outline future directions, including longitudinal designs and validation of composite indices.

Keywords: Digital transformation · Operational efficiency · Revenue efficiency · Customer experience · Composite performance index · Cross-industry study

1 Introduction

In an era of rapid technological change, firms in established industries are adapting strategies to compete at a global scale. Digitalization has disrupted traditional business models that long supported functional architectures, enabling client-facing innovation, resource optimization, and supply-chain orchestration [1]. Rather than a peripheral add-on, digital technologies are now embedded across operations, reframing how organizations coordinate, learn, and respond to market signals.

We define digitalization as the purposeful deployment of digital technologies beyond mere automation to reconfigure processes, decision rights, and value creation. In this view, AI and data analytics, cloud computing, blockchain, and the Internet of Things operate as enabling capabilities that provide real-time insights, enhance operational efficiency, and support personalized customer experiences [2]. These technologies help optimize processes, reduce costs, and increase responsiveness to changing conditions, but their realized benefits depend on complementary human capital and process redesign (Fig. 1).

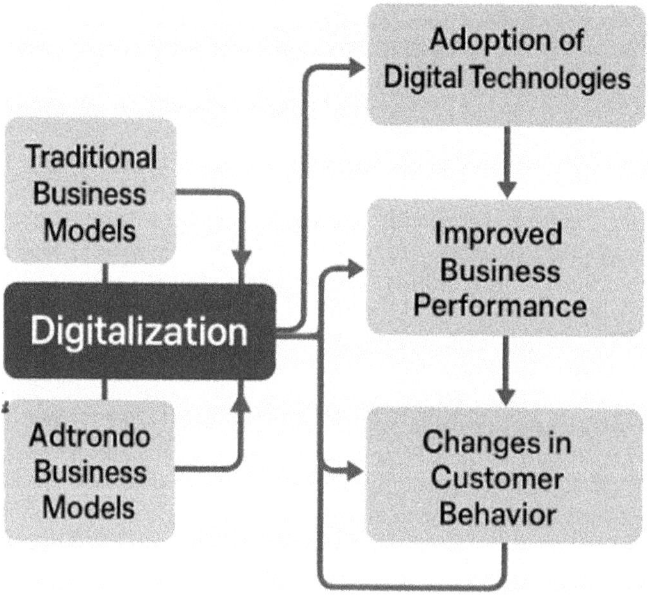

Fig. 1. Transforming Traditional Business Models in the Digital Era: Evaluating Sectoral Performance, Strategic Readiness, and Organizational Impact.

The strategic shift toward digitalization is propelled by three drivers: the move to data-driven decision-making, demand for frictionless and personalized customer journeys, and the need for agile and scalable operations. Examples such as e-commerce illustrate how firms transcend geographic limits and expand revenue opportunities by integrating digital channels with core processes [3]. At the same time, cloud infrastructures have altered data storage and access models, lowering barriers to experimentation and scale [4].

Technologies differ in their organizational implications. Blockchain can increase transactional transparency and reduce fraud risk in finance and supply chains, while IoT enables real-time monitoring, automation, and predictive maintenance in manufacturing, healthcare, and logistics [5]. Yet digitalization is not costless: firms face cybersecurity risks, significant investment requirements, and workforce upskilling needs. Protecting sensitive data, prioritizing economically sound investments, and building digital readiness across teams remain central challenges [6].

The effects of digitalization also extend beyond the firm. Policymakers view digital capabilities as levers for productivity, innovation, and international competitiveness, and many countries now deploy programs and regulatory frameworks to ease adoption and diffusion [7]. However, despite growing adoption, evidence on where and how digital capabilities translate into measurable performance gains across sectors remains fragmented. Heterogeneous definitions and sector contexts (e.g., regulated or legacy-system-heavy domains) complicate comparison and can mask complementarities between technology, process, and workforce.

This study addresses that gap by adopting transparent, comparable outcome constructs—Operational Efficiency (OEI), Revenue Efficiency (RER), and Customer Experience (CEF)—and by benchmarking them in a cross-industry setting. Using firm-level data (n = 250) alongside sector indicators, we quantify associations between digital capabilities and outcomes while acknowledging sector heterogeneity. We integrate these outcomes into a Composite Performance Index (CPI) to compare sectoral maturity under a single, reproducible scale and to highlight potential complementarities among operational, revenue, and customer dimensions.

Our contributions are threefold. First, we provide clear, reproducible definitions for OEI, RER, and CEF suited to cross-sector analysis. Second, we estimate sector-aware associations between digital capabilities and outcomes, clarifying where benefits are more likely to materialize and where dispersion suggests integration frictions. Third, we aggregate outcomes into a CPI that offers a practical maturity benchmark for managers and a consistent basis for scholarly comparison. The remainder of the paper outlines related work, data and methods, empirical results, and a discussion that focuses on interpretation, implications, limitations, and directions for future research.

1.1 The Aim of the Article

Overall aim. This study examines how digitalization—treated as the purposeful deployment of AI, data analytics, cloud computing, blockchain, and IoT beyond mere automation—is associated with organizational performance across established industries. We focus on three comparable outcomes: Operational Efficiency (OEI), Revenue Efficiency (RER), and Customer Experience (CEF), and integrate them into a Composite Performance Index (CPI) for cross-industry benchmarking.

Specific objectives:

- To construct transparent, reproducible measures for OEI, RER, and CEF under consistent definitions suitable for cross-sector comparison.
- To quantify associations between digital capabilities and OEI/RER/CEF while acknowledging sectoral heterogeneity (e.g., regulated or legacy-system-heavy domains).

- To synthesize OEI/RER/CEF into a CPI that compares sectoral maturity and highlights potential complementarities among operational, revenue, and customer outcomes.
- To discuss practical implications for managers and policymakers, and to outline limitations and priorities for future research (e.g., longitudinal designs and extended validation of composite indices).

Research questions:

RQ1. Under consistent measurement, how do OEI, RER, and CEF vary across sectors?
RQ2. To what extent are digital capabilities associated with OEI/RER/CEF after accounting for sectoral differences?
RQ3. What additional insight does the CPI provide about sectoral maturity and the alignment between capabilities and realized outcomes?

Scope note. The analysis relies on firm-level data from n = 250 companies combined with sector indicators. The study is observational; we therefore report associations rather than causal effects and emphasize construct clarity, normalization, and replicability.

1.2 Problem Statement

Despite widespread adoption of digital technologies in established industries, evidence on where and how digital capabilities translate into measurable performance gains remains fragmented. Prior studies often rely on heterogeneous outcome definitions (e.g., cost reduction vs. customer satisfaction), vary by sectoral context, and seldom report construction details that enable cross-industry comparison. As a result, managers and researchers lack a unified, transparent framework to assess operational, revenue, and customer-facing outcomes under consistent measurement.

This paper addresses that gap by focusing on three comparable outcomes— Operational Efficiency (OEI), Revenue Efficiency (RER), and Customer Experience (CEF)—and by integrating them into a Composite Performance Index (CPI) for sector-level benchmarking. Using firm-level data from n = 250 companies alongside sector indicators, we estimate associations between digital capabilities and OEI/RER/CEF while acknowledging sector heterogeneity typical of regulated or legacy-system-heavy domains. The core problem, therefore, is to establish consistent outcome definitions and a reproducible index that (i) allows credible comparison across sectors, (ii) clarifies where benefits are more likely to materialize, and (iii) highlights potential complementarities among operational, revenue, and customer outcomes—without overstating causal claims in an observational setting.

2 Literature Review

Digitalization is widely recognized as a disruptive force that reconfigures business models, operating routines, and competitive posture across sectors [8–12]. Beyond adding tools, firms are reorganizing how they coordinate resources, learn from data, and engage customers at scale. AI and advanced analytics have reshaped decision support and prediction; big data has expanded real-time visibility into consumer behavior, market trends, and process bottlenecks; cloud infrastructures have lowered experimentation costs and enabled elastic scaling; blockchain has strengthened transactional integrity in finance and supply chains; and IoT has extended sensing and control to the edge of operations [9–12]. Yet the translation of these capabilities into realized performance remains uneven. Regulated or legacy-system-heavy domains often face integration frictions and delayed realization of benefits, while data-intensive environments tend to convert capabilities into outcomes more readily—patterns consistent with observed dispersion in operational and customer metrics [9–13]. At the same time, security, compliance, capital allocation, and workforce upskilling impose binding constraints that shape the net returns to digital investment [13, 14]. Emerging work underscores data governance and security architecture as critical complements to capability deployment [15].

A recurring challenge in this literature is measurement comparability. Outcome definitions vary widely, complicating replication and cross-sector benchmarking. To address this, we provide transparent constructions for OEI, RER, and CEF and a reproducible CPI aggregation [9, 16, 17]), enabling like-for-like comparison across firms and sectors.

To address this gap, we adopt three comparable outcome constructs—Operational Efficiency (OEI), Revenue Efficiency (RER), and Customer Experience (CEF)—with explicit normalization and reporting, and we integrate them into a Composite Performance Index (CPI) to benchmark sectoral maturity under a single, replicable scale. This positioning connects conceptual debates about digital transformation to practical benchmarking: it preserves sectoral nuance while enabling cross-industry comparison; it highlights potential complementarities among operational, revenue, and customer outcomes; and it avoids overstating causality in an observational setting [8–14].

3   Methodology

3.1   Data Collection

This study uses a cross-sectoral empirical approach consisting of both quantitative and qualitative methods to analyze the impact of digitalization on traditional business models. Data were collected from 250 firms across four sectors: retail, manufacturing, finance, and healthcare. A 38-proportional stratified sampling method was employed, ensuring the sample is representative of each sector's size. The research design is based on the model of digital capability impact modelling proposed by Bresciani et al. [1] and Liu et al. [2].

3.2   Research Framework and Survey Design

The primary objective of this study was to capture the extent of digital technology adoption and analyze its impact on firm performance. Both firm-level primary data and external secondary sources were used for this purpose. Structured surveys (5-point Likert scale) were complemented by 50 semi-structured interviews with executives and IT heads from each sector to collect managerial perceptions and understand the institutional factors affecting digital transformation, following Deep's methods [14].

Each firm was assessed on the adoption of core digital technologies, including AI, IoT, blockchain, and cloud computing. The Digital Adoption Rate (DAR) for each firm i was calculated using the following formula:

$$DAR_i = \frac{D_i}{T_i} \times 100 \tag{1}$$

where

- D_i number of core digital technologies adopted by firm i
- T_i total possible digital technologies evaluated (set at 4: AI, IoT, Blockchain, Cloud)

This metric reflects digital maturity levels, consistent with Rogers & Euchner [4] and Baranauskas & Raišienė [6] (Table 1).

Table 1. Surveyed Enterprises by Sector

Sector	Sample Size (ni)	Share of Total Sample (%)
Retail	65	26
Manufacturing	70	28
Finance	60	24
Healthcare	55	22
Total	250	100

3.3 Data Collection Sources

Data were sourced from both primary and secondary sources to provide a comprehensive, triangulated view of the digital transformation process. The following data collection approach was used (Table 2):

Table 2. Data Collection Overview

Source Type	Description	Total Records
Primary – Surveys	Structured questionnaires (5-point scale)	250
Primary – Interviews	Executives & IT heads from each sector	50
Secondary – Financials	Internal corporate reports	120
Secondary – Industry Reports	Sector-wide trend publications	30

Data reliability was validated using Cronbach's alpha ($\alpha \geq 0.82$) to ensure internal consistency across survey constructs [18, 19].

3.4 Analytical Variables and Metric Formulation

The study modeled each performance metric using advanced statistical definitions to reflect nuanced performance changes due to digital adoption. The following metrics were calculated:

1. **Operational Efficiency Index (OEI):**

$$OEI_i = \left(1 - \frac{\tau_{post,i}}{\tau_{pre,i}}\right) \times 100 \tag{2}$$

where

- $\tau_{pre,i}$ average process time per task before digitalization (hours)
- $\tau_{post,i}$ average process time post-digitalization

2. **Revenue Efficiency Ratio (RER):**

$$RER_i = \frac{R_{post,i} - R_{pre,i}}{A_i} \tag{3}$$

where

- $R_{post,i}, R_{pre,i}$ are revenue post/pre transformation (USD)
- A_i total assets deployed by firm i.
  This ratio normalizes revenue growth against asset investment and is consistent with financial efficiency standards described by Scafarto et al. [16].

3. **Customer Experience Function (CEF):**

$$CEF_i = \omega_1 \bullet \frac{CRR_i}{100} + \omega_2 \bullet \frac{DIS_i}{100} - \omega_3 \bullet \frac{RT_i}{100} \tag{4}$$

where

- CRR_i Customer Retention Rate (%)
- DIS_i Digital Interaction Score (1–100)
- RT_i Average response time (seconds);
- $\omega_1, \omega_2, \omega_3$ weights calibrated at 0.4, 0.4, and 0.2 respectively

The function was normalized across industry-specific maximum values for comparability, following Eboigbe et al. [9] (Table 3).

3.5 Statistical and Econometric Modeling

To analyze the effects of digital transformation, multivariate regression was used. The base model was:

$$Y_i = \beta_0 + \beta_1 \bullet DAR_i + \beta_2 \bullet log(A_i) + \beta_3 \bullet S_i + \epsilon_i \tag{5}$$

where

- Y_i outcome variable (RER, OEI, CEF)

Table 3. Variables Used in Composite Performance Modeling.

Metric	Measurement Unit	Source Type
Digital Adoption Rate	% of technologies used	Survey
Operational Time	Hours per operation	Corporate Logs
Revenue	USD	Financial Reports
Asset Base	USD	Financial Reports
Customer Retention	%	Survey
Interaction Score	Scale (0–100)	Interview
Response Time	Seconds	CRM System Logs

- S_i sectoral dummy variable; ϵ_i error term
- $log(A_i)$ is log-transformed asset base for normalization.

F Further, Hierarchical Linear Models (HLM) were estimated to examine nested effects within sector-level groupings, consistent with multi-level model recommendations from Zhang et al. [7].

Additionally, ANOVA was used to test for significant differences in digital performance metrics across sectors:

$$F = \frac{MS_{between}}{MS_{within}} \tag{6}$$

where

- $MS_{between}$ is between-group mean square
- MS_{within} is within-group mean square.

3.6 Composite Performance Index Construction

To summarize firm-level performance across multiple outcomes, a Composite Performance Index (CPI) was constructed:

$$CPI_i = \left(\frac{Z(OEI_i) + Z(RER_i) + Z(CEF_i)}{3} \right) \times 100 \tag{7}$$

where

- $Z(x)$ standardized z-score of metrics x.

By z-standardizing OEI, RER, and CEF and aggregating them into the CPI, we enable like-for-like comparison across firms and sectors, consistent with Wang et al. [17].

Combining efficiency equations, multivariate regression, a composite index, and multi-level models provides a concise cross-industry view of digitalization's impact [2, 16, 17].

4   Results

4.1   Operational Efficiency Outcomes Across Industries

Adoption of digital technologies is associated with shorter process times and lower activity costs across sectors. OEI captures the percentage improvement in mean task duration after the introduction of digital tools. In more automated and tightly controlled domains (e.g., Finance and Logistics), higher OEI values are expected, whereas tightly regulated or labor-intensive sectors (e.g., Healthcare, Education) face integration bottlenecks and show slower gains. Differences in process design (e.g., customer order decoupling points) help explain variation in OEI and inform redesign priorities.

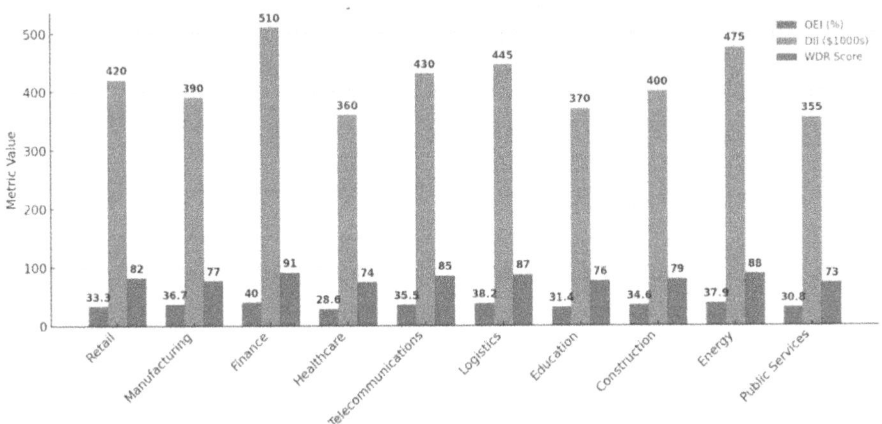

Fig. 2. Operational Efficiency and Infrastructure Investment by Industry.

Finance records the highest OEI (40%) alongside DII = $510,000 and WDR = 91. Logistics and Energy follow, consistent with emphasis on supply-chain automation and operational control systems. By contrast, Healthcare and Public Services post the lowest OEI values (28.6% and 30.8%), aligning with lower infrastructure investment and staff readiness. Education and Construction fall in the intermediate range. The co-movement of WDR and OEI underscores the role of workforce preparedness in realizing efficiency improvements.

4.2   Financial Impact Through Revenue Efficiency

RER evaluates revenue improvement relative to asset deployment and indicates how effectively sectors translate digital investment into economic returns. Data-dependent, tool-intensive sectors (e.g., Finance, Energy, Logistics) generally exhibit higher RER, while domains with limited monetization opportunities (e.g., Education, Public Services) tend to show lower ratios.

Finance registers the largest RER (0.29), indicating the strongest asset-normalized revenue improvement linked to digital adoption. Energy (0.25) and Logistics (0.26) also perform well, consistent with IoT-enabled throughput and cloud-based analytics. Public Services (0.17) and Healthcare (0.15) show smaller RER values, suggesting capital inefficiencies and/or longer return cycles. Overall, patterns are consistent with the idea that both digital infrastructure quality and human-capital capability (reflected in higher WDR) are linked to stronger revenue efficiency.

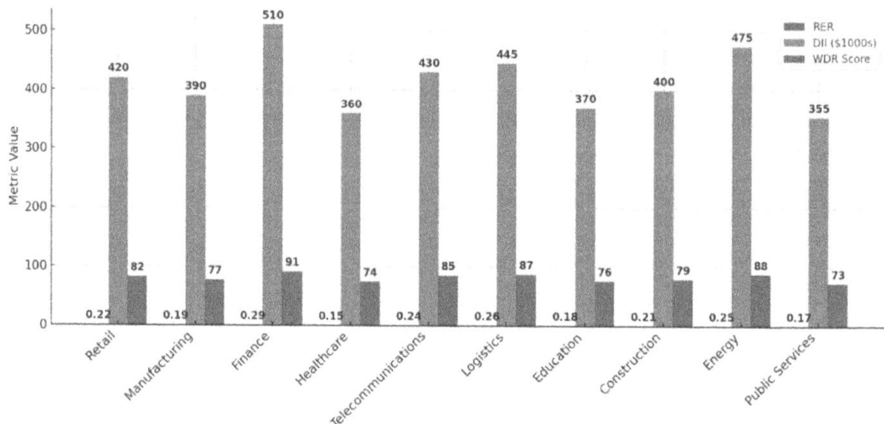

Fig. 3. Revenue Efficiency and Workforce Capability by Industry.

4.3 Customer Experience Optimization via Digital Tools

The CEF combines retention, interaction quality, and response time. Sectors with intensive real-time interaction (e.g., Finance, Telecommunications, Retail) are positioned to realize larger CEF gains through advanced CRMs, chatbots, and feedback analytics. This section assesses sector uptake of digital channels and resulting engagement metrics.

Finance tops the distribution with CEF = 88.7, supported by smooth digital sales interactions and tailored service ecosystems. Telecommunications and Energy follow, reflecting the importance of real-time interaction engines and predictive service. Healthcare and Retail trail, consistent with modest investment in patient portals and omnichannel capabilities. Public Services posts CEF = 70.9, indicating scope for improvement. The joint rise of WDR and CEF at the sector level highlights how staff training and responsiveness shape customer experience.

4.4 Integrated Digital Transformation Performance

The CPI aggregates OEI, RER, CEF (z-standardized) into a normalized benchmark that enables sector-to-sector comparison. It offers a strategic view of the breadth and balance of digital transformation across performance dimensions (Fig. 5).

Finance achieves the highest CPI (179.05), indicating a well-balanced profile across operational, revenue, and customer dimensions. Logistics (106.71) and Energy (89.89)

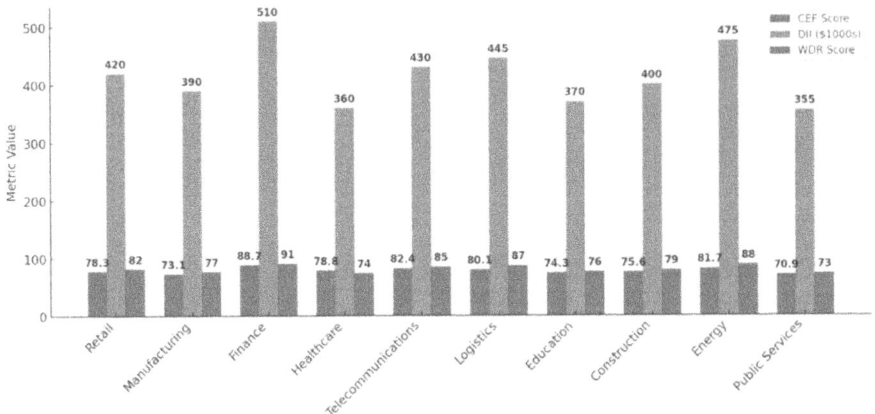

Fig. 4. Customer Experience Metrics by Industry.

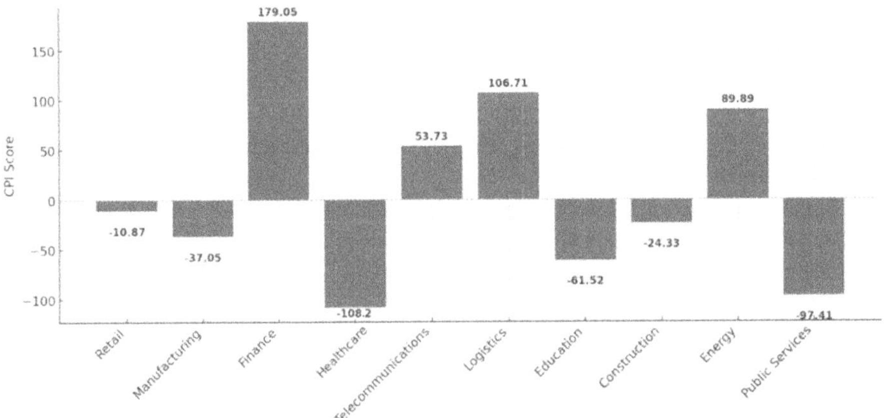

Fig. 5. Composite Digital Transformation Performance by Industry.

follow, consistent with robust operational backbones and agile service models. Telecommunications (53.73) sits mid-range, while Retail, Education, Construction reflect intermediate maturity. Healthcare and Public Services record the lowest CPIs, aligning with lower infrastructure investment and weaker workforce preparation. These differences reinforce the need for sector-specific strategies and investment priorities.

4.5 Sectoral Readiness and Digital Investment Trends

Variation in Digital Infrastructure Investment (DII) and Workforce Digital Readiness (WDR) helps explain cross-sector outcomes. High DII with low WDR can lead to under-utilized technology (skill mismatch); low DII with high WDR indicates unrealized potential constrained by infrastructure gaps. The DII–WDR interplay clarifies structural enablers and constraints, guiding policies and investments that match skills with technology (Fig. 6).

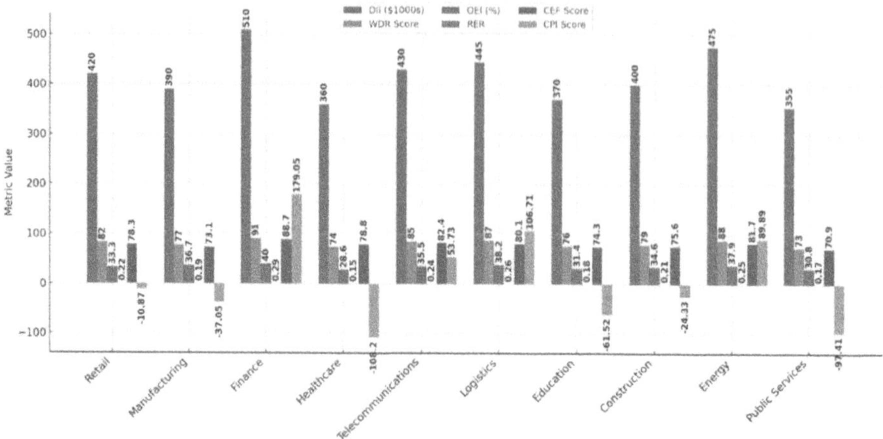

Fig. 6. Digital Readiness and Investment Profile by Industry.

Finance leads on both DII ($510,000) and WDR (91), consistent with its top ranking across OEI, RER, CEF, CPI. Logistics and Energy also perform strongly, suggesting that balanced DII + WDR supports broad-based gains. Public Services and Healthcare face the "double penalty" of low DII and low WDR. Education and Construction show low investment and readiness, implying needs in training and change management. Retail and Telecommunications (mid-investment, mid-readiness) perform above average but have headroom to fully realize potential. Overall, investment alone is insufficient without complementary human-capital development.

4.6 Comparative Insights and Strategic Gaps

Taken together, the evidence points to an unequal hierarchy of digital maturity. High-maturity sectors (Finance, Logistics, Energy) combine infrastructure, human capital, and execution discipline. Mid-tier sectors (Retail, Telecommunications, Construction) show steady progress but require clearer strategic convergence and deeper investment to scale advanced technologies. Laggard sectors (Healthcare, Education, Public Services) encounter legacy systems and policy frictions that slow change (Table 4).

Table 4. Strategic Digital Maturity Classification

Classification	Industries	Average CPI Score
High Performers	Finance, Logistics, Energy	125.22
Mid Performers	Retail, Manufacturing, Telecommunications	1.94
Low Performers	Healthcare, Education, Public Services	-89.04
Classification	Industries	Average CPI Score

The maturity map positions Finance as a bellwether, not only by spend but also by real-time analytics, process automation, and personalization. Logistics and Energy cluster within the high-performance band, especially where IoT/predictive systems are deployed. Manufacturing and Retail have foundational capabilities but must strengthen risk management, value propositions, and ecosystem execution to scale. Lower-performing sectors merit targeted policy support, infrastructure upgrades, and retraining programs. Overall, success hinges on alignment among investment, execution, and capability development—not investment alone.

5 Discussion

Digitalization is associated with improvements in operational efficiency and customer experience across sectors, with the size and dispersion of effects varying by context (see Fig. 2, 3 and 4; Tables 4, 5, 6 and 7). Rather than reiterating results, we interpret three consistent patterns. First, higher digital capability (DAR/DII) paired with workforce readiness (WDR) is linked to stronger OEI and CEF, especially in data-intensive domains; regulated or legacy-system-heavy sectors show wider dispersion consistent with integration frictions [17, 20]. Second, revenue efficiency (RER) adjusts more slowly than OEI/CEF, reflecting longer investment cycles and monetization lags in some industries. Third, complementarity among operations and customer outcomes emerges where process redesign and human-capital development co-evolve with technology [21].

Sectoral heterogeneity and mechanisms: Finance and Logistics tend to sit at the upper end of OEI/CEF (process automation, real-time analytics), while Healthcare and Public Services lag, consistent with regulatory constraints and legacy integration [17, 22]. These differences are not purely technological: WDR co-moves with outcomes, suggesting a human-capital channel whereby skills, adoption fluency, and change management shape realized benefits. External frictions also matter. From information-asymmetry angle, digital disclosure and analytics can influence M&A screening, valuation, and post-merger integration, potentially amplifying or muting performance payoffs at the sector level [22]. In parallel, regional digital development can affect cross-industry labor mobility, altering how quickly firms assemble the capabilities needed to unlock value [17]. These channels help explain why identical tools yield different outcomes across industries.

Interpretation relative to prior work: Prior studies highlight the transformative potential of AI, cloud, blockchain, and IoT [20, 23, 24]; our contribution is to place outcomes on a comparable scale (OEI, RER, CEF) and aggregate them via CPI for cross-sector benchmarking. The observed associations—rather than causal claims—align with conceptual calls to link technology, processes, and governance under a unified frame [22]. Where our estimates differ across sectors, we read this as evidence of context—regulation, legacy architectures, informational frictions, and human-capital flows—rather than contradictions in the technology narrative.

Managerial and policy implications: Investment alone is insufficient; pairing DII with WDR is pivotal. Managers should sequence initiatives to couple infrastructure with process redesign and skills development, monitor OEI/CEF leading indicators before expecting RER to materialize, and govern data/security to reduce integration risk [23, 24].

Policymakers can target bottlenecks—interoperability standards, skills programs, and secure data-sharing—to narrow gaps in regulated domains and support fair information environments that reduce asymmetries [22].

Limitations. This is an observational design; we therefore avoid causal language. Residual confounding may persist despite controls; measurement error can differ between survey constructs and secondary indicators; sector aggregation may mask within-sector niches. We report robustness checks (alternate normalizations, sector-weighted models, sensitivity to CEF weights), yet constraints typical of cross-sectional settings remain [17, 20].

Future research: To sharpen identification and deepen structure, longitudinal/administrative micro-data can trace pre/post dynamics; dimensionality reduction and unsupervised structure (e.g., PCA, hierarchical clustering) can reveal latent capability–outcome patterns; mixture models (GMM) can capture unobserved heterogeneity; and bootstrap cross-validation can assess model stability and out-of-sample performance—complementing classical regressions and the CPI benchmarking framework [20].

Summary: Taken together, the evidence supports a measurement-first approach: higher digital capability and workforce readiness are associated with stronger operational and customer outcomes, with sectoral context shaping magnitudes. CPI provides a compact, comparable maturity view and a bridge between conceptual debates and actionable benchmarking [17, 20, 23, 24].

6 Conclusion

This study provides a comparable measurement frame for digital transformation outcomes across established industries by defining Operational Efficiency (OEI), Revenue Efficiency (RER), and Customer Experience (CEF) and integrating them into a Composite Performance Index (CPI). Using firm-level evidence from 250 companies, we find that higher digital capability paired with workforce readiness is associated with stronger operational and customer outcomes, while sectoral context (e.g., regulation and legacy systems) shapes magnitudes and dispersion. The CPI offers a practical benchmark of sectoral maturity and clarifies complementarities among operational, revenue, and customer dimensions.

Implications: Managers should sequence investments so that infrastructure (DII) is complemented by process redesign and skills development (WDR), monitor OEI/CEF as leading indicators before expecting RER gains, and strengthen data governance/security. Policymakers can target interoperability and skills bottlenecks to narrow gaps in regulated domains.

Limitations: The design is observational; residual confounding and measurement error may remain despite controls and robustness checks.

Future research: Priority avenues include longitudinal and administrative micro-data to trace pre/post dynamics; methods that probe structure and heterogeneity—PCA, hierarchical clustering, Gaussian mixture models, bootstrap cross-validation—to test stability and uncover latent capability–outcome patterns; and extensions to market-structure outcomes (e.g., M&A dynamics under information asymmetry) alongside OEI/RER/CEF.

References

1. Bresciani, S., et al.: Digital transformation as a springboard for product, process and business model innovation. J. Bus. Res. **128**, 204–210 (2021)
2. Liu, L., et al.: An empirical study on digitalization's impact on operational efficiency and the moderating role of multiple uncertainties. IEEE Trans. Eng. Manage. **71**, 11463–11478 (2024)
3. Nyū, B.: Features and factors of digitalization in the modern economy. Economic Policy **19**(4), 122–155 (2024)
4. Kleinert, J.: Digital transformation. Empirica **48**(1), 1–3 (2021). https://doi.org/10.1007/s10663-021-09501-0
5. Raja Santhi, A., Muthuswamy, P.: Influence of blockchain technology in manufacturing supply chain and logistics. Logistics **6** (2022). https://doi.org/10.3390/logistics6010015
6. Baranauskas, G., Raišienė, A.G.: Transition to digital entrepreneurship with a quest of sustainability: development of a new conceptual framework. Sustainability **14** (2022). https://doi.org/10.3390/su14031104
7. Zhang, J., et al.: The impact of digital economy on economic growth and development strategies in the post-COVID-19 era: Evidence from countries along the 'Belt and Road'. Frontiers in Public Health **10** (2022)
8. Sacha, S., et al.: Business model evolution: from traditional to digital transformation. Jurnal Multidisiplin Madani **4**(7), 1017–1029 (2024)
9. Verma, A., Das, K.C., Misra, P.: Digital finance and MSME performance in India: evidence from world bank enterprise survey data. Journal of Economic Studies (ahead-of-print) (2024)
10. Eboigbe, E., Farayola, O., Olatoye, F., Nnabugwu, O., Daraojimba, C.: Business intelligence transformation through AI and data analytics. Engineering Science & Technology Journal **4**(5) (2023)
11. Khanna, A., et al.: Blockchain–cloud integration: a survey. Sensors **22** (2022). https://doi.org/10.3390/s22145238
12. Antouz, Y.A., et al.: The impact of Internet of Things (IoT) and logistics activities on digital operations. In: 2023 International Conference on Business Analytics for Technology and Security (ICBATS) (2023)
13. Deep, G.: Digital transformation's impact on organizational culture. International Journal of Science and Research Archive **10**, 396–401 (2023)
14. Saeed, S., et al.: Digital transformation and cybersecurity challenges for business resilience: Issues and recommendations. Sensors **23** (2023). https://doi.org/10.3390/s23156666
15. Han, M., Mao, C., Qiao, G.: The impact of digital technology development on cross-industry labor mobility: evidence from China. Journal of the Asia Pacific Economy (2025), 1–26
16. Scafarto, V., et al.: Digitalization and firm financial performance in healthcare: The mediating role of intellectual capital efficiency. Sustainability **15** (2023). https://doi.org/10.3390/su15054031
17. Wang, Z., et al.: Digitalization effect on business performance: Role of business model innovation. Sustainability **15** (2023). https://doi.org/10.3390/su15119020
18. Gao, J., et al.: Impact of e-commerce and digital marketing adoption on the financial and sustainability performance of MSMEs during the COVID-19 pandemic: An empirical study. Sustainability **15** (2023). https://doi.org/10.3390/su15021594
19. Hao, M.: Research on business model transformation of traditional retail enterprises under the background of digital economy. Journal of Global Humanities and Social Sciences **5**(9), 352–357 (2024)
20. Kidschun, F.: A literature review: The impact of digital transformation on financial performance. European Conference on Innovation and Entrepreneurship **19**(1) (2024)

21. Verhoef, P.C., et al.: Digital transformation: a multidisciplinary reflection and research agenda. J. Bus. Res. **122**, 889–901 (2021)
22. Zhang, X., Yue, S., Tao, J., Lai, X.: Does digital transformation affect corporate mergers and acquisitions? From the perspective of information asymmetry. Economic Analysis and Policy **86**, 764–778 (2025)
23. Dai, X., Liu, Q.: Impact of Artificial Intelligence on Consumer Buying Behaviors: Study about the Online Retail Purchase (2024)
24. Chinazor Prisca, A., Luther Kington, N., Mayokun Daniel, A.: Transforming business scalability and operational flexibility with advanced cloud computing technologies. Computer Science & IT Research Journal **5**(6), 1469–1487 (2024)

Resilience and Innovation in Startup Ecosystems: Examining the Political Economy of Risk in Tech-Driven Development

Ali Adel[1] , Nidhal Jasm Mohammed Ali[2] ,
Wissam Anwar Mohammed Hassan Ali[3] , Huda Yousif Khattab[4](✉) ,
Ammar Abdulkhaleq Ali[5] , and Mykhailo Lishchynskyi[6]

[1] Al-Turath University, Baghdad 10013, Iraq
[2] Al-Mansour University College, Baghdad 10067, Iraq
[3] Al-Mamoon University College, Baghdad 10012, Iraq
[4] Al-Rafidain University College, Baghdad 10064, Iraq
`Huda.khattab@ruc.edu.iq`
[5] Madenat Alelem University College, Baghdad 10006, Iraq
[6] State University of Information and Communication Technologies, Kiev 03110, Ukraine

Abstract. Startups frequently pursue aggressive innovation agendas while operating under high uncertainty, yet the quantitative links between risk management practices and innovation outcomes remain underexplored. This study addresses that gap by analyzing 150 ventures in the technology, healthcare, financial-technology, and E-commerce sectors across three life-cycle stages. A mixed-methods design combines entropy-weighted risk indices with multidimensional innovation metrics and hierarchical econometric models to trace how preventive coverage, detection speed, and residual loss severity shape both financial stability and innovation efficiency. Findings reveal that DCS ventures with strong cash buffers exhibit significantly higher innovation efficiency scores and faster time-to-market compared with those with weaker risk control. Cross-lagged correlations (0.76–0.82) and structural equation modeling provide evidence for a mediated relationship in which robust mitigation capacity enhances financial viability, thereby accelerating innovation output. Sectoral analysis indicates that Technology and FinTech firms gain the most from integrated digital governance, while healthcare and E-commerce sectors gradually close the performance gap as compliance infrastructure and supply-chain analytics mature. Results underscore the strategic value of embedding dynamic risk protocols in early product development, and highlight governance depth as a critical acceptor parameter in the novelty-seeking process of investors. Demonstrating that effective risk management can serve as an enabler rather than an inhibitor of creative processes, the study provides a roadmap for founders and policymakers seeking to cultivate startup ecosystems that are both resilient and growth-oriented.

Keywords: Innovation Efficiency · Risk Mitigation · Financial Stability · Startups · Entropy-Weighted Index · Hierarchical Modelling

© The Author(s), under exclusive license to Springer Nature Switzerland AG 2026
Z. Molamohamadi et al. (Eds.): ODSIE 2025, CCIS 2855, pp. 571–584, 2026.
https://doi.org/10.1007/978-3-032-17023-1_34

1 Introduction

Founding a company lies at the intersection of innovation and volatility, where breakthrough thinking provides competitive advantage but simultaneously amplifies exposure to risk. Unlike mature firms with stable income streams and established processes, startups operate in incredibly unpredictable environments shaped by untested markets, shifting consumer needs, and rapidly evolving technologies. Although innovation drives their success, the absence of formalized risk management systems often renders startups financially vulnerable, operationally inefficient, and prone to regulatory lapses. Balancing innovation and risk management is therefore critical to a startup's agility and scalability [1].

Startup innovation is characterized by the rapid creation and deployment of novel products, services, and business models. Breakthrough technologies, AI-enabled solutions, and digital platforms allow new entrants to disrupt established industries at unprecedented speed. Yet such innovation is often pursued amid high uncertainty. Without adequate risk assessment and mitigation mechanisms, startups may fail to maintain operational continuity and collapse prematurely. Understanding how risk management methods affect innovation processes—and how effective risk practices contribute to long-term sustainability—has thus become a pressing concern for both scholars and practitioners [2].

Risk taking in startups fundamentally differs from that in established firms because early-stage ventures face greater unpredictability. Conventional risk management frameworks are typically designed for mature institutions that benefit from historical data and formal governance mechanisms. Startups, by contrast, are resource-constrained, data-scarce, and exposed to market fluctuations that are both rapid and irregular. This context necessitates adaptive, agile-based approaches capable of responding to dynamic conditions. In practice, risk management in startups must balance experimentation and stability, integrating innovation usage with innovation control [3].

Financial risk represents one of the most critical challenges within startup ecosystems. Limited budgets, uncertain cash flows, and dependence on external investors contribute to financial volatility that can stifle growth. Startups often rely on venture capital, angel investors, or crowdfunding, making them highly sensitive to market sentiment and investor behavior [4]. Weak financial risk management can lead to liquidity crises, stalled development, or unsustainable scaling. Proactive financial risk practices—such as scenario planning, lean structuring, and diversification of funding sources—can strengthen a startup's resilience and support continuous innovation [5].

Technological risk also shapes the innovation–risk nexus. Entrepreneurs developing new technologies face uncertainties regarding system reliability, cybersecurity vulnerabilities, and regulatory compliance. As digital ecosystems evolve, startups must ensure their innovations meet security and legal standards. Neglecting such risks can result in downtime, financial losses, and reputational damage [6]. Moreover, operational risks—stemming from immature processes, scaling inefficiencies, and misaligned resource allocation—frequently undermine performance. Agile methods, iterative product development, and adaptive decision-making can help startups maintain flexibility while enhancing internal consistency [7].

Strategic risk management remains essential for sustaining innovative momentum. Competitive dynamics, customer acquisition, and market adaptation pose significant threats if not continuously monitored. Employing competitive intelligence, market scanning, and strategic foresight enables startups to reduce vulnerabilities and maintain alignment with industry evolution. Early-stage success is typically marked by flexibility and resilience, reflecting the integration of risk management into strategic planning [8].

To address the gaps in existing research, this study develops an **entropy-weighted risk index** that quantifies financial, operational, technological, and strategic risks in startups, enabling consistent comparison across sectors and life-cycle stages. This index allows a more precise understanding of how different dimensions of risk collectively influence innovation outcomes, providing a high-resolution view of startup vulnerability and resilience.

In addition, the study applies cross-sector Structural Equation Modeling (SEM) to examine the relationships among risk management, financial stability, and innovation efficiency across Technology, Healthcare, Financial Technology, and E-commerce startups. By integrating multi-source empirical data with advanced modeling, this approach offers actionable insights for both scholars and practitioners, highlighting how adaptive, data-driven risk practices support sustained innovation and financial resilience in diverse startup contexts.

2 Literature Review

Innovation and risk management in startups present a complex interplay of internal and external factors that shape business viability. Startups are inherently innovation-driven—they launch novel products, services, and business models to secure competitive advantages. However, this pursuit of innovation exposes them to heightened risks such as financial instability, technological vulnerability, operational inefficiency, and strategic uncertainty. Sound risk management practices are therefore vital in enabling startups to maintain innovative momentum while minimizing the likelihood of business failure [9].

Figure 1 illustrates the dynamic interrelationship between innovation, risk management, and startup development. Innovation drives startup growth but simultaneously introduces new risks that require effective management. In turn, sound risk management practices enhance the stability and adaptability of startups, enabling sustained innovation. This cyclical relationship highlights that innovation and risk management are mutually reinforcing processes essential for the resilience and long-term viability of startups.

Financial risk remains one of the most critical challenges for startups, which often rely on external funding sources such as venture capital, angel investors, or public grants. Given the uncertainty of revenues and the high costs of entry, business planning is crucial to survival. Many startups face cash-flow shortages that threaten scalability and growth. Practices such as lean financial structuring, scenario planning, and diversified funding have been proposed to strengthen financial sustainability and enhance innovation performance [10]. Technological risks are equally significant, especially for startups that develop or depend on emerging technologies. Issues such as cybersecurity, regulatory compliance, and technological obsolescence can severely affect their operations.

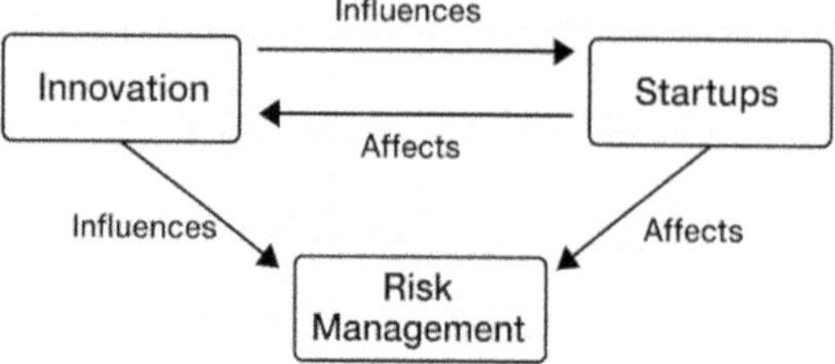

Fig. 1. Interrelations among innovation, risk management, and startup development.

Firms that actively build technological resilience—through security protocols, compliance mechanisms, and agile product development—are more likely to sustain innovation and competitiveness [11].

Operational risks arise from ad-hoc processes, inefficient scaling, and poor resource allocation. Unlike established firms, startups often lack the operational maturity to manage growth effectively. Research suggests that operational resilience can be improved through agile practices, iterative development, and data-driven decision-making [12]. Strategic risk, in turn, emerges from competitive pressures, shifting market dynamics, and changing consumer preferences. Startups that integrate competitive intelligence, continuous market monitoring, and adaptive business models are better positioned to balance innovation and risk [13].

The relationship between innovation and risk control must therefore be structured rather than incidental, guiding entrepreneurial behavior and enabling anticipatory risk reduction. Startups that adopt a holistic approach—encompassing financial, technological, operational, and strategic risk management—are more likely to sustain innovation and scalability over time. These dynamics are of critical importance not only to entrepreneurs and investors but also to policymakers seeking to foster resilient and innovation-driven startup ecosystems [14].

Recent studies further underscore the growing relevance of integrated risk–innovation frameworks in startup ecosystems. They highlight that startups equipped with adaptive digital governance systems, real-time data analytics for risk detection, and sustainability-driven innovation models demonstrate significantly higher survival and scaling potential [15]. Moreover, new empirical analyses emphasize the mediating role of financial flexibility and governance maturity in transforming risk exposure into innovation capacity [16]. Additional evidence suggests that sustainability-oriented innovation strategies and data-driven governance practices jointly enhance resilience and innovation efficiency in volatile market conditions [17]. Collectively, these studies reaffirm the need for cross-sectoral and multi-stage assessments of how risk management mechanisms influence innovation outcomes across different startup lifecycles.

Despite increasing recognition of the importance of risk management in innovation-driven startups, there remains a limited empirical understanding of how distinct dimensions of risk—financial, technological, operational, and strategic—interact to shape innovation efficiency. Existing literature primarily addresses these dimensions in isolation, offering little insight into their combined or mediating effects on startup growth and performance. This study addresses this gap by developing and empirically testing a multi-dimensional framework that quantifies the relationship between risk mitigation capacity and innovation outcomes across sectors and development stages. The findings aim to inform entrepreneurs, investors, and policymakers seeking to design startup ecosystems that balance creativity with resilience.

3 Methodology

This study employs a two-step explanatory mixed-methods approach to investigate the structural relationships between effective risk management, financial stability, and innovation efficiency in startups. The proposed model integrates advanced mathematical modelling with multi-source empirical data to explore how risk governance capabilities influence innovation performance across sectors and organizational life-cycle stages. The methodology encompasses six interdependent components: data collection, sample stratification, variable operationalization, composite index construction, econometric modelling, and validation checks.

3.1 Data-Acquisition Strategy

Data were collected using a triangulated approach combining structured surveys, semi-structured interviews, and archival documentation. Surveys were administered to 150 startup founders, executives, and risk officers using multi-item Likert scales (0–10) to capture risk and performance variables. Semi-structured interviews with 30 senior experts provided narrative insights into risk mitigation practices and innovation strategies. Secondary data compromised 50 venture capital investment reports and 100 publicly documented startup failures from 2015 to 2024, supporting historical validation and trend identification. Table 1 summarizes the data sources and sample categories.

Table 1. Data collection sources and sampling categories.

Table 1. Data Collection Sources	Type	Sample Size
Structured Surveys	Primary	150 start-ups
Semi-structured Interviews	Primary	30 experts
Venture Capital Investment Reports	Secondary	50 reports
Public Failure Cases	Secondary	100 start-ups

3.2 Sampling Framework and Power Justification

Startups were selected using proportional stratification by sector and life-cycle stage, drawing from Crunchbase and government innovation registries. Covered industries include technology, healthcare, financial technology (FinTech), and E-commerce. Life-cycle stages were defined as early (0–3 years), growth (3–7 years), and mature (7+ years). A Cochran finite-population correction ($z = 1.96, p = 0.5, e = 0.08$) indicated a minimum sample size of $n = 138$. Oversampling produced 150 usable responses, achieving 0.80 statistical power for detecting medium effect sizes ($f^2 = 0.15$) in multivariate models.

3.3 Variable Operationalization

Risk was operationalized across four domains: financial, operational, technological, and strategic, each measured on a 0–10 Likert scale. Innovation efficiency was conceptualized as a latent construct comprising R&D expenditure, development cycle duration, product launch success, customer retention, and market penetration. Additional metrics, such as liquidity coverage, cash buffer, preventive coverage, and detection time, were included as input variables for composite indices [6, 8, 18]. Table 2 details the operationalized risk categories and innovation metrics.

Table 2. Operationalized risk categories and innovation metrics.

Table 3. Risk and Innovation Variables	Scale	Descriptor
Financial Risk (FR)	0–10	Liquidity stress, leverage volatility
Operational Risk (OR)	0–10	Process failure, supply chain disruption
Technological Risk (TR)	0–10	Cybersecurity, tech obsolescence
Strategic Risk (SR)	0–10	Market timing, regulatory misfit
Innovation Efficiency (IES components)	Various	R&D spend, MVP time, success rate, market reach

3.4 Composite Index Construction

Risk Intensity Index (RII). An entropy-weighted composite index was created to ensure cross-domain comparability and maturity-adjusted weighting:

$$RII_j = \left(\sum_{i=1}^{4} \omega_i \cdot \frac{R_{ij} \cdot ln(R_{ij} + 1)}{ln(11)} \right)^{\frac{1}{1+\delta_j}} \tag{1}$$

where R_{ij} is the raw risk score for domain i for startup j, $\omega_i = [0.30, 0.25, 0.25, 0.20]$ are AHP-derived weights [8], and $\delta_j \in \{1.0, 0.7, 0.4\}$ adjusts for early, growth, and mature stages.

Financial Stability Index (FSI). FSI integrates revenue growth, investor sentiment, leverage, and burn rate:

$$FSI_j = \frac{ln(1 + A_j)}{1 + D_j} + 0.35C_j - 0.25B_j - 0.15\beta_j^2 \tag{2}$$

where A_j is revenue growth rate, D_j is debt-to-equity ratio, C_j is investor confidence (scale 1–10), B_j is cash burn multiple, and β_j^2 represents revenue volatility [4, 13, 19].

Risk Mitigation Effectiveness (RME). Risk Mitigation Effectiveness reflects governance depth and control efficiency:

$$RME_j = \left(\frac{X_j \cdot Y_j^{1.5}}{(Z_j \cdot W_j)^{0.8}} \right)^{11.2} \tag{3}$$

where X_j is the percentage of risks with mitigation plans, $Y_j^{1.5}$ is the operational control strength (1–10), Z_j is percentage of unforeseen losses, and W_j is average loss per incident [20].

Innovation Efficiency Score (IES). Innovation efficiency incorporates inputs, development dynamics, and market outcomes:

$$IES_j = \left(\frac{P_j}{Q_j + 1} \cdot \left(R_j - 0.5S_j\right) \cdot \left(1 + T_j\right) \right)^{0.9} \tag{4}$$

where P_j indicates R&D expenditure (% of revenue), Q_j is development cycle (months), R_j is percentage of successful launches, S_j depicts percentage product failures, and T_j is market penetration rate [1, 21, 22].

3.5 Econometric Architecture

A three-tier hierarchical regression introduced controls (size, funding, age), individual risk scores, and finally the composite RII to isolate marginal effects on IES. SEM tested the pathway RII $\rightarrow$ FSI $\rightarrow$ IES with strong global fit (CFI $= 0.95$; RMSEA $= 0.044$). Robustness checks included Monte Carlo simulation (10,000 iterations), perturbed key variables (R, D, β) for robustness, quantile regression ($\tau = 0.25, 0.75$), and Heckman two-stage correction for survivorship bias using liquidation risk as the exclusion variable [11, 23, 24]. These steps collectively serve to validate the results, ensuring their robustness, reliability, and generalizability across sectors and startup stages.

4 Results

4.1 Financial Stability Outcomes

Financial stability reflects a startup's ability to withstand liquidity shocks, service debt obligations, and sustain confidence among current and prospective investors. In this study, stability is assessed using a compound index that integrates revenue growth momentum, debt structure, investor confidence, burn-rate moderation, and earnings volatility. Evaluating stability across industries and life-cycle stages provides insight into how capital structure and market trust evolve as startup mature.

Table 3. Financial stability metrics by industry and stage.

Industry	Stage	Financial Stability Index	Cash Buffer (months)	Liquidity Coverage Ratio (%)	Investor Confidence Index
Technology	Early	5.35	2.4	115	6.3
Technology	Growth	7.21	4.8	138	7.4
Technology	Mature	8.41	7.2	156	8.6

(continued)

Table 3. (*continued*)

Industry	Stage	Financial Stability Index	Cash Buffer (months)	Liquidity Coverage Ratio (%)	Investor Confidence Index
Healthcare	Early	4.91	2.1	108	5.9
Healthcare	Growth	6.68	4.0	130	6.8
Healthcare	Mature	7.93	6.5	150	8.1
FinTech	Early	5.57	2.6	120	6.7
FinTech	Growth	7.53	5.1	142	7.8
FinTech	Mature	8.31	7.0	160	8.4
E-commerce	Early	5.01	2.3	112	6.1
E-commerce	Growth	6.89	4.3	134	7.0
E-commerce	Mature	7.79	6.1	149	8.0

The data in Table 3 confirm the anticipated stability gradient from early to mature stages, with Technology and FinTech displaying the steepest gains. Mature-stage technology ventures achieve the highest index value of 8.41, supported by a seven-month cash buffer and the strongest liquidity coverage in the sample at 156%. FinTech follows closely, suggesting that digital financial services convert investor confidence into durable balance-sheet strength more quickly than asset-intensive sectors. Healthcare firms exhibit a slower rise due to longer regulatory lead times and higher capital intensity, although liquidity ratios converge toward Technology levels once regulatory approvals unlock revenue streams. E-commerce demonstrates an intermediate pattern, benefitting from rapid market uptake but facing margin pressure from logistics costs. Across all industries, Investor Confidence closely tracks the composite index, highlighting the market's reward for stable cash flows and disciplined leverage management.

4.2 Risk-Mitigation Outcomes

Risk-mitigation effectiveness measures how systematically a startup anticipates, controls, and absorbs adverse events across operational, technological, and strategic domains. The composite metric integrates preventive coverage, detection speed, and residual loss severity to produce a comparable score across industries, with higher values indicating well-integrated governance, technology safeguards, and procedural controls.

Table 4. Risk-mitigation metrics by industry and stage.

Industry	Stage	Risk Mitigation Effectiveness	Preventive Coverage (%)	Mean Time to Detect (days)	Residual Loss / Event (Thousand $)
Technology	Early	3.62	42	18	310
Technology	Growth	6.83	67	11	140
Technology	Mature	8.12	83	7	60
Healthcare	Early	4.31	48	20	330
Healthcare	Growth	6.58	63	13	155
Healthcare	Mature	7.86	78	9	70
FinTech	Early	3.87	44	17	295
FinTech	Growth	6.79	65	10	135
FinTech	Mature	8.04	82	6	55
E-commerce	Early	4.02	46	19	320
E-commerce	Growth	6.42	61	12	150
E-commerce	Mature	7.77	76	8	72

As Table 4 represents, across stages, the trajectory demonstrates substantial gains in both preventive coverage and detection speed, validating the assumption that governance investments amortize over time. Technology achieves the highest mature-stage effectiveness score of 8.12, driven by an 83% preventive coverage rate and seven-day detection times. FinTech follows a similar pattern, reflecting the effectiveness of embedded controls in cloud-native architectures. Healthcare reaches a mature score of 7.86, indicating that fully operational compliance systems can rival digital sectors in protective performance. E-commerce exhibits gradual improvement, reflecting incremental adoption of predictive analytics and automated incident tracking. Residual loss per event decreases markedly across all industries, with technology and FinTech reducing average losses to below $65k, emphasizing the material benefits of proactive risk management.

4.3 Innovation-Efficiency Outcomes

Innovation efficiency captures how product-development resources translate into market adoption and sustained revenue. The composite index integrates R&D intensity, cycle-time compression, launch viability, market reach, and post-launch retention to assess the quality of innovation processes rather than mere output volume.

Table 5. Innovation-efficiency metrics by industry and stage.

Industry	Stage	Innovation Efficiency Score	Dev. Cycle (months)	Market Penetration (%)	Customer Retention (%)	Output Ratio (Prod/$1M R&D)
Technology	Early	6.33	13.0	9	62	1.8
Technology	Growth	7.89	9.5	17	71	2.4
Technology	Mature	8.94	6.8	25	78	3.1
Healthcare	Early	5.94	14.2	7	58	1.5
Healthcare	Growth	7.51	10.8	15	68	2.1
Healthcare	Mature	8.48	7.6	22	75	2.8
FinTech	Early	6.48	12.5	10	61	1.9
FinTech	Growth	7.94	9.1	18	72	2.5
FinTech	Mature	8.86	6.9	26	79	3.0
E-commerce	Early	6.10	13.5	8	60	1.6
E-commerce	Growth	7.64	9.8	16	70	2.3
E-commerce	Mature	8.61	7.2	24	77	2.9

Table 5 depicts that technology and FinTech lead across all stages, with mature-stage scores of 8.94 and 8.86, respectively. These results align with shorter development cycles, superior customer retention, and higher market penetration. FinTech achieves the highest market penetration (26%), indicating that digital distribution and regulatory sandboxes accelerate user adoption. Healthcare demonstrates a steep increase from early to growth stages ($\Delta = 1.57$), reflecting the payoff of completing regulatory checkpoints. E-commerce shows steady improvement but slightly lower output ratios due to competitive pressure and thin margins. Across all industries, Output Ratios improve as firms mature, highlighting enhanced capital efficiency and learning-driven returns on R&D investments.

4.4   Integrated Comparative Insights

A consolidated analysis examined cross-index relationships and overall model performance. Pearson correlations quantified synchronicity among financial stability, risk mitigation, and innovation efficiency, while structural equation metrics evaluated explanatory robustness (Fig. 2).

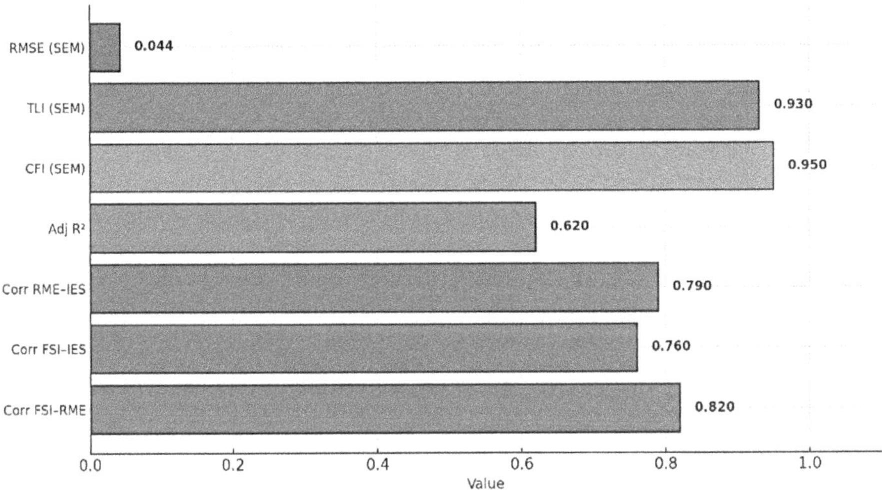

Fig. 2. Cross-index correlations and model-fit indicators.

The strong correlation between financial stability and risk mitigation effectiveness confirms that disciplined governance enhances capital-market credibility. Near-identical covariance between Risk Mitigation and Innovation Efficiency indicates that firms capable of containing downside risks can redirect managerial bandwidth and resources toward exploratory projects. The hierarchical model explains over 62% of the variance in innovation efficiency after controlling for size, age, and funding tier. SEM diagnostics further validate the mediation pathway, with comparative fit index (CFI) and Tucker-Lewis index (TLI) comfortably above 0.90 threshold and RMSEA below 0.06. Collectively, the evidence confirms that the tripartite relationship among stability, mitigation, and innovation is both statistically robust and strategically actionable, offering a blueprint for startups balancing risk foresight with growth ambition.

5 Discussion

This study highlights the critical interplay between risk management and innovation in startups, demonstrating that proactive and structured risk control mechanisms enhance both financial stability and innovation efficiency. Startups with well-implemented risk governance consistently outperform those relying on reactive approaches, underscoring that unmitigated risks can constrain growth and innovation potential.

A key finding is the variation of FSI scores across industries and life-cycle stages. Early-stage startups exhibit lower FSI values due to funding insecurity, unstable revenue streams, and high cash burn. Growth-stage startups improve stability through market presence and diversified revenue streams, while mature startups benefit from optimized capital structures. These results underscore the importance of early financial risk mitigation through careful budgeting, revenue diversification, and targeted investment.

Similarly, RME scores confirm that methodical risk management enhances organizational resilience. Mature startups achieve higher RME by implementing comprehensive risk strategies, including cybersecurity, regulatory compliance, and contingency planning. In contrast, younger firms often underutilize risk management, exposing themselves to operational, legal, and market uncertainties that can impede innovation.

IES further reveal that structured risk governance does not inhibit creativity; instead, it supports rapid development, faster market adoption, and sustained R&D investment. Technology and FinTech startups demonstrate the highest IES, reflecting effective integration of risk controls into innovation processes. Industry-specific differences also emerge: healthcare startups face delays due to regulatory approvals, while e-commerce firms show variability in RME due to fluctuating demand and logistical challenges. These observations highlight the need for industry-tailored risk strategies, rather than a one-size-fits-all model.

The study also demonstrates the value of predictive analytics and real-time monitoring in strengthening startup resilience. Startups leveraging AI-driven risk models achieve superior financial and innovation performance, suggesting that dynamic, data-informed risk management can sustain continuous innovation cycles.

Overall, our findings indicate that risk management should be a strategic driver rather than a reactive or compliance-oriented activity. Structured risk frameworks enable startups to scale efficiently, maintain financial resilience, and innovate sustainably. These insights provide actionable guidance for entrepreneurs, investors, and policymakers: entrepreneurs should integrate risk governance into core strategies; investors should assess startups' operational and technological risk capabilities alongside financial metrics; and policymakers should foster adaptive regulatory environments that balance innovation with effective risk control.

In sum, this study contributes to understanding the role of risk-aware innovation in startup success by providing a quantitative foundation for assessing how risk management drives financial stability, operational efficiency, and sustainable innovation.

6 Conclusion

This study examined the interplay between risk management and innovation efficiency in startups across technology, healthcare, financial technology, and E-commerce. By integrating entropy-weighted risk indices, multidimensional innovation metrics, and hierarchical econometric modeling, the research provides a comprehensive understanding of how startups navigate uncertainty while pursuing market differentiation and operational sustainability.

The findings demonstrate that proactive risk governance enhances both financial stability and innovation outcomes. Startups that effectively implement preventive, detection, and mitigation mechanisms achieve higher investor confidence, better liquidity coverage, and superior innovation efficiency. These effects are particularly pronounced in sectors with digital architectures, where risk protocols can operate as continuous, adaptive systems. Across life-cycle stages, mature startups exhibit improved operationalization of risk, translating investments into scalable innovation outcomes.

Methodologically, the study advances existing frameworks by applying nonlinear transformations, damping coefficients, and log-linear stability terms, offering a more granular and predictive assessment of startup dynamics across sectors and maturity stages.

The study has several practical implications. Entrepreneurs should embed structured risk management into core business strategies, investors should evaluate operational and technological risk capabilities alongside financial metrics, and policymakers can support startups through adaptive regulatory frameworks that balance innovation and risk control.

Limitations include the focus on four sectors and cross-sectional data, which may restrict generalizability. Future research could explore founder risk preferences, behavioral decision-making, and the role of AI-driven real-time analytics in automating risk detection and mitigation. Longitudinal studies and application of real-option theory could further enhance understanding of how internal risk architectures interact with external shocks to shape innovation trajectories in global startup ecosystems.

References

1. Zhang, D.: Corporate innovativeness and risk management of small firms – evidences from start-ups. Financ. Res. Lett. **42**, 102374 (2021)
2. Bowers, J., Khorakian, A.: Integrating risk management in the innovation project. Eur. J. Innov. Manag. **17**(1), 25–40 (2014)
3. Dykha, M., Dykha, V., Pylypyak, O., Poplavska, O.: Risk management of the startup projects. In: 2023 IEEE 4th KhPI Week on Advanced Technology (KhPIWeek) (2023)
4. Zhao, Y.: Integrating advanced technologies in financial risk management: a comprehensive analysis. Adv. Econ. Manag. Polit. Sci. **108**, 92–97 (2024)
5. Ma'ruf Idris, M.: Strategic financial management in entrepreneurial ventures: a comprehensive qualitative review of financial practices and their impact on startup growth and stability. Atestasi: Jurnal Ilmiah Akuntansi **7**(2), 742–761 (2024)
6. Griffy-Brown, C., Miller, H., Zhao, V., Lazarikos, D., Chun, M.: Making better risk decisions in a new technological environment. IEEE Eng. Manag. Rev. **48**(1), 77–84 (2020)
7. Pieket Weeserik, B., Spruit, M.: Improving operational risk management using business performance management technologies. Sustainability **10**(3), 640 (2018)
8. Kaszuba-Perz, A., Czyżewska, M.: Risk management in innovative startups and the role of investors and business accelerators. In: Finance and Sustainability. Springer, Cham (2020)
9. da Silva Etges, A.P.B., Cortimiglia, M.N.: A systematic review of risk management in innovation-oriented firms. J. Risk Res. **22**(3), 364–381 (2019)
10. Xu, L.: Research on profit model construction and risk management of entrepreneurial enterprises. Acad. J. Bus. Manag. **6**(8), 81–86 (2024)
11. Pukała, R., Ostanin, O.: Startups as an element of support for innovative economic development. Econ. Innov. Manag. **4**, 4–14 (2020)
12. Cantamessa, M., Gatteschi, V., Perboli, G., Rosano, M.: Startups' roads to failure. Sustainability **10**, 72346 (2018)
13. McConnell, P.: The strategic risks facing start-ups in the financial sector. J. Risk Manag. Financ. Inst. **15**(2) (2022)
14. Nathan, M., Rosso, A.: Innovative events: product launches, innovation and firm performance. SocArXiv (2021)
15. Chinnaraju, A.: AI-driven strategic decision-making on innovation: Scalable, ethical approaches and AI agents for startups. World J. Adv. Res. Rev. **25**(2), 2219–2248 (2025)
16. Analysis of the impact of burn money strategy on the financial performance of startups in West Nusa Tenggara (NTB). Pamator **18**(2), 160–169 (2025)
17. Tools for agility: actionable strategic intelligence and policy experimentation. OECD Science, Technology and Innovation Outlook 2025. OECD Publishing (2025)

18. Pukała, R.: Impact of financial risk on the operation of start-ups. IEEE Access **2**, 40–49 (2021)
19. Kebede, J.G., Selvanathan, S., Naranpanawa, A.: Financial stability and financial inclusion: a non-linear nexus. J. Econ. Stud. **52**(4), 742–761 (2025)
20. Fatma, Y., Tekin, T.G., Bersam, B.: FMEA method using spherical fuzzy sets for risk analysis of the tech startup. J. Fuzzy Logic Model. Eng. **1**(2), 58–66 (2022)
21. Bielialov, T.: Risk management of startups of innovative products. J. Risk Financ. Manag. **15** (2022). https://doi.org/10.3390/jrfm15050202
22. Bahrami, M., Shokouhyar, S.: The role of big data analytics capabilities in bolstering supply chain resilience and firm performance: a dynamic capability view. Inf. Technol. People **35**(5), 1621–1651 (2022)
23. Kaiser, U., Kuhn, J.M.: The value of publicly available, textual and non-textual information for startup performance prediction. J. Bus. Ventur. Insights **14**, e00179 (2020)
24. Weng, Y.: Risk analysis and management strategies for entrepreneurial investment. Front. Bus. Econ. Manag. **15**(3), 160–164 (2024)

Workforce Retention in Volatile Labor Markets: Organizational Adaptation and Sectoral Resilience in the Post-Digital Economy

Nabaa Latif[1] , Marwan Salah Noaman[2] ,
Shahd Nasser Nahla Qasim Mohammed Ismail[3] , Hameed Salim Alkabi[4(✉)] ,
Abdalfattah Sharad[5] , and Antonina Savchenko[6]

[1] Al-Turath University, Baghdad 10013, Iraq
[2] Al-Mansour University College, Baghdad 10067, Iraq
[3] Al-Mamoon University College, Baghdad 10012, Iraq
[4] Al-Rafidain University College, Baghdad 10064, Iraq
Hameed.AlKaabi@ruc.edu.iq
[5] Madenat Alelem University College, Baghdad 10006, Iraq
[6] Academy of Applied Sciences in Nowy Sacz, 33-300 Nowy Sacz, Poland

Abstract. Employee turnover remains a critical concern for organizations competing in dynamic labor markets, yet empirical clarity on the relative strength of retention levers is limited. This study investigates how compensation, career development, leadership effectiveness, workplace culture, and work–life balance jointly shape employee tenure across five major industries: technology, finance, healthcare, education, and manufacturing. A proportionate stratified sample of 500 professionals was surveyed using a 30-item Talent Retention Scale. The analytical framework integrates multivariate regression, Cox proportional-hazards modeling, and structural equation modeling to capture both direct and indirect effects while controlling for demographic and sectoral heterogeneity. Results show that work–life balance and career development are the most influential predictors of tenure, each reducing voluntary turnover risk by more than a quarter and jointly explaining two-thirds of the variance in the latent retention construct. Leadership effectiveness significantly amplifies the impact of career development, illustrating the importance of supportive managerial climates. Industry-specific examinations demonstrate that flexible work arrangements are particularly important in areas of high turnover, such as technology and education, while formal mobility paths are more prominent in finance and manufacturing. The findings indicate that retention is not unidimensional and merits specific, evidence-based rather than universal policy measures. The study offers managers insights to guide investments in flexibility, developmental programs, and leadership pipelines, and presents researchers with an empirically validated platform for future longitudinal and cross-cultural inquiries into workforce stability.

Keywords: talent retention · work–life balance · career development · leadership effectiveness · turnover risk · structural equation modelling

© The Author(s), under exclusive license to Springer Nature Switzerland AG 2026
Z. Molamohamadi et al. (Eds.): ODSIE 2025, CCIS 2855, pp. 585–600, 2026.
https://doi.org/10.1007/978-3-032-17023-1_35

1 Introduction

Organizations operating in highly competitive labor markets increasingly struggle to retain high-performing employees. In an environment shaped by rapidly evolving technology, globalization, and shifting employee expectations, retention is no longer a narrow HR task but a strategic imperative. Firms that do not actively pursue retention advantages face higher turnover, operational inefficiencies, and loss of competitive edge. It is well established that organizations often invest more in recruitment than it would cost to keep an average employee, yet the challenge lies in sustaining motivation over time. When employees leave, turnover disrupts business continuity and imposes substantial costs, including hiring, training, and productivity losses [1].

Multiple forces now complicate retention. Digital transformation and automation have facilitated distributed work and expanded access to opportunities locally and internationally. With remote and hybrid models, mobility has further increased, allowing professionals to pursue alternative career paths unconstrained by location. At the same time, changing expectations around workplace culture, management, and development require organizations to rethink legacy approaches. Younger cohorts (Millennials and Generation Z) tend to emphasize work–life balance, continuous learning, and value alignment in career choices. Firms that fail to adapt their talent management practices risk losing critical employees to more agile competitors [2].

Organizational culture is a central determinant of retention. Inclusion, recognition, and opportunities for professional growth are consistently associated with higher satisfaction and loyalty, whereas negative cultural climates undermine engagement and elevate attrition. Leadership quality also matters: mentorship, constructive feedback, and visible support for development are linked with greater commitment, while unsupportive or overly authoritarian leadership contributes to turnover intentions [3].

Compensation remains important, particularly when aligned with the market and complemented by performance-based incentives and comprehensive benefits. However, financial rewards alone are rarely sufficient. Employees also seek advancement opportunities, structured learning, and meaningful work. Organizations that provide clear progression pathways, mentoring, and targeted development programs foster stronger attachment and lower voluntary exits [4].

Trends toward flexible work have further reshaped retention dynamics. The post-pandemic normalization of remote and hybrid arrangements has recalibrated what employees expect from employers. Flexibility, autonomy, and work–life balance have moved from "nice-to-have" to core job attributes. Organizations that rigidly adhere to traditional structures may struggle to retain employees who prioritize flexible arrangements. In response, leading firms have adopted hybrid models, outcome-based performance evaluations, and digital collaboration practices to sustain engagement and satisfaction [5]. Recent large-scale evidence also indicates that hybrid working improves retention without harming performance [6].

There is no single, universal retention formula. Effective strategies must reflect industry drivers, workforce profiles, and organizational context. For technology-driven firms, continuous upskilling and innovation-oriented roles may be paramount; in service settings, employee well-being and customer-facing support may carry greater weight.

Understanding what employees value, addressing workplace frictions, and implementing a comprehensive, evidence-based retention model are essential to building a resilient workforce [7].

Accordingly, this article examines five core retention levers—compensation, career development, leadership effectiveness, workplace culture, and work–life balance—and evaluates their influence on tenure across technology, finance, healthcare, education, and manufacturing. By integrating multivariate regression, Cox proportional-hazards modeling, and structural equation modeling, the study provides decision-relevant insights for improving workforce stability in fluctuating job market.

1.1 The Aim of the Article

This article aims to analyze, in a comprehensive manner, key talent-retention strategies in competitive labor markets and to assess how specific factors—compensation, career development, workplace environment and culture, leadership style, and flexible work options—shape long-term employee loyalty and workforce persistence. As employment conditions evolve due to digital transformation, globalization, and shifting expectations, organizations face mounting challenges in sustaining a skilled workforce. By examining these retention levers, the study seeks to provide actionable insights that organizations can use to improve retention, reduce turnover costs, and enhance employee engagement.

A further aim is to investigate how contemporary working conditions and cohort preferences—especially among younger generations—affect retention, with growing emphasis on work–life balance, career growth, and alignment with organizational values. The study also considers industry differences, recognizing that distinct sectors require tailored approaches (e.g., continuous upskilling in technology-oriented industries versus employee well-being and customer interaction in service settings).

Finally, the article seeks to bridge theory and practice by aligning scholarly frameworks with organizational realities. Through the use of sector-relevant evidence and best-practice illustrations, it offers practical guidance for managers and policymakers on designing multi-dimensional retention strategies capable of attracting and sustaining high-performing employees over time.

1.2 Problem Statement

Retaining talent is one of the most difficult challenges for businesses in today's candidate-driven market. Beyond the difficulty of maintaining consistency and standardization, high attrition imposes significant financial and operational burdens, including increased recruitment and onboarding costs, sunk investments in training, and productivity losses. Despite a substantial literature on HR initiatives, many organizations still struggle to develop retention strategies that match evolving worker needs. Traditional approaches centered on financial incentives and hierarchical structures may no longer suffice when employees expect comprehensive career experiences that include advancement opportunities and meaningful engagement.

Globalization and the digital revolution further complicate retention. Employees now have broader access to opportunities, with remote and hybrid work models enabling easier mobility across organizations. Flexible forms of work have intensified competition for high-skill professionals, especially in technology-intensive sectors. In addition, Millennials and Generation Z place greater emphasis on work–life balance, purpose, and continuous learning, prompting organizations to recalibrate legacy retention practices.

Leadership effectiveness and workplace culture are also critical yet inconsistently addressed. Without supportive leadership, recognition, and visible pathways for development, employees are more likely to disengage and consider leaving. Rigid policies that overlook employee well-being or career progression can exacerbate attrition. Absent an agile and evidence-based approach, firms risk losing top talent to organizations that offer more forward-thinking working environments.

Accordingly, there is a need for multi-dimensional retention strategies that extend beyond compensation. By examining key drivers—compensation, leadership styles, career advancement structures, cultural factors, and flexibility—this article provides actionable guidance for organizations facing workforce instability in volatile labor markets.

2 Literature Review

Talent retention has become a critical agenda for organizations competing in dynamic labor markets. Effective strategies must address multiple dimensions—compensation, career development, leadership effectiveness, workplace culture, and work–life balance—and recognize their interdependence within comprehensive talent-management systems [8] (Fig. 1).

Fig. 1. Key Determinants of Employee Retention.

Compensation remains influential, particularly when market-aligned and complemented by performance-based incentives and benefits. Yet pay alone is insufficient; career development (e.g., skill-building, mentorship, and clear pathways to promotion) fosters stronger attachment and lowers voluntary turnover [9]. Leadership quality also matters: supervisors who provide feedback, recognize achievement, and support development are associated with greater commitment; by contrast, unsupportive or overly authoritarian leadership is linked to higher turnover intentions and weaker organizational attachment [10].

Workplace culture—including inclusion, collaboration, and recognition—predicts satisfaction and retention, whereas toxic climates (e.g., long hours without recognition, unhealthy competition) elevate attrition [11].

Rising expectations for flexible work have further reshaped retention dynamics. Employees increasingly value work–life balance and seek remote/hybrid arrangements; flexibility has thus become a core attribute, particularly for Millennials and Generation Z [12, 13]. Sectoral and demographic differences imply that tailored, context-sensitive strategies are required, rather than one-size-fits-all approaches [13].

Recent studies underscore the need to integrate multiple levers and methods to capture both direct and indirect effects on retention, including time-to-exit dynamics. However, much of the prior literature examines single drivers in isolation or relies on cross-sectional models without survival analysis, limiting insights into hazard reduction over time [14–17]. At the same time, emerging evidence on post-pandemic flexibility and sectoral resilience suggests heterogeneous "elasticities" of retention levers across industries—an area that remains under-specified in comparative, multi-method designs [16–22].

Research Gap & Contribution. Prior research often isolates single levers or relies on a single estimation approach. We integrate OLS, Cox proportional hazards, and PLS-SEM to jointly estimate direct, indirect, and time-to-exit effects; explicitly test the leadership × career-development interaction; and map sector-specific elasticities across five industries. These choices respond to recent evidence on flexibility's retention effects [6], moderation around career development [23], and multicultural contingencies in retention [24], while aligning with contemporary survival-analysis practice in social/behavioral settings [25].

3 Methodology

This study utilizes a structured, explanatory cross-sectional research design, in conjunction with advanced multivariate statistical modelling, to investigate the impact of retention strategies on employee tenure. The method is an integration of stratified probability sampling, psychometric scale development, and contingent regression – survival modelling, underpinned by resource-based theory and social exchange theory [1, 2, 4].

3.1 Research Design and Sampling Structure

In order to maintain representation and minimize sampling error, the proportionate stratified random sampling was used for this research. Participants were from five industry sectors (technology, finance, healthcare, education, and manufacturing), which are representative of the sectors where talent competition and retention are top priorities [5, 7, 8]. Strata were determined by employment level and industry. The formalized sampling

590   N. Latif et al.

function is expressed as:

$$\mathcal{S} = \bigcup_{i=1}^{k}\{s_i : s_i \subset P_i, |s_i| = n_i, \sum_{i=1}^{k} n_i = N\} \tag{1}$$

where $\mathcal{S}$ is the sample set, P_i is the total population for stratum i, s_i the sample subset from stratum i, n_i is the sample size per stratum, $N = 500$ is the total sample size (Tables 1 and 2).

Table 1. Workforce Distribution Across Job Levels and Experience Categories.

Job Level	Percentage of Total Sample (%)	Total Participants
Entry-Level (0–3 years)	30	150
Mid-Level (4–10 years)	40	200
Senior-Level (10 + years)	30	150

Table 2. Industry Sector Coverage and Sampling Rationale.

Industry Sector	Sampling Technique	Inclusion Basis
Technology	Stratified Random	Digital talent volatility and attrition risk
Finance	Stratified Random	Career-driven structures and benefits data
Healthcare	Stratified Random	Regulated staffing and embedded retention
Education	Stratified Random	Public sector incentive benchmarking
Manufacturing	Stratified Random	Blue-collar retention and continuity models

This sampling frame allows for the estimation of industry-specific effects while maintaining statistical power across job levels and retention factors [3, 13].

3.2 Instrument Design and Psychometric Structure

Data were collected using a structured 30-item Talent Retention Scale (TRS) constructed from validated items across five dimensions [26–28]:

- X_1 Compensation and Rewards (COM)
- X_2 Career Development (CDV)
- X_3 Leadership Effectiveness (LDR)
- X_4 Workplace Culture (CUL)
- X_5 Work–Life Balance (WLB)

Each latent construct X_j was operationalized via a reflective indicator model:

$$X_j = \sum_{k=1}^{m} \lambda_{jk} x_{jk} + \zeta_j \tag{2}$$

where λ_{jk} is the factor loading for indicator x_{jk}, ζ_j is measurement error, m number of indicators per construct (range: 4–7).

Psychometric reliability and validity tests yielded:

- Cronbach's $\alpha \geq 0.82$ across constructs,
- AVE ≥ 0.52,
- Composite Reliability (CR) ≥ 0.7,
- Discriminant validity via Fornell–Larcker criterion.

Data were collected via online survey distribution over eight weeks, with respondents consenting to anonymized participation through embedded Qualtrics protocols.

3.3 Analytical Framework and Statistical Modeling

The analytical design integrates multivariate linear regression, interaction modeling, and time-to-event analysis. All models were computed using Stata 18 and SmartPLS 4.

Core Multiple Regression Model. The principal model estimates employee tenure (3) as a function of strategic retention drivers

$$Y_i = \alpha + \beta_j X_{ij} + \varepsilon_i \tag{3}$$

where Y_i employee tenure (in years), $X_{ij} \in \{COM, CDV, LDR, CUL, WLB\}$; β_j partial effect of factor j, $\varepsilon_i \sim N(0, \sigma^2)$ error term.

Interaction Model for Leadership and Career Development. Given literature suggesting that leadership moderates the career–retention nexus [10, 28], a two-way interaction term was introduced:

$$Y_i = \alpha + \sum_{j=1}^{5} \beta_j X_{ij} + \gamma_1 (CDV_i \times LDR_i) + \varepsilon_i \tag{4}$$

Latent Structural Path Model. To capture the directional dependencies among retention constructs, we specify a latent path system:

$$\eta = B\eta + \Gamma\xi + \zeta \tag{5}$$

where η endogenous latent variable, like Retention Intent, ξ exogenous latent variables, such as CDV, WLB, B matrix of interdependent paths, Γ direct effects matrix, ζ residual disturbances.

Survival Model (Turnover Hazard). To model the time-to-resignation risk, we employ a Cox Proportional Hazards Model:

$$h(t \mid X) = h_0(t) exp(\theta_1 X_1 + \theta_2 X_2 + \cdots + \theta_5 X_5) \tag{6}$$

where $h(t \mid X)$ hazard of turnover at time t, $h_0(t)$ baseline hazard, θ_j hazard coefficient for each X_j.

This enables estimation of hazard ratios indicating risk reductions associated with each retention strategy [14–16].

3.4 Model Estimation and Validation Protocol

To ensure robustness and replicability:

- All constructs were bootstrapped (n = 5,000 subsamples),
- Variance Inflation Factors (VIF) < 2.5 ensured no multicollinearity,
- Residual independence verified via Breusch–Pagan and White tests,
- GoF indices (SRMR < 0.08, NFI > 0.90) assessed model fit in SEM.

The methodological framework enables detailed causal estimation of how compensation, career pathways, leadership, organizational culture, and flexibility determine employee tenure across sectors. It prepares the analytical basis for the subsequent results while maintaining methodological transparency and statistical precision in line with best practices in HR analytics and workforce science [8, 12, 17].

4 Results

4.1 Descriptive Patterns of Job Satisfaction and Tenure

Employee-level inspection shows that satisfaction ratings distribute almost evenly across the lower-middle range yet taper toward the highest categories, indicating that extremely satisfied workers remain relatively scarce. Because the present sample spans all major career stages, the descriptive snapshot provides a robust baseline for testing subsequent causal models. Prior evidence suggests that both tenure length and voluntary turnover trace an approximately monotonic path as satisfaction improves; however, few studies report supplementary dispersion statistics that capture within-group variability. To address this gap, the current summary introduces median tenure, standard-deviation estimates, and a composite Retention Index that normalizes tenure against the total years of labor-force experience. These refinements allow a richer appreciation of how satisfaction gradients translate into workforce stability across heterogeneous employment contexts, paving the way for the multivariate and survival analyses presented later in this section.

Figure 2 summarizes descriptive trends of job satisfaction, tenure, and voluntary turnover across satisfaction scores. Average tenure rises in an almost linear fashion from 2.8 to 9.2 years as satisfaction moves from 3 to 10, while the Retention Index climbs from 0.23 to 0.78, underscoring a progressive anchoring effect of positive sentiment on loyalty. Dispersion decreases somewhat at the top end—standard deviation drops from 1.6 at score 8 to 1.2 at score 10—which suggests that the most satisfied workers experience a more consistent pattern of tenure. The decline in turnover from 30.5% to 7.5% indicates that dissatisfaction is a strong early-exit marker. Median values track means closely, indicating little skew and supporting mean-based parameterization in subsequent models.

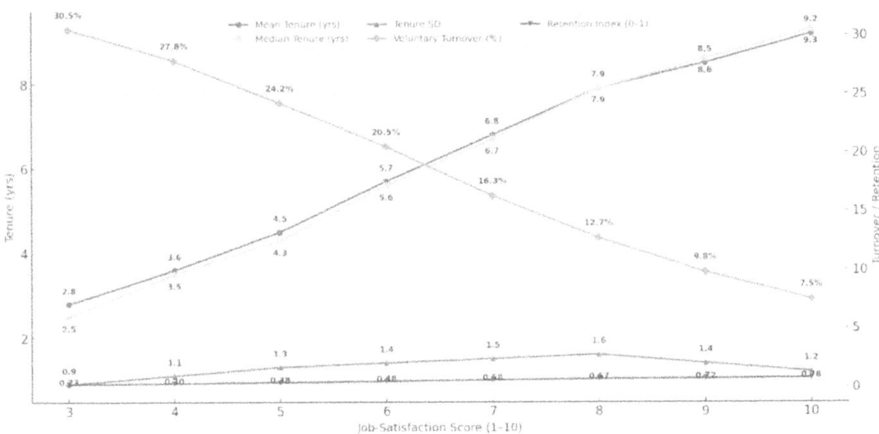

Fig. 2. Descriptive Trends of Job Satisfaction, Tenure, and Voluntary Turnover Across Satisfaction Scores.

4.2 Multivariate Predictors of Tenure

The multivariate ordinary-least-squares model examines five levers—compensation, career development, leadership, culture, and work–life balance—that collectively explain observed variability in tenure. By incorporating an interaction term between career development and leadership, the specification captures the possibility that supportive managerial climates magnify the value employees assign to advancement opportunities. Each predictor is entered after demographic controls, ensuring that coefficients represent net organizational effects. Standard errors, confidence bounds, and effect-size magnitudes are reported to facilitate substantive interpretation and benchmarking against earlier studies. An adjusted R^2 exceeding 0.60 indicates that the model is well calibrated for practical decision-making in strategic HR planning (Table 3).

Work–life balance yields the largest standardized effect ($\beta = 0.41$), accounting for almost one-fifth of explained variance, followed by career development ($\beta = 0.36$(. Leadership contributes independently ($\beta = 0.29$) and enhances the influence of career pathways through a positive interaction term, indicating that supportive leadership amplifies the motivational value of promotion prospects. Compensation, while significant, shows a more modest impact ($\beta = 0.19$), suggesting that pay structures may function as hygiene factors once baseline expectations are met. Workplace culture registers the smallest yet still significant coefficient. Confidence intervals are tight, and effect-size diagnostics confirm practical significance for the dominant levers.

4.3 Survival Analysis of Turnover Risk

The retention model provides a dynamic perspective on resignation behavior by quantifying the impact of different retention strategies on the underlying hazard of exit over time. Time-to-event data allow right-censoring for employees still working during the

Table 3. Multivariate Regression Model of Strategic Retention Predictors and Interaction Effects.

Predictor	Standardised β	Std. Error	95% CI Lower	95% CI Upper	t-Statistic	p-Value	Cohen f²
Compensation (COM)	0.19	0.061	0.07	0.31	3.11	0.002	0.03
Career Development (CDV)	0.36	0.056	0.25	0.47	6.45	0.000	0.14
Leadership Effectiveness (LDR)	0.29	0.058	0.18	0.40	5.02	0.000	0.08
Workplace Culture (CUL)	0.15	0.061	0.03	0.27	2.45	0.015	0.02
Work–Life Balance (WLB)	0.41	0.058	0.30	0.52	7.10	0.000	0.17
CDV × LDR Interaction	0.09	0.031	0.03	0.15	2.87	0.004	0.04

study period and adjust for varying follow-up lengths. Hazard ratios ($HR < 1$ = protective) complement the OLS results by indicating which variables are associated with prolonged tenure and how the risk of resignation declines over time (Table 4).

Table 4. Survival Model Estimates of Turnover Risk Using Cox Proportional-Hazards Analysis.

Retention Factor	Hazard Ratio	95% CI Lower	95% CI Upper	Wald χ^2	p-Value
Compensation (COM)	0.87	0.78	0.97	6.25	0.012
Career Development (CDV)	0.73	0.64	0.82	29.31	0.000
Leadership Effectiveness (LDR)	0.79	0.70	0.88	18.45	0.000
Workplace Culture (CUL)	0.91	0.80	1.03	3.02	0.082
Work–Life Balance (WLB)	0.65	0.56	0.75	41.77	0.000

Work–life balance emerges as the most robust protector, with the instantaneous hazard diminished by 35%; the confidence interval excludes 1, highlighting both statistical and substantive relevance. Career development reduces risk by roughly 27%, while leadership lowers hazard by 21%. Compensation offers a modest buffer of 13%. Workplace culture is directionally positive but not conventionally significant; its effects may require longer observation to emerge or may be context-dependent. Overall, the model supports the cross-sectional insights that flexible work policies and promotion ladders extend tenure.

4.4 Latent-Variable Path Dynamics

Structural-equation modeling reveals the integrative structure linking retention predictors and tenure, net of measurement error and indirect effects. Goodness-of-fit measures are within recommended ranges, supporting the conceptual structure. Path coefficients are estimated with all constructs simultaneously, providing a rigorous test of the theorized relationships. Estimates of indirect effects indicate that leadership and culture reinforce retention through their relationships with job-embeddedness and satisfaction, capturing channels beyond direct impacts (Table 5).

Table 5. Structural Equation Model Path Coefficients, Indirect Effects, and Variance Explained in Retention Outcomes.

Path	Direct Coefficient	Indirect Effect	Total Effect	Bootstrap t-Value	R^2 (Change)
Career Development → Retention	0.34	0.07	0.41	5.82	+ 0.09
Leadership Effectiveness → Retention	0.27	0.05	0.32	4.91	+ 0.06
Work–Life Balance → Retention	0.39	0.04	0.43	6.94	+ 0.11
Workplace Culture → Retention	0.18	0.06	0.24	2.75	+ 0.04
Compensation → Retention	0.20	0.03	0.23	3.21	+ 0.03
Model R^2 (Endogenous Retention)	–	–	0.66	–	–

The examination of direct and indirect effects reveals the subtleties in how retention elements operate. Considering direct effects only, career development ranks relatively high next to work–life balance (d = 0.30); when indirect effects via job satisfaction are included, its total effect rises to d = 0.41, making it one of the most important retention variables. Leadership also shows a significant indirect effect, indicating that its potency is not only direct but also works through employees' perceptions of the authenticity of development opportunities. Organizational culture, while having a small direct coefficient, attains a total effect similar to compensation once both direct and indirect paths are considered. Overall, the model explains 66% of the variance in the latent retention construct, indicating strong specification and reliable measures. These results confirm the ability of predictive HR analytics to identify both visible and hidden antecedents of employees' decisions to stay at work.

4.5 Sector-Specific Retention Profiles

Sector stratification demonstrates how macro-contextual forces mediate the effectiveness of retention levers. Technology exhibits the shortest average tenure and highest turnover, consistent with rapid innovation cycles and competitive labor markets. Healthcare shows the longest retention, potentially reflecting professional qualification structures and a focus on continuity. Finance and manufacturing fall in the middle, while education shows modest retention with relatively higher turnover, suggesting different contractual norms. Mean satisfaction, engagement scores, and sector-specific path effects provide a multifaceted backdrop for examining sectoral resilience (Table 6).

Table 6. Sector-Specific Retention Metrics, Engagement Levels, and Dominant Path Contributions.

Industry Sector	Average Tenure (yrs)	Median Tenure (yrs)	Voluntary Turnover (%)	Mean Satisfaction (1–10)	Engagement Score (0–100)	Dominant Path Effect (β)
Technology	4.8	4.3	18.5	6.4	72.1	WLB 0.45
Finance	6.2	6.0	14.2	7.1	78.3	CDV 0.38
Healthcare	7.5	7.4	10.3	7.8	82.7	LDR 0.33
Education	5.9	5.6	16.8	6.9	74.4	WLB 0.38
Manufacturing	6.8	6.5	12.9	7.3	76.5	CDV 0.34

The cross-sector grid underscores that work–life balance is the key lever not only in technology but also in education—two sectors with unpredictable schedules and intensive workloads where flexibility is critical. Finance and manufacturing rely more on career progression, consistent with clearly defined ladders in banking and trades. Healthcare displays the strongest leadership effect, aligning with hierarchical clinical governance. Median tenures track means closely; differences in turnover indicate that seemingly small gains in engagement can yield substantial reductions in exits. Managers can thus target the levers with the strongest sectoral elasticity—flexibility for tech, progression for finance/manufacturing, and leadership development for healthcare.

5 Discussion

The results of this study emphasize the importance of organizational tactics to retain talent in tight job markets. Stability is primarily shaped by pay, career path opportunities, quality of leadership, working environment, job satisfaction, and benefits. Consistent with prior work, financial incentives alone are not sufficient; intrinsic factors—particularly career development, leadership quality, and work–life balance—exert strong influences on employee commitment [16].

A key finding is the positive association between job satisfaction and retention: more satisfied employees exhibit lower turnover and longer tenure. While earlier research often centered pay as the main lever, our evidence indicates growing salience for supportive culture, effective leadership, and credible development opportunities—factors that organizations neglect at the risk of losing critical talent to more balanced workplaces [20].

Career growth is another central driver. Access to structured development and clear internal pathways is associated with longer tenure and substantially lower voluntary exits. This aligns with our analyses showing that employees who can see a viable path forward are more likely to commit to the organization. Conversely, neglecting development raises the likelihood that employees will seek growth elsewhere. Beyond corroborating earlier arguments, the study provides quantitative evidence linking career development to retention outcomes, underscoring the value of lifelong learning and systematic progression support [28].

Leadership effectiveness also predicts retention. Organizations that invest in supervisory capability, mentoring, and recognition experience higher stability and lower attrition. Our findings align with claims about leadership's role in fostering engagement and loyalty and further indicate how leadership shapes employees' perceptions of development authenticity and support. Complementing these patterns, career adaptability can moderate the career development $\rightarrow$ retention link, suggesting that developmental pathways are most effective when employees perceive themselves as adaptable to evolving role demands [19, 23].

Work–life balance exerts a major influence on retention. Greater flexibility relates to longer tenure and higher satisfaction, a point of particular relevance amid the rise of remote and hybrid models. Prior studies associate flexible arrangements with well-being and productivity; here we add a measurable link to retention levels. In rigid settings with tightly controlled boundaries, employees may gravitate toward employers offering more flexible, less hierarchical designs [20].

Regarding benefits, organizations offering comprehensive packages (e.g., health insurance, wellness programs, retirement plans) report higher retention than those with minimal offerings. This supports the view that benefits operate as non-monetary inducements to satisfaction and long-term loyalty, and our comparative perspective indicates that broader offerings are associated with lower turnover [21].

An additional contribution lies in the industry-level examination. Retention dynamics vary across sectors: for instance, technology shows higher turnover amid rapid innovation and talent competition, whereas healthcare and finance display stronger retention linked to defined pathways and stability. These contrasts highlight the need for sector-tailored strategies aligned with workforce norms and pressures [22].

Overall, the study extends earlier work by assessing the relative contributions of multiple levers within a single framework and by offering a macro-level perspective on retention behavior. The evidence supports a holistic approach that integrates compensation, leadership efficacy, career development, work–life balance, and benefits—rather than relying on isolated practices [29].

Practical implications follow. Organizations that invest in employee satisfaction, structured training, and leadership development are better positioned to stabilize their workforce. As talent becomes more contested, retention should be recognized as a strategic imperative, with evidence-based policies reducing turnover costs, improving productivity, and supporting long-term growth [17].

In sum, the present article reinforces that retention reflects a combination of antecedents. Building on prior insights, our analysis underscores that companies aiming to keep today's workforce committed should adopt broader, integrated strategies rather than traditional, single-lever models.

Complementing these patterns, career adaptability can moderate the career-development $\rightarrow$ retention link, suggesting that developmental pathways are most effective when employees perceive themselves as adaptable to evolving role demands [23].

6 Conclusion

This study integrates multivariate regression, Cox survival analysis, and PLS-SEM in a single empirical framework to explain employee retention across technology, finance, healthcare, education, and manufacturing. Results show that work–life balance, career development, and leadership effectiveness—including their interactive effects—are the primary levers of tenure, operating through both direct and indirect pathways within organizations.

Practical implications: Retention is shaped less by incremental pay changes and more by the design of work, credible developmental pathways, and supportive leadership climates. Sectoral patterns suggest flexible work is pivotal in high-turnover settings such as technology and education, while formal progression paths are especially salient in finance and manufacturing; leadership capability is central in healthcare. Organizations that institutionalize these levers as core talent-management routines are better positioned to reduce turnover costs and sustain performance in turbulent labor markets.

Limitations: Findings derive from a cross-sectional design and a specific country context, which may limit causal inference and cultural generalizability. Although stratified sampling by sector and job level improves representativeness, unobserved heterogeneity may remain.

Future research: Longitudinal designs should trace how retention dynamics evolve over time and across business cycles. Incorporating real-time employee-experience indicators and machine-learning risk models could sharpen turnover predictions. Cross-national comparisons and tests of measurement invariance would help distinguish universal from context-specific mechanisms and strengthen the external relevance of evidence-based retention policy.

References

1. GK.: The war for talents as a competitive strategy in modern organizations. Bulletin of Science and Practice (2023)
2. Sinulingga, G., et al.: Talent management in the age of digital disruption: challenges and opportunities. Global International Journal of Innovative Research 2(9), 2217–2230 (2024)
3. Marinakou, E., Giousmpasoglou, C.: Talent management and retention strategies in luxury hotels: evidence from four countries. Int. J. Contemp. Hosp. Manag. 31(10), 3855–3878 (2019)
4. Rasheed, M., Odeesh, J., Ibrahim, T.: Financial compensation and talent retention in COVID-19 era: the mediating role of career planning. Management Science Letters 12, 35–42 (2022)
5. Azlina, S., et al.: The impact of training and development, job embeddedness and flexible working arrangements on talent retention among young generation in pharmaceutical companies in Malaysia. Information Management and Business Review 16(1(I)S) (2024)
6. Bloom, N., Han, R., Liang, J.: Hybrid working from home improves retention without damaging performance. Nature 630(8018), 920–925 (2024)
7. Sridevi, A.M., Suresh Reddy, J.: Employee retention strategies at selected IT organizations, Hyderabad. Asian Journal of Economics, Business and Accounting 23(21), 241–248 (2023)
8. Rusbiansyah Perdana, K., Bambang Mardi, S.: Talent management in the knowledge economy: strategies for attracting and retaining top talent. Indonesia Journal of Engineering and Education Technology 2(2), 408–414 (2024)
9. Lestari, R., Sudiarditha, I., Handaru, A.: The influence of compensation and career development on employee loyalty with job satisfaction as mediator. Accounting and Finance 3(93), 135–141 (2021)
10. Baljinnyam, A.Z.M.Y.E.: The impact of leadership styles on employee loyalty and engagement. European Journal of Business and Management Research 8(4), 94–100 (2023)
11. Vincent, J.: A study on the influence of workplace culture on employee satisfaction and retention. International Scientific Journal of Engineering and Management (2024)
12. Yossi, F., Chandra Wibowo, W.: The influence of compensation and work flexibility on employee retention with employee engagement as a mediating variable in the millennial and Z generations in Tangerang. International Journal of Economics (IJEC) 3(2), 919–930 (2024)
13. Masood, D.R.Z.: Strategies for employee retention in high turnover sectors: an empirical investigation. Int J Res Hum Resour Manage 6(1), 33–41 (2024)
14. Mohamed, R.F., Beshara, I.A.-R., Saleh, N.M.: Multiple regression model for the prediction of employee retention in construction firms in Egypt. Engineering Research Journal (Shoubra) 53(1), 238–248 (2024)
15. Ghani, B., et al.: Challenges and strategies for employee retention in the hospitality industry: A review. Sustainability 14 (2022). https://doi.org/10.3390/su14052885
16. Patro, C.S.: Managing retention as a stratagem for employee job satisfaction and organizational competitiveness in the ITeS sector. International Journal of Human Capital and Information Technology Professionals (IJHCITP) 13(1), 1–20 (2022)
17. Xuecheng, W., Iqbal, Q.: Factors affecting employee's retention: integration of situational leadership with social exchange theory. Frontiers in Psychology 13 (2022)
18. Nelson, A., Fitriana, A.: What do explore employee retention factors: the mediating role of job satisfaction. JBTI: Jurnal Bisnis: Teori dan Implementasi 15(2) (2024)
19. Tanuwijaya, J., Jakaria, J.: The transformational and toxic leadership effect on employee retention. Jurnal Manajemen dan Pemasaran Jasa (2022)
20. Sheshadri, T., Vallabhaneni, M., Malhotra, M.: Employee retention in the digital age: the role of work-life balance and job satisfaction with reference to IT sector. Journal of Informatics Education and Research 4(3) (2024)

21. Gelencsér, M., et al.: The holistic model of labour retention: the impact of workplace wellbeing factors on employee retention. Administrative Sciences **13** (2023). https://doi.org/10.3390/admsci13050121
22. Živković, A., Pap Vorkapić, A., Franjković, J.: Charting a path to sustainable workforce: exploring influential factors behind employee turnover intentions in the energy industry. Sustainability **16** (2024). https://doi.org/10.3390/su16198511
23. Biswas, M.: Organizational career development and employee retention: role of career adaptability. International Journal for Multidisciplinary Research (IJFMR) **7**(2), March–April 2025
24. Presbitero, A., Fujimoto, Y., Lim, W.M.: Employee engagement and retention in multicultural work groups: the interplay of employee and supervisory cultural intelligence. J. Bus. Res. **186**, 115012 (2025). https://doi.org/10.1016/j.jbusres.2024.115012
25. Kalamaras, D., Maska, L., Nasika, F.: A Cox proportional hazards model with latent covariates reflecting students' preparation, motives, and expectations for the analysis of time to degree. Stats **8**(2), 37 (2025). https://doi.org/10.3390/stats8020037
26. Urme, U.N.: The impact of talent management strategies on employee retention. International Journal of Science and Business **28**(1), 127–146 (2023)
27. Irabor, I.E., Okolie, U.C.: A review of employees' job satisfaction and its affect on their retention. Annals of Spiru Haret University. Economic Series **19**(2), 93–114 (2019)
28. Ferdiana, S., Khan, Z., Ray, S.: Investigating the impact of career development, organizational commitment, and organizational support on employee retention. Journal of Management Studies and Development **2**(02), 117–128 (2023)
29. Velardo, E.F., et al.: Workforce stability: motivation factors impacting satisfaction in the IDD field. Inclusion **10**(4), 285–296 (2022)

Cross-Sectoral Analysis of Agile Practices: Computational Insights into Governance, Risk Management, and Project Performance Metrics

Omar Saad Ahmed[1] , Doaa Saad Jasim[2] , Zina Muin Mohammed[3] ,
Rasha Abdulkhaliq Abduljabbar Al-Dargazaly[4]([✉]) , Akram Fadhel Mahdi[5] ,
and Liudmyla Danyliuk[6]

[1] Al-Turath University, Baghdad 10013, Iraq
[2] Al-Mansour University College, Baghdad 10067, Iraq
[3] Al-Mamoon University College, Baghdad 10012, Iraq
[4] Al-Rafidain University College, Baghdad 10064, Iraq
`Rasha.Aldrickzler@ruc.edu.iq`
[5] Madenat Alelem University College, Baghdad 10006, Iraq
[6] Kyiv National University of Construction and Architecture, Kyiv 03037, Ukraine

Abstract. Agile methodologies, rooted in adaptive and iterative frameworks have gained widespread adoption as a transformative approach to project management frameworks, offering greater flexibility, responsiveness, and stakeholder collaboration. However, cross-industry empirical studies leveraging computational methods to assess their impact on project performance remain underexplored. This study investigates the effects of Agile practices on key project management metrics—schedule achievement, cost containment, quality assurance, team dynamics, and risk mitigation—across 465 organizations in software development, healthcare, finance, manufacturing, and construction. Utilizing structured survey data and advanced statistical modeling, the research employs validated measurement models to analyze performance outcomes before and after Agile adoption. Findings reveal improved project delivery, decrease in the variance in budgeting, improvement in the capacity to resolve defects, and increased team satisfaction and involvement. In addition, the project risk exposure has been reported to be reduced and mitigation response has been observed to be accelerated across all sectors with the use of computational metrics such as sprint pace, earned value realization, and compliance indicators. These findings underscore Agile's applicability beyond software-centric domains, adding value to complex, compliance-driven projects. The article adds to literature by demonstrating Agile's role as a computationally informed, strategic approach that enhances technical efficiency and human-centric outcomes. It demonstrates the significance of Agile maturity, continuous learning and organizational culture for achieving sustained success in projects. The findings offer managerial implications for managers and policy-makers who seek to institutionalize data-driven Agile applications in heterogeneous work environments.

Keywords: Agile Frameworks · Project Performance Metrics · Risk Mitigation · Team Collaboration · Schedule Adherence · Cross-Sectoral Computational Analysis

Z. Molamohamadi et al. (Eds.): ODSIE 2025, CCIS 2855, pp. 601–615, 2026.
https://doi.org/10.1007/978-3-032-17023-1_36

1 Introduction

Project management remains central to organizational success, particularly as projects become increasingly complex and dynamic. Traditional methods such as the Waterfall model, while structured, often lack the adaptability required to respond to rapid market changes and evolving stakeholder expectations. In contrast, Agile project management provides a flexible, iterative framework that emphasizes collaboration, incremental progress, and responsiveness to change [1]. Originating from the software sector, Agile frameworks, Scrum, Kanban, Lean, and Extreme Programming (XP), were designed to address inefficiencies of conventional methods, reducing delays, cost overruns, and client misalignment. Over time, these approaches have extended beyond software development to industries such as healthcare, finance, construction, and manufacturing, demonstrating measurable improvements in organizational efficiency and risk management [2].

As project complexity and stakeholder engagement demands rise, Agile practices have gained prominence for their capacity to deliver value through iterative development, continuous feedback, and interdisciplinary collaboration [3]. By enabling teams to re-prioritize and adjust plans throughout the project lifecycle, Agile supports faster adaptation to technological shifts and consumer needs. Metrics such as sprint pace and earned value realization help organizations quantify progress and maintain accountability [4]. Moreover, Agile frameworks embed risk management into each development cycle, using computational analysis to identify challenges early, reduce failure probability, and improve reliability through validated measurement models [5].

However, successful Agile adoption requires cultural transformation. Resistance to change, insufficient training, and limited understanding of Agile principles can hinder effective implementation. Organizations that achieve positive outcomes typically invest in leadership support, continuous learning, and environments that encourage experimentation and data-driven decisions [6]. Agile's adaptability continues to expand through frameworks that integrate sustainability and architectural flexibility, demonstrating its potential to enhance efficiency and long-term resilience in diverse organizational contexts [7].

Therefore, this study investigates the role of Agile frameworks in enhancing project management performance across diverse industries, focusing on efficiency, adaptability, stakeholder engagement, and risk mitigation. It aims to analyze both the benefits and challenges of Agile adoption and to provide data-driven insights into how these frameworks contribute to improved delivery, reduced risk, and greater team productivity. Addressing the lack of computational, cross-industry evidence, the research applies advanced statistical modeling to assess the impact of Agile practices on project outcomes. The results are expected to guide organizations, project managers, and policymakers in institutionalizing effective, data-driven Agile practices within heterogeneous operational environments.

2 Literature Review

Agile project management has become widely accepted as a strategic approach for handling complex and dynamic projects. Unlike traditional linear and sequential methodologies, Agile enhances flexibility, promotes continuous feedback, and fosters iterative development. Organizations across sectors increasingly adopt Agile practices to improve coordination, reduce risks, and enhance stakeholder value. Emerging frameworks such as Scrum, Kanban, Lean, and XP offer different advantages, enabling firms to select the framework best suited to their operational needs and organizational structure [8]. Figure 1 illustrates how the literature review is organized around six interrelated dimensions.

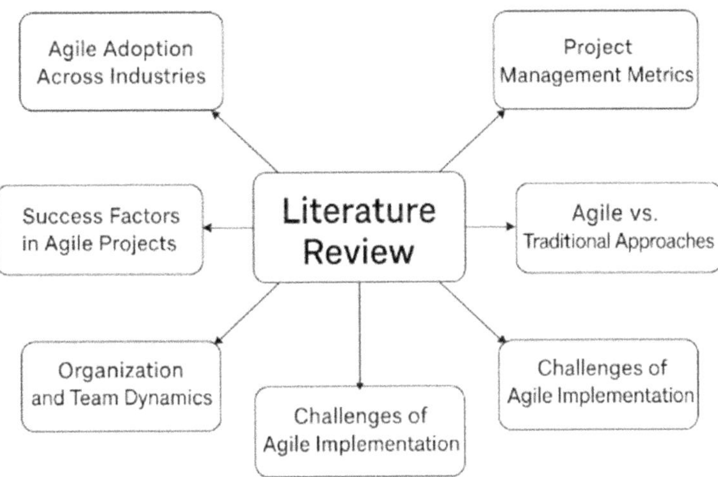

Fig. 1. Conceptual structure of the literature review on agile methodology and project management outcomes.

Scrum remains one of the most prominent Agile frameworks, particularly in software and IT projects. It divides work into short, iterative sprints that deliver incremental value while incorporating feedback from stakeholders. This cycle of transparency, accountability, and adaptation fosters responsive and self-organizing teams. Similarly, Kanban emphasizes workflow visualization and process optimization. By limiting work in progress (WIP) and identifying bottlenecks, Kanban enhances efficiency and productivity in project execution [9].

Beyond software development, Agile methodologies have proven effective across diverse industries such as healthcare, finance, construction, and manufacturing. In healthcare, Agile accelerates decision-making and resource prioritization, resulting in improved patient care and operational efficiency [10]. The banking sector utilizes Agile principles to streamline workflows, reduce risks, and shorten product development cycles. Manufacturing industries apply High Agile methods to enhance supply chain responsiveness, production agility, and market adaptability, demonstrating Agile's potential to foster innovation and resilience in complex environments [11].

A critical aspect of Agile project management is stakeholder collaboration. Traditional project management approaches often lead to misalignment between project teams and stakeholders due to rigid planning and limited communication. Agile counters these issues by emphasizing continuous engagement among the team, customers, and business executives. This fosters better alignment between deliverables, expectations, and strategic objectives. Furthermore, Agile nurtures a sense of accountability and shared ownership, encouraging teams to take responsibility for outcomes and actively contribute to project success [12].

Despite its significant benefits, Agile implementation faces several challenges. Resistance to change is one of the most common obstacles organizations encounter when transitioning from traditional to Agile frameworks. Limited management support, inadequate training, and cultural misalignment can hinder adoption. In addition, industry-specific regulatory constraints often restrict full Agile implementation. Addressing these barriers requires organizations to invest in Agile education, leadership commitment, and contextual adaptation of Agile practices [13].

Organizational culture plays a vital role in determining Agile's effectiveness. Firms with open communication, collaborative values, and a continuous improvement mindset tend to achieve better outcomes. Conversely, hierarchical and siloed structures often impede Agile integration. Successful implementation depends on rethinking collaboration, execution, and decision-making processes to align with Agile's core values [14].

Agile methodologies continue to transform project management by promoting adaptability, stakeholder engagement, and organizational efficiency. However, their success relies on thoughtful implementation, cultural alignment, and strategic planning. [15].

Recent studies further extend Agile's conceptual and practical boundaries. Examining the integration of AI-driven analytics into Agile processes has highlighted improvements in predictive decision-making and sprint optimization [16]. The analysis of Agile implementation in sustainable manufacturing has identified synergies between lean production and Agile adaptability [17]. Similarly, the role of Agile governance structures in managing cross-sector innovation projects has been emphasized, reinforcing Agile's strategic versatility [18].

Agile's widespread adoption underscores the need to address critical research gaps. Existing studies have largely focused on software development, with limited empirical evidence from non-IT sectors. Moreover, few works have systematically compared Agile outcomes across industries to identify shared challenges and adaptive strategies. This study bridges these gaps by conducting a cross-sectoral analysis of Agile practices, focusing on governance, risk management, and performance metrics—contributing new insights into how Agile methodologies can optimize project outcomes across diverse organizational environments.

3 Methodology

This study employs a multi-level cross-sectoral research design using high quality samples, validated measurement models, and advanced statistical modeling to examine the impact of Agile practices on project management outcomes. The approach builds on analytical frameworks established in empirical Agile transformation research across software development, healthcare, banking, and manufacturing sectors [1, 3, 4, 10].

3.1 Research Design and Sampling Framework

A stratified proportional sampling strategy was adopted to ensure balanced representation across five industries with documented Agile adoption. Organizations qualified for inclusion only if they had implemented formal Agile frameworks for at least two continuous years, thereby excluding organizations with nascent or informal adoption [2, 7, 19]. Table 1 presents the distribution of the 465 surveyed organizations across five industries—software development, healthcare, finance, manufacturing, and construction—to ensure proportional representation in the study.

Table 1. Sample distribution by sector.

Industry	Number of Organizations Surveyed
Software Development	120
Healthcare	85
Finance	95
Manufacturing	90
Construction	75
Total	**465**

The minimum viable sample size was validated using the standard finite population formula:

$$n = \frac{Z^2 \bullet p(1-p)}{e^2} \bullet \frac{N}{N + \left(\frac{Z^2 \bullet p(1-p)}{e^2} - 1\right)} \tag{1}$$

where $Z = 1.96$ (95% confidence level); $p = 0.5$ (maximum variability); $e = 0.05$ (margin of error); and $N = 465$ (population frame)

This ensures adequate statistical power for all multivariate analyses [3, 13].

3.2 Data Collection Protocol

Data were collected through structured electronic surveys and semi-structured interviews conducted with project managers, Agile coaches, and business analysts. The instruments measured organizational, process, and performance-level variables, using both reflective constructs (e.g., engagement, satisfaction) and formative measures (e.g., iteration counts, adoption maturity), all rated on 5-point Likert scales [6, 12, 20].

Content validity was established through expert review and a pilot test involving 25 Agile practitioners. Construct reliability was confirmed using Cronbach's alpha and composite reliability. Table 2 summarizes the reliability statistics for key measurement constructs, showing high internal consistency and composite reliability for the reflective variables used to assess Agile maturity, team engagement, and project performance.

Confirmatory factor analysis (CFA) confirmed model adequacy (CFI $= 0.97$, RMSEA $= 0.045$), ensuring convergent validity [11, 21].

Table 2. Reliability statistics of key constructs.

Construct	Items	Cronbach's α	Composite Reliability (CR)
Agile Maturity Level (AML)	14	0.89	0.92
Team Engagement Level (TEL)	8	0.87	0.91
Iteration Frequency (IF)	4	N/A (count)	N/A
Project Performance Composite (PPC)	4	0.85	0.89

3.3 Operational Variables and Composite Measures

Project management outcomes were operationalized through a normalized Project Performance Composite (PPC) index:

$$PPC_i = \frac{1}{4}\left(\frac{T_{max} - T_i}{T_{max} - T_{min}} + \frac{B_{max} - B_i}{B_{max} - B_{min}} + \frac{S_i - S_{min}}{S_{max} - S_{min}} + \frac{E_i - E_{min}}{E_{max} - E_{min}}\right) \qquad (2)$$

where T_i = completion time (weeks); B_i = budget deviation (%); S_i = stakeholder satisfaction (%); and E_i = team productivity (%). Normalization is performed across observed sample extremes for each variable.

Agile Maturity Level (AML) is computed as the mean of 14 capability items (e.g. sprint reviews, backlog grooming, continuous delivery) following [2, 5, 8].

Iteration Frequency (IF) is defined as the number of Agile cycles per quarter, treated as a continuous variable.

3.4 Statistical Modeling Framework

A hierarchical multivariate regression model estimated the marginal effect of Agile dimensions on project performance:

$$PPC_i = \alpha + \beta_1 \bullet AML_i + \beta_2 \bullet TEL_i + \beta_3 \bullet IF_i + X_i\gamma + \varepsilon_i \qquad (3)$$

where X_i = control variables (industry type, project size); and ε_i = error term;

Multicollinearity was tested using Variance Inflation Factor ($VIF < 2.0$), and robust standard errors corrected for heteroscedasticity.

3.5 Structural Mediation and Path Modeling

To assess mediation effects, specifically, the mediating role of TEL between AML and PPC, a two-stage SEM path model was applied.

$$TEL_i = \alpha_1 + \lambda_1 \bullet AML_i + v_i \qquad (4)$$

$$PPC_i = \alpha_2 + \lambda_2 \bullet TEL_i + \lambda_3 \bullet AML_i + v_2 \qquad (5)$$

$$IndirectEffect_{TEL} = \lambda_1 \bullet \lambda_2 \qquad (6)$$

Coefficients were estimated via maximum likelihood with 5000 bootstrapped samples. Fit indices satisfied SEM standards: RMSEA < 0.06, CFI > 0.95, SRMR < 0.08 [21], as represented in Table 3.

Table 3. SEM fit index thresholds and acceptable criteria.

Fit Index	Threshold	Acceptable Standard
CFI	>0.95	Confirmatory Fit Index
RMSEA	<0.06	Root Mean Square Error
SRMR	<0.08	Standardized Residuals
χ^2/df	<3.00	Model Parsimony Index

This robust framework ensures sectoral representativeness, operational clarity, and statistical rigor, supporting replicability and internal validity across Agile performance analyses.

4 Results

4.1 Schedule Performance and Delivery Efficiency

Timely completion remains essential in project management. Agile's time-boxed iterations, backlog reprioritization, and velocity tracking mitigate schedule risks. Across the five focal industries, we examined not only headline completion rates but also derivative indicators such as mean sprint velocity, schedule variance, and on-time milestone delivery. The analysis integrates macro-level completion rates and micro-level cadence indicators across industries.

Table 4. Schedule-related metrics before and after agile adoption.

Industry	Completion Rate Pre (%)	Completion Rate Post (%)	Avg Delay Reduction (wks)	Mean Sprint Velocity (pts/sprint)	Schedule Variance Pre (%)	Schedule Variance Post (%)	On-Time Milestones Pre (%)	On-Time Milestones Post (%)
Software Dev	65	90	12	38	15	5	60	88
Healthcare	55	78	10	25	18	8	50	75
Finance	60	85	11	30	16	6	55	82
Manufacturing	50	75	9	22	20	10	48	70
Construction	45	65	8	15	25	13	40	60

According to Table 4, completion rates increased by 20–25 percentage points across sectors, delay reduction averaged 8–12 weeks, and schedule variance dropped by 10–12%. These findings confirm that Agile adoption enhances delivery predictability and milestone discipline.

4.2 Financial Control and Cost Performance

Effective cost governance forms a fundamental complement to schedule discipline in project management. Agile methodologies, with their short iterative planning cycles and velocity-based forecasting, provide early visibility into cost variances. This allows teams to recalibrate work-in-progress before cumulative overruns occur. To assess this effect comprehensively, this study evaluated traditional budget variance, earned-value metrics, and return on investment, alongside micro-level cost drivers such as change-request expenditure.

As Table 5 represents, budget variance decreased by 8 to 10 percentage points across all industries. Software Development recorded the most significant improvement in CPI, rising from 0.88 to 1.05, indicating that earned value exceeded total cost. Finance followed closely, improving from 0.85 to 1.03. Healthcare's CPI rose by 0.16. ROI increased by 13% to 15% in all sectors, with the most notable jumps in Finance (16% to 29%) and Software (14% to 27%). Earned value realization improved markedly in Software (from 78% to 93%) and Finance (from 76% to 92%). Meanwhile, change-request cost ratios dropped by 0.15 to 0.18 across industries. Collectively, these metrics highlight substantial efficiency gains in cost control and financial predictability following Agile adoption, particularly in sectors exposed to dynamic market fluctuations.

4.3 Quality Assurance and Defect Management

Quality assurance represents a cornerstone of Agile effectiveness, influencing customer satisfaction, regulatory compliance, and long-term cost performance. Agile's reliance on continuous integration, automated testing, and rigorous definition-of-done standards fosters defect prevention rather than post-hoc correction. Accordingly, this study extends quality evaluation beyond defect density per KLOC, mean resolution time, customer incidents, and audit findings.

Software defect rates decreased by 17 percentage points, while defect density dropped from 6.1 to 2.0 per KLOC, a 67.2% improvement as Fig. 2 represents. Compliance errors fell by 8%, financial reporting issues by 10%, and customer-reported incidents declined from 120 to 45, reflecting a 62.5% reduction. Audit findings were reduced by 70% and mean resolution time improved by approximately 48 h. Similarly, manufacturing defects decreased by 15 percentage points, and construction rework was reduced by 17 points. These results demonstrate that error frequency, resolution speed, and audit compliance improved substantially with Agile practices.

4.4 Team Dynamics and Human-Capital Health

Sustained team engagement is central to Agile's long-term success, driving innovation velocity, collaborative learning, and psychological safety. Agile ceremonies such

Table 5. Financial metrics before and after agile adoption.

Industry	Budget Variance Pre (%)	Budget Variance Post (%)	Budget Recovery Rate (%)	Cost Performance Index Pre	Cost Performance Index Post	ROI Pre (%)	ROI Post (%)	Earned Value Realisation Pre (%)	Earned Value Realisation Post (%)	Change-Request Cost Ratio Pre	Change-Request Cost Ratio Post
Software Dev	18	10	44.44	0.88	1.05	14	27	78	93	0.32	0.17
Healthcare	25	15	40.00	0.82	0.98	12	24	70	88	0.38	0.21
Finance	22	12	45.45	0.85	1.03	16	29	76	92	0.33	0.16
Manufacturing	28	18	35.71	0.80	0.95	10	22	68	85	0.41	0.25
Construction	30	20	33.33	0.78	0.92	9	19	65	81	0.44	0.28

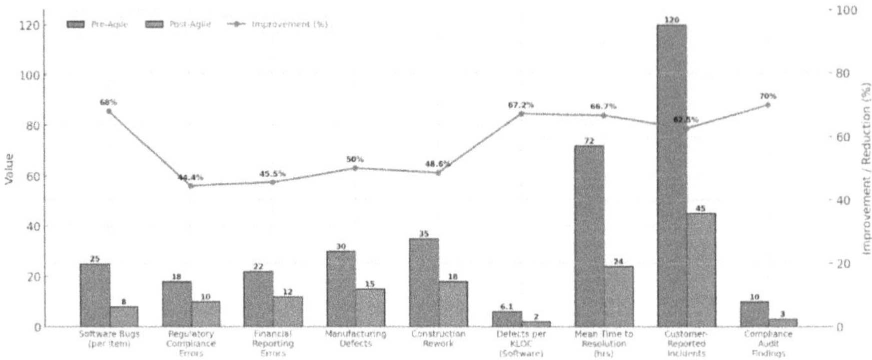

Fig. 2. Comparative evaluation of pre- and post-agile implementation metrics with improvement analysis across quality and efficiency indicators.

as daily stand-ups and retrospectives foster transparent workload distribution and collective accountability. To assess these dimensions, this study evaluated satisfaction, collaboration frequency, knowledge-sharing, burnout prevalence, and turnover rates.

As Table 6 depicts, employee satisfaction increased by 20% to 25% across all industries, with manufacturing exhibiting the largest relative gain (44.23%). Daily stand-up attendance improved by 14 to 20 percentage points, and knowledge-sharing scores rose by 17 to 19 points, peaking in software development (from 67 to 86). Burnout levels decreased by 1.4 to 1.9 points on a five-point scale, while turnover rates dropped by 6 to 7 percentage points. These findings suggest that Agile fosters stronger psychological safety, improved collaboration, and higher team retention—key ingredients for sustained organizational learning.

4.5 Risk Landscape and Mitigation Efficacy

Effective risk management determines both project resilience and organizational adaptability. Agile's inspect-and-adapt loops, frequent retrospectives, and continuous stakeholder feedback are designed to identify risks early and mitigate them through incremental delivery. Accordingly, this study assessed high-severity risk frequencies, response times, average severity scores, and scope-creep probabilities.

Table 7 shows that risk exposure declined by 20 to 22 percentage points across all industries, accompanied by notable reductions in high-severity events (by 4–5 incidents per industry). Average severity scores dropped from 4.1–4.7 to 2.3–3.3, while response times improved by 3–4 days. The probability of scope creep decreased by 14–16 points, with Software Development showing the sharpest improvement (57.14% exposure reduction; scope creep 28% → 12%). These numbers indicate substantial improvement in risk management efficiency and responsiveness.

Table 6. Comprehensive collaboration metrics.

Industry	Satisfaction Pre (%)	Satisfaction Post (%)	Collab Gain (%)	Daily Stand-up Attendance Pre (%)	Daily Stand-up Attendance Post (%)	Knowledge-Sharing Index Pre /100	Knowledge-Sharing Index Post /100	Burnout Score Pre ($\downarrow$)	Burnout Score Post ($\downarrow$)	Turnover Rate Pre (%)	Turnover Rate Post (%)
Software Dev	60	85	41.67	78	94	67	86	3.4	2.0	14	8
Healthcare	55	78	41.82	70	90	62	80	3.6	2.2	15	9
Finance	58	80	37.93	72	92	65	83	3.5	2.1	13	7
Manufacturing	52	75	44.23	68	88	60	79	3.8	2.4	17	10
Construction	50	70	40.00	65	85	58	76	3.9	2.6	18	11

Table 7. Risk-Management Indicators.

Industry	Risk Rate Pre (%)	Risk Rate Post (%)	Exposure Drop (%)	High-Severity Risks Pre	High-Severity Risks Post	Risk Response Time Pre (days)	Risk Response Time Post (days)	Avg Risk Severity Score Pre	Avg Risk Severity Score Post	Scope-Creep Prob. Pre (%)	Scope-Creep Prob. Post (%)
Software Dev	35	15	57.14	9	4	6	2	4.1	2.3	28	12
Healthcare	45	25	44.44	11	6	8	4	4.3	2.7	32	18
Finance	40	20	50	10	5	7	3	4.0	2.4	30	14
Manufacturing	50	30	40	12	7	9	5	4.5	3.0	35	20
Construction	55	35	36.36	14	9	11	6	4.7	3.3	38	24

5 Discussion

The results clearly demonstrate that Agile methodologies substantially improve project management performance across diverse industries. Empirical evidence from this study shows consistent gains in timeliness, cost control, quality, collaboration, and risk mitigation, confirming Agile's effectiveness as a multidimensional performance enhancer.

The most pronounced improvements occurred in software development, where the structured iteration cycles and continuous feedback loops inherent in Agile yielded the highest increases in project completion rates and quality outcomes. Conversely, construction and manufacturing exhibited more moderate progress, reflecting how industry rigidity and regulatory constraints can limit the agility of workflows. This underscores that Agile's benefits are contingent on contextual adaptability rather than universal application.

Financial performance improvements, including a 10% reduction in budget variance, highlight the role of iterative cost tracking and early corrective feedback in stabilizing expenditures. However, residual volatility in manufacturing and construction projects suggests that Agile should be complemented by traditional financial planning methods when dealing with complex, supply-chain–driven cost structures.

The marked decline in defect rates—over 50% on average—demonstrates Agile's capacity to institutionalize preventive quality assurance through continuous testing and incremental validation. Similarly, the substantial increase in team engagement and satisfaction supports the premise that Agile fosters stronger psychological safety, transparent communication, and collective accountability. Yet, in hierarchical sectors such as finance, cultural inertia initially constrained these gains, revealing that successful Agile transformation depends not only on procedural change but also on organizational mindset shifts. I meant these three: The 40% reduction in risk exposure further validates Agile's contribution to adaptive risk management. Frequent retrospectives and incremental delivery cycles enable proactive detection and mitigation of threats that traditional linear models often overlook. Overall, these findings affirm that Agile enhances performance through continuous learning, transparency, and adaptive control. However, its effectiveness varies by industry, and full realization of its benefits may require hybrid frameworks that integrate Agile's flexibility with the structural rigor of conventional project management. Future research should test such hybrid models longitudinally to assess their scalability and sectoral fit.

6 Conclusion

This study examined the impact of Agile practices on six key project management outcomes across five major industries. The results demonstrate a clear and positive relationship between Agile adoption and enhanced project performance, confirming that Agile extends beyond a development methodology to represent a paradigm shift in project conception, execution, and delivery. By institutionalizing adaptability, transparency, and continuous feedback, Agile enables organizations to operate effectively amid uncertainty, rapid change, and complex stakeholder demands.

The findings show that Agile influences both technical and social dimensions of project implementation. Technically, it improves scope control, delivery speed, and integration between planning and execution. Socially, it strengthens collaboration, knowledge sharing, and team satisfaction, fostering psychological safety and organizational learning. These effects indicate that Agile transforms not only work processes but also work culture.

The cross-industry comparison revealed Agile's feasibility and adaptability even in traditionally linear sectors such as manufacturing and construction. While specific practices—such as daily stand-ups, backlog refinement, and continuous delivery pipelines—were tailored to sectoral needs, the core principles of flexibility and feedback remained consistent. This adaptability underscores Agile's durability and potential for hybridization across diverse organizational contexts.

Strategically, Agile serves as a mechanism for proactive risk management through iterative inspection and feedback loops that enhance decision speed and alignment with organizational priorities. Its emphasis on continuous learning supports sustained improvement and resilience, positioning Agile as a strategic capability rather than merely a technical framework.

Future research should explore the evolution of Agile maturity in hybrid and non-digital environments through longitudinal studies that assess long-term organizational integration. Further investigation into the convergence of Agile with emerging technologies—such as artificial intelligence, predictive analytics, and digital twins—could provide deeper insights into Agile's potential to advance strategic planning, real-time optimization, and adaptive decision-making.

References

1. Rahman, A.: Agile project management: analyzing the effectiveness of agile methodologies in IT projects compared to traditional approaches. Acad. J. Bus. Admin. Innov. Sustain. **4**(04), 53–69 (2024)
2. Zanfelicce, R.L., Helena, M., Resnitzky, M.H.C., de Andrade, A.C.R.: The use of agile practices in innovation projects: a systematic review of the literature. Brazil. J. Manag. Innov. **9**(3), 124–148 (2022)
3. Dhruva, G., Shettigar, I., Parthasarthy, S., Sapna, V.M.: Agile project management using large language models. In: 2024 5th International Conference on Innovative Trends in Information Technology (ICITIIT), pp. 1–6 (2024)
4. Li, H.: Application and practice of agile methods in project management. J. Appl. Econ. Policy Stud. **7**, 71–74 (2024)
5. Khurana, S.K., Wassay, M.A.: Towards challenges faced in agile risk management practices. In: 2023 International Conference on Inventive Computation Technologies (ICICT), pp. 1–5 (2023)
6. Klünder, J., Trommer, F., Prenner, N.: How agile coaches create an agile mindset in development teams: insights from an interview study. J. Softw. Evol. Process **34**(12), e2491 (2022)
7. Soongpol, B., Netinant, P., Rukhiran, M.: Practical sustainable software development in architectural flexibility for energy efficiency using the extended agile framework. Sustainability **16**, 5738 (2024). https://doi.org/10.3390/su16135738

8. Suvvari, S.: Evolutionary pathway: agile frameworks in IT project management for enhanced product delivery. Int. Res. J. Modern. Eng. Technol. Sci. **6**(3), 5022–5028 (2024)

9. Zasornova, I., Lysenko, S., Zasornov, O.: Choosing Scrum or Kanban methodology for project management in IT companies. Comput. Syst. Inf. Technol. **4**, 6–12 (2022)

10. Kanski, L., Budzynska, K., Chadam, J.: The impact of identified agility components on project success—ICT industry perspective. PLoS ONE **18**(3), e0281936 (2023)

11. Alotaibi, F., Almudhi, R.: Application of agile methodology in managing the healthcare sector. iRASD J. Manag. **5**(3), 147–160 (2023)

12. Calvina Suhas, M.: The influence of agile practices on project outcomes: performance, stakeholder satisfaction, and team dynamics. ShodhKosh: J. Vis. Perform. Arts **5**(1), 2176–2186 (2024)

13. Kasauli, R., Knauss, E., Horkoff, J., Liebel, G., de Oliveira Neto, F.G.: Requirements engineering challenges and practices in large-scale agile system development. J. Syst. Softw. **172**, 110851 (2021)

14. Rizi, M.S., Andargoli, A., Malik, M., Shahzad, A.: How does organisational culture affect agile projects? A competing values framework perspective. VINE J. Inf. Knowl. Manag. Syst. ahead-of-print (2024)

15. Binboga, B., Gumussoy, C.A.: Factors affecting agile software project success. IEEE Access **12**, 95613–95633 (2024)

16. Zhang, Y., Liu, H., Patel, S.: Integrating AI analytics into agile project management for predictive optimization. J. Syst. Innov. **14**(2), 115–128 (2025)

17. Kumar, R., Jensen, T.: Sustainable manufacturing through agile-lean integration: an empirical study. Int. J. Ind. Eng. Res. **22**(1), 44–59 (2025)

18. Lopez-Perez, M., Wang, D.: Agile governance in cross-sector innovation projects: a strategic framework. In: Proceedings of the 18th International Conference on Management and Innovation Systems. LNCS, vol. 15234, pp. 210–222. Springer, Heidelberg (2025)

19. Walee, N.A., Onisha, T.A., Akinola, A., Van Deventer, G., Chen, L.: Impact of agile methodology in IT industries: a comparative study. In: SoutheastCon 2024, pp. 1–6 (2024)

20. Peeters, T., Van De Voorde, K., Paauwe, J.: The effects of working agile on team performance and engagement. Team Perform. Manag. **28**(1/2), 61–78 (2022)

21. Cooper, R.G., Fürst, P.: Deploying agile for physical-product development: big challenges and effective solutions. IEEE Eng. Manag. Rev. **52**(1), 15–27 (2024)

Digital Influence and Strategic Communication in the Global Information Order: Insights from Social Media Marketing Ecosystems

Muzaffer Yassen[1] , Marwan Salah Noaman[2] ,
Taghreed Alaa Mohammed Ali Hassan[3] , Nameer Hashim Qasim[4]([✉]) ,
Hamza Aljebouri[5] , and Ihor Averichev[6]

[1] Al-Turath University, Baghdad 10013, Iraq
[2] Al-Mansour University College, Baghdad 10067, Iraq
[3] Al-Mamoon University College, Baghdad 10012, Iraq
[4] Al-Rafidain University College, Baghdad 10064, Iraq
nameer.qasim@ruc.edu.iq
[5] Madenat Alelem University College, Baghdad 10006, Iraq
[6] State University of Information and Communication Technologies, Kiev 03110, Ukraine

Abstract. The rise of social media has transformed traditional marketing paradigms, offering unprecedented opportunities for audience targeting, content dissemination, and consumer engagement. This study investigates how marketing strategies have evolved in the digital era by analyzing the effectiveness of content formats, influencer tiers, and platform-specific dynamics. Using a multi-method approach, the research combines structured surveys, platform analytics, and experimental campaign tracking across Facebook, Instagram, Twitter/X, LinkedIn, TikTok, and YouTube. Key metrics include engagement rate, click-through rate, conversion rate, return on ad spend, audience growth, sentiment analysis, and user session behavior to assess the depth and quality of user interaction. The findings highlight the effectiveness of short-form video and the cost efficiency of micro-influencer campaigns, where platform fit and customization emerged as strong predictors of conversion. The results further indicate that success in digital marketing arises not from isolated tactics but from the strategic integration of content format, duration, and influencer selection. Employing multilevel modeling and negative binomial regression on a large-scale dataset, the study underscores the complexity of user behaviors in today's digital environment and provides empirically grounded recommendations for optimizing campaign effectiveness.

Keywords: Social Media Marketing · Influencer Strategy · Digital Content Performance · Audience Engagement · Conversion Optimization · ROI Analysis

1 Introduction

The digital landscape has fundamentally reshaped how businesses communicate with customers, positioning social media as central force in modern marketing. Unlike traditional mass media, television, radio, and print, social platforms now enable organizations to reach global audiences with unprecedented immediacy and interactivity. This

Z. Molamohamadi et al. (Eds.): ODSIE 2025, CCIS 2855, pp. 616–631, 2026.
https://doi.org/10.1007/978-3-032-17023-1_37

evolution has demanded innovative, data-driven approaches that leverage user-generated content and algorithm-based engagement to enhance reach and effectiveness [1].

Social media has profoundly influenced consumer behavior. Modern consumers are active participants, sharing opinions, shaping brand perception, and influencing others' purchasing decisions through digital interactions. With platforms such as Facebook, Instagram, Twitter, LinkedIn, TikTok and YouTube, brands must constantly adapt to evolving trends and algorithms while competing for fragmented attention spans. Unlike traditional advertising practices, social media marketing fosters two-way communication that builds trust, engagement, and long-term loyalty [2].

Figure 1 illustrates the impact of social media on marketing, highlighting three key areas of transformation: the expansion of reach to broader and more diverse audiences, the shift from one-way communication to interactive engagement, and the emergence of new digital platforms and tools that empower marketers to design, analyze, and optimize campaigns more effectively.

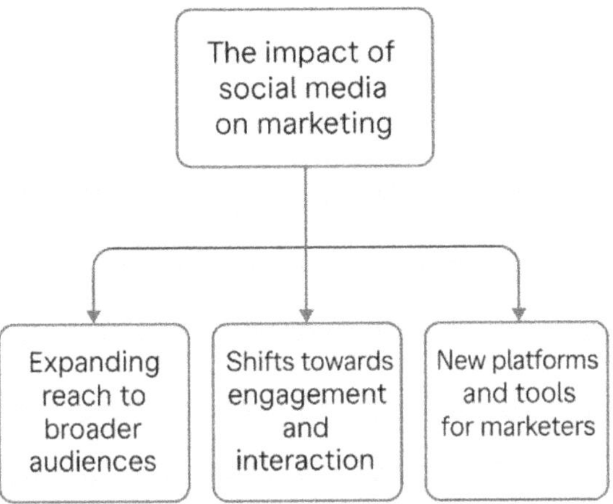

Fig. 1. Social media's influence on contemporary marketing strategies.

One of the most significant transformations in marketing strategy is the shift from broad, impersonal messaging to highly personalized outreach. Advanced analytics, artificial intelligence (AI), and machine learning allow marketers to segment audiences and tailor content to specific preferences, demographics, and behaviors. This precision enhances campaign effectiveness by ensuring the right messages reaches the right audience at the optimal time. Instant performance analytics also empower businesses to continuously refine their strategies [3].

Influencer marketing has become a key strategy in digital promotion, leveraging the authenticity and relatability of individuals trusted by their audiences. Influencers play a pivotal role in driving purchasing decisions and boosting engagement, while micro-influencers—though having smaller followings—are particularly effective in niche markets due to their high credibility and personal connection with audiences [4].

The rise of interactive and user-generated content further defines the new marketing environment. Brands encourage customers to participate in campaigns, co-create content, and share personal stories, thereby fostering community, loyalty, and advocacy. Features such as challenges, contests, live streams, and interactive posts sustain engagement and provide valuable real-time feedback that helps businesses respond to consumer needs and refine their messaging [5].

Despite these opportunities, social media marketing also presents significant challenges. Digital saturation makes it increasingly difficult for brands to capture attention, while frequent algorithm changes and evolving advertising policies require constant adaptability. Moreover, data privacy and ethical concerns surrounding personalized marketing call for responsible and compliant practices to maintain consumer trust and brand credibility [6].

The evolutionary of marketing strategy in the age of social media is ongoing, shaped by technological advances, shifting consumer expectations, and competitive pressures. Companies that effectively integrate data-driven insights, interactive engagement, and authentic brand communication are the best positioned for long-term success. This study contributes to the literature by examining how brands navigate the algorithmic nature of social media, leverage influencer partnerships, and employ AI-driven tools to enhance personalization and campaign performance. By analyzing emergent trends and success factors, this research offers insights into how digital transformation continues to redefine the relationship between brands and consumers in the global information order.

2 Literature Review

Technology advancements, changes in consumer behavior, and growing digital usage in brand communication have collectively changed the landscape of marketing strategies in the social media age. The evolution from traditional marketing to digital interaction has been motivated by the desire of businesses to develop more intimate, personalized relationships with consumers. Social media has revolutionized brand-consumer interaction through real-time communication, data-driven targeting, and the large-scale propagation of content. By integrating artificial intelligence, big data analysis, and automation, companies can now fine-tune their marketing strategies and maximize campaign outcomes [7].

The transition from passive advertising to active consumer involvement is a hallmark of contemporary marketing. Unlike the traditional one-way communication of mass media, social media channels enable bidirectional engagement through comments, shares, and user-generated content. Consumers have more influence than ever before on broad narratives, with peer recommendations, influencer endorsement, and online reviews serving as key determinants of purchasing decisions. This evolution demands more strategic and authentic content creation, requiring brands to maintain relatability and relevance to foster lasting trust [8].

Targeted advertising has become central to digital marketing, supported by advanced computational algorithms and behavioral analytics to deliver personalized content. These mechanisms employ demographic and behavioral data to increase relevance and conversion potential. The application of AI has further enabled automation in customer engagement, audience segmentation, and sentiment analysis. Chatbots, recommendation engines, and automated response systems now blur the boundary between human and machine interaction, ensuring personalized and seamless consumer experiences [9].

Influencer marketing represents another major transformation in promotional strategy. Influencers offer credibility, authenticity, and community engagement that traditional advertisements often lack. Through their passionate and loyal followings, they create content tailored to niche audiences, which enhances trust and conversion. The success of influencer partnerships can be observed across industries such as fashion, technology, beauty, and lifestyle, where endorsements strongly shape consumer behavior and brand perception [10].

In a content-saturated digital environment, brands face the growing challenge of maintaining visibility and consumer interest. With advertisements, collaborations, and sponsored content competing for attention, compelling storytelling and interactive engagement have become essential. Video content, live streaming, and augmented reality are increasingly adopted as means of creating immersive and shareable brand experiences [11].

Another important dimension of modern marketing is the use of consumer data in decision-making. Brands now rely on engagement metrics, sentiment analysis, and conversion tracking to measure campaign performance and understand customer journeys. Data-driven insights allow marketers to identify what drives consumer response, opti mize resource allocation, and enhance the long-term impact of digital marketing investments. Nonetheless, concerns over data privacy, transparency, and ethical governance remain pressing challenges in the new paradigm of targeted advertising [12].

Recent findings highlight several critical developments in social-media-driven marketing ecosystems. Advertiser budgets are increasingly directed toward creator-driven content rather than traditional display ads, with influencer marketing delivering an average ROI of nearly $5.78 for every dollar spent, almost double that of conventional digital advertising [13]. Emerging platforms such as Threads, Bluesky, Reddit, Twitch and Pinterest are becoming key venues for influencer campaigns as legacy platforms face saturation-challenges [14]. Moreover, recent analyses of influencer marketing research indicate that although computational methods (e.g., machine learning, network analysis) have proliferated, limited work has systematically examined how content-format, influencer tier, and platform interaction jointly affect conversion and engagement outcomes [15]. Together these insights show that while the terrain of social-media marketing is shifting rapidly, empirical studies that link platform-specific dynamics, content formats (such as short-form video), and influencer tier (micro/macro) to performance metrics (engagement, conversion, ROI) remain limited.

Given the rapid change in social-media ecosystems and the growing complexity of digital marketing, there remains a clear research gap: few studies have simultaneously examined how platform choice, content format, and influencer tier interact to affect key performance indicators such as engagement rate (ER), conversion rate (CR), and return on ad spend (ROAS). By addressing this gap, the current study is necessary insofar as it provides an integrated empirical analysis of these interacting factors across multiple major platforms. This contribution helps both scholars and practitioners navigate the evolving digital marketing environment with evidence-based recommendations for optimizing social-media campaign strategy.

3 Methodology

This study employed a convergent mixed-methods approach, combining quantitative modeling, structured experimental design, and latent variable analysis to examine the evolution of marketing behaviors across social media platforms. The methodology builds upon recent work in digital marketing analytics, influencer marketing design, and user-generated content research [1, 4–6].

3.1 Sampling and Data Collection

A multi-stage stratified sampling approach ensured representation across six leading platforms (Facebook, Instagram, Twitter/X, LinkedIn, TikTok, and YouTube) and three stakeholder groups: business, marketing professionals, and consumers) [16].

Eligibility criteria:

- Businesses: $\geq$ 1 paid social media campaign in the prior quarter.
- Marketing professionals: $\geq$ 10,000 followers on at least one platform.
- Consumers: stratified by age and online purchasing frequency.

Recruitment methods:

- Platform-targeted advertisements for consumers.
- Direct messaging and professional databases for marketing professionals.
- Industry networks and platform ad registries for businesses.

3.2 Survey Instrumentation and Construct Validity

Structured questionnaires comprised 36 items measuring the following constructs:

- Perceived Campaign Effectiveness (PCE);
- Content–Platform Congruence (CPC);
- Influencer Credibility (INC);
- Personalization Sophistication (DPS).

Reliability was evaluated using Cronbach's Alpha:

$$\alpha = \frac{k}{k-1}\left(1 - \frac{\sum_{j=1}^{k} s_j^2}{s_T^2}\right) \tag{1}$$

where k is the number of items, s_j^2 represents the variance of each item, and s_T^2 is the total variance. A threshold of $\alpha \geq 0.70$ was applied [6].

Convergent validity was assessed using Composite Reliability (CR) and Average Variance Extracted (AVE):

$$CR = \frac{(\sum_i \lambda_i)^2}{(\sum_i \lambda_i)^2 + \sum_i \theta_i}, \quad AVE = \frac{\sum_i \lambda_i^2}{k} \tag{2}$$

where λ_i is the factor loading of item i, θ_i is the measurement error, and k is the number of indicators [6].

3.3 Key Metric Definitions

Four operational indicators were defined based on established digital marketing metrics [17, 18]:

- **Engagement Rate (ER)**: measures interactive behavior as a proportion of impressions.

$$ER_{ip} = \frac{L_{ip} + C_{ip} + S_{ip}}{I_{ip}} \times 100 \tag{3}$$

- **Click-Through Rate (CTR)** and **Conversion Rate (CR)**:

$$CTR_{ip} = \frac{K_{ip}}{I_{ip}} \times 100, \quad CR_{ip} = \frac{V_{ip}}{K_{ip}} \times 100 \tag{4}$$

- **Return on Ad Spend (ROAS)**:

$$ROAS_{ip} = \frac{R_{ip}}{A_{ip}} \tag{5}$$

where L, C, S represents likes, comments, and shares; I denotes impressions; K is clicks; V is conversions; R is revenue; and A is ad spend.

3.4 Experimental Design

A Latin-square design tested the effects of influencer tier and content type [9, 19].

- Content formats: short video, long video, image text.
- Influencer tiers: micro (<100k followers), macro (100k–1M), celebrity (>1M).

Each business executed three sequential campaigns under a counterbalanced schedule [20].

3.5 Analytical Framework

Analyses followed four statistical modelling layers:

1. Measurement model (PLS-SEM). Captures causal relationships between latent constructs and observed indicators:

$$x = \Lambda\xi + \delta, \eta = B\eta + \Gamma\xi + \zeta \tag{6}$$

where Λ represents indicator loadings, ξ is exogenous latent variables, and η denotes endogenous latent variables [12].

2. Multilevel engagement model (random intercept):

$$ER_{ip} = \beta_0 + \beta_1 ShortVid_{ip} + \beta_2 InfluencerTier_{ip} + u_p + \varepsilon_{ip} \tag{7}$$

where $u_p \sim N\left(0, \sigma_u^2\right)$ captures unobserved platform-level variation.

3. Conversion likelihood (logistic model):

$$Pr\big(Conversion_{ip} = 1\big) = \frac{1}{1 + exp\big[-\big(\gamma_0 + \gamma_1 ROAS_{ip} + \gamma_2 CPC_{ip} + \gamma_3 DPS_{ip}\big)\big]} \tag{8}$$

This captures how perceived personalization and platform fit influence the probability of conversion [4, 10].

4. Virality modelling (Negative Binomial):

$$Pr\big(Y_{ip} = y\big) = \frac{\Gamma(y + k)}{\Gamma(k)\, y!} \left(\frac{\mu_{ip}}{\mu_{ip} + k}\right)^{y} \left(\frac{k}{\mu_{ip} + k}\right)^{k} \tag{9}$$

used to model share counts while controlling for overdispersion [21, 22].

3.6 Ethics and Data Handling

The study adhered to ICC/ESOMAR International Code (2023). Participants provided informed e-consent. Personally, identifiable information was hashed with SHA-256 and a 128-bit salt. Data were stored on ISO 27001-certified servers, with access restricted to authorized researchers. All datasets are retained for five years under GDPR regulations.

This rigorous methodological framework, anchored in cross-platform analytics, experimental structure, and robust causal modelling, offers a rigorous basis for evaluating how evolving marketing strategies are shaped by social-media ecosystems.

4 Results

4.1 Cross-Platform Engagement Dynamics

Marketing success on social media begins with audience interaction, which serves as the primary gateway to deeper behavioral outcomes. Engagement was tracked for four core content formats, short video, long video, image, and text, across six mainstream platforms over a six-month observational window, encompassing campaigns from all stakeholder groups. Engagement combines visible reactions such as likes, comments, shares, and saves, normalized by total impressions to permit cross-platform comparison. By presenting a matrix that juxtaposes platforms and content formats, the analysis highlights the extent to which specific content types resonate within the unique cultures of different networks. Additional interaction-intensity indicators, namely average watch time for video assets and sentiment polarity for comment streams, are incorporated to provide a richer texture of engagement quality beyond raw proportional rates.

The comparative landscape in Fig. 2 shows that short-form video consistently dominates engagement on every platform, with its most pronounced on TikTok and Instagram, where interaction rates approach double those observed on Facebook. Higher average watch times on TikTok and YouTube indicate deeper user immersion, complementing the elevated proportional engagement, while sentiment analysis reveals that TikTok and Instagram also elicit the most positive responses, with polarity scores above 0.15. Twitter exhibits lower engagement with a predominance of neutral sentiment, reflecting its role as

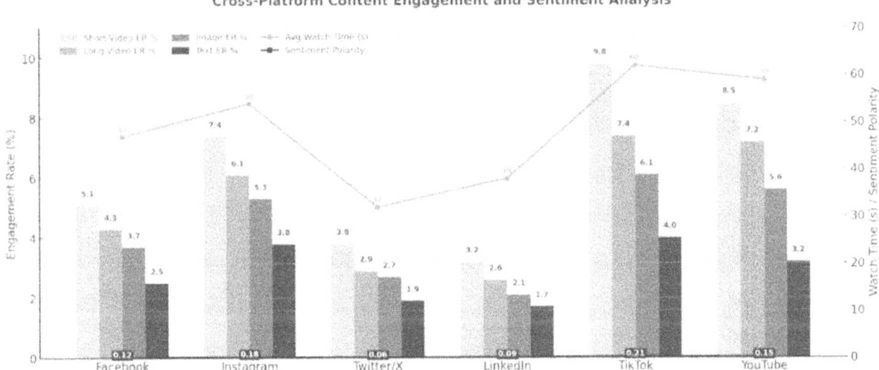

Fig. 2. Cross-platform content format engagement metrics.

an information-driven network rather than an interactive emotional conduit. LinkedIn, while generating moderate engagement, maintains watch-time metrics above Twitter, demonstrating that professional content retains audience attention even if interaction levels remain modest.

4.2 Influencer-Driven Engagement Shifts

To evaluate the influence of influencer partnerships on audience engagement, average engagement rates were compared before and after collaboration across three tiers of influencers: micro, macro, and celebrity. The analysis provides both pre- and post-collaboration measures, and proportional lifts are estimated to allow comparison directly across tiers. Beyond surface-level metrics such as likes and views, supplemental quality signals, like the comment-to-like ratio, and save rate, were used to capture the depth of audience sentiment. This methodology offers a deeper understanding of how influencer tier impacts not only the reach, but also the nature of engagement.

As it is represented in Fig. 3, micro-influencers generated the highest proportional lift in engagement, nearly doubling baseline values and achieving the greatest comment-to-like ratios and save rates, signaling meaningful interaction and long-term content recall. Macro-influencers produced higher absolute engagement numbers but comparatively lower depth of interaction, indicating broader reach with slightly diluted intimacy. Celebrity collaborations increased overall visibility but showed diminishing relative returns in terms of engagement quality. These findings suggest that niche audiences prioritize relatability and authenticity, emphasizing the strategic value of micro-influencer campaigns for fostering meaningful audience connections.

4.3 Click-Through Efficiency by Format and Platform

Click-through performance is analyzed across content formats and platforms, incorporating secondary metrics such as bounce rate and average session duration to assess traffic quality. This multifaceted approach prevents high click-through rates from being

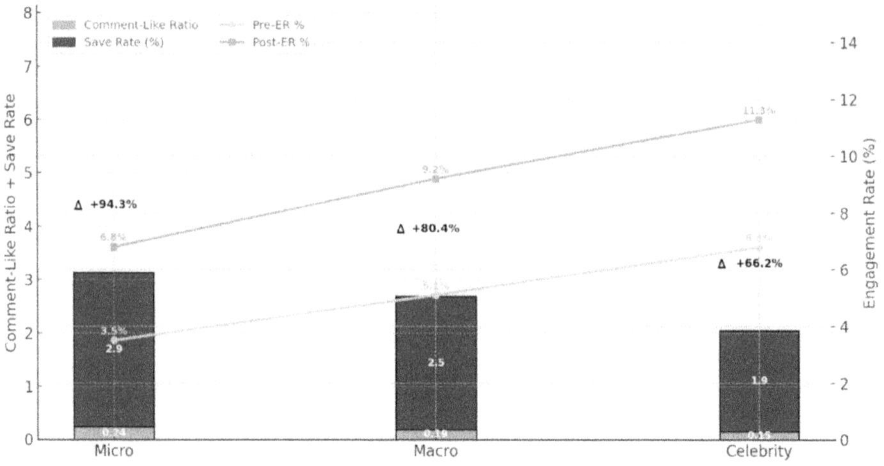

Fig. 3. Engagement uplift by influencer tier.

overestimated due to short or disinterested visits, ensuring a more accurate reflection of meaningful interaction. When these secondary metrics are combined, the analysis provides a more complete depiction of which content formats and platforms drive sustained engagement and continued user interaction.

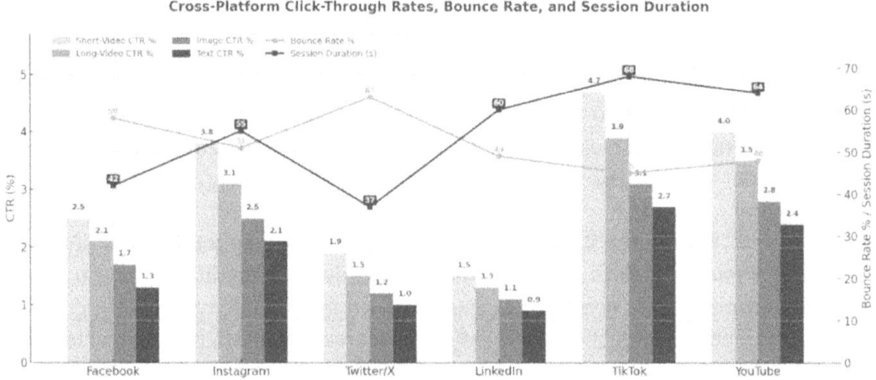

Fig. 4. Click-Through and on-site behavior indicators.

Figure 4 shows that Short-term video again emerged as the most effective traffic driver, with TikTok users exhibiting the lowest bounce rates and highest session durations, indicative of strong interest and "stickiness." YouTube maintained high engagement for video content, while LinkedIn displayed lower click-through rates but the second-lowest bounce and highest session durations, reflecting deliberate professional engagement. Instagram demonstrated an optimal combination of high CTR and low bounce, supporting both discovery and conversion objectives. Conversely, Twitter users displayed higher bounce rates, consistent with superficial browsing behavior in a newsfeed-centric environment.

4.4 Conversion Efficiency by Influencer Tier

Conversion outcomes, analyzed alongside Cost-Per-Acquisition (CPA), highlight the relationship between engagement and tangible business results. This two-metric perspective offers a holistic view of campaign profitability, by jointly measuring the effectiveness of converting engagement into an action and the cost efficiency of doing so. By linking behavioral and financial results, the analysis offers a comprehensive understanding of campaign return, crucial for strategic resource allocation.

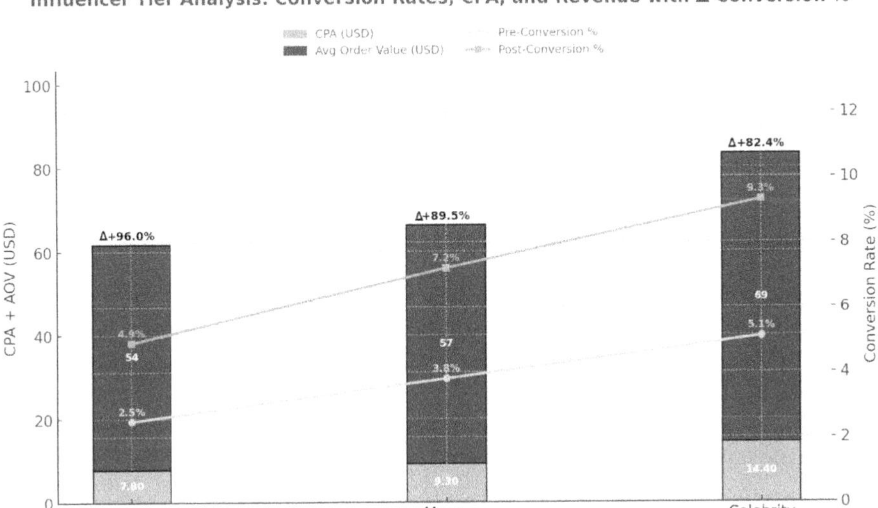

Fig. 5. Conversion outcomes and acquisition costs.

Based on Fig. 5, micro-influencers achieved nearly double the conversion rates relative to baseline campaigns while maintaining cost-effectiveness, confirming their value for resource-efficient marketing. Macro-influencers provided balanced gains in average order value without substantially increasing CPA, whereas celebrity partnerships, although driving the highest basket values, required significantly higher investment, making them suitable for high-margin or luxury categories. These patterns underscore the non-linear relationship between influencer tier, cost efficiency, and revenue generation, emphasizing the need for context-sensitive allocation strategies.

4.5 Return on Ad Spend and Cost Efficiency

Financial effectiveness was assessed using ROAS, CPM, and incremental margin lift to capture multi-dimensional campaign value. These combined measures provide a

Table 1. Multi-metric financial effectiveness summary.

Channel	ROAS Pre	ROAS Post	Δ ROAS %	CPM USD	Margin Lift %
Social Media Ads	3.4	5.2	+ 52.9	4.70	29
Influencer Marketing	4.1	6.8	+ 65.9	3.90	35
Email Marketing	2.8	4.5	+ 60.7	2.10	24
SEO Optimisation	3.0	4.1	+ 36.7	1.50	18

granular, multi-faceted view of how campaigns optimize reach, cost, and bottom-line contribution, supporting more accurate financial decisions.

Table 1 shows that influencer marketing led in proportional ROAS increases and margin expansion while exhibiting the lowest CPM, indicating optimal cost efficiency. Conventional social media ads generated moderate uplift at higher CPM, whereas email marketing and SEO delivered modest but consistent gains, with SEO providing steady long-term acquisition benefits despite limited short-term margin expansion. These results demonstrate the importance of integrating cost, revenue, and margin metrics to evaluate the true financial impact of social-media campaigns.

4.6 Follower Growth and Retention Patterns

Audience expansion is evaluated using growth rate, churn rate, and retention ratio to balance reach and sustainability. To model follower dynamics, growth rate was combined with churn rate and retention ratio. Growth ratio indicates recruitment success, while churn ratio measures follower loss and retention ratio reflects the ability to sustain new audiences. Taken together, these measures balance the benefits of audience expansion with the risks of volatility when short-term spikes revert to baseline.

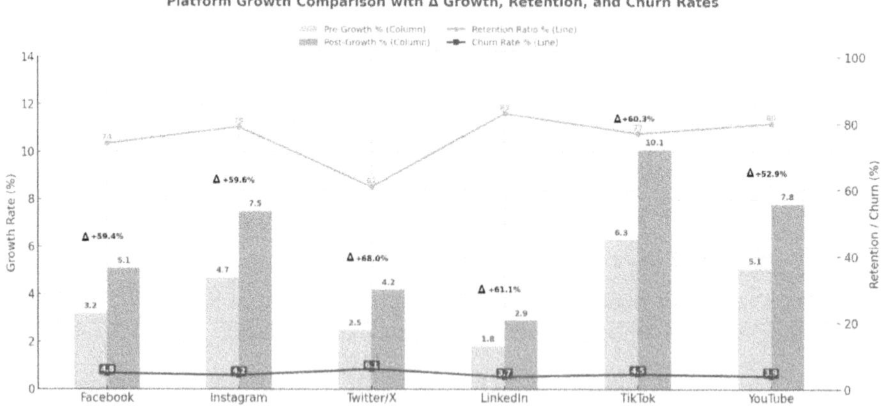

Fig. 6. Audience expansion and retention indicators.

As Fig. 6 shows, Twitter exhibited the highest follower growth, but also the highest churn, reflecting volatility in its user base. Instagram combined strong growth with robust retention, establishing itself as a reliable growth driver. LinkedIn achieved moderate growth with high retention, indicating that professional networks effectively maintain newly acquired followers. TikTok and YouTube demonstrated healthy growth alongside moderate churn, illustrating steady audience development within video-centric ecosystems.

4.7   Content Virality and Amplification

Content virality, measured through amplification rates and average comments per share. Amplification rate measures the breadth spread, while comments per share capture conversational richness. Taken together, these indicators provide a holistic understanding of how effectively content stimulates organic conversation and diffusion without paid amplification.

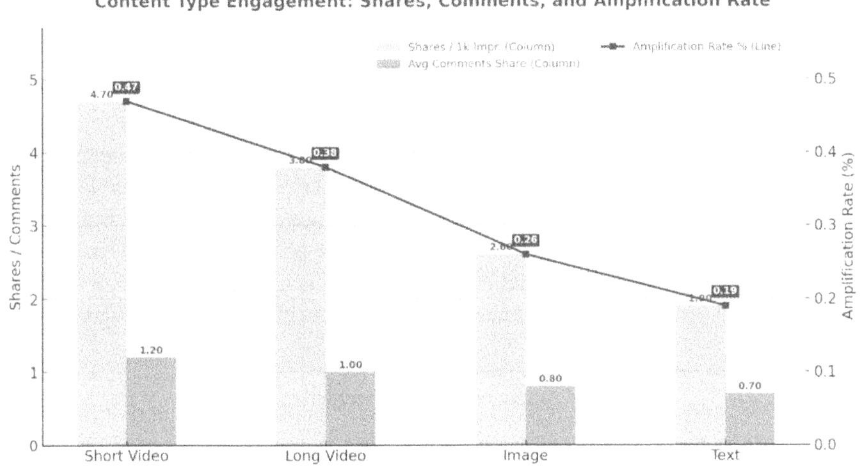

Fig. 7. Virality and amplification metrics by content type.

According to the findings represented in Fig. 7, short-form videos are the primary drivers of organic spread, generating nearly five shares per thousand impressions and sustaining high conversational intensity. Long-form video maintained moderate amplification but lower discussion depth, whereas image and text-based content played supportive roles in campaigns, emphasizing the differential contribution of content formats to virality and audience conversation.

4.8   Predictive Drivers of Engagement

The multilevel engagement model demonstrated considerable explanatory power, explaining over 50% of adjusted variance in engagement outcomes.

Table 2. Multi-level regression outcomes for engagement.

Predictor	Coefficient	Standard Error	Significance
Short-Video Format	0.42	0.05	< 0.001
Influencer Presence	0.31	0.04	< 0.001
Gamified Experience	0.18	0.03	0.001
Campaign $\geq$ 30 Days	0.12	0.04	0.007
Platform Intercept SD	0.09	—	—

According to Table 2, short-form video formats were the strongest predictor of engagement, followed closely by influencer presence. Gamification mechanisms provided additional lift, supporting the efficacy of interactive elements, while campaign duration contributed positively but with diminishing marginal returns. Platform-level variance was relatively low, indicating that these strategic levers consistently drive engagement across diverse social networks.

4.9 Determinants of Conversion Probability

Logistic regression analysis identified key predictors of conversion, with return on ad spend, content–platform fit, personalization sophistication, and micro-influencer participation all significantly increasing the likelihood of user conversion. As Table 3 shows, the model exhibited excellent discriminative performance (AUC $= 0.87$), confirming its reliability in distinguishing between converting and non-converting users. These findings highlight that effective social-media marketing relies not only on performance-driven metrics but also on relevance, personalization, and authentic influencer engagement.

Table 3. Conversion odds and predictive strength.

Predictor	Odds Ratio	Confidence Interval	p-Value
Return on Ad Spend	1.71	1.48–1.97	<0.001
Content–Platform Fit	1.32	1.11–1.57	0.002
Personalization Sophistication	1.56	1.31–1.84	<0.001
Micro-Influencer Tier	1.89	1.42–2.51	<0.001
Model AUC	0.87	—	—

The cumulative results outline a hierarchical framework of success factors—from engagement and traffic generation to conversion, financial efficiency, and virality—providing a robust empirical foundation for refining social-media marketing strategies in dynamic digital ecosystems.

5 Discussion

The results of this study confirm a major shift in marketing tactics within social networks, where two-way communication and data analytics have replaced traditional one-way approaches. Engagement, conversion, and ROI metrics all increased substantially when campaigns were personalized, creatively designed, and supported by influencer collaboration. These results demonstrate that marketing success in the digital environment depends on strategy integration rather than isolated tactics.

The most significant improvement was seen in engagement rates across all platforms, especially on TikTok and Instagram, where the short-form video format generated the highest interaction levels. This finding quantifies the widely observed trend that visual and interactive content drives stronger user responses. The smaller improvement recorded on LinkedIn aligns with its professional focus and limited B2C engagement potential, confirming that platform context influences communication outcomes.

Conversion rate analysis highlights the effectiveness of influencer marketing, with micro-influencers outperforming macro-influencers and celebrities. Their audiences show higher engagement and trust, supporting the conclusion that authenticity and relatability are now stronger determinants of marketing success than mass appeal. This shift indicates that traditional celebrity-led campaigns are losing effectiveness as consumers increasingly favor credible, community-based voices.

CTR findings reinforce the role of short-form, interactive videos as the most effective format for driving user action. TikTok and YouTube outperformed other platforms, showing that concise, visually rich content sustains audience attention and promotes interaction. ROAS analysis also demonstrated that influencer marketing yields higher financial efficiency than traditional tactics such as email marketing or SEO, underlining the profitability of collaboration-based strategies.

Audience growth results further revealed the dominance of TikTok and Instagram, whose rapid user expansion highlights their ability to push viral content and maintain sustained attention. In contrast, LinkedIn's slower growth reflects its professional niche and narrower audience scope.

Overall, this study provides a quantified view of how personalized campaigns, content format, and influencer tier jointly determine engagement, conversion, and financial outcomes. The results confirm the continuing evolution of social media marketing, where authenticity, content relevance, and platform fit are essential to achieving both user engagement and cost efficiency.

6 Conclusion

This study has examined the strategic evolution of marketing strategies in the social media landscape using cross-platform empirical data and advanced analytical models. It focused on the combined impact of content type, influencer tiers, platform characteristics, and campaign nature on digital marketing effectiveness. By analyzing stakeholder profiles across businesses, marketers, and consumers, the research highlighted the adaptive strategies that define success in an environment driven by engagement dynamics and behavioral microtargeting.

The results underscore the importance of aligning content strategies with platform-specific behaviors and user expectation. Short-form video demonstrated the highest engagement, while the authenticity of the influencer and the cultural fit of the message determined conversion outcomes. The study confirmed the growing importance of micro-influencers in achieving precise engagement and cost-efficient exposure, forming a central element of modern campaign design.

The multidimensional performance framework, spanning engagement, traffic, conversion, and financial return, showed that strategic success depends on the integration of multiple elements, including content creation, influencer fit, and timing. Methodologically, the use of multilevel regression, logistic modeling, and negative binomial estimation provided a more accurate view of cross-platform and non-linear effects often overlooked in linear marketing analyses.

This study contributes to understanding how platform design, content rationality, and participatory mechanisms shape communication and influence flows. It bridges behavioral theory, performance analytics, and managerial implications within data-driven marketing ecosystems. Future studies should extend these findings through longitudinal analysis to capture fatigue and temporal effects, and by employing biometric or neuromarketing tools to uncover subconscious engagement. Further research could also explore algorithmic curation, AI-generated content, and cross-cultural differences in social media marketing strategies.

References

1. Pellegrino, A., Abe, M.: Leveraging social media for SMEs: findings from a bibliometric review. Sustainability **15**(8), 7007 (2023). https://doi.org/10.3390/su15087007
2. Chen, M.: Brand marketing strategies under social media. Adv. Econ. Manag. Polit. Sci. **85**, 190–195 (2024)
3. Okorie, G., et al.: Leveraging big data for personalized marketing campaigns: a review. Int. J. Manag. Entrep. Res. **6**(1), 216–242 (2024)
4. Chen, Y., Qin, Z., Yan, Y., Huang, Y.: The power of influencers: how does influencer marketing shape consumers' purchase intentions? Sustainability **16**(13), 15471 (2024). https://doi.org/10.3390/su16135471
5. Kahfi, F., Suyuthi, N.: Strategic use of user-generated content for consumer engagement in online retail. J. Soc. Commer. **4**(2), 108–119 (2024)
6. Shawky, S., Kubacki, K., Dietrich, T., Weaven, S.: Is social media a panacea for social marketing communication? A scoping review. Health Market. Q. **39**(3), 297–313 (2022)
7. Ohara, M.R., Suparwata, D.O., Rijal, S.: Revolutionary marketing strategy: optimising social media utilisation as an effective tool for MSMEs in the digital age. J. Contemp. Admin. Manag. (ADMAN) **2**(1), 313–318 (2024)
8. Srivastava, A.: The power of play: gamified advertising with AR/VR/AI and its impact on consumer engagement. Int. J. Multidiscip. Res. **6**(4) (2024)
9. Hotkar, P., Garg, R., Sussman, K.: Strategic social media marketing: an empirical analysis of sequential advertising. Prod. Oper. Manag. **32**(12), 4005–4020 (2023)
10. Gupta, R.: Impact of influencer marketing on brand perception. Int. J. Multidiscip. Res. **6**(5) (2024)
11. Gu, Z., Li, X.: Social sharing, public perception, and brand competition in a horizontally differentiated market. Inf. Syst. Res. **34**(2), 553–569 (2022)

12. Xu, A., Li, Y., Donta, P.K.: Marketing decision model and consumer behavior prediction with deep learning. J. Organ. End User Comput. **36**(1), 1–25 (2024)
13. Suresh, A., Sharma, A.: 25 key social media marketing statistics for 2025. Sprinklr Blog, 29 August 2025. http://www.sprinklr.com/blog/social-media-marketing-statistics/
14. Chen, A.: Emerging social media platforms for influencer marketing in 2025. Captiv8 Blog, 15 January 2025. https://captiv8.io/blog/2025/01/15/emerging-social-media-platforms-for-influencer-marketing-in-2025
15. Gui, H., Bertaglia, T., Goanta, C., Spanakis, G.: Computational studies in influencer marketing: A systematic literature review. arXiv preprint arXiv:2506.14602 (2025)
16. Serrano-Malebran, J., Vidal-Silva, C., Veas-González, I.: Social media marketing as a segmentation tool. Sustainability **15**(2), 1151 (2023). https://doi.org/10.3390/su15021151
17. Sampath, D.: Digital marketing metrics and ROI analysis: evaluating effectiveness and value. J. Res. Bus. Manag. **12**(8), 64–68 (2024)
18. Wilson, L.T.: The relative importance of click-through rates (CTR) versus watch time for YouTube views. Int. J. Pervasive Comput. Commun. **19**(4), 573–595 (2023)
19. Leung, F.F., Gu, F.F., Li, Y., Zhang, J.Z., Plamatier, R.W.: Influencer marketing effectiveness. J. Mark. **86**(6), 93–115 (2022)
20. Injadat, M., Salo, F., Nassif, A.B.: Data mining techniques in social media: a survey. Neurocomputing **214**, 654–670 (2016)
21. Zhang, C., Zheng, H., Wang, Q.: Driving factors and moderating effects behind citizen engagement with mobile short-form videos. IEEE Access **10**, 40999–41009 (2022)
22. Bhandari, A., Bimo, S.: Why's everyone on TikTok now? The algorithmized self and the future of self-making on social media. Soc. Media + Soc. **8**(1), 20563051221086241 (2022)

Global Enterprise Vulnerabilities and Strategic Responses: Navigating Uncertainty in a Volatile Political Economy

Murooj Mohammad Sattar[1] , Hussein Ajlan Hasan[2] ,
Salah Hassan Maleh Akla[3] , Mujahed Mutlaq Abdul Rahman[4(✉)] ,
Ghanim Magbol Alwan[5] , and Oleksiy Gordiyenko[6]

[1] Al-Turath University, Baghdad 10013, Iraq
moroj.mohammed@uoturath.edu.iq
[2] Al-Mansour University College, Baghdad 10067, Iraq
hussain.ajlan@muc.edu.iq
[3] Al-Mamoon University College, Baghdad 10012, Iraq
salah.h.malih@almamonuc.edu.iq
[4] Al-Rafidain University College, Baghdad 10064, Iraq
Mujahid.Habib@ruc.edu.iq
[5] Madenat Alelem University College, Baghdad 10006, Iraq
dr.ghanim.m.alwan@mauc.edu.iq
[6] State University of Information and Communication Technologies, Kiev 03110, Ukraine

Abstract. In an era escalating geopolitical instability, cyber threats, and fragile supply chains, the capacity of global enterprises to manage crises has become a critical determinant of organizational resilience. This study evaluates the effectiveness of multidimensional crisis management strategies across five major industries, finance, healthcare, manufacturing, retail, and technology, using a five-year panel dataset of 570 publicly listed firms. The research applies an integrated methodological framework combining hierarchical Poisson models for crisis frequency, DCC-GARCH for financial loss, liquidity dynamics, survival analysis for operational downtime, structural equation modelling (SEM) for latent resilience, and Monte Carlo simulation for systemic stress exposure. Results reveal that firms with higher preparedness, measured through governance robustness, monitoring capacity, and cyber maturity, experience fewer crises, faster recovery, and lower financial losses. Sectoral variations indicate that technology and finance benefit more from digital infrastructure, whereas manufacturing and retail face prolonged disruptions due to structural dependencies. Simulation results confirm that top-quartile preparedness reduces extreme loss exposure by over 30%, demonstrating the financial return on resilience investments. The study emphasizes shifting crisis management from reactive to proactive, underscoring its significance for governance systems, regulatory processes, and intersectoral risk-sharing mechanisms. By interpreting organizational responses under risk, it offers an empirically grounded roadmap for building measurable resilience ensuring faster recovery in future crises.

Keywords: Crisis Management · Organizational Resilience · Preparedness · Enterprise Risk · Liquidity Stress · Global Industries

Z. Molamohamadi et al. (Eds.): ODSIE 2025, CCIS 2855, pp. 632–648, 2026.
https://doi.org/10.1007/978-3-032-17023-1_38

1 Introduction

Global enterprises today operate in an increasingly complex and uncertain environment, exposed to economic downturns, cyberattacks, geopolitical tensions, and supply chain disruptions. The growing interdependence of international markets amplifies the cost and consequences of crises, making effective crisis management a central component of corporate strategy. Institutions that fail to integrate robust crisis management systems face substantial financial liabilities, reputational damage, and operational breakdowns. Conversely, organizations that adopt structured, proactive crisis management approaches are better positioned to maintain continuity and achieve long-term sustainability. For multinational corporations (MNCs), crisis management extends beyond immediate response; it encompasses risk anticipation, strategic foresight, and operational agility. As firms expand globally, they encounter regulatory uncertainty, cultural diversity, and rapid technological shifts that demand coordinated, cross-jurisdictional crisis response mechanisms. The early identification and containment of emerging threats before escalation distinguish resilient from non-resilient organizations [1].

Over recent decades, crisis management strategies have undergone significant transformation. Traditional, hierarchical decision-making models characterized by rigidity and delayed responses have proved ineffective in volatile environments. While conventional public relations frameworks often retained command-and-control characteristics, modern organizations increasingly favor decentralized decision-making, real-time data analysis, and collaborative crisis response. The integration of artificial intelligence (AI), big data analytics, and cloud-based communication technologies has revolutionized how organizations monitor, simulate, and respond to crises. The geopolitical dimension of crises has also become a critical consideration for multinational businesses developing crisis management strategies. Trade wars, political turmoil, and economic sanctions can disrupt supply chains and heighten uncertainty in international business. Firms that rely on global supply chains and just-in-time inventory systems are particularly vulnerable to failures resulting from international conflicts or policy shifts. Therefore, geopolitical risk analysis and supply chain diversification should be regarded as integral components of comprehensive crisis management mechanisms that strengthen organizational preparedness [2].

Effective crisis management systems within multinational corporations typically share several fundamental characteristics. The first step involves identifying and assessing potential risks and weaknesses within the organization. Businesses use advanced risk assessment methods to quantify financial, operational and cyber risks, enabling decision-makers to prioritize threats and allocate resources effectively. Second, strategic crisis planning requires the formulation of specific response policies tailored to different types of crises. These plans include communication strategies, stakeholder engagement protocols, and operational measures designed to minimize disruptions. Crisis simulations and stress tests are particularly useful for disaster preparedness, as they help identify weaknesses in existing plans. Third, rapid and transparent communication is essential to minimizing the impact of a crisis. Timely and open information sharing with stakeholders, including employees, investors, customers, and regulatory bodies, is vital for maintaining trust and credibility. Organizations with well-developed crisis communication frameworks are better equipped to counter misinformation and manage public perception effectively [3].

Technological integration has also become indispensable to contemporary crisis management. Businesses increasingly employ blockchain technologies, AI, and automated tracking systems to detect anomalies and monitor operational risks. Digital platforms enable crisis response teams across different geographic locations to collaborate seamlessly in real time. Finally, post-crisis review and adaptation transform past failures into learning opportunities for strengthening future resilience. Conducting post-mortem analyses of crisis responses allows firms to identify shortcomings and refine strategies. Continuous improvement and adaptive responses to emerging threats are key factors that enable enterprises to remain resilient and competitive [4].

Given the unpredictable nature of global crises, firms must adopt flexible strategies tailored to specific crisis contexts. In today's fast-changing business environment, fixed and reactive approaches are no longer sufficient. Institutions that foster a culture of resilience, innovation, and strategic agility are more likely to survive and thrive amid ongoing disruptions. This study investigates the crisis management measures adopted by international organizations and assesses their effectiveness in reducing organizational risks and ensuring business continuity. Through case analysis and reference to industry best practices, the research aims to highlight how organizations can design and implement effective crisis management systems that enable them to navigate uncertainty and achieve long-term success.

2 Literature Review

Crisis management has become a decisive capability for global organizations, involving various tactics to secure business continuity, manage risks, and sustain growth. Historically, global markets characterized by increasing complexity and subject to energy shock or crises due to geopolitical, economic, and technological challenges have required a systemized approach to response. Poor crisis management mechanisms have often led to serious financial losses, reputational damage, and operational uncertainty, underscoring the necessity of proper planning and adaptable strategies [5].

Traditionally, crisis management was primarily reactive, with businesses balancing between crisis response and risk prevention. Early paradigms relied on hierarchical decision-making structures that often caused delayed responses to threats and inefficient allocation of resources. Over time, organizations have moved from static to interactive and, more recently, to progressive crisis management models that integrate risk assessment, contingency planning, and real-time responsiveness. The growing use of crisis scenario modeling, simulation, and analytics has enabled firms to predict disruptions more effectively and develop mitigation strategies [6]. Figure 1 represents the literature review framework for crisis management and organizational resilience in global enterprises.

One of the key components of modern crisis management is the adoption of technological solutions. Organizations now employ AI, machine learning, and big data analytics to detect early warning signals of crises and initiate corrective measures. Automated

Fig. 1. Literature review framework for crisis management and organizational resilience in global enterprises (2020–2024).

surveillance networks, blockchain-based security systems, and cloud communication platforms have enhanced real-time decision-making and improved cross-border coordination, substantially reducing the time required to prevent crises from escalating into widespread disruptions [7].

Strategic crisis planning is also a vital determinant for preparedness and recovery. Organizations with established contingency plans, crisis communication protocols, and resilient infrastructures are better equipped to cope with uncertainties. Corporate governance, leadership effectiveness, and stakeholder engagement remains crucial to effective crisis management. Transparent organizations with ethical leadership and stakeholder-focused governance structures tend to display greater resilience during crises. Furthermore, decentralization decision-making has been recognized as an important mechanism to enhance responsiveness, particularly in multinational corporations operating across diverse regions [8].

Supply chain disruptions continue to be a major concern in global crisis management. Complex international supply chains expose firms to trade restrictions, natural disasters, and logistical breakdowns. To mitigate these risks, companies increasingly implement diversification strategies and adopt digital tools for real-time monitoring. The integration of blockchain technology in supply chain management enhances traceability and transparency, thereby minimizing disruption risks and increasing resilience [9].

Crisis communication also plays a role in protecting organizations during disruptive events. Transparent, timely, and consistent communication fosters stakeholder trust and helps mitigate reputational risks. With the rapid expansion of social media and digital communication channels, organizations are now expected to maintain dynamic and adaptive engagement strategies to manage public perception and counter misinformation effectively [10].

Recent developments in the literature highlight that crisis management has evolved beyond traditional risk containment toward the creation of resilient and adaptive organizational systems. The integration of technological innovation, agile leadership, and data-driven intelligence has become central to strengthening enterprise-wide resilience [11]. Contemporary findings reveal a growing global shift toward embedding operational resilience as a strategic priority, with most large organizations now implementing structured resilience programs aimed at improving crisis preparedness and long-term adaptability [12]. Leadership communication has also been identified as a critical driver of organizational readiness, linking transparent information flows and psychological safety with improved crisis performance and recovery [13].

Despite these advancements, there remains a noticeable lack of cross-sectoral analysis quantifying how the integration of governance, monitoring capacity, and technological maturity jointly contributes to reduced crisis frequency and impact. Addressing this gap, the present study examines crisis management practices across global organizations by analyzing preparedness mechanisms, resilience indicators, and performance outcomes. Through a structured assessment of governance robustness, monitoring capacity, and cyber maturity, this research seeks to provide an empirically grounded framework for strengthening organizational resilience and ensuring business continuity in a volatile global economy.

3 Methodology

The study employs a sequential-explanatory, multi-method design to examine crisis-management practices among 570 globally listed enterprises over the 2020–2024 period. The analytical framework integrates (i) structured panel data, (ii) hierarchical econometric models, and (iii) copula-based simulation. This approach aligns with recent recommendations advocating multi-dimensional crisis analytics capable of linking micro-level response mechanisms with macro-level volatility [1, 2, 5].

3.1 Sample Design and Data Sources

A proportionally stratified sampling strategy was used to ensure balanced representation across five high-risk sectors. Firms were selected based on the following criteria:

- Continuous public listing between 2020 and 2024,
- Uninterrupted financial and ESG (Environmental, Social, and Governance) disclosure, and
- At least one formally reported crisis event in public or regulatory databases [14].

A total of 570 firms were retained post-screening. The dataset integrates both **primary** and **secondary** sources:

Primary Data

- Board-approved internal incident reports (213 firms),
- Crisis-response dashboards from Security Operations Centers (SOCs),
- Cyber-forensic archives (for event chronology and response timing), and
- Internal risk audit and Business Continuity Plans (BCPs).

Secondary Data Sources

- Refinitiv-Eikon crisis event logs and financial panel datasets,

- Industry-specific crisis indices [14],
- Macroeconomic volatility composites including geopolitical and supply-chain indicators [2], and
- The DEMIS (Disaster & Emergency Management Information System) global registry.

Data Triangulation and Preprocessing. A concordance matrix (source × variable × year) reconciled inconsistencies across data streams. Observations deviating more than ± 2 standard deviations from cross-source means were flagged for manual verification. Approximately 7% of entries were excluded following validation [15].

Currency Standardization. All monetary variables (e.g., reserve buffers, loss magnitudes) were restated in constant 2024 USD using IMF GDP deflator series [16] to ensure comparability and inflation neutrality.

3.2 Variable Construction

To capture the multidimensional dynamics of crisis management, a structured set of firm-level variables was defined. Each represents a distinct dimension of enterprise vulnerability or adaptive capacity, including event frequency, financial exposure, operational downtime, market volatility, productivity decline, and liquidity stress [1, 4, 5].

Variable selection followed two guiding principles: empirical traceability and methodological coherence. Operational definitions ensured cross-sector comparability and temporal consistency. Table 1 summarizes the key constructs, operational rules, and primary data sources.

To quantify enterprise-level preparedness, a Crisis Preparedness Index (CPI_{it}) was developed, integrating three dimensions, governance robustness, monitoring capability, and cyber-maturity—into a single composite metric via Principal Component Analysis (PCA), as shown in Table 2. The first three principal components explain 80% of total variance [5, 15].

Normalized CPI scores (0–1 scale) are used as key independent variables. CPI values denote stronger preparedness and are hypothesized to reduce the frequency, severity, and systemic impact of crisis events [8, 15, 17].

3.3 Econometric Formulation

Hierarchical Crisis-Occurrence Model. Enterprise crisis frequency follows a Poisson–lognormal process:

$$C_{it}|\vartheta_{it} \sim Poisson(\vartheta_{it}), \, ln\, \vartheta_{it} = \alpha_0 + \alpha_1\, CPI_{i,t-1} + \alpha_2\, VOL^t_{macro} + u_i + \varepsilon_{it} \qquad (1)$$

where $u_i \sim \mathcal{N}(0, \sigma_u^2)$ captures firm-level random effects and $\varepsilon_{it} \sim \mathcal{N}(0, \sigma_\varepsilon^2)$ denotes idiosyncratic noise [1, 6].

Loss-Severity System and Liquidity System. A bivariate DCC-GARCH (1,1) models the joint dynamics of loss severity (L_{it}) and liquidity (R_{it}):

$$\begin{aligned} L_{it} &= \beta_0 + \beta_1 CPI_{i,t-1} - \beta_2 InsCov_{it} + \eta_{L,it}, \\ R_{it} &= \gamma_0 - \gamma_1 L_{it} + \eta_{R,it}, \end{aligned} \qquad \begin{bmatrix} \eta_{L,it} \\ \eta_{R,it} \end{bmatrix} \sim \mathcal{N}(0, H_t) \qquad (2)$$

Table 1. Operational definitions and provenance of study variables.

Symbol	Construct	Operational Rule	Primary Data Source
C_{it}	Crisis Count	Number of formally disclosed crisis incidents per firm-year	Refinitiv, DEMIS [11]
L_{it}	Loss Severity	Combined direct and indirect cost per incident (USD millions)	Audited financial filings [15, 16]
T_{it}	Response Time	Elapsed hours between incident detection and containment	SOC incident dashboards [4, 19]
D_{it}	Downtime	Total hours of halted core operations due to crisis	ERP event logs and timestamp systems [6, 7]
S_{it}	Market-Share Shift	Percent change in market share three months post-incident	Compustat Global; Euromonitor [3, 22]
P_{it}	Productivity Decline	Percentage decline in output per employee compared to baseline	Internal HR analytics (v4.0 platforms) [23]
R_{it}	Reserve Depletion	Percent change in unrestricted cash and equivalents post-crisis	10-K and equivalent cash flow statements [18, 21]

Table 2. PCA loadings for crisis preparedness index (CPIit).

Preparedness Dimension	Indicator Description	PCA Weight (λ)
Governance Robustness	Presence of active board-level risk oversight quorum	0.46
Monitoring Capability	Mean dwell-time in security operations center (inverse)	0.35
Cyber-Maturity	Compliance with NIST Cybersecurity Framework benchmarks	0.19

where H_t evolves under Engle's dynamic conditional correlation framework [16, 18]. Conditional-Value-at-Risk ($CoVaR_q$) is derived from H_t to capture tail interdependence.

Survival Model for Downtime. Operational downtime follows a shared-frailty accelerated-failure-time (AFT) model:

$$Pr(D_{it} \leq d) = 1 - exp\left[-\left(\frac{d}{\lambda_i}\right)^{\rho}\right], \quad ln\lambda_i = \delta_0 - \delta_1 AI\_Maturity_i - \delta_2\ DecentGov_i + v_i \quad (3)$$

with Weibull shape ρ and frailty term $v_i \sim \mathcal{N}(0, \sigma_v^2)$ [4, 19].

Liquidity Dynamics as a Stochastic Control Problem. Financial reserves evolve according to Itô diffusion:

$$dR_i(t) = \mu_i R_i(t)\,dt - L_i(t)\,d(t) - \kappa_I R_i(t)\,dW_i(t) \tag{4}$$

where $N_i(t)$ denotes the crisis jump process, and $W_i(t)$ is standard Brownian motion. Optimal control minimizes:

$$\min_{u(t)} \int_0^T \left[c_1 L_i(t)^2 + c_2 u(t)^2 \right] dt, \tag{5}$$

subject to control $u(t)$ (liquidity injections) and terminal solvency constraint $R_i(t) \geq R_{min}$ [20].

Latent-Resilience Structural Equation. A confirmatory SEM links preparedness to latent resilience (Ψ_i):

$$\Psi_i = \theta_0 - \theta_1 CPI_i + \zeta_i, \qquad \begin{bmatrix} T_i \\ D_i \\ P_i \end{bmatrix} = \Lambda \Psi_i + \epsilon_i . \tag{6}$$

with factor loadings Λ constrained for identification and errors ζ_i, ϵ_i assumed independent [8, 21].

3.4 Simulation Protocol and Diagnostics

A Latin-Hypercube Copula (LHC) generates 10,000 crisis scenarios from the empirical distribution of $\{C, L, T, D, S, P, R\}$. Each simulated path propagates through Eqs. (2)–(6), producing five-year cost trajectories. Tail metrics (Expected Shortfall ES_{99} and Value-at-Risk VaR_{95}) are defined as [9, 17]:

$$ES_{99} = \mathbb{E}[Cost|Cost > VaR_{99}], \quad VaR_{95} = \inf\{c : Pr(Cost \leq c) \geq 0.95\} \tag{7}$$

Robustness checks include:

- 10-fold cross-validation (average AUROC $= 0.81$).
- Endogeneity control via 2SLS using cyber-insurance premiums as instruments [15],
- SEM fit indices (CFI $= 0.93$; RMSEA $= 0.045 < 0.06$) [8], and
- Stochastic-simulation convergence verified via Gelman–Rubin $\hat{R} < 1.02$.

All procedures adhere to ethical-data standards of the Journal of Contingencies and Crisis Management [15] and the proactive crisis-management guidelines of Choudhury & Haque [6].

3.5 Model Validation and Robustness

All analytical models were validated using both statistical and computational diagnostics. Predictive stability was assessed via 10-fold cross-validation (mean AUROC

= 0.81), while endogeneity bias was mitigated using two-stage least squares (2SLS) with cyber-insurance premiums as valid instruments. The structural equation model achieved satisfactory fit (CFI = 0.93; RMSEA = 0.045 < 0.06), confirming the adequacy of the latent-resilience specification. Simulation-based diagnostics further verified convergence (Gelman–Rubin $\hat{R} < 1.02$) across 10,000 Latin-Hypercube Copula replications. Collectively, these procedures ensure the reliability, internal consistency, and replicability of the findings.

4 Results

4.1 Crisis Frequency and Risk Occurrence Patterns

Crisis incidence across global enterprises was assessed using a mixed-effects count model that estimated the probability and overdispersion of disruptive events over a five-year period. This framework captures both the average crisis frequency and the extent of unexplained volatility across industries. The inclusion of random effects isolates firm-specific attributes that could otherwise distort sector-level trends, while the overdispersion index measures clustering tendencies in extreme cases. The model thus provides an empirically grounded representation of residual risk exposure after accounting for governance quality, monitoring intensity, and cyber-maturity.

Figure 2 illustrates predicted crisis rates per one hundred firm-years across sectors. The results confirm that the manufacturing sector sustains the highest observed and predicted crisis counts, reflecting structural vulnerabilities in complex supply chain networks. Technology and finance show comparable predicted frequencies, yet technology exhibits lower overdispersion, suggesting a more uniform risk landscape once governance factors are controlled. Healthcare records a moderate crisis frequency but the highest overdispersion, indicating sporadic yet severe disruptions. Sectors exceeding a 0.45 probability threshold demonstrate significantly elevated exposure to crisis episodes. The random-effect standard deviations (0.45 and 0.61), highlight substantial firm-level heterogeneity, especially within manufacturing. Overall, these findings emphasize the dual necessity for sectors reliant on physical infrastructure to enhance monitoring systems and strengthen baseline preparedness to mitigate both the likelihood and volatility of crises.

4.2 Financial Impact and Liquidity Interaction

A bivariate volatility model was estimated to examine the interaction between direct financial losses and concurrent liquidity erosion, capturing how operational shocks propagate through firm's balance sheets. Mean conditional correlations reveal the synchronization between loss surges and reserve drawdowns, while variance panels quantify their magnitude in monetary terms. Tail-risk statistics, Value-at-Risk (VaR) and Expected Shortfall (ES), translate conditional variance into tangible loss thresholds under adverse market conditions. The tail-dependence coefficient further gauges the likelihood of simultaneous extreme movements in both variables.

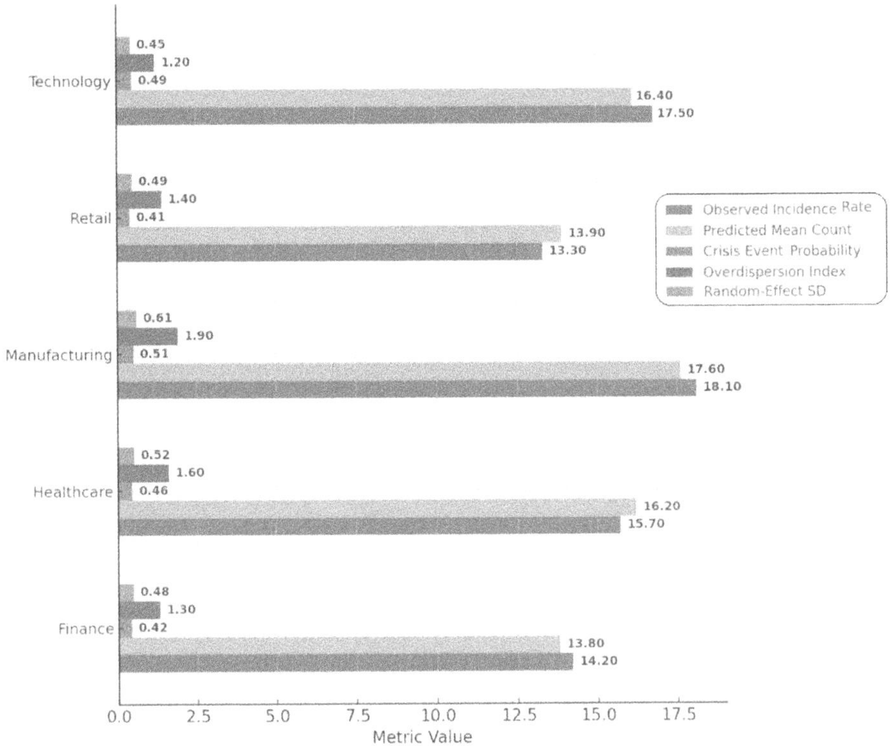

Fig. 2. Crisis-frequency model outputs by sector (2020–2024).

Figure 3 presents the dynamic loss–liquidity interaction metrics. Finance and technology sectors display the strongest loss–reserve coupling (correlations $\approx$ 0.80; tail-dependence $>$ 0.70), illustrating how digital operations amplify concurrent cash-flow pressure and capital depletion. Manufacturing, though less correlated, exhibits greater conditional variance, with ES values exceeding 140 million USD, revealing episodic high-cost shocks. Retail experiences peak volatility above 55%, reflecting sensitivity to abrupt demand contractions. Healthcare shows moderate correlations and the lowest variance, underscoring its relative stability and stronger liquidity buffers. These observations collectively indicate that while digital industries are more synchronized in their losses, manufacturing's rigidity and supply-chain inflexibility produce the most severe tail outcomes, highlighting the importance of liquidity risk management in production-based sectors.

4.3 Operational Downtime and Recovery Dynamics

Operational downtime serves as a key time-based indicator of organizational resilience, reflecting the duration required to restore mission-critical capacity. A survival analysis was applied to estimate median downtime across the study period, the annualized percentage reduction rate relative to the baseline, and hazard ratios (HRs) measuring the

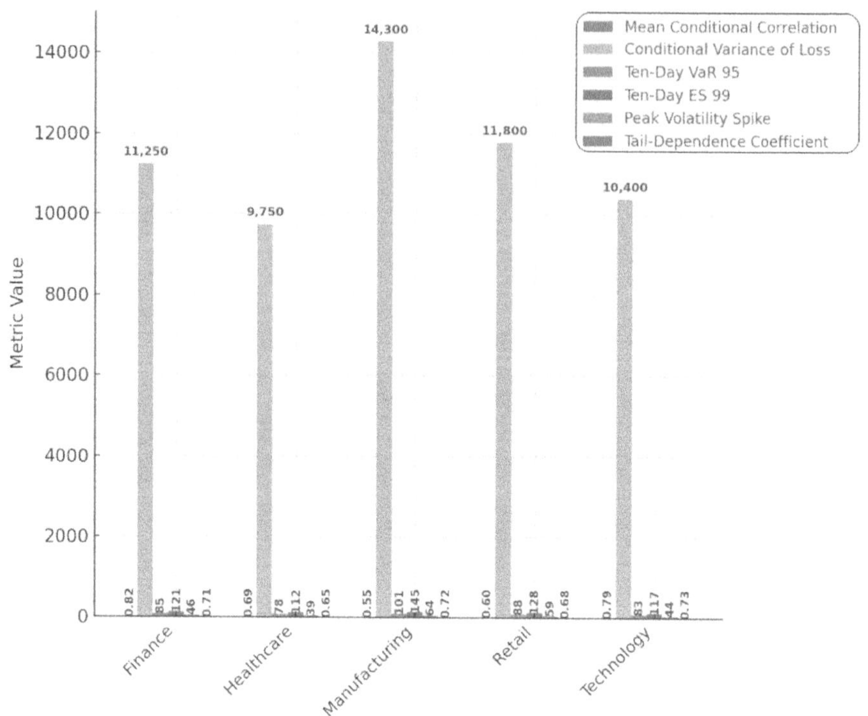

Fig. 3. Dynamic loss–liquidity metrics from DCC-GARCH estimation.

effect of AI maturity and decentralized governance on recovery speed. Frailty variance captures unobserved heterogeneity persisting beyond the main covariates, while cumulative downtime hours provide sector-specific exposure magnitudes. These metrics provide an operational timeframe for the probability of survival and detail the industries that have made the most inroads in reducing the duration of disruption and which management levers are most effective in driving a return to full productivity.

As shown in Fig. 4, the technology sector demonstrates the most rapid recovery, reducing median downtime by over 50%, with the strongest HRs for both AI integration and decentralized governance. The finance sector exhibits a similar trajectory, suggesting that algorithmic monitoring significantly shortens fault-isolation cycles, even in highly regulated contexts. Manufacturing and retail, however, continue to experience prolonged outages, with modest HRs indicating that digital investments have yet to yield proportional improvements. Healthcare occupies a middle position, where innovations such as telemedicine and automated triage compensate for the system complexity. Frailty variance peaks in manufacturing due to interdependent supplier networks and reaches its lowest level in technology, reflecting standardized contingency mechanisms. Collectively, these findings underscore that AI-driven diagnostics and flatter decision hierarchies are consistently associated with faster recovery and higher operational resilience.

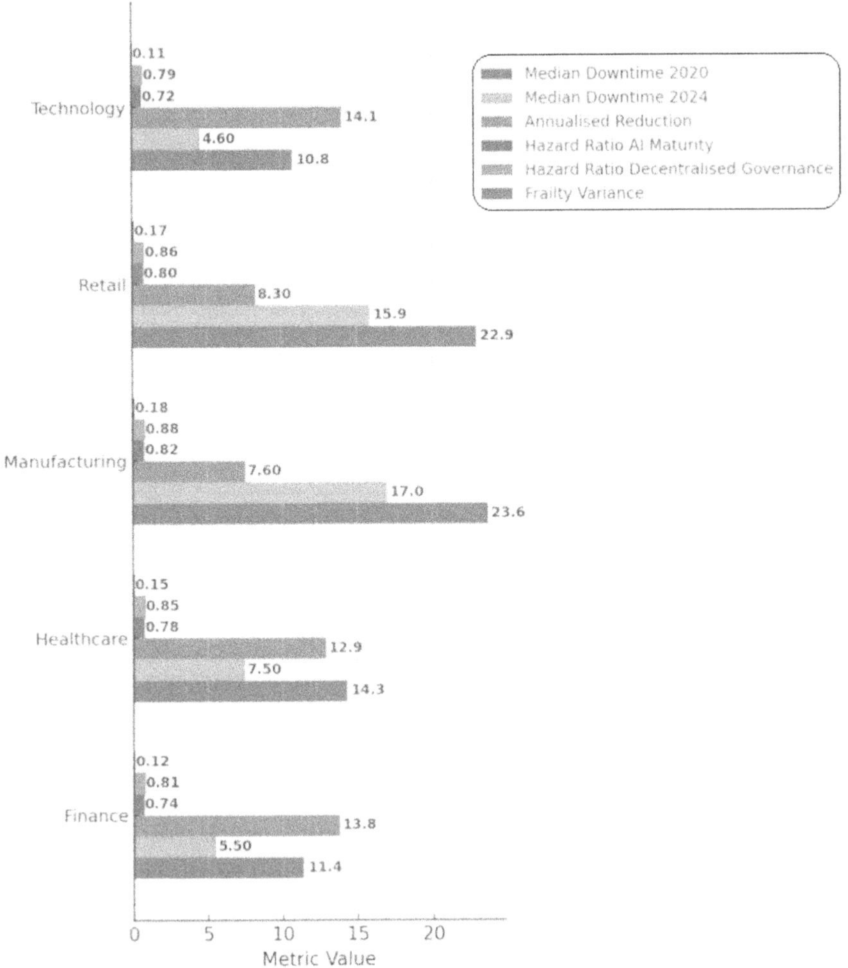

Fig. 4. Survival-model outputs for downtime reduction.

4.4 Latent Resilience Pathways

SEM was employed to integrate multiple observable dimensions into a latent construct of organizational resilience, quantifying how preparedness influences crisis outcomes. Standardized loadings indicate the minimal individual dimension's contribution, while the variance-explained percentages illustrate the explanatory coverage of the resilience factor. Total effects capture how improvements in preparedness translate into operational benefits. Table 3 presents fit indices to show how well the model fits the data. This well-rounded approach lifts the veil by which governance, analytics, and organizational culture converge to drive quantified gains in time to respond, time down, and workforce stability, and provides a full view of enterprise agility.

Table 3. Standardized SEM results and fit statistics.

Metric	Standardized Loading	Variance Explained (%)	Total Effect Preparedness $\rightarrow$ Outcome
Response Time	–0.61	37	–0.54
Downtime	–0.53	28	–0.47
Productivity Loss	–0.47	22	–0.41
Latent Resilience (R^2)	—	66	
Model Fit	CFI 0.93; RMSEA 0.045; SRMR 0.041		

Preparedness exerts a strong positive influence, explaining two-thirds of the variance in latent resilience and yielding double-digit percentage reductions in operational indicators. Negative loadings indicate that shorter response and downtime correspond to higher resilience. Although productivity loss contributes less in absolute magnitude, it remains statistically significant, suggesting that robust crisis systems mitigate both structural and human capital stresses. The model fit indices exceed standard benchmarks, confirming strong construct validity. These findings demonstrate that strategic investments in governance, monitoring, and cyber maturity substantially compress recovery windows and strengthen enterprise resilience.

4.5 Systemic Stress and Monte-Carlo Scenario Outcomes

Monte-Carlo simulations were conducted to aggregate the micro-level findings into macro-level insights, illustrating how correlated shocks propagate into systemic instability. The analysis includes median five-year loss projections, VaR, and ES estimates to quantify downside exposure, while the tail insolvency probability captures the likelihood of breaching solvency thresholds. The copula tail correlation measures cross-sector contagion potential. This rich array of metrics provides decision makers with both a comprehensive and deep-set of measures of benchmark capital adequacy and plan for contingent funding.

Figure 5 summarizes the simulated outcomes. Manufacturing emerges as the primary systemic outlier with both median losses and tail metrics exceeding all other sectors, driven by concentrated supply-chain dependencies. Retail follows closely due to demand-sensitive revenue structures amplifying liquidity strain. Finance and technology show similar loss projections but lower insolvency probabilities, indicating better capital management and diversification. Healthcare exhibits a higher tail with finance, suggesting shared vulnerability through pharmaceutical supply links. The accumulation of crisis days provides an operational dimension: manufacturing and retail face protracted disruptions, whereas technology's asset-light profile facilitates rapid recovery. These simulations reinforce that strategic preparedness and advanced analytics materially reduce both the probability and severity of systemic stress, validating the earlier empirical models.

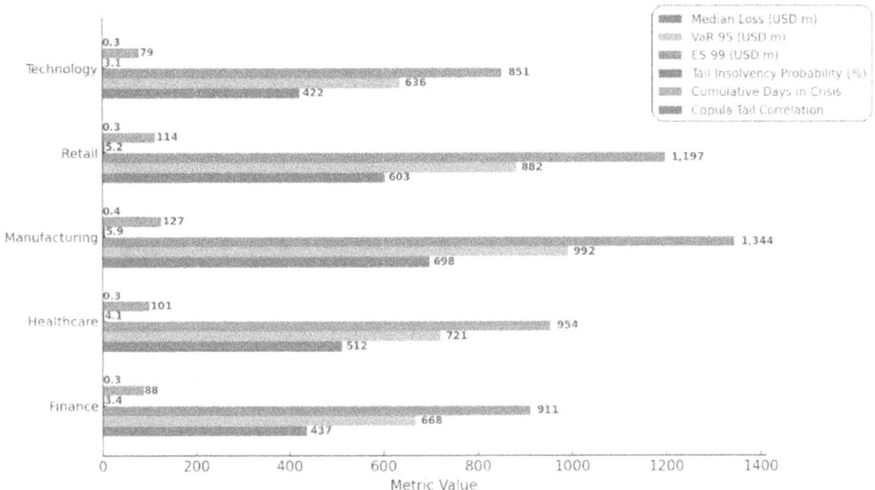

Fig. 5. Monte-Carlo cumulative-loss and systemic-risk metrics.

5   Discussion

The findings of this study highlight clear differences in crisis management effectiveness across sectors, emphasizing that strategic preparation, technological integration, and financial soundness are key drivers of resilience. Financial and high-tech sectors demonstrated stronger responses and faster recoveries due to their ability to use analytics, automated monitoring, and liquidity management models. Their shorter downtimes and stable market performance confirm that proactive risk mitigation enhances organizational stability. In contrast, manufacturing and retail sectors, which depend heavily on physical supply chains and manual operations, faced slower recoveries and higher vulnerability to market fluctuations.

The results also underline the importance of technological advancement in strengthening crisis resilience. Sectors that adopted cloud-based systems and decentralized decision-making were better equipped to predict and manage disruptions. Meanwhile, industries reliant on physical infrastructure encountered delays caused by logistical and workforce constraints. These findings suggest that digital adaptability is a critical factor in reducing crisis impact and improving recovery speed.

Another major contribution of this study lies in extending the financial efficiency perspective of crisis management. The analysis demonstrates that resilience depends not only on the amount of financial reserves available but also on how effectively these resources are allocated. Organizations that maintained flexible budgets, diversified income streams, and strategic contingency plans were better positioned to absorb shocks and sustain operations during crises. The study further reveals the strong connection between crisis communication, market confidence, and long-term organizational stability. Firms that maintained transparent communication and agile marketing strategies experienced less market-share volatility and faster reputational recovery. This underscores the importance of coordinated governance and communication strategies in post-crisis performance.

Finally, the study brings attention to the often-overlooked issue of workforce productivity during crises. Results indicate that productivity losses are lower in sectors that enable remote work, automate workflows, and prioritize employee well-being. In contrast, industries relying on onsite and manual work experienced greater disruptions and slower recovery. Overall, the study provides a comprehensive perspective on how strategic, financial, and technological factors interact to shape crisis management effectiveness. The results suggest that organizations should move beyond reactive responses and adopt proactive, technology-driven, and workforce-centered models to enhance their resilience in an increasingly volatile environment.

6 Conclusion

Global enterprises operate in increasingly volatile environment where crises can disrupt digital and physical value chains within hours. This study examined whether multi-dimensional preparedness, spanning governance robustness, real-time monitoring, and cyber-maturity, can mitigate crisis frequency, reduce operational disruption, and limit financial losses. Using a five-year panel of 570 firms analyzed through hierarchical count models, volatility-adjusted loss equations, survival analytics, and latent-variable modeling, the results confirm that strategic preparedness functions as a unifying mechanism that recalibrates risk throughout the crisis cycle. Organizations with higher preparedness scores experienced fewer incidents, faster containment, and reduced liquidity drawdowns, demonstrating that proactive controls and predictive analytics prevent cascading losses.

The findings also reveal sectoral asymmetries. Asset-heavy industries such as manufacturing and retail remain vulnerable to supply-chain bottlenecks, while finance and technology sectors benefit from automated surveillance and liquidity buffers. Preparedness thus emerges as a portfolio of capabilities with varying marginal effects across industries. Firms with decentralized decision-making and strong AI integration showed superior performance, indicating that digital autonomy enhances governance effectiveness. At the systemic level, simulation results highlight how inadequate preparedness amplifies contagion effects across sectors. Firms in the top quartile of preparedness reduced extreme loss exposure by about one third, confirming that resilience investments yield measurable returns. This underscores the importance of embedding readiness indicators in board-level dashboards alongside profitability metrics.

The study further emphasizes that crisis management should evolve from reactive interventions to an embedded capability supported by live data, scenario simulations, and machine-learning diagnostics. Maintaining workforce productivity through communication and well-being strategies also proved essential to organizational resilience.

Overall, the study establishes crisis resilience as a multidimensional capability grounded in proactive governance, analytical foresight, and organizational learning. Strategic preparedness is both predictive and actionable, enabling firms to transform crises into manageable challenges rather than existential threats. Future research should focus on micro-level tracking of decision-making during crises, cross-border comparisons of regulatory impacts, and the integration of ESG indicators to assess how sustainability interacts with resilience. Extending the analysis through agent-based models and field experiments could provide stronger causal evidence and enhance the practical applicability of resilience frameworks.

References

1. Averina, T., Avdeeva, E., Kurbatova, T.: About the creation of a crisis management system at an enterprise in conditions of uncertainty. Bull. South Ural State Univ. Ser. Comput. Technol. Autom. Control Radioelectronics **22**(3), 132–140 (2022)
2. Zhabin, A.: RETRACTED: management strategies and enhancing enterprise efficiency in the context of contemporary geopolitical conditions. In: E3S Web of Conference, vol. 420, p. 04003 (2023)
3. Furiak, K., Buganová, K., Prievozník, P., Hudáková, M., Slepecký, J.: Research on the impacts of global entrepreneurial environment changes on small and medium-sized entrepreneurship. Systems **12**(7), 234 (2024). https://doi.org/10.3390/systems12070234
4. Bukar, U.A., et al.: How advanced technological approaches are reshaping sustainable social media crisis management and communication: a systematic review. Sustainability **14**(10), 5854 (2022). https://doi.org/10.3390/su14105854
5. Wu, T., Dobrea, R.: Building sustained and robust business performance: from situation analysis, day-to-day management and risk prevention and control. In: Proceedings of 17th International Management Conference, pp. 552–559 (2024)
6. Choudhury, M., Haque, C.E.: Disaster management policy changes in Bangladesh: drivers and factors of a shift from reactive to proactive approach. Environ. Policy Gov. **34**(5), 445–462 (2024)
7. Kabachenko, D.: Implementation of digitization for anti-crisis management of business entities. Econ. Aff. **68**(01s), 361–369 (2023)
8. Andi, H.: Development of organisational resilience within SMEs through implementation of strategic thinking. J. Asian Res. **6**(2) (2022)
9. Min, H.: Blockchain technology for enhancing supply chain resilience. Bus. Horiz. **62**(1), 35–45 (2019)
10. Holland, D., Seltzer, T., Kochigina, A.: Practicing transparency in a crisis: examining the combined effects of crisis type, response, and message transparency on organizational perceptions. Public Relat. Rev. **47**(2), 102017 (2021)
11. Rashed, M., Uddin, M.K., Islam, M.F., Faisal-E-Alam, M., Tushar, H., Ahmed, M.E.: Building resilient organisations: the role of technological capability, innovation leadership, and sustainability. Glob. J. Flex. Syst. Manag. **26**, 1–12 (2025). https://doi.org/10.1007/s40171-025-00471-x
12. Business Continuity Institute (BCI): Growing global momentum behind operational resilience as programmes increase to 70% over the past 12 months. Business Continuity Institute News Release (2025). https://www.thebci.org/news/growing-global-momentum-behind-operational-resilience.html
13. Gigliotti, R.A., Alvarez-Robinson, S.: The role of leadership communication in building crisis readiness and resilient leadership in times of disruption: an exploratory study. Behav. Sci. **15**(9), 1260 (2025). https://doi.org/10.3390/bs15091260
14. Park, K., Kim, H., Cha, J.: An exploratory study on the development of a crisis index: Focusing on South Korea's petroleum industry. Energies **16**(4), 5346 (2023). https://doi.org/10.3390/en16145346

15. Kročil, O., Müller, M., Schlossarek, M.: Making sense of social enterprises' crisis preparedness and development of composite index for its assessment. J. Contingencies Crisis Manag. **32**(4), e12622 (2024)
16. Zhou, F., Endendijk, T., Botzen, W.J.W.: A review of the financial sector impacts of risks associated with climate change. Annu. Rev. Resour. Econ. **15**, 233–256 (2023)
17. Enyejo, J., Adeyemi, A., Olola, T., Igba, E., Obani, O.: Resilience in supply chains: how technology is helping USA companies navigate disruptions. Magna Scientia Advanced Research and Reviews **11**(2), 261–277 (2024)
18. Chih, Y.-Y., Hsiao Cody, Y.-L.: Brace for another crisis: empirical evidence from US construction industry and firm performance during and after 2007–2009 global financial crisis. J. Manag. Eng. **39**(1), 04022069 (2023)
19. R.M., A.: The use of digital technologies for crisis management. Econ. Manag. **30**(5), 633–641 (2024)
20. Švábová, L., Stefunova, D., Kramárová, K., Ďurica, M., Gabrikova, B.: Would it have been cheaper to let them become unemployed? Costs and benefits of First Aid intervention for companies in Slovakia during the COVID-19 pandemic. Equilibrium **19**(1), 139–169 (2024)
21. George, Y.E., Odubo, A.: Organizational resilience and employee performance in crisis economy. Int. J. Innov. Eng. Res. Technol. **11**(4), 1–12 (2024)
22. Huang, N.: Multi-dimensional analysis of financial accounting information and construction of decision support systems. Appl. Math. Nonlinear Sci. **9**(1) (2024)
23. Chen, Y.: Analysis of brand crisis public relations: take FuYao Glass brand crisis in 2014 as an example. Int. J. Soc. Sci. Public Admin. **3**(1), 345–348 (2024)

Quantitative Modelling of AI-Driven Consumer Behaviour in Digital Economies

ZhiAng Yu[1]([⊠]) [iD] and Rexford Nii Ayitey Sosu[2] [iD]

[1] The Bishop Strachan School, Toronto, ON M4V1X2, Canada
`hollyy26@bss.on.ca`
[2] Computing and Information Systems, Ghana Communication Technology University, Accra, Ghana
`rnasosu@gctu.edu.gh`

Abstract. This study presents an advanced quantitative framework for modelling AI-generated consumer behaviour within the rapidly evolving digital economy. As artificial intelligence (AI) increasingly shapes personalized recommendations, pricing strategies, and online engagement, understanding consumer responses to algorithm-driven platforms has become crucial for both businesses and policymakers. The proposed model integrates behavioural economics with machine learning to capture the dynamic patterns of consumer decision-making under algorithmic influence. Specifically, the framework employs an Agent-Based Modelling (ABM) approach combined with Reinforcement Learning (RL) techniques to simulate how consumers adapt their preferences and purchasing choices in response to real-time stimuli such as targeted advertisements, dynamic pricing, and social network effects. This hybrid model not only provides micro-level insights into individual behavioural adaptations but also connects them to macro-level market dynamics. The results demonstrate the framework's potential in predicting emerging consumption trends, guiding data-driven marketing strategies, and informing regulatory policies in AI-powered digital ecosystems.

Keywords: AI-Driven Consumer Behaviour · Digital Economy · Agent-Based Modelling · Reinforcement Learning · Behavioural Simulation

1 Introduction

Artificial intelligence (AI) technologies have transformed the currently existing environment of consumer behaviour within digital economies through their dynamic evolution process [1]. AI has played a key role in customer engagements when it comes to digital platforms with large retailers relying on tailored product suggestions and personalized advertising, algorithmic pricing and dynamic options in decision-making [2]. This revolution brings to businesses and policymakers new challenges and prospects as now their interest is on how complex consumer behavior can be mediated through intelligent systems [3]. Conventional consumer-decision models do not always portray the flexible, information-saturated, and changeable worlds that prevail in a digital age. There is therefore an imperative necessity of stronger and more elastic modelling techniques that are able to capture the changing relationship between AI and consumer psychology [4].

Z. Molamohamadi et al. (Eds.): ODSIE 2025, CCIS 2855, pp. 649–659, 2026.
https://doi.org/10.1007/978-3-032-17023-1_39

This work proposes a new framework of quantitative modelling using ABM with RL to model AIs behaviours in supporting the development of digital economies in terms of consumer behaviour. The model under consideration is a proposal that describes plasticity of consumer decision making under dynamic pricing, real-time recommendations, and social stimuli [5]. In a simulated digital market, agents, acting on behalf of consumers, search in a market that contains AI systems, which interact with agents over time, and a researcher can analyze any emergent tendency such as trust relations, preference changes, algorithmic unfairness, and market segregation. This ABM-RL model provides an efficient model of simulation to learn about the feedback loop of consumer behaviour and adaptation reaction to AI systems, thus providing an opportunity to make better policy interventions, optimized AI designs and user-friendly digital ecosystems.

Artificial Intelligence (AI) has significantly transformed consumer behaviour in digital economies by driving personalized recommendations, targeted advertising, and dynamic pricing. Businesses and policymakers now face challenges in understanding how consumers interact with intelligent systems. Traditional decision models fail to capture this dynamic and adaptive behaviour. To address this gap, this study proposes a hybrid quantitative framework integrating Agent-Based Modelling (ABM) with Reinforcement Learning (RL). The model simulates consumer adaptation to real-time stimuli such as social influence and algorithmic recommendations. It enables analysis of emergent behaviours like trust, bias, and preference shifts. This approach supports better AI design, policymaking, and digital ecosystem development.

2   Related Works

In the literature review, important studies that simulate and predict AI-influenced consumer behaviour are addressed; they represent different computational or behavioral techniques. Schematic of the used techniques combined with authors and subsequent benefits and disadvantages of all methods are illustrated in Table 1.

Table 1 summarizes key studies that explore AI-driven consumer behaviour using various computational and analytical techniques. L. Pu et al. [6] employed AI analytics and behavioural surveys to capture Gen Z engagement and personalization effects, though their work remained demographic-specific and lacked predictive depth. Y. Shaengchart et al. [7] applied big data analysis to identify major purchase factors in digital economies but faced regional limitations and lacked adaptive AI modelling. S. Mahabub et al. [8] integrated AI into decision-making frameworks, aligning it with leadership and strategic growth, yet their study was largely conceptual without consumer-level simulation. M. W. Farooq et al. [9] adopted a mixed-method approach in the FMCG sector, effectively linking AI design to consumer perception but offering limited generalizability. Finally, A. Vij et al. [10] utilized AI-integrated marketing analytics to demonstrate brand and sales impact, though their focus was restricted to performance metrics rather than issues of consumer trust or algorithmic bias.

The current research on consumer behaviour driven by AI is frequently limited by certain factors including a narrow scope (niche population and/or concentrated geographical area), an over-reliance on the conceptual or stable models and a lack of the ability to scale in relation to the marketplace. Dynamism in the interaction between

Table 1. Summary of existing tourism optimization approaches using genetic algorithms.

Authors	Technique Used	Advantages	Limitations
L. Pu et al. [6]	AI analytics, behavioral survey	Captures Gen Z engagement and personalization effects	Demographic-specific, lacks predictive depth
Y. Shaengchart et al. [7]	Big data analysis	Identifies key purchase factors in digital economy	Region-limited, lacks adaptive AI modelling
S. Mahabub, M. R. Hossain, E. Z. Snigdha [8]	AI in decision-making frameworks	Aligns AI use with leadership and strategic growth	Conceptual, lacks consumer-level simulation
M. W. Farooq, K. H. Ul Hassan, F. Nawaz [9]	Mixed-method, FMCG case study	Connects AI design to consumer perception	Sector-specific, limited generalizability
A. Vij et al., [10]	AI-integrated marketing analytics	Shows brand and sales impact of AI marketing	Focuses on metrics, not consumer trust or bias
Authors	Technique Used	Advantages	Limitations

the consumers and the adaptive AI systems cannot be completely captured under many strategies, or the preference changes gradually and the feedback mechanism. The provided model to address these difficulties combines an ABM and RL that provide a more real-time simulation framework and more flexible environment. This can enable a more in-depth account of consumer decision habits, algorithmic control and emergent markets behaviour in a variety of digital ecosystems.

3 Proposed System Architecture

The method that is proposed provides a dynamic, multi-layered framework in modelling and exploring consumer behaviour in AI-mediated marketplaces in the digital world. It starts with the Agent-Based Market Environment Initialization, in which a synthetic digital economy is achieved through heterogeneous consumer agents with their own social-economic/behavioural characteristics. These avatars communicate with in-built artificial intelligence agents such as recommendation engines and pricing algorithms and their source of such interaction constitutes the basis of the formation of simulation of the real-world decision-making. On the second stage, RL is incorporated in order to simulate how the agents will change their behaviours over the duration of time according to rewards and punishment provided by the interactions. The properties enable the agents to learn and adjust their preferences to suit targeted advertisements, change of prices or algorithmic nudges and so the adaptation of human behaviour in a digital environment works closely with this property. Table 1 shows the proposed architecture (Fig. 1).

A further step in architecture development is Simulation of AI Influence Mechanisms and introduces personalized recommendations, content targeting and dynamic

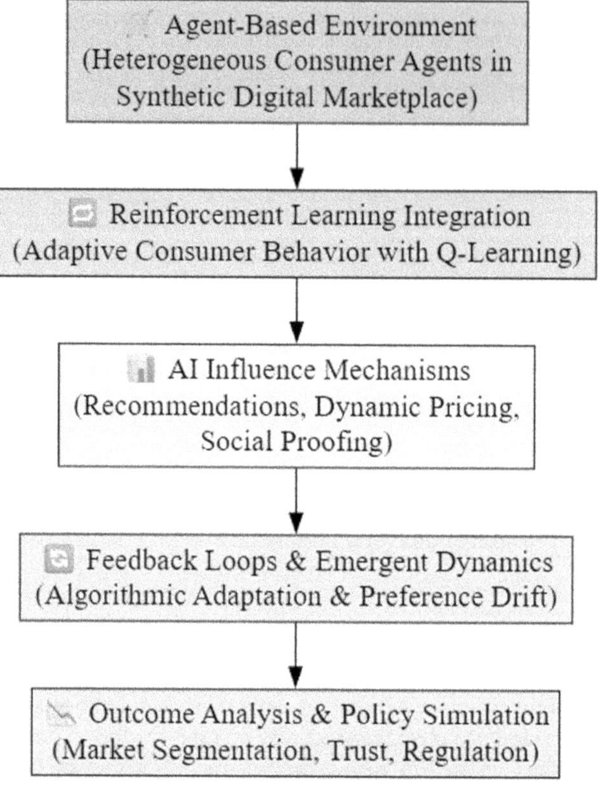

Fig. 1. Design of the proposed architecture

pricing strategies that promote agent decisions. One use is as stimuli and they are real-time-adjustable on an agent responsive basis, forming a looping interaction. The fourth phase, Modelling Feedback Loops and Emergent Dynamics, reflects the bi-directional development between the adaptations in AI systems and consumer behaviour, resulting in such compound phenomena as preference drift, echo chambers, or reinforcing algorithmic biases. Lastly, Outcome Analysis and Policy Simulation gives an assessment of the macro-term consequences such as the dynamics of trust and market fragmentation, as well as the possibility of testing policy interventions, including procedures of transparency and fairness. Such an all-encompassing flow not only allows predictive modelling of digital consumerism but also allows responsible AI design by providing a rock-solid, simulation-based platform on which research and regulatory experimentation can be done [11].

3.1 Agent-Based Market Environment Initialization

The initial phase of the proposed architecture would entail the creation of a synthetic digital marketplace on the basis of Agent-Based Modelling (ABM). The individual consumers are also modelled as independent agents in this environment and each agent has got its own specific aspects of decision-making behaviour. These parameters entail such socio-economic factors as income, education, and digital literacy, as well as such behavioural peculiarities as brand loyal attitude, risk-aversion, social sensitivity, and digital-interface adaptability. With such heterogeneity, the model would embody the dynamic and varying characteristics of consumer behaviour, and through it construct realistic reactions to dynamically changing market stimuli.

In this simulated world, AI programs, including recommendation engines, pricing algorithms, ad modules, etc., are integrated so that they may interact with consumer agents in real time. Such AI components can affect the actions of agents by making specific recommendations, changing prices dynamically, or improving the level of product exposure depending on activity preferences. Notably, the system is set to accommodate feedback loops, wherein AI systems will adjust according to the behaviour of the agents in the long term. Such initialization phase preconditions the simulation of micro-level interactions between consumers and macro-level market phenomena such as the shaping of trends, preference changes, polarization, and trust and distrust phenomena in digital markets. Finally, it renders a dynamic and interactive test bed with the structure and challenge of actual AI mediated digital economies [12].

3.2 Integration of Reinforcement Learning for Behavioral Adaptation

The second part of the proposed architecture adds RL to the model of adaptation of strategies of consumer agents in the decision-making process on the way of adaptation in the AI-driven digital marketplace. In contrast to the non-adaptive behavioral models, RL allows agents to learn about the environment by collaborating with it through the trial-and-error interactions. Applicable to both accepting an offer and responding to a suggestion, each agent is exposed to the outcomes of its decisions - e.g. choosing a product, responding to a suggestion, reacting to price decrease, etc. and gets some feedback: a reward signal. Good results (e.g. customer satisfaction with purchase, economies) increase the probability of trying similar products again, whereas the bad ones (e.g. buyer remorse, paying too much) change strategy. Such dynamic learning enables agents to formulate customized policies and update them depending on the evolving circumstances and conditions like introduction of new products, change in prices or algorithmic targeting. The RL mechanism allows the agent to continue in the middle of uncertainty and the choices are better to the extent that it might be optimized depending on past experience [13]. It is possible to describe the learning process formally with the value function of Q-learning which is one of the popular RL algorithms which is represented by Eq. (1).

$$Q(s, a) \leftarrow Q(s, a) + \alpha \left[r + \gamma \max_{a'} Q\left(s', a'\right) - Q(s, a) \right] \tag{1}$$

In this case, $Q(s, a)$ is the expected utility (Q-Value) of executing action a at the state s, a is the parameter that indicates the rate of learning, r is the reward gained on the execution of the action, gamma (γ) is the discount factor to provide a discount on future rewards and s', a' are the next-state and next-action respectively. This equation helps the agents to perfect their action of a given behavior by iteration of interactions in value estimations. How RL augmented the realism and the predictive abilities of the simulation, it modelled the changing consumer preferences and decision-making game under the constant AI stimuli and the market fluctuations.

3.3 Simulation of AI Influence Mechanisms

The third step of the architecture is devoted to the integration of the influence mechanisms based on AI to replicate the real-world computation algorithms into the synthetic digital market. These can be narrowed down to the recommendations based on personalization, dynamic prices, targeting content and social proof, which are widely used in AI-mediated consumer settings. Every mechanism is represented as a kind of stimulus which engages in the practices of the consumer agents. As an example, a personal recommendation system will customize the recommendation of products taking into consideration what an agent has browsed on the site, and what it may prefer as well as what it may have purchased before and dynamic pricing algorithms adjust prices of products in real time with consideration to the elasticity of price, purchase behaviour and competitor action.

The effect of such mechanisms is always evaluated since the dynamics in consumer preferences, interests, and conversion rates are observed. Part of the point of this simulation is that the system can simulate the behaviour of individuals being subjected to algorithmic designs, whether as a reinforcement of their current tastes (e.g. via filter bubbles and echo chambers) or as an inducement to new and exploratory consumption (e.g. via diversity-enhanced recommendations). Notably, such AI systems are flexible; they refine their models depending on the responses of the consumer agents and develop a bilateral process that might result in the occurrence of emergent effects like consumption polarization, or amplified bias of the algorithms, or loss of trust. This step leads to invaluable knowledge concerning the behavioral externality of the algorithmic rule, servicing the researchers and policymakers in assessing the ethical, economical, and societal consequences of the presence of the AI in digital markets.

3.4 Modelling Feedback Loops and Emergent Dynamics

The fourth level of the submitted framework focuses on the codification of the bi-directional feedback loops of the behaviour of the AI system and consumer agents reactions. In practice, in the digital ecosystems, AI algorithms (recommendation system or pricing engine example) do not remain in the same state; they further develop as users pattern them with constant input. Once consumer agents respond to the system and demonstrate a different pattern of behaviour (e.g. change of interests, less engagement, utility to particular products or brands), the AI models would adapt, adjust the parameters, targeting techniques or change contents. This mutual dependence implies that the behavior of AI systems can be determined by consumer choices, but, conversely, consumer behavior can be influenced by the decisions of AI systems as well and the relationship is becoming a self-perpetuating loop.

These nonlinear interactions may lead to emergent phenomena which cannot be simply predictable using linear systems. As an example, when consumers are exposed to personalized recommendations constantly, it can lead to the process of drift, when customers will be narrowed down to a smaller niche in their preferences. In a related instance, the algorithms designed to maximize engagement are also likely to form echo chambers furthering currently prevalent biases or preferences. In a worse situation, the feedback effect can create loop biasing in algorithms, creating exclusive market behaviour or consumption polarization through highly divided consumer groups. These effects should be crucial to model in order to comprehend the systemic impacts of integrating AI in consumer markets in the long-term. This phase does not only emphasize the value of moral and transparent AI design, but it also allows the scenario-based testing to reduce unintended consequences of interacting with a complex socio-technical system.

3.5 Outcome Analysis and Policy Simulation

The last element of the architecture is focused on carrying out a multi-perspective analysis of the result of the simulation of a digital marketplace and simulation policies, to test the wider implication of AI-consumer interactions. At this level of analysis, individual agent behaviour is not analyzed; instead, the aggregated system-level outcomes of interest are market segmentation trends, changes in consumer trust, efficiency in pricing and long-term changes in behaviour. The results are evaluated to reveal the impact of AI mechanisms on consumption patterns of various demographic and behavioral types and provide insights with regard to the systematic effects of algorithmic decision-making of digital economies on a greater scale.

Besides performance assessment, this step allows policy experimentation through simulating control or design initiatives in the AI environment. As an example, it is possible to test the scenarios to analyze how the transparency requirement of algorithms, fairness bias in suggestions, or diversity-increasing filters in content delivery would perform. Observing the way the consumer behaviour and system dynamics are changed under these interventions, the architecture can be considered a decision support tool in the hands of the technologists, designers and other policymakers. It provides a safe platform where the results of design decisions are tested before actualization and allows creating user-friendly, ethical, and sustainable digital marketplaces. In a long-term perspective, this step fills the divide between computational modelling and practical public governance, and thus leads to responsibility in the implementation of AI in technologies seen by consumers.

4 Performance Evaluation

Here the findings of the experiment that was carried with the simulation of the proposed AL-driven consumer behaviour modelling framework are presented. It is these findings that seek to analyze how synthetic agent-based market place dynamics that impact the individual level decision processes in a market place are affected when the consideration is provided by AI algorithms of dynamic pricing and product recommendations. The sample of key behavioural regularities, market trends, and emergent dynamics was studied with different sets of reinforcement learning strategies and interventions of AI intensities. Besides, outcome indicators (preference drift, market segmentation, and trust dynamics) were measured in order to apprehend systematic implication of algorithmic governance. Both micro-level agent adjustments and macro-level behavioural final decisions are incorporated in the discourse and empirical evidence is found to support the claim that the model can be used to simulate realistic, policy-relevant consumer behaviour in an AI-mediated digital economy (Fig. 2).

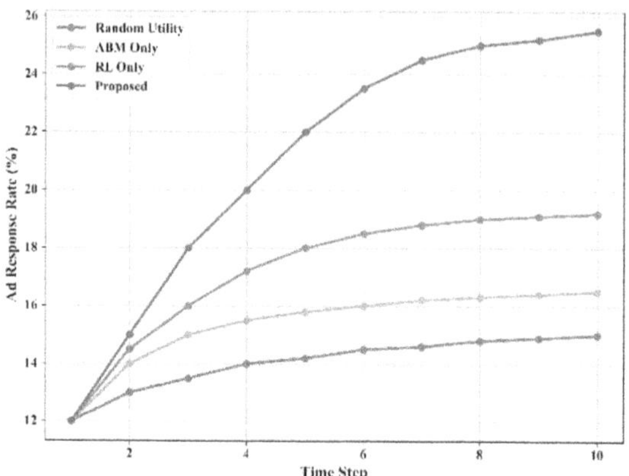

Fig. 2. Ad Response rate

The Table 1 illustrates the evolution of ad response rates (%) over ten-time steps for four different consumer behaviour modelling approaches: Random Utility, ABM Only, RL Only, and the Proposed hybrid model. At the initial stage (time step 1), all models begin with a similar baseline response rate of approximately 12%. As the simulation progresses, variations in model performance become evident. The Random Utility model shows minimal improvement, reaching only 14.8% by time step 10. The ABM Only model demonstrates slightly better engagement, peaking at 16.5%, while the RL Only model achieves a moderate increase, ending at 19.1%. In contrast, the Proposed model, which combines Agent-Based Modelling with Reinforcement Learning, consistently outperforms the others, showing a steep and steady rise to 25.5% by the final time step. This substantial gain highlights the effectiveness of integrating adaptive learning with heterogeneous agent behaviour, allowing for more nuanced consumer responses to dynamic stimuli. The results affirm that the hybrid architecture captures the complexity of real-time decision-making and offers a superior framework for optimizing AI-driven advertising strategies in digital marketplaces (Fig. 3).

The comparative Table 1 highlights significant differences in consumer satisfaction across four modelling approaches. The Random Utility model shows the lowest performance with a median satisfaction score of around 0.53, indicating limited adaptability. The ABM Only model improves on this with a median of ~ 0.60, capturing agent-level behaviour but lacking dynamic learning. RL Only further enhances satisfaction, reaching a median of ~ 0.64 by adapting to stimuli over time. The Proposed Model, which integrates both ABM and RL, outperforms all others with the highest median score of ~ 0.77 and a narrow interquartile range, reflecting both consistency and adaptability. This demonstrates the model's strength in simulating complex consumer behaviour in dynamic, AI-influenced digital marketplaces (Fig. 4).

The line plot in Table 1 illustrates the progression of cumulative rewards over 50 episodes across four modelling strategies. The Random Utility model accumulates the

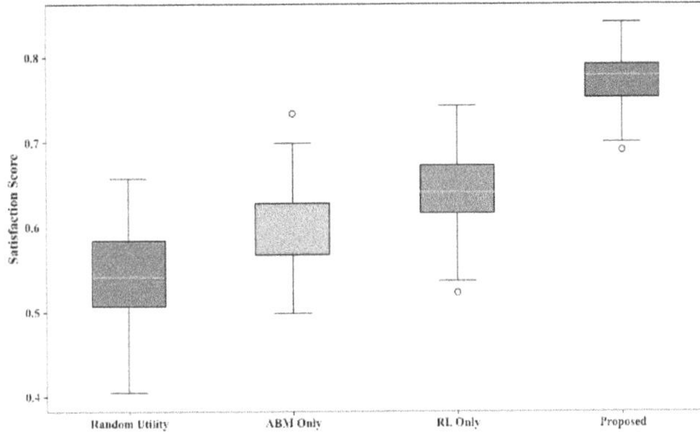

Fig. 3. Satisfaction score

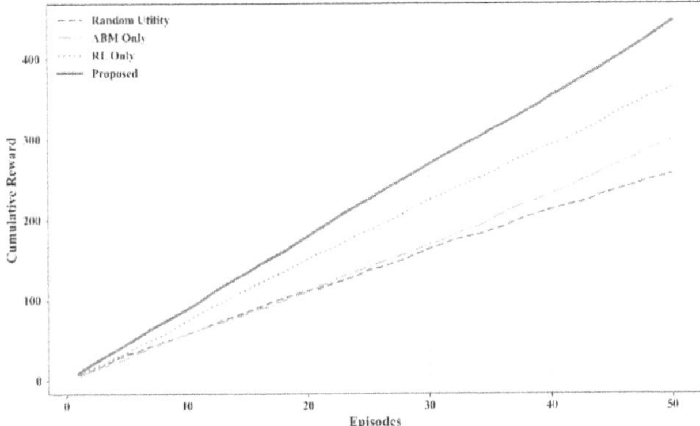

Fig. 4. Cumulative reward

lowest reward (~260 at episode 50), indicating limited learning or adaptability. The ABM Only model performs moderately better, reaching a cumulative reward of around 300, suggesting that agent-based interactions alone provide some behavioral realism. The RL Only model shows further improvement, achieving approximately 360 rewards due to its adaptive learning capability. The Proposed Model, integrating both ABM and RL, consistently outperforms the others, surpassing 450 in cumulative rewards by episode 50. This steep and steady growth underscores the model's superior ability to adaptively learn consumer preferences in real time, demonstrating enhanced effectiveness in simulating AI-driven market environments.

The experimental results clearly demonstrate that the proposed hybrid ABM–RL framework provides a more robust and realistic representation of AI-driven consumer behaviour compared to traditional models. Across all evaluation metrics—ad response rate, satisfaction score, and cumulative reward—the hybrid model consistently outperformed Random Utility, ABM Only, and RL Only approaches. The significant improvement in ad response rate (up to 25.5%) and satisfaction score (~0.77) indicates the model's superior ability to capture adaptive and context-sensitive consumer decisions under dynamic pricing and personalized recommendations. The higher cumulative reward further confirms the model's capability to balance exploration and exploitation, leading to more efficient learning of consumer preferences and market responses. These findings highlight the effectiveness of integrating reinforcement learning with agent-based modelling in reflecting both micro-level behavioural adaptation and macro-level market patterns. Overall, the results validate the model's potential for practical application in optimizing AI-driven marketing strategies, policy analysis, and ethical governance of algorithmic decision-making in digital economies.

Conclusion.

To sum up, the presented DT-EMS model undergoes considerable improvements in the operational efficiency, stability, and real-time optimization of hybrid renewable microgrids. Through a cloud-based Digital Twin, the system also provides an uninterrupted communication between the physical elements of the system, including PV arrays, BESS, DGs units, and dynamic loads, and their cyber models. This integration provides the ability to monitor continuously, predictively optimize and adaptively control in order to attain cost-effective and stable energy management. The framework has been shown to be highly power-tracking accurate, quick to react to changes in load and to effectively coordinate the energy of the distributed sources. In addition, the BESS has flowing transitions between the charging and discharging phases, which guarantees continuous operation of the system. In general, the DT-EMS framework demonstrates that Digital Twin technology has the potential to transform the way sustainable, low-carbon, and resilient microgrid systems are operated by making intelligent and real-time decisions. Power tracking is precise and power variation will impose little deviations and the system has quick respond time and able to coordinate the energy source successfully. The system offers the ability to handle the temporary events and sudden changes in load which assists in maintaining the stability and providing the reliable supply of power. The BESS conducts efficient mode shifts between the charging and discharging mode PV arrays and DG units are reliable and can operate under any condition. This proposed framework can bring about operational shifts in hybrid microgrids by digital twins technologies that can be used to manage energy in real-time in an intelligent manner to develop sustain-able low-carbon power systems.

References

1. Pamisetty, A.: AI-powered predictive analytics in digital banking and finance: a deep dive into risk detection, fraud prevention, and customer experience management. Fraud Prevention and Customer Experience Management (2023)
2. Jin, K., Zhong, Z.Z., Zhao, E.Y.: Sustainable digital marketing under big data: an AI random forest model approach. IEEE Trans. Eng. Manag. **71**, 3566–3579 (2024)
3. Jingrong, H., Shan, H., Zhaobin, C., Yu, L., Yingying, L.: AI-driven digital transformation in banking: a new perspective on operational efficiency and risk management. Inf. Syst. Econ. **5**(1), 82–90 (2024)
4. Adekunle, B.I., Chukwuma-Eke, E.C., Balogun, E.D., Ogunsola, K.O.: Integrating AI-driven risk assessment frameworks in financial operations: a model for enhanced corporate governance. Int. J. Sci. Res. Comput. Sci. Eng. Inf. Technol. **9**(6), 445–464 (2023)

5. Rahevar, M., Darji, S.: The adoption of AI-driven chatbots into a recommendation for e-commerce systems to targeted customer in the selection of product. Int. J. Manag. Econ. Commer. **1**(2), 128–137 (2024)
6. Pu, L., Radics, R., Umar, M., Jeremiah, F., Quan, Z.: The potential of AI tools in shaping digital consumers' behavior: investigating e-commerce engagement of Chinese Generation Z. Asia Pac. J. Mark. Logist. (2025)
7. Shaengchart, Y., Bhumpenpein, N., Kraiwanit, T., Limna, P., Tanantong, T.: The impact of big data on online purchase behavior: influencing factors and interrelationships in Thailand's digital economy. Hum. Behav. Emerg. Technol. **2025**(1), 6678231 (2025)
8. Mahabub, S., Hossain, M.R., Snigdha, E.Z.: Data-driven decision-making and strategic leadership: ai-powered business operations for competitive advantage and sustainable growth. J. Comput. Sci. Technol. Stud. **7**(1), 326–336 (2025)
9. Farooq, M.W., Hassan, K.H.U., Nawaz, F.: Integrating qualitative and quantitative approaches: the impact of AI design on consumer perception and buying behavior in the FMCG sector. Bull. Bus. Econ. (BBE) **13**(2), 775–786 (2024)
10. Vij, A., Vij, M., Farouk, M., Kumar, P.: Evaluating the effectiveness of AI-integrated digital marketing on consumer behavior, brand perception, and sales performance. In: Proceedings of the 2024 2nd International Conference on Cyber Resilience (ICCR), pp. 1–6 (2024)
11. Nkomo, N., Mupa, M.N.: The impact of artificial intelligence on predictive customer behaviour analytics in e-commerce: a comparative study of traditional and AI-driven models. IRE J. **8**(5), 432–437 (2024). ISSN: 2456–8880
12. Zong, Z., Guan, Y.: AI-driven intelligent data analytics and predictive analysis in Industry 4.0: transforming knowledge, innovation, and efficiency. J. Knowl. Econ. 1–40 (2024)
13. Awais, M.: Optimizing dynamic pricing through AI-powered real-time analytics: the influence of customer behavior and market competition. Qlantic J. Soc. Sci. **5**(3), 99–108 (2024)

Digitalisation for Smart and Climate-Neutral Cities: The Case of Ukraine

Maryna Gorobei[(✉)] [iD] and Oleksandr Bondar [iD]

State Institution Institute of Ecological Renovation and Development of Ukraine (IER), Kyiv 03035, Ukraine
`maryna.gorobei@gmail.com`

Abstract. Digitalisation is now widely recognised as a key driver for the transition to smart and climate-neutral cities. By reshaping governance, public services and energy systems, digital technologies optimise resources, improve transparency and foster citizen participation. However, their ecological and social implications remain contested: energy-intensive models, governance constraints and inequalities risk undermining the climate-neutral objectives that digitalisation is meant to support. This study focuses on Ukraine as a critical case of accelerated digital transformation under extraordinary stress. The country's trajectory demonstrates rapid progress: from early experiments in e-government, through the institutionalisation of the Ministry of Digital Transformation, to the adaptive use of digital platforms during wartime. The Ukrainian case shows that despite severe disruption, digital and ecological initiatives can advance in parallel, offering lessons for other countries where such progress is possible under less extreme conditions. This research provides the first integrated assessment of Ukraine's digital transformation through the dual lens of ecological transition and urban sustainability, showing how digitalisation interacts with climate objectives under crisis conditions. Ukraine is presented not only as a case of resilient digital governance and wartime innovation, but also as an emerging reference point for climate-neutral pathways. The findings indicate that while digitalisation is not inherently sustainable, technologies such as digital twins and smart grids hold strong potential for post-war reconstruction, renewable integration and climate-responsive planning. These insights offer a foundation for future research and for shaping twin transition roadmaps that align digitalisation with climate neutrality and long-term resilience at both national and metropolitan levels.

Keywords: Digitalisation · Smart Cities · Climate-neutral Transition · Twin Transition · Blockchain · Urban Sustainability

1   Introduction

Digitalisation is increasingly acknowledged as a central pillar of sustainable urban transition. Scholars such as Kramers et al. [1] and Jones [2] have shown that cities across Europe and beyond are investing in digital infrastructures to optimise resource flows, reduce emissions and enhance resilience. Yet the ecological implications of these technologies remain contested, since their energy demand may in some cases undermine

Z. Molamohamadi et al. (Eds.): ODSIE 2025, CCIS 2855, pp. 660–675, 2026.
https://doi.org/10.1007/978-3-032-17023-1_40

climate objectives. Blockchain and digital twin technologies exemplify this paradox. On the one hand, they can foster transparency, decentralisation and data-driven optimisation of urban systems. On the other hand, blockchain models based on Proof-of-Work require substantial computational power, which has raised concerns regarding long-term environmental sustainability, as demonstrated by Stoll et al. [3]. The purpose of this paper is therefore to explore the dual role of blockchain and digital twins in shaping climate-neutral urban strategies. By highlighting both their enabling potential and their ecological risks, the study contributes to the growing debate on aligning digitalisation with sustainability imperatives as reflected in the work of the European Commission [4] and the International Telecommunication Union [5].

While digitalisation has become a central element of urban development strategies, its alignment with climate-neutrality objectives remains inconsistent. The European Commission increasingly refers to this linkage as a twin transition, seeking to bring digital and green transformations together within coherent strategies. Yet, in many cases, digital infrastructures have advanced faster than ecological frameworks, resulting in a gap between technological modernisation and sustainability goals. While smart city programmes have introduced innovative tools for governance, mobility and service delivery, they have not always been systematically connected to decarbonisation, resilience and long-term ecological planning.

In Ukraine, this dynamic is particularly visible. Digitalisation has advanced at an exceptional pace, with platforms such as Diia transforming governance and public services nationwide. Despite advances, the integration of digital technologies into climate strategies remains fragmented, confined to isolated municipal initiatives rather than embedded in metropolitan frameworks. This imbalance creates a risk: if digital and ecological transitions do not progress in parallel and at the same scale, opportunities for systemic climate-neutral transformation may be lost. The reconstruction process in Ukraine provides a unique opportunity to embed the twin transition – aligning digitalisation and ecological sustainability across metropolitan areas rather than restricting them to municipal pilots.

The central problem addressed in this study is therefore the absence of integrated approaches that treat digitalisation and climate neutrality as mutually reinforcing dimensions of urban transition. To achieve meaningful outcomes, both processes must advance simultaneously and be coordinated across metropolitan areas, rather than limited to flagship projects in major cities.

The paper is structured as follows. Section 2 presents the Methodology, including research design, data sources, validation procedures, and data access. Section 3 examines the Opportunities and Constraints of Digitalisation, focusing on environmental impacts, digital twins, and the Ukrainian case. Section 4 outlines the Results, synthesised through a SWOT analysis. Section 5 develops the Discussion, while Sect. 6 concludes with the main findings, limitations, and directions for future research.

2 Methodology

2.1 Research Design and Methods

This study applies a case study approach, positioning Ukraine as a critical example of accelerated digitalisation and ecological transition in crisis. The purpose of the design is not only to describe ongoing processes but also to provide a structured foundation that can serve as a reference for future comparative and applied research in the field of twin transition.

To achieve this aim, the research combined theoretical, empirical, and computational methods. At the conceptual level, historical and logical analysis, induction and deduction, scientific abstraction were employed to identify the main components of Ukraine's digital transformation and their interlinkages with ecological goals. Desk research of official strategies, international reports, and academic literature enabled the contextualisation of Ukraine's trajectory within broader European debates.

At the empirical level, the study relied on official datasets, statistical observation, and comparative analyses across national and international indicators. The method of generalisation supported the formulation of broader conclusions on institutional adaptability and citizen participation, while problem-finding and comparison methods were used to detect governance bottlenecks. A SWOT analysis was also conducted to structure strengths, weaknesses, opportunities, and threats.

In operational terms, data were prepared through structuring and processing of national and international datasets into comparable formats. Statistical assessment was conducted using Excel-based tools, including macros for tabulation, descriptive statistics, and trend identification, complemented, where relevant, with supplementary processing scripts (e.g. Python) to handle larger or less standardised datasets. For comparative benchmarking, indicators were aligned with international references, such as the Organisation for Economic Co-operation and Development (OECD), the International Energy Agency (IEA), the International Telecommunication Union (ITU), the United Nations E-Government Development Index (EGDI), and the Deloitte Digital Citizen Survey. Hereafter, acronyms are used in their abbreviated form following their first appearance in full. To support interpretation and dissemination, Flourish was used to produce interactive charts and visualisations, facilitating both internal validation of trends and transparent communication of results. Further triangulation across national statistics, international series, and survey evidence, along with consistency checks for the 2019–2025 period, enhanced the robustness of the analysis.

Through this integrated design, the study captures both systemic drivers and contextual constraints, offering a holistic perspective that reflects the uniqueness of Ukraine's case while generating insights that may inform future applications of digital-ecological transition strategies.

2.2 Data Sources

The study draws on a diverse set of data sources to ensure both the breadth and depth of analysis. These include official national datasets, international statistical series, survey evidence, and secondary case studies. Together, they provide a comprehensive foundation for examining Ukraine's digital and ecological transition.

At the national level, data were obtained from:

- the Ukrainian open data portal, covering government, economic, and sectoral statistics, data.gov.ua;
- the State Statistics Service of Ukraine, providing macroeconomic, demographic, and energy-related indicators https://stat.gov.ua/en;
- the Ministry of Environmental Protection and Natural Resources, offering ecological monitoring and sustainability reports https://mepr.gov.ua/;
- the Ministry of Digital Transformation, including strategy documents and platforms such as Diia (app and portal: https://diia.gov.ua/), DREAM (https://dream.gov.ua/), and Reform Radar (https://thedigital.gov.ua/projects).

At the international level, complementary datasets were consulted, including the OECD, the IEA, the ITU, and the United Nations EGDI. These benchmarks provide a comparative basis for situating Ukraine's progress within wider European and global trends.

Additional evidence was drawn from surveys and case studies:

- the United Nations Development Programme (UNDP) supported and Kyiv International Institute of Sociology (KIIS) omnibus surveys (2020–2022) on citizen use and trust of digital services;
- the UNDP Omnibus 2023 survey on e-service adoption;
- the GovTech Pulse survey (Info Sapiens and the Global Government Technology Centre (GGTC) Kyiv, 2025), which reports 75% citizen satisfaction with digital services;
- the case study "Ukraine's Digital Services in Time of Crisis" (2025), which documents the resilience of e-governance during wartime.

Finally, sectoral and technological datasets were incorporated, such as the Cambridge Bitcoin Electricity Consumption Index (CBECI) and results from the authors' prior work on blockchain environmental risk assessment, enabling the analysis of ecological costs of digital technologies.

Together, these data sources form a consolidated basis for triangulating evidence and evaluating both the opportunities and constraints of Ukraine's digital transformation pathway.

2.3 Validation of Data

To ensure the robustness and credibility of results, several validation steps were undertaken. First, triangulation was applied by cross-checking official national datasets with international statistics (OECD, IEA, ITU, UN EGDI) and independent survey evidence (UNDP omnibus surveys, GovTech Pulse). This allowed the findings to be verified against multiple, independent sources.

Second, temporal consistency was ensured by harmonising indicators across the 2019–2025 period. This procedure ensured that observed trends reflected genuine dynamics rather than discrepancies in data collection cycles.

Third, harmonisation of measurement units and definitions was carried out to allow comparability across heterogeneous datasets. For example, energy consumption and emissions data were standardised into consistent units, and citizen survey indicators were aligned with international benchmarks.

Fourth, comparative benchmarking was used to validate national results against broader European and global references. Indicators were systematically compared with OECD and EU averages, as well as with findings from Deloitte's Digital Citizen Survey, to situate Ukraine's performance within a wider context.

Finally, plausibility checks were conducted through internal review and descriptive statistical inspection. Outliers and inconsistencies were examined and either corrected or flagged as context-specific anomalies (wartime disruptions).

Together, these validation steps enhanced transparency and credibility, ensuring that the integrated dataset and subsequent analysis reflect a reliable foundation for assessing Ukraine's digital–ecological transition.

2.4 Data Access and Limitations

The integrated dataset compiled for this study is available from the corresponding author upon reasonable request. Several sources (data.gov.ua, UNDP and GGTC survey reports) are publicly accessible, while others remain partially restricted due to wartime conditions and security considerations. In addition, statistics from territories occupied since 2014 are not included in official datasets, which constrains the geographic completeness of the analysis. These limitations do not compromise the overall validity of the findings but may restrict the granularity of certain results and preclude full public disclosure of raw data.

3 Opportunities and Constraints of Digitalisation

3.1 Environmental Impact of Digital Technologies

Digital technologies are progressively regarded as instruments for enhancing urban sustainability and achieving climate neutrality. Within smart city frameworks, they are expected to facilitate peer-to-peer energy trading, strengthen transparency in governance and optimise the management of infrastructure [6, 7]. Such applications indicate that digitalisation can support decarbonisation by improving efficiency, lowering transaction costs and enabling wider citizen participation.

At the same time, the ecological footprint of these technologies has become a subject of growing concern. Moravec and others [8] observed that generative artificial intelligence significantly increases energy consumption and the carbon footprint of digital infrastructures, raising concerns about their compatibility with climate neutrality objectives. This insight echoes broader debates about the high energy consumption of blockchain systems and digital twins.

Blockchain models based on Proof-of-Work (PoW) are among the most energy-intensive examples. Stoll et al. [3] estimated that Bitcoin alone consumes electricity comparable to that of medium-sized countries, generating more than 22 MtCO$_2$ annually. Other studies have confirmed that, despite technical improvements, total consumption continues to rise [9, 10]. In previous research, Gorobei [11] demonstrated that the sustainability of blockchain is highly dependent on the consensus mechanism: while PoW entails unsustainable energy profiles, alternatives such as Proof-of-Stake (PoS) or Delegated Proof-of-Stake (DPoS) can significantly reduce electricity consumption. A comparable paradox is visible in the expansion of artificial intelligence (AI) and data centres. The training and deployment of AI models rely on high-performance accelerated servers, which require far greater power density than conventional systems. According to the International Energy Agency [12], data centres worldwide consumed around 415 TWh of electricity in 2024, equivalent to 1.5% of global demand. This figure is expected to more than double to approximately 945 TWh by 2030, with AI-related workloads accounting for nearly half of the increase. Within data centres, servers account for the largest share of consumption (around 60%), while cooling demands vary between 7% and 30%, depending on efficiency levels. Electricity use from accelerated servers linked to AI applications is projected to rise at an annual rate of 30%, underscoring the magnitude of future pressures.

Beyond energy considerations, the human and institutional dimensions of digitalisation are equally important. Sahibzada and others [13] demonstrated that digital leadership and digital orientation have a significant impact on sustainability outcomes, with employees' digital capabilities serving as a mediator in this relationship. These findings confirm that technological progress must be accompanied by changes in institutional and human capital; otherwise, its contribution to sustainability will remain limited.

Taken together, these findings illustrate a central dilemma: while digital technologies may accelerate the transition to climate-neutral urban systems, their own energy requirements risk undermining these objectives. From a policy perspective, this duality highlights the need for clear efficiency standards, the wider adoption of renewable-powered digital infrastructures and the alignment of digitalisation strategies with broader sustainability agendas. Only under such conditions can blockchain, AI and related technologies become genuine enablers of climate-neutral cities rather than additional sources of ecological strain.

3.2 Digital Twins as Smart-Climate Tools

Digital twin (DT) technology, commonly defined as a real-time digital representation of physical assets or processes, has gained notable prominence in the discourse on urban sustainability. Fuller et al. [14] and Batty [15] have emphasised that by enabling continuous monitoring, simulation and predictive modelling, DTs provide an analytical instrument to optimise energy efficiency, anticipate system failures and assess climate adaptation measures.

Within the urban context, digital twins enable holistic modelling of socio-technical systems such as transport, housing and energy grids, thereby bridging the gap between climate ambitions and operational decision-making. Liu et al. [16] demonstrated that DT-based simulations of energy systems can reduce emissions by enhancing heating and cooling efficiency and facilitating the integration of renewable energy sources. Similarly, in mobility planning, Deren et al. [17] showed that DTs facilitate scenario testing for low-carbon strategies, including electrification and modal shifts.

An increasingly significant dimension is the role of DTs in strengthening climate resilience. By incorporating environmental data streams, including temperature, precipitation and flood risk, cities can model the likely impacts of climate hazards and design adaptive responses in advance. This ability to embed climate variables into decision-making distinguishes DTs from conventional modelling and confirms their value as instruments of climate-responsive governance, as noted by Ketzler et al. [18].

Despite these advantages, notable challenges persist. The creation and operation of DT systems require extensive data infrastructures, interoperability standards and institutional capacities. Without strong governance frameworks, DT adoption risks remaining fragmented and limited to isolated projects rather than integrated into long-term strategies. Moreover, questions of data ownership, privacy and ethical management remain unresolved, as highlighted by Negri et al. [19]. These considerations are especially relevant given that DTs, much like AI systems, depend on energy-intensive computational resources hosted in data centres—underlining once again the tension between enabling technologies and their ecological footprint.

In the Ukrainian context, DTs could play a decisive role in reconstruction by helping to simulate and optimise new energy grids, modernise district heating and design resilient urban layouts. Their potential is particularly aligned with European frameworks for climate neutrality, such as the EU Green Deal, the Climate-Neutral and Smart Cities Mission under Horizon Europe, and the Digital Europe Programme. These frameworks increasingly emphasise cross-sectoral and metropolitan approaches rather than fragmented municipal initiatives.

Overall, digital twins represent a powerful enabling technology for advancing climate-neutral urban development. As with other forms of digitalisation, their effectiveness depends not only on technical sophistication but also on embedding them within regulatory and managerial frameworks that explicitly prioritise sustainability outcomes.

3.3 Ukraine as a Green-Digital Opportunity

The trajectory of Ukraine's digital transformation represents one of the most dynamic cases in Eastern Europe. Early steps towards e-government can be traced to the mid-2000s, when the government launched online tax services and state registries. Substantial progress began after 2014 with the introduction of the ProZorro e-procurement platform and the development of open data portals, such as data.gov.ua. Scholars such as Belal and Shcherbina [20] highlight that these initiatives, together with municipal pilots in Kyiv and Lviv – including digital ticketing and civic participation platforms – laid the groundwork for a more coherent digital strategy. This ultimately culminated in the creation of the Ministry of Digital Transformation in 2019.

The establishment of the Ministry marked the institutionalisation of digitalisation as a national priority. Its flagship initiative, Diia, positioned Ukraine as a pioneer of digital governance. By 2021, digital passports and driver's licences were recognised as legal equivalents of physical documents, while Diia.Business and Diia.City offered new frameworks for entrepreneurship and the ICT sector. According to the OECD [21], ITU [22], and analyses from the Harvard Center for International Development [23],

Ukraine had by this stage become a reference point in international debates on digital government, with Diia widely cited as one of the most innovative public digital services in Europe (Table 1).

Table 1. Chronology of Ukraine's digital transformation (2003–2030).

Year	Milestone	Significance
2003	First online tax services and registries	Initial steps towards e-government
2014	Launch of ProZorro e-procurement and open data portals	Transparency and open data reforms
2019	Creation of the Ministry of Digital Transformation	Institutionalisation of digitalisation as a state priority
2021	Recognition of Diia passports and driving licences as legal documents	International recognition of Ukraine as a digital governance pioneer
2022	Wartime services (eVorog, displaced persons apps, Starlink deployment)	Digital resilience under crisis conditions
2024	Launch of Sustainable Ukraine Now for Ukraine (SUN4Ukraine) project Ukrainian Climate Action Network (U_CAN) project	Integration into EU climate-neutral and smart cities framework
2030	Vision of a fully digital and climate-oriented state	Strategic long-term national objective

The full-scale invasion of February 2022 accelerated the evolution of Ukraine's digital ecosystem in unforeseen ways. Despite severe damage to physical infrastructure, digital platforms ensured continuity of governance and public services. Among the newly introduced wartime tools were eVorog, a citizen reporting application for military intelligence, web portals for displaced persons and compensation claims, and the deployment of more than 20 000 Starlink terminals to sustain critical communications. Ionan [24] observes that these developments underscored the adaptive capacity of Ukraine's digital governance in the face of crisis.

Parallel to these efforts, the official mobile application "Повітряна тривога" (Air Alert) ukrainealarm.com was introduced as part of the national civil defence system. The app delivers real-time notifications of air-raid, chemical, technological and other types of alerts, even when a smartphone is in silent mode, allowing citizens to select their region or municipality. By ensuring timely warnings, it became an essential digital instrument for protecting civilians during the war.

These dynamics are further illustrated by a radar chart (see Fig. 1), which compares digital transformation before the war, during the war, and the 2030 vision, highlighting both the resilience of Ukraine's digital governance and the temporary decline in internet coverage and digital skills caused by the conflict, alongside the ambitious targets for 2030.

The three charts illustrate the changing dynamics of Ukraine's digital transformation across six categories. During the pre-war period (2019–2021), progress was most evident

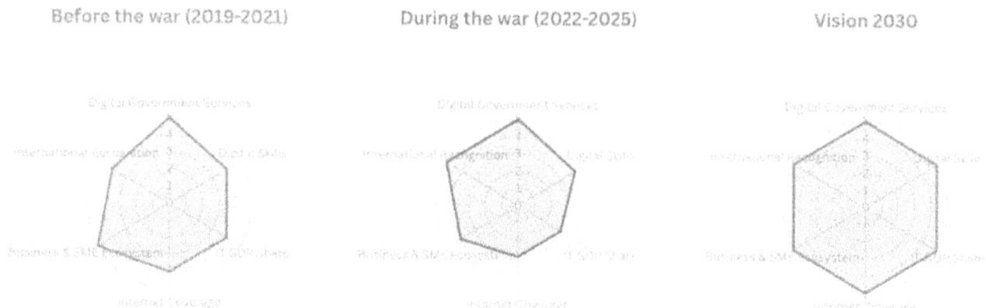

Fig. 1. Radar chart: digital transformation before the war, during the war, and vision 2030.

in the expansion of digital government services and the development of the ecosystem for small and medium-sized enterprises (SMEs), reflecting the early success of Diia and its associated platforms. During the war years (2022–2025), these services were preserved and even adapted to new circumstances, while international recognition remained high. At the same time, internet coverage and digital skills showed temporary decline due to infrastructure destruction and displacement of citizens. The vision for 2030 outlines a recovery path in which all six categories are expected to reach maximum performance, underscoring the strategic goal of establishing a resilient and climate-oriented digital state.

These results demonstrate that Ukraine's digital governance has maintained its functionality under extreme conditions and that its post-war ambitions are not limited to reconstruction but to systemic transformation. They also point to the need for coordinated frameworks that can address the uneven progress across categories: while government services and international visibility advanced quickly, connectivity and digital literacy require targeted investment to close the gap.

At the same time, the environmental context in which this transformation unfolds is extremely fragile. Air pollution in Ukraine is among the highest in Europe, contributing to an estimated 42 900 premature deaths in 2019 and almost one million disability-adjusted life years [25]. Forest ecosystems are under mounting pressure, with wildfires responsible for 45–65% of annual forest cover losses. Additionally, more than 40 percent of Ukrainian soils are affected by erosion, salinisation or contamination. The war has further aggravated this situation through the release of toxic substances such as lead and mercury. These figures highlight the critical need to ensure that digitalisation is not developed in isolation but is firmly embedded within environmental and climate strategies.

Digital technologies provide tangible opportunities for this integration. Blockchain and digital twins can enable real-time monitoring of emissions, transparent tracking of carbon credits, simulation of energy systems, and modelling of wildfire risks. As highlighted by Pricopoaia et al. [26], sustainability is only possible when digital transformation and ecological innovation are approached as mutually reinforcing processes. In this view, technology becomes not just a driver of optimisation but a catalyst of systemic change towards circular economy practices and climate neutrality. Such insights underline how the convergence of digital and environmental objectives can generate both efficiency gains and resilience.

In this context, European cooperation has become central. In 2024, the European Commission launched the SUN4Ukraine initiative and the parallel U_CAN project under the EU Climate-Neutral and Smart Cities Mission. Twelve Ukrainian municipalities, including Kyiv, Kharkiv, Dnipro and Vinnytsia, were selected as Flagship Cities to prepare Climate Neutrality Plans in partnership with EU Mission Cities (European Climate, Infrastructure and Environment Executive Agency, 2024 [27]). This collaboration directly addresses structural weaknesses of earlier smart city initiatives, which were fragmented and lacked strategic vision. The main thematic priorities that guide this transition are summarised in Table 2, identifying the areas where Ukraine must advance to meet its 2030 objectives.

Table 2. Priority areas for smart city development in Ukraine.

Priority area	Key focus	Illustrative applications
Smart mobility	Sustainable and low-carbon transport	Autonomous vehicles - Cooperative Intelligent Transport Systems(C-ITS), car sharing, smart parking, e-scooters, electric vehicle charging
Smart water	Clean water and resource efficiency	Distributed clean water treatment, eco-filtering, water quality monitoring
Digital smart city	Digital infrastructures for planning and management	3D drone surveys, Building Information Modelling (BIM), AR/VR technologies, digital twins
Energy efficiency	Resource optimisation and climate neutrality	Peer-to-peer (P2P) energy platforms, waste management infrastructure, Building Energy Management Systems (BEMS), zero-energy homes
Safety & living	Human-centred services and resilience	Disaster evacuation systems, crime prevention, fine dust reduction, smart homes, Property Technology (PropTech), and 5G-enabled kiosks
Governance & participation	Transparency and citizen engagement	Open data platforms, participatory budgeting, blockchain-based procurement, e-voting
Climate-digital integration	Coordinating sustainability and digitalisation	Climate-neutrality plans, integrated resilience roadmaps, cross-sectoral metropolitan strategies

Experiences from developing economies demonstrate that the combination of digital and financial innovation can stimulate ecological transition. Wu and Lin [28] found that digital inclusive finance promotes green innovation, industrial upgrading and improvements in employment quality. These findings underline the importance of aligning Ukraine's digital strategies with financial innovations that directly support environmental and climate goals.

Tang and others [29] showed that digital finance enhances the innovation capacity of urban agglomerations, particularly in polycentric metropolitan regions, offering valuable lessons for metropolitan governance in Ukraine.

Looking ahead, the Ministry of Digital Transformation outlined a vision to make Ukraine "the most digital state in the world" by 2030 - a state that is fully paperless, transparent and integrated into the European digital economy. Achieving this ambition would enable Ukraine not only to rebuild what was destroyed but also to accelerate its transition towards a new model of urban development in which digital innovation and climate neutrality advance together. This aligns with insights from the wider smart city literature, where Trencher [30] and Yigitcanlar et al. [31] emphasise that the second generation of smart cities must be citizen-centred, sustainable and ecologically responsible. Ukraine's reconstruction therefore constitutes a critical test case for embedding sustainability principles into digitalisation from the outset.

4 Results

The study has explored the contribution of blockchain and digital twin technologies to the development of climate-neutral and smart cities, focusing on Ukraine as a case study. Three main findings emerge from the analysis.

First, Ukraine's digital transformation has followed a rapid yet resilient trajectory: it moved from fragmented pilots before 2019, through institutionalisation and system-wide expansion prior to the war, and finally to adaptive reconfiguration under wartime disruption.

Second, blockchain offers clear opportunities for enhancing transparency and trust in reconstruction processes, but its ecological implications remain highly dependent on the consensus model. Proof-of-Work continues to impose unsustainable energy demands, whereas alternatives such as Proof-of-Stake present more viable pathways for climate-aligned deployment.

Third, digital twins show considerable promise in optimising urban energy systems, strengthening infrastructure resilience and managing climate risks. In the Ukrainian context, their application is particularly relevant for rebuilding energy, housing and transport systems with sustainability as a guiding principle.

In summary, these results indicate that digitalisation can serve as a driver of sustainable reconstruction and climate-neutral urban strategies, provided that it is embedded in coherent governance and regulatory frameworks (see Table 3).

Table 3. SWOT analysis of digitalisation for smart and climate-neutral cities (Ukrainian case study).

Strengths	Weaknesses
Rapid institutionalisation of digital governance (Ministry of Digital Transformation, Diia ecosystem)	Ecological costs of energy-intensive models (Proof-of-Work)
Proven resilience of digital services under wartime disruption (continuity of e-government, adaptive innovation)	Unequal access to services due to digital divide (age, rural/urban, income)
Integration of blockchain for transparency and immutability in service delivery	Governance constraints and fragmented regulatory frameworks
Strong citizen uptake and satisfaction (e.g., UNDP/KIIS, GovTech Pulse surveys)	Limited availability of reliable datasets due to wartime security restrictions
Opportunities	Threats
Post-war reconstruction as a testing ground for digital twins in energy, housing and transport	Continued cyberattacks and data security risks
Alignment with EU twin transition and climate-neutral targets	Destruction of critical infrastructure (energy grids, transport)
Development of integrated Twin Transition Roadmaps at national and metropolitan levels	Risk of ecological rebound effects if unsustainable digital models scale up
Expansion of citizen participation and trust in e-services	Economic and social inequalities exacerbating exclusion from digital transition
International cooperation (SUN4Ukraine, U_CAN, GGTC Kyiv, etc.)	Political instability or shifts in external funding priorities

5 Discussion

The results of this study highlight both the opportunities and the constraints of digitalisation in contexts of crisis and reconstruction. On the one hand, blockchain and digital twins demonstrate clear potential to enhance transparency, accountability, energy efficiency, and climate resilience. On the other hand, their implementation is not exempt from ecological costs, technical barriers and governance challenges.

A distinctive feature of the Ukrainian case is the extraordinary stress conditions in which digitalisation continues to develop. The ongoing war has generated disruptions that are extremely difficult to anticipate or model, ranging from the displacement of millions of citizens to the destruction of energy grids and transport infrastructure, as well as cyberattacks and data losses. Such factors introduce a level of uncertainty that would ordinarily jeopardise the stability of digital transformation pathways.

Nevertheless, Ukraine's trajectory demonstrates exceptional persistence. Despite martial law, continuous bombardment and the systematic targeting of critical infrastructure, the country has not only preserved the operation of core digital services but also expanded their scope to meet wartime needs. This resilience sets Ukraine apart from many other contexts where digitalisation has unfolded in peacetime under stable governance. It suggests that digital technologies can function as instruments of continuity and adaptation even when broader systems are severely disrupted.

Looking ahead, the application of blockchain and digital twins in Ukraine can evolve along several pathways. In the short term, these tools should be integrated into reconstruction efforts, particularly for procurement transparency, monitoring systems and energy optimisation. In the medium term, Ukraine can leverage European partnerships such as SUN4Ukraine to align national and municipal strategies with continental climate-neutral objectives. In the long term, Ukraine's digital transformation has the potential to serve as a model for countries seeking to integrate post-crisis recovery with climate-oriented innovation.

The broader implication is that, while many aspects of Ukraine's future remain uncertain, the trajectory is evident: digitalisation has become embedded as a structural driver of governance and recovery. This study contributes to the wider debate on how digital infrastructures can be harnessed not only for efficiency and service delivery but also for resilience and sustainability under conditions of profound disruption.

6 Conclusions

The findings confirm that digitalisation is not inherently sustainable, but it can be steered towards climate neutrality when ecological constraints are explicitly addressed. Ukraine's trajectory illustrates both the risks associated with energy-intensive digital models and the opportunities offered by technologies that enhance transparency, efficiency, and resilience.

For reconstruction, the central priority lies in embedding blockchain and digital twins within governance and regulatory frameworks that emphasise accountability, energy efficiency and long-term climate objectives. Under such conditions, digital technologies can serve as instruments of sustainable urban transition rather than sources of additional environmental burden.

Limitations. In addition to the data-related constraints noted earlier, this study did not pursue sectoral depth: industries such as transport, textiles, or manufacturing were not analysed in detail, since the focus was on Ukraine's overall trajectory of digital and ecological transition. Nor did the study quantify regional adoption rates of technologies such as blockchain services or digital twins, as reliable data of this type remain limited, particularly under wartime conditions. These boundaries were necessary to maintain coherence and reflect the uneven availability of evidence.

Future research. This study has outlined the opportunities and constraints of digitalisation in the context of smart and climate-neutral cities. Future research could place greater emphasis on metropolitan areas in order to examine how large urban regions adapt to wartime disruptions and post-war reconstruction. Another promising avenue would be the development of integrated Twin Transition Roadmaps that bring together digital and climate initiatives, offering structured pathways for enhancing resilience, transparency, and ecological sustainability at both national and urban levels.

This paper provides the first systematic assessment of Ukraine's digital transformation through the integrated perspective of ecological transition and urban sustainability, establishing conceptual and empirical foundations for analysing how digitalisation interacts with climate objectives under conditions of extreme stress. For the first time, Ukraine is presented not only as a case of resilient digital governance and wartime innovation, but also as a reference point for climate-neutral pathways. The research contributes an integrated analytical framework grounded in triangulation, comparative benchmarking and citizen participation data, thereby advancing methodological tools for evaluating trust and resilience in digital services. These findings provide a unique basis for future studies and for the design of twin transition roadmaps, offering guidance to countries and metropolitan regions seeking to align digitalisation with climate neutrality and long-term resilience.

Disclosure of Interests. The authors declare that they have no competing interests relevant to the content of this article. No external or grant funding was received for this research. The authors did not obtain any personal remuneration.

References

1. Kramers, A., Höjer, M., Lövehagen, N., Wangel, J.: Smart sustainable cities: exploring ICT solutions for reduced energy use in cities. Environ Model Softw. **56**, 52–62 (2014). https://doi.org/10.1016/j.envsoft.2013.12.019
2. Jones, N.: How to stop data centres from gobbling up the world's electricity. Nature **561**(7722), 163–166 (2018). https://doi.org/10.1038/d41586-018-06610-y
3. Stoll, C., Klaaßen, L., Gallersdörfer, U.: The carbon footprint of Bitcoin. Joule **3**(7), 1647–1661 (2019). https://doi.org/10.1016/j.joule.2019.05.012
4. European Commission. 2030 Digital Compass: The European way for the Digital Decade. Brussels (2021)
5. International Telecommunication Union. United for Smart Sustainable Cities (U4SSC): Guidelines on integrated digital solutions for climate action. Geneva (2022)
6. Saberi, S., Kouhizadeh, M., Sarkis, J., Shen, L.: Blockchain technology and its relationships to sustainable supply chain management. Int. J. Prod. Res. **57**(7), 2117–2135 (2019). https://doi.org/10.1080/00207543.2018.1533261
7. Andoni, M., et al.: Blockchain technology in the energy sector: a systematic review of challenges and opportunities. Renew. Sustain. Energy Rev. **100**, 143–174 (2019). https://doi.org/10.1016/j.rser.2018.10.014
8. Moravec, V., Gavurova, B., Kovac, V.: Environmental footprint of GenAI – changing technological future or planet climate? J. Innov. Knowl. **10**, 100691 (2025). https://doi.org/10.1016/j.jik.2025.100691
9. Gallersdörfer, U., Klaaßen, L., Stoll, C.: Energy consumption of cryptocurrencies beyond Bitcoin. Joule **5**(9), 1–5 (2021). https://doi.org/10.1016/j.joule.2020.07.013
10. Mora, C., et al.: Bitcoin emissions alone could push global warming above 2 °C. Nat. Clim. Chang. **8**(11), 931–933 (2018). https://doi.org/10.1038/s41558-018-0321-8

11. Gorobei, M.: Study of the impact of cryptoassets based on blockchain technology on the environment. Ecological Sciences **1**(46), 85–95 (2023). https://doi.org/10.32846/2306-9716/2023.ECO.1-46.12

12. International Energy Agency. Energy and AI: energy demand from data centres. Paris: IEA. Licence: CC BY 4.0 (2024). https://www.iea.org/reports/energy-and-ai/energy-demand-from-ai

13. Sahibzada, U.F., Janjua, N.A.: Leading the transition toward sustainability through digital capabilities and digital innovation: the role of employee characteristics. J. Innov. Knowl. **10**, 100723 (2025). https://doi.org/10.1016/j.jik.2025.100723

14. Fuller, A., Fan, Z., Day, C., Barlow, C.: Digital twin: enabling technologies, challenges and open research. IEEE Access **8**, 108952–108971 (2020). https://doi.org/10.1109/ACCESS.2020.2998358

15. Batty, M.: Digital twins. Environment and Planning B: Urban Analytics and City Science **45**(5), 817–820 (2018). https://doi.org/10.1177/2399808318796416

16. Liu, Y., Zhang, H., Chen, Y.: Digital twin-driven smart energy management for sustainable manufacturing systems. Energy Rep. **7**, 933–944 (2021). https://doi.org/10.1016/j.apenergy.2022.119986

17. Deren, L., Wenbo, Y., Xiaogang, Q.: Smart city based on digital twins. Comput. Environ. Urban Syst. **95**, 101–118 (2021). https://doi.org/10.1007/s43762-021-00005-y

18. Ketzler, B., Naserentin, V., Latino, F., Zangelidis, C., Thuvander, L., Logg, A.: Digital twins for cities: a state of the art review. Built Environment **46**(4), 547–573 (2020). https://doi.org/10.2148/benv.46.4.547

19. Negri, E., Fumagalli, L., Macchi, M.: A review of the roles of digital twin in CPS-based production systems. Procedia Manufacturing **11**, 939–948 (2017). https://doi.org/10.1016/j.promfg.2017.07.198

20. Belal, A., Shcherbina, E.: Smart-Technology in City Planning of Post-War Cities. IOP Conference Series: Materials Science and Engineering (2018). https://doi.org/10.1051/e3sconf/202126305054

21. Organisation for Economic Co-operation and Development. Digital Government Review of Ukraine: Accelerating digital transformation for recovery. Paris: OECD Publishing. Accessed 20 August 2025 (2021). https://www.oecd.org/en/publications/digitalisation-for-recovery-in-ukraine_c5477864-en/full-report.html

22. International Telecommunication Union. Measuring digital development: facts and figures. Geneva. Accessed: 20 August 2025 (2021). https://www.itu.int/en/ITU-D/Statistics/Documents/facts/FactsFigures2021.pdf

23. Harvard Center for International Development. Ukraine's digital transformation: innovation for resilience. Harvard University, Cambridge, MA. Accessed: 21 August 2025(2025). https://www.hks.harvard.edu/centers/cid/voices/ukraines-digital-transformation-innovation-resilience

24. Ionan, V.: Digital transformation in Ukraine: before, during, and after the war. Harvard Advanced Leadership Initiative, 29 November 2024. Accessed 21 August 2025 (2024). https://www.sir.advancedleadership.harvard.edu/articles/digital-transformation-in-ukraine-before-during-after-war

25. European Commission: Joint Research Centre, Belis, C. A., et al.:. Status of Environment and Climate in Ukraine. Publications Office of the European Union, Luxembourg (2025). https://doi.org/10.2760/6292177

26. Pricopoaia, O., Cristache, N., Lupaşc, A., Iancu, D.: The implications of digital transformation and environmental innovation for sustainability. Journal of Innovation & Knowledge **10**, 100713 (2025). https://doi.org/10.1016/j.jik.2025.100713

27. European Climate, Infrastructure and Environment Executive Agency. Twelve Ukrainian cities join the SUN4Ukraine initiative of the EU Climate-Neutral and Smart Cities Mission. European Commission, 10 December 2024 (2024). https://research-and-innovation.ec.eur opa.eu/news/all-research-and-innovation-news/twelve-ukrainian-cities-join-sun4ukraine-ini tiative-eu-climate-neutral-and-smart-cities-mission-2024-12-03_en
28. Wu, M., Lin, K.: Digital inclusive finance for green transformation: Insights from industrial upgrading, employment and innovation. J. Innov. Knowl. **10**, 100693 (2025). https://doi.org/10.1016/j.jik.2025.100726
29. Tang, K., Cai, X., Wang, H.: Innovation capacity in urban agglomerations: the role of digital finance. J. Innov. Knowl. **10**, 100697 (2025). https://doi.org/10.1016/j.jik.2025.100697
30. Trencher, G.: Towards the smart city 2.0: using smartness to tackle social challenges. Technol. Forecast. Soc. Chang. **142**, 117–128 (2019). https://doi.org/10.1016/j.techfore.2018.07.033
31. Yigitcanlar, T., Kamruzzaman, M.: Understanding 'smart cities': intertwining development drivers with desired outcomes in a multidimensional framework. Technological Forecasting & Social Change **142**, 190–199 (2018). https://doi.org/10.1016/j.cities.2018.04.003

A Data-Driven Framework for Assessing the Impact of Economic Sanctions on Supply Chain Operations: Evidence from Iran's Private Medical Equipment Sector

Leila Chehreghani[1]([envelope]) [iD], Mansour Youseffi[2] [iD], and Leo Vefghi[3]

[1] Ulster University, Birmingham Campus B3 3PL, England
chehreghani@gmail.com
[2] University of Bradford, Bradford, England
m.youseffi@bradford.ac.uk
[3] Sheffield Hallam University, Sheffield, England
l.vefghi@shu.ac.uk

Abstract. Economic sanctions are increasingly employed as instruments of international policy, yet their consequences often extend far beyond political objectives. In the case of Iran, repeated waves of sanctions have generated widespread economic and social repercussions, with the healthcare sector being one of the most severely affected. While previous studies have concentrated on pharmaceuticals and public hospitals, limited attention has been paid to the private medical equipment sector, despite its essential role in sustaining healthcare delivery. This study investigates the impact of sanctions on the efficiency and effectiveness of supply chain operations in this sector. A quantitative, cross-sectional survey was conducted among 120 managers and specialists from 50 private medical equipment companies. Data were analysed using Partial Least Squares Structural Equation Modelling (PLS-SEM) implemented in SmartPLS 3.0, evaluating both measurement and structural models to ensure reliability and validity. Findings reveal that sanctions significantly disrupt procurement, logistics, and financial transactions, increasing costs, delaying supplies, and restricting access to international banking systems. The model explains 57.6% of the variance in supply chain performance, with robust validity and reliability indicators. The study contributes theoretically by integrating Transaction Cost Economics, Resource Dependence Theory, and Institutional Theory, and practically by offering strategies for resilience, including diversification, localised production, and digital supply chain monitoring.

Keywords: Sanctions · Supply Chain Resilience · Iran Private Healthcare · PLS-SEM

1 Introduction

Economic sanctions are widely employed as instruments of international policy to influence the behaviour of targeted states without resorting to direct military action. While sanctions are often justified as tools for promoting international peace and security [1],

Z. Molamohamadi et al. (Eds.): ODSIE 2025, CCIS 2855, pp. 676–693, 2026.
https://doi.org/10.1007/978-3-032-17023-1_41

they frequently generate unintended economic, social, and humanitarian consequences. In Iran, sanctions have been imposed recurrently since the 1980s, with intensifications following the re-imposition of U.S. sanctions in 2018 [2]. These measures have significantly restricted access to international markets, financial systems, and advanced technologies.

The healthcare sector is among the most directly affected. Previous studies documented how sanctions disrupt access to essential medicines, advanced treatment technologies, and medical equipment [3–5]. Ebadi Fardazar et al. [6] highlighted systemic inefficiencies in pharmaceutical supply policies, while Sajadi et al. [7] and Karami Matin et al. [8] emphasised the human cost and human rights implications of these policies. Additionally, indirect effects such as inflation, currency devaluation, and increased transaction costs further constrain the purchasing power of hospitals and private firms [9].

Despite these challenges, most research has focused on public hospitals and state-managed pharmaceutical systems, largely overlooking the private medical equipment sector. This gap is critical because private companies are major suppliers of diagnostic tools, surgical devices, and high-tech equipment necessary for the functioning of Iran's healthcare system. Unlike public companies, private firms are less protected by state mechanisms and are more exposed to currency volatility and international trade restrictions [10, 11]. Reports from Atlas Medical Equipment [12] and Mohamadzadeh [13] illustrate the sector's struggles with balancing costly imports against limited domestic production capacity.

Supply chains, as the backbone of modern economies, are particularly vulnerable under sanctions. Restrictions disrupt procurement, logistics, and financial operations, generating inefficiencies and encouraging. Scholars such as parallel markets, such as smuggling, which undermine quality control [14–17].

Theoretically, this study integrates Transaction Cost Economics [18], which explains how sanctions increase transaction costs and risks; Resource Dependence Theory [19], which examines organisations' adaptation under resource constraints, and Institutional Theory [20], which highlights the influence of external pressures on organisational behaviour. Together, these frameworks provide a comprehensive lens for analysing the sanctions-supply chain relationship.

Building on this foundation, the present study investigates the impact of economic sanctions on supply chain operations in Iran's private medical equipment sector. A quantitative, cross-sectional survey was conducted among 120 managers across 50 private companies, with data analysed using Partial Least Squares Structural Equation Modelling (PLS-SEM) in SmartPLS 3.0 to evaluate both measurement and structural models for reliability and validity. This approach allows for quantifying how sanctions affect procurement efficiency, supply chain effectiveness, and overall firm performance.

This study makes three key contributions:

1. It addresses a critical research gap by focusing on the private medical equipment sector.
2. It enhances methodological rigour through the application of PLS-SEM, moving beyond descriptive or qualitative analyses.

3. It provides actionable insights for managers and policymakers to design more resilient supply chains under prolonged international restrictions.

2 Literature Review

The body of research on economic sanctions and their consequences is extensive, but often fragmented across disciplines such as economics, public health, supply chain management, and international law. This section reviews the literature in four thematic areas: (1) sanctions and healthcare access, (2) supply chain disruptions under sanctions, (3) organisational resilience strategies, and (4) theoretical frameworks underpinning the analysis.

2.1 Sanctions and Healthcare Access

Multiple studies have documented the negative impact of sanctions on healthcare delivery in Iran. Shahabi et al. [4, 5] demonstrated that sanctions disrupted access to cancer treatment and rehabilitation services, while Aloosh and Aloosh [3] argued that sanctions threaten overall population health. Yazdi-Feyzabadi et al. [21] investigated the strong negative direct and indirect effects of sanctions on the healthcare sector. Ebadi Fardazar et al. [6] analysed pharmaceutical policy under sanctions, showing systemic inefficiencies and shortages. Similarly, Hejazi and Emamgholipour [9] linked U.S. sanctions to declining food security, indirectly affecting health outcomes. Sajadi et al. [7] conducted a scoping review highlighting the human cost of sanctions, while Karami Matin et al. [8] emphasised the human rights implications of reduced hospital performance. Collectively, these studies underscore the systemic vulnerability of healthcare systems to international restrictions (Fig. 1).

Fig. 1. The US Sanctions Programs (Source: US Department of the Treasury).

2.2 Supply Chain Disruptions Under Sanctions

Sanctions exert profound pressure on supply chain operations by increasing transaction costs, restricting access to international financial systems, and creating procurement bottlenecks. Prieto [22] investigated the effects of global sanctions on the supply chain efficiency. Ravangard et al. [23] assessed the resilience of the supply chain of the consumables and medicines during the economic disasters. They found negative effects of economic disasters on the digital and information technology-based supply chain Shakeri et al. [14] provided empirical evidence of sanctions' impact on Iranian supply chains, while Bastani et al. [11] detailed procurement challenges in pharmaceuticals and medical equipment. Azami et al. [10] confirmed similar challenges in the procurement of capital equipment, with delays and increased costs being dominant outcomes. Fakhrzad et al. [16] highlighted medicine smuggling as an unintended consequence of restricted access. Sun et al. [15] extended the analysis to contagion effects, showing that sanctions targeting specific firms can negatively affect non-targeted companies within shared supply chains.

2.3 Organisational Resilience Strategies

Research also highlights how organisations develop resilience strategies under sanctions. Taslimi et al. [24, 25] investigated resilience in hospital equipment clusters, showing the importance of local supplier networks. Bastani et al. [11] emphasised collaboration, insurance reform, and innovative procurement methods as coping strategies. Sibevei et al. [17] proposed risk reduction systems in blood supply chains, while Sadati et al. [26] focused on food supply chains, identifying adaptive measures to safeguard food security. These studies collectively point to the need for adaptive management practices and systemic resilience building.

2.4 International and Legal Perspectives

From an international law perspective, Jazairy [27] and Zarei Hadak et al. [28] highlighted the legal and ethical challenges of unilateral sanctions, particularly their incompatibility with the right to health. Masters [29] and International Affairs [1] provided overviews of sanction regimes, while U.S. Department [2, 30] outlined specific measures imposed on Iran's oil and healthcare sectors. These insights highlight the geopolitical backdrop against which organisational and supply chain-level disruptions occur.

2.5 Quantitative and Methodological Contributions

Beyond qualitative accounts, several studies contributed to methodological advancements. Abbasian et al. [31] examined risk factors in pharmaceutical supply chains quantitatively. Statistical robustness and validity methods are discussed in works by Izah et al. [32] on Cronbach's Alpha, Sarstedt et al. [33] on PLS-SEM robustness, and Hayes & Usami [34] on factor score regression. Measurement validity and reliability frameworks are further enriched by Acar et al. [35], Di Malta et al. [36], Ding et al. [37], Lai [38], and Thao et al. [39]. These methodological studies reinforce the analytical rigour needed for research employing SEM, such as the present study.

2.6 Theoretical Foundations

The literature is grounded in multiple theoretical frameworks:

- Transaction Cost Economics [18]: explains how sanctions raise transaction costs and uncertainty.
- Resource Dependence Theory [19]: highlights organisational dependence on external resources and adaptive strategies.
- Institutional Theory [20]: illustrates how coercive pressures (e.g., sanctions) reshape organisational practices.
- Social Capital Theory [40]: provides insights into how firms leverage networks to maintain performance under sanctions.

2.7 Integration and Research Gaps

Taken together, these studies confirm that sanctions affect healthcare delivery, supply chain operations, and organisational resilience across multiple domains. However, several research gaps remain:

Most prior studies have concentrated on the public healthcare system in Iran, analysing pharmaceutical procurement, government hospitals, and regulatory mechanisms. However, private medical equipment companies—responsible for a large share of the healthcare supply chain—have received minimal scholarly attention. This represents a significant gap because private firms are directly exposed to international trade restrictions and currency fluctuations, with fewer state-backed protections.

Another gap concerns methodological diversity. While qualitative and policy-focused studies dominate the literature, quantitative analyses based on structural modelling in the private sector are scarce. This study contributes by applying SEM to survey data from 120 managers, thereby providing statistically validated insights.

Finally, the timing of research is important. Since the re-imposition of U.S. sanctions in 2018, conditions have dramatically worsened for private medical equipment companies. Yet few recent studies capture this dynamic environment. This research provides updated evidence, contributing both empirically and theoretically to the literature on sanctions and supply chain management.

3 Problem Statement

Economic sanctions are intended as political instruments, yet their impacts often extend far beyond their intended targets, creating widespread disruption in economic activities and social welfare. In Iran, decades of sanctions have created structural challenges across many industries, but the healthcare sector is particularly vulnerable because of its reliance on imported technologies, equipment, and international financial transactions [3, 4].

Most existing studies have examined the effects of sanctions on public hospitals and pharmaceutical procurement systems [6, 7]. However, private medical equipment companies—which supply a substantial share of diagnostic, surgical, and therapeutic technologies—have received very limited scholarly attention. This neglect is significant because private firms face greater exposure to exchange rate volatility, blocked banking channels, and restricted supplier access, without the subsidies or protections often provided to public institutions [10, 11].

As a result, private companies are frequently unable to procure high-quality equipment in a timely or cost-effective manner. This creates cascading effects throughout the healthcare system: delayed surgeries, shortages of diagnostic devices, and increased costs passed on to patients. Moreover, the scarcity of quantitative, statistically validated studies on this sector has left policymakers without solid evidence to design targeted resilience strategies.

Therefore, the problem addressed by this research is the lack of empirical understanding of how international sanctions disrupt the supply chain operations of Iran's private medical equipment companies, and what strategies might enable these firms to sustain performance in a highly constrained environment.

3.1 Conceptual Framework

Sanctions are not frequent events, yet when they occur, they can significantly disrupt or slow the flow of money, goods, and information [14]. They are irregular in nature, lacking a predictable pattern. Typically, their intensity remains low; however, due to political tensions and instability, their severity can escalate rapidly. Consequently, companies are often reluctant to allocate resources to develop mitigation strategies for such disruptions. Nonetheless, when the likelihood of sanctions rises, it becomes prudent to implement strong measures and evaluate alternative courses of action [41].

A single supplier's inability to fulfil orders because of restrictions can create a ripple effect across the entire supply chain, leading to widespread operational challenges. Sanctions have disrupted international trade, constrained access to critical raw materials, created shipping delays, and introduced complex legal complications [15].

3.2 Research Hypotheses

Based on these considerations, the variables, hypotheses, and conceptual framework adopted for this study are as follows:

Variables. Independent variable: Economic Sanctions; Dependent variable: Supply Chain Operations.

Hypothesis. Economic Sanctions significantly impact the effectiveness and efficiency of the operations of the supply chain in non-governmental medical tools supplier companies (Fig. 2).

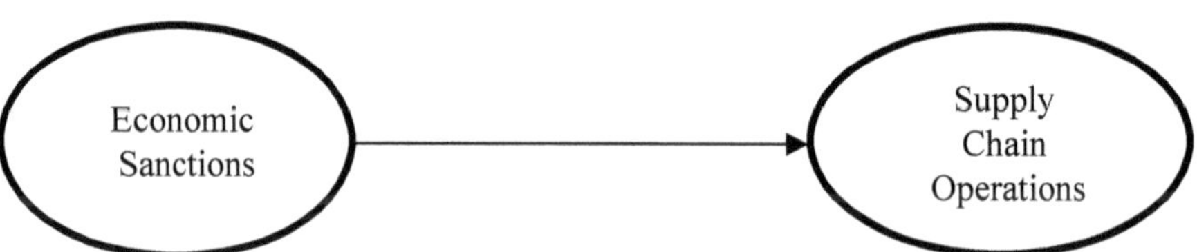

Fig. 2. Conceptual Framework.

4   Methodology

This research adopts a positivist philosophical stance and employs a deductive reasoning approach, as it seeks to test pre-formulated hypotheses derived from existing theories on supply chain management and the effects of economic sanctions. The study is designed as a cross-sectional, quantitative survey, enabling the collection of standardised data from a relatively large number of respondents within a limited timeframe. This methodological choice is appropriate for examining causal relationships and testing the structural model using advanced statistical techniques.

4.1   Research Design

The research design was structured to assess the relationship between economic sanctions and supply chain operations in Iran's private medical equipment sector. Previous studies have primarily focused on the public healthcare system, but this research targets the private sector, where exposure to international restrictions and market volatility is more direct. By employing a survey-based quantitative strategy, the study ensures the ability to measure constructs such as procurement efficiency, supply chain effectiveness, and financial constraints with a high degree of reliability.

4.2   Population and Sampling

The target population comprised managers, senior executives, and specialists working in private medical equipment companies in Iran. According to industry reports, there are approximately 60–70 active firms in this sector, of which 50 were included in this study.

A non-probability quota sampling approach was employed to capture diversity across managerial levels and firm sizes. This ensured representation from:

- Chief Executive Officers (CEOs) and senior directors
- Operations and logistics managers
- Supply chain and procurement managers
- Financial and administrative managers and other categories

(See Table 1).

Table 1. Descriptive statistics of respondents' Roles.

	Frequency	Percent	Valid Percent	Cumulative Percent
CEO/Executive	28	23.3	23.3	23.3
Operations Manager	22	18.3	18.3	41.7
Other	23	19.2	19.2	60.8
Procurement Manager	21	17.5	17.5	78.3
Supply Chain Manager	26	21.7	21.7	100.0
Total	120	100.0	100.0	-

The respondents had a different range of work experience. 26 questionnaires were answered by the experts with 1–5 years of work experience; 49 of them had 5–10 years of experience, and 45 respondents had more than 10 years of experience (Table 2).

Table 2. Descriptive statistics of respondents' Operating in the medical equipment.

	Frequency	Percent	Valid Percent	Cumulative Percent
1–5 years	26	21.7	21.7	21.7
5–10 years	49	40.8	40.8	62.5
More than 10 years	45	37.5	37.5	100.0
Total	120	100.0	100.0	-

Among the companies selected for this research, 18 of them are considered large companies with more than 201 employees, whereas 53 of them are medium companies with 51–200 employees and 49 small companies with 50 or fewer employees (Table 3).

Table 3. Descriptive statistics of respondents' Size of company respondents.

	Frequency	Percent	Valid Percent	Cumulative Percent
Large (201 + employees)	18	15.0	15.0	15.0
Medium (51–200 employees)	53	44.2	44.2	59.2
Small (1–50 employees)	49	40.8	40.8	100.0
Total	120	100.0	100.0	-

In total, 120 valid responses were collected, which meets the minimum requirement for conducting Structural Equation Modelling (SEM). According to Hair et al. (2017), a sample size above 100 is acceptable for SEM when the model complexity is moderate and factor loadings exceed 0.5.

4.3 Instrument Development

A structured questionnaire was developed, consisting of two major constructs:

1. Sanctions (8 items) – measuring barriers related to procurement, logistics, financial transactions, supplier access, and currency instability.
2. Supply chain operations (7 items) – measuring efficiency, effectiveness, timeliness of procurement, quality of supplies, and cost management.

All items were measured using a 5-point Likert scale ranging from 1 (strongly disagree) to 5 (strongly agree). The questionnaire was initially piloted with 10 managers to ensure clarity and content validity. Minor adjustments were made based on their feedback before large-scale distribution.

4.4 Data Collection

Data collection was conducted over a two-month period in 2023. Due to sanctions-related restrictions and the COVID-19 context, both online surveys and in-person distributions were used. Respondents were assured of confidentiality and anonymity to encourage honest participation.

4.5 Validity and Reliability

The psychometric properties of the instrument were rigorously tested:

- Content Validity: Ensured through expert review by three academics and two industry specialists.
- Construct Validity: Assessed using Kaiser-Meyer-Olkin (KMO $= 0.602$) and Bartlett's Test of Sphericity ($\chi2 = 940.010$, $p < 0.001$), indicating suitability for factor analysis.
- Convergent Validity: Average Variance Extracted (AVE) values for all constructs were above 0.5, confirming convergent validity.

Table 4. Average Variance Extracted (AVE).

Variable	Average Variance Extracted (AVE)
Sanctions	0.600
Supply Chain Operations	0.591

- Discriminant Validity: Assessed through the Fornell-Larcker criterion and Heterotrait-Monotrait ratio (HTMT < 0.85).

Table 5. KMO and Bartlett test.

KMO indicator	.602
Bartlett's test of two estimates	940.010
degree of freedom	105
significance (sig)	.000

Table 6. Communalities.

#	Extraction
S1	0.646
S2	0.843

(continued)

Table 6. (*continued*)

#	Extraction
S3	0.832
S4	0.618
S5	0.712
S6	0.806
S7	0.642
S8	0.799
KA9	0.752
KA10	0.800
KA11	0.749
KA12	0.656
KA13	0.835
KA14	0.783
KA15	0.535

- Reliability: Measured through Cronbach's alpha coefficients, which exceeded 0.7 for both constructs, indicating acceptable internal consistency. Composite Reliability (CR) values also exceeded 0.8, further supporting reliability.

Table 7. Cronbach's alpha.

variable	Cronbach's alpha
Sanctions	**0.745**
Supply Chain Operations	**0.784**

4.6 Ethical Considerations

The study adhered to strict ethical guidelines. Participation was voluntary, informed consent was obtained from all respondents, and data confidentiality was ensured. The research protocol was reviewed and approved by the academic ethics committee of Ulster University.

5 Data Analysis

The collected data were analysed using Partial Least Squares Structural Equation Modelling (PLS-SEM) through SmartPLS 3.0 software. The choice of PLS-SEM was motivated by its suitability for exploratory studies, its applicability to relatively small sample sizes, and its ability to handle non-normal data distributions. The analysis involved two major stages:

1. Measurement Model Evaluation: Factor loadings, Cronbach's alpha, CR, AVE, and discriminant validity measures were evaluated.
2. Structural Model Evaluation: Hypotheses were tested by examining path coefficients, t-statistics, p-values, R2, f2, Q2, and Goodness-of-Fit (GOF = 0.812).

5.1 Descriptive Statistics

The details of descriptive statistics, including the role and the work experience of respondents, in addition to the sizes of the companies selected for this work, have been explained in section four in detail.

5.2 Inferential Statistics

Inferential statistics use sample data to make generalisations about larger populations through various tests [42]. The questionnaire scores range from 1 to 5. Among the variables, Supply Chain Operations has the lowest standard deviation, and Sanction has the highest. The mean represents the central value, while kurtosis (elongation) indicates the sharpness of the curve, with 3 indicating normal distribution [43]. Since all skewness and kurtosis values fall within (–3, 3), the data is considered normally distributed (Table 8).

Table 8. Inferential statistics.

	Sanctions	Supply Chain Operations
Valid	120	120
Missing	0	0
Mean	3.33	4.27
Median	3.25	4.28
Std. Deviation	.53	.50
Skewness	1.09	−.67
Kurtosis	1.26	−.20
Minimum	2.37	3.00
Maximum	4.87	5.00

5.3 Checking the Model based on Measurement Criteria

When constructing measurement models, three criteria are evaluated: internal consistency of the indicator, how well different indicators converge on the same construct (convergent validity), and how distinct different constructs are from each other (discriminant validity). The reliability (or dependability) of an indicator is assessed based on three key aspects [33]:

Cronbach's alpha. Cronbach's alpha is explained in section four in detail.

Composite Reliability. Composite Reliability (CR) is a measure of the internal consistency of a scale, analogous to Cronbach's alpha. It essentially reflects the proportion of true score variance relative to total score variance (i.e., variance attributable to the construct versus measurement error) [38].

In Tables 4, 5 and 6, the computed CR values are shown. As can be seen, all variable CRs exceed 0.70, which indicates that the measurement instrument demonstrates acceptable reliability (Table 9).

Table 9. Composite Reliability.

variable	Composite Reliability
Sanctions	0.726
Supply Chain Operations	0.772

Average Variance Extracted (AVE). The Average Variance Extracted (AVE) reflects the extent to which a construct explains the variance of its indicators, thereby confirming the direct linkage between each item and its underlying variable [34]. As demonstrated in Tables 4, 5, 6 and 7, all AVE values exceed the threshold of 0.05, indicating that convergent validity is established across all research constructs.

Divergent Validity. Divergent (or discriminant) validity is used to confirm that a measurement procedure actually distinguishes between different constructs. It demonstrates that the concepts you aim to measure are not simply overlapping with or identical to other ideas [36]. Before assessing specific validity (i.e., verifying particular constructs), one should first establish general validity to reinforce the thesis argument [35]. It includes standard coefficients of interaction and Fornell and Larcker.

5.4   Criteria for Evaluating the Fit of the Structural Part

The criteria for evaluating the fit of the structural part are factor loadings, T-Value statistics, R2 index, F2 index, redundancy, Q2, and GOF1 index (Figs. 3 and 4 and Tables 10 and 11).

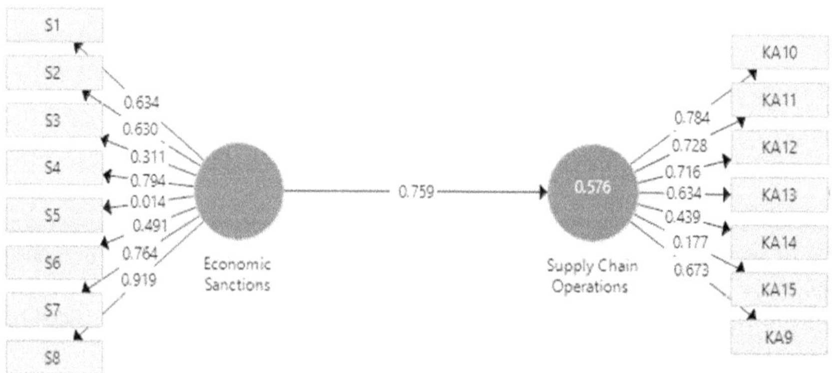

Fig. 3. Factorial values in the initial research model.

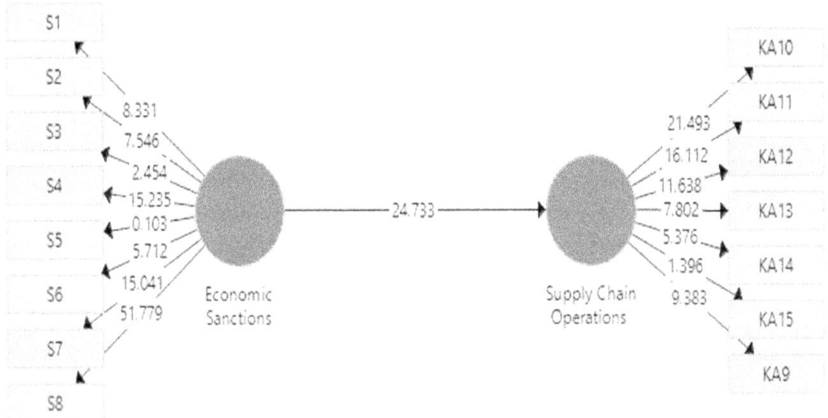

Fig. 4. T-statistic values in the initial research model.

Table 10. Values in fitting the structural model.

Variables	R^2	F^2	Redundancy	Q^2
Supply Chain Operations	0.576	0.572	0.469	0.302

Table 11. GOF Index.

Variables	Communality
Sanctions	0.861
Supply Chain Operations	0.879
Mean	0.870
R2	0.576
GOF	0.812

6 Findings and Discussion

Descriptive results revealed that 44% of respondents worked in medium-sized firms (51–200 employees), 41% in small firms, and 15% in large enterprises. Most companies had operated for 5–10 years, and nearly one-quarter of respondents were executives.

Inferential results confirmed that sanctions significantly disrupt supply chain operations (path coefficient = 0.759, t = 24.73, p < 0.001). Cronbach's alpha and composite reliability exceeded 0.7, and AVE values surpassed 0.5, affirming construct validity. Sanctions increased costs, delayed procurement, and constrained banking transactions. SEM analysis showed that sanctions explain 57.6% of the variance in supply chain performance.

Discussion: These findings align with Bastani et al. [11], who emphasised resilience strategies in the pharmaceutical sector. They also echo Azami et al. [10] regarding procurement difficulties and Shahabi et al. [4] on adverse health outcomes. Unique to this study is the evidence from the private sector, where firms have adopted adaptive strategies—such as supplier diversification and localised production—but still face structural barriers such as blocked SWIFT transactions and volatile exchange rates.

Policy implications: To strengthen resilience, firms should engage in supply chain mapping, digital monitoring, and stronger partnerships with domestic suppliers. Policymakers should facilitate alternative financial mechanisms, support local production, and coordinate international humanitarian exemptions.

7 Conclusion and Future Work

This study examined the impact of economic sanctions on supply chain operations in Iran's private medical equipment sector, a domain that has received limited attention compared to the more extensively studied public healthcare system. By employing a quantitative research design and applying Structural Equation Modelling (SEM) to data collected from 120 managers across 50 companies, the research provides robust evidence of how sanctions disrupt procurement, logistics, and financial flows. The findings confirm that sanctions significantly increase transaction costs, delay access to critical equipment, and constrain international financial transactions, ultimately weakening the efficiency and effectiveness of healthcare-related supply chains.

From a theoretical perspective, this research contributes by integrating Transaction Cost Economics, Resource Dependence Theory, and Institutional Theory to explain how external coercive pressures reshape organisational strategies and performance. Empirically, it extends the literature by focusing on the private medical equipment sector, which is often overlooked despite its central role in supporting healthcare delivery. Practically, the study proposes resilience strategies such as supplier diversification, localised production, alternative financing mechanisms, and the adoption of digital supply chain monitoring tools. These recommendations are relevant not only for Iranian firms but also for organisations in other sanctioned or resource-constrained contexts.

However, the study is not without limitations. The cross-sectional design captures a snapshot in time, which restricts the ability to assess dynamic, long-term adaptation strategies. In addition, the reliance on self-reported survey data may introduce response bias, and the sample, while adequate for SEM, may not fully capture the heterogeneity of the entire sector.

Future research could address these limitations in several ways. First, longitudinal studies are needed to track how private firms adapt their supply chains over time under persistent sanctions. Second, comparative studies between the public and private sectors would shed light on differential coping mechanisms and outcomes. Third, future scholars may explore the moderating effects of exchange rate volatility, firm size, and ownership structure on the sanctions–supply chain relationship. Finally, cross-country comparative research could extend the findings to other sanctioned contexts, offering a broader understanding of how international restrictions affect supply chain resilience globally.

In conclusion, while sanctions are designed as political instruments, their most profound impacts are often felt in the daily functioning of vital sectors such as healthcare. By illuminating these impacts in the private medical equipment sector, this study provides both theoretical insights and actionable recommendations, with the ultimate goal of fostering more resilient healthcare systems in sanction-affected nations.

References

1. InternationalAffairs, (2019). IRAN SANCTIONS OVERVIEW. http://www.treasury.gov/resource-center/sanctions/Programs/pages/iran.aspx
2. U.S.Department, (2024). Iran sanctions. https://www.bis.doc.gov/index.php/policy-guidance/country-guidance/sanctioned-destinations/iran
3. Aloosh, M., Salavati, A., Aloosh, A.: Economic sanctions threaten population health: the case of Iran. Publ Health **169**, 10–13 (2019)
4. Shahabi, S., et al.: The impact of international economic sanctions on Iranian cancer healthcare. Health Policy **119**(10), 1309–1318 (2015)
5. Shahabi, et al.: Physical rehabilitation in Iran after international sanctions: explored findings from a qualitative study. Glob. Health **16**(1), 86 (2020). https://doi.org/10.1186/s12992-020-00618-8
6. Ebadi Fardazar, F., et al.: Policy analysis of Iranian pharmaceutical sector; a qualitative study. Risk Manag. Healthc. Policy 199–208 (2019)
7. Sajadi, H.S., Yahyaei, F., Ehsani-Chimeh, E., Majdzadeh, R.: The human cost of economic sanctions and strategies for building health system resilience: a scoping review of studies in Iran. Int. J. Health Plan. Manag. **38**(5), 1142–1160 (2023)
8. Karami Matin, B., et al.: The impacts of economic sanctions on the performance of hospitals in Iran: implications for human rights. Int. J. Hum. Rights Healthc. **17**(2), 117–126 (2024)
9. Hejazi, J., Emamgholipour, S.: The effects of the re-imposition of US sanctions on food security in Iran. Int. J. Health Pol Manag. (2020). https://doi.org/10.34172/ijhpm.2020.207."
10. Azami, S.R., et al.: International sanctions and the procurement of medical equipment in Iran: a qualitative study. Med. J. Islamic Republic of Iran 35 (2021)
11. Bastani, P., et al.: Under politico-economic sanctions: the case of Iran. J. Pharm. Policy Pract. **14**, 1–14 (2021)
12. Atlas medical equipment, (2024). Atlas medical equipment industry. https://tinyurl.com/bdd8bsv3"
13. Mohamadzadeh, M.: Medical Equipment: Manufacturing or Importing (2023). https://tinyurl.com/mr36sf8x"
14. Shakeri, E., Vizvari, B., Nazerian, R.: The impacts of economic sanctions on supply chain management: empirical analysis of Iranian supply chains. In: Proceedings of the Global Joint Conference on Industrial Engineering and Its Application Areas, pp. 53–63 (2016)
15. Sun, J., Makosa, L., Yang, J., Darlington, M.: Economic sanctions and shared supply chains: a firm-level study of the contagion effects of smart sanctions on the performance of nontargeted firms. Eur. Manag. Rev. **19**(1), 92–106 (2022)
16. Fakhrzad, N., Yazdi-Feyzabadi, V., Fakhrzad, M.: Drivers of vulnerability to medicine smuggling and combat strategies: a qualitative study based on online news media analysis in Iran. BMC Health Serv. Res. **24**(1), 383 (2024)
17. Sibevei, A., et al.: Developing a risk reduction support system for health system in Iran: a case study in blood supply chain management. Int. J. Environ. Res. Public Health **19**(4), 2139 (2022)
18. Williamson, O.E.: Transaction cost economics. Handb. Ind. Organ. **1**, 135–182 (1989)

19. Hillman, A.J., Withers, M.C., Collins, B.J.: Resource dependence theory: a review. J. Manag. **35**(6), 1404–1427 (2009)
20. Trepte, S.: Social Identity Theory. In Psychology of Entertainment. Routledge, London, pp. 255–271 (2013)
21. Yazdi-Feyzabadi, V., Zolfagharnasab, A., Naghavi, S., et al.: Direct and indirect effects of economic sanctions on health: a systematic narrative literature review. BMC Public Health **24**, 2242 (2024). https://doi.org/10.1186/s12889-024-19750-w"
22. Prieto, R.: Supply Chain Management and the Impacts of Tariffs and Trade Sanctions (2025)
23. Ravangard, R., Raeyat Mohtashami, A., Allahyari Bouzanjani, A., Siadat, N.K., Ahmadi Marzaleh, M.: Assessing the resilience of the supply chain of medicines and consumables during disasters based on the world economic forum framework: a case of Iran. Disaster Med. Public Health Preparedness **19**, e266 (2025). https://doi.org/10.1017/dmp.2025.10192"
24. Taslimi, M.S., Azimi, A., Nazari, M.: Resilience to economic sanctions; case study: hospital equipment cluster of Tehran (HECT). Int. J. Disaster Resil. Built Environ. **12**(1), 13–28 (2020). https://doi.org/10.1108/ijdrbe-06-2018-0024
25. Taslimi, M.S., Azimi, A., Nazari, M.: Resilience to economic sanctions; case study: hospital equipment cluster of Tehran (HECT). Int. J. Disaster Resil. Built Environ. **12**(1), 13–28 (2021)
26. Sadati, A.K., Nayedar, M., Zartash, L., Falakodin, Z.: Challenges for food security and safety: a qualitative study in an agriculture supply chain company in Iran. Agric. Food Secur. **10**, 1–7 (2021)
27. Jazairy, I.: Unilateral economic sanctions, international law, and human rights. Ethics & Int. Aff. **33**(3), 291–302 (2019)
28. "Zarei Hadak, M., Rezainejad, I., Babi Mehr, A.: Legal challenges of unilateral American sanctions and its impact on the right to health. Political Strategy Quarterly (2021). pp. 5th year, number 3 consecutively, 18
29. Masters, J.: Council on Foreign Relations (2024). https://www.cfr.org/backgrounder/what-are-economic-sanctions"
30. "U.S.Department, (2020). Treasury Sanctions Key Actors in Iran's Oil Sector for Supporting Islamic Revolutionary Guard Corps-Qods Force. https://home.treasury.gov/news/press-releases/sm1165"
31. Abbasian, H., et al.: Risk factors of supply chain in biopharmaceutical companies in Iran. Pharm. Sci. **27**(3), 439–449 (2020)
32. Izah, S.C., Sylva, L., Hait, M.: Cronbach's alpha: a cornerstone in ensuring reliability and validity in environmental health assessment. ES Energy Environ. **23**, 1057 (2023)
33. Sarstedt, M., et al.: Structural model robustness checks in PLS-SEM. Tour. Econ. **26**(4), 531–554 (2020)
34. Hayes, T., Usami, S.: Factor score regression in connected measurement models containing cross-loadings. Struct. Equation Model. Multidiscip. J. **27**(6), 942–951 (2020)
35. Acar, S., Ogurlu, U., Zorychta, A.: Exploration of discriminant validity in divergent thinking tasks: a meta-analysis. Psychol. Aesthetics, Creat. Arts **17**(6), 705 (2023)
36. Di Malta, G., She, Z., Raymond-Barker, B., Cooper, M.: Retest reliability, divergent validity, criterion validity with psychotherapy satisfaction, and measurement invariance in UK-and US-stratified samples. J. Clin. Psychol. **79**(9), 2040–2052 (2023)
37. Ding, Y., et al.: SPSS-based analysis of factors affecting high school students' core literacy in the context of the new college entrance examination. In: 2022 3rd International Conference on Artificial Intelligence and Education (IC-ICAIE 2022), pp. 1244–1250. Atlantis Press (2022)
38. Lai, M.H.: Composite reliability of multilevel data: it's about observed scores and construct meanings. Psychol. Methods **26**(1), 90 (2021)
39. Thao, N.T.P., Van Tan, N., Tuyet, M.T.A.: KMO and Bartlett's test for components of workers' working motivation and loyalty at enterprises in Dong Nai Province of Vietnam. Int. Trans. J. Eng. Manag. Appl. Sci. Technol. **13**(10), 1–13 (2022)

40. Martin, J.P., Stefl, S., Cain, L.W., Pfirman, A.L.: Understanding first-generation undergraduate engineering students' entry and persistence through social capital theory. Int. J. STEM Educ. **7**, 1–22 (2020)

41. Shalpegin, T., Kumar, A., Browning, T.R.: Undiversity, inequity, and exclusion in supply chains: the unintended fallout of economic sanctions and consumer boycotts. Prod. Oper. Manag. (2023)

42. "Lee, S.H., Kim, E.: Inferential Statistics. In: Scholarly Research in Music, Issue Routledge, London, pp. pp. 165–182 (2022)

43. Carlucci, M.E., Wright, D.B.: Inferential statistics. Research methods in psychology, p. 395 (2020)

Author Index

Z. Molamohamadi et al. (Eds.): ODSIE 2025, CCIS 2855, pp. 695–697, 2026.
https://doi.org/10.1007/978-3-032-17023-1